GW00482032

Towers' International Opamp Linear-IC Selector

Towers' International Opamp Linear-IC Selector

Specification data for the identification,
selection and substitution of opamp linear-ICs

by **T D Towers,** MBE, MA, BSc, C Eng, MIERE &
N S Towers, BA(Cantab)

BPB PUBLICATIONS

B-14, CONNAUGHT PLACE, NEW DELHI-110001

FIRST INDIAN EDITION 1982

REPRINTED 2007

Distributors:

MICRO BOOK CENTRE
2, City Centre, CG Road,
Near Swastic Char Rasta,
AHMEDABAD-380009 Phone: 6421611

COMPUTER BOOK CENTRE
12, Shrungar Shopping Centre, M.G. Road,
BANGALORE-560001 Phone: 5587923, 5584641

MICRO BOOKS
Shanti Niketan Building, 8, Camac Street,
KOLKATTA-700017 Phone: 2826518, 2826519

BUSINESS PROMOTION BUREAU
8/1, Ritchie Street, Mount Road,
CHENNAI-600002 Phone: 8534796, 8550491

DECCAN AGENCIES
4-3-329, Bank Street,
HYDERABAD-400001 Phone: 4756400, 4756967

MICRO MEDIA
Shop No. 5, Mahendra Chambers, 150 D.N. Road,
Next to Capital Cinema V.T. (C.S.T.) Station,
MUMBAI-400001 Ph.: 2078296, 2078297, 2002732

BPB PUBLICATIONS
B-14, Connaught Place, **NEW DELHI-110001**
Phone: 3325760, 3723393, 3737742

INFO TECH
G-2, Sidhartha Building, 96 Nehru Place,
NEW DELHI-110019
Phone: 6438245, 6415092, 6234208

INFO TECH
Shop No. 2, F-38, South Extension Part-1
NEW DELHI-110049
Phone: 4691288, 4641941

INFO TECH
B-11, Vardhman Plaza, Sector-16,
Electronics Nagar, **NOIDA-201301**
Phone: 914-512329, 515917, 515918

BPB BOOK CENTRE
376, Old Lajpat Rai Market,
DELHI-110006 PHONE: 3861747

Copyright © W. Foulsham

All rights reserved. No part of this book shall be reproduced, stored in a retrieval system, or transmitted by any means, electronic, mechanical, photocopying, recording, or otherwise, without written permission from the publisher except for the inclusion of brief quotation in a review.

All of the corporation represented in this book make no warranty, express or implied, that information on the products contained herein is still applicable. Specifications and availability of products herein are subject to change without notice.

Printed in India by arrangement with

W. FOULSHAM & CO. LTD.
Yeovil Road, Slough, Berks, England

Price : Rs. 150/-

ISBN 81-7656-625-X

Published by Manish Jain for BPB Publications, B-14, Connaught Place
New Delhi-110 001 and Printed by him at Pressworks, Delhi.

Preface

In the 1950s, the vacuum tube was the workhorse of the electronic circuit designer. In the 1960s, the circuit man had come to use the transistor as his basic active element. Finally, by the 1970s the silicon integrated circuit had taken over much of the work formerly done by the transistor and its vacuum tube predecessor.

From the mid 1970s, a new generation of circuit design engineers grew up, for whom the opamp linear-IC is the basic circuit element, with transistors and other 'discrete' semiconductor devices relegated to the position of ancillaries to the basic opamp.

If you deal with opamp linear-ICs — whether as a student, a hobbyist, a circuit engineer, a buyer, a teacher or a serviceman — you often want data on a specific opamp of which you know only the type number.

Specifications apart, you may be even more interested in where you can get the device in question. And perhaps more important still (particularly with obsolete devices) you may want guidance on a readily available possible substitute.

This opamp linear-IC compendium, a comprehensive tabulation of basic specifications for over four thousand opamps, offers information on:

1. Ratings
2. Characteristics
3. Case details
4. Terminal identifications
5. Applications use
6. Manufacturers
7. Substitution equivalents (both European and American)

This work covers not only classical 'pure' opamps, but also several classes of 'quasi-opamps' (i.e. linear-ICs with opamp-like characteristics) such as dc-comparators, operational-transconductance-amplifiers, differential-output amplifiers, current-difference amplifiers, and voltage-follower amplifiers.

The more than 4,000 opamps covered in this self-contained volume are a selection of the more common current and widely-used obsolete types.

The coverage is international in scope and includes opamps not only from the USA and Continental Europe, but also from the United Kingdom and the Far East (Japan).

The tabulation format used is similar to that used in the well-known companion publications, *Towers' International Transistor Selector and Fet Selector*.

Every reasonable care has been taken to ensure accuracy of information in the tables, but no responsibility can be accepted for inaccuracies that may have arisen.

COMBATING MENACE OF PIRACY

Dear Readers, Publishers, Booksellers, all College & University Administrations, & Librarians,

The Federation of Publishers' & Book Sellers' Associations in India (FPBAI) would like to bring a matter of great importance to your urgent notice. Though possession of pirated copies is a cognizable and non-bailable offence, punishable by law and carries stiff penalties including imprisonment, it continues to thrive in markets, bookshops and photocopy shops. This is a matter of grave concern for all of us in the education and publishing community. Let us all do our bit in combating this crime against creation and spread of knowledge. If you come across a pirated book, photocopied pages from a book or are in any kind of doubt about the genuineness of a copy, please contact Chairman, Anti Piracy Committee, FPBAI, 84, Second Floor (Opp. Cambridge Primary School), Daryaganj, New Delhi-110 002.

Your co-operation and help in bringing the culprits to book will help the publishing industry combat piracy and facilitate the continued availability of good books of quality production at reasonable prices.

RELEVANT PORTIONS OF THE COPYRIGHT ACT ARE GIVEN BELOW:

INFRINGEMENT OF COPYRIGHT

The owner of the copyright has the exclusive right in respect of the reproduction of the work and such other acts which enables the owner to get the financial benefits by exercising such rights. If any of these acts relating to the work is carried out by a person other than the owner without a license from the owner or competent person/authority under the Copyright Act 1957, it constitutes infringement of copyright in the work.

PIRACY

It is a kind of illegal activity which has been caused by rapid technological advancement. Latest techniques of photocopying and printing have made it easy to produce unauthorized copies of a book within a short span of time at a relatively low cost on a large scale. This offence deprives the author of the work from getting his legitimate due and ultimately hampers the growth of original and creative work by the pursuit of hard work and intellectual skill and national economy as well.

COPYRIGHT LEGISLATION

The Principal Act (Copyright Act, 1957) was amended in 1984 to incorporate anti-piracy legislation to check widespread piracy of books, etc. and it has been made a cognizable and non-bailable offence.

The punishment for various offences have been enhanced by amending sections 63 and 65 and by inserting new sections 63A & 63B.

SECTION 63:

Any person who knowingly infringes or abets the infringement of the copyright work shall be punishable with:

    (i)   Imprisonment for a term which shall not be less than 6 months but which may be extended up to 3 years;

    (ii)   Fine of not less than Rs. 50,000/- but up to Rs. 2 lakhs.

SECTION 63A:

The quantum of enhanced penalty on second or subsequent conviction shall be:

    (i)   Imprisonment for not less than 1 year but up to 3 years; and

    (ii)   Fine of not less than rupees 1 lakh but may be extended up to Rs. 2 lakhs.

SECTION 63B:

The quantum penalty for the offence of knowing use of infringing copy of computer program shall be:

    (i)   Imprisonment for not less than 7 days but up to 3 years; and

    (ii)   Fine of not less than Rs. 50,000/- but may be extended up to Rs. 2 lakhs.

More powers have been given to the police for prompt action and speedy apprehension of the offender by amending section 64, as any police officer not below the rank of Sub-Inspector may seize without warrant all infringing copies or the work if he is satisfied that the offence is under section 63 in respect of the infringement of copyright. The Economic Offence (Inapplicability of Limitation) Act, 1974 was amended by incorporating in the Schedule the clause (a) of section 63 of the Copyright Act, 1957 which declared infringement of copyright as an economic offence.

Contents

An Introduction to the Opamp

Formally, an *opamp* (abbreviation for operational amplifier') is usually taken to be a multi-purpose, very-high-gain, dc amplifier with differential input (two balanced input leads) and single-ended output (one output lead), which uses external feedback for control of response characteristics. Within its limitations, the opamp can be considered to be truly a 'universal' amplifier.

Nowadays, opamps are readily commercially available off-the-shelf in pre-assembled, pretested, packaged 'integrated circuit' form. Such opamps may be *modular* (an interconnection of subminiature conventional cased components embedded in a closed protective package), *hybrid* (interconnection of uncased components in a closed package), or *monolithic* (all components fabricated by gaseous diffusion within a single tiny chip of silicon).

Originally the opamp was developed for analog computer mathematical operations such as addition and subtraction, but it has recently extended its use very widely to become one of the most versatile electronic circuitry tools. It can fill most economically a wide diversity of needs in signal conditioning, signal generation, active filters, electronic measurement and control, as well as in the traditional mathematical computing functions.

More and more engineers are facing problems that can best be solved with the prepackaged gain block available in the commercial opamp. As a result, the design of circuits with individual transistors, etc., is tending to give way to design with opamps as the basic building blocks.

Opamp Symbols

As a circuit element, the complex conglomerate of devices in the opamp is nowadays usually represented in circuit diagrams by the single simple triangular symbol of Figure 1(a). This has two input terminals on the left, designated (−) and (+), and one output terminal on the right without a designation. Only a single output terminal is shown because it is usually assumed that the output voltage is measured with reference to ground, and thus there is no need to show a second ('ground') output terminal.

The (−) input terminal is known as the *inverting input* and the (+) input terminal as the *non-inverting input*. This is because a +ve voltage applied to the (−) input with respect to the (+) input will cause the output to go negative, i.e. *inverting* the input signal. Conversely, a +ve voltage on the (+) input (with respect to the (−) input) causes the output to go +ve, i.e. *non-inverting*, or producing the same polarity as, the input.

Sometimes the inverting input is known as the 'virtual earth', 'virtual ground' or 'summing point', and the non-inverting input as the 'reference input'.

The 'A' notation inside the opamp symbol is optional and represents the 'open-loop voltage gain' or 'open-loop transfer function'. This voltage gain is the output voltage (E_o) divided by the difference ($E_+ - E_-$) between the non-inverting and inverting input voltages, measured under open-loop conditions, i.e. with no feedback from output to input. Open-loop voltage gain is sometimes more precisely symbolised as 'A_{VOL}' (= voltage gain A_V under OL = open-loop conditions).

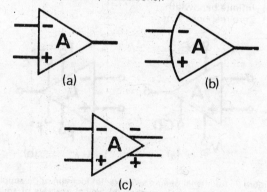

Figure 1. Opamp circuit symbols: (a) modern preferred, (b) older, obsolescent, (c) balanced ('differential') output.

The alternative segment-shaped symbol of Figure 1(b) may sometimes be found used for the opamp, particularly in older literature, but it has now been virtually completely superseded by the triangle of Figure 1(a)

Although single-ended output is the norm for an opamp, a special type, the 'differential-output' one, will sometimes be met with, and for this the two separate balanced output terminals are indicated as in the symbol of Figure 1(c).

In general, power supply and/or ground (earth) connections are not included in the opamp symbol. However, if they are required for any reason, it is usual to site the +ve dc supply terminal indication on the upper side and the −ve dc supply and ground terminals on the lower side as in Figure 2(a).

Apart from the input, output, power supply and ground terminals, practical opamps frequently have other special terminals for such things as dc biasing or frequency-compensation component attachment. Such extra terminals are normally shown spaced out at convenient drawing points round the symbol circumference as in the illustrative example of Figure 2(b)

Real v Ideal Opamps

The main characteristics of an ideal opamp are:
(a) Infinite open-loop voltage gain,
(b) Infinite input resistance,
(c) Zero output resistance,
(d) Zero offset voltage,
(e) Infinite bandwidth,
(f) Zero response time,

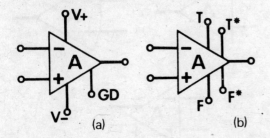

(a)   (b)

Figure 2. Additional opamp symbols: (a) location of dc supply and ground terminals in symbol, (b) typical example of additional special-purpose terminals with F, F* = frequency compensation and T, T* = dc biasing offset voltage trimming or balancing terminals.

(g) Zero variation of characteristics with common-mode input voltage change,
(h) Zero variation of characteristics with power supply voltage change,
(i) Zero variation of characteristics with temperature.

Real-life opamps obviously cannot produce these infinity and zero characteristics. However, nowadays types are readily commercially available in which the departure from the ideal limits are of little significance in most applications.

Open-loop Voltage Gain (A_{VOL})

Most opamps on the market provide a guaranteed minimum open-loop voltage gain of at least 80dB

($\times 10,000$). High-gain types with A_{VOL} greater than 100dB ($\times 100,000$) are readily and cheaply available. Extra-high-gain opamps with A_{VOL} greater than 120dB ($\times 1,000,000$) are not uncommon, though comparatively rather costly.

For most practical ordinary circuit purposes, opamp open-loop gains of more than 80dB can be regarded as infinite, so that the exact value of this characteristic is usually relatively unimportant.

Instead of open-loop voltage gain, some opamp data sheets specify 'large-signal voltage gain'. This is the open-loop voltage gain measured while driving the opamp so that its output voltage swing is relatively large but not clipped. Large-signal voltage gain can be taken as equal to open-loop voltage gain for all practical purposes.

Input Resistance (R_{IN})

The input resistance, R_{IN}, of an opamp is normally taken to be the *differential* input resistance, i.e. the resistance seen looking in across the two input terminals under open-loop condition.

Some opamp data sheets also specify the *common-mode* input resistance, i.e. the resistance with respect to ground or a common point looking into both inputs tied together, but this is not normally a very important characteristic. Usually it can be taken to be 100 to 1,000 times greater than the differential input resistance.

The differential input resistance of an opamp depends mainly on the fabrication of its input stages. Commercial opamps nowadays tend to use symmetrical balanced differential input stages in which the input resistance is largely decided by the amplifying device type used therein.

The ranges of input resistance characteristic of the different commercial opamps are:

Conventional (bipolar) transistor input 300K–3M min.
Darlington-connected transistor input 1–10M min.
'Superbeta' transistor input   10–30M min.
Fet-input   100–1,000G min.

The input resistance of an opamp when actually used with negative feedback (as is normal in most applications) is much greater than the open-loop input resistance. The feedback causes the input resistance to be multiplied by the 'loop-gain', i.e. the ratio of the open-loop gain to the feedback factor.

Output Resistance (R_{OUT})

The output resistance of an opamp is usually measured under open-loop conditions and is generally not regarded as sufficiently significant to be included in manufacturers' data. Most commonly, commercial opamps have typical output resistance around 150–200 ohms, although values down to around 10 ohms and up to around 1–5 kilohms may be met with.

Output resistance is not normally regarded as an important characteristic, because, with the usual level of feedback employed, the opamp output resistance is divided by the loop-gain . . . which is not usually less than 30dB ($\times 30$), and may be as high as 60dB ($\times 1,000$).

Input Offset Voltage (V_{IO})

Ideally, when both input terminals of an open-loop opamp are grounded (i.e. set at zero potential with $E_- = E_+ = 0$), the output too is zero (i.e. $E_o = 0$). In a real opamp, when the inputs are shorted to ground, the output voltage will not be zero, however, and a small dc 'offset voltage', V_{IO}, must be applied to one of the inputs to bring the output to zero.

In commercial opamps, the offset voltage is usually specified as a maximum at room temperature, and can range from as low as 10µV max. up to as high as 100mV max. For most general-purpose opamps, military types tend to have maximum offset of around 1mV, industrial types around 3mV and consumer types around 10mV, with the exception that fet-input opamps tend to have maximum offset voltages of about five times this.

In practical opamp circuits, the zeroing (trimming, or nulling) of the output voltage under no input signal can always be effected by a suitable dc bias applied to the input terminals. Some opamps, however, also have special separate extra terminals giving access to internal bias points which can be used to balance out the input offset voltage without having to make balance connections to the sensitive input terminals. Where no separate trimming (balancing, nulling) terminals are provided the opamp may be termed a *pre-input-nulling* or *external-nulling* type. Where separate nulling terminals are provided, it may be termed a *post-input-nulling* or *internal-nulling* type.

In use, it is often unnecessary to trim, i.e. balance out, the opamp input offset voltage, if it is sufficiently low in the first instance. This is because the resulting output offset, being equal to the input offset voltage multiplied only by the closed-loop gain, can be sufficiently small to be ignored. As an example, in a 20dB ($\times 10$) amplifier connection of an opamp with 1mV max. input offset voltage, the output offset will not be more than 10mV when the input terminals are connected together.

Input Bias Current (I_B)

In real opamps, bias currents flow in both the inputs. The *input bias current* (I_B) is defined as the average (i.e. half the sum) of the currents flowing into the two input terminals, when the output is at zero voltage, measured with no load on the output and at room temperature (nominally 25C ambient).

In commercial opamps, I_B may range from 1pA max. (with some specialist fet-input types) up to 100µA max. Maximum input bias current levels vary considerably with the device type used in the opamp balanced input circuits as follows:

	Input bias current levels		
	Military	*Industrial*	*Consumer*
Fet-input	50pA max.	100pA max.	200pA max
Superbeta-transistor-input	2nA max.	5nA max.	10nA max.
Ordinary-transistor-input	100nA max.	200nA max.	500nA max.

Input Offset Current (I_{IO})

As the opamp is a balanced amplifier, any unbalance on the bias currents into the two inputs can be important. Because of this, most data sheets specify a max. value for the *input offset current* (I_{IO}), i.e. the difference between the input currents when the output is at zero.

In commercial opamps, I_{IO} tends to come out about 1/5th of the corresponding I_B, with max. values typically varying with input device type as follows:

	Input offset current levels		
	Military	*Industrial*	*Consumer*
Fet-input	10pA max.	20pA max.	50pA max.
Superbeta-transistor-input	0.2nA max.	0.5nA max.	1nA max.
Ordinary-transistor-input	10nA max.	25nA max.	50nA max.

Opamp Ac Characteristics (GBP)

As noted earlier, the ideal opamp is regarded as having infinite bandwidth. This means that its small-signal open-loop voltage gain, A_{VOL}, remains constant as the signal frequency is varied from dc (zero frequency) up through to the highest frequencies. In real opamps, as the frequency increases, A_{VOL} tends to drop off until at some higher frequency it drops to unity (0dB). The manner in which A_{VOL} changes with frequency determines whether an opamp will be stable in the widest range of feedback configurations.

Manufacturers may be found specifying one or other of three different frequency response characteristics for their opamps:

(a) BW = 'bandwidth', which is the frequency at which A_{VOL} drops to 0.707 (3dB down) of its dc level for a constant amplitude input.
(b) GBP = 'gain-bandwidth product', which is the product of gain and frequency at any point on the A_{VOL} v. frequency curve of the opamp (assumed usually to be constant).
(c) f_T = 'unity-gain frequency', which is the frequency at which A_{VOL} drops to 0dB (×1) and which can often be taken for practical purposes as equal to GBP.

The open-loop bandwidth, BW, cannot be readily used to establish the 3dB bandwidth of a feedback-controlled amplifier using the opamp. Most manufacturers, therefore, do not usually specify BW. As a rough approximation, it can be taken that for any individual opamp, BW = GBP/A_{VOL}. For one well known opamp, the '741', for example, a typical A_{VOL} of 106dB (×200,000) combined with a typical GBP of 1MHz, gives a 3dB open-loop bandwidth of only 1,000,000/200,000 = 5Hz.

In currently-available commercial opamps, the range of the gain-bandwidth product minimum (GBP$_{min}$) is wide, varying from as low as 0.075MHz up to around 75MHz. In standard general-purpose opamps, GBP$_{min}$ varies typically from around 0.3MHz up to 3MHz. Taking the typical GBP, GBP$_{typ}$, as some three times GBP$_{min}$, gives a range of GBP$_{typ}$ of 1–10MHz.

Earlier, we had a look at the input and output resistances of an opamp. For ac use, manufacturers sometimes specify impedances rather than resistances, but these are usually stated ohmically as the modulus of a complex impedance. For most practical purposes, the impedance values given can be taken as resistances.

Frequency-compensation of Opamps

At higher frequencies, in a real opamp, not only does the voltage gain decrease with rising frequency, but the output signal tends to lag by more than the 180° phase-reversal from the inverting input implied in an ideal opamp. This lag, called phase-shift, can be a source of instability in practical circuits. It is customary, therefore, to add to the basic real opamp some form of frequency-compensation network.

The frequency-compensation network may be included within the opamp package to give 'internally-compensated', 'frequency-compensated', or, simply, 'compensated' types. (In the data tables later such types are coded 'INT' = INTernally-compensated.)

By contrast, in some practical opamps, no internal frequency-compensation is provided, but separate terminals are included in the package for attaching external compensation networks to, giving 'uncompensated' or 'tailored-response' types. (In the tables these are coded 'EXT' = EXTernal-compensation required.)

Frequency-compensation can be carried out on the input stages of an opamp, and is then known as 'input' compensation. (In the tables, input-compensation terminals are indicated by an 'F' = Frequency-compensation.) Compensation can also be applied to later amplifier stages, and this is known as 'output' compensation. (Terminals for output compensation are indicated in the data tables by the sign 'Φ', i.e. the Greek letter 'phi' corresponding to the English letter, F.)

Frequency-compensation external networks vary with different uncompensated opamps, and may range from a single capacitor up to a network of as many as seven resistors and capacitors.

Opamp Switching Characteristics (SR)

Ideally an opamp should have a zero response time. This means that the output voltage should respond instantly to any change in the input.

In a real opamp, the speed of response is limited by the rate at which it can charge and discharge circuit capacitances. In practice, these capacitances can be internal to the opamp (either parasitics or internal frequency compensation additions) or external (either external compensation additions or load capacitances).

Response times for opamps are usually specified in terms of slew rate (sometimes called slewing rate). Slew rate is, in simple terms, the rate of change of the opamp output voltage. Usually denoted by 'SR', it is normally expressed in units of volts per microsecond, V/μS.

Slew rate varies with compensating capacitance values, and manufacturers normally specify it for the 'voltage follower' configuration, i.e. as a unity-gain, non-inverting amplifier, because this calls for the largest compensation capacitance.

Where high speed switching is important, the *minimum* limit of the slew rate, SR_{min}, is normally given by the manufacturer. Where only the typical value is given in the data sheet, it can be taken that the minimum value lies around 1/3rd to 1/6th of the typical.

In commercial opamps, SR_{min} may lie between 0·02V/μS and 250V/μS. In general-purpose units, it tends to lie between 0·3 and 3·0V/μS. For switching applications, manufacturers also produce specialised types which can be classified as 'high slew rate' (with SR_{min} between 3 and 30V/μS and coded HSR = *H*igh *S*lew *R*ate in the tables) or 'extra-high slew rate' (with SR_{min} over 30V/μS and coded XSR = e*X*tra-high *S*lew *R*ate in the tables).

In some applications, an important characteristic of an opamp is its peak-to-peak output capability with higher frequencies. This is defined by its 'power bandwidth' or 'full-power response', f_p. Where an opamp is designed mainly for amplifying an audio frequency sinewave, the power bandwidth is stated in the data sheet, but usually it is not specified and has to be computed from SR_{min} through the relation $f_p = SR/(2\pi E_{op})$, where E_{op} is the peak amplifier output voltage.

Another switching characteristic of the opamp is its settling time, i.e. the time required for the output to settle within a given percentage of final value in response to an input voltage step. Data sheets do not usually specify limits to the settling time, and engineers have tended to use slew rate and unity-gain bandwidth as approximate indicators of relative settling time performance when comparing or choosing amplifiers. The difficulty is that settling time is really a closed-loop parameter (all other opamp specifications being open-loop) and therefore depends on the closed-loop configuration and gain.

Opamp Sensitivity to Common-mode Voltages (CMRR)

Ideal opamps respond only to differential input signals and ignore 'common-mode' input signals, i.e. signals common to both inputs. Real opamps exhibit real, if small, outputs ('errors') arising from their non-zero response to common-mode signals.

The common-mode rejection ratio, CMRR, may be defined in several different, but essentially equivalent, ways, by various manufacturers.

It can be defined as the ratio of the change in the input common-mode voltage, dV_{ICM}, to the resulting input offset voltage change, dV_{IO} (i.e. the input balancing voltage change required to return the output to zero). It

can also be shown to be equal to the ratio of the closed-loop gain to the common-mode gain.

CMRR is most commonly specified in decibel units (dB), where $CMRR(dB) = 20 \log_{10}(dV_{ICM}/dV_{IO})$.

CMRR is important to non-inverting or differential amplifier configurations, because these see a common-mode voltage.

CMRR is usually specified at dc or very low frequency. Generally, the data specification includes a minimum limit to this characteristic. Few commercial opamps have a $CMRR_{min}$ less than 60dB. Most good quality opamps have a minimum of 80dB, and special instrumentation amplifiers (coded PIA = *P*recision *I*nstrumentation *A*mplifier in the data tables) specify $CMRR_{min}$ of 100dB and even 120dB.

In general, $CMRR_{min}$ is higher in transistor-input opamps than in fet-input types, since fets have inherently poor common-mode rejection.

Commercial opamps frequently specify an 'operating common-mode range' or 'common-mode input voltage range'. This is the maximum range of common-mode voltage swing that the input stage can tolerate and still operate within the specification limits of the data sheet. This is effectively the input voltage that can be applied to either input separately because in normal opamp circuitry applications, the two input terminals are held at virtually the same potential.

Opamp Sensitivity to Power-supply Voltage Levels (PSRR)

The characteristics of an ideal opamp are independent of the supply voltage levels. In a real opamp, the electrical characteristics vary with the supply voltage levels, and, because the opamp is capable of amplifying dc voltages, it is inherently sensitive to changes in its own dc supply voltages, V_S+ and V_S-. Also, if the supply voltages are poorly filtered and vary at some ripple frequency, the opamp characteristics may vary at the same frequency. It is essential therefore that a real opamp is, as far as practicable, insensitive to supply voltage changes.

The insensitivity of a commercial opamp to variations in its supply voltages is usually specified in one of a number of essentially equivalent ways. The commonest specification is of the 'power supply rejection ratio', PSRR, which is usually expressed in decibels or volts per microvolt. In decibel form, $PSRR = 20 \log_{10}(dV_S/dV_{IO})$, where dV_S = change in the power supply voltage and dV_{IO} = resulting change in input offset voltage. In volts per microvolt, $PSRR = dV_S$ (volts)/dV_{IO} (microvolts). Opamp data sheets generally specify a minimum for this parameter, $PSRR_{min}$.

Instead of PSRR, data sheets sometimes specify 'power supply sensitivity' or 'supply voltage sensitivity', expressed as microvolts per volt $= dV_{IO}$ (microvolts)/dV_S (volts), which can be seen to be the numerical inverse of the logarithmic PSRR. Sometimes you will find this parameter specified as microvolts per percentage change in supply voltage, $\mu V/\%$. However specified, supply voltage sensitivity normally has a maximum laid down in the data sheet.

In commercial opamps, $PSRR_{min}$ will be found to vary typically from some 60dB min. for consumer types, through 80–100dB min. for industrial types, up to 100–120dB min. for specialist instrumentation types. Corresponding maximum limits for supply voltage sensitivity are 1,000$\mu V/V$ ($= 60$dB PSRR), 100$\mu V/V$ ($= 80$dB), 10$\mu V/V$ ($= 100$dB) and 1$\mu V/V$ ($= 120$dB).

Temperature Drift of Opamp Characteristics (dV_{IO}/dT)

In the ideal opamp, the characteristics do not change with the temperature. In real opamps, all characteristics change or 'drift' from their initial values with temperature, and temperature drift is the main source of error in most precision applications.

For most commercial opamp parameters, the temperature drift is not important. Data sheets do not usually specify such drift, except for three special characteristics . . . input offset voltage, input bias current and input offset current. Of these, by far the most important (and always specified) is the offset voltage drift.

The input offset voltage drift (temperature coefficient) is the ratio of change in input offset voltage to the change of circuit temperature for a constant output voltage, and can be denoted by dV_{IO}/dT. It is usually specified in microvolts per degree centigrade ($\mu V/C$) of offset-voltage change, and is computed as an average value over the temperature range of the opamp. Data sheets normally specify a maximum value for the offset voltage temperature drift.

In commercial opamps, the maximum value of dV_{IO}/dT, i.e. dV_{IO}/dT_{max}, can range from as low as 0.1$\mu V/C$ up to 150$\mu V/C$. For general-purpose, run-of-the-mill opamps, maximum voltage drifts range from about 10 to 50$\mu V/C$. It is only in special-purpose instrumentation amplifiers (coded LVD = *Low Voltage*

Drift in the tables) that maximum drifts of the order of 1$\mu V/C$ will be found.

Input bias current, I_B, is another opamp characteristic that drifts with temperature, and its drift, dI_B/dT, is a measure of how stable the input bias currents remain over the operating temperature range. It is often regarded as a secondary characteristic and omitted from data sheets, but some indication of its magnitude can be deduced from a knowledge of the input-stage devices used. For example, in bipolar-transistor-input opamps, the input bias current is the base current of a high-gain transistor operated at a low collector current (usually at most tens of microamps), and falls with temperature rise. A typical I_B drift specification is around 1nA/C for such transistor-input opamps. In fet-input opamps, by contrast, the input bias current, although low (being essentially the reverse-biased silicon diode leakage current) approximately doubles with every 10C temperature rise.

The other temperature drift characteristic sometimes to be found in opamp data sheets is the input offset current temperature coefficient, dI_{IO}/dT. This is an average over the specified operational range of the opamp, and is a measure of the change in offset current over the range. Some commercial opamps are carefully designed for very low offset current drift, and these may be found coded LCD (= *Low Current Drift*) in the tables.

Noise in Opamps

An ideal opamp is 'noiseless', i.e. adds no noise of its own to the signal it amplifies. A real opamp, containing active and passive components that generate and add noise to its output, is 'noisy'. Noise in this context is taken to describe random ac voltages and currents generated in the amplifier which limit its signal sensitivity. Where very low voltage signals are to be amplified, very high closed-loop gain must be used to bring the input signals up to usable levels. With a very high gain, however, the noise is amplified along with the signals to the point where nearly as much noise as signal appears at the output. There are three recognised main types of noise phenomena associated with solid state opamps. These are (a) Schottky or 'shot', (b) Johnson or 'thermal', and (c) 1/f or 'flicker' noise.

For most general-purpose applications, opamp noise is not a significant characteristic, and it is not usually found specified in the manufacturers' data. For some applications (e.g. high-gain instrumentation amplifiers), noise levels in the opamp are significant, and may be found specified. The data sheet will then be found to specify one or more of three different noise parameters: (a) Input-referred voltage noise (nV/Hz$^{\frac{1}{2}}$); (b) Input-referred current noise (pA/Hz$^{\frac{1}{2}}$); and (c) Popcorn noise transition amplitude (µV).

Where specified, noise is often expressed as a typical value only, but some specialist amplifiers have maximum limits placed on noise in their data sheets. Opamps specially intended for low noise applications will be found coded LNA (=Low Noise Amplifier) in the data tables.

Opamp Power Supplies (V_S+, V_S-)

In many applications the opamp's output voltage must be capable of swinging in both positive and negative directions. For such, the opamp requires two source voltages: one positive (V_S+) and the other negative (V_S-) with respect to ground or common point. Often in practice the positive and negative rail voltages are made symmetrical, i.e. equal. In general-purpose commercial opamps, +15V, −15V rails are most commonly used.

Although designed for nominal ±15V supplies as standard, most commercial opamps will operate satisfactorily over a wide range of supply voltages, some from as low as ±1V, and some up to ±22V. High voltage opamps (denoted HVO = High Voltage Output in the tables) work with rails of not less than ±40V, and some work up to around ±150V supplies.

In the tables the rated maximum permissible supply voltages are specified (V_S+_{max} and V_S-_{max}) for each opamp.

Some opamp dc rails are not symmetrical. As an example, the old well-known '702' has dc supply maximum voltage ratings of +14V, −7V only.

Some more recent commercial opamps are designed for 'single-supply' operation, as for example on an automobile 12V battery. In these, of course, the output cannot swing about zero volts, but they are designed to swing on both sides of the half-rail voltage. For such types, the tables show only a positive rated supply voltage, V_S+_{max}.

Quiescent Current and Power Characteristics (I_Q, P_Q)

A circuit designer normally needs to know the maximum bias current that the opamp will consume under no-signal, 'quiescent' conditions. In data sheets this may be found stated either directly as a maximum to the supply current drain (I_{Qmax}) or indirectly as a maximum to the dc power consumption (P_{Qmax}).

For typical commercial general-purpose opamps, I_{Qmax} will be found to range from as low as 1mA up to as high as 30mA, although it usually lies in the range of 3–10mA.

The maximum dc power consumption, P_{Qmax}, is the power consumed for biasing purposes. It is not usable output power, because none of it is delivered to the load. (It should not be confused with the maximum permissible package power dissipation, P_{TOTmax}, dealt with later below.) In commercial general-purpose opamps, P_{Qmax} is normally specified for rail voltages of ±15V.

If only one of the two characteristics, I_{Qmax} and P_{Qmax} is given, the other can be derived through the equation (assuming 15V supplies): $P_{Qmax} = 30 \times I_{Qmax}$.

With some opamps, quiescent current and power consumption can be controlled externally over a wide range with a master bias terminal (denoted B = Bias in the tables). This terminal may be used to adjust the level of bias current, which in turn, of course, varies the quiescent dc power consumption of the opamp. Opamps of this type are usually referred to as programmable (coded PRA = PRogrammable Amplifier). In some programmable types, the external terminal bias control can even be used for switching the opamp completely off or on, i.e. strobing.

Rated Output Voltage and Current Swings (V_{OUTmin}, I_{OUTmin})

An opamp data sheet usually specifies a minimum output voltage, V_{OUTmin}. This is the lowest value of the peak output voltage swing, referred to zero (or to half-supply-voltage level for single-supply types) that is guaranteed can be obtained without clipping for that type. V_{OUTmin} is a measure of the ability to deliver its rated output voltage across a specified value of load resistance. A symmetrical output swing is usually implied, although not necessarily true in all cases. The output can become limited due to loading effects, power supply levels, frequency effects and output resistance of the opamp. Usually opamps can supply peak voltage swings to within a few volts of the supply voltages used. In general-purpose opamps, V_{OUTmin} tends to range from about 10V to 15V.

Circuit designers may also be interested in the opamp's rated current output, I_{OUTmin}. This is the minimum peak swing of output current, referred to zero, that the data sheet guarantees the opamp will provide.

For general-purpose opamps, I_{OUTmin} will be found usually to lie in the range of 1mA to 5mA. Special opamps are also available (denoted in the tables by HCO=High Current Output) to provide guaranteed minimum output current swings of up to 50mA peak. (I_{OUTmin} should not be confused with the rated maximum permissible output current, discussed in the next section below.)

Opamp Maximum Ratings

Opamp data sheets normally specify a number of maximum ratings, i.e. values which must not be exceeded by the user.

For dc supply rails, discussed earlier above, the opamp data sheet normally specifies V_S+_{max} and V_S-_{max}, the maximum allowable supply voltages that can safely be applied to the amplifier. Sometimes these are stated merely as a single overall total positive-to-negative permissible maximum supply voltage. If the rated supply voltages are exceeded, there is a considerable danger of catastrophic breakdown in the semiconductor devices incorporated in the opamp, and, even if this does not occur, there is the possibility of internal overheating due to excessive power dissipation that can lead to deterioration in performance.

Opamp data sheets normally specify a maximum permissible internal power dissipation, P_{TOTmax}, which is the power that a particular device is capable of dissipating safely on a continuous basis while operating within a specified temperature range. This rating varies according to the type of package used for the opamp. In general, ceramic packages allow the highest power dissipation, metal packages the next highest, and plastic-encapsulated packages the lowest. For run-of-the-mill general-purpose opamps, P_{TOTmax} will be found to lie in the range of 300–900mW for free-air mounting of the package (denoted in mWF=milliwatts free-air in the tables).

Manufacturers normally specify for their opamps an operating ambient temperature range over which they will perform within their specifications. Military-grade devices typically operate from −55C to +125C, industrial-grade from −25C to +85C, and commercial (consumer) grades from 0C to +75C. For most users, only the maximum permissible operating temperature is generally of interest, and in the tables this will be found shown as T_{opmax}. Standard jargon is to identify 125C devices as military, 85C as industrial (including automotive), and 75C as 'entertainment' or 'consumer'.

A secondary temperature rating to be found in opamp data sheets is the maximum permissible lead temperature, relating to precautions against damaging by overheating of leads in soldering packages into circuit. Conventional specification is to limit hand-soldering time to less than 10 seconds for an iron tip temperature less than 245C and to less than 5 seconds for a tip temperature between 245C and 400C. For dip or flow soldering, it is conventional to limit the solder temperature to 245C and the time of immersion to not more than 5 seconds.

Earlier we took a look at the rated output current, I_{OUTmin}, i.e. the minimum output current that the data sheet guarantees the opamp can produce. This should not be confused with the rated maximum permissible output current, I_{OUTmax}, which will be found in some data sheets. Typically this tends to be about 5 to 10 times I_{OUTmin}, and, not being usually required by the circuit designer, is not included in the tabular data of this Selector.

Most data sheets include a rated maximum permissible value for the common-mode input voltage, V_{ICMmax}. This is the maximum voltage that can safely be applied between both input terminals together and circuit common. There are usually positive and negative limits to this rating, although often these will be found to be equal. In the tables, therefore, only one voltage rating will be found for V_{ICMmax}, without a sign indication. As a rule of thumb, it can be taken that with modern opamps V_{ICMmax} can be taken as approximately equal to V_S+_{max}, but care must be taken with some of the older types, because V_{ICMmax} can be much lower than the rail voltage rating.

V_{ICMmax} ratings should not be confused with the corresponding input voltage 'ranges', sometimes quoted in data sheets, which represent the limits of common-mode input voltage within which the opamp characteristics will remain inside the data sheet specification limits, i.e. the opamps will operate linearly.

Data sheets normally also include a rated maximum permissible value for the differential input voltage, V_{IDmax}. This figure, representing the maximum voltage that can safely be applied between the differential input terminals without excessive input current flow. For many general-purpose opamps, it will be found that V_{IDmax} equals twice the supply voltage rating, V_{Smax}. However, this is not always true, because some opamps have protective clamping diodes connected back-to-back across the input terminals. Such types are limited to a maximum differential-input voltage of less than 1V. Without clamping diodes, some opamp types are limited to about 5V or less in their differential input voltage to prevent reverse emitter-base breakdown in an input stage transistor. The highest V_{IDmax} ratings will, however, generally be found in devices using a combination pnp-npn cascode input.

Opamp Output Protection

Earlier opamps often had a specification of the maximum permissible time for holding the output short circuited, giving a rating t_{SCmax}. For example, the archetypal μA709 had only a 5-second output short-circuit rating. More recent opamps generally have built-in current-limiting short-circuit protection, which allows the output to be short-circuited indefinitely without damage to the device.

Quasi-opamps

A large number of commercial devices have opamp-like characteristics without being precisely 'classical' opamps, and such quasi-opamps have been included in the tabular data.

Many manufacturers provide opamps with two balanced differential outputs instead of the single output of the conventional opamp. These quasi-opamps will be found denoted BDO (=Balanced Differential Output). Probably the best known balanced output opamp is the μA733.

Some opamps are packaged by the manufacturer with no external inverting input terminal, but with the inverting input internally connected to the output terminal to give a voltage-follower amplifier (denoted VFA=Voltage Follower Amplifier in the tables). Such 'committed' opamps are also known as buffers.

The opamp's ability to provide both positive and negative output voltage swings is not always necessary or desirable. In such cases, the opamp's need for both positive and negative dc rail voltages can be a costly inconvenience. Standard double-supply opamps can be wired to work with a single dc supply, with the output swinging about approximately half the single rail voltage, but this technique requires additional relatively costly components. Manufacturers have developed a cheaper alternative. This is the current-differencing-amplifier (denoted CDA in the tables) or Norton amplifier with the circuit symbol shown in Figure 3. The CDA is designed to operate from low-voltage, single-ended dc power supplies. CDAs are not one-to-one direct replacements

of opamps, for they require special external wiring to set them up, and are not intended for dc voltage amplification. Typical commercial CDA types are the MC3401 and LM3900.

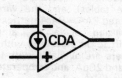

Figure 3. Circuit symbol for current-differencing amplifier.

Some commercial operational amplifiers are specifically designed to work efficiently as dc comparators, a comparator being one that provides an output indication of the relative state of two input terminals. These specialised opamps will be found denoted CPR (=ComPaRator) in the tables.

Another type of quasi-opamp readily commercially available is the 'Operational Transconductance Amplifier' (usually abbreviated OTA). This is like a standard opamp in many respects, and many of the characteristics of the opamp also apply to the OTA. The major difference from the standard opamp is that the OTA gives a current, rather than a voltage, output. Ideally an OTA has infinite output resistance, in contrast to the zero output resistance of the ideal opamp. Instead of an open-loop voltage gain, the OTA data sheet gives a transconductance, g_m, which provides, for an input differential voltage E_{in}, an output single-ended current $g_m E_{in}$. The archetypal OTA is the CA3080.

Multiple-opamp Packages

General-purpose opamps, particularly internally compensated types, require relatively few external terminals (for V_s+, V_s-, E_-, E_+, and output, with possibly a few nulling and frequency-compensation connections). Because of this, it has been possible to fabricate multi-element opamps with 2, 3 and even 4 identical isolated circuits in a single industry-standard package. In the tables, multi-element opamps will be found characterised by an applications coding starting with D (for Dual elements), T (for Triple) and Q (for Quadruple).

Low-drift Opamps

Commercial opamps have attractive features like low cost and small size. However, where extremely low drift and bias currents are required for some stringent applications, specialised-technique opamps are available. The two main specialised opamps of this type are the *chopper-stabilized* (denoted CHP = *CH*o*P*per-stabilized in the tables), and the *varactor-bridge* or *parametric* (denoted PAA = *PArA*metric).

Chopper-stabilizing of a dc amplifier enables microvolts of drift to be measured with high accuracy. Chopper amplifiers are available with initial offsets of less than 20μV and 20pA, and temperature drifts of less than 0·5μV/C and 1μV/week.

Parametric (varactor-bridge) amplifiers feature very low input bias and offset current and drift, low noise at very low frequencies, high input impedance and high common-mode rejection and isolation characteristics. They go a step beyond the performance of fet amplifiers and offer a solid-state alternative to the electrometer vacuum tube opamp. They are particularly suited to four basic jobs: measuring very low level currents; measuring voltages from very high impedance source; accommodating very wide input signal variations (using logarithmic or other nonlinear feedback); and charging/discharging capacitors in integrator, differentiator, charge amplifier and sample/hold circuits.

Some comparison of performance between the different opamp types can be seen in the typical characteristics values set out below:

In the DIL package it will be seen that the leads emerge on the sides and turn downwards. In one variant, the leads still lie in two parallel rows in line, but emerge as pins from the bottom of the package. This is known as the 'modified dual-in-line' type and is coded DIM (= *D*ual *I*n-line, *M*odified) in the tables.

Another variant of the DIL is one with a heat-sink incorporated as an integral part of the package. This will be found denoted HIL (= *H*eatsinked *I*n *L*ine package) in the tables.

DIL packages vary considerably in size, but miniature versions will be found denoted MDL (= *M*iniature *D*ual In *L*ine) in the tables. These should not be confused with the shortest of the standard DIL packages, the 8-lead 'minidip', denoted DIL-8 in the tables.

A final DIL variant met with is one in which the leads are staggered out from two into four parallel rows, to provide greater interlead spacings between the lead holes in the printed wiring board on which it is to be mounted. This variant is known as a 'quad-in-line' (= QIL) package.

The next commonest style of package used for opamps is the 'transistor-can' type, which is derived from conventional metal package outlines used for single transistors, and adapted for use with linear ICs. Nowadays the commonest opamp metal can packages of this type are:

	Typical values			
Parameter	Transistor-input	Fet-input	Chopper-stabilized	Parametric (Varactor)
Current drift (25–35C)	30pA/C	1pA/C	0·3pA/C	0·05pA/C
Voltage drift	0·75μV/C	2μV/C	0·5μV/C	60μV/C
Current noise (DC–1Hz)	5pA	0·1pA	10pA	0·01pA
Voltage noise (DC–1Hz)	1μV	3μV	5μV	2μV

Opamp Packages

Commercial opamps will be found supplied in a variety of different package configurations.

The commonest package is by far the 'dual-in-line' package, coded DIL (= *D*ual *I*n *L*ine) in the tables, and with a number of different possible lead configurations as detailed in Appendix F. Abbreviations other than DIL may be found used for dual-in-line packages, such as DIP (= *D*ual *I*n-line *P*ackage).

(a) *Multilead TO5* low-power package (coded TO5 in the tables and Appendix F, but also commonly appearing in the literature under EIA standard numbers such as TO99 (≐ 8-lead) and TO100 (= 10-lead);

(b) *Multilead TO8* medium-power package (coded TO8 in tables);

(c) *Multilead TO66* small 'diamond' power package (coded T66 in tables);

and

(d) *Multilead TO3* diamond' high-power package (coded TO3 in tables).

The type of package that was popular with early opamps, but is now much less often met with, is the *'flatpack'* (coded FLP in the tables), in which the leads emerge horizontally from two opposite edges of the package.

Low-cost packages developed recently and increasing in popularity are the 'single-in-line' (coded SIL), in which the leads emerge in a single straight line from only one edge of the package.

As with DILs, the SIL may be provided with an integral heat sink. Such a heat-sinked single-in-line package will be found coded SIH (= *S*ingle-*I*n-line, *H*eatsinked).

In recent years, manufacturers have been providing their opamps in 'naked chip' form, i.e. without a protective packaging. These generally come in one or other of three basic forms:

(a) *'Chip'* or *'Dice'* (denoted CHP = *CH*i*P* in the tables), available as separate individual single chips, or in 'wafer' form with many chips in a silicon slice.
(b) *'Flip Chip'* (denoted CFL = *C*hip, *FL*ip) with solder bump terminals on the chip face.
(c) *'Beam Lead'* (denoted BML = *B*ea*M* *L*ead) with terminal metal strips, or 'beams' deposited along the face of the chip to project beyond the edge.

Leadout Assignments

There is no standard assignment of leadouts in the various opamp packages met with. The only common factor is that the leads are usually taken to number sequentially upwards in an anticlockwise direction from an indicating tab, dot, mark, etc.,.when looked down from above at the package mounted in its printed wiring board or socket.

The lead numbering sequence of the different packages will be found set out in Appendix F to the tables. The terminal assignments, i.e. the actual connections to be made to the different lead numbers will be found set out in the body of the tables.

In general, once you have an idea of the shape and lead numbering sequence of the different package styles, you will find that you can identify the lead connections from the opamp entry line in the body of the tables without having to look up the outlines in Appendix F. The connections are coded in the table according to a listing shown in detail in Appendix G and repeated for convenience on each page of the tabular data.

Applications of the Opamp

As a general-purpose gain block, the opamp has a multitude of applications. It is impossible to bring into the tabular data the precise details of specific applications for which each opamp is designed. However, it has been possible to characterise each one in the tabular data by a mnemonic-type applications coding for the major applications areas detailed in Appendix E.

Opamp Type Numbers

Opamps are produced by semiconductor manufacturing companies in all the major industrialised countries of the world. As yet there is no global industry-standard type numbering system in operation, but fortunately many manufacturers operating multinationally retain their own numbering system in every outlet. As a result, it is often possible to tell from an opamp type number at least who is the manufacturer. To help in this, Appendix D sets out the initial letters of the numerical series used by different manufacturers.

Apart from such house-code' numbers for opamps, some national agencies have introduced standard numbering systems. Examples of these are the JEDEC system of the EIA in the USA, the 'Proelectron' system of the Association Internationale Pro Electron in Europe, and the JIS system of the Japanese Industrial Standards organisation. Regrettably the various 'standard' numbering systems do not indicate by their type number sufficiently precise details of the opamp to enable it to be used without reference to other data information.

Some particular house-code numbers have become so well known that it is common to drop the alphabetic prefix to the number and use only the numerical part. For example, the µA709 in the jargon of the electronic engineer is usually referred to simply as the '709' or the LM101A as the '101'.

Opamp Manufacturers

The electronic engineer often wants to identify the manufacturer of an opamp, of which he knows only the device type. To meet this situation, the tables include a 'manufacturer' column, in which a major manufacturer is indicated by a coding. At Appendix C will be found details of the different manufacturers, together with the codings assigned to them.

Where more than one manufacturer makes a particular device, the maker assigned to that opamp in the table is generally the one whose house-code forms part of the type number.

'Workhorse' Opamps

A number of opamp types have become so widely used that they have become virtually industry standards.

These 'workhorses' may appear under different type numbers from different manufacturers, but the numerical part of the type code is usually sufficient to identify it. For example, the original LM*101A*H (National Semiconductors) may also appear as μA*101A*H (Fairchild), AMLM*101A*H (Advanced Microdevices), CA*101A*T (RCA), RC*101A*T (Raytheon), SFC2*101A* (Thompson –CSF), SG*101A*T (Silicon General), SN52*101A*L (Texas Instruments), TOA*101A*V (Transitron), and even U5B*101A*312.(obsolete, Fairchild).

The more widely-used of the workhorse opamps are set out below, grouped into classes of application. Under each application, the opamps are distinguished first as 'comp' (= internally frequency-compensated) and 'uncomp' (uncompensated, i.e. with no internal frequency-compensation and requiring frequency compensation through special terminals provided in the package). A subsidiary distinction is then made between 'ext. trim' (= with special terminals available for input offset voltage trim adjustment) and 'no ext. trim' (= no special trim terminals).

(a) General-purpose amplifiers

Uncomp., ext. trim	101	(LM101/201/301/ 101A/201A/301A)
	748	(μA748/748C)
Uncomp., no ext. trim	709	(μA709/709A/ 709C)
Comp., ext. trim	741	(μA741/741A/ 741C)
	799	(μA799/799C)
Comp., no ext. trim	107	(LM107/207/307)
	101H	(LH101/201)

(b) Superbeta-input, low-input-current amplifiers

Uncomp., ext. trim	112	(LM112/212/312)
	216	(LM216/316/ 216A/316A)
Uncomp., no ext. trim	108	(LM108/208/308/ 108A/208A/308A)
Comp., ext. trim	1536	(MC1536/1436/ 1556/1456)

	143	(LM143/343)
	2645	(HA2645)

(c) High-voltage-output amplifiers

Uncomp., ext. trim	144	(LM144/344)
Comp., ext. trim	1556	(MC1556)

(d) Voltage-follower amplifiers

Comp., ext. trim	110	(LM110/210/310)
	102	(LM102/202/302)

(e) Extra-high-slew-rate amplifiers

Uncomp., ext. trim	118	(LM118/218/318)
	531	(NE531/SE531)
	2005	(HA2005/2055/ 2065)
	2505	(HA2505/2515/ 2525)
Uncomp., no ext. trim	715	(μA715/715C)
	2535	(HA2535)

(f) Precision instrument amplifiers

Uncomp., ext. trim	152F	(LF152/252/352)
	725	(μA725/725A/ 725C)
	777	(μA777C)
Comp., ext. trim	121	(LM121/221/321)

(g) Fet-input amplifiers

Uncomp., ext. trim	3130	(CA3130/3130A/ 3130B)
Comp., ext. trim	740	(μA740/740C)
	536	(NE536/SU536)
	13741	(LF13741)
	155F	(LF155/255/355/ 156/256/356/ 157/257/357/ 155A/355A/156A/ 356A/157A/357A)
	3140	(CA3140/3140A/ 3140B)
	8007	(ICL8007/8007C)

(h) Wideband amplifiers

Uncomp., no ext. trim	702	(μA702/702A/ 702C)

(i) Low-offset-voltage-drift amplifiers

Comp., ext. trim	714	(μA714)
	OP–07	(OP–07/07A/07C)

(j) *Low-noise amplifiers*
Uncomp., no ext. trim   *911*   (HA909/911)

(k) *Differential-output amplifiers*
Uncomp., no ext. trim   *727*   (μA727/727C)
                               730   (μA730/730C)
                               733   (μA733/733C)

(l) *Programmable amplifiers*
Uncomp., ext. trim.   *2778*   (SFC2778/2778C)
                               3078   (CA3078/3078A)
Comp., ext. trim   *776*   (μA776/776C)
                               4250   (LM4250/4250C)
                               8021   (ICL8021M/
                                          8021C)
                               2725   (HA2725)

(m) *Operational-transconductance amplifiers*
Uncomp., no ext. trim   *3094*   (CA3094/3094A/
                                          3094B)
Comp., no ext. trim   *3080*   (CA3080/3080A)

(n) *Dc-comparators, single-output*
No-strobe, ext. trim   *734*   (μA734/734C)
                               8001   (ICL8001M/
                                          8001C)
No-strobe, no ext. trim   *710*   (μA710/710C)
1-strobe, ext. trim   *111*   (LM111/211/311)
                               111F   (LF111/211/311)
2-strobe, no ext. trim   *106*   (LM106/206/306)

(o) *Dc-comparators, double-output*
No-strobe, no ext. trim   *760*   (μA760/760C)
                               160   (LM160/260/360)
2-strobe, no ext. trim   *161*   (LM161/261/361)

(p) *Dual general-purpose amplifiers*
Uncomp., ext. trim   *2101*   (LH2101A/2201A/
                                          2301A)
Uncomp., no ext. trim   *1537*   (MC1537/1437)
Comp., ext. trim   *747*   (μA747/747A/
                                          747C)
Comp., no ext. trim   *1558*   (MC1558/1458)
                               158   (LM158/258/358/
                                          158A/258A/358A)
                               798   (μA798/798C)
                               4558   (MC4558/4558C)
                               2904   (LM2904)

(q) *Dual superbeta-transistor-input amplifiers*
Uncomp., no ext. trim   *2108*   (LH2108/2208/
                                          2308/2108A/
                                          2208A/2308A)

(r) *Dual voltage-follower amplifiers*
Comp., ext. trim   *2110*   (LH2110/2210/
                                          2310)

(s) *Dual fet-input amplifiers*
Comp., ext. trim   *8043*   (ICL8043M/
                                          ICL8043C)

(t) *Dual programmable amplifiers*
Comp., ext. trim   *8022*   (ICL8022M/
                                          8022C)
                               2735   (HA2735)
                               24250   (LH24250/
                                          24250C)

(u) *Dual comparators*
1–o/p, 2-strobe, no ext. trim *711*   (μA711/711C)
2–o/p, no-strobe, no ext.   *193*   (LM193/293/393/
   trim   193A/293A/393A)
                               2903   (LM2903)
                               119   (LM119/219/319)
2–o/p, 2-strobe, ext. trim   *2111*   (LH2111/2211/
                                          2311)
2–o/p, 2-strobe, no ext. trim *1514*   (MC1514/1414)

(v) *Dual low-noise amplifiers*
Uncomp., no ext. trim   *739*   (μA739C)
                               749   (μA749/749C/
                                          749D)
                               381   (LM381/381A)
                               1303   (MC1303)
Comp. no ext. trim   *382*   (LM382)
                               387   (LM387/387A)

(w) *Quad (quadruple) current-difference amplifiers*
Comp., no ext. trim   *2900*   (LM1900/2900/
                                          3900)
                               3301   (MC3301/3401)

(x) *Quad general-purpose amplifiers*
Comp., no ext. trim   *124*   (LM124/224/324/
                                          124A/224A/324A)
                               148   (LM148/248/348)
                               149   (LM149/249/349)
                               2902   (LM2902)

	3303	(MC3303/3403/3503)
	4136	(RC4136/RM4136/RV4136)
	4741	(HA4741)

(y) *Quad dc-comparators*

4–o/p. no-strobe. no ext. trim	139	(LM139/239/339/139A/239A/339A)
	775	(μA775/775C)
	2901	(LM2901)
	3302	(MC3302)

(z) *Triple low-power, programmable amplifiers*

Comp., no ext. trim	8023	(ICL8023M/8023C)

Substitute Opamps

Opamp users often want to know of a suitable substitute when a direct replacement device is not available. Substitutes can be *exact* or *functional*.

An exact substitute is a device which is sufficiently near in package body size and shape and leadout configuration to be able to be directly wired or plugged into the circuit from which the defective opamp has been removed. At the same time its ratings should be at least equal to those of the device being substituted and its characteristics within the limits of that device. An exact substitute or replacement device may also be found referred to as a 'drop-in', 'plug-in' or 'pin-for-pin' replacement.

To help people looking for substitutes for opamp devices that have failed or become obsolete the Selector tables include two special columns providing for most opamps listed an exact mechanical and electrical substitute, one indicating an equivalent USA device and the other a European. Where no specific European substitute is readily available, the Euro-substitute column is used to offer a second independent USA alternative.

Functional substitutes or equivalents are those which differ materially in mechanical or electrical characteristics from the opamp being replaced, to such an extent that the functional substitute cannot be just dropped into circuit as a direct replacement. Scrutinizing the tables will be of assistance in identifying an opamp of near electrical performance to the item being replaced. The package and leadout assignment information of the proposed functional equivalent given in the table will enable the necessary wiring modifications to be carried out for the circuit to accept the functional substitute.

Opamp Glossary

To some engineers, particularly of the older school, the various special terms used in specifying opamps may not be entirely familiar. For this reason, a special glossary, bringing together in one place in alphabetical order a listing of opamp terminology and its explanation has been included at Appendix B for easy reference.

Opamp Circuit Design Guidelines

The degree of engineering skill called for in using opamps effectively is complicated because the performance capability of commercial off-the-shelf units spans an enormous dynamic range. Thus, while many problems can be solved by opamps with little or no circuit experience, others require considerable efforts from very experienced circuit designers.

Failure to recognise the difficulty of a design problem accounts for many of the disappointments and frustrations experienced by the opamp beginner. The data sheets, applications notes and other literature from manufacturers can lead an engineer to think that with an opamp and an hour or two at the bench he can do anything. However, while a sophisticated opamp is a very powerful circuit design tool, it very often takes much engineering skill and many days of hard design and experimental work to apply the device in some cases.

The table below sets some guidelines as to the relative difficulties of different opamp application areas:

Of course this table is largely a matter of personal opinion, but it is probably accurate at least to an order of magnitude. The commentary applies, however, to only one variable at a time. For example, it is virtually impossible to achieve a 10V/μS slew rate at an impedance level of 100meg., or to measure 1mV signals with 0 01% accuracy. This compounding of problems is why greater experience and skills are required at the extremes of performance.

Opamp Data Sources

Every reasonable care has been taken to ensure accuracy of information in the tables, but no responsibility can be accepted for inaccuracies that may have arisen.

Readers seeking more detailed information than it is

Circuit operation area	Safe for beginners	Some experience required	Considerable experience required	Strictly for practised expert
Accuracy	Worse than 1%	1.0–0.1%	0.1–0.01%	Better than 0.01%
Voltage-signals	Above 100mV	100–5mV	5–0.1mV	Below 0.1mV
Current-signals	Above 100nA	100–5nA	5–0.1nA	Below 0.1nA
Impedance-levels	Below 1meg	1–30meg	30–1,000meg	Above 1,000meg
Slew-rate	Below 1V/μS	1–10V/μS	10–100V/μS	Above 100V/μS
Frequency-response	Below 0.1MHz	0.1–1.0MHz	1–100MHz	Above 100MHz

possible to include within the limitations of this tabular data presentation can always consult the data information provided by the manufacturer indicated in the opamp table entry.

Individual data sheets can sometimes be obtained through the local office or agent of the manufacturer, but more commonly, data is available either in short-form catalogue published by the manufacturer and setting out only the main control specification of the devices in his range or in a comprehensive data book with detailed specifications for a large number of devices. For many manufacturers nowadays, the data books have become quite large volumes and for cost reasons have now generally to be paid for. Indeed good technical bookshops nowadays often carry the data books of at least the major manufacturers.

Tabulations

TYPE NUMBER	MFR	APP	CMP	GBP MIN	SLEW RATE MIN	V_S^+ MAX	V_S^- MAX	T_{op} MAX	A_{VOL} MIN	V_{IO} MAX	I_B MAX	I_{IO} MAX	P_{TOT} MAX	I_{OUT} MIN	V_{OUT} MIN	V_{ICM} MAX	V_{IDF} MAX	dV_{IO}/dT MAX	P_Q MAX	I_O MAX	CMRR MIN	PSRR MIN	R_{IN} MIN
20-007A1	BLU	LVD	INT	2MHZ	2V/uS	+22V	-22V	85C	100dB	0.2MV	150NA	.	.	5MA	10V	15V	30V	1uV/C	.	7MA	90dB	.	200K
20-007A2	BLU	LVD	INT	2MHZ	2V/uS	+22V	-22V	85C	100dB	0.2MV	150NA	.	.	5MA	10V	15V	30V	2uV/C	.	7MA	90dB	.	200K
20-007B1	BLU	LVD	INT	2MHZ	2V/uS	+22V	-22V	85C	100dB	0.5MV	150NA	.	.	5MA	10V	15V	30V	5uV/C	.	7MA	90dB	.	200K
20-007C1	BLU	LVD	INT	2MHZ	2V/uS	+22V	-22V	85C	100dB	1MV	200NA	.	.	5MA	10V	15V	30V	10uV/C	.	7MA	90dB	.	200K
20-008A1	BLU	FET	INT	2MHZ	3V/uS	+22V	-22V	85C	100dB	1MV	5pA	.	.	5MA	10V	15V	30V	5uV/C	.	7MA	80dB	.	10G
20-008A2	BLU	FET	INT	2MHZ	3V/uS	+22V	-22V	85C	100dB	1MV	5pA	.	.	5MA	10V	15V	30V	10uV/C	.	7MA	80dB	.	10G
20-008B1	BLU	FET	INT	2MHZ	3V/uS	+22V	-22V	85C	100dB	1MV	5pA	.	.	5MA	10V	15V	30V	25uV/C	.	7MA	80dB	.	10G
20-008B2	BLU	FET	INT	2MHZ	3V/uS	+22V	-22V	85C	100dB	1MV	5pA	.	.	5MA	10V	15V	30V	50uV/C	.	7MA	80dB	.	10G
20-008B3	BLU	FET	INT	2MHZ	3V/uS	+22V	-22V	85C	100dB	2MV	5pA	.	.	5MA	10V	15V	30V	75uV/C	.	MA	80dB	.	10G
20-107A1	BLU	LVD	INT	2MHZ	2V/uS	+22V	-22V	85C	100dB	0.2MV	150NA	.	.	5MA	10V	15V	30V	1uV/C	.		90dB	.	200K
20-107A2	BLU	LVD	INT	2MHZ	2V/uS	+22V	-22V	85C	100dB	0.2MV	150NA	.	.	5MA	10V	15V	30V	2uV/C	.	7MA	90dB	.	200K
20-107B1	BLU	LVD	INT	2MHZ	2V/uS	+22V	-22V	85C	100dB	0.5MV	150NA	.	.	5MA	10V	15V	30V	5uV/C	.	7MA	90dB	.	200K
20/107C1	BLU	LVD	INT	2MHZ	2V/uS	+22V	-22V	85C	100dB	1MV	200NA	.	.	5MA	10V	15V	30V	10uV/C	.	7MA	90dB	.	200K
20-108A1	BLU	FET	INT	2MHZ	3V/uS	+22V	-22V	85C	100dB	1MV	5pA	.	.	5MA	10V	15V	30V	5uV/C	.	7MA	80dB	.	10G
20-108A2	BLU	FET	INT	2MHZ	3V/uS	+22V	-22V	85C	100dB	1MV	5pA	.	.	5MA	10V	15V	30V	10uV/C	.	7MA	80dB	.	10G
20-108B1	BLU	FET	INT	2MHZ	3V/uS	+22V	-22V	85C	100dB	1MV	5pA	.	.	5MA	10V	15V	30V	25uV/C	.	7MA	80dB	.	10G
20-108B2	BLU	FET	INT	2MHZ	3V/uS	+22V	-22V	85C	100dB	1MV	5pA	.	.	5MA	10V	15V	30V	50uV/C	.	7MA	80dB	.	10G
20/108B3	BLU	FET	INT	2MHZ	3V/uS	+22V	-22V	85C	100dB	2MV	5pA	.	.	5MA	10V	15V	30V	75uV/C	.	7MA	80dB	.	10G
20-247A1	BLU	LVD	INT	2MHZ	2V/uS	+22V	-22V	125C	100dB	0.2MV	100NA	.	.	5MA	10V	15V	30V	1uV/C	.	7MA	90dB	.	200K
20-247A2	BLU	LVD	INT	2MHZ	2V/uS	+22V	-22V	125C	100dB	0.2MV	100NA	.	.	5MA	10V	15V	30V	2uV/C	.	7MA	90dB	.	200K
20-247B1	BLU	GPK	INT	2MHZ	2V/uS	+22V	-22V	125C	100dB	0.5MV	100NA	.	.	5MA	10V	15V	30V	5uV/C	.	7MA	90dB	.	200K
20-247C1	BLU	GPK	INT	2MHZ	2V/uS	+22V	-22V	125C	100dB	1MV	100NA	.	.	5MA	10V	15V	30V	10uV/C	.	7MA	90dB	.	200K
20-248A1	BLU	FET	INT	2MHZ	6V/uS	+22V	-22V	125C	100dB	1MV	5pA	.	.	5MA	10V	15V	30V	5uV/C	.	7MA	62dB	.	10G
20-248A2	BLU	FET	INT	2MHZ	6V/uS	+22V	-22V	125C	100dB	1MV	5pA	.	.	5MA	10V	15V	30V	10uV/C	.	7MA	62dB	.	10G
20-248B1	BLU	FET	INT	2MHZ	6V/uS	+22V	-22V	125C	100dB	1MV	5pA	.	.	5MA	10V	15V	30V	25uV/C	.	7MA	62dB	.	10G
20-248B2	BLU	FET	INT	2MHZ	6V/uS	+22V	-22V	125C	100dB	1MV	5pA	.	.	5MA	10V	15V	30V	50uV/C	.	7MA	62dB	.	10G
20-248B3	BLU	FET	INT	2MHZ	6V/uS	+22V	-22V	125C	100dB	2MV	5pA	.	.	5MA	10V	15V	30V	75uV/C	.	7MA	62dB	.	10G
40J	ANU	FET	INT	1MHZ	6V/uS	+18V	-18V	70C	94dB	2MV	50pA	20pA	.	5MA	10V	8V	15V	50uV/C	.	.	70dB	70dB	10G
40K	ANU	FET	INT	.	6V/uS	-18V	-18V	70C	94dB	10V	20pA	10pA	.	5MA	10V	8V	15V	20uV/C	.	.	70dB	70dB	10G
41J	ANU	FET	INT	.3MHZ	3V/uS	18V	-18V	70C	100dB	2MV	0.5pA	0.2pA	.	5MA	10V	10V	15V	25uV/C	.	.	84dB	84dB	1T
41K	ANU	FET	INT	.3MHZ	3V/uS	+18V	-18V	70C	100dB	2MV	.25pA	0.1pA	.	5MA	10V	10V	15V	10uV/C	.	.	84dB	84dB	1T
41L	ANU	FET	INT	.3MHZ	3V/uS	+18V	-18V	70C	100dB	2MV	.15pA	0.1pA	.	5MA	10V	10V	15V	25uV/C	.	.	84dB	84dB	1T
42J	ANU	FET	INT	.2MHZ	.25V/uS	+18V	-18V	70C	110dB	2MV	.35pA	.	.	5MA	10V	10V	15V	50uV/C	.	5MA	56dB	80dB	1T
42K	ANU	FET	INT	.2MHZ	.25V/uS	+18V	-18V	70C	110dB	2MV	0.1pA	.	.	5MA	10V	10V	15V	15uV/C	.	5MA	56dB	80dB	1T
42L	ANU	FET	INT	.2MHZ	.25V/uS	+18V	-18V	70C	110dB	2MV	75FA	.	.	5MA	10V	10V	15V	15uV/C	.	5MA	56dB	80dB	1T
43J	ANU	FET	INT	1MHZ	6V/uS	+18V	-18V	70C	94dB	2MV	10pA	3pA	.	5MA	10V	8V	15V	30uV/C	.	.	80dB	70dB	10G
43K	ANU	FET	INT	1MHZ	12V/uS	+18V	-18V	70C	94dB	2MV	20pA	3pA	.	5MA	10V	8V	15V	5uV/C	.	.	80dB	84dB	10G
44J	ANU	FET	INT	2MHZ	50V/uS	+18V	-18V	70C	100dB	10MV	50pA	.	.	20MA	10V	11V	15V	50uV/C	.	9MA	80dB	.	10G
44K	ANU	FET	INT	2MHZ	50V/uS	+18V	-18V	70C	100dB	10MV	25pA	.	.	20MA	10V	11V	15V	15uV/C	.	9MA	80dB	.	10G
45J	ANU	FET	INT	2MHZ	20V/uS	+18V	-18V	70C	94dB	10MV	50pA	.	.	20MA	10V		15V	50uV/C	.	7MA	64dB	72dB	10G
45K	ANU	FET	INT	2MHZ		+18V	-18V	70C	94dB	10MV	25pA	.	.	20MA	10V		15V	15uV/C	.	7MA	64dB	72dB	10G
47A	ANU	FET	INT	2MHZ	50V/uS	+18V	-18V	85C	100dB	10MV	50pA	.	.	20MA	10V	11V	15V	50uV/C	.	9MA	80dB	.	10G
47B	ANU	FET	INT	2MHZ	50V/uS	+18V	-18V	85C	100dB	10MV	25pA	.	.	20MA	10V	11V	15V	15uV/C	.	9MA	80dB	.	10G
48J	ANU	XSR	INT	3MHZ	90V/uS	+18V	-18V	70C	100dB	10MV	50pA	.	.	20MA	10V	11V	15V	50uV/C	.	9MA	80dB	.	10G
48K	ANU	XSR	INT	3MHZ	90V/uS	+18V	-18V	70C	100dB	10MV	25pA	.	.	20MA	10V	11V	15V	15uV/C	.	9MA	80dB	.	10G
52J	ANU	FET	INT	.1MHZ	.25V/uS	+18V	-18V	70C	120dB	500uV	3pA	1pA	.	5MA	10V	10V	15V	3uV/C	.	.	100dB	96dB	100G
52K	ANU	FET	INT	.1MHZ	.25V/uS	+18V	-18V	70C	120dB	500uV	3pA	1pA	.	5MA	10V	10V	15V	1uV/C	.	.	100dB	96dB	100G
101A(CHP)	RAU	GPU	EXT	.	.	+22V	-22V	125C	94dB	2MV	75NA	10NA	.	5MA	12V	15V	30V	15uV/C	.	3MA	80dB	80dB	1.5M
101A(DIL14)	ING	GPU	EXT	.	.	+22V	-22V	125C	94dB	2MV	75NA	10NA	500MWF	5MA	12V	15V	30V	15uV/C	.	3MA	80dB	80dB	1.5M
101A(FLP)	ING	GPU	EXT	.	.	+22V	-22V	125C	94dB	2MV	75NA	10NA	500MWF	5MA	12V	15V	30V	15uV/C	.	3MA	80dB	80dB	1.5M
101A-LN DIL	ING	LNA	EXT	.2MHZ	.15V/uS	+20V	-20V	125C	88dB	3MV	100NA	20NA	500MWF	5MA	12V	15V	30V	15uV/C	.	3MA	80dB	.	.
101A-LN-FLP	ING	LNA	EXT	.2MHZ	.15V/uS	+20V	-20V	125C	88dB	3MV	100NA	20NA	500MWF	5MA	12V	15V	30V	15uV/C	.	3MA	80dB	.	.
101A-LN-TO5	ING	LNA	EXT	.2MHZ	.15V/uS	+20V	-20V	125C	88dB	3MV	100NA	20NA	500MWF	5MA	12V	15V	30V	15uV/C	.	3MA	80dB	.	.
101A(TO5)	ING	GPU	EXT	.	.	+22V	-22V	125C	94dB	2MV	75NA	10NA	500MWF	5MA	12V	15V	30V	15uV/C	.	3MA	80dB	80dB	1.5M
102(FLP)	ING	VFA	INT	.	.	+18V	-18V	125C	0dB	5MV	10NA	.	500MWF	1MA	10V	.	.	30uV/C	.	6MA	.	60dB	10G
102(TO5)	ING	VFA	INT	.	.	+18V	-18V	125C	0dB	5MV	10NA	.	500MWF	1MA	10V	.	.	30uV/C	.	6MA	.	60dB	10G
106(CHP)	RAU	CPR	INT	.	.	+15V	-15V	125C	84dB	2MV	20uA	3uA	.	50MA	2.5V	.	.	10uV/C	163MW	.	.	.	.
107(CHP)	RAU	GPK	INT	.	.	+22V	-22V	125C	94dB	2MV	75NA	10NA	.	5MA	12V	15V	30V	15uV/C	.	3MA	80dB	80dB	1.5M
107(TO5)	ING	GPK	INT	.	.	+22V	-22V	125C	94dB	2MV	75NA	10NA	500MWF	5MA	12V	15V	30V	15uV/C	.	3MA	80dB	80dB	1.5M
108A(DIL)	ING	SBA	EXT	.	.	+20V	-20V	125C	98dB	0.5MV	2NA	0.2NA	500MWF	1MA	13V	15V	1V	5uV/C	.	.6MA	96dB	96dB	30M

For detailed explanations of column heading notations, see App. A

Also for ready references the more important abbreviations used in the column headings are listed below

LEFT HAND PAGE

- APP = application (codes at APP.E.)
- CMRR = common mode rejection ratio
- CMP = compensation (frequency)
- dV_{in}/dT = input offset voltage temperature drift
- GB^P = gain bandwidth product
- I_B = input bias current
- I_{io} = input bias offset current
- I_q = quiescent supply current
- MFR = manufacturer (codes at App.C.)
- P_q = quiescent power consumer
- PSRR = power supply rejection ratio
- V_{ICM} = common mode input voltage rating
- V_{idi} = differential input voltage rating
- V_{io} = input offset voltage
- V_s = dc supply voltage

RIGHT HAND PAGE

Lead out coding summary (details at APP.G.) for different cases (APP.F.)

- A = gain adjust
- B = bias adjust
- C = case
- E− = inverting input
- E+ = non-inverting input
- F,F* = input frequency compensation
- G = ground
- J = high level input
- K = output, open collector
- L = output, open emitter
- M = metal case
- N = not connected
- Q = special terminal
- R,R* = outputs
- S = strobe
- T,T* = offset balance
- V+ = +ve dc supply
- V− = −ve dc supply
- W = guard ring
- X = blank position, no lead
- ++ = +ve supplementary dc supply
- −− = −ve supplementary dc supply
- ◊,◊* = output frequency compensation

CASE (APP F)	LD 1	LD 2	LD 3	LD 4	LD 5	LD 6	LD 7	LD 8	LD 9	LD 10	LD 11	LD 12	LD 13	LD 14	LD 15	LD 16	EUROPE SUBSTITUTE	USA SUBSTITUTE	I S	TYPE NUMBER
FLP-5/6P	E+	E-	V+	V-	R														0	20-007A1
FLP-5/6P	E+	E-	V+	V-	R													20-007A1	0	20-007A2
FLP-5/6P	E+	E-	V+	V-	R													20-007A2	0	20-007B1
FLP-5/6P	E+	E-	V+	V-	R													20-007B1	0	20-007C1
FLP-5/6P	E+	E-	V+	V-	R														0	20-008A1
FLP-5/6P	E+	E-	V+	V-	R													20-008A1	0	20-008A2
FLP-5/6P	E+	E-	V+	V-	R													20-008A2	0	20-008B1
FLP-5/6P	E+	E-	V+	V-	R													20-008B1	0	20-008B2
FLP-5/6P	E+	E-	V+	V-	R													20-008B2	0	20-008B3
DIM-5/4P	E+	E-	V+	V-	R														0	20-107A1
DIM-5/4P	E+	E-	V+	V-	R													20-107A1	0	20-107A2
DIM-5/4P	E+	E-	V+	V-	R													20-107A2	0	20-107B1
DIM-5/4P	E+	E-	V+	V-	R													20-107B1	0	20-107C1
DIM-5/4P	E+	E-	V+	V-	R														0	20-108A1
DIM-5/4P	E+	E-	V+	V-	R													20-108A1	0	20-108A2
DIM-5/4P	E+	E-	V+	V-	R													20-108A2	0	20-108B1
DIM-5/4P	E+	E-	V+	V-	R													20-108B1	0	20-108B2
DIM-5/4P	E+	E-	V+	V-	R													20-108B2	0	20-108B3
T08-12/1M	E+	E-	N	V+	G	V-	M	R	N	N	N	N							0	20-247A1
T08-12/1M	E+	E-	N	V+	G	V-	M	R	N	N	N	N						20-247A1	0	20-247A2
T08-12/1M	E+	E-	N	V+	G	V-	M	R	N	N	N	N						20-247A2	0	20-247B1
T08-12/1M	E+	E-	N	V+	G	V-	M	R	N	N	N	N						20-247B1	0	20-247C1
T08-12/1M	E+	E-	N	V+	G	V-	M	R	N	N	N	N							0	20-248A1
T08-12/1M	E+	E-	N	V+	G	V-	M	R	N	N	N	N						20-248A1	0	20-248A2
T08-12/1M	E+	E-	N	V+	G	V-	M	R	N	N	N	N						20-248A2	0	20-248B1
T08-12/1M	E+	E-	N	V+	G	V-	M	R	N	N	N	N						20-248B1	0	20-248B2
T08-12/1M	E+	E-	N	V+	G	V-	M	R	N	N	N	N						20-248B2	0	20-248B3
DIM-7/5P	E+	E-	V+	X	V-	R	T											43J	0	40J
DIM-7/5P	E+	E-	V+	X	V-	R	T											43K	0	40K
DIM-9/5P	T	E+	E-	T*	V+	N	V-	R	N									41K	0	41J
DIM-9/5P	T	E+	E-	T*	V+	N	V-	R	N									41L	0	41K
DIM-9/5P	T	E+	E-	T*	V+	N	V-	R	N										0	41L
DIM-7/5P	E+	E-	V+	G	V-	R	T											42K	0	42J
DIM-7/5P	E+	E-	V+	G	V-	R	T											42L	0	42K
DIM-7/5P	E+	E-	V+	G	V-	R	T												0	42L
DIM-7/5P	E+	E-	V+	X	V-	R	T											43K	0	43J
DIM-7/5P	E+	E-	V+	X	V-	R	T												0	43K
DIM-7/5P	E+	E-	V+	G	V-	R	T											44K	0	44J
DIM-7/5P	E+	E-	V+	G	V-	R	T												0	44K
DIM-7/5P	E+	E-	V+	G	V-	R	T											45K	0	45J
DIM-7/5P	E+	E-	V+	G	V-	R	T												0	45K
DIM-7/5P	E+	E-	V+	G	V-	R	T											47B	0	47A
DIM-7/5P	E+	E-	V+	G	V-	R	T												0	47B
DIM-7/5P	E+	E-	V+	G	V-	R	T											48K	0	48J
DIM-7/5P	E+	E-	V+	G	V-	R	T												0	48K
DIM-9/5P	T	E+	E-	T*	V+	X	V-	R	T1									52K	0	52J
DIM-9/5P	T	E+	E-	T*	V+	X	V-	R	T1										0	52K
CHP																			0	101A(CHP)
DIL-14/1C	N	N	FT	E-	E+	V-	N	N	T*	R	V+	F*	N	N			UA101AD	LM101AD	0	101A (DIL14
FLP-10/3C	N	FT	E-	E+	V-	T*	R	V+	F*	N							SFC2101APM	LM101AF	0	101A(FLP)
DIL-14/1C	N	N	TF	E-	E+	V-	N	N	T*	R	V+	F*	N	N					0	101A-LN-DIL
FLP-10/3C	N	TF	E-	E+	V-	T*	R	V+	F*	N									0	101A-LN-FLP
T05-8/1M	TF	E-	E+	V-	T*	R	V+	F*											0	101A-LN-T05
T05-8/1M	FT	E-	E+	V-M	T*	R	V+	F*									SFC2101A	LM101AH	0	101A(T05)
FLP-10/3C	N	T	N	E+	V-	L	R	V+	T*	N							AMLM102F	LM102F	0	102(FLP)
T05-8/1M	T	N	E+	V-	L	R	V+	T*									UA102M	LM102H	0	102(T05)
CHP																			0	106(CHP)
CHP																			0	107(CHP)
T05-8/1M	N	E-	E+	V-M	N	R	V+										SFC2107M	LM107H	0	107(T05)
DIL-14/1C	N	F	N	E-	E+	N	V-	N	N	R	V+	F*	N	N			UA108AD	LM108AD	0	108A(DIL)

TYPE NUMBER	MFR	APP	CMP	GBP MIN	SLEW RATE MIN	V_S^+ MAX	V_S^- MAX	T_{op} MAX	A_{VOL} MIN	V_{IO} MAX	I_B MAX	I_{IO} MAX	P_{TOT} MAX	I_{OUT} MIN	V_{OUT} MIN	V_{ICM} MAX	V_{IDF} MAX	dV_{IO}/dT MAX	P_Q MAX	I_O MAX	CM RR MIN	PS RR MIN	R_{IN} MIN
108A(T05)	ING	SBA	EXT	.	.	+20V	-20V	125C	98dB	0.5MV	2NA	0.2NA	500MWF	1MA	13V	15V	1V	5uV/C	.	.6MA	96dB	96dB	30M
108(CHP)	RAU	SBA	EXT	.	.	+20V	-20V	125C	96dB	2MV	2NA	0.2NA	.	1MA	13V	15V	1V	15uV/C	.	.6MA	85dB	80dB	30M
108(DIL)	ING	SBA	EXT	.	.	+20V	-20V	125C	96dB	2MV	2NA	0.2NA	500MWF	1MA	13V	15V	1V	15uV/C	.	.6MA	85dB	80dB	30M
108-LN-T05	ING	LNA	EXT	.1MHZ	0.1V/uS	+20V	-20V	125C	88dB	3MV	3NA	0.4NA	500MWF	1MA	13V	15V	1V	15uV/C	.	1MA	85dB	80dB	.
108(T05)	ING	SBA	EXT	.	.	+20V	-20V	125C	96dB	2MV	2NA	0.2NA	500MWF	1MA	13V	15V	1V	15uV/C	.	.6MA	85dB	80dB	30M
110(DIL)	ING	VFA	INT	.	15V/uS	+18V	-18V	125C	0dB	4MV	3NA	.	500MWF	1MA	10V	15V	15V	50uV/C	.	6MA	.	70dB	10G
110(FLP)	ING	VFA	INT	.	15V/uS	+18V	-18V	125C	0dB	4MV	3NA	.	500MWF	1MA	10V	15V	15V	50uV/C	.	6MA	.	70dB	10G
110(T05)	ING	VFA	INT	.	15V/uS	+18V	-18V	125C	0dB	4MV	3NA	.	500MWF	1MA	10V	15V	15V	50uV/C	.	6MA	.	70dB	10G
111(CHP)	RAU	CPR	EXT	.	.	+18V	-18V	125C	100dB	3MV	100NA	10NA	.	.	.	15V	30V	.	.	6MA	.	.	.
111(DIL)	ING	CPR	EXT	.	.	+18V	-18V	125C	100dB	3MV	100NA	10NA	500MWF	.	.	15V	30V	.	.	6MA	.	.	.
111(FLP)	ING	CPR	EXT	.	.	+18V	-18V	125C	100dB	3MV	100NA	10NA	500MWF	.	.	15V	30V	.	.	6MA	.	.	.
111(T05)	ING	CPR	EXT	.	.	+18V	-18V	125C	100dB	3MV	100NA	10NA	500MWF	.	.	15V	30V	.	.	6MA	.	.	.
112(CHP)	RAU	SBA	INT	.	.	+20V	-20V	125C	94dB	2MV	2NA	0.2NA	.	1MA	13V	14V	14V	15uV/C	.	.6MA	85dB	80dB	30M
118A	ANU	GPK	INT	.3MHZ	6V/uS	+18V	-18V	85C	108dB	5MV	35NA	3NA	.	5MA	10V	10V	15V	20uV/C	.	4MA	70dB	63dB	100K
118(CHP)	RAU	XSR	INT	.	50V/uS	+20V	-20V	125C	94dB	4MV	250NA	50NA	.	6MA	12V	15V	1V	.	.	8MA	80dB	70dB	1M
118K	ANU	GPK	INT	.3MHZ	6V/uS	+18V	-18V	70C	108dB	5MV	35NA	3NA	.	4MA	10V	10V	15V	5uV/C	.	4MA	70dB	63dB	100K
119A	ANU	GPK	INT	.3MHZ	6V/uS	+18V	-18V	85C	114dB	5MV	35NA	3NA	.	5MA	10V	10V	15V	20uV/C	.	2MA	70dB	64dB	100K
119K	ANU	GPK	INT	.3MHZ	6V/uS	+18V	-18V	70C	114dB	5MV	35NA	3NA	.	5MA	10V	10V	15V	5uV/C	.	2MA	70dB	64dB	100M
120A	ANU	XSR	INT	20MHZ	250V/uS	+18V	-18V	85C	114dB	.	55NA	.	.	25MA	10V	.	15V	15uV/C	.	20MA	.	80dB	.
120B	ANU	XSR	INT	20MHZ	250V/uS	+18V	-18V	85C	114dB	.	55NA	.	.	25MA	10V	.	15V	8uV/C	.	20MA	.	80dB	.
136	ZEU	HVO	INT	.3MHZ	6V/uS	125V	125V	70C	100dB	1MV	100pA	.	.	20MA	100V	120V	120V	50uV/C	.	10MA	90dB	.	10G
146J	ANU	FET	INT	1M5HZ	10V/uS	+18V	-18V	70C	100dB	0.7MV	30pA	10pA	.	20MA	10V	10V	15V	7uV/C	.	.	80dB	80dB	10G
146K	ANU	FET	INT	1M5HZ	10V/uS	+18V	-18V	70C	100dB	0.7MV	20pA	10pA	.	20MA	10V	10V	15V	2uV/C	.	.	80dB	80dB	10G
153J	ANU	LVD	INT	30KHZ	.02V/uS	+18V	-18V	70C	94dB	1MV	3NA	3NA	.	1MA	1V	1V	10V	5uV/C	.	.	80dB	80dB	0.2M
153K	ANU	LVD	INT	30KHZ	.02V/uS	+18V	-18V	70C	94dB	.25MV	3NA	3NA	.	1MA	1V	1V	10V	2uV/C	.	.	80dB	80dB	0.2M
163A	ANU	HVO	INT	.3MHZ	6V/uS	+26V	-26V	85C	114dB	5MV	35NA	3NA	.	20MA	20V	20V	40V	20uV/C	.	2MA	76dB	74dB	100K
163K	ANU	HVO	INT	.3MHZ	6V/uS	+26V	-26V	70C	114dB	5MV	35NA	3NA	.	20MA	20V	20V	40V	5uV/C	.	2MA	76dB	74dB	100K
165A	ANU	HVO	INT	.3MHZ	6V/uS	+26V	-26V	85C	108dB	5MV	35NA	3NA	.	5MA	20V	20V	40V	20uV/C	.	4MA	70dB	68dB	100K
	ANU	HVO	INT	.3MHZ	6V/uS	+26V	-26V	70C	108dB	5MV	35NA	3NA	.	5MA	20V	20V	40V	5uV/C	.	4MA	70dB	68dB	100K
	ANU	LVD	INT	.2MHZ	0.6V/uS	+18V	-18V	70C	110dB	250uV	4NA	.	.	2M5A	10V	10V	15V	1.5uV/C	.	6MA	100dB	84dB	400K
180K	ANU	LVD	INT	.2MHZ	0.6V/uS	+18V	-18V	70C	110dB	100uV	4NA	.	.	2M5A	10V	10V	15V	0.5uV/C	.	6MA	100dB	84dB	400K
184J	ANU	LVD	INT	.2MHZ	0.3V/uS	+18V	-18V	70C	110dB	250uV	25NA	2NA	.	5MA	10V	10V	15V	1.5uV/C	.	9MA	90dB	80dB	1M
184K	ANU	LVD	INT	.2MHZ	0.3V/uS	+18V	-18V	70C	110dB	100uV	25NA	2NA	.	5MA	10V	10V	15V	0.5uV/C	.	9MA	90dB	80dB	1M
184L	ANU	LVD	INT	.2MHZ	0.3V/uS	+18V	-18V	70C	110dB	100uV	25NA	2NA	.	5MA	10V	10V	15V	.25uV/C	.	9MA	90dB	80dB	1M
202(T05)	ING	VFA	INT	.	.	+18V	-18V	85C	0dB	10MV	15NA	.	500MWF	1MA	10V	.	.	60uV/C	.	6MA	.	60dB	10G
207(T05)	ING	GPK	INT	.	.	+22V	-22V	85C	94dB	2MV	75NA	10NA	500MWF	5MA	12V	15V	30V	15uV/C	.	3MA	80dB	80dB	1.5M
208A(DIL)	ING	SBA	EXT	.	.	+20V	-20V	85C	98dB	0.5MV	2NA	0.2NA	500MWF	1MA	13V	15V	1V	5uV/C	.	.6MA	96dB	96dB	30M
208A(T05)	ING	SBA	EXT	.	.	+20V	-20V	85C	98dB	0.5MV	2NA	0.2NA	500MWF	1MA	13V	15V	1V	5uV/C	.	.6MA	96dB	96dB	30M
208(DIL)	ING	SBA	EXT	.	.	+20V	-20V	85C	96dB	2MV	2NA	0.2NA	500MWF	1MA	13V	15V	1V	15uV/C	.	.6MA	85dB	80dB	30M
208(T05)	ING	SBA	EXT	.	.	+20V	-20V	85C	96dB	2MV	2NA	0.2NA	500MWF	1MA	13V	15V	1V	15uV/C	.	.6MA	85dB	80dB	30M
210(DIL)	ING	VFA	INT	.	15V/uS	+18V	-18V	85C	0dB	4MV	3NA	.	500MWF	1MA	10V	15V	15V	50uV/C	.	6MA	.	70dB	10G
210(T05)	ING	VFA	INT	.	15V/uS	+18V	-18V	85C	0dB	4MV	3NA	.	500MWF	1MA	10V	15V	15V	50uV/C	.	6MA	.	70dB	10G
211(DIL)	ING	CPR	EXT	.	.	+18V	-18V	85C	100dB	3MV	100NA	10NA	500MWF	.	.	15V	30V	.	.	6MA	.	.	.
211(FLP)	ING	CPR	EXT	.	.	+18V	-18V	85C	100dB	3MV	100NA	10NA	500MWF	.	.	15V	30V	.	.	6MA	.	.	.
211(T05)	ING	CPR	EXT	.	.	+18V	-18V	85C	100dB	3MV	100NA	10NA	500MWF	.	.	15V	30V	.	.	6MA	.	.	.
216(CHP)	RAU	LBC	INT	.	.	+20V	-20V	85C	86dB	10MV	150pA	50pA	.	1MA	13V	15V	14V	.	.	.8MA	80dB	80dB	300M
231J	ANU	CHP	INT	.1MHZ	0.2V/uS	+18V	-18V	70C	140dB	15uV	100pA	.	.	25MA	10V	.	15V	.25uV/C	.	10MA	.	.	60K
231K	ANU	CHP	INT	.1MHZ	0.2V/uS	+18V	18V	70C	140dB	10uV	50pA	.	.	25MA	10V	.	15V	0.1uV/C	.	10MA	.	.	60K
233J	ANU	CHP	EXT	.1MHZ	.25V/uS	+18V	-18V	70C	140dB	50uV	50pA	.	.	5MA	10V	.	15V	1uV/C	.	5MA	.	120dB	120K
233K	ANU	CHP	EXT	.1MHZ	.25V/uS	+18V	-18V	70C	140dB	20uV	50pA	.	.	5MA	10V	.	15V	0.3uV/C	.	5MA	.	120dB	120K
233L	ANU	CHP	EXT	.1MHZ	.25V/uS	+18V	-18V	70C	140dB	20uV	50pA	.	.	5MA	10V	.	15V	0.1uV/C	.	5MA	.	120dB	120K
234J	ANU	CHP	EXT	.5MHZ	30V/uS	+18V	-18V	70C	140dB	50uV	100pA	.	.	5MA	10V	.	15V	1uV/C	.	5MA	.	120dB	60K
234K	ANU	CHP	EXT	.5MHZ	30V/uS	+18V	-18V	70C	140dB	25uV	100pA	.	.	5MA	10V	.	15V	0.3uV/C	.	5MA	.	120dB	60K
234L	ANU	CHP	EXT	.5MHZ	30V/uS	+18V	-18V	70C	140dB	25uV	100pA	.	.	5MA	10V	.	15V	0.1uV/C	.	5MA	.	120dB	60K
260J	ANU	CHP	EXT	20HZ	.1MV/uS	+18V	-18V	70C	134dB	25uV	300pA	.	.	5MA	10V	0.5V	20V	0.3uV/C	.	7MA	100dB	120dB	15K
260K	ANU	CHP	EXT	20HZ	.1MV/uS	+18V	-18V	70C	134dB	25uV	300pA	.	.	5MA	10V	1V	20V	0.1uV/C	.	7MA	100dB	120dB	15K
261J	ANU	CHP	EXT	20HZ	.1MV/uS	+16V	-16V	70C	140dB	25uV	300pA	.	.	5MA	10V	0.5V	20V	0.3uV/C	.	7MA	100dB	120dB	8K
261K	ANU	CHP	EXT	20HZ	.1MV/uS	+16V	-16V	70C	140dB	25uV	300pA	.	.	5MA	10V	1V	20V	0.1uV/C	.	7MA	100dB	120dB	8K
301A(DIL8)	ING	GPU	EXT	.	.	+18V	-18V	70C	88dB	7.5MV	250NA	50NA	500MWF	5MA	12V	15V	30V	30uV/C	.	3MA	70dB	70dB	500K
301A-LN-DIL8	ING	LNA	EXT	.2MHZ	.15V/uS	+15V	-15V	70C	84dB	10MV	300NA	70NA	500MWF	5MA	12V	15V	30V	30uV/C	.	3MA	70dB	70dB	.

For detailed explanations of column heading notations, see App. A

Also for ready references the more important abbreviations used in the column headings are listed below.

LEFT HAND PAGE

APP = application (codes at APP.E.)
CMRR = common mode rejection ratio
CMP = compensation (frequency)
dV_{io}/dT = input offset voltage temperature drift
GBP = gain bandwidth product
I_e = input bias current
I_{io} = input bias offset current
I_q = quiescent supply current
MFR = manufacturer (codes at App.C.)
P = quiescent power consumer
PSRR = power supply rejection ratio
V_{ICM} = common mode input voltage rating
V = differential input voltage rating
V = input offset voltage
V = dc supply voltage

RIGHT HAND PAGE

Lead out coding summary (details at APP.G.) for different cases (APP.F.)

A = gain adjust
B = bias adjust
C = case
E- = inverting input
E+ = non-inverting input
F,F* = input frequency compensation
G = ground
J = high level input
K = output, open collector
L = output, open emitter
M = metal case
N = not connected
Q = special terminal
R,R* = outputs
S = strobe
T,T* = offset balance
V+ = +ve dc supply
V- = -ve dc supply
W = guard ring
X = blank position, no lead
++ = +ve supplementary dc supply
-- = -ve supplementary dc supply
φ,φ* = output frequency compensation

CASE (APP F)	LD 1	LD 2	LD 3	LD 4	LD 5	LD 6	LD 7	LD 8	LD 9	LD 10	LD 11	LD 12	LD 13	LD 14	LD 15	LD 16	EUROPE SUBSTITUTE	USA SUBSTITUTE	IS	TYPE NUMBER
T05-8/1M	F	E-	E+	V-M	N	R	V+	F*									SFC2108A	LM108AH	O	108A(T05)
CHP																			C	108(CHP)
DIL-14/1C	N	F	N	E-	E+	N	V-	N	N	R	V+	F*	N	N			UA108D	LM108D	O	108(DIL)
T05-8/1M	F	E-	E+	V-	N	R	V+	F*											O	108-LN-T05
T05-8/1M	F	E-	E+	V-M	N	R	V+	F*									SFC2108M	LM108H	O	108(T05)
DIL-14/1C	N	N	T	N	E+	V-	N	N	L	R	V+	T*	N	N			SN52110JA	LM110D	O	110(DIL)
FLP-10/3C	N	T	N	E-	V-	L	R	V+	T*	N							MLM110F	LM110F	O	110(FLP)
T05-8/1M	T	N	E+	V-	L	R	V+	T*									SFC2110M	LM110H	O	110(T05)
CHP																			O	111(CHP)
DIL-14/1C	N	G	E+	E-	N	V-	T	T*S	R	N	V+	N	N	N			SN52111J	LM111D	O	111(DIL)
FLP-10/3G	G	E+	E-	N	V-	T	T*S	N	R	V+							SN52111FA	LM111F	O	111(FLP)
T05-8/1M	G	E+	E-	V-	T	T*S	R	V+									SFC2111M	LM111H	O	111(T05)
CHP																			O	112(CHP)
DIM-7/5P	E+	E-	V+	G	V-	R	T												O	118A
CHP																			O	118(CHP)
DIM-7/5P	E+	E-	V+	G	V-	R	T												O	118K
DIM-7/5P	E+	E-	V+	G	V-	R	T												O	119A
DIM-7/5P	E+	E-	V+	G	V-	R	T												O	119K
DIM-9/5P	N	Q	E-	ø	V+	G	V-	R	T									120B	O	120A
DIM-9/5P	N	Q	E-	ø	V+	G	V-	R	T										O	120B
DIM-9/5P	T	E+	E-	T*	V+	G	V-	R	T1										O	136
DIM-7/5P	E+	E-	V+	G	V-	R	T											146K	O	146J
DIM-7/5P	E+	E-	V+	G	V-	R	T												O	146K
DIM-7/5P	E+	E-	V+	G	V-	R	T											153K	O	153J
DIM-7/5P	E+	E-	V+	G	V-	R	T												O	153K
DIM-7/5P	E+	E-	V+	G	V-	R	T											163K	O	163A
DIM-7/5P	E+	E-	V+	G	V-	R	T												O	163K
DIL-7/5P	E+	E-	V+	G	V-	R	T											165K	O	165A
DIL-7/5P	E+	E-	V+	G	V-	R	T												O	165K
DIM-7/5P	E+	E-	V+	G	V-	R	T											180K	O	180J
DIM-7/5P	E+	E-	V+	G	V-	R	T												O	180K
DIM-7/5P	E+	E-	V+	G	V-	R	T											184K	O	184J
DIM-7/5P	E+	E-	V+	G	V-	R	T											184L	O	184K
DIM-7/5P	E+	E-	V+	G	V-	R	T												O	184L
T05-8/1M	T	N	E+	V-	L	R	V+	T*									UA102M	LM202H	O	202(T05)
T05-8/1M	N	E-	E+	V-M	N	R	V+	N									SFC2207	LM207H	O	207(T05)
DIL-14/1C	N	F	N	E-	E+	N	V-	N	N	R	V+	F*	N	N			UA208AD	LM208AD	O	208A(DIL)
T05-8/1M	F	E-	E+	V-M	N	R	V+	F*									SFC2208A	LM208AH	O	208A(T05)
DIL-14/1C	N	F	N	E-	E+	N	V-	N	N	R	V+	F*	N	N			UA208D	LM208D	O	208(DIL)
T05-8/1M	F	E-	E+	V-M	N	R	V+	F*									SFC2208	LM208H	O	208(T05)
DIL-14/1C	N	N	T	N	E+	V-	N	N	L	R	V+	T*	N	N			SN52110JA	LM210D	O	210(DIL)
T05-8/1M	T	N	E+	V-	L	R	V+	T*									SFC2210	LM210H	O	210(T05)
DIL-14/1C	N	G	E+	E-	N	V-	T	T*S	R	N	V+	N	N	N			SN52111J	LM211D	O	211(DIL)
FLP-10/3C	G	E+	E-	N	V-	T	T*S	N	R	V+							SN52111FA	LM211F	O	211(FLP)
T05-8/1M	G	E+	E-	V-	T	T*5	R	V+									SFC2211	LM211H	Q	211(T05)
CHP																				216(CHP)
DIM-7/5P	Q	E-	V+	G	V-	R	T											231K	O	231J
DIM-7/5P	Q	E-	V+	G	V-	R	T												O	231K
DIM-9/5P	N	Q	E-	ø	V+	G	V-	R	T									233K	O	233J
DIM-9/5P	N	Q	E-	ø	V+	G	V-	R	T									233L	O	233K
DIM-9/5P	X	E+	E-	ø	V+	G	V-	R	T										O	233L
DIM-9/5P	N	Q	E-	ø	V+	G	V-	R	T									234K	O	234J
DIM-9/5P	N	Q	E-	ø	V+	G	V-	R	T									234L	O	234K
DIM-9/5P	N	Q	E-	ø	V+	G	V-	R	T										O	234L
DIM-9/5P	X	E+	E-	ø	V+	G	V-	R	T									260K	O	260J
DIM-9/5P	X	E+	E-	ø	V+	G	V-	R	T										O	260K
DIM-9/5P	X	E+	E-	ø	V+	G	V-	R	T									261K	O	261J
DIM-9/5P	X	E+	E-	ø	V+	G	V-	R	T										O	261K
DIL-8/1P	FT	E-	E+	V-	T*	R	V+	F*									SFC2301ADC	LM301AN	O	301A(DIL8)
DIL-8/1P	TF	E-	E+	V-	T*	R	V+	F*											O	301A-LN-DIL

TYPE NUMBER	MFR	APP	CMP	GBP MIN	SLEW RATE MIN	V_S^+ MAX	V_S^- MAX	T_{op} MAX	A_{VOL} MIN	V_{IO} MAX	I_B MAX	I_{IO} MAX	P_{TOT} MAX	I_{OUT} MIN	V_{OUT} MIN	V_{ICM} MAX	V_{IDF} MAX	dV_{IO}/dT MAX	P_Q MAX	I_Q MAX	CM RR MIN	PS RR MIN	R_{IN} MIN
301A-LN-T05	ING	LNA	EXT	.2MHZ	.15V/uS	+15V	-15V	70C	84dB	10MV	300NA	70NA	500MWF	5MA	12V	15V	30V	30uV/C		3MA	70dB	70dB	
301A(T05)	ING	GPU	EXT	.	.	+18V	-18V	70C	88dB	7.5MV	250NA	50NA	500MWF	5MA	12V	15V	30V	30uV/C		3MA	70dB	70dB	500K
302(T05)	ING	VFA	INT			+18V	-18V	70C	0dB	15MV	30NA		500MWF	1MA	10V			90uV/C		6MA		60dB	10G
307(DIL)	ING	GPK	INT			+18V	-18V	70C	84dB	7.5MV	250NA	50NA	500MWF	5MA	12V	15V	30V	30uV/C			70dB	70dB	0.5M
307(T05)	ING	GPK	INT			+18V	-18V	70C	84dB	7.5MV	250NA	50NA	500MWF	5MA	12V	15V	30V	30uV/C			70dB	70dB	0.5M
308A(DIL)	ING	SBA	EXT			+18V	-18V	70C	98dB	0.5MV	7NA	1NA	500MWF	1MA	13V	15V	1V	5uV/C		.6MA	96dB	96dB	10M
308A(T05)	ING	SBA	EXT			+18V	-18V	70C	98dB	0.5MV	7NA	1NA	500MWF	1MA	13V	15V	1V	5uV/C		.6MA	96dB	96dB	10M
308(DIL)	ING	SBA	EXT			+18V	-18V	70C	88dB	7.5MV	7NA	1NA	500MWF	1MA	13V	15V	1V	5uV/C		.6MA	80dB	80dB	10M
308-LN-T05	ING	LNA	EXT	.1MHZ	0.1V/uS	+20V	-20V	70C	84dB	10MV	10NA	1.5NA	500MWF	1MA	13V	15V	1V	30uV/C		1MA	80dB	80dB	10M
308(T05)	ING	SBA	EXT			+18V	-18V	70C	88dB	7.5MV	7NA	1NA	500MWF	1MA	13V	15V	1V	30uV/C		.6MA	80dB	80dB	10M
310(DIL)	ING	VFA	INT		15V/uS	+18V	-18V	70C	0dB	7.5MV	7NA		500MWF	1MA	10V	15V	15V	50uV/C		6MA		70dB	10G
310(T05)	ING	VFA	INT		15V/uS	+18V	-18V	70C	0dB	7.5MV	7NA		500MWF	1MA	10V	15V	15V	50uV/C		6MA		70dB	10G
311(DIL)	ING	CPR	EXT			+18V	-18V	70C	100dB	7.5MV	250NA	50NA	500MWF			15V	30V			8MA			
311(FLP)	ING	CPR	EXT			+18V	-18V	70C	100dB	7.5MV	250NA	50NA	500MWF			15V	30V			8MA			
311(T05)	ING	CPR	EXT			+18V	-18V	70C	100dB	7.5MV	250NA	50NA	500MWF			15V	30V			8MA			
350A	ANU	CPR	INT			+16V	-16V	85C	94dB	2MV	50pA			7MA	10V	10V	30V	75uV/C		3M5A	50dB		10G
350B	ANU	CPR	INT			+16V	-16V	85C	94dB	2MV	30pA			7MA	10V	10V	30V	40uV/C		3M5A	50dB		10G
350C	ANU	CPR	INT			+16V	-16V	85C	94dB	2MV	30pA			7MA	10V	10V	30V	25uV/C		3M5A	50dB		10G
435	OAU	GPK	INT	1MHZ	2V/uS	+24V	-24V	85C	80dB	50MV	1uA			.24A	20V	20V		50uV/C		15MA	70dB		10K
702(CHP)	RAU	WBA	EXT		0.5V/uS	+14V	-7V	125C	63dB	5MV	10uA	2uA		.3MA	5V	1.5V	5V	20uV/C	120MW	7MA	70dB	70dB	8K
709AE	TDG	GPU	EXT	.3MHZ	.15V/uS	+18V	-18V	125C	88dB	2MV	200NA	50NA	500MWF	5MA	12V	10V	5V	10uV/C	108MW		80dB	80dB	350K
709AH	TDG	GPU	EXT	.3MHZ	.15V/uS	+18V	-18V	125C	88dB	2MV	200NA	50NA	570MWF	5MA	12V	10V	5V	10uV/C	108MW		80dB	80dB	350K
709AL	TDG	GPU	EXT	.3MHZ	.15V/uS	+18V	-18V	125C	88dB	2MV	200NA	50NA	670MWF	5MA	12V	10V	5V	10uV/C	108MW		80dB	80dB	350K
709BE	TDG	GPU	EXT			+18V	-18V	125C	88dB	5MV	500NA	200NA	500MWF	5MA	12V	10V	5V	15uV/C	165MW		70dB	76dB	150K
709BH	TDG	GPU	EXT			+18V	-18V	125C	88dB	5MV	500NA	200NA	500MWF	5MA	12V	10V	5V	15uV/C	165MW		70dB	76dB	150K
709BL	TDG	GPU	EXT			+18V	-18V	125C	88dB	5MV	500NA	200NA	500MWF	5MA	12V	10V	5V	15uV/C	165MW		70dB	76dB	150K
709CE	TDG	GPU	EXT	.3MHZ	.15V/uS	+18V	-18V	125C	84dB	7.5MV	1.5uA	500NA	500MW	5MA	12V	10V	5V	15uV/C	165MW		65dB	76dB	150K
709(CHP)	RAU	GPU	EXT			+18V	-18V	125C	88dB	5MV	500NA	200NA		5MA	12V	10V	5V	15uV/C	200MW	6MA	70dB	76dB	150K
709CJ	TDG	GPU	EXT	.3MHZ	.15V/uS	+18V	-18V	70C	84dB	7.5MV	1.5uA	500NA	670MWF	5MA	12V	10V	5V		200MW		65dB	74dB	50K
709CL	TDG	GPU	EXT	.3MHZ	.15V/uS	+18V	-18V	70C	84dB	7.5MV	1.5uA	500NA	670MWF	5MA	12V	10V	5V		200MW		65dB	74dB	50K
710BE	TDG	CPR	EXT			+14V	-7V	125C	62dB	2MV	20uA	3uA	500MWF	5MA	2.5V	7V	5V	10uV/C	150MW		80dB		
710BH	TDG	CPR	EXT			+14V	-6V	125C	62dB	2MV	20uA	3uA	570MWF	5MA	2.5V	7V	5V	10uV/C	150MW		80dB		
710BL	TDG	CPR	EXT			+14V	-6V	125C	62dB	2MV	20uA	3uA	670MWF	5MA	2.5V	7V	5V	10uV/C	150MW		80dB		
710CE	TDG	CPR	EXT			+14V	-7V	125C	62dB	2MV	20uA	3uA	500MWF	5MA	2.5V	7V	5V	10uV/C	150MW		80dB		
710(CHP)	RAU	CPR	EXT			+14V	-7V	75C	60dB	5MV	25uA	5uA		5MA	2.5V	5V	5V	20uV/C	150MW	9MA	70dB		
710CJ	TDG	CPR	EXT			+14V	-6V	70C	60dB	5MV	25uA	5uA	670MWF	5MA	2.5V	7V	5V	20uV/C	150MW		70dB		
710CL	TDG	CPR	EXT			+14V	-6V	70C	60dB	5MV	25uA	5uA	670MWF	5MA	2.5V	7V	5V	20uV/C	150MW		70dB		
710CP	TDG	CPR	EXT			+14V	-7V	70C	60dB	5MV	25uA	5uA	530MWF	5MA	1V	7V	5V	20uV/C	150MW	9MA	70dB		
711BE	TDG	DCP	EXT			+14V	-7V	125C	58dB	3.5MV	75uA	10uA	500MWF	5MA	2.5V	7V	5V	20uV/C	200MW				
711BH	TDG	DCP	EXT			+14V	-7V	125C	58dB	3.5MV	75uA	10uA	570MWF	5MA	2.5V	7V	5V	20uV/C	200MW				
711BL	TDG	DCP	EXT			+14V	-7V	125C	58dB	3.5MV	75uA	10uA	670MWF	5MA	2.5V	7V	5V	20uV/C	200MW				
711CE	TDG	DCP	EXT			+14V	-7V	70C	57dB	5MV	100uA	15uA	500MWF	5MA	2.5V	7V	5V	20uV/C	230MW				
711(CHP)	RAU	DCP	EXT			+14V	-7V	70C	57dB	5MV	100uA	25uA				7V	5V	20uV/C	180MW		70dB		
711CJ	TDG	DCP	EXT			+14V	-7V	70C	57dB	5MV	100uA	15uA	670MWF	5MA	2.5V	7V	5V	20uV/C	230MW				
711CL	TDG	DCP	EXT			+14V	-7V	70C	57dB	5MV	100uA	15uA	670MWF	5MA	2.5V	7V	5V	20uV/C	230MW				
715DC	ADU	HSR	EXT		10V/uS	+18V	-18V	70C	80dB	7.5MV	1.5uA	250NA	670MWF	5MA	10V	15V	15V		300MW	10MA	74dB	68dB	300K
715DM	ADU	HSR	EXT		15V/uS	+18V	-18V	125C	84dB	5MV	750NA	250NA	670MWF	5MA	10V	15V	15V		210MW	7MA	74dB	70dB	300K
715FM	ADU	HSR	EXT		15V/uS	+18V	-18V	125C	84dB	5MV	750NA	250NA		5MA	10V	15V	15V		210MW	7MA	74dB	70dB	300K
715HC	ADU	HSR	EXT		10V/uS	+18V	-18V	70C	80dB	7.5MV	1.5uA	250NA	500MWF	5MA	10V	15V	15V		300MW	10MA	74dB	68dB	300K
715HM	ADU	HSR	EXT		15V/uS	+18V	-18V	125C	84dB	5MV	750NA	250NA	500MWF	5MA	10V	15V	15V		210MW	7MA	74dB	70dB	300K
715XC	ADU	HSR	EXT		10V/uS	+18V	-18V	70C	80dB	7.5MV	1.5uA	250NA		5MA	10V	15V	15V		300MW	10MA	74dB	68dB	300K
715XM	ADU	HSR	EXT		15V/uS	+18V	-18V	125C	84dB	5MV	750NA	250NA		5MA	10V	15V	15V		210MW	7MA	74dB	70dB	300K
725(CHP)	RAU	PIA	EXT			+22V	-22V	125C	120dB	1MV	100NA	20NA		5MA	12V	22V	5V	5uV/C	105MW		110dB	100dB	500K
725CN	ADU	PIA	EXT			+22V	-22V	70C	106dB	2.5MV	125NA	35NA	500MWF	5MA	12V	22V	5V	10uV/C	150MW		94dB	90dB	500K
725DC	ADU	PIA	EXT			+22V	-22V	70C	106dB	2.5MV	125NA	35NA	500MWF	5MA	12V	22V	5V	10uV/C	150MW		94dB	90dB	500K
725DM	ADU	PIA	EXT			+22V	-22V	125C	120dB	1MV	100NA	20NA	500MWF	5MA	12V	22V	5V	5uV/C	105MW		110dB	100dB	500K
725HC	ADU	PIA	EXT			+22V	-22V	70C	108dB	2.5MV	125NA	35NA	500MWF	5MA	12V	22V	5V	5uV/C	150MW		94dB	90dB	500K
725HM	ADU	PIA	EXT			+22V	-22V	125C	120dB	1MV	100NA	20NA	500MW	5MA	12V	22V	5V	5uV/C	150MW		110dB	100dB	500K
725XC	ADU	PIA	EXT			+22V	-22V	70C	108dB	2.5MV	125NA	35NA		5MA	12V	22V	5V	5uV/C	150MW		94dB	90dB	500K
725XM	ADU	PIA	EXT			+22V	-22V	125C	120dB	1MV	100NA	20NA		5MA	12V	22V	5V	5uV/C	150MW		110dB	100dB	500K

For detailed explanations of column heading notations, see App. A.

Also for ready references the more important abbreviations used in the column headings are listed below

LEFT HAND PAGE

APP = application (codes at APP.E.)

CMRR = common mode rejection ratio

CMP = compensation (frequency)

dV_{10}/dT = input offset voltage temperature drift

GBP = gain bandwidth product

I_B = input bias current

I_{10} = input bias offset current

I_0 = quiescent supply current

MFR = manufacturer (codes at App.C.)

P_0 = quiescent power consumer

PSRR = power supply rejection ratio

V_{ICM} = common mode input voltage rating

V_{IDF} = differential input voltage rating

V_{10} = input offset voltage

V_S = dc supply voltage

RIGHT HAND PAGE

Lead out coding summary (details at APP.G.) for different cases (APP.F.)

A = gain adjust

B = bias adjust

C = case

E— = inverting input

E+ = non-inverting input

F,F* = input frequency compensation

G = ground

J = high level input

K = output, open collector

L = output, open emitter

M = metal case

N = not connected

Q = special terminal

R,R* = outputs

S = strobe

T,T* = offset balance

V+ = +ve dc supply

V— = —ve dc supply

W = guard ring

X = blank position, no lead

++ = +ve supplementary dc supply

—— = —ve supplementary dc supply

∅,∅* = output frequency compensation

CASE (APP F)	LD 1	LD 2	LD 3	LD 4	LD 5	LD 6	LD 7	LD 8	LD 9	LD 10	LD 11	LD 12	LD 13	LD 14	LD 15	LD 16	EUROPE SUBSTITUTE	USA SUBSTITUTE	IS	TYPE NUMBER
TO5-8/1M	TF	E-	E+	V-	T*	R	V+	F*							·	·		101ALN-TO5	0	301A-LN-TO5
TOS-8/1M	FT	E-	E+	V-M	T*	R	V+	F*							·	·	SFC2301AH	LM301AH	0	301A(TO5)
TO5-8/1M	T	N	E+	V-	L	R	V+	T*							·	·	UA302C	LM302H	0	302(TO5)
DIL-8/1P	N	E-	E+	V-	N	R	V+	N							·	·	SFC2307DC	LM307N	0	307(DIL)
TO5-8/1M	N	E-	E+	V-M	N	R	V+	N							·	·	SFC2307	LM307H	0	307(TO5)
DIL-14/1C	N	F	N	E-	E+	N	V-	N	N	R	V+	F*	N	N			SN72308AJA	LM308AD	0	308A(DIL)
TO5-8/1M	F	E-	E+	V-M	N	R	V+	F*							·	·	SFC2308A	LM308AH	0	308A(TO5)
DIL-14/1C	F	F	N	E-	E+	N	V-	N	N	R	V+	F*	N	N			SN72308JA	LM308D	0	308(DIL)
TO5-8/1M	F	E-	E+	V-	N	R	V+	N							·	·		108-LN-TO5	0	308-LN-TO5
TO5-8/1M	F	E-	E+	V-M	N	R	V+	F*							·	·	SFC2308	LM308H		308(TO5)
DIL-14/1C	N	N	T	N	E+	N	V-	N	N	L	R	V+	T*	N	N		SFC2310EC	LM310D	0	310(DIL)
TO5-8/1M	N	T	N	E+	V-	L	R	V+	T*						·	·	SFC2310EC	LM310H	0	310(TO5)
DIL-14/1C	N	G	E+	E-	N	V-	T	T*S	R	R	N	V+	N	N			SFC2311EC	LM311D	0	311(DIL)
FLP-10/3G	G	E+	E-	N	V-	T	T*S	N	R	R	V+				·	·	SN72311FA	LM311F	0	311(FLP)
TO5-8/1M	G	E+	E-	V-	T	T*S	R	V+							·	·	SFC2311	LM311H	0	311(TO5)
DIM-7/5P	E+	E-	V+	G	V-	R	T								·	·		350B	0	350A
DIM-7/5P	E+	E-	V+	G	V-	R	T								·	·		350C	0	350B
DIM-7/5P	E+	E-	V+	G	V-	R	T								·	·			0	350C
TO3-5/2M	V+	R	V-	E+	E-										·	·			0	435
CHP															·	·			0	702(CHP)
TO5-8/1M	F	E-	E+	V-	∅	∅*R	V+	F*							·	·	TAA522	UA709AHM	0	709AE
FLP-10/3C	N	F	E-	E+	V-	∅	R	V+	F*	N					·	·	SN52709AFA	UA709AFM	0	709AH
DIL-14/1C	N	N	F	E-	E+	V-	N	N	∅	R	V+	F*	N	N			LM709AJ	UA709ADM	0	709AL
TO5-8/1M	F	E-	E+	V-M	∅	R∅*	V+	F*							·	·	MC1709AG	UA709AHM	0	709BE
FLP-10/3C	N	F	E-	E+	V-	∅	R	V+	F*	N					·	·	SN52709AFA	UA709AFM	0	709BH
DIL-14/1C	N	N	F	E-	E+	V-	N	N	∅	R	V+	F*	N	N			LM709AJ	UA709ADM	0	709BL
TO5-8/1M	F	E-	E+	V-	∅	∅*R	V	F*							·	·	TAA521	UA709HC	0	709CE
CHP															·	·			0	709(CHP)
DIL-14/1P	N	N	F	E-	E+	N	N	N	∅	R	V+	F*	N	N			TAA521A	UA709DC	0	709CJ
DIL-14/1C	N	N	F	E-	E+	V-	N	N	∅	R	V+	F*	N	N			TAA521A	UA709DC	0	709CL
TO5-8/1M	G	E+	E-	V-	N	N	R	V+							·	·	SFC2710M	UA710HM	0	710BE
FLP-10/3C	G	E+	E-	N	V-	R	N	V+	N	N					·	·	SFC2710PM	UA710FM	0	710BH
DIL-14/1C	N	G	E+	E-	N	V-	N	N	R	N	V+	N	N		·	·	SFC2710KM	UA710DM	0	710BL
TO5-8/1M	G	E+	E-	V-	N	N	R	V+							·	·	SFC2710M	UA710HM	0	710CE
CHP															·	·			0	710(CHP)
DIL-14/1P	N	G	E+	E-	N	V-	N	N	R.	N	V+	N	N		·	·	SFC2710EC	UA710DC	0	710CJ
DIL-14/1C	N	G	E+	E-	N	V-	N	N	R	N	V+	N	N		·	·	SFC2710EC	UA710DC	0	710CL
DIL-8/1P	G	E+	E-	V-	N	N	N	V+							·	·			0	710CP
TO5-10/1M	G	S1	E-1	E+1	V-	E+2	E-2	S2	R	V+					·	·	SFC2711M	UA711HM	0	711BE
FLP-10/3C	E-1	E+1	V-	E+2	E-2	S2	R	V+	G	S1					·	·	SFC2711PM	UA711FM	0	711BH
DIL-14/1C	N	E-1	E+1	V-	E+2	E-2	N	N	S2	R	V+	G	S1	N			SFC2711KM	UA711DM	0	711BL
TO5-10/1M	G	S1	E-1	E+1	V-	E+2	E-2	S2	R	V+					·	·	SFC2711C	UA711HC	0	711CE
CHP															·	·			0	711(CHP)
DIL-14/1P	N	E-1	E+1	V-	E+2	E-2	N	N	S2	R	V+	G	S1	N			SFC2711EC	UA711DC	0	711CJ
DIL-14/1C	N	E-1	E+1	V-	E+2	E-2	N	N	S2	R	V+	G	S1	N			SFC2711EC	UA711CN	0	711CL
DIL-14/1C	F	F*	Q	E-	E+	N	N	N	N	V-	R	∅	V+	∅*				UA715DC	0	715DC
DIL-14/1C	F	F*	Q	E-	E+	N	N	N	N	V-	R	∅	V+	∅*				UA715DM	0	715DM
FLP-10/3C	F	F	Q	E-	E+	V-	R	∅	V+	∅*					·	·		AM715FM	0	715FM
TO5-8/1M	F	Q	E-	E+	V-	∅	V+	∅*	F*						·	·		UA715HC	0	715HC
TO5-10/1M	F	Q	E-	E+	V-	∅	V+	∅*	F*						·	·		UA715HM	0	715HM
CHP															·	·			0	715XC
CHP															·	·			0	715XM
CHP															·	·			0	725(CHP)
DIL-8/1P	T	E-	E+	V-	∅	∅*R		T*							·	·		LM725CN	0	725CN
DIL-14/1C	N	N	T	E-	E+	V-	N	N	∅	∅*R		T*	N	N				LM725CJ-14	0	725DC
DIL-14/1C	N	N	T	E-	E+	V-	N	N	∅	∅*R	V	T*	N	N				LM725D	0	725DM
TO5-8/1M	T	E-	E+	V-	∅	∅*		T*							·	·	RC725T	UA725HC	0	725HC
TO5-8/1M	T	E-	E+	V-	∅	∅*	V+	T*							·	·	RM725T	UA725HM	0	725HM
CHP															·	·			0	725XC
CHP															·	·			0	725XM

TYPE NUMBER	MFR	APP	CMP	GBP MIN	SLEW RATE MIN	Vs+ MAX	Vs- MAX	Tmin MAX	Avol MIN	Vio MAX	ie MAX	Iio MAX	Ptot MAX	Iout MIN	Vout MIN	Vicm MAX	Vide MAX	dVio/dT MAX	Po MAX	Io MAX	CMRR MIN	PSRR MIN	Rin MIN
733(CHP)	RAU	BDO	EXT	20MHZ	.	+8V	-8V	125C	50dB	5MV	20uA	3uA		2MA	3V	6V	5V		.	24MA	60dB	50dB	2K
733DC	ING	BDO	EXT	20MHZ	.	+8V	-8V	70C	48dB	6MV	30uA	5uA	670MWF	2MA	1.5V	6V	5V			24MA	60dB	50dB	2K
733DCDD	ING	BDO	EXT	20MHZ	.	+8V	-8V	70C	48dB	6MV	30uA	5uA	670MWF	2MA	1.5V	6V	5V			24MA	60dB	50dB	2K
733DM	ING	BDO	EXT	20MHZ	.	+8V	-8V	125C	50dB	5MV	20uA	3uA	670MWF	2MA	1.5V	6V	5V			24MA	60dB	50dB	2K
733DMDD	ING	BDO	EXT	20MHZ	.	+8V	-8V	125C	50dB	5MV	20uA	3uA	670MWF	2MA	1.5V	6V	5V			24MA	60dB	50dB	2K
733FM	ADU	BDO	EXT	20MHZ	.	+8V	-8V	125C	50dB	5MV	20uA	3uA	570MWF	2MA	1.5V	6V	5V			24MA	60dB	50dB	2K
733HC	ADU	BDO	EXT	20MHZ	.	+8V	-8V	70C	48dB	6MV	30uA	5uA	500MWF	2MA	1.5V	6V	5V			24MA	60dB	50dB	2K
733HCTB	ING	BDO	EXT	20MHZ	.	+8V	-8V	70C	48dB	6MV	30uA	5uA	500MWF	2MA	1.5V	6V	5V			24MA	60dR	50dB	2K
733HM	ADU	BDO	EXT	20MHZ	.	+8V	-8V	125C	50dB	5MV	20uA	3uA	500MWF	2MA	1.5V	6V	5V			24MA	60dB	50dB	2K
733HMTB	ING	BDO	EXT	20MHZ	.	+8V	-8V	125C	50dB	5MV	20uA	3uA	500MWF	2MA	1.5V	6V	5V			24MA	60dB	50dB	2K
733XC	ADU	BDO	EXT	20MHZ	.	+8V	-8V	70C	48dB	6MV	30uA	5uA	.	2MA	1.5V	6V	5V			24MA	60dB	50dB	2K
733XM	ADU	BDO	EXT	20MHZ	.	+8V	-8V	125C	50dB	5MV	20uA	3uA	.	2MA	1.5V	6V	5V			24MA	60dB	50dB	2K
741ADM	ADU	GPK	INT	.4MHZ	0.3V/uS	+22V	-22V	125C	94dB	3MV	80NA	30NA	670MWF	10MA	16V	15V	30V	15uV/C	150MW	.	80dB	86dB	1M
741AFM	ADU	GPK	INT	.4MHZ	0.3V/uS	+22V	-22V	125C	94dB	3MV	80NA	30NA	570MWF	10MA	16V	15V	30V	15uV/C	150MW	.	80dB	86dB	1M
741AHM	ADU	GPK	INT	.4MHZ	0.3V/uS	+22V	-22V	125C	94dB	3MV	80NA	30NA	500MWF	10MA	16V	15V	30V	15uV/C	150MW	.	80dB	86dB	1M
741BE	TDG	GPK	INT	.	0.2V/uS	+22V	-22V	125C	94dB	5MV	500NA	200NA	500MWF	7MA	12V	15V	30V		85MW	3MA	70dB	76dB	300K
741BH	TDG	GPK	INT	.	0.2V/uS	+22V	-22V	125C	94dB	5MV	500NA	200NA	570MWF	5MA	12V	15V	30V		85MW	3MA	70dB	76dB	300K
741BL	TDG	GPK	INT	.4MHZ	0.3V/uS	+22V	-22V	125C	94dB	3MV	80NA	30NA	670MWF	5MA	12V	15V	30V	15uV/C	150MW	.	80dB	86dB	1M
741C(DIL)	ING	GPK	INT	.	0.2V/uS	+18V	-18V	70C	86dB	6MV	500NA	200MA	670MWF	5MA	12V	15V	30V		85MW	.	70dB	76dB	300K
741CE	TDG	GPK	INT	.	0.2V/uS	+18V	-18V	70C	86dB	6MV	500NA	200NA	500MWF	5MA	12V	15V	30V			3MA	70dB	76dB	300K
741(CHP)	RAU	GPK	INT	.	0.3V/uS	+22V	-22V	125C	94dB	5MV	500NA	200NA		5MA	12V	15V	30V		85MW	.	70dB	76dB	300K
741CHS-DIL8	ING	GPK	INT	.	0.7V/uS	+18V	-18V	70C	94dB	5MV	500NA	200NA	500MWF	5MA	12V	15V	30V			3MA	70dB	77dB	300K
741CHSPA	ING	GPK	INT	.	0.7V/uS	+18V	-18V	70C	94dB	5MV	500NA	200NA	500MWF	5MA	12V	15V	30V			3MA	70dB	77dB	300K
741CHS-T05	ING	GPK	INT	.	0.7V/uS	+18V	-18V	70C	88dB	6MV	500NA	200NA	500MWF	5MA	12V	15V	30V			3MA	70dB	77dB	300K
741CHSTY	ING	GPK	INT	.	0.7V/uS	+18V	-18V	70C	88dB	6MV	500NA	200NA	500MWF	5MA	12V	15V	30V			3MA	70dB	77dB	300K
741CJ	TDG	GPK	INT	.	0.2V/uS	+18V	-18V	70C	86dB	6MV	500NA	200NA	670MWF	5MA	12V	15V	30V		85MW	3MA	70dB	76dB	300K
741CL	TDG	GPK	INT	.	0.2V/uS	+18V	-18V	70C	86dB	6MV	500NA	200NA	670MWF	5MA	12V	15V	30V		85MW	3MA	70dB	76dB	300K
741C-LN-DIL8	ING	LNA	INT	.2MHZ	.15V/uS	+15V	-15V	70C	84dB	7.5MV	0.8uA	0.3uA	310MWF	5MA	12V	15V	30V	50uV/C		3MA	70dB	76dB	
741C-LN-T05	ING	LNA	INT	.2MHZ	.15V/uS	+15V	-15V	70C	84dB	7.5MV	0.8uA	0.3uA	500MWF	5MA	12V	15V	30V	50uV/C		3MA	70dB	76dB	
741CP	TDG	GPK	INT	.	0.2V/uS	+18V	-18V	70C	86dB	6MV	500NA	200NA	310MWF	5MA	12V	15V	30V			3MA	70dB	76dB	300K
741C(T05)	ING	GPK	INT	.	0.2V/uS	+18V	-18V	70C	86dB	6MV	500NA	200NA	500MWF	5MA	12V	15V	30V			3MA	70dB	76dB	300K
741DC	ADU	GPK	INT	.	0.2V/uS	+18V	-18V	70C	86dB	6MV	500NA	200NA	670MWF	5MA	12V	15V	30V		85MW	3MA	70dB	76dB	300K
741DM	ADU	GPK	INT	.4MHZ	0.3V/uS	+22V	-22V	125C	94dB	3MV	80NA	30NA	670MWF	5MA	12V	15V	30V	15uV/C	150MW	.	80dB	86dB	1M
741EDC	ADU	GPK	INT	.4MHZ	0.3V/uS	+22V	-22V	70C	94dB	3MV	80NA	30NA	670MWF	7MA	16V	15V	30V	15uV/C	150MW	.	80dB	86dB	1M
741EHC	ADU	GPK	INT	.4MHZ	0.3V/uS	+22V	-22V	70C	94dB	3MV	80NA	30NA	500MWF	7MA	16V	15V	30V	15uV/C	150MW	.	80dB	86dB	1M
741FM	ADU	GPK	INT	.	0.2V/uS	+22V	-22V	125C	94dB	5MV	500NA	200NA	570MWF	5MA	12V	15V	30V		85MW	3MA	70dB	76dB	300K
741HC	ADU	GPK	INT	.	0.2V/uS	+18V	-18V	70C	86dB	6MV	500NA	200NA	500MWF	5MA	12V	15V	30V			3MA	70dB	76dB	300K
741HM	ADU	GPK	INT	.	0.2V/uS	+22V	-22V	125C	94dB	5MV	500NA	200NA	500MWF	7MA	12V	15V	30V		85MW	3MA	70dB	76dB	300K
741-LN-DIL	ING	LNA	INT	.2MHZ	.15V/uS	+20V	-20V	125C	88dB	6MV	1.5uA	500NA	670MWF	5MA	12V	15V	30V	30uV/C		3MA	70dB	76dB	
741-LN-FLP	ING	LNA	INT	.2MHZ	.15V/uS	+20V	-20V	125C	88dB	6MV	1.5uA	500NA	570MWF	5MA	12V	15V	30V	30uV/C		3MA	70dB	76dB	
741-LN-T05	ING	LNA	INT	.2MHZ	.15V/uS	+24V	-20V	125C	88dB	6MV	1.5uA	500NA	500MWF	5MA	12V	15V	30V	30uV/C		3MA	70dB	76dB	
741MHSDD	ING	GPK	INT	.	0.7V/uS	+18V	-18V	125C	94dB	5MV	500NA	200NA	500MWF	5MA	12V	15V	30V			3MA	70dB	77dB	300K
741MHS-DIL	ING	GPK	INT	.	0.7V/uS	+18V	-18V	125C	94dB	5MV	500NA	200NA	500MWF	5MA	12V	15V	30V			3MA	70dB	77dB	300K
741MHSFD	ING	BDO	INT	.	0.7V/uS	+18V	-18V	125C	94dB	5MV	500NA	200NA	500MWF	5MA	12V	15V	30V			3MA	70dB	77dB	300K
741MHS-FLP	ING	GPK	INT	.	0.7V/uS	+18V	-18V	125C	94dB	5MV	500NA	200NA	500MWF	5MA	12V	15V	30V			3MA	70dB	77dB	300K
741MHS-T05	ING	GPK	INT	.	0.7V/uS	+18V	-18V	125C	94dB	5MV	500NA	200NA	500MWF	5MA	12V	15V	30V			3MA	70dB	77dB	300K
741MHSTY	ING	GPK	INT	.	0.7V/uS	+18V	-18V	125C	94dB	5MV	500NA	200NA	500MWF	5MA	12V	15V	30V			3MA	70dB	77dB	300K
741(T05)	ING	GPK	INT	.	0.2V/uS	+22V	-22V	125C	94dB	5MV	500NA	200NA	500MWF	7MA	12V	15V	30V		85MW	3MA	70dB	76dB	300K
741XC	ADU	GPK	INT	.	0.2V/uS	+18V	-18V	70C	86dB	6MV	500NA	200NA		5MA	12V	15V	30V			3MA	70dB	76dB	300K
741XM	ADU	GPK	INT	.	0.2V/uS	+22V	-22V	125C	94dB	5MV	500NA	200NA		7MA	12V	15V	30V		85MW	3MA	70dB	76dB	300K
747ADM	ADU	DGK	INT	.4MHZ	0.3V/uS	+22V	-22V	125C	94dB	3MV	80NA	30NA	670MWF	10MA	16V	15V	30V	15uV/C	150MW	3MA	80dB	86dB	1M
747AFM	ADU	DGK	INT	.4MHZ	0.3V/uS	+22V	-22V	125C	94dB	3MV	80NA	30NA	500MWF	10MA	16V	15V	30V	15uV/C	150MW	3MA	80dB	86dB	1M
747AHM	ADU	DGK	INT	.4MHZ	0.3V/uS	+22V	-22V	125C	94dB	3MV	80NA	30NA	500MWF	10MA	16V	15V	30V	15uV/C	150MW	3MA	80dB	86dB	1M
747BE	TDG	DGK	INT	.	0.2V/uS	+22V	-22V	125C	94dB	5MV	500NA	200NA	500MWF	5MA	12V	15V	30V			3MA	70dB	76dB	300K
747BL	TDG	DGK	INT	.	0.2V/uS	+22V	-22V	125C	94dB	5MV	500NA	200NA	670MWF	5MA	12V	15V	30V		85MW	3MA	70dB	76dB	300K
747CE	TDG	DGK	INT	.	0.2V/uS	+18V	-18V	70C	88dB	6MV	500NA	200NA	500MWF	5MA	12V	15V	30V		85MW	3MA	70dB	76dB	300K
747(CHP)	RAU	DGK	INT	.	0.2V/uS	+22V	-22V	125C	94dB	5MV	500NA	200NA	.	5MA	12V	15V	30V		85MW	3MA	70dB	76dB	300K
747CJ	TDG	DGK	INT	.	0.2V/uS	+18V	-18V	70C	88dB	6MV	500NA	200NA	670MWF	5MA	12V	15V	30V		85MW	3MA	70dB	76dB	300K
747DC	ADU	DGK	INT	.	0.2V/uS	+18V	-18V	70C	86dB	6MV	500NA	200NA	670MWF	5MA	12V	15V	30V		85MW	3MA	70dB	76dB	300K
747DM	ADU	DGK	INT	.	0.2V/uS	+22V	-22V	125C	94dB	5MV	500NA	200NA	670MWF	5MA	12V	15V	30V		85MW	3MA	70dB	76dB	300K

For detailed explanations of column heading notations, see App. A.

Also for ready references the more important abbreviations used in the column headings are listed below:

LEFT HAND PAGE

- APP = application (codes at APP.E.)
- CMRR = common mode rejection ratio
- CMP = compensation (frequency)
- dV_{IO}/dT = input offset voltage temperature drift
- GBP = gain bandwidth product
- I_B = input bias current
- I_{IO} = input bias offset current
- I_Q = quiescent supply current
- MFR = manufacturer (codes at App.C.)
- P_Q = quiescent power consumer
- PSRR = power supply rejection ratio
- V_{ICM} = common mode input voltage rating
- V_{IDF} = differential input voltage rating
- V_{IO} = input offset voltage
- V_S = dc supply voltage

RIGHT HAND PAGE

Lead out coding summary (details at APP.G.) for different cases (APP.F.)

- A = gain adjust
- B = bias adjust
- C = case
- E-- = inverting input
- E+ = non-inverting input
- F,F* = input frequency compensation
- G = ground
- J = high level input
- K = output, open collector
- L = output, open emitter
- M = metal case
- N = not connected
- O = special terminal
- R,R*. = outputs
- S = strobe
- T,T* = offset balance
- V+ = +ve dc supply
- V- = -ve dc supply
- W = guard ring
- X = blank position, no lead
- + + = +ve supplementary dc supply
- --- = -ve supplementary dc supply
- ♦,♦* = output frequency compensation

CASE (APP F)	LD 1	LD 2	LD 3	LD 4	LD 5	LD 6	LD 7	LD 8	LD 9	LD 10	LD 11	LD 12	LD 13	LD 14	LD 15	LD 16	EUROPE SUBSTITUTE	USA SUBSTITUTE	IS	TYPE NUMBER
CHP																			0	733(CHP)
DIL-14/1C	E+	N	A2	A*2	V-	N	R	R*	N	V+	A1	A*1	N	E-			LM733CD	UA733DC	0	733DC
DIL-14/1C	E+	N	A2	A*2	V-	N	R	R*	N	V+	A1	A*1	N	E-			LM733CD	UA733DC	0	733DCDD
DIL-14/1C	E+	N	A2	A*2	V-	N	R	R*	N	V+	A1	A*1	N	E-			LM733D	UA733DM	0	733DM
DIL-14/1C	E+	N	A2	A*2	V-	N	R	R*	N	V+	A1	A*1	N	E-			LM733D	UA733DM	0	733DMDD
FLP-10/3C	E+	A2	A*2	V-	R	R*	V+	A1	A*1	E-							SN52733A	UA733FM	0	733FM
TO5-10/1M	E-	E+	A2	A*2	V-	R	R*	V+	A1	A*1							LM733CH	UA733HC	0	733HC
TO5-10/1M	E-	E+	A2	A*2	V-	R	R*	V+	A1	A*1							LM733CH	UA733HC	0	733HCTB
TO5-10/1M	E-	E+	A2	A*2	V-	R	R*	V+	A1	A*1							LM733H	UA733HM	0	733HM
TO5-10/1M	E-	E+	A2	A*2	V-	R	R*	V+	A1	A*1							LM733H	UA733HM	0	733HMTB
CHP																			0	733XC
CHP																			0	733XM
DIL-14/1C	N	N	T	E-	E+	V-	N	N	T*	R	V+	N	N	N			LM741AD	UA741ADM	0	741ADM
FLP-10/3C	N	T	E-	E+	V-	T*	R	V+	N	N							SFC2741PM	UA741AFM	0	741AFM
TO5-8/1M	T	E-	E+	V-M	T*	R	V+	N									TBA222	UA741AHM	0	741AHM
TO5-8/1M	T	E-	E+	V-M	T*	R	V+	N									TBA222	UA741HM	0	741BE
FLP-10/3C	N	T	E-	E+	V-	T*	R	V+	N	N							SFC2741PM	UA741FM	0	741BH
DIL-14/1C	N	N	T	E-	E+	V-	N	N	T*	R	V+	N	N	N			LM741D	UA741DM	0	741BL
DIL-8/1P	T	E-	E+	V-	T*	R	V+	N									TBA221B	UA741TC	0	741C(DIL)
TO5-8/1M	T	E-	E+	V-M	T*	R	V+	N									TBA221	UA741HC	0	741CE
CHP																			0	741(CHP)
DIL-8/1P	T	E-	E+	V-	T*	R	V+	N											0	741CHS-DIL8
DIL-8/1P	T	E-	E+	V-	T*	R	V+	N										741CHSDIL8	0	741CHSPA
TO5-8/1M	T	E-	E+	V-M	T*	R	V+	N										741MHS-TO5	0	741CHS-TO5
TO5-8/1M	T	E-	E+	V-M	T*	R	V+	N										741CHS-TO5	0	741CHSTY
DIL-14/1P	N	N	T	E-	E+	V-	N	N	T*	R	V+	N	N	N			TBA221A	UA741DC	0	741CJ
DIL-14/1C	N	N	T	E-	E+	V-	N	N	T*	R	V+	N	N	N			TBA221A	UA741DC	0	741CL
DIL-8/1P	T	E-	E+	V-	T*	R	V+	N											0	741C-LN-DIL8
TO5-8/1M	T	E-	E+	V-M	T*	R	V+	N										741-LN-TO5	0	741C-LN-TO5
DIL-8/1P	T	E-	E+	V-	T*	R	V+	N									TBA221B	UA741TC	0	741CP
TO5-8/1M	T	E-	E+	V-M	T*	R	V+	N									TBA221	UA741HC	0	741C(TO5)
DIL-14/1C	N	N	T	E-	E+	V-	N	N	T*	R	V+	N	N	N			TBA221A	UA741DC	0	741DC
DIL-14/1C	N	N	T	E-	E+	V-	N	N	T*	R	V+	N	N	N			LM741D	UA741DM	0	741DM
DIL-14/1C	N	N	T	E-	E+	V-	N	N	T*	R	V+	N	N	N			TBA221A	UA741EDC	0	741EDC
TO5-8/1M	T	E-	E+	V-M	T*	R	V+	N									TBA221	UA741EHC	0	741EHC
FLP-10/3C	N	T	E-	E+	V	T*	R	V+	N	N							SFC2741PM	UA741FM	0	741FM
TO5-8/1M	T	E-	E+	V-M	T*	R	V+	N									TBA221	UA741HC	0	741HC
TO5-8/1M	T	E-	E+	V-M	T*	R	V+	N									TBA222	UA741HM	0	741HM
DIL-14/1C	N	N	T	E-	E+	V-	N	N	T	R	V+	N	N	N					0	741-LN-DIL
FLP-10/3C	N	T	E-	E+	V-	T*	R	V+	N	N									0	741-LN-FLP
TO5-8/1M	T	E-	E+	V-M	T*	R	V+	N											0	741-LN-TO5
DIL-14/1C	N	N	T	E-	E+	V-	N	N	T*	R	V+	N	N	N				741MHS-DIL	0	741MHSDD
DIL-14/1C	N	N	T	E-	E+	V-	N	N	T*	R	V+	N	N	N				741MHS-DIL	0	741MHS-DIL
FLP-10/3C	N	T	E-	E+	V-	T*	R	V+	N	N								741MHS-FLP	0	741MHSFD
FLP-10/3C	N	T	E-	E+	V-	T*	R	V+	N	N									0	741MHS-FLP
TO5-8/1M	T	E-	E+	V-M	T*	R	V+	N											0	741-LN-TO5
TO5-8/1M	T	E-	E+	V-M	T*	R	V+	N										741MHS-TO5	0	741MHSTY
TO5-8/1M	T	E-	E+	V-M	T*	R	V+	N									TBA222	UA741HM	0	741(TO5)
CHP																			0	741XC
CHP																			0	741XM
DIL-14/1C	E-1	E+1	T1	V-	T2	E+2	E-2	T*2	V+2	R2	N	R1	V+	T*1			SFC2747KM	UA747ADM	0	747ADM
FLP-14/3C	E-1	E+1	T1	V-	T2	E+2	E-2	T*2	V+2	R2	N	R1	V+1	T*1				AM747AFM	0	747AFM
TO5-10/1M	R1	V+1	E-1	E+1	V-	E+2	E-2	V+2	R2	N							TBC0747	UA747AHM	0	747AHM
TO5-10/1M	R1	V+1	E-1	E+1	V-	E+2	E-2	V+2	R2	N							TBC0747	UA747HM	0	747BE
DIL-14/1C	E-1	E+1	T1	V-	T2	E+2	E-2	T*2	V+2	R2	N	R1	V+1	T*1			SFC2747KM	UA747DM	0	747BL
TO5-10/1M	R1	V+1	E-1	E+1	V-	E+2	E-2	V+2	R2	N							TBB0747	UA747HC	0	747CE
CHP																			0	747(CHP)
DIL-14/1P	E-1	E+1	T1	V-	T2	E+2	E-2	T*2	V+2	R2	N	R1	V+1	T*1			TBB0747A	UA747DC	0	747CJ
DIL-14/1C	E-1	E+1	T1	V-	T2	E+2	E-2	T*2	V+2	R2	N	P1	V+1	T*1			TBB0747A	UA747DC	0	747DC
DIL-14/1C	E-1	E+1	T1	V-	T2	E+2	E-2	T*2	V+2	R2	N	R1	V+1	T*1			SFC2747KM	UA747DM	0	747DM

TYPE NUMBER	MFR	APP	CMP	GBP MIN	SLEW RATE MIN	V_S^+ MAX	V_S^- MAX	T_{op} MAX	A_{VOL} MIN	V_{IO} MAX	I_B MAX	I_{IO} MAX	P_{TOT} MAX	I_{OUT} MIN	V_{OUT} MIN	V_{ICM} MAX	V_{IDF} MAX	dV_{IO}/dT MAX	P_Q MAX	I_Q MAX	CMRR MIN	PSRR MIN	R_{IN} MIN
747EDC	ADU	DGK	INT	.4MHZ	0.3V/uS	+22V	-22V	70C	94dB	3MV	80NA	30NA	670MWF	10MA	16V	15V	30V	15uV/C	150MW	3MA	80dB	86dB	1M
747EHC	ADU	DGK	INT	.4MHZ	0.3V/uS	+22V	-22V	70C	94dB	3MV	80NA	30NA	500MWF	10MA	16V	15V	30V	15uV/C	150MW	3MA	80dB	86dB	1M
747FM	ADU	DGK	INT	.	0.2V/uS	+22V	-22V	125C	94dB	5MV	500NA	200NA	500MWF	5MA	12V	15V	30V		85MW	3MA	70dB	76dB	300K
747HC	ADU	DGK	INT	.	0.2V/uS	+18V	-18V	70C	88dB	6MV	500NA	200NA	500MWF	5MA	12V	15V	30V		85MW	3MA	70dB	76dB	300K
747HM	ADU	DGK	INT	.	0.2V/uS	+22V	-22V	125C	94dB	5MV	500NA	200NA	500MWF	5MA	12V	15V	30V		85MW	3MA	70dB	76dB	300K
747PC	ADU	DGK	INT	.	0.2V/uS	+18V	-18V	70C	88dB	6MV	500NA	200NA	800MWF	5MA	12V	15V	30V		85MW	3MA	70dB	76dB	300K
747XC	ADU	DGK	INT	.	0.2V/uS	+18V	-18V	70C	88dB	6MV	500NA	200NA	.	5MA	12V	15V	30V		85MW	3MA	70dB	76dB	300K
747XM	ADU	DGK	INT	.	0.2V/uS	+22V	-22V	125C	94dB	5MV	500NA	200NA		5MA	12V	15V	30V		85MW	3MA	70dB	76dB	300K
748BE	TDG	GPU	EXT	.	0.2V/uS	+22V	-22V	125C	94dB	5MV	500NA	200NA	500MWF	5MA	12V	15V	30V		85MW	3MA	70dB	76dB	300K
748BH	TDG	GPU	EXT	.	0.2V/uS	+22V	-22V	125C	94dB	5MV	500NA	200NA	570MWF	5MA	12V	15V	30V		85MW	3MA	70dB	76dB	300K
748BL	TDG	GPU	EXT	.	0.2V/uS	+22V	-22V	125C	94dB	5MV	500NA	200NA	670MWF	5MA	12V	15V	30V		85MW	3MA	70dB	76dB	300K
748CE	TDG	GPU	EXT	.	0.2V/uS	+18V	-18V	70C	86dB	6MV	500NA	200NA	500MWF	5MA	12V	15V	30V		85MW	3MA	70dB	76dB	300K
748(CHP)	RAU	GPU	EXT	.	.25V/uS	+22V	-22V	125C	94dB	5MV	500NA	200NA	.	5MA	10V	15V	30V		85MW	3MA	70dB	76dB	300K
748CJ	TDG	GPU	EXT	.	0.2V/uS	+22V	-22V	70C	86dB	6MV	500NA	200NA	670MWF	5MA	12V	15V	30V		85MW	3MA	70dB	76dB	300K
748CL	TDG	GPU	EXT	.	0.2V/uS	+22V	-22V	70C	86dB	6MV	500NA	200NA	670MWF	5MA	12V	15V	30V		85MW	3MA	70dB	76dB	300K
748CP	TDG	GPU	EXT	.	0.2V/uS	+18V	-18V	70C	86dB	6MV	500NA	200NA	310MWF	5MA	12V	15V	30V		85MW	3MA	70dB	76dB	300K
748DC	ADU	GPU	EXT	.	0.2V/uS	+22V	-22V	70C	86dB	6MV	500NA	200NA	670MWF	5MA	12V	15V	30V		85MW	3MA	70dB	76dB	300K
748DM	ADU	GPU	EXT	.	0.2V/uS	+22V	-22V	125C	94dB	5MV	500NA	200NA	670MWF	5MA	12V	15V	30V		85MW	3MA	70dB	76dB	300K
748HC	ADU	GPU	EXT	.	0.2V/uS	+18V	-18V	70C	86dB	6MV	500NA	200NA	500MWF	5MA	12V	15V	30V		85MW	3MA	70dB	76dB	300K
748HM	ADU	GPU	EXT	.	0.2V/uS	+22V	-22V	125C	94dB	5MV	500NA	200NA	500MWF	5MA	12V	15V	30V		85MW	3MA	70dB	76dB	300K
748XC	ADU	GPU	EXT	.	0.2V/uS	+18V	-18V	70C	86dB	6MV	500NA	200NA		5MA	12V	15V	30V		85MW	3MA	70dB	76dB	300K
748XM	ADU	GPU	EXT	.	0.2V/uS	+22V	-22V	125C	94dB	5MV	500NA	200NA		5MA	12V	15V	30V		85MW	3MA	70dB	76dB	300K
835BJ	TDG	QGK	INT	.3MHZ	0.2V/uS	+18V	-18V	125C	94dB	6MV	300NA	30NA	540MWF	5MA	10V	18V	30V		.	4MA	70dB	76dB	
835BL	TDG	QGK	INT	.3MHZ	0.2V/uS	+18V	-18V	125C	94dB		300NA	30NA	900MWF	5MA	10V	18V	30V		.	4MA	70dB	76dB	
835CJ	TDG	QGK	INT	.3MHZ	0.2V/uS	+16V	-16V	70C	88dB	6MV	500NA	100NA	540MWF	5MA	10V	16V	30V		.	5MA	70dB	76dB	
835CL	TDG	QGK	INT	.3MHZ	0.2V/uS	+16V	-16V	70C	88dB	6MV	500NA	100NA	900MWF	5MA	10V	16V	30V		.	5MA	70dB	76dB	
836BJ	TDG	QGK	INT	.5MHZ	0.3V/uS	+18V	-18V	125C	94dB	5MV	300NA	30NA	540MWF	5MA	10V	18V	30V		.	4MA	70dB	76dB	
836BL	TDG	QGK	INT	.5MHZ	0.3V/uS	+18V	-18V	125C	94dB	5MV	300NA	30NA	900MWF	5MA	10V	18V	30V		.	4MA	70dB	76dB	
836CJ	TDG	QGK	INT	.5MHZ	0.3V/uS	+16V	-16V	70C	88dB	8MV	500NA	100NA	540MWF	5MA	10V	16V	30V		.	5MA	70dB	76dB	
836CL	TDG	QGK	INT	.5MHZ	0.3V/uS	+16V	-16V	70C	88dB	8MV	500NA	100NA	900MWF	5MA	10V	16V	30V		.	5MA	70dB	76dB	
844BE	TDG	GPK	EXT	.	1V/uS	+22V	-22V	125C	100dB	2MV	30NA	5NA	500MWF	5MA	12V	15V	30V	15uv/C	.	3MA	80dB	80dB	25
844BL	TDG	GPK	EXT	.	1V/uS	+22V	-22V	125C	100dB	2MV	30NA	5NA	670MWF	5MA	12V	15V	30V	15uv/C	.	3MA	80dB	80dB	25
844CE	TDG	GPK	INT	.	1V/uS	+18V	-18V	70C	94dB	5MV	75NA	10NA	500MWF	5MA	12V	15V	30V	50uv/C	.	3MA	70dB	70dB	10
844CL	TDG	GPK	INT	.	1V/uS	+18V	-18V	70C	94dB	5MV	75NA	10NA	670MWF	5MA	12V	15V	30V	50uv/C	.	3MA	70dB	70dB	10
844CP	TDG	GPK	INT	.	1V/uS	+18V	-18V	70C	94dB	5MV	75NA	10NA	310MWF	5MA	12V	15V	30V	50uv/C	.	3MA	70dB	70dB	10
845BE	TDG	GPK	INT	.	.15V/uS	+22V	-22V	125C	100dB	2MV	30NA	5NA	500MWF	5MA	12V	15V	30V	15uv/C	.	3MA	80dB	80dB	25
845BL	TDG	GPK	INT	.	.15V/uS	+22V	-22V	125C	100dB	2MV	30NA	5NA	670MWF	5MA	12V	15V	30V	15uv/C	.	3MA	80dB	80dB	25
845CE	TDG	GPK	INT	.	.15V/uS	+18V	-18V	70C	94dB	5MV	75NA	10NA	500MWF	5MA	12V	15V	30V	50uv/C	.	3MA	70dB	70dB	10
845CL	TDG	GPK	INT	.	.15V/uS	+18V	-18V	70C	94dB	5MV	75NA	10NA	670MWF	5MA	12V	15V	30V	50uv/C	.	3MA	70dB	70dB	10
845CP	TDG	GPK	INT	.	.15V/uS	+18V	-18V	70C	94dB	5MV	75NA	10NA	310MWF	5MA	12V	15V	30V	50uv/C	.	3MA	70dB	70dB	10
846BE	TDG	GPU	EXT	.	1V/uS	+22V	-22V	125C	100dB	2MV	30NA	5NA	500MWF	5MA	12V	15V	30V	15uv/C	.	3MA	80dB	80dB	25
846BL	TDG	GPU	EXT	.	1V/uS	+22V	-22V	125C	100dB	2MV	30NA	5NA	670MWF	5MA	12V	15V	30V	15uv/C	.	3MA	80dB	80dB	25
846CE	TDG	GPU	EXT	.	1V/uS	+18V	-18V	70C	94dB	5MV	75NA	10NA	500MWF	5MA	12V	15V	30V	50uv/C	.	3MA	70dB	70dB	10
846CL	TDG	GPU	EXT	.	1V/uS	+18V	-18V	70C	94dB	5MV	75NA	10NA	670MWF	5MA	12V	15V	30V	50uv/C	.	3MA	70dB	70dB	10
846CP	TDG	GPU	EXT	.	1V/uS	+18V	-18V	70C	94dB	5MV	75NA	10NA	310MWF	5MA	12V	15V	30V	50uv/C	.	3MA	70dB	70dB	10
1017	TEU	HCO	INT	1MHZ	0.3V/uS	+18V	-18V	85C	86dB	6MV	100NA		.	.12A	10V	15V	30V	45uV/C	.	15MA	60dB	70dB	40
1020	TEU	LVD	INT	.5MHZ	0.3V/uS	+18V	-18V	85C	120dB	3MV	25NA	5NA	.	5MA	10V	15V	30V	5uV/C	.	12MA	86dB	.	2
1020-01	TEU	LVD	INT	.5MHZ	0.3V/uS	+18V	-18V	85C	120dB	0.5MV	25NA	5NA	.	5MA	10V	15V	30V	1.5uV/C	.	12MA	86dB	.	2
1020-02	TEU	LVD	INT	.5MHZ	0.3V/uS	+18V	-18V	85C	120dB	0.5MV	25NA	5NA	.	5MA	10V	15V	30V	0.5uV/C	.	12MA	86dB	.	2
1020-03	TEU	LVD	INT	.5MHZ	0.3V/uS	+18V	-18V	85C	120dB	0.5MV	25NA	5NA	.	5MA	10V	15V	30V	.25uV/C	.	12MA	86dB	.	2
1021	TEU	FET	INT	2MHZ	6V/uS	+18V	-18V	85C	100dB	10MV	10pA	.		20MA	10V	15V	30V	50uv/C	.	5MA	100dB	.	10
1023	TEU	FET	INT	2MHZ	6V/uS	+18V	-18V	60C	100dB	0.7MV	5pA	.		20MA	10V	15V	30V	5uV/C	.	6MA	100dB	.	10
1023-01	TEU	FET	INT	2MHZ	6V/uS	+18V	-18V	60C	100dB	0.7MV	2pA	.		20MA	10V	15V	30V	2uV/C	.	6MA	100dB	.	10
1024	TEU	HCO	INT	.5MHZ	6V/uS	+18V	-18V	85C	94dB	30MV	50NA	50NA	.	20MA	10V	15V	5V	20uV/C	.	9MA	80dB	80dB	300
1025	TEU	XSR	INT	50MHZ	500V/uS	+18V	-18V	85C	100dB	10MV	20pA	20pA	.	50MA	10V	15V	30V	50uV/C	.	40MA	70dB	80dB	10
1026	TEU	FET	INT	1MHZ	10V/uS	+18V	-18V	85C	100dB	10MV	50pA	.		5MA	10V	15V	30V	50uv/C	.	10MA	86dB	.	10
1026-01	TEU	FET	INT	1MHZ	10V/uS	+18V	-18V	85C	100dB	10MV	20pA	.		5MA	10V	15V	30V	20uV/C	.	10MA	86dB	.	10
1027	TEU	XSR	INT	10MHZ	60V/uS	+18V	-18V	85C	100dB	150MV	50pA	20pA	.	20MA	10V	15V	30V	50uv/C	.	12MA	80dB	70dB	100
1027-01	TEU	HSR	INT	10MHZ	60V/uS	+18V	-18V	85C	100dB	150MV	50pA	20pA	.	20MA	10V	15V	30V	15uV/C	.	12MA	80dB	70dB	100
1028	TEU	XHG	INT	1MHZ	6V/uS	+18V	-18V	85C	108dB	10MV	35NA	10NA	.	5MA	10V	15V	15V	20uV/C	.	5MA	80dB	80dB	300

For detailed explanations of column heading notations, see App. A.

Also for ready references the more important abbreviations used in the column headings are listed below:

LEFT HAND PAGE

APP = application (codes at APP.E.)

CMRR = common mode rejection ratio

CMP = compensation (frequency)

dV_{io}/dT = input offset voltage temperature drift

GBP = gain bandwidth product

I_B = input bias current

I_{io} = input bias offset current

I_Q = quiescent supply current

MFR = manufacturer (codes at App.C.)

P_G = quiescent power consumer

PSRR = power supply rejection ratio

V_{ICM} = common mode input voltage rating

V_{iDF} = differential input voltage rating

V_{io} = input offset voltage

V_S = dc supply voltage

RIGHT HAND PAGE

Lead out coding summary (details at APP G.) for different cases (APP.F.)

A = gain adjust

B = bias adjust

C = case

E— = inverting input

E+ = non-inverting input

F,F* = input frequency compensation

G = ground

J = high level input

K = output, open collector

L = output, open emitter

M = metal case

N = not connected

Q = special terminal

R,R* = outputs

S = strobe

T,T* = offset balance

V* = +ve dc supply

V— = —ve dc supply

W = guard ring

X = blank position, no lead

++ = +ve supplementary dc supply

—— = —ve supplementary dc supply

∅,∅* = output frequency compensation

CASE (APP F)	LD 1	LD 2	LD 3	LD 4	LD 5	LD 6	LD 7	LD 8	LD 9	LD 10	LD 11	LD 12	LD 13	LD 14	LD 15	LD 16	EUROPE SUBSTITUTE	USA SUBSTITUTE	ISS	TYPE NUMBER
DIM-9/5P	T	E+	E-	T*	V+	G	V-	R	T1										0	1030
DIM-7/5P	E+	E-	V+	G	V-	R	T												0	1034
DIM-7/5P	E+	E-	V+	G	V-	R	T												0	1034-01
DIM-9/5P	T	E+	E-	T*	V+	G	V-	R	T1									1035-01	0	1035
DIM-9/5P	T	E+	E-	T*	V+	G	V-	R	T1									1035-02	0	1035-01
DIM-9/5P	T	E+	E-	T*	V+	G	V-	R	T1										0	1035-02
TO5-8/1M	T	E-	E+	V-	T*	R	V+	N									TBA222	UA741HM	0	1319
TO5-8/1M	T	E-	E+	V-	T*	R	V+	N									LM741AH	UA741AHM	0	1319-01
TO5-8/1M	T	E-	E+	V-	T*	R	V+	∅										1321-01	0	1321
TO5-8/1M	T	E-	E+	V-	T*	R	V+	∅											0	1321-01
TO5-8/1M	T	E-	E+	V-	T*	R	V+	∅										1322-01	0	1322
TO5-8/1M	T	E-	E+	V-	T*	R	V+	∅											0	1322-01
TO5-8/1M	T	E-	E+	V-M	N	R	V+	T*										1323-01	0	1323
TO5-8/1M	T	E-	E+	V-M	N	R	V+	T*										1323-02	0	1323-01
TO5-8/1M	T	E-	E+	V-M	N	R	V+	T*											0	1323-02
TO5-8/1M	T	E-	E+	V-	T*	R	V+	∅										LM144H	0	1332
TO5-8/1M	F	E-	E+	V-	∅	R	V+	F*										MC1439G	0	1339
TO5-8/1M	F	E-	E+	V-	∅	R	V+	F*										MC1539G	0	1339-01
TO5-8/1M	F	E-	E+	V-	∅	R	V+	F*										MC1539G	0	1339-02
TO5-8/1M	Q	E-	E+	V-	Q*	R	V+	Q1											0	1340
TO5-8/1M	T	E-	E+	V-	T*	R	V+	N									UA740HM	SU536T	0	1421
TO5-8/1M	T	E-	E+	V-	T*	R	V+	N										ICL8007MTV	0	1421-01
TO5-8/1M	T	E-	E+	V-	T*	R	V+	N											0	1421-02
TO5-8/1M	T	E-	E+	V-	T*	R	V+	N										1421-25	0	1421-24
TO5-8/1M	T	E-	E+	V-	T*	R	V+	N											0	1421-25
TO5-8/1M	T	E-	E+	V-	T*	R	V+	N									ICL8007CTV	SU536T	0	1424
TO5-8/1M	T	E-	E+	V-	T*	R	V+	N									SU536T	ICL8007MTV	0	1425
TO5-8/1M	T	E-	E+	V-	T*	R	V+	N											0	1425-01
TO5-8/1M	T	E-	E+	V-	T*	R	V+	N											0	1425-02
TO5-8/1M	T	E-	E+	V-	T*	R	V+	N									ICL8007MTV	SU536T	0	1426
TO5-8/1M	T	E-	E+	V-	T*	R	V+	N										1426-02	0	1426-01
TO5-8/1M	T	E-	E+	V-	T*	R	V+	N										ICL8007MTV	0	1426-02
TO5-8/1M	T	E-	E+	V-	T*	R	V+	N											0	1426-03
DIM-14/1M	N	T	T*	E-	E+	V-	N	G	N	R	V+	T1	N	N					0	1430
TO5-8/1M	T	E-	E+	V-	T*	R	V+	∅										1433-01	0	1433
TO5-8/1M	T	E-	E+	V-	T*	R	V+	∅											0	1433-01
TO5-8/1M	T	E-	E+	V-	T*	R	V+	∅										1434-01	0	1434
TO5-8/1M	T	E-	E+	V-	T*	R	V+	∅											0	1434-01
DIM-14/1M	∅	R	∅*	V+	T	T*	E-	E+	N	N	N	V-	R*	M					0	1435
DIM-14/1M	∅	R	∅*	V+	T	T*	E-	E+	N	N	N	V-	R*	M					0	1435-83
CHP																			0	1437(CHP)
TO5-8/1M	T	E-	E+	V-	T*	R	V+	M										ICL8007ATV	0	1439
TO5-8/1M	R1	E-1	E+1	V-	E+2	E-2	R2	V+									TBC1458	MC1458G	0	1458CE
CHP																			0	1458(CHP)
DIL-8/1P	R1	E-1	E+1	V-	E+2	E-2	R2	V+									TBB1458B	MC1458U	0	1458CP
TO5-8/1M	R1	E-1	E+1	V-	E+2	E-2	R2	V+									TBB1458	MC1458G	0	1458E
DIL-8/1P	R1	E-1	E+1	V-	E+2	E-2	R2	V+									TBB1458B	MC1458U	0	1458P
CHP																			0	1514(CHP)
CHP																			0	1537(CHP)
CHP																			0	1556A(CHP)
CHP																			0	1558(CHP)
TO5-8/1M	R1	E-1	E+1	V-	E+2	E-2	R2	V+									TBC1458	MC1558G	0	1558E
DIM-7/5P	E+	E-	V+	G	V-	R	T											1701-01	0	1701
DIM-7/5P	E+	E-	V+	G	V-	R	T												0	1701-01
DIM-7/5P	E+	E-	V+	G	V-	R	T											1702-01	0	1702
DIM-7/5P	E+	E-	V+	G	V-	R	T												0	1702-01
DIM-7/5P	E+	E-	V+	G	V-	R	T											1703-01	0	1703
DIM-7/5P	E+	E-	V+	G	V-	R	T												0	1703-01
FLP-14/3G	N	∅	T	E-	E+	N	N	N	N	M	V-	T*	R	V+	N			HA9-2500	0	2500(FLP)
TO5-8/1M	T	E-	E+	V-	T*	R	V+	∅										HA2-2500	0	2500(TOS)

For detailed explanations of column heading notations, see App. A.

Also for ready references the more important abbreviations used in the column headings are listed below:

LEFT HAND PAGE

APP = application (codes at APP.E.)

CMRR = common mode rejection ratio

CMP = compensation (frequency)

dV_{IO}/dT = input offset voltage temperature drift

GBP = gain bandwidth product

I_B = input bias current

I_{IO} = input bias offset current

I_Q = quiescent supply current

MFR = manufacturer (codes at App.C.)

P_Q = quiescent power consumer

PSRR = power supply rejection ratio

V_{ICM} = common mode input voltage rating

V_{IDF} = differential input voltage rating

V_{IO} = input offset voltage

V_S = dc supply voltage

RIGHT HAND PAGE

Lead out coding summary (details at APP.G.) for different cases (APP.F.)

A = gain adjust

B = bias adjust

C = case

E– = inverting input

E+ = non-inverting input

F,F* = input frequency compensation

G = ground

J = high level input

K = output, open collector

L = output, open emitter

M = metal case

N = not connected

Q = special terminal

R,R* = outputs

S = strobe

T,T* = offset balance

V+ = +ve dc supply

V– = –ve dc supply

W = guard ring

X = blank position, no lead

++ = +ve supplementary dc supply

–– = –ve supplementary dc supply

ø,ø* = output frequency compensation

CASE (APP F)	LD 1	LD 2	LD 3	LD 4	LD 5	LD 6	LD 7	LD 8	LD 9	LD 10	LD 11	LD 12	LD 13	LD 14	LD 15	LD 16	EUROPE SUBSTITUTE	USA SUBSTITUTE	ISS	TYPE NUMBER
FLP-14/3G	N	ø	T	E-	E+	N	N	N	M	V-	T*	R	V+	N				HA9-2502	0	2502(FLP)
T05-8/1M	T	E-	E+	V-	T*	R	V+	ø										HA2-2502	0	2502(T05)
FLP-14/3G	N	ø	T	E-	E+	N	N	N	M	V-	T*	R	V+	N				HA9-2505	0	2505(FLP)
T05-8/1M	T	E-	E+	V-	T*	R	V+	ø										HA2-2505	0	2505(T05)
FLP-14/3G	N	ø	T	E-	E+	N	N	N	M	V-	T*	R	V+	N				HA9-2520	0	2520(FLP)
T05-8/1M	T	E-	E+	V-	T*	R	V+	ø										HA2-2520	0	2520(T05)
FLP-14/3G	N	ø	T	E-	E+	N	N	N	W	V-	T*	R	V+	N				HA9-2522	0	2522(FLP)
T05-8/1M	T	E-	E+	V-	T*	R	V+	ø										HA2-2522	0	2522(T05)
FLP-14/3G	N	ø	T	E-	E+	N	N	N	C	V-	T*	R	V+	N				HA9-2525	0	2525(FLP)
T05-8/1M	T	E-	E+	V-	T*	R	V+	ø										HA2-2525	0	2525(T05)
T05-8/1M	T	E-	E+	V-	T*	R	V+	MW											0	2740BE
T05-8/1M	T	E-	E+	V-	T*	R	V+	MW											0	2740CE
T05-8/1M	T	E-	E+	V-	T*	R	V+	MW											0	2740DE
T05-8/1M	F	E-	E+	V-M	N	R	V+	F*									SFC2108A	LM108AH	0	3051
T05-8/1M	F	E-	E+	V-M	N	R	V+	F*									SFC2108A	LM108AH	0	3052
T05-8/1M	F	E-	E+	V-M	N	R	V+	F*									SFC2108A	LM108AH	0	3053
T05-8/1M	F	E-	E+	V-M	N	R	V+	F*									SFC2108M	LM108H	0	3055
T05-8/1M	F	E-	E+	V-M	N	R	V+	F*									SFC2108M	LM108H	0	3056
T05-8/1M	F	E-	E+	V-M	N	R	V+	F*									SFC2108M	LM108H	0	3057
T05-8/1M	T	E-	E+	V-	T*	R	V+	N											0	3268/14
T05-8/1M	T	E-	E+	V-	T*	R	V+	N										3268/14	0	3269/14
DIM-7/5P	E+	E-	V+	G	V-	R	T											3291/14	0	3291/14
DIM-7/5P	E+	E-	V+	G	V-	R	T											3292/14	0	3292/14
DIM-7/5P	E+	E-	V+	G	V-	R	T											3293/14	0	3293/14
DIL-14/1P	R1	R2	V+	E-	E+	E-1	E+1	E-4	E+4	E-3	E+3	G	R3	R4			LM3302J	MC3302L	0	3302J
DIM-14/1P	X	X	X	X	E+	X	V-	X	X	R	X	X	X	V+					0	3329/03
DIM-7/5P	E+	E-	V+	G	V-	R	T												0	3400A
DIM-7/5P	E+	E-	V+	G	V-	R	T												0	3400B
T05-8/1M	T	E-	E+	V-M	T*	R	V+	N										RC1556AT	0	3500A
DIL-8/1P	T	E-	E+	V-	T*	R	V+	N										RC1556ANB	0	3500AN
T05-8/1M	T	E-	E+	V-M	T*	R	V+	N										RM1556AT	0	3500B
DIL-8/1P	T	E-	E+	V-	T*	R	V+	N										RC1556ANB	0	3500BN
T05-8/1M	T	E-	E+	V-M	T*	R	V+	N										RM1556AT	0	3500C
DIL-8/1P	T	E-	E+	V-	T*	R	V+	N										RC1556ANB	0	3500CN
T05-8/1M	T	E-	E+	V-M	T*	R	V+	N											0	3500E
																			0	3500MP(PAIR)
T05-8/1M	T	E-	E+	V-M	T*	R	V+	N										RM1556AT	0	3500R
T05-8/1M	T	E-	E+	V-M	T*	R	V+	N										RC1556AT	0	3500S
T05-8/1M	T	E-	E+	V-M	T*	R	V+	N										RM1556AT	0	3500T
T05-8/1M	T	E-	E+	V-M	T*	R	V+	N										RC4131T	0	3501A
																		RC4131T	0	3501B
T05-8/1M	T	E-	E+	V-M	T*	R	V+	N										RM4131T	0	3501C
T05-8/1M	T	E-	E+	V-M	T*	R	V+	N										RM4131T	0	3501R
T05-8/1M	T	E-	E+	V-M	T*	R	V+	N										RM4131T	0	3501S
T05-8/1M	T	E-	E+	V-M	T*	R	V+	N										3503B	0	3503A
																			0	3503B
T05-8/1M	T	E-	E+	V-M	T*	R	V+	N										3503S	0	3503R
T05-8/1M	T	E-	E+	V-M	T*	R	V+	N											0	3503S
T05-8/1M	T	E-	E+	V-M	T*	R	V+	N											0	3505J
T05-8/1M	T	E-	E+	V-	T*	R	V+	N											0	3506J
T05-8/1M	T	E-	E+	V-	T*	R	V+	ø											0	3507J
T05-8/1M	T	E-	E+	V-	T*	R	V+	ø											0	3508J
T05-8/1M	T	E-	E+	V-	T*	R	V+	M											0	3521H
T05-8/1M	T	E-	E+	V-	T*	R	V+	M											0	3521J
T05-8/1M	T	E-	E+	V-	T*	R	V+	M											0	3521K
T05-8/1M	T	E-	E+	V-	T*	R	V+	M											0	3521L
T05-8/1M	T	E-	E+	V-	T*	R	V+	M											0	3521R
T05-8/1M	T	E-	E+	V-	T*	R	V+	M										3522K	0	3522J
T05-8/1M	T	E-	E+	V-	T*	R	V+	M										3522L	0	3522K
T05-8/1M	T	E-	E+	V-	T*	R	V+	M											0	3522L

TYPE NUMBER	MFR	APP	CMP	GBP MIN	SLEW RATE MIN	Vs+ MAX	Vs- MAX	Top MAX	AVOL MIN	VIO MAX	IB MAX	IIO MAX	PTOT MAX	IOUT MIN	VOUT MIN	VICM MAX	VIDF MAX	dVIO/dT MAX	PQ MAX	IQ MAX	CMRR MIN	PSRR MIN	RIN MIN
3522S	BUU	LBC	INT	.3MHZ	0.6V/uS	+20V	-20V	125C	94dB	500uV	5pA	1pA		10MA	10V	15V	20V	25uV/C		4MA	80dB	80dB	10G
3523J	BUU	LBC	INT	.3MHZ	0.6V/uS	+20V	-20V	70C	94dB	1MV	0.5pA	0.2pA		10MA	10V	15V	20V	50uV/C		4MA	70dB	80dB	100G
3523K	BUU	LBC	INT	.3MHZ	0.6V/uS	+20V	-20V	70C	94dB	500uV	.25pA	0.1pA		10MA	10V	15V	20V	25uV/C		4MA	70dB	80dB	100G
3523L	BUU	LBC	INT	.3MHZ	0.6V/uS	+20V	-20V	70C	94dB	500uV	0.1pA	.05pA		10MA	10V	15V	20V	25uV/C		4MA	70dB	80dB	100G
3540J	BUU	FET	INT	.3MHZ	6V/uS	+18V	-18V	70C	86dB	50MV	50pA	0.5pA		5MA	10V	13V	18V	75uV/C		6MA	70dB	60dB	1G
3542J	BUU	FET	INT	.3MHZ	.15V/uS	+20V	-20V	70C	88dB	20MV	25pA	2pA		10MA	10V	15V	20V	50uV/C		4MA	70dB	74dB	10G
3542S	BUU	FET	INT	.3MHZ	.15V/uS	+20V	-20V	125C	88dB	20MV	25pA	2pA		10MA	10V	15V	20V	50uV/C		4MA	70dB	74dB	10G
3550J	BUU	XSR	INT	3MHZ	65V/uS	+20V	-20V	70C	90dB	1MV	100pA			10MA	10V	15V	20V	50uV/C		11MA	70dB	56dB	10G
3550K	BUU	XSR	INT	6MHZ	100V/uS	+20V	-20V	70C	90dB	1MV	100pA			10MA	10V	15V	20V	50uV/C		11MA	70dB	56dB	10G
3550S	BUU	XSR	INT	3MHZ	65V/uS	+20V	-20V	125C	90dB	1MV	100pA			10MA	10V	15V	20V	50uV/C		11MA	70dB	56dB	10G
3551J	BUU	WBA	EXT	20MHZ	100V/uS	+20V	-20V	70C	90dB	1MV	100pA			10MA	10V	15V	20V	50uV/C		11MA	60dB	56dB	10G
3551S	BUU	WBA	EXT	20MHZ	100V/uS	+20V	-20V	125C	90dB	1MV	100pA			10MA	10V	15V	20V	50uV/C		11MA	60dB	56dB	10G
3553AM	BUU	VFA	INT	10MHZ		+20V	-20V	85C	0dB					0.2A	10V	20V							10G
3554AM	BUU	XSR	EXT	.3GHZ	1KV/uS	+18V	-18V	70C	96dB	1MV	100pA			50MA	10V	13V	18V	25uV/C		30MA	50dB	56dB	10G
3554BM	BUU	XSR	EXT	.3GHZ	1KV/uS	+18V	-18V	70C	96dB	0.5MV	100pA			50MA	10V	13V	18V	5uV/C		30MA	50dB	56dB	10G
3554SM	BUU	XSR	EXT	.3GHZ	1KV/uS	+18V	-18V	125C	96dB	1MV	50pA			50MA	10V	13V	18V	25uV/C		30MA	50dB	56dB	10G
3571AM	BUU	HPO	EXT	.2MHZ	1V/uS	+40V	-40V	85C	94dB	2MV	100pA			2A	30V	30V	40V	40uV/C		25MA	80dB		10G
3572AM	BUU	HPO	EXT	.2MHZ	1V/uS	+40V	-40V	85C	94dB	2MV	100pA			5A	30V	30V	40V	40uV/C		25MA	80dB		10G
3580J	BUU	HVO	INT	2MHZ	5V/uS	+35V	-35V	70C	96dB	10MV	50pA			60MA	30V	30V	35V	30uV/C		10MA	76dB		10G
3581J	BUU	HVO	INT	2MHZ	6V/uS	+75V	-75V	70C	102dB	3MV	20pA			30MA	70V	65V	75V	25uV/C		8MA	100dB		10G
3582J	BUU	HVO	INT	2MHZ	6V/uS	150V	150V	70C	108dB	3MV	20pA			15MA	145V	140V	150V	25uV/C		7MA	100dB		10G
3583AM	BUU	HVO	INT	2MHZ	10V/uS	150V	150V	85C	108dB	3MV	20pA			75MA	145V	150V	150V	25uV/C		9MA	100dB		10G
3583JM	BUU	HVO	INT	2MHZ	10V/uS	150V	150V	70C	108dB	3MV	20pA			75MA	145V	140V	150V	25uV/C		9MA	100dB		10G
3621J	BUU	FET	INT	.	0.1V/uS	+18V	-18V	85C	66dB	2MV	10pA			10MA	10V	8V		50uV/C		5MA	100dB	80dB	1G
3621K	BUU	FET	INT	.	0.1V/uS	+18V	-18V	85C	66dB	2MV	10pA			10MA	10V	8V		15uV/C		5MA	100dB	80dB	1G
3621L	BUU	FET	INT	.	0.1V/uS	+18V	-18V	85C	66dB	2MV	10pA			10MA	10V	8V		5uV/C		5MA	100dB	80dB	1G
3625A	BUU	PIA	INT	.	0.3V/uS	+18V	-18V	85C	60dB	.25MV	60NA			5MA	10V	10V		4uV/C		8MA	74dB	100dB	1G
3625B	BUU	PIA	INT	.	0.3V/uS	+18V	-18V	85C	60dB	.25MV	60NA			5MA	10V	10V		2uV/C		8MA	74dB	100dB	1G
3625C	BUU	PIA	INT	.	0.3V/uS	+18V	-18V	85C	60dB	.25MV	60NA			5MA	10V	10V				8MA	74dB	100dB	1G
4009	OAU	HVO	INT	10MHZ	3V/uS	+36V	-36V	125C	50dB	100MV	1uA			5MA	30V	30V		1.5uV/C					
4131(CHP)	RAU	GPK	INT	1MHZ	.	+22V	-22V	125C	94dB	2MV	50NA	10NA		7MA	16V	15V	30V	500uV/C	500MW	12MA	70dB		10K
4136(CHP)	RAU	QGK	INT	2MHZ	0.5V/uS	+22V	-22V	125C	94dB	5MV	500NA	200NA		5MA	12V	15V	30V	15uV/C		2MA	80dB	80dB	2M2
4250C(TO5)	ING	PRA	INT	.	.	+18V	-18V	70C	95dB	6MV	75NA	20NA	500MWF	1MA	12V	15V	30V		340MW	.	70dB	76dB	300K
4250(TO5)	ING	PRA	INT	.	.	+18V	-18V	125C	100dB	5MV	75NA	10NA	500MWF	1MA	12V	15V	30V		3MW	.1MA	70dB	74dB	
4253	TEU	LVD	INT	.	.	+18V	-18V	85C	80dB	1MV	10pA	.		5MA	10V	18V	36V	2uV/C	2.7MW	90UA	70dB	76dB	1T
4253-01	TEU	LVD	INT	.	.	+18V	-18V	85C	80dB	0.5MV	10pA	.		5MA	10V	18V	36V	1uV/C		16MA	76dB	110dB	1T
4531(CHP)	RAU	HSR	EXT	.	10V/uS	+22V	-22V	125C	94dB	5MV	500NA	200NA		5MA	10V	12V	15V		210MW	7MA	80dB	110dB	1T
4558(CHP)	RAU	DGK	INT	.	0.2V/uS	+22V	-22V	125C	94dB	5MV	500NA	200NA		5MA	12V	15V	30V		170MW		70dB	76dB	300K
4739(CHP)	RAU	DLN	EXT	.	0.3V/uS	+18V	-18V	70C	86dB	6MV	500NA	200NA		5MA	12V	15V	30V		170MW		70dB	74dB	300K
8001C(TO5)	ING	CPR	EXT	.	.	+18V	-18V	70C	84dB	5MV	250NA	50NA	500MWF		3V	18V	15V	30uV/C	60MW		70dB	70dB	3M
8001(TO5)	ING	CPR	EXT	.	.	+18V	-18V	125C	84dB	3MV	100NA	20NA	500MWF		3V	18V	15V	20uV/C	60MW		70dB	70dB	3M
8007AC	ING	FET	INT	.3MHZ	2.5V/uS	+18V	-18V	70C	86dB	30MV	1pA	0.5pA	500MWF	5MA	12V	15V	30V	50uV/C	180MW	6MA	86dB	74dB	0.5T
8007AM	ING	FET	INT	.3MHZ	2.5V/uS	+18V	-18V	125C	86dB	30MV	1pA	0.5pA	500MWF	5MA	12V	15V	30V	50uV/C	156MW	5MA	86dB	74dB	0.5T
8007C	ING	FET	INT	.3MHZ	2V/uS	+18V	-18V	70C	86dB	50MV	50pA	5pA	500MWF	5MA	12V	15V	30V	75uV/C	180MW	6MA	70dB	64dB	0.5T
8007C-1	ING	FET	INT	.3MHZ	3V/uS	+18V	-18V	70C	94dB	2MV	10pA	5pA	500MWF	5MA	12V	15V	30V	5uV/C	180MW	6MA	70dB	70dB	0.1T
8007C-2	ING	FET	INT	.3MHZ	3V/uS	+18V	-18V	70C	94dB	2MV	10pA	5pA	500MWF	5MA	12V	15V	30V	15uV/C	180MW	6MA	70dB	70dB	0.1T
8007C-3	ING	FET	INT	.3MHZ	3V/uS	+18V	-18V	70C	94dB	4MV	20pA	5pA	500MWF	5MA	12V	15V	30V	30uV/C	180MW	6MA	70dB	70dB	0.1T
8007C-5	ING	FET	INT	.3MHZ	3V/uS	+18V	-18V	70C	94dB	10MV	10pA	5pA	500MWF	5MA	12V	15V	30V	10uV/C	180MW	6MA	70dB	70dB	0.1T
8007M	ING	FET	INT	.3MHZ	2V/uS	+18V	-18V	125C	94dB	20MV	20pA	2pA	500MWF	5MA	12V	15V	30V	15uV/C	180MW	5MA	70dB	70dB	0.5T
8007M-2	ING	FET	INT	.3MHZ	3V/uS	+18V	-18V	125C	94dB	2MV	10pA	5pA	500MWF	5MA	12V	15V	30V	15uV/C	156MW	6MA	70dB	70dB	100G
8007M-5	ING	FET	INT	.3MHZ	3V/uS	+18V	-18V	125C	94dB	10MV	10pA	5pA	500MWF	5MA	12V	15V	30V	15uV/C	156MW	6MA	70dB	70dB	100G
8008C(DIL8)	ING	GPK	INT	.	0.1V/uS	+15V	-15V	70C	86dB	6MV	25NA	20NA	500MWF	5MA	12V	15V	30V	75uV/C	85MW	3MA	70dB	76dB	5M
8008C(TO5)	ING	GPK	INT	.	0.1V/uS	+15V	-15V	70C	86dB	6MV	25NA	20NA	500MWF	5MA	12V	15V	30V	75uV/C	85MW	3MA	70dB	76dB	5M
8008M(TO5)	ING	GPK	INT	.	0.1V/uS	+15V	-15V	125C	86dB	5MV	10NA	5NA	500MWF	5MA	12V	15V	30V	35uV/C	85MW	3MA	70dB	76dB	5M
8021C	ING	PRA	INT	.1MHZ	0.3V/uS	+18V	-18V	70C	94dB	6MV	30NA	10NA	300MWF	1MA	12V	15V	15V	25uV/C	0.6MW	.	70dB	76dB	3M
8021M	ING	PRA	INT	.1MHZ	.03V/uS	+18V	-18V	125C	94dB	3MV	20NA	7.5NA	300MWF	1MA	12V	15V	15V	25uV/C	.48MW	.	70dB	76dB	3M
8022C	ING	DPR	INT	.1MHZ	.03V/uS	+18V	-18V	70C	94dB	6MV	30NA	10NA	300MWF	1MA	12V	15V	15V	25uV/C	0.6MW	.	70db	76dB	3M
8022M	ING	DPR	INT	.1MHZ	.03V/uS	+18V	-18V	125C	94dB	3MV	20NA	7.5NA	300MWF	1MA	12V	15V	15V	25uV/C	.48MW	.	70dB	76dB	3M
8023C	ING	TPR	INT	.1MHZ	.03V/uS	+18V	-18V	70C	94dB	6MV	30NA	10NA	300MWF	1MA	12V	15V	15V	25uV/C	0.6MW	.	70dB	76dB	3M

For detailed explanations of column heading notations, see App. A.
Also for ready references the more important abbreviations used in the column headings are listed below:

LEFT HAND PAGE
APP = application (codes at APP.E.)
CMRR = common mode rejection ratio
CMP = compensation (frequency)
dV_{IO}/dT = input offset voltage temperature drift
GBP = gain bandwidth product
I_B = input bias current
I_{IO} = input bias offset current
I_Q = quiescent supply current
MFR = manufacturer (codes at App.C.)
P_Q = quiescent power consumer
PSRR = power supply rejection ratio
V_{ICM} = common mode input voltage rating
V_{IDF} = differential input voltage rating
V_{IO} = input offset voltage
V_S = dc supply voltage

RIGHT HAND PAGE
Lead out coding summary (details at APP.G.) for different cases (APP.F.)
A = gain adjust
B = bias adjust
C = case
E– = inverting input
E+ = non-inverting input
F,F* = input frequency compensation
G = ground
J = high level input
K = output, open collector
L = output, open emitter
M = metal case
N = not connected
Q = special terminal
R,R* = outputs
S = strobe
T,T* = offset balance
V+ = +ve dc supply
V– = –ve dc supply
W = guard ring
X = blank position, no lead
++ = +ve supplementary dc supply
–– = –ve supplementary dc supply
Ø,Ø* = output frequency compensation

CASE (APP F)	LD 1	LD 2	LD 3	LD 4	LD 5	LD 6	LD 7	LD 8	LD 9	LD 10	LD 11	LD 12	LD 13	LD 14	LD 15	LD 16	EUROPE SUBSTITUTE	USA SUBSTITUTE	I S	TYPE NUMBER
T05-8/1M	T	E-	E+	V-	T*	R	V+	M											0	3522S
T05-8/1M	T	E-	E+	V-	T*	R	V+	M											0	3523J
T05-8/1M	T	E-	E+	V-	T*	R	V+	M										3523K	0	3523K
T05-8/1M	T	E-	E+	V-	T*	R	V+	M										3523L	0	3523L
T05-8/1M	T	E-	E+	V-	T*	R	V+	N											0	3540J
T05-8/1M	T	E-	E+	V-	T*	R	V+	M										3542S	0	3542J
T05-8/1M	T	E-	E+	V-	T*	R	V+	M											0	3542S
T05-8/1M	T	E-	E+	V-	T*	R	V+	M											0	3550J
T05-8/1M	T	E-	E+	V-	T*	R	V+	M											0	3550K
T05-8/1M	T	E-	E+	V-	T*	R	V+	M											0	3550S
T05-8/1M	T	E-	E+	V-	T*	RØ	V+	Ø*											0	3551J
T05-8/1M	T	E-	E+	V-	T*	RØ	V+	Ø*											0	3551S
T03-8/2M	N	V+	R	E+	N	N	V-	N											0	3553AM
T03-8/2M	RØ	V+	Ø*	T	E-	E+	V-	T*											0	3554AM
T03-8/2M	RØ	V+	Ø*	T	E-	E+	V-	T*											0	3554BM
T03-8/2M	RØ	V+	Ø*	T	E-	E+	V-	T*											0	3554SM
T03-8/2M	R	Q	V+	E+	E-	V-	Ø	Q*										3572AM	0	3571AM
T03-8/2M	R	Q	V+	E+	E-	V-	Ø	Q*											0	3572AM
T03-8/2M	R	V+	T	T*	E-	E+	V-	N										3581J	0	3580J
T03-8/2M	R	V+	T	T*	E-	E+	V-	N										3582J	0	3581J
T03-8/2M	R	V+	T	T*	E-	E+	V-	N											0	3582J
T03-8/2M	R	V+	T	T*	E-	E+	V-	N											0	3583AM
T03-8/2M	R	V+	T	T*	E-	E+	V-	N										3583AM	0	3583JM
DIM-9/5P	A	E+	E-	T	V+	G	V-	R	A*									3621K	0	3621J
DIM-9/5P	A	E+	E-	T	V+	G	V-	R	A*									3621L	0	3621K
DIM-9/5P	A	E+	E-	T	V+	G	V-	R	A*										0	3621L
DIM-9/5P	A	E+	E-	T	V+	G	V-	R	A*									3625C	0	3625A
DIM-9/5P	A	E+	E-	T	V+	G	V-	R	A*									3625A	C	3625B
DIM-9/5P	A	E+	E-	T	V+	G	V-	R	A*										0	3625C
T05-8/1M	E-	V+	R	N	N	V-	N	E+											0	4009
CHP																			0	4131(CHP)
CHP																			0	4136(CHP)
T05-8/1M	T	E-	E+	V-	T*	R	V+	B									SG4250CT	LM4250CH	0	4250C(T05)
T05-8/1M	T	E-	E+	V-	T*	R	V+	B									SG4250T	LM4250H	0	4250(T05)
DIM-11/5P	A	E+	E-	A*	V+	G	V-	R	T	Q	Q*							4253-01	0	4253
DIM-11/5P	A	E+	E-	A*	V+	G	V-	R	T	Q	Q*								0	4253-01
CHP																			0	4531(CHP)
CHP																			0	4558(CHP)
CHP																			0	4739(CHP)
T05-10/1M	E-	N	T	T*	V-M	G	R	V+	++	E+								ICL8001CTZ	0	8001C(T05)
T05-10/1M	E-	N	T	T*	V-M	G	R	V+	++	E+								ICL8001MTZ	0	8001(T05)
T05-8/1M	T	E-	E+	V-	T*	R	V+	MW										ICL8007AC	0	8007AC
T05-8/1M	T	E-	E+	V-	T*	R	V+	MW										ICL8007AM	0	8007AM
T05-8/1M	T	E-	E+	V-	T*	R	R	N										ICL8007CTV	0	8007C
T05-8/1M	N	E-	E+	V-M	N	R	V+	N										ICL8007C1	0	8007C-1
T05-8/1M	N	E-	E+	V-M	N	R	V+	N										ICL8007C-2	0	8007C-2
T05-8/1M	N	E-	E+	V-M	N	R	V+	B										ICL8007C-3	0	8007C-3
T05-8/1M	T	E-	E+	V-M	T*	R	V+	N										ICL8007C-4	0	8007C-4
T05-8/1M	T	E-	E+	V-M	T*	R	V+	N										ICL8007C-5	0	8007C-5
T05-8/1M	T	E-	E+	V-	T*	R	V+	N										ICL8007MTV	0	8007M
T05-8/1M	N	E-	E+	V-M	N	R	V+	N										ICL8007M-2	0	8007M-2
T05-8/1M	T	E-	E+	V-M	T*	R	V+	N										ICL8007M-5	0	8007M-5
DIL-8/1P	T	E-	E+	V-	T*	R	V+	N									RC1556NB	ICL8008CPA	0	8008C(DIL8)
T05-8/1M	T	E-	E+	V-	T*	R	V+	N									MC1456G	ICL8008CTY	0	8008C(T05)
T05-8/1M	T	E-	E+	V-	T*	R	V+	N									MC1556G	ICL8008MTY	0	8008M(T05)
T05-8/1M	T	E-	E+	V-	T*	R	V+	B										LM4250CH	ICL8021CTA	0	8021C
T05-8/1M	T	E-	E+	V-	T*	R	V+	B										LM4250CH	ICL8021MTA	0	8021M
DIL-14/1C	E+1	V-	T1	R1	R2	B2	T2	E-2	E+2	T*2	V+	B1	T*1	E-1				ICL8022CDD	0	8022C
DIL-14/1C	E+1	V-	T1	R1	R2	B2	T2	E-2	E+2	T*2	V+	B1	T*1	E-1				ICL8022MDD	0	8022M
DIL-16/1C	N	E-1	E+1	R2	V23	B3	E-3	E+3	V-	R3	B2	E-2	E+2	R1	V+1	B1		ICL8023CDE	0	8023C

TYPE NUMBER	MFR	APP	CMP	GBP MIN	SLEW RATE MIN	V_S^+ MAX	V_S^- MAX	T_{op} MAX	A_{VOL} MIN	V_{IO} MAX	I_B MAX	I_{IO} MAX	P_{TOT} MAX	I_{OUT} MIN	V_{OUT} MIN	V_{ICM} MAX	V_{IDF} MAX	dV_{IO}/dT MAX	P_Q MAX	I_Q MAX	CMRR MIN	PSRR MIN	R_{IN} MIN
8023M	ING	TPR	INT	.1MHZ	.03V/uS	+18V	-18V	125C	94dB	3MV	20NA	7.5NA	300MWF	1MA	12V	15V	15V	25uV/C	48MW	.	70dB	76dB	3M
8043C-DIL/C	ING	DFE	INT	.3MHZ	2V/uS	+18V	-18V	70C	86dB	50MV	50pA	5pA	500MWF	5MA	12V	15V	30V	75uV/C	204MW	7MA	70dB	64dB	0.5T
8043C-DIL/P	ING	DFE	INT	.3MHZ	2V/uS	+18V	-18V	70C	86dB	50MV	50pA	5pA	500MWF	5MA	12V	15V	30V	75uV/C	204MW	7MA	70dB	64dB	0.5T
8043M	ING	DFE	INT	.3MHZ	2V/uS	+18V	-18V	125C	94dB	20MV	20pA	2pA	500MWF	5MA	12V	15V	30V	75uV/C	180MW	6MA	70dB	70dB	0.5T
9905	OTU	VFA	INT	15MHZ	1KVV/uS	+18V	-18V	125C	0dB	.	100pA	.	.	0.1A	10V	.	10V	300uV/C	240MW	8MA	.	54dB	100G
9906	OTU	XSR	EXT	.1GHZ	250V/uS	+18V	-18V	125C	57dB	10MV	10uA	3uA	.	5MA	10V	10V	5V	150uV/C	600MW	20MA	50dB	50dB	50K
9908	OTU	FET	INT	1GHZ	200V/uS	+18V	-18V	125C	50dDB	.	300pA	100pA	.	5MA	10V	11V	15V	3uV/C	360MW	12MA	500dB	600dB	100G
9910	OTU	VFA	INT		2KVV/uS	+18V	-18V	125C	50dB	20uA	20UA	.	.	0.1A	10V	12V	.	100uV/C	150MW	5MA	.	.	1M D
9916	OTU	XSR	INT	.2GHZ	300V/uS	+20V	-20V	125C	60dB	30MV	30uA	10uA	.	5MA	5V	1V	5V	150uV/C	450MW	15MA	50dB	50dB	10K
AD0042C	ANU	FET	INT	.2MHZ	1.2V/uS	+18V	-18V	70C	88dB	20MV	50pA	.	.	10MA	12V	10V	20V	25uV/C	.	4MA	70dB	68dB	0.1T
AD101A(FLP)	ANU	GPU	EXT	.	.	+22V	-22V	125C	94dB	2MV	75NA	10NA	500MWF	5MA	12V	15V	30V	15uV/C	.	3MA	80dB	80dB	1.5M
AD101A(T099)	ANU	GPU	EXT	.	.	+22V	-22V	125C	94dB	2MV	75NA	10NA	500MWF	5MA	12V	15V	30V	15uV/C	.	3MA	80dB	80dB	1.5M
AD108	ANU	SBA	EXT	.	.	+20V	-20V	125C	96dB	2MV	2NA	0.2NA	500MWF	1MA	13V	15V	1V	15uV/C	.	.6MA	85dB	80dB	30M
AD108A	ANU	SBA	EXT	.	.	+20V	-20V	125C	98dB	0.5MV	2NA	0.2NA	500MWF	1MA	13V	15V	1V	5uV/C	.	.6MA	96dB	96dB	30M
AD111	ANU	CPR	EXT	.	.	+18V	-18V	125C	100dB	3MV	100NA	10NA	500MWF	.	.	15V	30V	.	.	6MA	.	.	.
AD201A(FLP)	ANU	GPU	EXT	.	.	+22V	-22V	85C	94dB	2MV	75NA	10NA	500MWF	5MA	12V	15V	30V	15uV/C	.	3MA	80dB	80dB	500K
AD201A(MDP)	ANU	GPU	EXT	.	.	+22V	-22V	70C	94dB	2MV	75NA	10NA	500MWF	5MA	12V	15V	30V	30uV/C	.	3MA	70dB	70dB	500K
AD301AL-MDIP	ANU	GPU	EXT	.2MHZ	.05V/uS	+18V	-18V	70C	98dB	0.5MV	30NA	5NA	.	5MA	12V	15V	15V	5uV/C	.	3MA	90dB	90dB	1M5
AD301AL-T099	ANU	GPU	EXT	.2MHZ	.05V/uS	+18V	-18V	70C	98dB	0.5MV	30NA	5NA	.	5MA	12V	15V	15V	5uV/C	.	3MA	90dB	90dB	1M5
AD201A(T099)	ANU	GPU	EXT	.	.	+22V	-22V	85C	94dB	2MV	75NA	10NA	500MWF	5MA	12V	15V	30V	15uV/C	.	3MA	80dB	80dB	500K
AD208	ANU	SBA	EXT	.	.	+20V	-20V	85C	96dB	2MV	2NA	0.2NA	500MWF	1MA	13V	15V	1V	15uV/C	.	.6MA	85dB	80dB	30M
AD208A	ANU	SBA	EXT	.	.	+20V	-20V	85C	98dB	0.5MV	2NA	0.2NA	500MWF	1MA	13V	15V	1V	5uV/C	.	.6MA	96dB	96dB	30M
AD211	ANU	CPR	EXT	.	.	+18V	-18V	85C	100dB	3MV	100NA	10NA	500MWF	.	.	15V	30V	.	.	6MA	.	.	.
AD301A(MDIP)	ANU	GPU	EXT	.	.	+18V	-18V	70C	88dB	7.5MV	250NA	50NA	500MWF	5MA	12V	15V	30V	30uV/C	.	3MA	70dB	70dB	500K
AD301A(T099)	ANU	GPU	EXT	.	.	+18V	-18V	70C	88dB	7.5MV	250NA	50NA	500MWF	5MA	12V	15V	30V	30uV/C	.	3MA	70dB	70dB	500K
AD308	ANU	SBA	EXT	.	.	+18V	-18V	70C	88dB	7.5MV	7NA	1NA	500MWF	1MA	13V	15V	1V	30uV/C	.	.6MA	80dB	80dB	10M
AD308A	ANU	SBA	EXT	.	.	+18V	-18V	70C	98dB	0.5MV	7NA	1NA	500MWF	1MA	13V	15V	1V	5uV/C	.	.6MA	96dB	96dB	10M
AD311(DIL-8)	ANU	CPR	EXT	.	.	+18V	-18V	70C	100dB	7.5MV	250NA	50NA	500MWF	.	.	15V	30V	.	.	8MA	.	.	.
AD311(T099)	ANU	CPR	EXT	.	.	+18V	-18V	70C	100dB	7.5MV	250NA	50NA	500MWF	.	.	15V	30V	.	.	8MA	.	.	.
AD351J	ANU	CPR	INT	.	.	+18V	-18V	70C	84dB	6MV	250NA	.	.	.	3V	10V	30V	20uV/C	.	.	70dB	.	1M
AD351K	ANU	CPR	INT	.	.	+18V	-18V	70C	84dB	6MV	250NA	.	.	.	3V	10V	30V	5uV/C	.	.	70dB	.	1M
AD351S	ANU	CPR	INT	.	.	+18V	-18V	125C	84dB	6MV	250NA	.	.	.	3V	10V	30V	10uV/C	.	.	70dB	.	1M
AD501A	ANU	FET	INT	.8MHZ	3V/uS	+18V	-18V	85C	88dB	2MV	25pA	.	.	5MA	12V	10V	.	75uV/C	.	9MA	70dB	60dB	10G
AD501B	ANU	FET	INT	.8MHZ	3V/uS	+18V	-18V	85C	88dB	1MV	10pA	.	.	5MA	12V	10V	.	25uV/C	.	9MA	70dB	60dB	10G
AD501C	ANU	FET	INT	.8MHZ	3V/uS	+18V	-18V	85C	88dB	1MV	5pA	.	.	5MA	12V	10V	.	25uV/C	.	9MA	70dB	60dB	10G
AD502J	ANU	GPK	INT	.2MHZ	0.1V/uS	+18V	-18V	125C	86dB	6MV	25NA	12NA	.	5MA	10V	10V	18V	40uV/C	.	3MA	70dB	76dB	25M
AD502K	ANU	GPK	INT	.2MHZ	0.1V/uS	+18V	-18V	125C	86dB	5MV	7NA	4NA	.	5MA	10V	10V	18V	20uV/C	.	3MA	70dB	76dB	25M
AD502L	ANU	GPK	INT	.2MHZ	0.1V/uS	+18V	-18V	125C	86dB	5MV	4NA	1NA	.	5MA	10V	10V	18V	10uV/C	.	3MA	70dB	76dB	25M
AD503J	ANU	FET	INT	.3MHZ	3V/uS	+18V	-18V	70C	86dB	50MV	15pA	.	.	5MA	12V	10V	3V	75uV/C	.	7MA	70dB	68dB	10G
AD503K	ANU	FET	INT	.3MHZ	3V/uS	+18V	-18V	70C	94dB	20MV	10pA	.	.	5MA	12V	10V	3V	25uV/C	.	7MA	80dB	74dB	10G
AD503S	ANU	FET	INT	.3MHZ	3V/uS	+22V	-22V	125C	94dB	20MV	10pA	.	.	5MA	12V	10V	3V	50uV/C	.	7MA	80dB	74dB	10G
AD504J	ANU	LNA	EXT	60KHZ	.02V/uS	+18V	-18V	70C	108dB	2.5MV	200NA	40NA	.	10MA	10V	18V	18V	5uV/C	.	4MA	94dB	92dB	100K
AD504K	ANU	LNA	EXT	60KHZ	.02V/uS	+18V	-18V	70C	114dB	1.5MV	100NA	15NA	.	10MA	10V	18V	18V	3uV/C	.	3MA	100dB	96dB	200K
AD504L	ANU	LNA	EXT	60KHZ	.02V/uS	+18V	-18V	70C	120dB	0.5MV	80NA	10NA	.	10MA	10V	18V	18V	1uV/C	.	3MA	110dB	100dB	300K
AD504M	ANU	LNA	EXT	60KHZ	.02V/uS	+18V	-18V	70C	120dB	0.5MV	80NA	10NA	.	10MA	10V	18V	18V	0.5uV/C	.	3MA	110dB	100dB	300K
AD504S	ANU	LNA	EXT	60KHZ	.02V/uS	+18V	-18V	125C	120dB	0.5MV	80NA	10NA	.	10MA	10V	18V	18V	1uV/C	.	3MA	110dB	100dB	300K
AD505J	ANU	XSR	INT	3MHZ	120V/uS	+20V	-20V	85C	100dB	5MV	75NA	.	.	5MA	10V	.	10V	15uV/C	.	8MA	.	60dB	400K
AD505K	OTU	XSR	INT	3MHZ	120V/uS	+20V	-20V	85C	108dB	2.5MV	75NA	.	.	5MA	10V	.	10V	15uV/C	.	8MA	.	60dB	400K
AD505S	ANU	XSR	INT	3MHZ	120V/uS	+20V	-20V	125C	108dB	2.5MV	75NA	.	.	5MA	10V	.	10V	20uV/C	.	8MA	.	60dB	400K
AD506J	ANU	FET	INT	.2MHZ	3V/uS	+18V	-18V	70C	86dB	3.5MV	15pA	.	.	5MA	12V	10V	4V	75uV/C	.	7MA	70dB	74dB	10G
AD506K	ANU	FET	INT	.2MHZ	3V/uS	+18V	-18V	70C	94dB	1.5MV	10pA	.	.	5MA	12V	10V	4V	25uV/C	.	7MA	80dB	80dB	10G
AD506L	ANU	FET	INT	.2MHZ	3V/uS	+18V	-18V	70C	98dB	1MV	5pA	.	.	5MA	12V	10V	4V	10uV/C	.	7MA	80dB	80dB	10G
AD506S	ANU	FET	INT	.2MHZ	3V/uS	+22V	-22V	125C	94dB	1.5MV	10pA	.	.	5MA	12V	10V	4V	50uV/C	.	7MA	80dB	80dB	10G
AD507J	ANU	WBA	EXT	10MHZ	20V/uS	+20V	-20V	70C	98dB	5MV	25NA	25NA	.	10MA	10V	11V	12V	15uV/C	.	4MA	74dB	74dB	40M
AD507K	ANU	WBA	EXT	10MHZ	25V/uS	+20V	-20V	70C	100dB	3MV	15NA	15NA	.	10MA	10V	11V	12V	15uV/C	.	4MA	80dB	80dB	40M
AD507S	ANU	WBA	EXT	10MHZ	20V/uS	+20V	-20V	125C	100dB	4MV	15NA	15NA	.	15MA	10V	11V	12V	20uV/C	.	4MA	80dB	80dB	40M
AD508J	ANU	LVD	T	60KHZ	.02V/uS	+18V	-18V	70C	108dB	2.5MV	25NA	5NA	.	10MA	10V	18V	18V	3uV/C	.	4MA	94dB	92dB	1M
AD509J	ANU	XSR	EXT	4MHZ	80V/uS	+20V	-20V	70C	78dB	10MV	250NA	50NA	.	10MA	10V	10V	15V	100uV/C	.	6MA	74dB	74dB	40M
AD509K	ANU	XSR	EXT	4MHZ	100V/uS	+20V	-20V	70C	80dB	8MV	200NA	25NA	.	10MA	10V	10V	15V	30uV/C	.	6MA	80dB	80dB	50M
AD509S	ANU	XSR	EXT	4MHZ	100V/uS	+20V	-20V	125C	80dB	8MV	200NA	25NA	.	10MA	10V	10V	15V	30uV/C	.	6MA	80dB	80dB	50M

For detailed explanations of column heading notations, see App. A.

Also for ready references the more important abbreviations used in the column headings are listed below:

LEFT HAND PAGE

APP := application (codes at APP.E.)
CMRR = common mode rejection ratio
CMP = compensation (frequency)
dV_{io}/dT = input offset voltage temperature drift
GBP = gain bandwidth product
I_B = input bias current
I_{io} = input bias offset current
I_Q = quiescent supply current
MFR = manufacturer (codes at App.C.)
P_Q = quiescent power consumer
PSRR = power supply rejection ratio
V_{ICM} = common mode input voltage rating
V_{IDF} = differential input voltage rating
V_{io} = input offset voltage
V_S = dc supply voltage

RIGHT HAND PAGE

Lead out coding summary (details at APP.G.) for different cases (APP.F.)

A = gain adjust
B = bias adjust
C = case
E— = inverting input
E+ = non-inverting input
F,F* = input frequency compensation
G = ground
J = high level input
K = output, open collector
L = output, open emitter
M = metal case
N = not connected
Q = special terminal
R,R* = outputs
S = strobe
T,T* = offset balance
V+ = +ve dc supply
V- = —ve dc supply
W = guard ring
X = blank position, no lead
++ = +ve supplementary dc supply
—— = —ve supplementary dc supply
ø,ø* = output frequency compensation

CASE (APP F)	LD 1	LD 2	LD 3	LD 4	LD 5	LD 6	LD 7	LD 8	LD 9	LD 10	LD 11	LD 12	LD 13	LD 14	LD 15	LD 16	EUROPE SUBSTITUTE	USA SUBSTITUTE	I S	TYPE NUMBER
DIL-16/1C	N	E-1	E+1	R2	V23	B3	E-3	E+3	V-	R3	B2	E-2	E+2	R1	V+1	B1		ICL8023MDE	O	8023M
DIL-16/1C	E+1	E-1	N	T1	T*1	V-	R1	N	N	R2	V+	T2	T*2	N	E-2	E+2		ICL8043CDE	O	8043C-DIL/C
DIL-16/1P	E+1	E-1	N	T1	T*1	V-	R1	N	N	R2	V+	T2	T*2	N	E-2	E+2		ICL8043CPE	O	8043C-DIL/P
DIL-16/1C	E+1	E-1	N	T1	T*1	V-	R1	N	N	R2	V+	T2	T*2	N	E-2	E+2		ICL8043MDE	O	8043M
DIL-8/1P	V-2	N	E+	V-1	T	R	V+1	V+2											O	9905
DIL-8/1P	N	E-	E+	V-	T	R	V+	N	*	*	*	*	*	*	*	*		*	O	9908
DIL-8/1P	B	N	E+	V-	B*	R	V+	N											O	9910
DIL-14/1P	N	N	T	E-	E+	V-	N	N	B	R	V+	T*	N	N					O	9916
T05-8/1M	T	E-	E+	V-	T*	R	V+	N										LH740A	O	AD0042C
FLP-10/3C	N	FT	E-	E+	V-	T*	R	V+	F*	N							SFC2101APM	LM101AF	O	AD101A(FLP)
T05-8/1M	FT	E-	E+	V-M	T*	R	V+	F*									SFC2101A	LM101AH	O	AD101A(T099)
T05-8/1M	F	E-	E+	V-M	N	R	V+	F*									SFC2108M	LM108H	O	AD108
T05-8/1M	F	E-	E+	V-M	N	R	V+	F*									SFC2108A	LM108AH	O	AD108A
T05-8/1M	G	E+	E-	V-	T	T*S	R	V+									SFC2111M	LM111H	O	AD111
FLP-10/3C	N	FT	E-	E+	V-	T*	R	V+	F*	N							SFC2201APM	LM201AF	O	AD201A(FLP)
DIL-8/1C	FT	E-	E+	V-	T*	R	V+	F*											O	AD201A(MDP)
DIL-8/1P	TF	E-	E+	V-	T*	R	V+	F*											O	AD301AL-MDIP
T05-8/1M	TF	E-	E+	V-	T*	R	V+	F*											O	AD301AL-T099
T05-8/1M	FT	E-	E+	V-M	T*	R	V+	F*									SFC2101A	LM201AH	O	AD201A(T099)
T05-8/1M	F	E-	E+	V-M	N	R	V+	F*									SFC2208	LM208H	O	AD208
T05-8/1M	F	E-	E+	V-M	N	R	V+	F*									SFC2208A	LM208AH	O	AD208A
T05-8/1M	G	E+	E-	V-	T	T*S	R	V+									SFC2211	LM211H	O	AD211
DIL-8/1P	FT	E-	E+	V-	T*	R	V+	F*									SFC2301ADC	LM301AJ	O	AD301A(MDIP)
T05-8/1M	FT	E-	E+	V-M	T*	R	V+	F*									SFC2301A	LM301AH	O	AD301A(T099)
T05-8/1M	F	E-	E+	V-M	N	R	V+	F*									SFC2308	LM308H	O	AD308
T05-8/1M	F	E-	E+	V-M	N	R	V+	F*									SFC2308A	LM308AH	O	AD308A
DIL-8/1P	G	E+	E-	V-	T	T*S	R	V+									SFC2311DC	LM311N	O	AD311(DIL-8)
T05-8/1M	G	E+	E-	V-	T	T*S	R	V+									SFC2311	LM311H	O	AD311(T099)
T05-10/1M	E-	N	T	T*	V-	G	R	Q	V+	E+								ICL8001M	O	AD351J
T05-10/1M	E-	N	T	T*	V-	G	R	Q	V+	E+								ICL8001M	O	AD351K
T05-10/1M	E-	N	T	T*	V-	G	R	Q	V+	E+								AD351K	O	AD351S
FLP-5/6P	E+	E-	V+	V-	R													AD501B	O	AD501A
FLP-5/6P	E+	E-	V+	V-	R													AD501C	O	AD501B
FLP-5/6P	E+	E-	V+	V-	R														O	AD501C
T05-8/1M	T	E-	E+	V-	T*	R	V+	N										SSS741J	O	AD502J
T05-8/1M	T	E-	E+	V-	T*	R	V+	N										SSS741J	O	AD502K
T05-8/1M	T	E-	E+	V-	T*	R	V+	N										SSS741J	O	AD502L
T05-8/1M	T	E-	E+	V-	T*	R	V+	N										AD503K	O	AD503J
T05-8/1M	T	E-	E+	V-	T*	R	V+	N										AD503S	O	AD503K
T05-8/1M	T	E-	E+	V-	T*	R	V+	N										ICL8007MTV	O	AD503S
T05-8/1M	T	E-	E+	V-	ø	Rø*	V+	T*										AD504K	O	AD504J
T05-8/1M	T	E-	E+	V-	ø	Rø*	V+	T*										AD504L	O	AD504K
T05-8/1M	T	E-	E+	V-	ø	Rø*	V+	T*										AD504M	O	AD504L
T05-8/1M	T	E-	E+	V-	ø	Rø*	V+	T*										AD504S	O	AD504M
T05-8/1M	T	E-	E+	V-	ø	Rø*	V+	T*											O	AD504S
T05-10/1M	F	GE+	E-	ø	V-	R	N	V+	F*	F1								AD505K	O	AD505J
T05-10/1M	F	GE+	E-	ø	V-	R	N	V+	F*	F1								AD505S	O	AD505K
T05-10/1M	F	GE+	E-	ø	V-	R	N	V+	F*	F1									O	AD505S
T05-8/1M	T	E-	E+	V-	T*	R	V+	N										AD506K	O	AD506J
T05-8/1M	T	E-	E+	V-	T*	R	V+	N										AD506L	O	AD506K
T05-8/1M	T	E-	E+	V-	T*	R	V+	N											O	AD506L
T05-8/1M	T	E-	E+	V-	T*	R	V+	N											O	AD506S
T05-8/1M	T	E-	E+	V-	T*	Rø	V+	ø*										AD507K	O	AD507J
T05-8/1M	T	E-	E+	V-	T*	Rø	V+	ø*										AD507S	O	AD507K
T05-8/1M	T	E-	E+	V-	T*	Rø	V+	ø*											O	AD5075
T05-8/1M	T	E-	E+	V-	ø	Rø*	V+	T*											O	AD508J
T05-8/1M	T	E-	E+	V-	T*	R	V+	ø										AD509K	O	AD509J
T05-8/1M	T	E-	E+	V-	T*	R	V+	ø										AD509S	O	AD509K
T05-8/1M	T	E-	E+	V-	T*	R	V+	ø											O	AD509S
FLP-5/6P	E+	E-	V+	V-	R													AD511B	O	AD511

TYPE NUMBER	MFR	APP	CMP	GBP MIN	SLEW RATE MIN	V$_S^+$ MAX	V$_S^-$ MAX	T$_{OP}$ MAX	A$_{VOL}$ MIN	V$_{IO}$ MAX	I$_B$ MAX	I$_{IO}$ MAX	P$_{TOT}$ MAX	I$_{OUT}$ MIN	V$_{OUT}$ MIN	V$_{ICM}$ MAX	V$_{IDF}$ MAX	dV$_{IO}$/dT MAX	P$_O$ MAX	I$_O$ MAX	CMRR MIN	PSRR MIN	R$_{IN}$ MIN
AD511A	ANU	FET	INT	.3MHZ	3V/uS	+18V	-18V	85C	88dB	3.5MV	25pA	.	.	5MA	12V	10V	.	75uV/C	.	7MA	70dB	70dB	10G
AD511B	ANU	FET	INT	.3MHZ	3V/uS	+18V	-18V	85C	88dB	1.5MV	10pA	.	.	5MA	12V	10V	.	25uV/C	.	7MA	70dB	70dB	10G
AD511C	ANU	FET	INT	.3MHZ	3V/uS	+18V	-18V	85C	88dB	1MV	5pA	.	.	5MA	12V	10V	.	25uV/C	.	7MA	70dB	70dB	10G
AD512K	ANU	HCO	INT	.2MHZ	0.1V/uS	+18V	-18V	70C	94dB	3MV	100NA	50NA	.	10MA	10V	15V	30V	20uV/C	.	3M3A	80dB	80dB	200K
AD512S	ANU	HCO	INT	.2MHZ	0.1V/uS	+22V	-22V	125C	94dB	3MV	100NA	50NA	.	10MA	10V	15V	30V	25uV/C	.	3M3A	80dB	80dB	200K
AD513J	ANU	HSR	EXT	.2MHZ	10V/uS	+18V	-18V	70C	86dB	50MV	30pA	.	.	5MA	12V	10V	36V	75uV/C	.	7MA	70dB	70dB	10G
AD513K	ANU	HSR	EXT	.2MHZ	10V/uS	+18V	-18V	70C	94dB	20MV	20pA	.	.	5MA	12V	10V	36V	25uV/C	.	7MA	70dB	74dB	10G
AD513S	ANU	HSR	EXT	.2MHZ	10V/uS	+18V	-18V	125C	94dB	20MV	20pA	.	.	5MA	12V	10V	36V	50uV/C	.	7MA	70dB	74dB	10G
AD514J	ANU	FET	INT	.2MHZ	0.1V/uS	+18V	-18V	70C	86dB	50MV	50pA	.	.	5MA	12V	10V	20V	75uV/C	.	7MA	70dB	68dB	1G
AD514K	ANU	FET	INT	.2MHZ	0.1V/uS	+18V	-18V	70C	94dB	20MV	20pA	.	.	5MA	12V	10V	20V	25uV/C	.	7MA	70dB	70dB	1G
AD514L	ANU	FET	INT	.2MHZ	0.1V/uS	+18V	-18V	70C	94dB	20MV	20pA	.	.	5MA	12V	10V	20V	25uV/C	.	7MA.	70dB	70dB	1G
AD514S	ANU	FET	INT	.2MHZ	0.1V/uS	+18V	-18V	125C	94dB	20MV	10pA	.	.	5MA	12V	10V	20V	50uV/C	.	7MA	70dB	70dB	1G
AD516J	ANU	HSR	EXT	.2MHZ	10V/uS	+18V	-18V	70C	86dB	3.5MV	30pA	.	.	5MA	12V	10V	36V	75uV/C	.	7MA	70dB	70dB	10G
AD516K	ANU	HSR	EXT	.2MHZ	10V/uS	+18V	-18V	70C	94dB	1.5MV	20pA	.	.	5MA	12V	10V	36V	25uV/C	.	7MA	70dB	74dB	10G
AD516S	ANU	HSR	EXT	.2MHZ	10V/uS	+18V	-18V	125C	94dB	1.5MV	20pA	.	.	5MA	12V	10V	36V	50uV/C	.	7MA	70dB	74dB	10G
AD518J	ANU	XSR	EXT	3MHZ	50V/uS	+20V	-20V	70C	88dB	10MV	500NA	200NA	.	10MA	12V	10V	15V	50uV/C	.	10MA	70dB	65dB	500K
AD518K	ANU	XSR	EXT	10MHZ	50V/uS	+20V	-20V	70C	94dB	4MV	200NA	50NA	.	10MA	12V	10V	15V	10uV/C	.	7MA	80dB	70dB	500K
AD518S	ANU	XSR	EXT	10MHZ	50V/uS	+20V	-20V	125C	94dB	4MV	200NA	50NA	.	10MA	12V	10V	15V	10uV/C	.	7MA	80dB	70dB	500K
AD523J	ANU	FET	INT	.1MHZ	3V/uS	+18V	-18V	125C	88dB	50MV	1pA	.	.	5MA	10V	8V	10V	90uV/C	.	7MA	70dB	74dB	0.1T
AD523K	ANU	FET	INT	.1MHZ	3V/uS	+18V	-18V	125C	92dB	50MV	0.5pA	.	.	5MA	10V	8V	10V	30uV/C	.	7MA	80dB	80dB	0.1T
AD523L	ANU	FET	INT	.1MHZ	3V/uS	+18V	-18V	125C	92dB	20MV	.25pA	.	.	5MA	10V	8V	10V	60uV/C	.	7MA	80dB	80dB	0.1T
AD528J	ANU	FET	EXT	2MHZ	50V/uS	+20V	-20V	70C	88dB	3MV	30pA	5pA	.	10MA	10V	20V	20V	50uV/C	.	7MA	70dB	70dB	0.1T
AD528K	ANU	FET	EXT	2MHZ	50V/uS	+20V	-20V	70C	94dB	1MV	15pA	2pA	.	10MA	10V	20V	20V	25uV/C	.	7MA	80dB	80dB	0.1T
AD528S	ANU	FET	EXT	2MHZ	50V/uS	+20V	-20V	125C	94dB	1MV	15pA	2pA	.	10MA	10V	20V	20V	25uV/C	.	7MA	80dB	80dB	0.1T
AD540J	ANU	FET	INT	.2MHZ	2V/uS	+18V	-18V	70C	86dB	50MV	50pA	.	.	5MA	12V	10V	20V	75uV/C	.	7MA	70dB	68dB	1G
AD540K	ANU	FET	INT	.2MHZ	2V/uS	+18V	-18V	70C	94dB	20MV	25pA	.	.	5MA	12V	10V	20V	25uV/C	.	7MA	70dB	70dB	1G
AD540S	ANU	FET	INT	.2MHZ	2V/uS	+18V	-18V	125C	94dB	20MV	25pA	.	.	5MA	12V	10V	20V	50uV/C	.	7MA	70dB	70dB	1G
AD741	ANU	GPK	INT	.	0.2V/uS	+22V	-22V	125C	94dB	5MV	500NA	200NA	500MWF	7MA	12V	15V	30V	.	85MW	3MA	70dB	76dB	300K
AD741C(MDIP)	ANU	GPK	INT	.	0.2V/uS	+18V	-18V	70C	86dB	6MV	500NA	200NA	310MWF	.	12V	15V	30V	.	.	3MA	70dB	76dB	300K
AD741C(TO99)	ANU	GPK	INT	.	0.2V/uS	+18V	-18V	70C	86dB	6MV	500NA	200NA	500MWF	5MA	12V	15V	30V	.	.	3MA	70dB	76dB	300K
AM741HM	ADU	GPK	INT	.	0.2V/uS	+22V	-22V	125C	94dB	5MV	500NA	200NA	500MWF	7MA	12V	15V	30V	.	85MW	3MA	70dB	76dB	300K
AD741J(MDIP)	ANU	GPK	INT	.2MHZ	0.1V/uS	+18V	-18V	70C	94dB	3MV	200NA	50NA	.	10MA	10V	15V	30V	20uV/C	.	3M3A	80dB	80dB	200K
AD741J(TO99)	ANU	GPK	INT	.2MHZ	0.1V/uS	+18V	-18V	70C	94dB	3MV	200NA	50NA	.	10MA	10V	15V	30V	20uV/C	.	3M3A	80dB	80dB	200K
AD741K(MDIP)	ANU	GPK	INT	.2MHZ	0.1V/uS	+22V	-22V	70C	94dB	2MV	75NA	10NA	.	5MA	10V	15V	30V	15uV/C	.	2M8A	90dB	96dB	400K
AD741K(TO99)	ANU	GPK	INT	.2MHZ	0.1V/uS	+22V	-22V	70C	94dB	2MV	75NA	10NA	.	5MA	10V	15V	30V	15uV/C	.	2M8A	90dB	96dB	400K
AD741L(MDIP)	ANU	GPK	INT	.2MHZ	0.1V/uS	+22V	-22V	70C	94dB	0.5MV	50NA	5NA	.	5MA	10V	15V	30V	5uV/C	.	2M8A	90dB	96dB	400K
AD741L(TO99)	ANU	GPK	INT	.2MHZ	0.1V/uS	+22V	-22V	70C	94dB	0.5MV	50NA	5NA	.	5MA	10V	15V	30V	5uV/C	.	2M8A	90dB	96dB	400K
AD741S(MDIP)	ANU	GPK	INT	.2MHZ	0.1V/uS	+22V	-22V	125C	94dB	2MV	75NA	10NA	.	10MA	10V	15V	30V	15uV/C	.	2M8A	90dB	96dB	400K
AD741S(TO99)	ANU	GPK	INT	.2MHZ	0.1V/uS	+22V	-22V	125C	94dB	2MV	75NA	10NA	.	10MA	10V	15V	30V	15uV/C	.	2M8A	90dB	96dB	400K
AM741XM	ADU	GPK	INT	.	0.2V/uS	+22V	-22V	125C	94dB	5MV	500NA	200NA	.	7MA	12V	15V	30V	.	85MW	3MA	70dB	76dB	300K
AM747ADM	ADU	DGK	INT	.4MHZ	0.3V/uS	+22V	-22V	125C	94dB	3MV	80NA	30NA	670MWF	10MA	16V	15V	30V	15uV/C	150MW	3MA	80dB	86dB	1M
AM747AFM	ADU	DGK	INT	.4MHZ	0.3V/uS	+22V	-22V	125C	94dB	3MV	80NA	30NA	500MWF	10MA	16V	15V	30V	15uV/C	150MW	3MA	80dB	86dB	1M
AM747AHM	ADU	DGK	INT	.4MHZ	0.3V/uS	+22V	-22V	125C	94dB	3MV	80NA	30NA	500MWF	10MA	16V	15V	30V	15uV/C	150MW	3MA	80dB	86dB	1M
AM747DC	ADU	DGK	INT	.	0.2V/uS	+18V	-18V	70C	88dB	6MV	500NA	200NA	670MWF	5MA	12V	15V	30V	.	85MW	3MA	70dB	76dB	300K
AM747DM	ADU	DGK	INT	.	0.2V/uS	+22V	-22V	125C	94dB	5MV	500NA	200NA	670MWF	5MA	12V	15V	30V	.	85MW	3MA	70dB	76dB	300K
AM747EDC	ADU	DGK	INT	.4MHZ	0.3V/uS	+22V	-22V	70C	94dB	3MV	80NA	30NA	670MWF	10MA	16V	15V	30V	15uV/C	150MW	3MA	80dB	86dB	1M
AM747EHC	ADU	DGK	INT	.4MHZ	0.3V/uS	+22V	-22V	70C	94dB	3MV	80NA	30NA	500MWF	10MA	16V	15V	30V	15uV/C	150MW	3MA	80dB	86dB	1M
AM747HC	ADU	DGK	INT	.	0.2V/uS	+18V	-18V	70C	88dB	6MV	500NA	200NA	500MWF	5MA	12V	15V	30V	.	85MW	3MA	70dB	76dB	300K
AM747HM	ADU	DGK	INT	.	0.2V/uS	+22V	-22V	125C	94dB	5MV	500NA	200NA	500MWF	5MA	12V	15V	30V	.	85MW	3MA	70dB	76dB	300K
AM747PC	ADU	DGK	INT	.	0.2V/uS	+18V	-18V	70C	88dB	6MV	500NA	200NA	800MWF	5MA	12V	15V	30V	.	85MW	3MA	70dB	76dB	300K
AM747XC	ADU	DGK	INT	.	0.2V/uS	+18V	-18V	70C	88dB	6MV	500NA	200NA	.	5MA	12V	15V	30V	.	85MW	3MA	70dB	76dB	300K
AM747XM	ADU	DGK	INT	.	0.2V/uS	+22V	-22V	125C	94dB	5MV	500NA	200NA	.	5MA	12V	15V	30V	.	85MW	3MA	70dB	76dB	300K
AD801A	ANU	GPU	EXT	.1MHZ	2V/uS	+18V	-18V	125C	84dB	5MV	4NA	2NA	.	5MA	10V	8V	10V	40uV/C	.	6MA	65dB	74dB	25M
AD801B	ANU	GPU	EXT	.1MHZ	2V/uS	+18V	-18V	125C	84dB	5MV	4NA	2NA	.	5MA	10V	8V	10V	10uV/C	.	6MA	65dB	74dB	25M
AD801S	ANU	GPU	EXT	.1MHZ	2V/uS	+18V	-18V	125C	84dB	5MV	4NA	2NA	.	5MA	10V	8V	10V	20uV/C	.	6MA	65dB	74dB	25M
AD3542J	ANU	FET	INT	.2MHZ	0.1V/uS	+20V	-20V	70C	88dB	20MV	25pA	.	.	10MA	12V	10V	15V	50uV/C	.	6MA	80dB	68dB	10G
AD8007C	ANU	FET	INT	.2MHZ	1.2V/uS	+18V	-18V	70C	86dB	50MV	50pA	.	.	10MA	12V	10V	20V	75uV/C	.	6MA	70dB	64dB	0.1T
ADM501A	ANU	FET	INT	.8MHZ	3V/uS	+18V	-18V	85C	88dB	2MV	25pA	.	.	5MA	12V	10V	.	75uV/C	.	9MA	70dB	60dB	10G
ADM501B	ANU	FET	INT	.8MHZ	3V/uS	+18V	-18V	85C	88dB	1MV	10pA	.	.	5MA	12V	10V	.	25uV/C	.	9MA	70dB	60dB	10G
ADM501C	ANU	FET	INT	.8MHZ	3V/uS	+18V	-18V	85C	88dB	1MV	5pA	.	.	5MA	12V	10V	.	25uV/C	.	9MA	70dB	60dB	10G

LEFT HAND PAGE

APP = application (codes at APP.E.)
CMRR = common mode rejection ratio
CMP = compensation (frequency)
dV_{io}/dT = input offset voltage temperature drift
GBP = gain bandwidth product
I_B = input bias current
I_{io} = input bias offset current
I_q = quiescent supply current
MFR = manufacturer (codes at App.C.)
P_q = quiescent power consumer
PSRR = power supply rejection ratio
V_{icm} = common mode input voltage rating
V_{idr} = differential input voltage rating
V_{io} = input offset voltage
V_S = dc supply voltage

RIGHT HAND PAGE

Lead out coding summary (details at APP.G.) for different cases (APP.F.)

A = gain adjust
B = bias adjust
C = case
E− = inverting input
E+ = non-inverting input
F,F* = input frequency compensation
G = ground
J = high level input
K = output, open collector
L = output, open emitter
M = metal case
N = not connected
Q = special terminal
R,R* = outputs
S = strobe
T T* = offset balance
∴ = +ve dc supply
V− = −ve dc supply
W = guard ring
X = blank position, no lead
+ + = +ve supplementary dc supply
− − = −ve supplementary dc supply
♦♦* = output frequency compensation

CASE (APP F)	LD 1	LD 2	LD 3	LD 4	LD 5	LD 6	LD 7	LD 8	LD 9	LD 10	LD 11	LD 12	LD 13	LD 14	LD 15	LD 16	EUROPE SUBSTITUTE	USA SUBSTITUTE	IS	TYPE NUMBER
FLP-5/6P	E+	E-	V+	V-	R	.	.											AD511C	0	AD511B
FLP-5/6P	E+	E-	V+	V-	R													AD511C	0	AD511C
T05-8/1M	T	E-	E+	V-	T*	R	V+	N										AD512S	0	AD512K
T05-8/1M	T	E-	E+	V-	T*	R	V+	N											0	AD512S
T05-8/1M	TF	E-	E+	V-	T*	R	V+	F*										CA3130T	0	AD513J
T05-8/1M	TF	E-	E+	V-	T*	R	V+	F*										CA3130AT	0	AD513K
T05-8/1M	TF	E-	E+	V-	T*	R	V+	F*										CA3130AT	0	AD513S
T05-8/1M	T	E-	E+	V-	T*	R	V+	N										NE536T	0	AD514J
T05-8/1M	T	E-	E+	V-	T*	R	V+	N										ICL8007MTV	0	AD514K
T05-8/1M	T	E-	E+	V-	T*	R	V+	N										ICL8007MTV	0	AD514L
T05-8/1M	T	E-	E+	V-	T*	R	V+	N										ICL8007MTV	0	AD514S
T05-8/1M	TF	E-	E+	V-	T*	R	V+	F*										AD516K	0	AD516J
T05-8/1M	TF	E-	E+	V-	T*	R	V+	F*										AD516S	0	AD516K
T05-8/1M	TF	E-	E+	V-	T*	R	V+	F*											0	AD516S
T05-8/1M	F1T	E-	E+	V-	F3T	R	V+	F2										AD518K	0	AD518J
T05-8/1M	F1T	E-	E+	V-	F3T	R	V+	F2										AD518S	0	AD518K
T05-8/1M	F1T	E-	E+	V-	F3T	R	V+	F2											0	AD518S
T05-8/1M	T	E-	E+	V-	T*	R	V+	W										8007AM	0	AD523J
T05-8/1M	T	E-	E+	V-	T*	R	V+	W										8007AM	0	AD523K
T05-8/1M	T	E-	E+	V-	T*	R	V+	W										8007AM	0	AD523L
T05-8/1M	F1T	E-	E+	V-	F3T	R	V+	F2T										AD528K	0	AD528J
T05-8/1M	F1T	E-	E+	V-	F3T	R	V+	F2T										AD528S	0	AD528K
T05-8/1M	F1T	E-	E+	V-	F3T	R	V+	F2T											0	AD528S
T05-8/1M	T	E-	E+	V-	T*	R	V+	N										ICL8007CTV	0	AD540J
T05-8/1M	T	E-	E+	V-	T*	R	V+	N										ICL8007MTV	0	AD540K
T05-8/1M	T	E-	E+	V-	T*	R	V+	N										ICL8007MTV	0	AD540S
T05-8/1M	T	E-	E+	V-M	T*	R	V+	N									TBA222	UA741HM	0	AD741
DIL-8/1C	T	E-	E+	V-	T*	R	V+	N									TBA221B	UA741TC	0	AD741C(MDIP)
T05-8/1M	T	E-	E+	V-M	T*	R	V+	N									TBA221	UA741HC	0	AD741C(T099)
T05-8/1M	T	E-	E+	V-M	T*	R	V+	N									TBA222	UA741HM	0	AM741HM
DIL-8/1P	T	E	E+	V-	T*	R	V+	N										LM741EJ	0	AD741J(MDIP)
T05-8/1M	T	E-	E+	V-	T*	R	V+	N									UA741EHC	RC4131T	0	AD741J(T099)
DIL-8/1P	T	E-	E+	V-	T*	R	V+	N										LM741EJ	0	AD741K(MDIP)
T05-8/1M	T	E-	E+	V-	T*	R	V+	N										RM4131T	0	AD741K(T099)
DIL-8/1P	T	E-	E+	V-	T*	R	V+	N											0	AD741L(MDIP)
T05-8/1M	T	E-	E+	V-	T*	R	V+	N											0	AD741L(T099)
DIL-8/1P	T	E-	E+	V-	T*	R	V+	N											0	AD741S(MDIP)
T05-8/1M	T	E-	E+	V-	T*	R	V+	N											0	AD741S(T099)
CHP	.	.	.	.															0	AM741XM
DIL-14/1C	E-1	E+1	T1	V-	T2	E+2	E-2	T*2	V+2	R2	N	R1	V+	T*1			SFC2747KM	UA747ADM	0	AM747ADM
FLP-14/3C	E-1	E+1	T1	V-	T2	E+2	E-2	T*2	V+2	R2	N	R1	V+1	T*1				747AFM	0	AM747AFM
T05-10/1M	R1	V+1	E-1	E+1	V-	E+2	E-2	V+2	R2	N							TBC0747	UA747AHM	0	AM747AHM
DIL-14/1C	E-1	E+1	T1	V-	T2	E+2	E-2	T*2	V+2	R2	N	R1	V+1	T*1			TBB0747A	UA747DC	0	AM747DC
DIL-14/1C	E-1	E+1	T1	V-	T2	E+2	E-2	T*2	V+2	R2	N	R1	V+1	T*1			SFC2747KM	UA747DM	0	AM747DM
DIL-14/1C	E-1	E+1	T1	V-	T2	E+2	E-2	T*2	V+2	R2	N	R1	V+1	T*1			SFC2747KM	UA747EDC	0	AM747EDC
T05-10/1M	R1	V+1	E-1	E+1	V-	E+2	E-2	V+2	R2	N							LM747EH	UA747EHC	0	AM747EHC
T05-10/1M	R1	V+1	E-1	E+1	V-	E+2	E-2	V+2	R2	N							TBB0747	UA747HC	0	AM747HC
T05-10/1M	R1	V+1	E-1	E+1	V-	E+2	E-2	V+2	R2	N							TBC0747	UA747HM	0	AM747HM
DIL-14/1P	E-1	E+1	T1	V-	T2	E+2	E-2	T*2	V+2	R2	N	R1	V+1	T*1			TBB0747A	UA747DC	0	AM747PC
CHP	.	.																	0	AM747XC
CHP	.	.																	0	AM747XM
T05-8/1M	F	E-	E+	V-	∅*	R∅	V+	F*										UA709AHM	0	AD801A
T05-8/1M	F	E-	E+	V-	∅*	R∅	V+	F*										UA709AHM	0	AD801B
T05-8/1M	F	E-	E+	V-	∅*	R∅	V+	F*										UA709AHM	0	AD801S
T05-8/1M	T	E-	E+	V-	T*	R	V+	N										8007M	0	AD3542J
T05-8/1M	T	E-	E+	V-	T*	R	V+	N										8007C	0	AD8007C
T08-12/1M	E+	E-	N	V+	N	V-	N	R	N	N	N	N						ADM501B	0	ADM501A
T08-12/1M	E+	E-	N	V+	N	V-	N	R	N	N	N	N						ADM501C	0	ADM501B
T08-12/1M	E+	E-	N	V+	N	V-	N	R	N	N	N								0	ADM501C
DIM-5/4P	E+	E-	V+	V-	R													ADP501B	0	ADP501A

TYPE NUMBER	MFR	APP	CMP	GBP MIN	SLEW RATE MIN	V_S+ MAX	V_S- MAX	T_{op} MAX	A_{VOl} MIN	V_{IO} MAX	I_B MAX	I_{IO} MAX	P_{TO1} MAX	I_{OUT} MIN	V_{OUT} MIN	V_{ICM} MAX	V_{IDF} MAX	dV_{IO}/dT MAX	P_Q MAX	I_O MAX	CM RR MIN	PS RR MIN	R_{IN} MIN
ADP501A	ANU	FET	INT	.8MHZ	3V/uS	+18V	-18V	85C	88dB	2MV	25pA	.		5MA	12V	10V	.	75uV/C		9MA	70dB	60dB	10G
ADP501B	ANU	FET	INT	.8MHZ	3V/uS	+18V	-18V	85C	88dB	1MV	10pA			5MA	12V	10V	.	25uV/C		9MA	70dB	60dB	10G
ADP501C	ANU	FET	INT	.8MHZ	3V/uS	+18V	-18V	85C	88dB	1MV	5pA			5MA	12V	10V	.	25uV/C		9MA	70dB	60dB	10G
ADP511A	ANU	FET	INT	.3MHZ	3V/uS	+18V	-18V	85C	88dB	3.5MV	25pA			5MA	12V	10V	.	75uV/C		7MA	70dB	70dB	10G
ADP511B	ANU	FET	INT	.3MHZ	3V/uS	+18V	-18V	85C	88dB	1.5MV	10pA			5MA	12V	10V	.	25uV/C		7MA	70dB	70dB	10G
ADP511C	ANU	FET	INT	.3MHZ	3V/uS	+18V	-18V	85C	88dB	1MV	5pA			5MA	12V	10V		25uV/C		7MA	70dB	70dB	10G
ADX118	ANU	XSR	INT	.	50V/uS	+20V	-20V	125C	94dB	4MV	250NA	50NA	500MWF	6MA	12V	15V	1V			8MA	80dB	70dB	1M
ADX218	ANU	XSR	INT	.	50V/uS	+20V	-20V	85C	94dB	4MV	250NA	50NA	500MWF	6MA	12V	15V	1V			8MA	80dB	70dB	1M
ADX318	ANU	XSR	INT	.	50V/uS	+20V	-20V	70C	88dB	10MV	500NA	200NA	500MWF	6MA	12V	15V	1V			8MA	70dB	65dB	500K
AM101A-DICE	ADU	GPU	EXT	.	.	+22V	-22V	125C	94dB	2MV	75NA	10NA		5MA	12V	15V	30V	15uV/C		3MA	80dB	80dB	1.5M
AM101A-DIP	ADU	GPU	EXT	.	.	+22V	-22V	125C	94dB	2MV	75NA	10NA	500MWF	5MA	12V	15V	30V	15uV/C		3MA	80dB	80dB	1.5M
AM101A-FLP	ADU	GPU	EXT	.	.	+22V	-22V	125C	94dB	2MV	75NA	10NA	500MWF	5MA	12V	15V	30V	15uV/C		3MA	80dB	80dB	1.5M
AM101A-TO5	ADU	GPU	EXT	.	.	+22V	-22V	125C	94dB	2MV	75NA	10NA	500MWF	5MA	12V	15V	30V	15uV/C		3MA	80dB	80dB	1.5M
AM101-DICE	ADU	GPU	EXT	.	.	+22V	-22V	125C	94dB	5MV	500NA	200NA		5MA	12V	15V	30V	15uV/C		3MA	70dB	70dB	300K
AM101-DIP	ADU	GPU	EXT	.	.	+22V	-22V	125C	94dB	5MV	1.5uA	0.5uA	500MWF	5MA	12V	15V	30V	15uV/C		3MA	70dB	70dB	300K
AM101-FLP	ADU	GPU	EXT	.	.	+22V	-22V	125C	.94dB	5MV	1.5uA	0.5uA	5007WF	5MA	12V	15V	30V	15uV/C		3MA	70dB	70dB	300K
AM101-TO5	ADU	GPU	EXT	.	.	+22V	-22V	125C	94dB	5MV	500NA	200NA	500MWF	5MA	12V	15V	30V	15uV/C		3MA	70dB	70dB	300K
AM102-DICE	ADU	VFA	INT	.	.	+18V	-18V	125C	0dB	5MV	10NA	.		1MA	10V	.	.	30uV/C		6MA	.	60dB	10G
AM102-DIP	ADU	VFA	INT	.	.	+18V	-18V	125C	0dB	5MV	10NA	.	500MWF	1MA	10V	.	.	30uV/C		6MA	.	60dB	10G
AM102-FLP	ADU	VFA	INT	.	.	+18V	-18V	125C	0dB	5MV	10NA	.	500MWF	1MA	10V	.	.	30uV/C		6MA	.	60dB	10G
AM102-TO5	ADU	VFA	INT	.	.	+18V	-18V	125C	0dB	5MV	10NA	.	500MWF	1MA	10V	.	.	30uV/C		6MA	.	60dB	10G
AM106-DICE	ADU	CPR	EXT	.	.	+15V	-15V	125C	84dB	2MV	20uA	3uA		50MA	2.5V	.	.	10uV/C	163MW	.	.	.	.
AM106-FLP	ADU	CPR	EXT	.	.	+15V	-15V	125C	84dB	2MV	20uA	3uA	600MWF	50MA	2.5V	.	.	10uV/C	163MW	.	.	.	.
AM106-TO5	ADU	CPR	EXT	.	.	+15V	-15V	125C	84dB	2MV	20uA	3uA	600MWF	50MA	2.5V	.	.	10uV/C	163MW	.	.	.	.
AM107-DICE	ADU	GPK	INT	.	.	+22V	-22V	125C	94dB	2MV	75NA	10NA		5MA	12V	15V	30V	15uV/C		3MA	80dB	80dB	1.5M
AM107-DIP	ADU	GPK	INT	.	.	+22V	-22V	125C	94dB	2MV	75NA	10NA	500MWF	5MA	12V	15V	30V	15uV/C		3MA	80dB	80dB	1.5M
AM107-FLP	ADU	GPK	INT	.	.	+22V	-22V	125C	94dB	2MV	75NA	10NA	500MWF	5MA	12V	15V	30V	15uV/C		3MA	80dB	80dB	1.5M
AM107-TO5	ADU	GPK	INT	.	.	+22V	-22V	125C	94dB	2MV	75NA	10NA	500MWF	5MA	12V	15V	30V	15uV/C		3MA	80dB	80dB	1.5M
AM108A-DICE	ADU	SBA	EXT	.	.	+20V	-20V	125C	98dB	0.5MV	2NA	0.2NA			13V	15V	1V	5uV/C		.6MA	96dB	96dB	30M
AM108A-DIP	AUU	SBA	EXT	.	.	+20V	-20V	125C	98dB	0.5MV	2NA	0.2NA	500MWF	1MAA	13V	15V	1V	5uV/C		.6MA	96dB	96dB	30M
AM108A-TO5	ADU	SBA	EXT	.	.	+20V	-20V	125C	98dB	0.5MV	2NA	0.2NA	500MWF	1MA	13V	15V	1V	5uV/C		.6MA	96dB	96dB	30M
AM108-DICE	ADU	SBA	EXT	.	.	+20V	-20V	125C	96dB	2MV	2NA	0.2NA		1MA	13V	15V	1V	15uV/C		.6MA	85dB	80dB	30M
AM108-DIP	ADU	SBA	EXT	.	.	+20V	-20V	125C	96dB	2MV	2NA	0.2NA	500MWF	1MA	13V	15V	1V	15uV/C		.6MA	85dB	80dB	30M
AM108-TO5	ADU	SBA	EXT	.	.	+20V	-20V	125C	96dB	2MV	2NA	0.2NA	500MWF	1MA	13V	15V	1V	15uV/C		.6MA	85dB	80dB	30M
AM110-DICE	ADU	VFA	INT	.	15V/uS	+18V	-18V	125C	0dB	4MV	3NA	.		1MA	10V	15V	15V	50uV/C		6MA	.	70dB	10G
AM110-DIP	ADU	VFA	INT	.	15V/uS	+18V	-18V	125C	0dB	4MV	3NA	.	500MWF	1MA	10V	15V	15V	50uV/C		6MA	.	70dB	10G
AM110-FLP	ADU	VFA	INT	.	15V/uS	+18V	-18V	125C	0dB	4MV	3NA	.	500MWF	1MA	10V	15V	15V	50uV/C		6MA	.	70dB	10G
AM110-TO5	ADU	VFA	INT	.	15V/uS	+18V	-18V	125C	0dB	4MV	3NA	.	500MWF	1MA	10V	15V	15V	50uV/C		6MA	.	70dB	10G
AM111-DICE	ADU	CPR	EXT	.	.	+18V	-18V	125C	100dB	3MV	100NA	10NA				15V	30V			6MA	.	.	.
AM111-DIP	ADU	CPR	EXT	.	.	+18V	-18V	125C	100dB	3MV	100NA	10NA	500MWF			15V	30V			6MA	.	.	.
AM111-FLP	ADU	CPR	EXT	.	.	+18V	-18V	125C	100dB	3MV	100NA	10NA	500MWF			15V	30V			6MA	.	.	.
AM111-TO5	ADU	CPR	EXT	.	.	+18V	-18V	125C	100dB	3MV	100NA	10NA	500MWF			15V	30V			6MA	.	.	.
AM112-DICE	ADU	SBA	INT	.	.	+20V	-20V	125C	94dB	2MV	2NA	0.2NA		1MA	13V	14V	14V	15uV/C		.6MA	85dB	80dB	30M
AM112-DIP	ADU	SBA	INT	.	.	+20V	-20V	125C	94dB	2MV	2NA	0.2NA	500MWF	1MA	13V	14V	14V	15uV/C		.6MA	85dB	80dB	30M
AM112-FLP	ADU	SBA	INT	.	.	+20V	-20V	125C	94dB	2MV	2NA	0.2NA	500MWF	1MA	13V	14V	14V	15uV/C		.6MA	85dB	80dB	30M
AM112-TO5	ADU	SBA	INT	.	.	+20V	-20V	125C	94dB	2MV	2NA	0.2NA	500MWF	1MA	13V	14V	14V	15uV/C		.6MA	85dB	80dB	30M
AM118-DICE	ADU	XSR	INT	.	50V/uS	+20V	-20V	125C	94dB	4MV	250NA	50NA		6MA	12V	15V	1V			8MA	80dB	70dB	1M
AM118-DIP	ADU	XSR	INT	.	50V/uS	+20V	-20V	125C	94dB	4MV	250NA	50NA	500MWF	6MA	12V	15V	1V			8MA	80dB	70dB	1M
AM118-FLP	ADU	XSR	INT	.	50V/uS	+20V	-20V	125C	94dB	4MV	250NA	50NA	500MWF	6MA	12V	15V	1V			8MA	80dB	70dB	1M
AM118-TO5	ADU	XSR	INT	.	50V/uS	+20V	-20V	125C	94dB	4MV	250NA	50NA	500MWF	6MA	12V	15V	1V			8MA	80dB	70dB	1M
AM119-DICE	ADU	DCP	INT	.	.	+18V	-18V	125C	80dB	4MV	500NA	75NA				15V	5V			12MA	.	.	.
AM119-DIP	ADU	DCP	INT	.	.	+18V	-18V	125C	80dB	4MV	500NA	75NA	500MWF			15V	5V			12MA	.	.	.
AM119-FLP	ADU	DCP	INT	.	.	+18V	-18V	125C	80dB	4MV	500NA	75NA	500MWF			15V	5V			12MA	.	.	.
AM119-TO5	ADU	DCP	INT	.	.	+18V	-18V	125C	80dB	4MV	500NA	75NA	500MWF			15V	5V			12MA	.	.	.
AM124A-DICE	ADU	OGK	INT	.	.	+16V	-16V	125C	94dB	2MV	50NA	10NA				16V	16V	20uV/C		2MA	70dB	65dB	.
AM124A-DIP	ADU	OGK	INT	.	.	+16V	-16V	125C	94dB	2MV	50NA	10NA	900MWF			16V	16V	20uV/C		2MA	70dB	65dB	.
AM124A-FLP	ADU	OGK	INT	.	.	+16V	-16V	125C	94dB	2MV	50NA	10NA	800MWF			16V	16V	20uV/C		2MA	70dB	65dB	.
AM124-DICE	ADU	OGK	INT	.	.	+16V	-16V	125C	94dB	5MV	150NA	30NA				16V	16V	35uV/C		2MA	70dB	65dB	.
AM124-DIP	ADU	OGK	INT	.	.	+16V	-16V	125C	94dB	5MV	150NA	30NA				16V	16V	35uV/C		2MA	70dB	65dB	.
AM124-FLP	ADU	OGK	INT	.	.	+16V	-16V	125C	94dB	5MV	150NA	30NA	800MWF			16V	16V	35uV/C		2MA	70dB	65dB	.

For detailed explanations of column heading notations, see App. A.

Also for ready references the more important abbreviations used in the column headings are listed below:

LEFT HAND PAGE

APP = application (codes at APP.E.)

CMRR = common mode rejection ratio

CMP = compensation (frequency)

dV_{10}/dT = input offset voltage temperature drift

GBP = gain bandwidth product

I_B = input bias current

I_{10} = input bias offset current

I_C = quiescent supply current

MFR = manufacturer (codes at App.C.)

P_0 = quiescent power consumer

PSRR = power supply rejection ratio

V_{ICM} = common mode input voltage rating

V_{IDF} = differential input voltage rating

V_{IO} = input offset voltage

V_S = dc supply voltage

RIGHT HAND PAGE

Lead out coding summary (details at APP.G.) for different cases (APP.F.)

A = gain adjust

B = bias adjust

C = case

E— = inverting input

E+ = non-inverting input

F,F* = input frequency compensation

G = ground

J = high level input

K = output, open collector

L = output, open emitter

M = metal case

N = not connected

Q = special terminal

R,R* = outputs

S = strobe

T,T* = offset balance

V+ = +ve dc supply

V— = —ve dc supply

W = guard ring

X = blank position, no lead

++ = +ve supplementary dc supply

—— = —ve supplementary dc supply

ϕ,ϕ^* = output frequency compensation

CASE (APP F)	LD 1	LD 2	LD 3	LD 4	LD 5	LD 6	LD 7	LD 8	LD 9	LD 10	LD 11	LD 12	LD 13	LD 14	LD 15	LD 16	EUROPE SUBSTITUTE	USA SUBSTITUTE	I S	TYPE NUMBER
DIM-5/4P	E+	E-	V+	V-	R	.	.	.	.	.	.	.	.	.	.	.	.	ADP501C	O	ADP501B
DIM-5/4P	E+	E-	V+	V-	R	.	.	.	.	.	.	.	.	.	.	.	.	.	O	ADP501C
DIM-5/4P	E+	E-	V+	V-	R	.	.	.	.	.	.	.	.	.	.	.	.	ADP511B	O	ADP511A
DIM-5/4P	E+	E-	V+	V-	R	.	.	.	.	.	.	.	.	.	.	.	.	ADP511C	O	ADP511B
DIM-5/4P	E+	E-	V+	V-	R	.	.	.	.	.	.	.	.	.	.	.	.	.	O	ADP511C
TO5-8/1M	T*F	E-	E+	V-	F*T	R	V+	ø	.	.	.	.	.	.	.	.	TDC0118CM	LM118H	O	ADX118
TO5-8/1M	T*F	E-	E+	V-	F*T	R	V+	ø	.	.	.	.	.	.	.	.	TDB0118CM	LM218H	O	ADX218
TO5-8/1M	T*F	E-	E+	V-	F*T	R	V+	ø	.	.	.	.	.	.	.	.	TDE0118CM	LM318H	O	ADX318
CHP	.	.	.	.	.	.	.	.	.	.	.	.	.	.	.	.	.	.	O	AM101A-DICE
DIL-14/1P	N	N	FT	E-	E+	V-	N	N	T*	R	V+	F*	N	N	.	.	UA101AD	LM101AD	O	AM101A-DIP
FLP-10/3C	N	FT	E-	E+	V-	T*	R	V+	F*	N	.	.	.	.	.	.	SFC2101APM	LM101AF	O	AM101A-FLP
TO5-8/1M	FT	E-	E+	V-M	T*	R	V+	F*	.	.	.	.	.	.	.	.	SFC2101A	LM101AH	O	AM101A-TO5
CHP	.	.	.	.	.	.	.	.	.	.	.	.	.	.	.	.	AMLD101	LD101	O	AM101-DICE
DIL-14/1C	N	N	FT	E-	E+	V-	N	N	T*	R	V+	F*	N	N	.	.	UA101AD	LM101D	O	AM101-DIP
FLP-10/3M	N	FT	E-	E+	V-	T*	R	V+	F*	N	.	.	.	.	.	.	SFC2101APM	LM101F	O	AM101-FLP
TO5-8/1M	FT	E-	E+	V-M	T*	R	V+	F*	.	.	.	.	.	.	.	.	SFC2101A	LM101H	O	AM101-TO5
CHP	.	.	.	.	.	.	.	.	.	.	.	.	.	.	.	.	.	LM102D	O	AM102-DICE
DIL-14/1C	N	N	T	N	E+	V-	N	N	L	R	V+	T*	N	N	.	.	.	LM102D	O	AM102-DIP
FLP-10/3C	N	T	N	E+	V-	L	R	V+	T*	N	.	.	.	.	.	.	102(FLP)	LM102F	O	AM102-FLP
TO5-8/1M	T	N	E+	V-	L	R	V+	T*	.	.	.	.	.	.	.	.	UA102M	LM102H	O	AM102-TO5
CHP	.	.	.	.	.	.	.	.	.	.	.	.	.	.	.	.	.	LD106	C	AM106-DICE
FLP-14/3C	N	G	E+	E-	N	V-	S1	S2	R	N	V+	N	N	N	.	.	SN52106FA	LM106F	O	AM106-FLP
TO5-8/1M	G	E+	E-	V-M	S1	S2	R	V+	.	.	.	.	.	.	.	.	SN52106L	LM106H	O	AM106-TO5
CHP	.	.	.	.	.	.	.	.	.	.	.	.	.	.	.	.	AMLD107	LD107	O	AM107-DICE
DIL-14/1C	N	N	N	E-	E+	N	N	N	R	N	V+	N	N	N	.	.	SN52107JA	LM107D	O	AM107-DIP
FLP-10/3C	N	N	E-	E+	V-	N	R	V+	N	N	.	.	.	.	.	.	SFC2107PM	LM107F	O	AM107-FLP
TO5-8/1M	N	E-	E+	V-M	N	R	V+	.	.	.	.	.	.	.	.	.	SFC2107M	LM107H	O	AM107-TO5
CHP	.	.	.	.	.	.	.	.	.	.	.	.	.	.	.	.	.	LD108A	O	AM108A-DICE
DIL-14/1C	N	F	N	E-	E+	N	V-	N	N	R	V+	F*	N	N	.	.	UA108AD	LM108AD	O	AM108A-DIP
TO5-8/1M	F	E-	E+	V-M	N	R	V+	F*	.	.	.	.	.	.	.	.	SFC2108A	LM108AH	O	AM108A-TO5
CHP	.	.	.	.	.	.	.	.	.	.	.	.	.	.	.	.	.	LD108	O	AM108-DICE
DIL-14/1C	N	F	N	E-	E+	N	V-	N	N	R	V+	F*	N	N	.	.	UA108D	LM108D	O	AM108-DIP
TO5-8/1M	F	E-	E+	V-M	N	R	V+	F*	.	.	.	.	.	.	.	.	SFC2108M	LM108H	O	AM108-TO5
CHP	.	.	.	.	.	.	.	.	.	.	.	.	.	.	.	.	AMLD110	LD110	O	AM110-DICE
DIL-14/1C	N	N	T	N	E+	V-	N	N	L	R	V+	T*	N	N	.	.	SN52110JA	LM110D	O	AM110-DIP
FLP-10/3C	N	T	N	E+	V-	L	R	V+	T*	N	.	.	.	.	.	.	MLM110F	LM110F	O	AM110-FLP
TO5-8/1M	T	N	E+	V-	L	R	V+	T*	.	.	.	.	.	.	.	.	SFC2110M	LM110H	O	AM110-TO5
CHP	.	.	.	.	.	.	.	.	.	.	.	.	.	.	.	.	AM111-DICE	AMLD111	O	AM111-DICE
DIL-14/1C	N	G	E+	E-	N	V-	T	T*S	R	N	V+	N	N	N	.	.	SN52111J	LM111D	O	AM111-DIP
FLP-10/3C	G	E+	E-	N	V-	T	T*S	N	R	V+	.	.	.	.	.	.	SN52111FA	LM111F	O	AM111-FLP
TO5-8/1M	G	E+	E-	V-	T	T*S	R	V+	.	.	.	.	.	.	.	.	SFC211M	LM111H	O	AM111-TO5
CHP	.	.	.	.	.	.	.	.	.	.	.	.	.	.	.	.	AMLD112	LD112	O	AM112-DICE
DIL-14/1C	N	T	W	E-	E+	W*	V-	N	F	R	V+	T*	N	N	.	.	.	LM112D	O	AM112-DIP
FLP-10/3C	N	W	E-	E+	W*	V-	R	V+	T	T*	.	.	.	.	.	.	MLM112F	LM112F	O	AM112-FLP
TO5-8/1M	T	E-	E+	V-	R	V+	T*	.	.	.	.	.	.	.	.	.	.	LM112H	O	AM112-TO5
CHP	.	.	.	.	.	.	.	.	.	.	.	.	.	.	.	.	AMLD118	LD118	O	AM118-DICE
DIL-14/1C	N	N	T*F	E-	E+	V-	N	N	F*T	R	V+	ø	N	N	.	.	SN52118JA	LM118DP	O	AM118-DIP
FLP-10/3C	N	T*F	E-	E+	V-	F*T	R	V+	ø	N	.	.	.	.	.	.	SN52118FA	LM118F	O	AM118-FLP
TO5-8/1M	T*F	E-	E+	V-	F*T	R	V+	ø	.	.	.	.	.	.	.	.	TDC0118CM	LM118H	O	AM118-TO5
CHP	.	.	.	.	.	.	.	.	.	.	.	.	.	.	.	.	AMLD119	LD119	O	AM119-DICE
DIL-14/1C	N	N	G1	E+1	E-1	V-	R2	G2	E+2	E-2	V+	R1	N	N	.	.	TDC0119DC	LM119D	O	AM119-DIP
FLP-10/3C	R1	G1	E+1	E-1	V-	R2	G2	E+2	E-2	V+	.	.	.	.	.	.	.	LM119F	O	AM119-FLP
TO5-10/1M	R1	G1	E+1	E-1	V-	R2	G2	E+2	E-2	V+	.	.	.	.	.	.	TDC0119CM	LM119H	O	AM119-TO5
CHP	.	.	.	.	.	.	.	.	.	.	.	.	.	.	.	.	AMLD124A	LD124A	O	AM124A-DICE
DIL-14/1C	R1	E-1	E+1	V+	E+2	E-2	R2	R3	E-3	E+3	G	E+4	E-4	R4	.	.	LM124AD	AMLM124AD	O	AM124A-DIP
FLP-14/3C	R1	E-1	E+1	V+	E+2	E-2	R2	R3	E-3	E+3	G	E+4	E-4	R4	.	.	.	LM124AF	O	AM124A-FLP
CHP	.	.	.	.	.	.	.	.	.	.	.	.	.	.	.	.	AMLD124	LD124	O	AM124-DICE
DIL-14/1C	R1	E-1	E+1	V+	E+2	E-2	R2	R3	E-3	E+3	G	E+4	E-4	R4	.	.	MLM124L	LM124D	O	AM124-DIP
FLP-14/3C	R1	E-1	E+1	V+	E+2	E-2	R2	R3	E-3	E+3	G	E+4	E-4	R4	.	.	AMLM124F	LM124F	O	AM124-FLP
CHP	.	.	.	.	.	.	.	.	.	.	.	.	.	.	.	.	.	.	O	AM139A-DICE

TYPE NUMBER	M F H	A P P	C M P	GBP MIN	SLEW RATE MIN	V_S^+ MAX	V_S^- MAX	T_{OP} MAX	A_{VOL} MIN	V_{IO} MAX	I_B MAX	I_{IO} MAX	P_{TOT} MAX	I_{OUT} MIN	V_{OUT} MIN	V_{ICM} MAX	V_{IDF} MAX	dV_{IO}/dT MAX	P_O MAX	I_O MAX	CM RR MIN	PS RR MIN	R_{IN} MIN
AM139A-DICE	ADU	QCP	EXT	.	.	+18V	-18V	125C	94dB	2MV	100NA	25NA	.	.	.	18V	18V	.	.	2MA	.	.	.
AM139A-DIP	ADU	QCP	EXT	.	.	+18V	-18V	125C	94dB	2MV	100NA	25NA	900MWF	.	.	18V	18V	.	.	2MA	.	.	.
AM139A-FLP	ADU	QCP	EXT	.	.	+18V	-18V	125C	94dB	2MV	100NA	25NA	800MWF	.	.	18V	18V	.	.	2MA	.	.	.
AM139-DICE	ADU	QCP	EXT	.	.	+18V	-18V	125C	88dB	5MV	100NA	25NA	.	.	.	18V	18V	.	.	2MA	.	.	.
AM139-DIP	ADU	QCP	EXT	.	.	+18V	-18V	125C	88dB	5MV	100NA	25NA	900MWF	.	.	18V	18V	.	.	2MA	.	.	.
AM139-FLP	ADU	QCP	EXT	.	.	+18V	-18V	125C	88dB	5MV	100NA	25NA	800MWF	.	.	18V	18V	.	.	2MA	.	.	.
AM148-DICE	ADU	QGK	INT	.3MHZ	0.2V/uS	+22V	-22V	125C	94dB	5MV	100NA	25NA	.	5MA	12V	22V	44V	.	.	1MA	70dB	77dB	800K
AM148-DIP	ADU	QGK	INT	.3MHZ	0.2V/uS	+22V	-22V	125C	94dB	5MV	100NA	25NA	.	5MA	12V	22V	44V	.	.	1MA	70dB	77dB	800K
AM149-DICE	ADU	QGK	INT	1MHZ	0.5V/uS	+22V	-22V	125C	94dB	5MV	100NA	25NA	.	5MA	12V	22V	44V	.	.	1MA	70dB	77dB	800K
AM149-DIP	ADU	QGK	INT	1MHZ	0.5V/uS	+22V	-22V	125C	94dB	5MV	100NA	25NA	900MWF	5MA	12V	22V	44V	.	.	1MA	70dB	77dB	800K
AM201A-DIP	ADU	GPU	EXT	.	.	+22V	-22V	85C	94dB	2MV	75NA	10NA	500MWF	5MA	12V	15V	30V	15uV/C	.	3MA	80dB	80dB	500K
AM201A-FLP	ADU	GPW	EXT	.	.	+22V	-22V	85C	94dB	2MV	75NA	10NA	500MWF	5MA	12V	15V	30V	15uV/C	.	3MA	80dB	80dB	500K
AM201A-T05	ADU	GPW	EXT	.	.	+22V	-22V	85C	94dB	2MV	75NA	10NA	500MWF	5MA	12V	15V	30V	15uV/C	.	3MA	80dB	80dB	500K
AM201-DIP	ADU	GPU	EXT	.	.	+22V	-22V	85C	86dB	7.5MV	1.5uA	0.5uA	500MWF	5MA	12V	15V	30V	30uV/C	.	3MA	65dB	70dB	100K
AM201-FLP	ADU	GPU	EXT	.	.	+22V	-22V	85C	86dB	7.5MV	1.5uA	0.5uA	500MWF	5MA	12V	15V	30V	30uV/C	.	3MA	65dB	70dB	100K
AM201-T05	ADU	GPU	EXT	.	.	+22V	-22V	85C	86dB	7.5MV	1.5uA	0.5uA	500MWF	5MA	12V	15V	30V	30uV/C	.	3MA	65dB	70dB	100K
AM202-DIP	ADU	VFA	INT	.	.	+18V	-18V	85C	0dB	10MV	15NA	.	500MWF	1MA	10V	.	.	60uV/C	.	6MA	.	60dB	10G
AM202-T05	ADU	VFA	INT	.	.	+18V	-18V	85C	0dB	10MV	15NA	.	500MWF	1MA	10V	.	.	60uV/C	.	6MA	.	60dB	10G
AM206-T05	ADU	CPR	EXT	.	.	+15V	-15V	85C	84dB	2MV	20uA	3uA	600MWF	50MA	2.5V	.	.	10uV/C	163MW	.	.	.	.	.
AM207-DIP	ADU	GPK	INT	.	.	+22V	-22V	85C	94dB	2MV	75NA	10NA	500MWF	5MA	12V	15V	30V	15uV/C	.	3MA	80dB	80dB	1.5M
AM207-FLP	ADU	GPK	INT	.	.	+22V	-22V	85C	94dB	2MV	75NA	10NA	500MWF	5MA	12V	15V	30V	15uV/C	.	3MA	80dB	80dB	1.5M
AM207-T05	ADU	GPK	INT	.	.	+22V	-22V	85C	94dB	2MV	75NA	10NA	500MWF	5MA	12V	15V	30V	15uV/C	.	3MA	80dB	80dB	1.5M
AM208A-DIP	ADU	SBA	EXT	.	.	+20V	-20V	85C	98dB	0.5MV	2NA	0.2NA	500MWF	1MA	13V	15V	1V	5uV/C	.	.6MA	96dB	96dB	30M
AM208A-T05	ADU	SBA	EXT	.	.	+20V	-20V	85C	98dB	0.5MV	2NA	0.2NA	500MWF	1MA	13V	15V	1V	5uV/C	.	.6MA	96dB	96dB	30M
AM208-DIP	ADU	SBA	EXT	.	.	+20V	-20V	85C	96dB	2MV	2NA	0.2NA	500MWF	1MA	13V	15V	1V	15uV/C	.	.6MA	85dB	80dB	30M
AM208-T05	ADU	SBA	EXT	.	.	+20V	-20V	85C	96dB	2MV	2NA	0.2NA	500MWF	1MA	13V	15V	1V	15uV/C	.	.6MA	85dB	80dB	30M
AM210-DIP	ADU	VFA	INT	.	15V/uS	+18V	-18V	85C	0dB	4MV	3NA	.	500MWF	1MA	10V	15V	15V	50uV/C	.	6MA	.	70dB	10G
AM210-FLP	ADU	VFA	INT	.	15V/uS	+18V	-18V	85C	0dB	4MV	3NA	.	500MWF	1MA	10V	15V	15V	50uV/C	.	6MA	.	70dB	10G
AM210-T05	ADU	VFA	INT	.	15V/uS	+18V	-18V	85C	0dB	4MV	3NA	.	500MWF	1MA	10V	15V	15V	50uV/C	.	6MA	.	70dB	10G
AM211-DIP	ADU	CPR	EXT	.	.	+18V	-18V	85C	100dB	3MV	100NA	10NA	500MWF	.	.	15V	30V	.	.	6MA	.	.	.
AM211-T05	ADU	CPR	EXT	.	.	+18V	-18V	85C	100dB	3MV	100NA	10NA	500MWF	.	.	15V	30V	.	.	6MA	.	.	.
AM212-DIP	ADU	SBA	INT	.	.	+20V	-20V	85C	94dB	2MV	2NA	0.2NA	500MWF	1MA	13V	14V	14V	15uV/C	.	.6MA	85dB	80dB	30M
AM212-FLP	ADU	SBA	INT	.	.	+20V	-20V	85C	94dB	2MV	2NA	0.2NA	500MWF	1MA	13V	14V	14V	15uV/C	.	.6MA	85dB	80dB	30M
AM212-T05	ADU	SBA	INT	.	.	+20V	-20V	85C	94dB	2MV	2NA	0.2NA	500MWF	1MA	13V	14V	14V	15uV/C	.	.6MA	85dB	80dB	30M
AM216A-DICE	ADU	LBC	INT	.	.	+20V	-20V	85C	92dB	3MV	50pA	15pA	.	1MA	13V	15V	14V	.	.	.6MA	80dB	80dB	2G
AM216A-DIP	ADU	LBC	INT	.	.	+20V	-20V	85C	92dB	3MV	50pA	15pA	500MWF	1MA	13V	15V	14V	.	.	.6MA	80dB	80dB	2G
AM216A-FLP	ADU	LBC	INT	.	.	+20V	-20V	85C	92dB	3MV	50pA	15pA	500MWF	1MA	13V	15V	14V	.	.	.6MA	80dB	80dB	2G
AM216A-T05	ADU	LBC	INT	.	.	+20V	-20V	85C	92dB	3MV	50pA	15pA	500MWF	1MA	13V	15V	14V	.	.	.6MA	80dB	80dB	2G
AM216-DICE	ADU	LBC	INT	.	.	+20V	-20V	85C	86dB	10MV	150pA	50pA	.	1MA	13V	15V	14V	.	.	.8MA	80dB	80dB	300M
AM216-DIP	ADU	LBC	INT	.	.	+20V	-20V	85C	86dB	10MV	150pA	50pA	500MWF	1MA	13V	15V	14V	.	.	.8MA	80dB	80dB	300M
AM216-FLP	ADU	LBC	INT	.	.	+20V	-20V	85C	86dB	10MV	150pA	50pA	500MWF	1MA	13V	15V	14V	.	.	.8MA	80dB	80dB	300M
AM216-T05	ADU	LBC	INT	.	.	+20V	-20V	85C	86dB	10MV	150pA	50pA	500MWF	1MA	13V	15V	14V	.	.	.8MA	80dB	80dB	300M
AM218-DIP	ADU	XSR	INT	.	50V/uS	+20V	-20V	85C	94dB	4MV	250NA	50NA	500MWF	6MA	12V	15V	1V	.	.	8MA	80dB	70dB	1M
AM218-FLP	ADU	XSR	INT	.	50V/uS	+20V	-20V	85C	94dB	4MV	250NA	50NA	500MWF	6MA	12V	15V	1V	.	.	8MA	80dB	70dB	1M
AM218-T05	ADU	XSR	INT	.	50V/uS	+20V	-20V	85C	94dB	4MV	250NA	50NA	500MWF	6MA	12V	15V	1V	.	.	8MA	80dB	70dB	1M
AM219-DIP	ADU	DCP	INT	.	.	+18V	-18V	85C	80dB	4MV	500NA	75NA	500MWF	.	.	15V	5V	.	.	12MA	.	.	.
AM219-FLP	ADU	DCP	INT	.	.	+18V	-18V	85C	80dB	4MV	500NA	75NA	500MWF	.	.	15V	5V	.	.	12MA	.	.	.
AM219-T05	ADU	DCP	INT	.	.	+18V	-18V	85C	80dB	4MV	500NA	75NA	500MWF	.	.	15V	5V	.	.	12MA	.	.	.
AM224A-DIP	ADU	QGK	INT	.	.	+16V	-16V	85C	94dB	3MV	80NA	15NA	900MWF	.	.	16V	16V	20uV/C	.	2MA	70dB	65dB	.
AM224-DIP	ADU	QGK	INT	.	.	+16V	-16V	85C	94dB	5MV	150NA	30NA	900MWF	.	.	16V	16V	35uV/C	.	2MA	70dB	65dB	.
AM239A-DIP	ADU	QCP	EXT	.	.	+18V	-18V	85C	94dB	2MV	250NA	50NA	900MWF	6MA	.	18V	18V	.	.	2MA	.	.	.
AM239-DIP	ADU	QCP	EXT	.	.	+18V	-18V	85C	88dB	5MV	250NA	50NA	900MWF	6MA	.	18V	18V	.	.	2MA	.	.	.
AM248-DIP	ADU	QGK	INT	.3MHZ	0.2V/uS	+18V	-18V	85C	88dB	6MV	200NA	50NA	900MWF	5MA	12V	18V	36V	.	.	1MA	70dB	77dB	800K
AM249-DIP	ADU	QGK	INT	1MHZ	0.5V/uS	+18V	-18V	85C	88dB	6MV	200NA	50NA	900MWF	5MA	12V	18V	36V	.	.	1MA	70dB	77dB	800K
AM301A-DICE	ADU	GPU	EXT	.	.	+18V	-18V	70C	88dB	7.5MV	250NA	50NA	.	5MA	12V	15V	30V	30uV/C	.	3MA	70dB	70dB	500K
AM301A-DIL8	ADU	GPU	EXT	.	.	+18V	-18V	70C	88dB	7.5MV	250NA	50NA	500MWF	5MA	12V	15V	30V	30uV/C	.	3MA	70dB	70dB	500K
AM301A-DIP	ADU	GPU	EXT	.	.	+18V	-18V	70C	88dB	7.5MV	250NA	50NA	500MWF	5MA	12V	15V	30V	30uV/C	.	3MA	70dB	70dB	500K
AM301A-T05	ADU	GPU	EXT	.	.	+18V	-18V	70C	88dB	7.5MV	250NA	50NA	500MWF	5MA	12V	15V	30V	30uV/C	.	3MA	70dB	70dB	500K
AM301-DICE	ADU	GPU	EXT	.	.	+18V	-18V	70C	83dB	10MV	2uA	.75uA	.	.	12V	15V	30V	30uV/C	.	3MA	65dB	70dB	100K
AM301-DIP	ADU	GPU	EXT	.	.	+18V	-18V	70C	83dB	10MV	2uA	.75uA	500MWF	5MA	12V	15V	30V	30uV/C	.	3MA	65dB	70dB	100K

48

For detailed explanations of column heading notations, see App. A.

Also for ready references the more important abbreviations used in the column headings are listed below:

LEFT HAND PAGE

APP = application (codes at APP.E.)
CMRR = common mode rejection ratio
CMP = compensation (frequency)
dV_{IO}/dT = input offset voltage temperature drift
GBP = gain bandwidth product
I_B = input bias current
I_{IO} = input bias offset current
I_Q = quiescent supply current
MFR = manufacturer (codes at App.C.)
P_Q = quiescent power consumer
PSRR = power supply rejection ratio
V_{ICM} = common mode input voltage rating
V_{IDF} = differential input voltage rating
V_{IO} = input offset voltage
V_S = dc supply voltage

RIGHT HAND PAGE

Lead out coding summary (details at APP.G.) for different cases (APP.F.)

A = gain adjust
B = bias adjust
C = case
E– = inverting input
E+ = non-inverting input
F,F* = input frequency compensation
G = ground
J = high level input
K = output, open collector
L = output, open emitter
M = metal case
N = not connected
] = special terminal
R,R* = outputs
S = strobe
T,T* = offset balance
V+ = +ve dc supply
V– = –ve dc supply
W = guard ring
X = blank position, no lead
+ + = +ve supplementary dc supply
– – = –ve supplementary dc supply
* = output frequency compensation

CASE (APP F)	LD 1	LD 2	LD 3	LD 4	LD 5	LD 6	LD 7	LD 8	LD 9	LD 10	LD 11	LD 12	LD 13	LD 14	LD 15	LD 16	EUROPE SUBSTITUTE	USA SUBSTITUTE	ISS	TYPE NUMBER
DIL-14/1C	R2	R1	V+	E-1	E+1	E-2	E+2	E+3	E-3	E-4	E+4	G	R4	R3			MLM139AL	LM139AD	0	AM139A-DIP
FLP-14/3C	R2	R1	V+	E-1	E+1	E-2	E+2	E+3	E-3	E-4	E+4	G	R4	R3				LM139AF	0	AM139A-FLP
CHP																	AMLD139	LD139	0	AM139-DICE
DIL-14/1C	R2	R1	V+	E-1	E+1	E-2	E+2	E+3	E-3	E-4	E+4	G	R4	R3			LM139DDD	LM139D	0	AM139-DIP
FLP-14/3C	R2	R1	V+	E-1	E+1	E-2	E+2	E+3	E-3	E-4	E+4	G	R4	R3				LM139F	0	AM139-FLP
CHP																	AMLD148	LD148	0	AM148-DICE
DIL-14/1C	R1	E-1	E+1	V+	E+2	E-2	R2	R3	E-3	E+3	V-	E+4	E-4	R4			AMLM148Q	LM148D	0	AM148-DIP
CHP																	AMLD149	LD149	0	AM149-DICE
DIL-14/1C	R1	E-1	E+1	V+	E+2	E-2	R2	R3	E-3	E+3	V-	E+4	E-4	R4			LM149D	AMLM149D	0	AM149-DIP
DIL-14/1C	N	N	FT	E-	E+	V-	N	N	T*	R	V+	F*	N	N			UA201AD	LM201AD	0	AM201-DIP
FLP-10/3C	N	FT	E-	E+	V-	T*	R	V+	F*	N							SFC2201APT	LM201AF	0	AM201A-FLP
T05-8/1M	FT	E-	E+	V-M	T*	R	V+	F*									SFC2101A	LM201AH	0	AM201A-T05
DIL-14/1C	N	N	FT	E-	E+	V-	N	N	T*	R	V+	F*	N	N			UA201AD	LM201D	0	AM201-DIP
FLP-10/3M	N	FT	E-	E+	V-	T*	R	V+	F*	N							SFC2201APM	LM201F	0	AM201-FLP
T05-8/1M	FT	E-	E+	V-M	T*	R	V+	F*									SFC2101A	LM201H	0	AM201-T05
DIL-14/1C	N	N	T	N	E+	V-	N	N	L	R	V+	T*	N	N				LM202D	0	AM202-DIP
T05-8/1M	T	N	E+	V-	L	R	V+	T*									UA102M	LM202H	0	AM202-T05
T05-8/1M	G	E+	E-	V-M	S1	S2	R	V+									SN52106L	LM206H	0	AM206-T05
DIL-14/1C	N	N	E-	E+	V-	N	N	N	R	V+	N	N	N	N			SN52107JA	LM207D	0	AM207-DIP
FLP-10/3C	N	N	E-	E+	V-	N	R	V+	N								SFC2207PT	LM207F	0	AM207-FLP
T05-8/1M	N	E-	E+	V-M	R	V+	N										SFC2207	LM207H	0	AM207-T05
DIL-14/1C	N	F	N	E-	E+	N	V-	N	R	V+	F*	N	N				UA208AD	LM208AD	0	AM208A-DIP
T05-8/1M	F	E-	E+	V-M	N	R	V+	F*									SFC2208A	LM208AH	0	AM208A-T05
DIL-14/1C	N	F	N	E-	E+	N	V-	N	R	V+	F*	N	N				UA208D	LM208D	0	AM208-DIP
T05-8/1M	F	E-	E+	V-M	N	R	V+	F*									SFC2208	LM208H	0	AM208-T05
DIL-14/1C	N	N	T	N	E+	V-	N	N	L	R	V+	T*	N	N			SN52110JA	LM210D	0	AM210-DIP
FLP-10/3C	N	T	N	E+	V-	L	R	V+	T*	N								LM210F	0	AM210-FLP
T05-8/1M	T	N	E+	V-	L	R	V+	T*									SFC2210	LM210H	0	AM210-T05
DIL-14/1C	N	G	E+	E-	N	V-	T	T*S	R	N	N	V+	N	N			SN52111J	LM211D	0	AM211-DIP
T05-8/1M	G	E+	E-	V-	T	T*S	R	V+									SFC2211	LM211H	0	AM211-T05
DIL-14/1C	N	T	W	E-	E+	W*	V-	N	F	R	V+	T*	N	N				LM212D	0	AM212-DIP
FLP-10/3C	N	W	E-	E+	W*	V-	R	V+	T	T*								LM212F	0	AM212-FLP
T05-8/1M	T	E-	E+	V-	F	R	V+	T*										LM212H	0	AM212-T05
CHP																		LD216A	0	AM216A-DICE
DIL-14/1C	N	T	W	E-	E+	W*	V-	N	F	R	V+	T*	N	N				LM216AD	0	AM216A-DIP
FLP-10/3C	N	W	E-	E+	W*	V-	R	V+	T	T*								LM216AF	0	AM216A-FLP
T05-8/1M	T	E-	E+	V-	F	R	V+	T*										LM216AH	0	AM216A-T05
CHP																		LD216	0	AM216-DICE
DIL-14/1C	N	T	W	E-	E+	W*	V-	N	F	R	V+	T*	N	N				LM216D	0	AM216-DIP
FLP-10/3C	N	W	E-	E+	W*	V-	R	V+	T	T*								LM216F	0	AM216-FLP
T05-8/1M	T	E-	E+	V-	F	R	V+	T*										LM216H	0	AM216-T05
DIL-14/1C	N	N	T*F	E-	E+	V-	N	F*T	R	V+	ø	N	N					LM218D	0	AM218-DIP
FLP-10/3C	N	T*F	E-	E+	V-	F*T	R	V+	ø	N								LM218F	0	AM218-FLP
T05-8/1M	T*F	E-	E+	V-	F*T	R	V+	ø									TDB0118CM	LM218H	0	AM218-T05
DIL-14/1C	N	N	G1	E+1	E-1	V-	R2	G2	E+2	E-2	V+	R1	N	N			TDE0119DP	LM219D	0	AM219-DIP
FLP-10/3C	R1	G1	E+1	E-1	V-	R2	G2	E+2	E-2	V+								LM219F	0	AM219-FLP
T05-10/1M	R1	G1	E+1	E-1	V-	R2	G2	E+2	E-2	V+							TDE0119CM	LM219H	0	AM219-T05
DIL-14/1C	R1	E-1	E+1	V+	E+2	E-2	R2	R3	E-3	E+3	G	E+4	E-4	R4				LM224AD	0	AM224A-DIP
DIL-14/1C	R1	E-1	E+1	V+	E+2	E-2	R2	R3	E-3	E+3	G	E+4	E-4	R4				LM224D	0	AM224-DIP
DIL-14/1C	R2	R1	V+	E-1	E+1	E-2	E+2	F+3	E-3	E-4	E+4	G	R4	R3			MLM239AL	LM239AD	0	AM239A-DIP
DIL-14/1C	R2	R1	V+	E-1	E+1	E-2	E+2	E+3	E-3	E-4	E+4	G	R4	R3			MLM239L	LM239D	0	AM239-DIP
DIL-14/1C	R1	E-1	E+1	V+	E+2	E-2	R2	R3	E-3	E+3	V-	E+4	E-4	R4				LM248D	0	AM248-DIP
DIL-14/1C	R1	E-1	E+1	V+	E+2	E-2	R2	R3	E-3	E+3	V-	E+4	E-4	R4			LM248D	LM249D	0	AM249-DIP
CHP																		LD301A	0	AM301A-DICE
DIL-8/1P	FT	E-	E+	V-	T*	R	V+	F*									SFC2301ADC	LM301AN	0	AM301A-DIL8
DIL-14/1C	N	N	FT	E-	E+	V-	N	N	T*	R	V+	F*	N	N			UA301AD	LM301AJ14	0	AM301A-DIP
T05-8/1M	FT	E-	E+	V-M	T*	R	V+	F*									SFC2301AH	LM301AH	0	AM301A-T05
CHP																		LD301	0	AM301-DICE
DIL-14/1C	N	N	FT	E-	E+	V-	N	N	T*	R	V+	F*	N	N				LM301D	0	AM301-DIP
FLP-10/3C	N	FT	E-	E+	V-	T*	R	V+	F*	N							SFC2201APT	LM301F	0	AM301-FLP

TYPE NUMBER	MFR	APP	CMP	GBP MIN	SLEW RATE MIN	V_S^+ MAX	V_S^- MAX	T_{OP} MAX	A_{VOL} MIN	V_{IO} MAX	I_B MAX	I_{IO} MAX	P_{TOT} MAX	I_{OUT} MIN	V_{OUT} MIN	V_{ICM} MAX	V_{IDF} MAX	dV_{IO}/dT MAX	P_Q MAX	I_Q MAX	CMRR MIN	PSRR MIN	R_{IN} MIN
AM301-FLP	ADU	GPU	EXT	.	.	+18V	-18V	70C	83dB	10MV	2uA	.75uA	500MWF	.	12V	15V	30V	30uV/C	.	3MA	65dB	70dB	100K
AM301-TO5	ADU	GPU	EXT	.	.	+18V	-18V	70C	83dB	10MV	2uA	.75uA	500MWF	.	12V	15V	30V	30uV/C	.	3MA	65dB	70dB	100K
AM302-DIP	ADU	VFA	INT	.	.	+18V	-18V	85C	0dB	5MV	30NA	.	500MWF	1MA	10V	.	.	90uV/C	.	6MA	.	.	10G
AM302-TO5	ADU	VFA	INT	.	.	+18V	-18V	70C	0dB	15MV	30NA	.	500MWF	1MA	10V	.	.	90uV/C	.	6MA	.	60dB	10G
AM302-DICE	ADU	VFA	INT	.	.	+18V	-18V	70C	0dB	15MV	30NA	.	.	1MA	10V	.	.	90uV/C	.	6MA	.	60dB	10G
AM306-DICE	ADU	CPR	EXT	.	.	+15V	-15V	70C	84dB	5MV	25uA	5uA	.	50MA	2.5V	.	.	20uV/C	163MW	.	.	.	.
AM306-TO5	ADU	CPR	EXT	.	.	+15V	-15V	70C	84dB	5MV	25uA	5uA	600MWF	50MA	2.5V	.	.	20uV/C	163MW	.	.	.	.
AM307-DICE	ADU	GPK	INT	.	.	+18V	-18V	70C	84dB	7.5MV	250NA	50NA	.	5MA	12V	15V	30V	30uV/C	.	.	.	70dB	70dB	0.5M
AM307-DIP	ADU	GPK	INT	.	.	+18V	-18V	70C	84dB	7.5MV	250NA	50NA	500MWF	5MA	12V	15V	30V	30uV/C	.	.	.	70dB	70dB	0.5M
AM307-TO5	ADU	GPK	INT	.	.	+18V	-18V	70C	84dB	7.5MV	250NA	50NA	.	5MA	12V	15V	30V	30uV/C	.	.	.	70dB	70dB	0.5M
AM308A-DICE	ADU	SBA	EXT	.	.	+18V	-18V	70C	98dB	0.5MV	7NA	1NA	.	1MA	13V	15V	1V	5uV/C	.	.6MA	96dB	96dB	10M
AM308A-DIL8	ADU	SBA	EXT	.	.	+18V	-18V	70C	98dB	0.5V	7NA	1NA	500MWF	1MA	13V	15V	1V	5uV/C	.	.6MA	96dB	96dB	10M
AM308A-DIP	ADU	SBA	EXT	.	.	+18V	-18V	70C	98dB	0.5MV	7NA	1NA	.	1MA	13V	15V	1V	5uV/C	.	.6MA	96dB	96dB	10M
AM308A-TO5	ADU	SBA	EXT	.	.	+18V	-18V	70C	98dB	0.5MV	7NA	1NA	500MWF	1MA	13V	15V	1V	5uV/C	.	.6MA	96dB	96dB	10M
AM308-DICE	ADU	SBA	EXT	.	.	+18V	-18V	70C	88dB	7.5MV	7NA	1NA	.	1MA	13V	15V	1V	30uV/C	.	.6MA	80dB	80dB	10M
AM308-DIL8	ADU	SBA	EXT	.	.	+18V	-18V	70C	88dB	7.5MV	7NA	1NA	500MWF	1MA	13V	15V	1V	30uV/C	.	.6MA	80dB	80dB	10M
AM308-DIP	ADU	SBA	EXT	.	.	+18V	-18V	70C	88dB	7.5MV	7NA	1NA	500MWF	1MA	13V	15V	1V	30uV/C	.	.6MA	80dB	80dB	10M
AM308-TO5	ADU	SBA	EXT	.	.	+18V	-18V	70C	88dB	7.5MV	7NA	1NA	500MWF	1MA	13V	15V	1V	30uV/C	.	.6MA	80dB	80dB	10M
AM310-DICE	ADU	VFA	INT	15V/uS	.	+18V	-18V	70C	0dB	7.5MV	7NA	.	.	1MA	10V	15V	15V	50uV/C	.	6MA	.	70dB	10G
AM310-DIL8	ADU	VFA	INT	15V/uS	.	+18V	-18V	70C	0dB	7.5MV	7NA	.	500MWF	1MA	10V	15V	15V	50uV/C	.	6MA	.	70dB	10G
AM310-DIP	ADU	VFA	INT	15V/uS	.	+18V	-18V	70C	0dB	7.5MV	7NA	.	500MWF	1MA	10V	15V	15V	50uV/C	.	6MA	.	70dB	10G
AM310-FLP	ADU	VFA	INT	15V/uS	.	+18V	-18V	70C	0dB	7.5MV	7NA	.	500MWF	1MA	10V	15V	15V	50uV/C	.	6MA	.	70dB	10G
AM310-TO5	ADU	VFA	INT	15V/uS	.	+18V	-18V	70C	0dB	7.5MV	7NA	.	500MWF	1MA	10V	15V	15V	50uV/C	.	6MA	.	70dB	10G
AM311-DICE	ADU	CPR	EXT	.	.	+18V	-18V	70C	100dB	7.5MV	250NA	50NA	.	.	.	15V	30V	.	.	8MA	.	.	.
AM311-DIP	ADU	CPR	EXT	.	.	+18V	-18V	70C	100dB	7.5MV	250NA	50NA	500MWF	.	.	15V	30V	.	.	8MA	.	.	.
AM311-TO5	ADU	CPR	EXT	.	.	+18V	-18V	70C	100dB	7.5MV	250NA	50NA	500MWF	.	.	15V	30V	.	.	8MA	.	.	.
AM312-DICE	ADU	SBA	INT	.	.	+18V	-18V	70C	88dB	7.5MV	7NA	1NA	.	1MA	13V	14V	14V	15uV/C	.	.8MA	80dB	80dB	10M
AM312-DIP	ADU	SBA	INT	.	.	+18V	-18V	70C	88dB	7.5MV	7NA	1NA	500MWF	1MA	13V	14V	14V	15uV/C	.	.8MA	80dB	80dB	10M
AM312-TO5	ADU	SBA	INT	.	.	+18V	-18V	70C	88dB	7.5MV	7NA	1NA	500MWF	1MA	13V	14V	14V	15uV/C	.	.8MA	80dB	30dB	10M
AM316A-DICE	ADU	LBC	INT	.	.	+20V	-20V	70C	92dB	3MV	50pA	15pA	.	1MA	13V	15V	14V	.	.	.6MA	80dB	80dB	2G
AM316A-DIF	ADU	LBC	INT	.	.	+20V	-20V	70C	92dB	3MV	50pA	15pA	500MWF	1MA	13V	.15V	14V	.	.	.6MA	80dB	80dB	2G
AM316A-FLP	ADU	LBC	INT	.	.	+20V	-20V	70C	92dB	3MV	50pA	15pA	500MWF	1MA	13V	15V	14V	.	.	.6MA	80dB	80dB	2G
AM316-DICE	ADU	LBC	INT	.	.	+20V	-20V	70C	86dB	10MV	150pA	50pA	.	1MA	13V	15V	14V	.	.	.8MA	80dB	80dB	300M
AM316-DIP	ADU	LBC	INT	.	.	+20V	-20V	70C	86dB	10MV	150pA	50pA	500MWF	1MA	13V	15V	14V	.	.	.8MA	80dB	80dB	300M
AM316-FLP	ADU	LBC	INT	.	.	+20V	-20V	70C	86dB	10MV	150pA	50pA	500MWF	1MA	13V	15V	14V	.	.	.8MA	80dB	80dB	300M
AM316-TO5	ADU	LBC	INT	.	.	+20V	-20V	70C	86dB	10MV	150pA	50pA	500MWF	1MA	13V	15V	14V	.	.	.8MA	80dB	80dB	300M
AM318-DICE	ADU	XSR	INT	50V/uS	.	+20V	-20V	70C	88dB	10MV	500NA	200NA	.	6MA	12V	15V	1V	.	.	8MA	70dB	65dB	500K
AM318-DIL8	ADU	XSR	INT	50V/uS	.	+20V	-20V	70C	88dB	10MV	500NA	200NA	500MWF	6MA	12V	15V	1V	.	.	8MA	70dB	65dB	500K
AM318-DIP	ADU	XSR	INT	50V/uS	.	+20V	-20V	70C	88dB	10MV	500NA	200NA	500MWF	6MA	12V	15V	1V	.	.	8MA	70dB	65dB	500K
AM318-FLP	ADU	XSR	INT	50V/uS	.	+20V	-20V	70C	88dB	10MV	500NA	200NA	500MWF	6MA	12V	15V	1V	.	.	8MA	70dB	65dB	500K
AM318-TO5	ADU	XSR	INT	50V/uS	.	+20V	-20V	70C	88dB	10MV	500NA	200NA	500MWF	6MA	12V	15V	1V	.	.	8MA	70dB	65dB	500K
AM319-DICE	ADU	DCP	INT	.	.	+18V	-18V	70C	78dB	8MV	1uA	0.2uA	.	.	.	15V	5V	.	.	12MA	.	.	.
AM319-DILP	ADU	DCP	INT	.	.	+18V	-18V	70C	78dB	8MV	1uA	0.2uA	500MWF	.	.	15V	5V	.	.	12MA	.	.	.
AM319-DIP	ADU	DCP	INT	.	.	+18V	-18V	70C	78dB	8MV	1uA	0.2uA	500MWF	.	.	15V	5V	.	.	12MA	.	.	.
AM319-TO5	ADU	DCP	INT	.	.	+18V	-18V	70C	78dB	8MV	1uA	0.2uA	500MWF	.	.	15V	5V	.	.	12MA	.	.	.
AM324A-DICE	ADU	QGK	INT	.	.	+16V	-16V	70C	88dB	3MV	100NA	30NA	.	.	.	16V	16V	30uV/C	.	2MA	65dB	65dB	.
AM324A-DILP	ADU	QGK	INT	.	.	+16V	-16V	70C	88dB	3MV	100NA	30NA	900MWF	.	.	16V	16V	30uV/C	.	2MA	65dB	65dB	.
AM324A-DIP	ADU	QGK	INT	.	.	+16V	-16V	70C	88dB	3MV	100NA	30NA	900MWF	.	.	16V	16V	30uV/C	.	2MA	65dB	65dB	.
AM324-DICE	ADU	QGK	INT	.	.	+16V	-16V	70C	88dB	7MV	250NA	50NA	.	.	.	16V	16V	35uV/C	.	2MA	65dB	65dB	.
AM324-DILP	ADU	QGK	INT	.	.	+16V	-16V	70C	88dB	7MV	250NA	50NA	900MWF	.	.	16V	16V	35uV/C	.	2MA	65dB	65dB	.
AM324-DIP	ADU	QGK	INT	.	.	+16V	-16V	70C	88dB	7MV	250NA	50NA	900MWF	.	.	16V	16V	35uV/C	.	2MA	65dB	65dB	.
AM339A-DICE	ADU	QCP	EXT	.	.	+18V	-18V	70C	94dB	2MV	250NA	50NA	.	6MA	.	18V	18V	.	.	2MA	.	.	.
AM339A-DILP	ADU	QCP	EXT	.	.	+18V	-18V	70C	94dB	2MV	250NA	50NA	900MWF	6MA	.	18V	18V	.	.	2MA	.	.	.
AM339A-DIP	ADU	QCP	EXT	.	.	+18V	-18V	70C	94dB	2MV	250NA	50NA	900MWF	6MA	.	18V	18V	.	.	2MA	.	.	.
AM339-DICE	ADU	QCP	EXT	.	.	+18V	-18V	70C	88dB	5MV	250NA	50NA	.	6MA	.	18V	18V	.	.	2MA	.	.	.
AM339-DILP	ADU	QCP	EXT	.	.	+18V	-18V	70C	88dB	5MV	250NA	50NA	900MWF	6MA	.	18V	18V	.	.	2MA	.	.	.
AM339-DIP	ADU	QCP	EXT	.	.	+18V	-18V	70C	88dB	5MV	250NA	50NA	900MWF	6MA	.	18V	18V	.	.	2MA	.	.	.
AM348-DICE	ADU	QGK	INT	.3MHZ	0.2V/uS	+18V	-18V	70C	88dB	6MV	200NA	50NA	.	5MA	12V	18V	36V	.	.	1MA	70dB	77dB	800K
AM348-DILP	ADU	QGK	INT	.3MHZ	0.2V/uS	+18V	-18V	70C	88dB	6MV	200NA	50NA	900MWF	5MA	12V	18V	36V	.	.	1MA	70dB	77dB	800K

50

For detailed explanations of column heading notations, see App. A.

Also for ready references the more important abbreviations used in the column headings are listed below:

LEFT HAND PAGE

- APP = application (codes at APP.E.)
- CMRR = common mode rejection ratio
- CMP = compensation (frequency)
- dV_{io}/dT = input offset voltage temperature drift
- GBP = gain bandwidth product
- I_8 = input bias current
- I_{io} = input bias offset current
- I_Q = quiescent supply current
- MFR = manufacturer (codes at App.C.)
- P_Q = quiescent power consumer
- PSRR = power supply rejection ratio
- V_{icm} = common mode input voltage rating
- V_{idf} = differential input voltage rating
- V_{io} = input offset voltage
- V_S = dc supply voltage

RIGHT HAND PAGE

Lead out coding summary (details at APP.G.) for different cases (APP.F.)

- A = gain adjust
- B = bias adjust
- C = case
- E− = inverting input
- E+ = non-inverting input
- F,F* = input frequency compensation
- G = ground
- J = high level input
- K = output, open collector
- L = output, open emitter
- M = metal case
- N = not connected
- Q = special terminal
- R,R* = outputs
- S = strobe
- T,T* = offset balance
- V+ = +ve dc supply
- V− = −ve dc supply
- W = guard ring
- X = blank position, no lead
- ++ = +ve supplementary dc supply
- −− = −ve supplementary dc supply
- ◖,◗* = output frequency compensation

CASE (APP F)	LD 1	LD 2	LD 3	LD 4	LD 5	LD 6	LD 7	LD 8	LD 9	LD 10	LD 11	LD 12	LD 13	LD 14	LD 15	LD 16	EUROPE SUBSTITUTE	USA SUBSTITUTE	I/S	TYPE NUMBER
T05-8/1M	FT	E-	E+	V-M	T*	R	V+	F*									SFC2301A	LM301H	O	AM301-T05
DIL-14/1C	N	N	T	N	E+	V-	N		L	R	V+	T*	N	N				LM302D	O	AM302-DIP
T05-8/1M	T	N	E+	V-	L	R	V+	T*									UA302C	LM302H	O	AM302-T05
CHP																		LD302	O	AM302-DICE
CHP																		LD306	O	AM306-DICE
T05-8/1M	G	E+	E-	V-M	S1	S2	R	V+									SN72306L	LM306H	O	AM306-T05
CHP																			O	AM307-DICE
DIL-14/1C	N	N	N	E-	E+	V-	N	N	N	R	V+	N	N	N			SN72307JA	LM307D	O	AM307-DIP
T05-8/1M	N	E-	E+	V-M	N	R	V+	N									SFC2307	LM307H	O	AM307-T05
CHP																			O	AM308A-DICE
DIL-8/1P	F	E-	E+	V-	N	R	V+	F*										LM308AN	O	AM308A-DIL8
DIL-14/1C	N	F	N	E-	E+	N	V-	N	N	R	V+	F*	N	N			SN72308AJA	LM308AD	O	AM308A-DIP
T05-8/1M	F	E-	E+	V-M	N	R	V+	F*									SFC2308A	LM308AH	O	AM308A-T05
CHP																			O	AM308-DICE
DIL-8/1P	F	E-	E+	V-	N	R	V+	F*									SFC2308DC	LM308N	O	AM308-DIL8
DIL-14/1C	N	F	N	E-	E+	N	V-	N	N	R	V+	F*	N	N			SN72308JA	LM308D	O	AM308-DIP
T05-8/1M	F	E-	E+	V-M	N	R	V+	F*									SFC2308	LM308H	O	AM308-T05
CHP																			O	AM310-DICE
DIL-8/1P	T	N	E+	V-	L	R	V+	T*									SFC2310DC	LM310N	O	AM310-DIL8
DIL-14/1C	N	N	T	N	E+	V-	N	N	L	R	V+	T*	N	N			SFC2310EC	LM310D	O	AM310-DIP
FLP-10/3C	N	T	N	E+	V-	L	R	V+	T*	N								LM310F	O	AM310-FLP
T05-8/1M	T	N	E+	V-	L	R	V+	T*									SFC2310EC	LM310H	O	AM310-T05
CHP																			O	AM311-DICE
DIL-14/1P	N	G	E+	E-	N	V-	T	T*S	R	N	V+	N	N	N			SFC2311EC	LM311D	O	AM311-DIP
T05-8/1M	G	E+	E-	V-	T	T*	R	V+									SFC2311	LM311H	O	AM311-T05
CHP																			O	AM312-DICE
DIL-14/1C	N	T	W	E-	E+	W*	V-	N	F	R	V+	T*	N	N				LM312D	O	AM312-DIP
T05-8/1M	T	E-	E+	V-	F	R	V+	T*										LM312H	O	AM312-T05
CHP																			O	AM316A-DICE
DIL-14/1C	N	T	W	E-	E+	W*	V-	N	F	R	V+	T*	N	N				LM316AD	O	AM316A-DIP
FLP-10/3C	N	W	E-	E+	W*	V-	R	V+	T	T*								LM316AF	O	AM316A-FLP
T05-8/1M	T	E-	E+	V-	F	R	V+	T*										LM316AH	O	AM316A-T05
CHP																			O	AM316-DICE
DIL-14/1C	N	T	W	E-	E+	W*	V-	N	F	R	V+	T*	N	N			MLM316D	LM316D	O	AM316-DIP
FLP-10/3C	N	W	E-	E+	W*	V-	R	V+	T	T*								LM316F	O	AM316-FLP
T05-8/1M	T	E-	E+	V-	F	R	V+	T*										LM316H	O	AM316-T05
CHP																			O	AM318-DICE
DIL-8/1P	T*F	E-	E+	V-	F*	R	V-	ø									SN72318JP	LM318N	O	AM318-DIL8
DIL-14/1C	N	N	T*F	E-	E+	V-	N	N	F*T	R	V+	ø	N	N			SN72318JA	LM318D	O	AM318-DIP
FLP-10/3C	N	T*F	E-	E+	V-	F*T	R	V+	ø	N								LM318F	O	AM318-FLP
T05-8/1M	T*F	E-	E+	V-	F*T	R	V+	ø									TDE0118CM	LM318H	O	AM318-T05
CHP																			O	AM319-DICE
DIL-14/1P	N	N	G1	E+1	E-1	V-	R2	G2	E+2	E-2	V+	R1	N	N			TDB0119DP	LM319N	O	AM319-DILP
DIL-14/1P	N	N	G1	E+1	E-1	V-	R2	G2	E+2	E-2	V+	R1	N	N			TDB0119DP	LM319D	O	AM319-DIP
T05-10/1M	R1	G1	E+1	E-1	V-	R2	G2	E+2	E-2	V+							TDB0119CM	LM319H	O	AM319-T05
CHP																			O	AM324A-DICE
DIL-14/1P	R1	E-1	E+1	V+	E+2	E-2	R2	R3	E-3	E+3	G	E+4	E-4	R4				LM324AN	O	AM324A-DILP
DIL-14/1C	R1	E-1	E+1	V+	E+2	E-2	R2	R3	E-3	E+3	G	E+4	E-4	R4				LM324AD	O	AM324A-DIP
CHP																			O	AM324-DICE
DIL-14/1P	R1	E-1	E+1	V+	E+2	E-2	R2	R3	E-3	E+3	G	E+4	E-4	R4			MLM324J	LM324N	O	AM324-DILP
DIL-14/1C	R1	E-1	E+1	V+	E+2	E-2	R2	R3	E-3	E+3	G	E+4	E-4	R4			MLM324L	LM324D	O	AM324-DIP
CHP																			O	AM339A-DICE
DIL-14/1P	R2	R1	V+	E-1	E+1	E-2	E+2	E+3	E-3	E-4	E+4	G	R4	R3			MLM339AL	LM339AN	O	AM339A-DILP
DIL-14/1C	R2	R1	V+	E-1	E+1	E-2	E+2	E+3	E-3	E-4	E+4	G	R4	R3			MLM339AL	LM339AD	O	AM339A-DIP
CHP																			O	AM339-DICE
DIL-14/1P	R2	R1	V+	E-1	E+1	E-2	E+2	E+3	E-3	E-4	E+4	G	R4	R3			MLM339L	LM339D	O	AM339-DILP
DIL-14/1C	R2	R1	V+	E-1	E+1	E-2	E+2	E+3	E-3	E-4	E+4	G	R4	R3			MLM339L	LM339D	O	AM339-DIP
CHP																			O	AM348-DICE
DIL-14/1P	R1	E-1	E+1	V+	E+2	E-2	R2	R3	E-3	E+3	V-	E+4	E-4	R4				LM348N	O	AM348-DILP
DIL-14/1C	R1	E-1	E+1	V+	E+2	E-2	R2	R3	E-3	E+3	V-	E+4	E-4	R4				LM348D	O	AM348-DIP

TYPE NUMBER	M F R	A P P	C M P	GBP MIN	SLEW RATE MIN	V_S^+ MAX	V_S^- MAX	T_{OP} MAX	A_{VOL} MIN	V_{IO} MAX	I_B MAX	I_{IO} MAX	P_{TOT} MAX	I_{OUT} MIN	V_{OUT} MIN	V_{ICM} MAX	V_{IDF} MAX	dV_{IO}/dT MAX	P_Q MAX	I_Q MAX	CM RR MIN	PS RR MIN	R_{IN} MIN
AM348-DIP	ADU	QGK	INT	.3MHZ	0.2V/uS	+18V	-18V	70C	88dB	6MV	200NA	50NA	900MWF	5MA	12V	18V	36V	.	.	1MA	70dB	77dB	800K
AM349-DICE	ADU	QGK	INT	1MHZ	0.5V/uS	+18V	-18V	70C	88dB	6MV	200NA	50NA		5MA	12V	18V	36V			1MA	70dB	77dB	800K
AM349-DILP	ADU	QGK	INT	1MHZ	0.5V/uS	+18V	-18V	70C	88dB	6MV	200NA	50NA	900MWF	5MA	12V	18V	36V			1MA	70dB	77dB	800K
AM349-DIP	ADU	QGK	INT	1MHZ	0.5V/uS	+18V	-18V	70C	88dB	6MV	200NA	50NA	900MWF	5MA	12V	18V	36V			1MA	70dB	77dB	800K
AM500GC	DAU	XSR	INT	.1GHZ	500V/uS	+18V	-18V	70C	100dB	3MV	4NA	0.5NA	.	50MA	10V	18V	5V	5uV/C		30MA		80dB	10M
AM500MM	DAU	XSR	INT	.1GHZ	500V/uS	+18V	-18V	125C	100dB	3MV	4NA	0.5NA		50MA	10V	18V	5V	5uV/C		30MA		80dB	10M
AM500MR	DAU	XSR	INT	.1GHZ	500V/uS	+18V	-18V	85C	100dB	3MV	4NA	0.5NA		50MA	10V	18V	5V	5uV/C		30MA		80dB	10M
AM592DC	ADU	BDO	EXT	20MHZ		+8V	-8V	70C	48dB	6MV	30uA	5uA	670MWF	2MA	1.5V	6V	5V			24MA	60dB	50dB	2K
AM592DM	ADU	BDO	EXT	20MHZ		+8V	-8V	125C	50dB	5MV	20uA	3uA	500MWF	2MA	1.5V	6V	5V			24MA	60dB	50dB	2K
AM592HC	ADU	BDO	EXT	20MHZ		+8V	-8V	70C	48dB	6MV	30uA	5uA	500MWF	2MA	1.5V	6V	5V			24MA	60dB	50dB	2K
AM592HM	ADU	BDO	EXT	20MHZ		+8V	-8V	125C	50dB	5MV	20uA	3uA	500MWF	2MA	1.5V	6V	5V			24MA	60dB	50dB	2K
AM592PC	ADU	BDO	EXT	20MHZ		+8V	-8V	70C	48dB	6MV	30uA	5uA	670MWF	2MA	1.5V	6V	5V			24MA	60dB	50dB	2K
AM685DL	ADU	CPR	EXT	.		+7V	-7V	85C		2MV	10uA	1uA	500MWF			4V	6V	10uV/C	300MW	26MA	80dB	70dB	6K
AM685DM	ADU	CPR	EXT	.		+7V	-7V	125C		2MV	10uA	1uA	500MWF			4V	6V	10uV/C	300MW	26MA	80dB	70dB	6K
AM685HL	ADU	CPR	EXT	.		+7V	-7V	85C		2MV	10uA	1uA	500MWF			4V	6V	10uV/C	300MW	26MA	80dB	70dB	6K
AM685HM	ADU	CPR	EXT	.		+7V	-7V	125C		2MV	10uA	1uA	500MWF			4V	6V	10uV/C	300MW	26MA	80dB	70dB	6K
AM685XL	ADU	CPR	EXT			+7V	-7V	85C		2MV	10uA	1uA				4V	6V	10uV/C	300MW	26MA	80dB	70dB	6K
AM685XM	ADU	CPR	EXT			+7V	-7V	125C		2MV	10uA	1uA				4V	6V	10uV/C	300MW	26MA	80dB	70dB	6K
AM686DC	ADU	CPR	EXT			+7V	-7V	70C		3MV	10uA	1uA	600MWF			4V	6V	10uV/C	415MW	42MA	80dB	70dB	
AM686DM	ADU	CPR	EXT			+7V	-7V	125C		2MV	10uA	1uA	600MWF			4V	6V	10uV/C	400MW	40MA	80dB	70dB	
AM686HC	ADU	CPR	EXT			+7V	-7V	70C		3MV	10uA	1uA	600MWF			4V	6V	10uV/C	415MW	42MA	80dB	70dB	
AM686HM	ADU	CPR	EXT			+7V	-7V	125C		2MV	10uA	1uA	600MWF			4V	6V	10uV/C	400MW	40MA	80dB	70dB	
AM686XC	ADU	CPR	EXT			+7V	-7V	70C		3MV	10uA	1uA				4V	6V	10uV/C	415MW	42MA	80dB	70dB	
AM686XM	ADU	CPR	EXT			+7V	-7V	125C		2MV	10uA	1uA				4V	6V	10uV/C	400MW	40MA	80dB	70dB	
AM687ADL	ADU	DCP	EXT			+7V	-7V	85C		3MV	10uA	1uA	600MWF			4V	6V	10uV/C	485MW	48MA	80dB	70dB	
AM687ADM	ADU	DCP	EXT			+7V	-7V	125C		2MV	10uA	1uA	600MWF			4V	6V	10uV/C	450MW	44MA	80dB	70dB	
AM687DL	ADU	DCP	EXT			+7V	-7V	85C		3MV	10uA	1uA	600MWF			4V	6V	10uV/C	485MW	48MA	80dB	70dB	
AM687DM	ADU	DCP	EXT			+7V	-7V	125C		2MV	10uA	1uA	600MWF			4V	6V	10uV/C	450MW	44MA	80dB	70dB	
AM687XL	ADU	DCP	EXT			+7V	-7V	85C		3MV	10uA	1uA				4V	6V	10uV/C	485MW	48MA	80dB	70dB	
AM687XM	ADU	DCP	EXT			+7V	-7V	125C		2MV	10uA	1uA				4V	6V	10uV/C	450MW	44MA	80dB	70dB	
AM715DC	ADU	HSR	EXT		10V/uS	+18V	-18V	70C	80dB	7.5MV	1.5uA	250NA	670MWF	5MA	10V	15V	15V	.	300MW	10MA	74dB	68dB	300K
AM715DM	ADU	HSR	EXT		15V/uS	+18V	-18V	125C	84dB	5MV	750NA	250NA	670MWF	5MA	10V	15V	15V		210MW	7MA	74dB	70dB	300K
AM715FM	ADU	HSR	EXT	.	15V/uS	+18V	-18V	125C	84dB	5MV	750NA	250NA	500MWF	5MA	10V	15V	15V		210MW	7MA	74dB	70dB	300K
AM715HC	ADU	HSR	EXT		10V/uS	+18V	-18V	70C	80dB	7.5MV	1.5uA	250NA	500MWF	5MA	10V	15V	15V		300MW	10MA	74dB	68dB	300K
AM715HM	ADU	HSR	EXT		15V/uS	+18V	-18V	125C	84dB	5MV	750NA	250NA	500MWF	5MA	10V	15V	15V		210MW	7MA	74dB	70dB	300K
AM715XC	ADU	HSR	EXT		10V/uS	+18V	-18V	70C	80dB	7.5MV	1.5uA	250NA		5MA	10V	15V	15V		300MW	10MA	74dB	68dB	300K
AM715XM	ADU	HSR	EXT		15V/uS	+18V	-18V	125C	84dB	5MV	750NA	250NA		5MA	10V	15V	15V		210MW	7MA	74dB	70dB	300K
AM725CN	ADU	PIA	EXT			+22V	-22V	70C	106dB	2.5MV	125NA	35NA	500MWF	5MA	12V	22V	5V	10uV/C	150MW		94dB	90dB	500K
AM725DM	ADU	PIA	EXT			+22V	-22V	125C	120dB	1MV	100NA	20NA	500MWF	5MA	12V	22V	5V	5uV/C	105MW		110dB	100dB	500K
AM725DC	ADU	PIA	EXT			+22V	-22V	70C	106dB	2.5MV	125NA	35NA	500MWF	5MA	12V	22V	5V	10uV/C	150MW		94dB	90dB	500K
AM725HC	ADU	PIA	EXT			+22V	-22V	70C	108dB	2.5MV	125NA	35NA	500MWF	5MA	12V	22V	5V	5uV/C	150MW		94dB	90dB	500K
AM725HM	ADU	PIA	EXT			+22V	-22V	125C	120dB	1MV	100NA	20NA	500MWF	5MA	12V	22V	5V	5uV/C	150MW		110dB	100dB	500K
AM725XC	ADU	PIA	EXT			+22V	-22V	70C	108dB	2.5MV	125NA	35NA		5MA	12V	22V	5V	5uV/C	150MW		94dB	90dB	500K
AM725XM	ADU	PIA	EXT			+22V	-22V	125C	120dB	1MV	100NA	20NA		5MA	12V	22V	5V	5uV/C	150MW		110dB	100dB	800K
AM733DC	ADU	BDO	EXT	20MHZ		+8V	-8V	70C	48dB	6MV	30uA	5uA	670MWF	2MA	1.5V	6V	5V			24MA	60dB	50dB	2K
AM733DM	ADU	BDO	EXT	20MHZ		+8V	-8V	125C	50dB	5MV	20uA	3uA	670MWF	2MA	1.5V	6V	5V			24MA	60dB	50dB	2K
AM733FM	ADU	BDO	EXT	20MHZ		+8V	-8V	125C	50dB	5MV	20uA	3uA	570MWF	2MA	1.5V	6V	5V			24MA	60dB	50dB	2K
AM733HC	ADU	BDO	EXT	20MHZ		+8V	-8V	70C	48dB	6MV	30uA	5uA	500MWF	2MA	1.5V	6V	5V			24MA	60dB	50dB	2K
AM733HM	ADU	BDO	EXT	20MHZ		+8V	-8V	125C	50dB	5MV	20uA	3uA	500MWF	2MA	1.5V	6V	5V			24MA	60dB	50dB	2K
AM733XC	ADU	BDO	EXT	20MHZ		+8V	-8V	70C	48dB	6MV	30uA	5uA		2MA	1.5V	6V	5V			24MA	60dB	50dB	2K
AM733XM	ADU	BDO	EXT	20MHZ		+8V	-8V	125C	50dB	5MV	20uA	3uA		2MA	1.5V	6V	5V			24MA	60dB	50dB	2K
AM741ADM	ADU	GPK	INT	.4MHZ	0.3V/uS	+22V	-22V	125C	94dB	3MV	80NA	30NA	670MWF	10MA	16V	15V	30V	15uV/C	150MW		80dB	86dB	1M
AM741AFM	ADU	GPK	INT	.4MHZ	0.3V/uS	+22V	-22V	125C	94dB	3MV	80NA	30NA	570MWF	10MA	16V	15V	30V	15uV/C	150MW		80dB	86dB	1M
AM741AHM	ADU	GPK	INT	.4MHZ	0.3V/uS	+22V	-22V	125C	94dB	3MV	80NA	30NA	500MWF	10MA	16V	15V	30V	15uV/C	150MW		80dB	86dB	1M
AM741DC	ADU	GPK	INT		0.2V/uS	+18V	-18V	70C	86dB	6MV	500NA	200NA	670MWF	5MA	12V	15V	30V		85MW	3MA	70dB	76dB	300K
AM741DM	ADU	GPK	INT	.4MHZ	0.3V/uS	+22V	-22V	125C	94dB	3MV	80NA	30NA	670MWF	5MA	12V	15V	30V	15uV/C	150MW		80dB	86dB	1M
AM741EDC	ADU	GPK	INT	.4MHZ	0.3V/uS	+22V	-22V	70C	94dB	3MV	80NA	30NA	670MWF	7MA	16V	15V	30V	15uV/C	150MW		80dB	86dB	1M
AM741EHC	ADU	GPK	INT	.4MHZ	0.3V/uS	+22V	-22V	70C	94dB	3MV	80NA	30NA	500MWF	7MA	16V	15V	30V	15uV/C	150MW		80dB	86dB	1M
AM741FM	ADU	GPK	INT		0.2V/uS	+22V	-22V	125C	94dB	5MV	500NA	200NA	570MWF	5MA	12V	15V	30V		85MW	3MA	70dB	76dB	300K
AM741HC	ADU	GPK	INT		0.2V/uS	+18V	-18V	70C	86dB	6MV	500NA	200NA	500MWF	5MA	12V	15V	30V			3MA	70dB	76dB	300K

For detailed explanations of column heading notations, see App. A.

Also for ready references the more important abbreviations used in the column headings are listed below

LEFT HAND PAGE

APP = application (codes at APP.E.)
CMRR = common mode rejection ratio
CMP = compensation (frequency)
dV_{io}/dT = input offset voltage temperature drift
GBP = gain bandwidth product
I_B = input bias current
I_{io} = input bias offset current
I_0 = quiescent supply current
MFR = manufacturer (codes at App.C.)
P_0 = quiescent power consumer
PSRR = power supply rejection ratio
V_{ICM} = common mode input voltage rating
V_{IDF} = differential input voltage rating
V_{IO} = input offset voltage
V_S = dc supply voltage

RIGHT HAND PAGE
Lead out coding summary (details at APP.G.) for different cases (APP.F.)

A = gain adjust
B = bias adjust
C = case
E— = inverting input
E+ = non-inverting input
F,F* = input frequency compensation
G = ground
J = high level input
K = output, open collector
L = output, open emitter
M = metal case
N = not connected
Q = special terminal
R,R* = outputs
S = strobe
T,T* = offset balance
V+ = +ve dc supply
V— = —ve dc supply
W = guard ring
X = blank position, no lead
++ = +ve supplementary dc supply
—— = —ve supplementary dc supply
ø,ø* = output frequency compensation

CASE (APP F)	LD1	LD2	LD3	LD4	LD5	LD6	LD7	LD8	LD9	LD10	LD11	LD12	LD13	LD14	LD15	LD16	EUROPE SUBSTITUTE	USA SUBSTITUTE	IS	TYPE NUMBER
CHP																			0	AM349-DICE
DIL-14/1P	R1	E-1	E+1	V+	E+2	E-2	R2	R3	E-3	E+3	V-	E+4	E-4	R4				LM349N	0	AM349-DILP
DIL-14/1C	R1	E-1	E+1	V+	E+2	E-2	R2	R3	E-3	E+3	V-	E+4	E-4	R4				LM349D	0	AM349-DIP
DIM-14/1G	N	N	N	E-	E+	V-	N	G	N	R	V+	N	N	N					0	AM500GC
DIM-14/1M	N	N	N	E-	E+	V-	N	G	N	R	V+	N	N	N					0	AM500MM
DIM-14/1M	N	N	N	E-	E+	V-	N	G	N	R	V+	N	N	N					0	AM500MR
DIL-14/1C	E+	N	A2	A*2	V-	N	R	R*	N	V+	A1	A*1	N	E-			UA733DC	NE592F	0	AM592DC
DIL-14/1C	E+	N	A2	A*2	V-	N	R	R*	N	V+	A1	A*1	N	E-			UA733DM	SE592F	0	AM592DM
T05-10/1M	E-	E+	A2	A*2	V-	R	R*	V+	A1	A*1							UA733HC	NE592K	0	AM592HC
T05-10/1M	E-	E+	A2	A*2	V-	R	R*	V+	A1	A*1							UA733HM	SE592K	0	AM592HM
DIL-14/1P	E+	N	A2	A*2	V-	N	E-										UA733DC	NE592A	0	AM592PC
DIL-16/1C	G	V+	E+	E-	N	Q	N	V-	N	N	L	L*	N	N	N	G*		AM685DM	0	AM685DL
DIL-16/1C	G	V+	E+	E-	N	Q	N	V-	N	N	L	L*	N	N	N	G*			0	AM685DM
T05-10/1M	V+	E+	E-	Q	V-	N	L	L*	G*	G								AM685HM	0	AM685HL
T05-10/1M	V+	E+	E-	Q	V-	N	L	L*	G*	G									0	AM685HM
CHP																		AM685XM	0	AM685XL
CHP																			0	AM685XM
DIL-16/1C	N	N	V+	E+	E-	V-	N	N	N	N	Q	G	R	R*	N	N		AM686DM	0	AM686DC
DIL-16/1C	N	N	V+	E+	E-	V-	N	N	N	N	Q	G	R	R*	N	N			0	AM686DM
T05-10/1M	V+	N	E+	E-	V-	Q	G	R	R*	N								AM686HM	0	AM686HC
T05-10/1M	V+	N	E+	E-	V-	Q	G	R	R*	N									0	AM686HM
CHP																		AM686XM	0	AM686XC
CHP																			0	AM686XM
DIL-16/1C	L1	L*1	G1	Q1	Q*1	V-	E-1	E+1	E+2	E-2	V+	Q*2	Q2	G2	L*2	L2		AM687ADM	0	AM687ADL
DIL-16/1C	L1	L*1	G1	Q1	Q*1	V-	E-1	E+1	E+2	E-2	V+	Q*2	Q2	G2	L*2	L2			0	AM687ADM
DIL-16/1C	L1	L*1	G1	Q1	Q*1	V-	E-1	E+1	E+2	E-2	V+	Q*2	Q2	G2	L*2	L2		AM687DM	0	AM687DL
DIL-16/1C	L1	L*1	G1	Q1	Q*1	V-	E-1	E+1	E+2	E-2	V+	Q*2	Q2	G2	L*2	L2			0	AM687DM
CHP																		AM687XM	0	AM687XL
CHP																			0	AM687XM
DIL-14/1C	F	F*	Q	E-	E+	N	N	N	N	V-	R	ø	V+	ø*				UA715DC	0	AM715DC
DIL-14/1C	F	F*	Q	E-	E+	N	N	N	N	V-	R	ø	V+	ø*				UA715DM	0	AM715DM
FLP-10/3C	F	Q	E-	E+	V-	R	ø	V+	ø*	F*								715FM	0	AM715FM
T05-10/1M	F	Q	E-	E+	V-	R	ø	V+	ø*	F*								UA715HC	0	AM715HC
T05-10/1M	F	Q	E-	E+	V-	R	ø	V+	ø*	F*								UA715HM	0	AM715HM
CHP																			0	AM715XC
CHP																			0	AM715XM
DIL-8/1P	T	E-	E+	V-	ø	ø*R	V+	T*										LM725CN	0	AM725CN
DIL-14/1C	N	N	T	E-	E+	V-	N	N	ø	ø*R	V+	T*	N	N				LM725D	0	AM725DM
DIL-14/1C	N	N	T	E-	E+	V-	N	N	ø	ø*R	V+	T*	N	N				LM725CJ14	0	AM725DC
T05-8/1M	T	E-	E+	V-	ø	ø*	V+	T*									RC725T	UA725HC	0	AM725HC
T05-8/1M	T	E-	E+	V-	ø	ø*	V+	T*									RM725T	UA725HM	0	AM725HM
CHP																			0	AM725XC
CHP																			0	AM725XM
DIL-14/1C	E+	N	A2	A*2	V-	N	R	R*	N	V+	A1	A*1	N	E-			LM733CD	UA733DC	0	AM733DC
DIL-14/1C	E+	N	A2	A*2	V-	N	R	R*	N	V+	A1	A*1	N	E-			LM733D	UA733DM	0	AM733DM
FLP-10/3C	E+	A2	A*2	V-	R	R*	V+	A1	A*1	E-							SN52733FA	UA733FM	0	AM733FM
T05-10/1M	E-	E+	A2	A*2	V-	R	R*	V+	A1	A*1							LM733CH	UA733HC	0	AM733HC
T05-10/1M	E-	E+	A2	A*2	V-	R	R*	V+	A1	A*1							LM733H	UA733HM	0	AM733HM
CHP																			0	AM733XC
CHP																			0	AM733XM
DIL-14/1C	N	N	T	E-	E+	V-	N	N	T*	R	V+	N	N	N			LM741AD	UA741ADM	0	AM741ADM
FLP-10/3C	N	T	E-	E+	V-	T*	V+	N	N								SFC2741PM	UA741AFM	0	AM741AFM
T05-8/1M	T	E-	E+	V-M	T*	R	V+	N									TBA222	LM741AH	0	AM741AHM
DIL-14/1C	N	N	T	E-	E+	V-	N	N	T*	R	V+	N	N	N			TBA221A	UA741DC	0	AM741DC
DIL-14/1C	N	N	T	E-	E+	V-	N	N	T*	R	V+	N	N	N			LM741D	UA741DM	0	AM741DN
DIL-14/1C	N	N	T	E-	E+	V-	N	N	T*	R	V+	N	N	N			TBA221A	UA741EDC	0	AM741EDC
T05-8/1M	T	E-	E+	V-M	T*	R	V+	N									TBA221	UA741EHC	0	AM741EHC
FLP-10/3C	N	T	E-	E+	V-	T*	R	V+	N	N							SFC2741PM	UA741FM	0	AM741FM
T05-8/1M	T	E-	E+	V-M	T*	R	V+	N									TBA221	UA741HC	0	AM741HC
T05-8/1M	T	E-	E+	V-M	T*	R	V+	N									TBA222	UA741HM	0	AM741HM

TYPE NUMBER	MFR	APP	CMP	GBP MIN	SLEW RATE MIN	V_S^+ MAX	V_S^- MAX	T_{op} MAX	A_{VOL} MIN	V_{IO} MAX	I_B MAX	I_{IO} MAX	P_{TOT} MAX	I_{OUT} MIN	V_{OUT} MIN	V_{ICM} MAX	V_{IDF} MAX	dV_{IO}/dT MAX	P_Q MAX	I_Q MAX	CMRR MIN	PSRR MIN	R_{IN} MIN
AM741HM	ADU	GPK	INT	.	0.2V/uS	+22V	-22V	125C	94dB	5MV	500NA	200NA	500MWF	7MA	12V	15V	30V		85MW	3MA	70dB	76dB	3005
AM741XC	ADU	GPK	INT	.	0.2V/uS	+18V	-18V	70C	86dB	6MV	500NA	200NA	.	5MA	12V	15V	30V		.	3MA	70dB	76dB	300K
AM747DC	ADU	DGK	INT	.	0.2V/uS	+18V	-18V	70C	88dB	6MV	500NA	200NA	670MWF	5MA	12V	15V	30V		85MW	3MA	70dB	76dB	300K
AM747DM	ADU	DGK	INT	.	0.2V/uS	+22V	-22V	125C	94dB	5MV	500NA	200NA	670MWF	5MA	12V	15V	30V		85MW	3MA	70dB	76dB	300K
AM747FM	ADU	DGK	INT	.	0.2V/uS	+22V	-22V	125C	94dB	5MV	500NA	200NA	500MWF	5MA	12V	15V	30V		85MW	3MA	70dB	76dB	300K
AM747HC	ADU	DGK	INT	.	0.2V/uS	+18V	-18V	70C	88dB	6MV	500NA	200NA	500MWF	5MA	12V	15V	30V		85MW	3MA	70dB	76dB	300K
AM747HM	ADU	DGK	INT	.	0.2V/uS	+22V	-22V	125C	94dB	5MV	500NA	200NA	500MWF	5MA	12V	15V	30V		85MW	3MA	70dB	76dB	300K
AM748DC	ADU	GPU	EXT	.	0.2V/uS	+22V	-22V	70C	86dB	6MV	500NA	200NA	670MWF	5MA	12V	15V	30V		85MW	3MA	70dB	76dB	300K
AM748DM	ADU	GPU	EXT	.	0.2V/uS	+22V	-22V	125C	94dB	5MV	500NA	200NA	670MWF	5MA	12V	15V	30V		85MW	3MA	70dB	76dB	300K
AM748HC	ADU	GPU	EXT	.	0.2V/uS	+18V	-18V	70C	86dB	6MV	500NA	200NA	500MWF	5MA	12V	15V	30V		85MW	3MA	70dB	76dB	300K
AM748HM	ADU	GPU	EXT	.	0.2V/uS	+22V	-22V	125C	94dB	5MV	500NA	200NA	500MWF	5MA	12V	15V	30V		85MW	3MA	70dB	76dB	300K
AM748XC	ADU	GPU	EXT	.	0.2V/uS	+18V	-18V	70C	86dB	6MV	500NA	200NA		5MA	12V	15V	30V		85MW	3MA	70dB	76dB	300K
AM748XM	ADU	GPU	EXT	.	0.2V/uS	+22V	-22V	125C	94dB	5MV	500NA	200NA		5MA	12V	15V	30V		85MW	3MA	70dB	76dB	300K
AM1458-DICE	ADU	DGK	INT	.5MHZ	0.3V/uS	+18V	-18V	75C	86dB	6MV	0.5uA	0.2uA		5MA	12V	15V	30V	50uV/C	170MW	6MA	70dB	76dB	300K
AM1458H	ADU	DGK	INT	.5MHZ	0.3V/uS	+18V	-18V	75C	86dB	6MV	0.5uA	0.2uA	680MWF	5MA	12V	15V	30V	50uV/C	170MW	6MA	70dB	76dB	300K
AM1500DC	ADU	DCP	EXT	.	.	+18V	-18V	70C	100dB	7.5MV	250NA	50NA	500MWF	8MA	.	15V	30V		.	8MA	.	.	.
AM1500DL	ADU	DCP	EXT	.	.	+18V	-18V	85C	100dB	3MV	100NA	10NA	500MWF	8MA	.	15V	30V			8MA			
AM1500DM	ADU	DCP	EXT	.	.	+18V	-18V	125C	100dB	3MV	100NA	10NA	500MWF	8MA	.	15V	30V			6MA			
AM1500FL	ADU	DCP	EXT	.	.	+18V	-18V	85C	100dB	3MV	100NA	10NA	500MWF	8MA	.	15V	30V			6MA			
AM1500FM	ADU	DCP	EXT	.	.	+18V	-18V	125C	100dB	3MV	100NA	10NA	500MWF	8MA	.	15V	30V			6MA			
AM1500FC	ADU	DCP	EXT	.	.	+18V	-18V	70C	100dB	7.5MV	250NA	50NA	500MWF	8MA	.	15V	30V			8MA			
AM1501DC	ADU	DGU	EXT	.	.	+22V	-22V	70C	88dB	7.5MV	250NA	50NA	500MWF	5MA	12V	15V	30V	30uV/C		3MA	70dB	70dB	0.5M
AM1501DL	ADU	DGU	EXT	.	.	+22V	-22V	85C	94dB	3MV	75NA	10NA	500MWF	5MA	12V	15V	30V	15uV/C		4MA	80dB	80dB	1.5M
AM1501DM	NAU	DGU	EXT	.	.	+22V	-22V	125C	94dB	2MV	75NA	10NA	500MWF	5MA	12V	15V	30V	15uV/C		3MA	80dB	80dB	1.5M
AM1501FC	ADU	DGU	EXT	.	.	+22V	-22V	70C	88dB	7.5MV	250NA	50NA	500MWF	5MA	12V	15V	30V	30uV/C		3MA	70dB	70dB	0.5M
AM1501FL	ADU	DGU	EXT	.	.	+22V	-22V	85C	94dB	2MV	75NA	10NA	500MWF	5MA	12V	15V	30V	15uV/C		4MA	80dB	80dB	1.5M
AM1501FM	ADU	DGU	EXT	.	.	+22V	-22V	125C	94dB	2MV	75NA	10NA	500MWF	5MA	12V	15V	30V	15uV/C		3MA	80dB	80dB	1.5M
AM1558-DICE	ADU	DGK	INT	5MH7	0.3V/uS	+22V	-22V	125C	94dB	5MV	0.5uA	0.2uA		5MA	12V	15V	30V	50uV/C	150MW	5MA	70dB	76dB	300K
AM1558H	ADU	DGK	INT	.5MHZ	0.3V/uS	+22V	-22V	125C	94dB	5MV	0.5uA	0.2uA	680MWF	5MA	12V	15V	30V	50uV/C	150MW	5MA	70dB	76dB	300K
AMLD101	ADU	GPU	EXT	.	.	+22V	-22V	125C	94dB	5MV	500NA	200NA		5MA	12V	15V	30V	15uV/C		3MA	70dB	70dB	300K
AMLD101A	ADU	GPU	EXT	.	.	+22V	-22V	125C	94dB	2MV	75NA	10NA		5MA	12V	15V	30V	15uV/C		3MA	80dB	80dB	1.5M
AMLD102	ADU	VFA	INT	.	.	+18V	-18V	125C	0dB	5MV	10NA	.		1MA	10V	.	30V	30uV/C		6MA	.	60dB	10G
AMLD106	ADU	CPR	EXT	.	.	+15V	-15V	125C	84dB	2MV	20uA	3uA		50MA	2.5V	.	.	10uV/C	163MW	.	.	.	
AMLD107	ADU	GPK	INT	.	.	+22V	-22V	125C	94dB	2MV	75NA	10NA		5MA	12V	15V	30V	15uV/C		3MA	80dB	80dB	1.5M
AMLD108	ADU	SBA	EXT	.	.	+20V	-20V	125C	96dB	2MV	2NA	0.2NA		1MA	13V	15V	1V	15uV/C		.6MA	85dB	80dB	30M
AMLD108A	ADU	SBA	EXT	.	.	+20V	-20V	125C	98dB	0.5MV	2NA	0.2NA		.	13V	15V	1V	5uV/C		.6MA	96dB	96dB	30M
AMLD110	ADU	VFA	INT	.	15V/uS	+18V	-18V	125C	0dB	4MV	3NA	.		.	10V	15V	15V	50uV/C		6MA	.	70dB	10G
AMLD111	ADU	CPR	EXT	.	.	+18V	-18V	125C	100dB	3MV	100NA	10NA		.	.	15V	30V			6MA			
AMLD112	ADU	SBA	EXT	.	.	+20V	-20V	125C	94dB	2MV	2NA	0.2NA		1MA	13V	14V	14V	15uV/C		.6MA	85dB	80dB	30M
AMLD118	ADU	XSR	INT	.	50V/uS	+20V	-20V	125C	94dB	4MV	250NA	50NA		6MA	12V	15V	1V			8MA	80dB	70dB	1M
AMLD119	ADU	DCP	INT	.	.	+18V	-18V	125C	80dB	4MV	500NA	75NA		.	.	15V	5V			12MA			
AMLD124	ADU	QGK	INT	.	.	+16V	-16V	125C	94dB	5MV	150NA	30NA		.	.	16V	16V	35uV/C		2MA	70dB	65dB	
AMLD124A	ADU	QGK	INT	.	.	+16V	-16V	125C	94dB	2MV	50NA	10NA		.	.	16V	16V	20uV/C		2MA	70dB	65dB	
AMLD139	ADU	QCP	EXT	.	.	+18V	-18V	125C	88dB	5MV	100NA	25NA		.	.	18V	18V			2MA			
AMLD139A	ADU	QCP	EXT	.	.	+18V	-18V	125C	94dB	2MV	100NA	25NA		.	.	18V	18V			2MA			
AMLD148	ADU	QGK	INT	.3MHZ	0.2V/uS	+22V	-22V	125C	94dB	5MV	100NA	25NA		5MA	12V	22V	44V			1MA	70dB	77dB	800K
AMLD149	ADU	QGK	INT	1MHZ	0.5V/uS	+22V	-22V	125C	94dB	5MV	100NA	25NA		5MA	12V	22V	44V			1MA	70dB	77dB	800K
AMLD155	ADU	FET	INT	.5MHZ	2V/uS	+22V	-22V	125C	94dB	5MV	100pA	20pA		5MA	12V	20V	40V	20uV/C		4MA	85dB	85dB	0.1T
AMLD155A	ADU	FET	INT	.5MHZ	3V/uS	+22V	-22V	125C	94dB	2MV	50pA	10pA		5MA	12V	20V	40V	5uV/C		4MA	85dB	85dB	0.1T
AMLD156	ADU	HSR	INT	1MHZ	7.5V/uS	+22V	-22V	125C	94dB	5MV	100pA	20pA		5MA	12V	20V	40V	20uV/C		7MA	85dB	85dB	0.1T
AMLD156A	ADU	HSR	INT	4MHZ	10V/uS	+22V	-22V	125C	94dB	2MV	50pA	10pA		5MA	12V	20V	40V	5uV/C		4MA	85dB	85dB	0.1T
AMLD157	ADU	XSR	INT	15MHZ	6V/uS	+22V	-22V	125C	94dB	5MV	100pA	20pA		5MA	12V	20V	40V	20uV/C		7MA	85dB	85dB	0.1T
AMLD157A	ADU	XSR	INT	15MHZ	8V/uS	+22V	-22V	125C	94dB	2MV	50pA	10pA		5MA	12V	20V	40V	5uV/C		7MA	85dB	85dB	0.1T
AMLD216	ADU	LBC	INT	.	.	+20V	-20V	85C	86dB	10MV	150pA	50pA		1MA	13V	15V	14V			.8MA	80dB	80dB	300M
AMLD216A	ADU	LBC	INT	.	.	+20V	-20V	85C	92dB	3MV	50pA	15pA		1MA	13V	15V	14V			.6MA	80dB	80dB	2G
AMLD301	ADU	GPU	EXT	.	.	+18V	-18V	70C	83dB	10MV	2uA	.75uA		.	12V	15V	30V	30uV/C		3MA	65dB	70dB	100K
AMLD301A	ADU	GPU	EXT	.	.	+18V	-18V	70C	88dB	7.5MV	250NA	50NA		5MA	12V	15V	30V	30uV/C		3MA	70dB	70dB	500K
AMLD302	ADU	VFA	INT	.	.	+18V	-18V	70C	0dB	15MV	30NA	.		1MA	10V	.	.	90uV/C		6MA	.	60dB	10G
AMLD306	ADU	CPR	EXT	.	.	+15V	-15V	70C	84dB	5MV	25uA	5uA		50MA	2.5V	.	.	20uV/C	163MW	.	.	.	
AMLD307	ADU	GPK	INT	.	.	+18V	-18V	70C	84dB	7.5MV	250NA	50NA		5MA	12V	15V	30V	30uV/C		.	70dB	70dB	0.5M

For detailed explanations of column heading notations, see App. A.

Also for ready references the more important abbreviations used in the column headings are listed below:

LEFT HAND PAGE

APP = application (codes at APP.E.)

CMRR = common mode rejection ratio

CMP = compensation (frequency)

dV_{IO}/dT = input offset voltage temperature drift

GBP = gain bandwidth product

I_B = input bias current

I_{IO} = input bias offset current

I_Q = quiescent supply current

MFR = manufacturer (codes at App.C.)

P_Q = quiescent power consumer

PSRR = power supply rejection ratio

V_{ICM} = common mode input voltage rating

V_{IDF} = differential input voltage rating

V_{IO} = input offset voltage

V_S = dc supply voltage

RIGHT HAND PAGE

Lead out coding summary (details at APP.G.) for different cases (APP.F.)

A = gain adjust

B = bias adjust

C = case

E— = inverting input

E+ = non-inverting input

F,F* = input frequency compensation

G = ground

J = high level input

K = output, open collector

L = output, open emitter

M = metal case

N = not connected

Q = special terminal

R,R* = outputs

S = strobe

T,T* = offset balance

V+ = +ve dc supply

V- = —ve dc supply

W = guard ring

X = blank position, no lead

++ = +ve supplementary dc supply

—— = —ve supplementary dc supply

ϕ,ϕ^* = output frequency compensation

CASE (APP F)	LD 1	LD 2	LD 3	LD 4	LD 5	LD 6	LD 7	LD 8	LD 9	LD 10	LD 11	LD 12	LD 13	LD 14	LD 15	LD 16	EUROPE SUBSTITUTE	USA SUBSTITUTE	I S	TYPE NUMBER
CHP																			O	AM741XC
DIL-14/1P	E-1	E+1	T1	V-	T2	E+2	E-2	T*2	V+2	R2	N	R1	V+1	T*1			TBB0747A	UA747DC	O	AM747DC
DIL-14/1P	E-1	E+1	T1	V-	T2	E+2	E-2	T*2	V+2	R2	N	R1	V+	T*1			SFC2747KM	UA747DM	O	AM747DM
FLP-14/3C	E-	E+1	T1	V-	T2	E+2	E-2	T*2	V+2	R2	N	R1	V+1	T*1				LM747F	O	AM747FM
T05-10/1M	R1	V+1	E-1	E+1	V-	E+2	E-2	V+2	R2	N							TBB0747	UA747HC	O	AM747HC
T05-10/1M	R1	V+1	E-1	E+1	V-	E+2	E-2	V+2	R2	N							TBC0747	UA747HM	O	AM747HM
DIL-14/1C	N	N	FT	E-	E+	V-	N	N	T*	R	V+	F*	N	N			SN72748J	UA748DC	O	AM748DC
DIL-14/1C	N	N	FT	E-	E+	V-	N	N	T*	R	V+	F*	N	N			SN52748JA	UA748DM	O	AM748DM
T05-8/1M	FT	E-	E+	V-	T*	R	V+	F*									TBB0748	UA748HC	O	AM748HC
T05-8/1M	FT	E-	E+	V-	T*	R	V+	F*									TBC0748	UA748HM	O	AM748HM
CHP																			O	AM748XC
CHP																			O	AM748XM
CHP																	TBB1458	MC1458G	O	AM1458-DICE
T05-8/1M	R1	E-1	E+1	V-	E+2	E-2	R2	V+											O	AM1458H
DIL-16/1C	V+1	L1	E+1	E-1	V-	T*2	TS2	K2	V+2	L2	E+2	E-2	T*1	TS1	K1		AMLH2311D	LH2311D	O	AM1500DC
DIL-16/1C	V+1	L1	E+1	E-1	V-	T*2	TS2	K2	V+2	L2	E+2	E-2	T*1	TS1	K1	N	AMLH2211D	LH2211D	O	AM1500DL
DIL-16/1C	V+1	L1	E+1	E-1	V-	T*2	TS2	K2	V+2	L2	E+2	E-2	T*1	TS1	K1	N	AMLH2111D	LH2111D	O	AM1500DM
FLP-16/3C	V+1	L1	E+1	E-1	V-	T*2	TS2	K2	V+2	L2	E+2	E-2	T*1	TS	K1	N	AMLH2211F	LH2211F	O	AM1500FL
FLP-16/3C	V+1	L1	E+1	E-1	V-	LT*2	TS2	K2	V+2	L2	E+2	E-2	T*1	TS1	K1	N	AMLH2111F	LH2111F	O	AM1500FM
FLP-16/3M	V+1	L1	E+1	E-1	V-	T*2	TS2	K2	V+2	L2	E+2	E-2	T*1	TS1	K1	N	AMLH2311F	LH2311F	O	AM1500FC
DIL-16/1C	V+1	ϕ*1	Tϕ1	E-1	E+1	V-	T*2	R2	V+2	ϕ*2	Tϕ2	E-2	E+2	T*1	N	R1	AMLH2301AD	LH2201AD	O	AM1501DC
DIL-16/1C	V+1	ϕ*1	Tϕ1	E-1	E+1	V-	T*2	R2	V+2	ϕ*2	Tϕ2	E-2	E+2	T*1	N	R1	AMLH2201AD	LH2201AD	O	AM1501DL
DIL-16/1C	V+1	ϕ*1	Tϕ1	E-1	E+1	V-	T*2	R2	V+2	ϕ*2	Tϕ2	E-2	E+2	T*1	N	R1		LH2101AD	O	AM1501DM
FLP-16/3C	V+1	ϕ*1	Tϕ1	E-1	E+1	V-	T*2	R2	V+2	ϕ*2	Tϕ2	E-2	E+2	T*1	N	R1	AMLH2301AF	LH2301AF	O	AM1501FC
FLP-16/3C	V+1	ϕ*1	Tϕ1	E-1	E+1	V-	T*2	R2	V+2	ϕ*2	Tϕ2	E-2	E+2	T*1	N	R1	AMLH2201AF	LH2201AF	O	AM1501FL
FLP-16/3C	V+1	ϕ*1	Tϕ1	E-1	E+1	V-	T*2	R2	V+2	ϕ*2	Tϕ2	E-2	E+2	T*1	N	R1	AMLH2101AF	LH2101AF	O	AM1501FM
CHP																			O	AM1558-DICE
T05-8/1M	R1	E-1	E+1	V-	E+2	E-2	R2	V+									TBC1458	MC1558G	O	AM1558H
CHP																	AM101-DICE	LD101	O	AMLD101
CHP																		LD101A	O	AMLD101A
CHP																		LD102	O	AMLD102
CHP																		LD106	O	AMLD106
CHP																	AM107-DICE	LD107	O	AMLD107
CHP																		LD108	O	AMLD108
CHP																		LD108A	O	AMLD108A
CHP																	AM110-DICE	LD110	O	AMLD110
CHP																	AM111-DICE	LD111	O	AMLD111
CHP																	AM112-DICE	LD112	O	AMLD112
CHP																	AM118-DICE	LD118	O	AMLD118
CHP																	AM119-DICE	LD119	O	AMLD119
																	AM124-DICE	LD124	O	AMLD124
CHP																	AM124ADICE	LD124A	O	AMLD124A
CHP																	AM139-DICE	LD139	O	AMLD139
CHP																		LD139A	O	AMLD139A
CHP																		LD148	O	AMLD148
CHP																	AM149-DICE	LD149	O	AMLD149
CHP																		LD155	O	AMLD155
CHP																		LD155A	O	AMLD155A
CHP																		LD156	O	AMLD156
CHP																		LD156A	O	AMLD156A
CHP																		LD157	O	AMLD157
CHP																		LD157A	O	AMLD157A
CHP																		LD216	O	AMLD216
CHP																		LD216A	O	AMLD216A
CHP																		LD301	O	AMLD301
CHP																		LD301A	O	AMLD301A
CHP																		LD302	O	AMLD302
CHP																		LD306	O	AMLD306
CHP																		LD307	O	AMLD307
CHP																		LD308	O	AMLD308

TYPE NUMBER	MFR	APP	CMP	GBP MIN	SLEW RATE MIN	V_S^+ MAX	V_S^- MAX	T_{op} MAX	A_{VOL} MIN	V_{IO} MAX	I_B MAX	I_{IO} MAX	P_{TOT} MAX	I_{OUT} MIN	V_{OUT} MIN	V_{ICM} MAX	V_{IDF} MAX	dV_{IO}/dT MAX	P_Q MAX	I_O MAX	CM RR MIN	PS RR MIN	R_{IN} MIN
AMLD308	ADU	SBA	EXT	.	.	+18V	-18V	70C	88dB	7.5MV	7NA	1NA	.	1MA	13V	15V	1V	30uV/C	.	.6MA	80dB	80dB	10M
AMLD308A	ADU	SBA	EXT	.	.	+18V	-18V	70C	98dB	0.5MV	7NA	1NA	.	1MA	13V	15V	1V	5uV/C	.	.6MA	96dB	96dB	10M
AMLD310	ADU	VFA	INT	.	15V/uS	+18V	-18V	70C	0dB	7.5MV	7NA	.	.	1MA	10V	15V	15V	50uV/C	.	6MA	.	70dB	10G
AMLD311	ADU	CPR	EXT	.	.	+18V	-18V	70C	100dB	7.5MV	250NA	50NA	.	.	15V	30V	.	.	.	8MA	.	.	.
AMLD312	ADU	SBA	INT	.	.	+18V	-18V	70C	88dB	7.5MV	7NA	1NA	.	1MA	13V	14V	14V	15uV/C	.	.8MA	80dB	80dB	10M
AMLD316	ADU	LBC	INT	.	.	+20V	-20V	70C	86dB	10MV	150pA	50pA	.	1MA	13V	15V	14V	.	.	.8MA	80dB	80dB	300M
AMLD316A	ADU	LBC	INT	.	.	+20V	-20V	70C	92dB	3MV	50pA	15pA	.	1MA	13V	15V	14V	.	.	.6MA	80dB	80dB	2G
AMLD318	ADU	XSR	INT	.	50V/uS	+20V	-20V	70C	88dB	10MV	500NA	200NA	.	6MA	12V	15V	1V	.	.	8MA	70dB	65dB	500K
AMLD319	ADU	DCP	INT	.	.	+18V	-18V	70C	78dB	8MV	1uA	0.2uA	.	.	15V	5V	.	.	.	12MA	.	.	.
AMLD324	ADU	QGK	INT	.	.	+16V	-16V	70C	88dB	7MV	250NA	50NA	.	.	16V	16V	35uV/C	.	.	2MA	65dB	65dB	.
AMLD324A	ADU	QGK	INT	.	.	+16V	-16V	70C	88dB	3MV	100NA	30NA	.	.	16V	16V	30uV/C	.	.	2MA	65dB	65dB	.
AMLD339	ADU	QCP	EXT	.	.	+18V	-18V	70C	88dB	5MV	250NA	50NA	.	6MA	.	18V	18V	.	.	2MA	.	.	.
AMLD339A	ADU	QCP	EXT	.	.	+18V	-18V	70C	94dB	2MV	250NA	50NA	.	6MA	.	18V	18V	.	.	2MA	.	.	.
AMLD348	ADU	QGK	INT	.3MHZ	0.2V/uS	+18V	-18V	70C	88dB	6MV	200NA	50NA	.	5MA	12V	18V	36V	.	.	1MA	70dB	77dB	800K
AMLD349	ADU	QGK	INT	1MHZ	0.5V/uS	+18V	-18V	70C	88dB	6MV	200NA	50NA	.	5MA	12V	18V	36V	.	.	1MA	70dB	77dB	800K
AMLD355	ADU	FET	INT	.5MHZ	2V/uS	+18V	-18V	70C	88dB	10MV	100pA	20pA	.	5MA	12V	16V	30V	20uV/C	.	4MA	80dB	80dB	0.1T
AMLD355A	ADU	FET	INT	.5MHZ	3V/uS	+18V	-18V	70C	94dB	2MV	50pA	10pA	.	5MA	12V	16V	30V	5uV/C	.	4MA	85dB	85dB	0.1T
AMLD356	ADU	HSR	INT	1MHZ	7.5V/uS	+18V	-18V	70C	88dB	10MV	200pA	50pA	.	5MA	12V	16V	30V	20uV/C	.	10MA	80dB	80dB	0.1T
AMLD356A	ADU	HSR	INT	4MHZ	10V/uS	+18V	-18V	70C	94dB	2MV	50pA	10pA	.	5MA	12V	16V	30V	5uV/C	.	10MA	85dB	85dB	0.1T
AMLD357	ADU	XSR	INT	4MHZ	6V/uS	+18V	-18V	70C	88dB	10MV	200pA	50pA	.	5MA	12V	16V	30V	20uV/C	.	10MA	80dB	80dB	0.1T
AMLD357A	ADU	XSR	INT	15MHZ	8V/uS	+18V	-18V	70C	94dB	2MV	50pA	10pA	.	5MA	12V	16V	30V	5uV/C	.	10MA	85dB	85dB	0.1T
AMLD592	ADU	BDO	EXT	20MHZ	.	+8V	-8V	125C	50dB	5MV	20uA	3uA	.	2MA	1.5V	6V	5V	.	.	24MA	60dB	50dB	2K
AMLD592C	ADU	BDO	EXT	20MHZ	.	+8V	-8V	70C	48dB	6MV	30uA	5uA	.	2MA	1.5V	6V	5V	.	.	24MA	60dB	50dB	2K
AMLF111D	ADU	CPR	EXT	.	.	+18V	-18V	125C	100dB	4MV	50pA	25pA	500MWF	8MA	.	15V	30V	.	.	6MA	.	.	.
AMLF111F	NAU	CPR	EXT	.	.	+18V	-18V	125C	100dB	4MV	50pA	25pA	500MWF	8MA	.	15V	30V	.	.	6MA	.	.	.
AMLF111H	ADU	CPR	EXT	.	.	+18V	-18V	125C	100dB	4MV	50pA	25pA	500MWF	8MA	.	15V	30V	.	.	6MA	.	.	.
AMLF155AH	ADU	FET	INT	5MHZ	3V/uS	+22V	-22V	125C	94dB	2MV	50pA	10pA	670MWF	5MA	12V	20V	40V	5uV/C	.	4MA	85dB	85dB	0.1T
AMLF155H	ADU	FET	INT	5MHZ	2V/uS	+22V	-22V	125C	94dB	5MV	100pA	20pA	670MWF	5MA	12V	20V	40V	20uV/C	.	4MA	85dB	85dB	0.1T
AMLF156AH	ADU	HSR	INT	4MHZ	10V/uS	+22V	-22V	125C	94dB	2MV	50pA	10pA	670MWF	5MA	12V	20V	40V	5uV/C	.	4MA	85dB	85dB	0.1T
AMLF156H	ADU	HSR	INT	1MHZ	7.5V/uS	+22V	-22V	125C	94dB	5MV	100pA	20pA	670MWF	5MA	12V	20V	40V	20uV/C	.	7MA	85dB	85dB	0.1T
AMLF157AH	ADU	XSR	INT	15MHZ	8V/uS	+22V	-22V	125C	94dB	2MV	50pA	10pA	670MWF	5MA	12V	20V	40V	5uV/C	.	7MA	85dB	85dB	0.1T
AMLF157H	ADU	XSR	INT	4MHZ	6V/uS	+22V	-22V	125C	94dB	5MV	100pA	50pA	670MWF	5MA	12V	20V	40V	20uV/C	.	7MA	85dB	85dB	0.1T
AMLF211D	ADU	CPR	EXT	.	.	+18V	-18V	85C	100dB	4MV	50pA	25pA	500MWF	8MA	.	15V	30V	.	.	6MA	.	.	.
AMLF211F	ADU	CPR	EXT	.	.	+18V	-18V	85C	100dB	4MV	50pA	25pA	500MWF	8MA	.	15V	30V	.	.	6MA	.	.	.
AMLF211H	ADU	CPR	EXT	.	.	+18V	-18V	85C	100dB	4MV	50pA	25pA	500MWF	8MA	.	15V	30V	.	.	6MA	.	.	.
AMLF255H	ADU	FET	INT	5MHZ	2V/uS	+22V	-22V	85C	94dB	5MV	100pA	20pA	570MWF	5MA	12V	20V	40V	20uV/C	.	4MA	85dB	85dB	0.1T
AMLF256H	ADU	HSR	INT	1MHZ	7.5V/uS	+22V	-22V	85C	94dB	5MV	100pA	20pA	570MWF	5MA	12V	20V	40V	20uV/C	.	7MA	85dB	85dB	0.1T
AMLF257H	ADU	XSR	INT	4MHZ	6V/uS	+22V	-22V	85C	94dB	5MV	100pA	20pA	570MWF	5MA	12V	20V	40V	20uV/C	.	7MA	85dB	85dB	0.1T
AMLF311D	ADU	CPR	EXT	.	.	+18V	-18V	70C	100dB	10MV	150pA	75pA	500MWF	8MA	.	15V	30V	.	.	8MA	.	.	.
AMLF311F	ADU	CPR	EXT	.	.	+18V	-18V	70C	100dB	10MV	150pA	75pA	500MWF	8MA	.	15V	30V	.	.	8MA	.	.	.
AMLF311H	ADU	CPR	EXT	.	.	+18V	-18V	70C	100dB	10MV	150pA	75pA	500MWF	8MA	.	15V	30V	.	.	8MA	.	.	.
AMLF355AH	ADU	FET	INT	.5MHZ	3V/uS	+18V	-18V	70C	94dB	2MV	50pA	10pA	500MWF	5MA	12V	16V	30V	5uV/C	.	4MA	85dB	85dB	0.1T
AMLF355H	ADU	FET	INT	.5MHZ	2V/uS	+18V	-18V	70C	88dB	10MV	100pA	20pA	500MWF	5MA	12V	16V	30V	20uV/C	.	4MA	80dB	80dB	0.1T
AMLF355N	ADU	FET	INT	.5MHZ	2V/uS	+18V	-18V	70C	88dB	10MV	100pA	20pA	500MWF	5MA	12V	16V	30V	20uV/C	.	4MA	80dB	80dB	0.1T
AMLF356AH	ADU	HSR	INT	4MHZ	10V/uS	+18V	-18V	70C	94dB	2MV	50pA	10pA	.	5MA	12V	16V	30V	5uV/C	.	10MA	85dB	85dB	0.1T
AMLF356H	ADU	HSR	INT	1MHZ	7.5V/uS	+18V	-18V	70C	88dB	10MV	200pA	50pA	500MWF	5MA	12V	16V	30V	20uV/C	.	10MA	80dB	80dB	0.1T
AMLF356N	ADU	HSR	INT	1MHZ	7.5V/uS	+18V	-18V	70C	88dB	10MV	200pA	50pA	500MWF	5MA	12V	16V	30V	20uV/C	.	10MA	80dB	80dB	0.1T
AMLF357AH	ADU	XSR	INT	15MHZ	8V/uS	+18V	-18V	70C	94dB	2MV	50pA	10pA	500MWF	5MA	12V	16V	30V	5uV/C	.	10MA	85dB	85dB	0.1T
AMLF357H	ADU	XSR	INT	4MHZ	6V/uS	+18V	-18V	70C	88dB	10MV	200pA	50pA	500MWF	5MA	12V	16V	30V	20uV/C	.	10MA	80dB	80dB	0.1T
AMLF357N	ADU	XSR	INT	4MHZ	6V/uS	+18V	-18V	70C	88dB	10MV	200pA	50pA	500MWF	5MA	12V	16V	30V	20uV/C	.	10MA	80dB	80dB	0.1T
AMLFD111	ADU	CPR	EXT	.	.	+18V	-18V	125C	100dB	4MV	50pA	25pA	.	8MA	.	15V	30V	.	.	6MA	.	.	.
AMLFD311	ADU	CPR	EXT	.	.	+18V	-18V	70C	100dB	10MV	150pA	75pA	.	8MA	.	15V	30V	.	.	8MA	.	.	.
AMLH2111D	ADU	DCP	EXT	.	.	+18V	-18V	125C	100dB	3MV	100NA	10NA	500MWF	8MA	.	15V	30V	.	.	6MA	.	.	.
AMLH2111F	ADU	DCP	EXT	.	.	+18V	-18V	125C	100dB	3MV	100NA	10NA	500MWF	8MA	.	15V	30V	.	.	6MA	.	.	.
AMLH2101AD	ADU	DGU	EXT	.	.	+22V	-22V	125C	94dB	2MV	75NA	10NA	500MWF	5MA	12V	15V	30V	15uV/C	.	3MA	80dB	80dB	1.5M
AMLH2101AF	ADU	DGU	EXT	.	.	+22V	-22V	125C	94dB	2MV	75NA	10NA	500MWF	5MA	12V	15V	30V	15uV/C	.	3MA	80dB	80dB	1.5M
AMLH2201AD	ADU	DGU	EXT	.	.	+22V	-22V	85C	94dB	2MV	75NA	10NA	500MWF	5MA	12V	15V	30V	15uV/C	.	4MA	80dB	80dB	1.5M
AMLH2201AF	ADU	DGU	EXT	.	.	+22V	-22V	85C	94dB	2MV	75NA	10NA	500MWF	5MA	12V	15V	30V	15uV/C	.	4MA	80d9	80dB	1.5M
AMLH2211D	ADU	DCP	EXT	.	.	+18V	-18V	85C	100dB	3MV	100NA	10NA	500MWF	8MA	.	15V	30V	.	.	6MA	.	.	.
AMLH2211F	ADU	DCP	EXT	.	.	+18V	-18V	85C	100dB	3MV	100NA	10NA	500MWF	8MA	.	15V	30V	.	.	6MA	.	.	.

For detailed explanations of column heading notations, see App. A. Also for ready references the more important abbreviations used in the column headings are listed below:

LEFT HAND PAGE
APP = application (codes at APP.E.)
CMRR = common mode rejection ratio
CMP = compensation (frequency)
dV_{IO}/dT = input offset voltage temperature drift
GBP = gain bandwidth product
I_B = input bias current
I_{IO} = input bias offset current
I_O = quiescent supply current
MFR = manufacturer (codes at App.C.)
P_O = quiescent power consumer
PSRR = power supply rejection ratio
V_{ICM} = common mode input voltage rating
V_{IDF} = differential input voltage rating
V_{IO} = input offset voltage
V_S = dc supply voltage

RIGHT HAND PAGE
Lead out coding summary (details at APP.G.) for different cases (APP.F.)
A = gain adjust
B = bias adjust
C = case
E- = inverting input
E+ = non-inverting input
F,F* = input frequency compensation
G = ground
J = high level input
K = output, open collector
L = output, open emitter
M = metal case
N = not connected
Q = special terminal
R,R* = outputs
S = strobe
T,T* = offset balance
V+ = +ve dc supply
V- = -ve dc supply
W = guard ring
X = blank position, no lead
++ = +ve supplementary dc supply
-- = -ve supplementary dc supply
ø.ø* = output frequency compensation

CASE (APP F)	LD1	LD2	LD3	LD4	LD5	LD6	LD7	LD8	LD9	LD10	LD11	LD12	LD13	LD14	LD15	LD16	EUROPE SUBSTITUTE	USA SUBSTITUTE	ISS	TYPE NUMBER
CHP																		LD308A	0	AMLD308A
CHP																		LD310	0	AMLD310
CHP																		LD311	0	AMLD311
CHP																		LD312	0	AMLD312
CHP																		LD316	0	AMLD316
CHP																		LD316A	0	AMLD316A
CHP																		LD318	0	AMLD318
CHP																		LD319	0	AMLD319
CHP																		LD324	0	AMLD324
CHP																		LD324A	0	AMLD324A
CHP																		LD339	0	AMLD339
CHP																		LD339A	0	AMLD339A
CHP																		LD348	0	AMLD348
CHP																		LD349	0	AMLD349
CHP																		LD355	0	AMLD355
CHP																		LD355A	0	AMLD355A
CHP																		LD356	0	AMLD356
CHP																		LD356A	0	AMLD356A
CHP																		LD357	0	AMLD357
CHP																		LD357A	0	AMLD357A
CHP																		LD592	0	AMLD592
CHP																		LD592C	0	AMLD592C
DIL-14/1C	N	G	E+	E-	N	V-	T	T*S	R	N	V+	N	N	N			UAF111D	LF111D	0	AMLF111D
FLP-10/3C	G	G	E+	E-	N	V-	T	T*S	N	R	V+							LF111F	0	AMLF111F
TO5-8/1M	G		E+	E-	V-	T		T*S	R	V+							UAF111H	LF111H	0	AMLF111H
TQ5-8/1M	T	E-	E+	V-	T*	R	V+	N									UAF155AHM	LF155AH	0	AMLF155AH
TO5-8/1M	T	E-	E+	V-	T*	R	V+	N									UAF155HM	LF155H	0	AMLF155H
TO5-8/1M	T	E-	E+	V-	T*	R	V+	N									UAF156AHM	LF156AH	0	AMLF156AH
TO5-8/1M	T	E-	E+	V-	T*	R	V+	N									UAF156HM	LF156H	0	AMLF156H
TO5-8/1M	T	E-	E+	V-	T*	R	V+	N									UA157AHM	LF157AH	0	AMLF157AH
TO5-8/1M	T	E-	E+	V-	T*	R	V+	N									UAF157HM	LF157H	0	AMLF157H
DIL-14/1C	N	G	E+	E-	N	V-	T	T*S	R	N	V+	N	N	N				LF211D	0	AMLF211D
FLP-10/3C	G	G	E+	E-	N	V-	T	T*S	N	R	V+							LF211F	0	AMLF211F
TO5-8/1M	G		E+	E-	V-	T		T*S	R	V+							UAF111H	LF211H	0	AMLF211H
TO5-8/1M	T	E-	E+	V-	T*	R	V+	N										LF255H	0	AMLF255H
TO5-8/1M	T	E-	E+	V-	T*	R	V+	N										LF256H	0	AMLF256H
TO5-8/1M	T	E-	E+	V-	T*	R	V+	N										LF257H	0	AMLF257H
DIL-14/1C	N	G	E+	E-	N	V-	T	T*S	R	N	V+	N	N	N			UAF311D	LF311D	0	AMLF311D
FLP-10/3C	G	G	E+	E-	N	V-	T	T*S	N	R	V+							LF311F	0	AMLF311F
TO5-8/1M	G		E+	E-	V-	T		T*S	R	V+							UAF311H	LF311H	0	AMLF311H
TO5-8/1M	T	E-	E+	V-	T*	R	V+	N									UA355AHC	LF355AH	0	AMLF355AH
TO5-8/1M	T	E-	E+	V-	T*	R	V+	N									UAF355HC	LF355H	0	AMLF355H
DIL-8/1P	T	E-	E+	V-	T*	R	V+	N										LF355N	0	AMLF355N
TO5-8/1M	T	E-	E+	V-	T*	R	V+	N									UA356AHC	LF356AH	0	AMLF356AH
TO5-8/1M	T	E-	E+	V-	T*	R	V+	N									UA356HC	LF356H	0	AMLF356H
DIL-8/1P	T	E-	E+	V-	T*	R	V+	N										LF356N	0	AMLF356N
TO5-8/1M	T	E-	E+	V-	T*	R	V+	N									UA357AHC	LF357AH	0	AMLF357AH
TO5-8/1M	T	E-	E+	V-	T*	R	V+	N									UA357HC	LF357H	0	AMLF357H
DIL-8/1P	T	E-	E+	V-	T*	R	V+	N										LF357N	0	AMLF357N
CHP																			0	AMLFD111
CHP																		LFD311	0	AMLFD311
DIL-16/1C	V+1	L1	E+1	E-1	V-	T*2	TS2	K2	V+2	L2	E+2	E-2	T*1	TS1	K1	N	AM1500DM	LH2111D	0	AMLH2111D
DIL-16/1C	V+1	L1	E+1	E-1	V-	T*2	TS2	K2	V+2	L2	E+2	E-2	T*1	TS1	K1	N	AM1500FM	LH2111F	0	AMLH2111F
DIL-16/1C	V+1	ø*1	Tø1	E-1	E+1	V-	T*2	R2	V+2	ø*2	Tø2	E-2	E+2	T*1	N	R1		LH2101AD	0	AMLH2101AD
FLP-16/3C	V+1	ø*1	Tø1	E-1	E+1	V-	T*2	R2	V+2	ø*2	Tø2	E-2	E+2	T*1	N	R1	AM1501FM	LH2101AF	0	AMLH2101AF
DIL-16/1C	V+1	ø*1	Tø1	E-1	E+1	V-	T*2	R2	V+2	ø*2	Tø2	E-2	E+2	T*1	N	R1	AM1501DL	LH2201AD	0	AMLH2201AD
FLP-16/3C	V+1	ø*1	Tø1	E-1	E+1	V-	T*2	R2	V+2	ø*2	Tø2	E-2	E+2	T*1	N	R1	AM1501FL	LH2201AF	0	AMLH2201AF
DIL-16/1C	V+1	L1	E+1	E-1	V-	T*2	TS2	K2	V+2	L2	E+2	E-2	T*1	TS1	K1	N	AM1500DL	LH2211D	0	AMLH2211D
FLP-16/3C	V+1	L1	E+1	E-1	V-	T*2	TS2	K2	V+2	L2	E+2	E-2	T*1	TS1	K1	N	AM1500FL	LH2211F	0	AMLH2211F
DIL-16/1C	V+1	ø*1	Tø1	E-1	E+1	V-	T*2	R2	V+2	ø*2	Tø2	E-2	E+2	T*1	N	R1		LH2301AD	0	AMLH2301AD

TYPE NUMBER	MFR	APP	CMP	GBP MIN	SLEW RATE MIN	V_S^+ MAX	V_S^- MAX	T_{op} MAX	A_{VOL} MIN	V_{IO} MAX	I_B MAX	I_{IO} MAX	P_{TOT} MAX	I_{OUT} MIN	V_{OUT} MIN	V_{ICM} MAX	V_{IDF} MAX	dV_{IO}/dT MAX	P_Q MAX	I_Q MAX	CM RR MIN	PS RR MIN	R_{IN} MIN
AMLH2301AD	ADU	DGU	EXT	.	.	+22V	-22V	70C	88dB	7.5MV	250NA	50NA	500MWF	5MA	12V	15V	30V	30uV/C	.	3MA	70dB	70dB	0.5M
AMLH2301AF	ADU	DGU	EXT	.	.	+22V	-22V	70C	88dB	7.5MV	250NA	50NA	500MWF	5MA	12V	15V	30V	30uV/C	.	3MA	70dB	70dB	0.5M
AMLH2311D	ADU	DCP	EXT	.	.	+18V	-18V	70C	100dB	7.5MV	250NA	50NA	500MWF	8MA	.	15V	30V	.	.	8MA	.	.	.
AMLH2311F	ADU	DCP	EXT	.	.	+18V	-18V	70C	100dB	7.5MV	250NA	50NA	500MWF	8MA	.	15V	30V	.	.	8MA	.	.	.
AMLM101AD	ADU	GPU	EXT	.	.	+22V	-22V	125C	94dB	2MV	75NA	10NA	500MWF	5MA	12V	15V	30V	15uV/C	.	3MA	80dB	80dB	1.5M
AMLM101AF	ADU	GPU	EXT	.	.	+22V	-22V	125C	94dB	2MV	75NA	10NA	500MWF	5MA	12V	15V	30V	15uV/C	.	3MA	80dB	80dB	1.5M
AMLM101AH	ADU	GPU	EXT	.	.	+22V	-22V	125C	94dB	2MV	75NA	10NA	500MWF	5MA	12V	15V	30V	15uV/C	.	3MA	70dB	70dB	300K
AMLM101D	ADU	GPU	EXT	.	.	+22V	-22V	125C	94dB	5MV	1.5uA	0.5uA	500MWF	5MA	12V	15V	30V	15uV/C	.	3MA	70dB	70dB	300K
AMLM101F	ADU	GPU	EXT	.	.	+22V	-22V	125C	94dB	5MV	1.5uA	0.5uA	500MWF	5MA	12V	15V	30V	15uV/C	.	3MA	70dB	70dB	300K
AMLM101H	ADU	GPU	EXT	.	.	+22V	-22V	125C	94dB	5MV	500NA	200NA	500MWF	5MA	12V	15V	30V	15uV/C	.	3MA	70dB	70dB	300K
AMLM102D	ADU	VFA	INT	.	.	+18V	-18V	125C	0dB	5MV	10NA	.	500MWF	1MA	10V	.	.	30uV/C	.	6MA	.	60dB	10G
AMLM102F	ADU	VFA	INT	.	.	+18V	-18V	125C	0dB	5MV	10NA	.	500MWF	1MA	10V	.	.	30uV/C	.	6MA	.	60dB	10G
AMLM102H	ADU	VFA	INT	.	.	+18V	-18V	125C	0dB	5MV	10NA	.	500MWF	1MA	10V	.	.	30uV/C	.	6MA	.	60dB	10G
AMLM106F	ADU	CPR	EXT	.	.	+15V	-15V	125C	84dB	2MV	20uA	3uA	600MWF	50MA	2.5V	.	.	10uV/C	163MW	.	.	.	.
AMLM106H	ADU	CPR	EXT	.	.	+15V	-15V	125C	84dB	2MV	20uA	3uA	600MWF	50MA	2.5V	.	.	10uV/C	163MW	.	.	.	.
AMLM107D	ADU	GPK	INT	.	.	+22V	-22V	125C	94dB	2MV	75NA	10NA	500MWF	5MA	12V	15V	30V	15uV/C	.	3MA	80dB	80dB	1.5M
AMLM107F	ADU	GPK	INT	.	.	+22V	-22V	125C	94dB	2MV	75NA	10NA	500MWF	5MA	12V	15V	30V	15uV/C	.	3MA	80dB	80dB	1.5M
AMLM107H	ADU	GPK	INT	.	.	+22V	-22V	125C	94dB	2MV	75NA	10NA	500MWF	5MA	12V	15V	30V	15uV/C	.	3MA	80dB	80dB	1.5M
AMLM108AD	ADU	SBA	EXT	.	.	+20V	-20V	125C	98dB	0.5MV	2NA	0.2NA	500MWF	1MA	13V	15V	1V	5uV/C	.	.6MA	96dB	96dB	30M
AMLM108AF	ADU	SBA	EXT	.	.	+20V	-20V	125C	98dB	0.5MV	2NA	0.2NA	500MWF	1MA	13V	15V	1V	5uV/C	.	.6MA	96dB	96dB	30M
AMLM108AH	ADU	SBA	EXT	.	.	+20V	-20V	125C	98dB	0.5MV	2NA	0.2NA	500MWF	1MA	13V	15V	1V	5uV/C	.	.6MA	96dB	96dB	30M
AMLM108D	ADU	SBA	EXT	.	.	+20V	-20V	125C	96dB	2MV	2NA	0.2NA	500MWF	1MA	13V	15V	1V	15uV/C	.	.6MA	85dB	80dB	30M
AMLM108F	ADU	SBA	EXT	.	.	+20V	-20V	125C	96dB	2MV	2NA	0.2NA	500MWF	1MA	13V	15V	1V	15uV/C	.	.6MA	85dB	80dB	30M
AMLM108H	ADU	SBA	EXT	.	.	+20V	-20V	125C	96dB	2MV	2NA	0.2NA	500MWF	1MA	13V	15V	1V	15uV/C	.	.6MA	85dB	80dB	30M
AMLM110D	ADU	VFA	INT	15V/uS	.	+18V	-18V	125C	0dB	4MV	3NA	.	500MWF	1MA	10V	15V	15V	50uV/C	.	6MA	.	70dB	10G
AMLM110F	ADU	VFA	INT	15V/uS	.	+18V	-18V	125C	0dB	4MV	3NA	.	500MWF	1MA	10V	15V	15V	50uV/C	.	6MA	.	70dB	10G
AMLM110H	ADU	VFA	INT	15V/uS	.	+18V	-18V	125C	0dB	4MV	3NA	.	500MWF	1MA	10V	15V	15V	50uV/C	.	6MA	.	70dB	10G
AMLM111D	ADU	CPR	EXT	.	.	+18V	-18V	125C	100dB	3MV	100NA	10NA	500MWF	.	.	15V	30V	.	.	6MA	.	.	.
AMLM111F	ADU	CPR	EXT	.	.	+18V	-18V	125C	100dB	3MV	100NA	10NA	500MWF	.	.	15V	30V	.	.	6MA	.	.	.
AMLM111H	ADU	CPR	EXT	.	.	+18V	-18V	125C	100dB	3MV	100NA	10NA	500MWF	.	.	15V	30V	.	.	6MA	.	.	.
AMLM112D	ADU	SBA	INT	.	.	+20V	-20V	125C	94dB	2MV	2NA	0.2NA	500MWF	1MA	13V	14V	14V	15uV/C	.	.6MA	85dB	80dB	30M
AMLM112F	ADU	SBA	INT	.	.	+20V	-20V	125C	94dB	2MV	2NA	0.2NA	500MWF	1MA	13V	14V	14V	15uV/C	.	.6MA	85dB	80dB	30M
AMLM112H	ADU	SBA	INT	.	.	+20V	-20V	125C	94dB	2MV	2NA	0.2NA	500MWF	1MA	13V	14V	14V	15uV/C	.	.6MA	85dB	80dB	30M
AMLM118D	ADU	XSR	INT	50V/uS	.	+20V	-20V	125C	94dB	4MV	250NA	50NA	500MWF	6MA	12V	15V	1V	.	.	8MA	80dB	70dB	1M
AMLM118F	ADU	XSR	INT	50V/uS	.	+20V	-20V	125C	94dB	4MV	250NA	50NA	500MWF	6MA	12V	15V	1V	.	.	8MA	80dB	70dB	1M
AMLM118H	ADU	XSR	INT	50V/uS	.	+20V	-20V	125C	94dB	4MV	250NA	50NA	500MWF	6MA	12V	15V	1V	.	.	8MA	80dB	70dB	1M
AMLM119D	ADU	DCP	INT	.	.	+18V	-18V	125C	80dB	4MV	500NA	75NA	500MWF	.	.	15V	5V	.	.	12MA	.	.	.
AMLM119F	ADU	DCP	INT	.	.	+18V	-18V	125C	80dB	4MV	500NA	75NA	500MWF	.	.	15V	5V	.	.	12MA	.	.	.
AMLM119H	ADU	DCP	INT	.	.	+18V	-18V	125C	80dB	4MV	500NA	75NA	500MWF	.	.	15V	5V	.	.	12MA	.	.	.
AMLM124AD	ADU	QGK	INT	.	.	+16V	-16V	125C	94dB	2MV	50NA	10NA	900MWF	.	.	16V	16V	20uV/C	.	2MA	70dB	65dB	.
AMLM124D	ADU	QGK	INT	.	.	+16V	-16V	125C	94dB	5MV	150NA	30NA	.	.	.	16V	16V	35uV/C	.	2MA	70dB	65dB	.
AMLM124AF	ADU	QGK	INT	.	.	+16V	-16V	125C	94dB	2MV	50NA	10NA	800MWF	.	.	16V	16V	20uV/C	.	2MA	70dB	65dB	.
AMLM124F	ADU	QGK	INT	.	.	+16V	-16V	125C	94dB	5MV	150NA	30NA	800MWF	.	.	16V	16V	35uV/C	.	2MA	.	.	.
AMLM139AD	ADU	QCP	EXT	.	.	+18V	-18V	125C	94dB	2MV	100NA	25NA	900MWF	.	.	18V	18V	.	.	2MA	.	.	.
AMLM139AF	ADU	QCP	EXT	.	.	+18V	-18V	125C	94dB	2MV	100NA	25NA	800MWF	.	.	18V	18V	.	.	2MA	.	.	.
AMLM139D	ADU	QCP	EXT	.	.	+18V	-18V	125C	88dB	5MV	100NA	25NA	900MWF	.	.	18V	18V	.	.	2MA	.	.	.
AMLM139F	ADU	QCP	EXT	.	.	+18V	-18V	125C	88dB	5MV	100NA	25NA	800MWF	.	.	18V	18V	.	.	2MA	.	.	.
AMLM148D	ADU	QGK	INT	.3MHZ	0.2V/uS	+22V	-22V	125C	94dB	5MV	100NA	25NA	.	5MA	12V	22V	44V	.	.	1MA	70dB	77dB	800K
AMLM149D	ADU	QGK	INT	1MHZ	0.5V/uS	+22V	-22V	125C	94dB	5MV	100NA	25NA	900MWF	5MA	12V	22V	44V	.	.	1MA	70dB	77dB	800K
AMLM201AD	ADU	GPU	EXT	.	.	+22V	-22V	85C	94dB	2MV	75NA	10NA	.	5MA	12V	15V	30V	15uV/C	.	3MA	80dB	80dB	500K
AMLM201AF	ADU	GPU	EXT	.	.	+22V	-22V	85C	94dB	2MV	75NA	10NA	500MWF	5MA	12V	15V	30V	15uV/C	.	3MA	80dB	80dB	500K
AMLM201AH	ADU	GPU	EXT	.	.	+22V	-22V	85C	94dB	2MV	75NA	10NA	500MWF	5MA	12V	15V	30V	30uV/C	.	3MA	65dB	70dB	100K
AMLM201D	ADU	GPU	EXT	.	.	+22V	-22V	85C	86dB	7.5MV	1.5uA	0.5uA	500MWF	5MA	12V	15V	30V	30uV/C	.	3MA	65dB	70dB	100K
AMLM201F	ADU	GPU	EXT	.	.	+22V	-22V	85C	86dB	7.5MV	1.5uA	0.5uA	500MWF	5MA	12V	15V	30V	30uV/C	.	3MA	65dB	70dB	100K
AMLM201H	ADU	GPU	EXT	.	.	+22V	-22V	85C	86dB	7.5MV	1.5uA	0.5uA	500MWF	5MA	12V	15V	30V	30uV/C	.	3MA	65dB	70dB	100K
AMLM202D	ADU	VFA	INT	.	.	+18V	-18V	85C	0dB	10MV	15NA	.	500MWF	1MA	10V	.	.	60uV/C	.	6MA	.	60dB	10G
AMLM202H	ADU	VFA	INT	.	.	+18V	-18V	85C	0dB	10MV	15NA	.	500MWF	1MA	10V	.	.	60uV/C	.	6MA	.	60dB	10G
AMLM206H	ADU	CPR	EXT	.	.	+15V	-15V	85C	84dB	2MV	20uA	3uA	600MWF	50MA	2.5V	.	.	10uV/C	163MW	.	.	.	.
AMLM207D	ADU	GPK	INT	.	.	+22V	-22V	85C	94dB	2MV	75NA	10NA	500MWF	5MA	12V	15V	30V	15uV/C	.	3MA	80dB	80dB	1.5M
AMLM207F	ADU	GPK	INT	.	.	+22V	-22V	85C	94dB	2MV	75NA	10NA	500MWF	5MA	12V	15V	30V	15uV/C	.	3MA	80dB	80dB	1.5M

For detailed explanations of column heading notations, see App. A.

Also for ready references the more important abbreviations used in the column headings are listed below:

LEFT HAND PAGE

APP = application (codes at APP.E.)
CMRR = common mode rejection ratio
CMP = compensation (frequency)
dV_{in}/dT = input offset voltage temperature drift
GBP = gain bandwidth product
I_B = input bias current
I_{iO} = input bias offset current
I_Q = quiescent supply current
MFR = manufacturer (codes at App.C.)
P_Q = quiescent power consumer
PSRR = power supply rejection ratio
V_{icm} = common mode input voltage rating
V_{idf} = differential input voltage rating
V_{io} = input offset voltage
V_S = dc supply voltage

RIGHT HAND PAGE

Lead out coding summary (details at APP.G.) for different cases (APP.F.)

A = gain adjust
B = bias adjust
C = case
E— = inverting input
E+ = non-inverting input
F,F* = input frequency compensation
G = ground
J = high level input
K = output, open collector
L = output, open emitter
M = metal case
N = not connected
O = special terminal
R,R* = outputs
S = strobe
T,T* = offset balance
V+ = +ve dc supply
V- = —ve dc supply
W = guard ring
X = blank position, no lead
+ + = +ve supplementary dc supply
— — = —ve supplementary dc supply
∅,∅* = output frequency compensation

CASE (APP F)	LD 1	LD 2	LD 3	LD 4	LD 5	LD 6	LD 7	LD 8	LD 9	LD 10	LD 11	LD 12	LD 13	LD 14	LD 15	LD 16	EUROPE SUBSTITUTE	USA SUBSTITUTE	I S S	TYPE NUMBER
FLP-16/3C	V+1	∅*1	T∅1	E-1	E+1	V-	T*2	R2	V+2	∅*2	T∅2	E-2	E+2	T*1	N	R1	AM1501FC	LH2301AF	0	AMLH2301AF
DIL-16/1C	V+1	L1	E+1	E-1	V-	T*2	TS2	K2	V+2	L2	E+2	E-2	T*1	TS1	K1		AM1500DC	LH2311D	0	AMLH2311D
FLP-16/3C	V+1	L1	E+1	E-1	V-	T*2	TS2	K2	V+2	L2	E+2	E-2	T*1	TS1	K1	N	AM1500FC	LH2311F	0	AMLH2311F
DIL-14/1C	N	N	FT	E-	E+	V-	N	N	T*	R	V+	F*	N	N			UA101AD	LM101AD	0	AMLM101AD
FLP-10/3G	N	FT	E-	E+	V-	T*	R	V+	F*	N							SFC2101APM	LM101AF	0	AMLM101AF
TO5-8/1M	FT	E-	E+	V-M	T*	R	V+	F*									SFC2101A	LM101AH	0	AMLM101AH
DIL-14/1C	N	N	FT	E-	E+	V-	N	N	T*	R	V+	F*	N	N			UA101D	LM101D	0	AMLM101D
FLP-10/3G	N	FT	E-	E+	V-	T*	R	V+	F*	N							SFC2101APM	LM101F	0	AMLM101F
TO5-8/1M	FT	E-	E+	V-M	T*	R	V+	F*									SFC2101A	LM101H	0	AMLM101H
DIL-14/1P	N	N	T	N	E+	V-	N	N	L	R	V+	T*	N	N				LM102D	0	AMLM102D
FLP-10/3C	N	T	N	E+	V-	L	R	V+	T*	N							102(FLP)	LM102F	0	AMLM102F
TO5-8/1M	T	N	E+	V-	L	R	V+	T*									102(TO5)	LM102H	0	AMLM102H
FLP-14/3C	N	G	E+	E-	N	V-	S1	S2	R	N	V+	N	N	N			MLM106F	LM106F	0	AMLM106F
TO5-8/1M	G	E+	E-	V-M	S1	S2	R	V+									SN52106L	LM106H	0	AMLM106H
DIL-14/1C	N	N	N	E-	E+	V-	N	N	N	N	R	V+	N	N			MLM107D	LM107D	0	AMLM107D
FLP-10/3G	N	N	E-	E+	V-	N	R	V+	N	N							MLM107F	LM107F	0	AMLM107F
TO5-8/1M	N	E-	E+	V-M	N	R	V+	N									MLM107H	LM107H	0	AMLM107H
DIL-14/1C	N	F	N	E-	E+	N	V-	N	N	N	R	V+	F*	N	N		MLM108AD	LM108AD	0	AMLM108AD
FLP-10/3G	N	N	E-	E+	N	V-	R	V+	F*	F							MLM108AF	LM108AF	0	AMLM108AF
TO5-8/1M	F	E-	E+	V-M	N	R	V+	F*									MLM108AH	LM108AH	0	AMLM108AH
DIL-14/1C	N	F	N	E-	E+	N	V-	N	N	R	V+	F*	N	N			MLM108D	LM108D	0	AMLM108D
FLP-10/3G	N	N	E-	E+	N	V-	R	V+	F*	N							MLM108F	LM108F	0	AMLM108F
TO5-8/1M	F	E-	E+	V-M	N	R	V+	F*									MLM108H	LM108H	0	AMLM108H
DIL-14/1C	N	N	T	N	E+	V-	N	N	L	R	V+	T*	N	N			MLM110D	LM110D	0	AMLM110D
FLP-10/3G	N	T	N	E+	V-	L	R	V+	T*	N							MLM110F	LM110F	0	AMLM110F
TO5-8/1M	T	N	E+	V-	L	R	V+	T*									UA110M	LM110H	0	AMLM110H
DIL-14/1C	N	G	E+	E-	N	V-	T	T*S	R	N	N	V+	N	N			MLM111D	LM111D	0	AMLM111D
FLP-10/3G	G	E+	E-	N	V-	T	T*S	N	R	V+							MLM111F	LM111F	0	AMLM111F
TO5-8/1M	G	E+	E-	V-	T	T*S	R	V+									MLM111H	LM111H	0	AMLM111H
DIL-14/1C	N	T	W	E-	E+	W*	V-	N	F	R	V+	T*	N	N			MLM112D	LM112D	0	AMLM112D
FLP-10/3G	N	W	E-	E+	W*	V-	R	V+	T	T*							MLM112F	LM112F	0	AMLM112F
TO5-8/1M	T	E-	E+	V-	F	R	V+	T*										LM112H	0	AMLM112H
DIL-14/1C	N	N	T*F	E-	E+	V-	N	N	F*T	R	V+	∅	N	N			MLM118D	LM118D	0	AMLM118D
FLP-10/3G	N	T*F	E-	E+	V-	F*T	R	V+	∅	N							MLM118F	LM118F	0	AMLM118F
TO5-8/1M	T*F	E-	E+	V-	F*T	R	V+	∅									MLM118H	LM118H	0	AMLM118H
DIL-14/1C	N	N	G1	E+1	E-1	V-	R2	G2	E+2	E-2	V+	R1	N	N			TDC0119DC	LM119D	0	AMLM119D
FLP-10/3G	N	N	G1	E+1	E-1	V-	R2	G2	E+2	E-2	V+	R1	N	N			TDC0119DC	LM119F	0	AMLM119F
TO5-10/1M	R1	G1	E+1	E-1	V-	R2	G2	E+2	E-2	V+							TDC0119CM	LM119H	0	AMLM119H
DIL-14/1P	R1	E-1	E+1	V+	E+2	E-2	R2	R3	E-3	E+3	G	E+4	E-4	R4				LM124AD	0	AMLM124AD
DIL-14/1C	R1	E-1	E+1	V+	E+2	E-2	R2	R3	E-3	E+3	G	E+4	E-4	R4			MLM124L	LM124D	0	AMLM124D
FLP-14/3C	R1	E-1	E+1	V+	E+2	E-2	R2	R3	E-3	E+3	G	E+4	E-4	R4				LM124AF	0	AMLM124AF
FLP-14/1C	R1	E-1	E+1	V+	E+2	E-2	R2	R3	E-3	E+3	G	E+4	E-4	R4				LM124F	0	AMLM124F
DIL-14/1C	R2	R1	V+	E-1	E+1	E-2	E+2	E+3	E-3	E-4	E+4	G	R4	R3			MLM139AD	LM139AD	0	AMLM139AD
FLP-14/3C	R2	R1	V+	E-1	E+1	E-2	E+2	E+3	E-3	E-4	E+4	G	R4	R3				LM139AF	0	AMLM139AF
DIL-14/1C	R2	R1	V+	E-1	E+1	E-2	E+2	E+3	E-3	E-4	E+4	G	R4	R3				LM139D	0	AMLM139D
FLP-14/3G	R2	R1	V+	E-1	E+1	E-2	E+2	E+3	E-3	E-4	E+4	G	R4	R3				LM139F	0	AMLM139F
DIL-14/1C	R1	E-1	E+1	V+	E+2	E-2	R2	R3	E-3	E+3	V-	E+4	E-4	R4				LM148D	0	AMLM148D
DIL-14/1C	R1	E-1	E+1	V+	E+2	E-2	R2	R3	E-3	E+3	V-	E+4	E-4	R4				LM149D	0	AMLM149D
DIL-14/1C	N	N	FT	E-	E+	V-	N	N	T*	R	V+	F*	N	N			UA201AD	LM201AD	0	AMLM201AD
FLP-10/3G	N	FT	E-	E+	V-	T*	R	V+	F*	N							SFC2201APT	LM201AF	0	AMLM201AF
TO5-8/1M	FT	E-	E+	V-M	T*	R	V+	F*									UA201AH	LM201AH	0	AMLM201AH
DIL-14/1C	N	N	FT	E-	E+	V-	N	N	T*	R	V+	F*	N	N			UA201AD	LM201D	0	AMLM201D
FLP-10/3G	N	FT	E-	E+	V-	T*	R	V+	F*	N							MLM201F	LM201F	0	AMLM201F
TO5-8/1M	FT	E-	E+	V-M	T*	R	V+	F*									SFC2101A	LM201H	0	AMLM201H
DIL-14/1C	N	N	N	E-	E+	V-	N	N	N	N	R	V+	T*	N	N			LM202D	0	AMLM202D
TO5-8/1M	T	N	E+	V-	L	R	V+	T*									UA102M	LM202H	0	AMLM202H
TO5-8/1M	G	E+	E-	V-M	S1	S2	R	V+									SN52106L	LM206H	0	AMLM206H
DIL-14/1C	N	N	N	E-	E+	V-	N	N	N	N	R	V+	N	N			SN52107JA	LM207D	0	AMLM207D
FLP-10/3G	N	N	E-	E+	V-	N	R	V+	N	N							SFC2207PT	LM207F	0	AMLM207F
TO5-8/1M	N	E-	E+	V-M	N	R	V+	N									SFC2207	LM207H	0	AMLM207H

TYPE NUMBER	M F R	A P P	C M P	GBP MIN	SLEW RATE MIN	Vs+ MAX	Vs− MAX	Top MAX	Avol MIN	Vio MAX	Ib MAX	Iio MAX	Ptot MAX	Iout MIN	Vout MIN	Vicm MAX	Vidf MAX	dVio/dT MAX	Po MAX	Io MAX	CMRR MIN	PSRR MIN	Rin MIN
AMLM207H	ADU	GPK	INT	.	.	+22V	−22V	85C	94dB	2MV	75NA	10NA	500MWF	5MA	12V	15V	30V	15uV/C	.	3MA	80dB	80dB	1.5M
AMLM208AD	ADU	SBA	EXT	.	.	+20V	−20V	85C	98dB	0.5MV	2NA	0.2NA	500MWF	1MA	13V	15V	1V	5uV/C	.	.6MA	96dB	96dB	30M
AMLM208AF	ADU	SBA	EXT	.	.	+20V	−20V	85C	98dB	0.5MV	2NA	0.2NA	500MWF	1MA	13V	15V	1V	5uV/C	.	.6MA	96dB	96dB	30M
AMLM208AH	ADU	SBA	EXT	.	.	+20V	−20V	85C	98dB	0.5MV	2NA	0.2NA	500MWF	1MA	13V	15V	1V	5uV/C	.	.6MA	96dB	96dB	30M
AMLM208D	ADU	SBA	EXT	.	.	+20V	−20V	85C	96dB	2MV	2NA	0.2NA	500MWF	1MA	13V	15V	1V	15uV/C	.	.6MA	85dB	80dB	30M
AMLM208F	ADU	SBA	EXT	.	.	+20V	−20V	85C	96dB	2MV	2NA	0.2NA	500MWF	1MA	13V	15V	1V	15uV/C	:	.6MA	85dB	80dB	30M
AMLM208H	ADU	SBA	EXT	.	.	+20V	−20V	85C	96dB	2MV	2NA	0.2NA	500MWF	1MA	13V	15V	1V	15uV/C	.	.6MA	85dB	80dB	30M
AMLM210D	ADU	VFA	INT	.	15V/uS	+18V	−18V	85C	0dB	4MV	3NA	.	500MWF	1MA	10V	15V	15V	50uV/C	.	6MA	.	70dB	10G
AMLM210F	ADU	VFA	INT	.	15V/uS	+18V	−18V	85C	0dB	4MV	3NA	.	500MWF	1MA	10V	15V	15V	50uV/C	.	6MA	.	70dB	10G
AMLM210H	ADU	VFA	INT	.	15V/uS	+18V	−18V	85C	0dB	4MV	3NA	.	500MWF	1MA	10V	15V	15V	50uV/C	.	6MA	.	70dB	10G
AMLM211D	ADU	CPR	EXT	.	.	+18V	−18V	85C	100dB	3MV	100NA	10NA	500MWF	.	.	15V	30V	.	.	6MA	.	.	.
AMLM211H	ADU	CPR	EXT	.	.	+18V	−18V	85C	100dB	3MV	100NA	10NA	500MWF	.	.	15V	30V	.	.	6MA	.	.	.
AMLM212D	ADU	SBA	INT	.	.	+20V	−20V	85C	94dB	2MV	2NA	0.2NA	500MWF	1MA	13V	14V	14V	15uV/C	.	.6MA	85dB	80dB	30M
AMLM212H	ADU	SBA	INT	.	.	+20V	−20V	85C	94dB	2MV	2NA	0.2NA	500MWF	1MA	13V	14V	14V	15uV/C	.	.6MA	85dB	80dB	30M
AMLM212F	ADU	SBA	INT	.	.	+20V	−20V	85C	94dB	2MV	2NA	0.2NA	500MWF	1MA	13V	14V	14V	15uV/C	.	.6MA	85dB	80dB	30M
AMLM216AD	ADU	LBC	INT	.	.	+20V	−20V	85C	92dB	3MV	50pA	15pA	500MWF	1MA	13V	15V	14V	.	.	.6MA	80dB	80dB	2G
AMLM216AF	ADU	LBC	INT	.	.	+20V	−20V	85C	92dB	3MV	50pA	15pA	500MWF	1MA	13V	15V	14V	.	.	.6MA	80dB	80dB	2G
AMLM216AH	ADU	LBC	INT	.	.	+20V	−20V	85C	92dB	3MV	50pA	15pA	500MWF	1MA	13V	15V	14V	.	.	.6MA	80dB	80dB	2G
AMLM216D	ADU	LBC	INT	.	.	+20V	−20V	85C	86dB	10MV	150pA	50pA	500MWF	1MA	13V	15V	14V	.	.	.8MA	80dB	80dB	300M
AMLM216F	ADU	LBC	INT	.	.	+20V	−20V	85C	86dB	10MV	150pA	50pA	500MWF	1MA	13V	15V	14V	.	.	.8MA	80dB	80dB	300M
AMLM216H	ADU	LBC	INT	.	.	+20V	−20V	85C	86dB	10MV	150pA	50pA	500MWF	1MA	13V	15V	14V	.	.	.8MA	80dB	80dB	300M
AMLM218D	ADU	XSR	INT	.	50V/uS	+20V	−20V	85C	94dB	4MV	250NA	50NA	500MWF	6MA	12V	15V	1V	.	.	.8MA	80dB	70dB	1M
AMLM218F	ADU	XSR	INT	.	50V/uS	+20V	−20V	85C	94dB	4MV	250NA	50NA	500MWF	6MA	12V	15V	1V	.	.	.8MA	80dB	70dB	1M
AMLM218H	ADU	XSR	INT	.	50V/uS	+20V	−20V	85C	94dB	4MV	250NA	50NA	500MWF	6MA	12V	15V	1V	.	.	.8MA	80dB	70dB	1M
AMLM219D	ADU	QCP	EXT	.	.	+18V	−18V	85C	94dB	2MV	250NA	50NA	900MWF	6MA	.	18V	18V	.	.	2MA	.	.	.
AMLM219F	ADU	DCP	INT	.	.	+18V	−18V	85C	80dB	4MV	500NA	75NA	500MWF	.	.	15V	5V	.	.	12MA	.	.	.
AMLM219H	ADU	DCP	INT	.	.	+18V	−18V	85C	80dB	4MV	500NA	75NA	500MWF	.	.	15V	5V	.	.	12MA	.	.	.
AMLM224AD	ADU	QQK	INT	.	.	+16V	−16V	85C	94dB	3MV	80NA	15NA	900MWF	.	.	16V	16V	20uV/C	.	2MA	70dB	65dB	.
AMLM224D	ADU	QGK	INT	.	.	+16V	−16V	85C	94dB	5MV	150NA	30NA	900MWF	.	.	16V	16V	35uV/C	.	2MA	70dB	65dB	.
AMLM239AD	ADU	DCP	INT	.	.	+18V	−18V	85C	80dB	4MV	500NA	75NA	500MWF	.	.	15V	5V	.	.	12MA	.	.	.
AMLM239D	ADU	QCP	EXT	.	.	+18V	−18V	85C	88dB	5MV	250NA	50NA	900MWF	6MA	.	18V	18V	.	.	2MA	.	.	.
AMLM248D	ADU	QGK	INT	.3MHZ	0.2V/uS	+18V	−18V	85C	88dB	6MV	200NA	50NA	900MWF	5MA	12V	18V	36V	.	.	1MA	70dB	77dB	800K
AMLM249D	ADU	QGK	INT	1MHZ	0.5V/uS	+18V	−18V	85C	88dB	6MV	200NA	50NA	900MWF	5MA	12V	18V	36V	.	.	1MA	70dB	77dB	800K
AMLM301AD	ADU	GPU	EXT	.	.	+18V	−18V	70C	88dB	7.5MV	250NA	50NA	500MWF	5MA	12V	15V	30V	30uV/C	.	3MA	70dB	70dB	500K
AMLM301AH	ADU	GPU	EXT	.	.	+18V	−18V	70C	88dB	7.5MV	250NA	50NA	500MWF	5MA	12V	15V	30V	30uV/C	.	3MA	70dB	70dB	500K
AMLM301AN	ADU	GPU	EXT	.	.	+18V	−18V	70C	88dB	7.5MV	250NA	50NA	500MWF	5MA	12V	15V	30V	30uV/C	.	3MA	70dB	70dB	500K
AMLM301D	ADU	GPU	EXT	.	.	+18V	−18V	70C	83dB	10MV	2uA	.75uA	500MWF	5MA	12V	15V	30V	30uV/C	.	3MA	65dB	70dB	100K
AMLM301F	ADU	GPU	EXT	.	.	+18V	−18V	70C	83dB	10MV	2uA	.75uA	.	.	12V	15V	30V	30uV/C	.	3MA	65dB	70dB	100K
AMLM301H	ADU	GPU	EXT	.	.	+18V	−18V	70C	83dB	10MV	2uA	.75uA	500MWF	.	12V	15V	30V	30uV/C	.	3MA	65dB	70dB	100K
AMLM302D	ADU	VFA	INT	.	.	+18V	−18V	85C	0dB	5MV	30NA	.	500MWF	1MA	10V	.	.	90uV/C	.	6MA	.	.	10G
AMLM302H	ADU	VFA	INT	.	.	+18V	−18V	70C	0dB	15MV	30NA	.	500MWF	1MA	10V	.	.	90uV/C	.	6MA	.	60dB	10G
AMLM306H	ADU	CPR	EXT	.	.	+15V	−15V	70C	84dB	5MV	25uA	5uA	600MWF	50MA	2.5V	.	.	20uV/C	163MW	.	.	.	.
AMLM307D	ADU	GPK	INT	.	.	+18V	−18V	70C	84dB	7.5MV	250NA	50NA	500MWF	5MA	12V	15V	30V	30uV/C	.	.	70dB	70dB	0.5M
AMLM307H	ADU	GPK	INT	.	.	+18V	−18V	70C	84dB	7.5MV	250NA	50NA	500MWF	5MA	12V	15V	30V	30uV/C	.	.	70dB	70dB	0.5M
AMLM308AD	ADU	SBA	EXT	.	.	+18V	−18V	70C	98dB	0.5MV	7NA	1NA	500MWF	1MA	13V	15V	1V	5uV/C	.	.6MA	96dB	96dB	10M
AMLM308AH	ADU	SBA	EXT	.	.	+18V	−18V	70C	98dB	0.5MV	7NA	1NA	500MWF	1MA	13V	15V	1V	5uV/C	.	.6MA	96dB	96dB	10M
AMLM308AN	ADU	SBA	EXT	.	.	+18V	−18V	70C	98dB	0.5MV	7NA	1NA	500MWF	1MA	13V	15V	1V	5uV/C	.	.6MA	96dB	96dB	10M
AMLM308D	ADU	SBA	EXT	.	.	+18V	−18V	70C	88dB	7.5MV	7NA	1NA	500MWF	1MA	13V	15V	1V	30uV/C	.	.6MA	80dB	80dB	10M
AMLM308H	ADU	SBA	EXT	.	.	+18V	−18V	70C	88dB	7.5MV	7NA	1NA	500MWF	1MA	13V	15V	1V	30uV/C	.	.6MA	80dB	80dB	10M
AMLM308N	ADU	SBA	EXT	.	.	+18V	−18V	70C	88dB	7.5MV	7NA	1NA	500MWF	1MA	13V	15V	1V	30uV/C	.	.6MA	80dB	80dB	10M
AMLM310D	ADU	VFA	INT	.	15V/uS	+18V	−10V	70C	0dB	7.5MV	7NA	.	500MWF	1MA	10V	15V	15V	50uV/C	.	6MA	.	70dB	10G
AMLM310F	ADU	VFA	INT	.	15V/uS	+18V	−18V	70C	0dB	7.5MV	7NA	.	500MWF	1MA	10V	15V	15V	50uV/C	.	6MA	.	70dB	10G
AMLM310H	ADU	VFA	INT	.	15V/uS	+18V	−18V	70C	0dB	7.5MV	7NA	.	500MWF	1MA	10V	15V	15V	50uV/C	.	6MA	.	70dB	10G
AMLM310N	ADU	VFA	INT	.	15V/uS	+18V	−18V	70C	0dB	7.5MV	7NA	.	500MWF	1MA	10V	15V	15V	50uV/C	.	6MA	.	70dB	10G
AMLM311D	ADU	CPR	EXT	.	.	+18V	−18V	70C	100dB	7.5MV	250NA	50NA	500MWF	.	.	15V	30V	.	.	8MA	.	.	.
AMLM311H	ADU	CPR	EXT	.	.	+18V	−18V	70C	100dB	7.5MV	250NA	50NA	500MWF	.	.	15V	30V	.	.	8MA	.	.	.
AMLM312D	ADU	SBA	INT	.	.	+18V	−18V	70C	88dB	7.5MV	7NA	1NA	500MWF	1MA	13V	14V	14V	15uV/C	.	.8MA	80dB	80dB	10M
AMLM312H	ADU	SBA	INT	.	.	+18V	−18V	70C	88dB	7.5MV	7NA	1NA	500MWF	1MA	13V	14V	14V	15uV/C	.	.8MA	80dB	80dB	10M
AMLM316AD	ADU	LBC	INT	.	.	+20V	−20V	70C	92dB	3MV	50pA	15pA	500MWF	1MA	13V	15V	14V	.	.	.6MA	80dB	80dB	2G
AMLM316AF	ADU	LBC	INT	.	.	+20V	−20V	70C	92dB	3MV	50pA	15pA	500MWF	1MA	13V	15V	14V	.	.	.6MA	80dB	80dB	2G

For detailed explanations of column heading notations, see App. A.

Also for ready references the more important abbreviations used in the column headings are listed below:

LEFT HAND PAGE

APP = application (codes at APP.E.)
CMRR = common mode rejection ratio
CMP = compensation (frequency)
dV_{io}/dT = input offset voltage temperature drift
GBP = gain bandwidth product
I_B = input bias current
I_{io} = input bias offset current
I_Q = quiescent supply current
MFR = manufacturer (codes at App.C.)
P_G = quiescent power consumer
PSRR = power supply rejection ratio
V_{ICM} = common mode input voltage rating
V_{ID} = differential input voltage rating
V_{IO} = input offset voltage
V_S = dc supply voltage

RIGHT HAND PAGE

Lead out coding summary (details at APP.G.) for different cases (APP.F.)

A = gain adjust
B = bias adjust
C = case
E− = inverting input
E+ = non-inverting input
F.F* = input frequency compensation
G = ground
J = high level input
K = output, open collector
L = output, open emitter
M = metal case
N = not connected
Q = special terminal
R,R* = outputs
S = strobe
T,T* = offset balance
V+ = +ve dc supply
V− = −ve dc supply
W = guard ring
X = blank position, no lead
++ = +ve supplementary dc supply
−− = −ve supplementary dc supply
ø,ø* = output frequency compensation

CASE (APP F)	LD 1	LD 2	LD 3	LD 4	LD 5	LD 6	LD 7	LD 8	LD 9	LD 10	LD 11	LD 12	LD 13	LD 14	LD 15	LD 16	EUROPE SUBSTITUTE	USA SUBSTITUTE	ISS	TYPE NUMBER
DIL-14/1C	N	F	N	E−	E+	N	V−	N	N	R	V+	F*	N	N			UA208AD	LM208AD	0	AMLM208AD
FLP-10/3G	N	N	E−	E+	N	V−	R	V+	F*	F								LM208AF	0	AMLM208AF
T05-8/1M	F	E−	E+	V−M	N	R	V+	F*									SFC2208A	LM208AH	0	AMLM208AH
DIL-14/1C	N	F	N	E−	E+	N	V−	N	N	R	V+	F*	N	N			UA208D	LM208D	0	AMLM208D
FLP-10/3G	N	N	E−	E+	N	V−	R	V+	F*	F							SFC2208PT	LM208F	0	AMLM208F
T05-8/1M	F	E−	E+	V−M	N	R	V+	F*									SFC2208	LM208H	0	AMLM208H
DIL-14/1C	N	N	T	N	E+	V−	N	N	L	R	V+	T*	N	N			SN52110JA	LM210D	0	AMLM210D
FLP-10/3C	N	T	N	E+	V−	L	R	V+	T*	N								LM210F	0	AMLM210F
T05-8/1M	T	N	E+	V−	L	R	V+	T*									MLM210G	LM210H	0	AMLM210H
DIL-14/1C	N	G	E+	E−	N	V−	T	T*S	R	N	N	V+	N	N			SN52111J	LM211D	0	AMLM211D
T05-8/1M	G	E+	E−	V−	T	T*S	R	V+									SFC2211	LM211H	0	AMLM211H
DIL-14/1P	N	T	W	E−	E+	W*	V−	N	F	R	V+	T*	N	N				LM212D	0	AMLM212D
T05-8/1M	T	E−	E+	V−	F	R	V+	T*										LM212H	0	AMLM212H
FLP-10/3C	N	W	E−	E+	W*	V−	R	V+	T	T*								LM212F	0	AM212-FLP
DIL-14/1C	N	T	W	E−	E+	W*	V−	N	F	R	V+	T*	N	N			MLM216AD	LM216AD	0	AMLM216AD
FLP-10/3C	N	W	E−	E+	W*	V−	R	V+	T	T*								LM216AF	0	AMLM216AF
T05-8/1M	T	E−	E+	V−	F	R	V+	T*										LM216AH	0	AMLM216AH
DIL-14/1C	N	T	W	E−	E+	W*	V−	N	F	R	V+	T*	N	N				LM216D	0	AMLM216D
FLP-10/3C	N	W	E−	E+	W*	V−	R	V+	T	T*								LM216F	0	AMLM216F
T05-8/1M	T	E−	E+	V−	F	R	V+	T*										LM216H	0	AMLM216H
DIL-14/1C	N	N	T*F	E−	E+	V−	N	N	F*T	R	V+	ø					SN52118JA	LM218D	0	AMLM218D
FLP-10/3C	N	T*F	E−	E+	V−	F*T	R	V+	ø	N								LM218F	0	AMLM218F
T05-8/1M	T*F	E−	E+	V−	F*T	R	V+	ø									TDB0118CM	LM218H	0	AMLM218H
DIL-14/1C	R2	R1	V+	E−1	E+1	E−2	E+2	E+3	E−3	E−4	E+4	G	R4	R3			TDE0119DP	LM219D	0	AMLM219D
FLP-10/3C	R1	G1	E+1	E−1	V−	R2	G2	E+2	E−2	V+								LM219F	0	AMLM219F
T05-10/1M	R1	G1	E+1	E−1	V−	R2	G2	E+2	E−2	V+							TDE0119CM	LM219H	0	AMLM219H
DIL-14/1C	R1	E−1	E+1	V+	E+2	E−2	R2	R3	E−3	E+3	G	E+4	E−4	R4				LM224AD	0	AMLM224AD
DIL-14/1C	R1	E−1	E+1	V+	E+2	E−2	R2	R3	E−3	E+3	G	E+4	E−4	R4				LM224D	0	AMLM224D
DIL-14/1C	N	N	G1	E+1	E−1	V−	R2	G2	E+2	E−2	V+	R1	N	N			MLM239AL	LM239AD	0	AMLM239AD
DIL-14/1C	R2	R1	V+	E−1	E+1	E−2	E+2	E+3	E−3	E−4	E+4	G	R4	R3			MLM239L	LM239D	0	AMLM239D
DIL-14/1C	R1	E−1	E+1	V+	E+2	E−2	R2	R3	E−3	E+3	V−	E+4	E−4	R4				LM248D	0	AMLM248D
DIL-14/1P	R1	E−1	E+1	V+	E+2	E−2	R2	R3	E−3	E+3		E+4	E−4	R4				LM249D	0	AMLM249D
DIL-14/1P	FT	E−	E+	V−	T*	R	V+	F*									TDA0301D	LM301AD	0	AMLM301AD
T05-8/1M	FT	E−	E+	V−M	T*	R	V+	F*									SFC2301AH	LM301AH	0	AMLM301AH
DIL-8/1P	FT	E−	E+	V−	T*	R	V+	F*									SFC2301ADC	LM301AN	0	AMLM301AN
DIL-14/1P	N	N	FT	E−	E+	V−	N	N	T*	R	V+	F*	N	N				LM301D	0	AMLM301D
FLP-10/3C	N	FT	E−	E+	V−	T*	R	V+	F*	N								LM301F	0	AMLM301F
T05-8/1M	FT	E−	E+	V−M	T*	R	V+	F*									SFC2301A	LM301H	0	AMLM301H
DIL-14/1C	N	N	T	N	E+	V−	N	N	L	R	V+	T*	N	N				LM302D	0	AMLM302D
T05-8/1M	T	N	E+	V−	L	R	V+	T*									UA302C	LM302H	0	AMLM302H
T05-8/1M	G	E+	E−	V−M	S1	S2	R	V+									SN72306L	LM306H	0	AMLM306H
DIL-14/1P	N	N	N	E−	E+	N	V−	N	N	R	V+	N	N	N			SN72307JA	LM307D	0	AMLM307D
T05-8/1M	N	E−	E+	V−M	N	R	V+	N									SFC2307	LM307H	0	AMLM307H
DIL-14/1P	N	F	N	E−	E+	N	V−	N	N	R	V+	F*	N	N			SN72308AJA	LM308AD	0	AMLM308AD
T05-8/1M	F	E−	E+	V−M	N	R	V+	F*									SFC2308A	LM308AH	0	AMLM308AH
DIL-8/1P	F	E−	E+	V−	N	R	V+	F*										LM308AN	0	AMLM308AN
DIL-14/1P	N	F	N	E−	E+	N	V−	N	N	R	V+	F*	N	N			UA308D	LM308D	0	AMLM308D
T05-8/1M	F	E−	E+	V−M	N	R	V+	F*									SFC2308	LM308H	0	AMLM308H
DIL-8/1P	F	E−	E+	V−	N	R	V+	F*									SFC23080C	LM308N	0	AMLM308N
DIL-14/1P	N	N	T	N	E+	V−	N	N	L	R	V+	T*	N	N			SFC2310EC	LM310D	0	AMLM310D
FLP-10/3C	N	T	N	E+	V−	L	R	V+	T*	N								LM310F	0	AMLM310F
T05-8/1M	T	N	E+	V−	L	R	V+	T*									SFC2310EC	LM310H	0	AMLM310H
DIL-8/1P	T	N	E+	V−	L	R	V+	T*									SFC2310C	LM310N	0	AMLM310N
DIL-14/1P	N	G	E+	E−	N	V−	T	T*S	R	N	N	V+	N	N			SFC2311EC	LM311D	0	AMLM311D
T05-8/1M	G	E+	E−	V−	T	T*S	R	V+									SFC2311	LM311H	0	AMLM311H
DIL-14/1P	N	T	C	E−	E+	C*	V−	N	F	R	V+	T*	N	N				LM312D	0	AMLM312D
T05-8/1M	T	E−	E+	V−	F	R	V+	T*										LM312H	0	AMLM312H
DIL-14/1P	N	T	W	E−	E+	W*	V−	N	F	R	V+	T*	N	N			MLM316AD	LM316AD	0	AMLM316AD
FLP-10/3C	N	W	E−	E+	W*	V−	R	V+	T	T*								LM316AF	0	AMLM316AF
T05-8/1M	T	E−	E+	V−	F	R	V+	T*										LM316AH	0	AMLM316AH

TYPE NUMBER	MFR	APP	CMP	GBP MIN	SLEW RATE MIN	V_S^+ MAX	V_S^- MAX	T_{OP} MAX	A_{VOL} MIN	V_{IO} MAX	I_B MAX	I_{IO} MAX	P_{TOT} MAX	I_{OUT} MIN	V_{OUT} MIN	V_{ICM} MAX	V_{IDF} MAX	dV_{IO}/dT MAX	P_O MAX	I_O MAX	CM RR MIN	PS RR MIN	R_{IN} MIN
AMLM316AH	ADU	LBC	INT	.	.	+20V	-20V	70C	92dB	3MV	50pA	15pA	500MWF	1MA	13V	15V	14V	.	.	.6MA	80dB	80dB	2G
AMLM316D	ADU	LBC	INT	.	.	+20V	-20V	70C	86dB	10MV	150pA	50pA	500MWF	1MA	13V	15V	14V	.	.	.8MA	80dB	80dB	300M
AMLM316D	ADU	LBC	INT	.	.	+20V	-20V	70C	86dB	10MV	150pA	50pA	500MWF	1MA	13V	15V	14V	.	.	.8MA	80dB	80dB	300M
AMLM316F	ADU	LBC	INT	.	.	+20V	-20V	70C	86dB	10MV	150pA	50pA	500MWF	1MA	13V	15V	14V	.	.	.8MA	80dB	80dB	300M
AMLM316H	ADU	LBC	INT	.	.	+20V	-20V	70C	86dB	10MV	150pA	50pA	500MWF	1MA	13V	15V	14V	.	.	.8MA	80dB	80dB	300M
AMLM318D	ADU	XSR	INT	.	50V/uS	+20V	-20V	70C	88dB	10MV	500NA	200NA	500MWF	6MA	12V	15V	1V	.	.	8MA	70dB	65dB	500K
AMLM318F	ADU	XSR	INT	.	50V/uS	+20V	-20V	70C	88dB	10MV	500NA	200NA	500MWF	6MA	12V	15V	1V	.	.	8MA	70dB	65dB	500K
AMLM318H	ADU	XSR	INT	.	50V/uS	+20V	-20V	70C	88dB	10MV	500NA	200NA	500MWF	6MA	12V	15V	1V	.	.	8MA	70dB	65dB	500K
AMLM318N	ADU	XSR	INT	.	50V/uS	+20V	-20V	70C	88dB	10MV	500NA	200NA	500MWF	6MA	12V	15V	1V	.	.	8MA	70dB	65dB	500K
AMLM319D	ADU	DCP	INT	.	.	+18V	-18V	70C	78dB	8MV	1uA	0.2uA	500MWF	.	.	15V	5V	.	.	12MA	.	.	.
AMLM319H	ADU	DCP	INT	.	.	+18V	-18V	70C	78dB	8MV	1uA	0.2uA	500MWF	.	.	15V	5V	.	.	12MA	.	.	.
AMLM319N	ADU	DCP	INT	.	.	+18V	-18V	70C	78dB	8MV	1uA	0.2uA	500MWF	.	.	15V	5V	.	.	12MA	.	.	.
AMLM324AD	ADU	OGK	INT	.	.	+16V	-16V	70C	88dB	3MV	100NA	30NA	900MWF	.	.	16V	16V	30uV/C	.	.	2MA	65dB	65dB	.
AMLM324AN	ADU	OGK	INT	.	.	+16V	-16V	70C	88dB	3MV	100NA	30NA	900MWF	.	.	16V	16V	30uV/C	.	.	2MA	65dB	65dB	.
AMLM324D	ADU	OGK	INT	.	.	+16V	-16V	70C	88dB	7MV	250NA	50NA	900MWF	.	.	16V	16V	35uV/C	.	.	2MA	65dB	65dB	.
AMLM324N	ADU	OGK	INT	.	.	+16V	-16V	70C	88dB	7MV	250NA	50NA	900MWF	.	.	16V	16V	35uV/C	.	.	2MA	65dB	65dB	.
AMLM339AD	ADU	QCP	EXT	.	.	+18V	-18V	70C	94dB	2MV	250NA	50NA	900MWF	6MA	.	18V	18V	.	.	.	2MA	.	.	.
AMLM339AN	ADU	QCP	EXT	.	.	+18V	-18V	70C	94dB	2MV	250NA	50NA	900MWF	6MA	.	18V	18V	.	.	.	2MA	.	.	.
AMLM339D	ADU	QCP	EXT	.	.	+18V	-18V	70C	88dB	5MV	250NA	50NA	900MWF	6MA	.	18V	18V	.	.	.	2MA	.	.	.
AMLM339N	ADU	QCP	EXT	.	.	+18V	-18V	70C	88dB	5MV	250NA	50NA	800MWF	.	.	18V	18V	.	.	.	2MA	.	.	.
AMLM348D	ADU	OGK	INT	.3MHZ	0.2V/uS	+18V	-18V	70C	88dB	6MV	200NA	50NA	900MWF	5MA	12V	18V	36V	.	.	1MA	70dB	77dB	300K
AMLM348N	ADU	OGK	INT	.3MHZ	0.2V/uS	+18V	-18V	70C	88dB	6MV	200NA	50NA	900MWF	5MA	12V	18V	36V	.	.	1MA	70dB	77dB	300K
AMLM349D	ADU	OGK	INT	1MHZ	0.5V/uS	+18V	-18V	70C	88dB	6MV	200NA	50NA	900MWF	5MA	12V	18V	36V	.	.	1MA	70dB	77dB	300K
AMLM349N	ADU	OGK	INT	1MHZ	0.5V/uS	+18V	-18V	70C	88dB	6MV	200NA	50NA	900MWF	5MA	12V	18V	36V	.	.	1MA	70dB	77dB	300K
AMSSS725BJ	ADU	PIA	EXT	.	.	+22V	-22V	70C	120dB	.75MV	80NA	5NA	500MWF	11MA	12V	22V	30V	2.8uV/C	120MW	.	.	110dB	106dB	700K
AMSSS725EJ	ADU	PIA	EXT	.	.	+22V	-22V	70C	120dB	0.5MV	80NA	5NA	500MWF	11MA	12V	22V	30V	2uV/C	120MW	.	.	120dB	106dB	700K
AMSSS725J	ADU	PIA	EXT	.	.	+22V	-22V	125C	120dB	0.5MV	80NA	5NA	500MWF	11MA	12V	22V	30V	2uV/C	120MW	.	.	120dB	106dB	700K
AMSSS741CJ	ADU	GPK	INT	.	.	+18V	-18V	70C	86dB	6MV	100NA	25NA	500MWF	5MA	12V	18V	30V	.	85MW	.	70dB	76dB	1M
AMSSS741J	ADU	GPK	INT	.	.	+22V	-22V	125C	100dB	2MV	50NA	5NA	500MWF	5MA	12V	22V	30V	.	85MW	.	80dB	80dB	2M
AMSSS747CK	ADU	DGK	INT	.	.	+22V	-22V	70C	94dB	5MV	100NA	25NA	500MWF	5MA	12V	22V	30V	.	85MW	.	70dB	76dB	1M
AMSSS747CP	ADU	DGK	INT	.	.	+22V	-22V	70C	94dB	5MV	100NA	25NA	500MWF	5MA	12V	22V	30V	.	85MW	.	70dB	76dB	1M
AMSSS747K	ADU	DGK	INT	.	.	+22V	-22V	125C	100dB	2MV	80NA	5NA	500MWF	5MA	12V	22V	30V	.	85MW	.	80dB	80dB	2M
AMSSS747M	ADU	DGK	INT	.	.	+22V	-22V	125C	100dB	2MV	80NA	5NA	500MWF	5MA	12V	22V	30V	.	85MW	.	80dB	80dB	2M
AMSSS747P	ADU	DGK	INT	.	.	+22V	-22V	125C	100dB	2MV	80NA	5NA	500MWF	5MA	12V	22V	30V	.	85MW	.	80dB	80dB	2M
AMU3F7733312	ADU	BDO	EXT	20MHZ	.	+8V	-8V	125C	50dB	5MV	20uA	3uA	570MWF	2MA	1.5V	6V	5V	.	.	24MA	60dB	50dB	2K
AMU3F7748312	ADU	GPU	EXT	.	0.2V/uS	+22V	-22V	125C	94dB	5MV	500NA	200NA	570MWF	5MA	12V	15V	30V	.	85MW	3MA	70dB	76dB	300K
AMU5B7741312	ADU	GPK	INT	.	0.2V/uS	+22V	-22V	125C	94dB	5MV	500NA	200NA	500MWF	7MA	12V	15V	30V	.	85MW	3MA	70dB	76dB	300K
AMU5B7741393	ADU	GPK	INT	.	0.2V/uS	+18V	-18V	70C	86dB	6MV	500NA	200NA	500MWF	5MA	12V	15V	30V	.	.	3MA	70dB	76dB	300K
AMU5B7748312	ADU	GPU	EXT	.	0.2V/uS	+22V	-22V	125C	94dB	5MV	500NA	200NA	500MWF	5MA	12V	15V	30V	.	85MW	3MA	70dB	76dB	300K
AMU5B7748393	ADU	GPU	EXT	.	0.2V/uS	+18V	-18V	70C	86dB	6MV	500NA	200NA	500MW	5MA	12V	15V	30V	.	85MW	3MA	70dB	76dB	300K
AMU6A7733312	ADU	BDO	EXT	20MHZ	.	+8V	-8V	125C	50dB	5MV	20uA	3uA	670MWF	2MA	1.5V	6V	5V	.	.	24MA	60dB	50dB	2K
AMU6A7733393	ADU	BDO	EXT	20MHZ	.	+8V	-8V	70C	48dB	6MV	30uA	5uA	670MWF	2MA	1.5V	6V	5V	.	.	24MA	60dB	50dB	2K
AMU6A7741312	ADU	GPK	INT	.4MHZ	0.3V/uS	+22V	-22V	125C	94dB	3MV	80NA	30NA	670MWF	5MA	12V	15V	30V	15uV/C	150MW	.	80dB	86dB	1M
AMU6A7741393	ADU	GPK	INT	.	0.2V/uS	+18V	-18V	70C	86dB	6MV	500NA	200NA	670MWF	5MA	12V	15V	30V	.	85MW	3MA	70dB	76dB	300K
AMU6A7748312	ADU	GPU	EXT	.	0.2V/uS	+22V	-22V	125C	94dB	5MV	500NA	200NA	670MWF	5MA	12V	15V	30V	.	85MW	3MA	70dB	76dB	300K
AMU6A7748393	ADU	GPU	EXT	.	0.2V/uS	+22V	-22V	70C	86dB	6MV	500NA	200NA	670MWF	5MA	12V	15V	30V	.	85MW	3MA	70dB	76dB	300K
AMU6W7747312	ADU	DGK	INT	.	0.2V/uS	+22V	-22V	125C	94dB	5MV	500NA	200NA	670MWF	5MA	12V	15V	30V	.	85MW	3MA	70dB	76dB	300K
AMU6W7747393	ADU	DGK	INT	.	0.2V/uS	+18V	-18V	70C	88dB	6MV	500NA	200NA	670MWF	5MA	12V	15V	30V	.	85MW	3MA	70dB	76dB	300K
B100	ANU	VFA	INT	.	65V/uS	+16V	-16V	85C	-2dB	200MV	500NA	.	.	.	0.1A	10V	.	15V	1V/C	.	8MA	.	.	2K
C118A	BLU	FET	INT	.5MHZ	0.5/uS	+18V	-18V	85C	100dB	1MV	5pA	.	.	5MA	10V	15V	30V	10uV/C	.	4MA	76dB	.	10G
C118B	BLU	FET	INT	.5MHZ	0.5/uS	+18V	-18V	85C	100dB	1MV	5pA	.	.	5MA	10V	15V	30V	25uV/C	.	4MA	76dB	.	10G
C118C	BLU	FET	INT	.5MHZ	0.5/uS	+18V	-18V	85C	100dB	2MV	10pA	.	.	5MA	10V	15V	30V	50uV/C	.	4MA	76dB	.	10G
C218A	BLU	FET	INT	.5MHZ	0.5/uS	+18V	-18V	85C	100dB	1MV	5pA	.	.	5MA	10V	15V	30V	10uV/C	.	4MA	76dB	.	10G
C218B	BLU	FET	INT	.5MHZ	0.5/uS	+18V	-18V	85C	100dB	1MV	5pA	.	.	5MA	10V	15V	30V	25uV/C	.	4MA	76dB	.	10G
C218C	BLU	FET	INT	.5MHZ	0.5/uS	+18V	-18V	85C	100dB	2MV	10pA	.	.	5MA	10V	15V	30V	50uV/C	.	4MA	76dB	.	10G
C228A	BLU	FET	INT	7MHZ	25V/uS	+18V	-18V	125C	100dB	1MV	5pA	.	.	5MA	10V	15V	30V	10uV/C	.	10MA	80dB	.	10G
C228B	BLU	FET	INT	7MHZ	25V/uS	+18V	-18V	125C	100dB	1MV	5pA	.	.	5MA	10V	15V	30V	25uV/C	.	10MA	80dB	.	10G
C228C	BLU	FET	INT	7MHZ	25V/uS	+18V	-18V	125C	100dB	2MV	10pA	.	.	5MA	10V	15V	30V	50uV/C	.	10MA	80dB	.	10G
C238A	BLU	FET	INT	10MHZ	50V/uS	+22V	-22V	125C	.	1MV	5pA	.	.	20MA	10V	15V	30V	10uV/C	.	12MA	86dB	70dB	10G
C238B	BLU	FET	INT	10MHZ	50V/uS	+22V	-22V	125C	.	1MV	5pA	.	.	20MA	10V	15V	30V	25uV/C	.	12MA	86dB	70dB	10G

For detailed explanations of column heading notations, see App. A.

Also for ready references the more important abbreviations used in the column headings are listed below:

LEFT HAND PAGE

APP = application (codes at APP.E.)

CMRR = common mode rejection ratio

CMP = compensation (frequency)

dV_{IO}/dT = input offset voltage temperature drift

GBP = gain bandwidth product

I_B = input bias current

I_{IO} = input bias offset current

I_Q = quiescent supply current

MFR = manufacturer (codes at App.C.)

P_Q = quiescent power consumer

PSRR = power supply rejection ratio

V_{ICM} = common mode input voltage rating

V_{IDF} = differential input voltage rating

V_{IO} = input offset voltage

V_S = dc supply voltage

RIGHT HAND PAGE

Lead out coding summary (details at APP.G.) for different cases (APP.F.)

A = gain adjust

B = bias adjust

C = case

E— = inverting input

E+ = non-inverting input

F,F* = input frequency compensation

G = ground

J = high level input

K = output, open collector

L = output, open emitter

M = metal case

N = not connected

Q = special terminal

R,R* = outputs

S = strobe

T,T* = offset balance

V+ = +ve dc supply

V— = —ve dc supply

W = guard ring

X = blank position, no lead

++ = +ve supplementary dc supply

—— = —ve supplementary dc supply

ϕ,ϕ^* = output frequency compensation

CASE (APP F)	LD 1	LD 2	LD 3	LD 4	LD 5	LD 6	LD 7	LD 8	LD 9	LD 10	LD 11	LD 12	LD 13	LD 14	LD 15	LD 16	EUROPE SUBSTITUTE	USA SUBSTITUTE	ISS	TYPE NUMBER
DIL-14/1C	N	T	W	E-	E+	W*	V-	N	F	R	V+	T*	N	N				LM316D	0	AMLM316D
DIL-14/1P	N	T	W	E-	E+	W*	V-	N	F	R	V+	T*	N	N				LM316D	0	AMLM316D
FLP-10/3C	N	W	E-	E+	W*	V-	R	V+	T	T*								LM316F	0	AMLM316F
T05-8/1M	T	E-	E+	V-	F	R	V+	T*										LM316H	0	AMLM316H
DIL-14/1C	N	N	T*F	E-	E+	V-	N	N	F*T	R	V+	ø	N	N			SFC2318EC	LM318D	0	AMLM318D
FLP-10/3C	N	T*F	E-	E+	V-	F*T	R	V+	ø	N								LM318F	0	AMLM318F
T05-8/1M	T*F	E-	E+	V-	F*T	R	V+	ø									TDE0118CM	LM318H	0	AMLM318H
DIL-8/1P	T*F	E-	E+	V-	F*	R	V-	ø									SN72318JP	LM318N	0	AMLM318N
DIL-14/1C	N	N	G1	E+1	E-1	V-	R2	G2	E+2	E-2	V+	R1	N	N			TDB0119DP	LM319D	0	AMLM319D
T05-10/1M	R1	G1	E+1	E-1	V-	R2	G2	E+2	E-2	V+							TDB0119CM	LM319H	0	AMLM319H
DIL-14/1P	N	N	G1	E+1	E-1	V-	R2	G2	E+2	E-2	V+	R1	N	N			TDB0119DP	LM319N	0	AMLM319N
DIL-14/1C	R1	E-1	E+1	V+	E+2	E-2	R2	R3	E-3	E+3	G	E+4	E-4	R4				LM324AD	0	AMLM324AD
DIL-14/1P	R1	E-1	E+1	V+	E+2	E-2	R2	R3	E-3	E+3	G	E+4	E-4	R4				LM324AN	0	AMLM324AN
DIL-14/1C	R1	E-1	E+1	V+	E+2	E-2	R2	R3	E-3	E+3	G	E+4	E-4	R4			MLM324L	LM324D	0	AMLM324D
DIL-14/1P	R1	E-1	E+1	V+	E+2	E-2	R2	R3	E-3	E+3	G	E+4	E-4	R4			MLM324L	LM324N	0	AMLM324N
DIL-14/1C	R2	R1	V+	E-1	E+1	E-2	E+2	E+3	E-3	E-4	E+4	G	R4	R3			MLM339AL	LM339AD	0	AML339AD
DIL-14/1P	R2	R1	V+	E-1	E+1	E-2	E+2	E+3	E-3	E-4	E+4	G	R4	R3			MLM339AN	LM339AN	0	AMLM339AN
DIL-14/1C	R2	R1	V+	E-1	E+1	E-2	E+2	E+3	E-3	E-4	E+4	G	R4	R3			MLM339L	LM339D	0	AMLM339D
DIL-14/1P	R2	R1	V+	E-1	E+1	E-2	E+2	E+3	E-3	E-4	E+4	G	R4	R3			MLM339L	LM339N	0	AMLM339N
DIL-14/1C	R1	E-1	E+1	V+	E+2	E-2	R2	R3	E-3	E+3	V-	E+4	E-4	R4				LM348D	0	AMLM348D
DIL-14/1P	R1	E-1	E+1	V+	E+2	E-2	R2	R3	E-3	E+3	V-	E+4	E-4	R4				LM348N	0	AMLM348N
DIL-14/1C	R1	E-1	E+1	V+	E+2	E-2	R2	R3	E-3	E+3	V-	E+4	E-4	R4				LM349D	0	AMLM349D
DIL-14/1P	R1	E-1	E+1	V+	E+2	E-2	R2	R3	E-3	E+3	V-	E+4	E-4	R4				LM349N	0	AMLM349N
T05-8/1M	T	E-	E+	V-M	ø	R	V+	T*									UA725AHM	SSS725BJ	0	AMSSS725BJ
T05-8/1M	T	E-	E+	V-M	ø	R	V+	T*									UA725EHC	SSS725EJ	0	AMSSS725EJ
T05-8/1M	T	E-	E+	V-M	ø	R	V+	T*									UA725AHM	SSS725J	0	AMSSS725J
T05-8/1M	T	E-	E+	V-M	T*	R	V+	N									UA741EHC	SSS741CJ	0	AMSSS741CJ
T05-8/1M	T	E-	E+	V-M	T*	R	V+	N									RM4131T	SSS741J	0	AMSSS741J
T05-10/1M	R1	V+	E-1	E+1	V-	E+2	E-2	V+2	R2	N							SFC2747M	SSS747CK	0	AMSSS747CK
DIL-14/1C	E-1	E+1	T1	V-	T2	E+2	E+2	T*2	V+2	R2	N	R1	V+1	T*1			SSS747CY	SSS747CP	0	AMSSS747CP
T05-10/1M	R1	V+1	E-1	E+1	V-	E+2	E-2	V+2	R2	N								SSS747K	0	AMSSS747K
FLP-14/3G	E-1	E+1	T1	V-	T2	E+2	E+2	T*2	V+2	R2	N	R1	V+1	T*1				SSS747M	0	AMSSS747M
DIL-14/1C	E-1	E+1	T1	V-	T2	E+2	E-2	T*2	V+2	R2	N	R1	V+1	T*1			SSS747Y	SSS747P	0	AMSSS747P
FLP-10/3C	E+	A2	A*2	V-	R	R*	V+	A1	A*1	E-							SN52733FA	UA733FM	0	AMU3F7733312
FLP-10/3C	N	FT	E-	E+	V-	T*	R	V+	F*	N							SN52748FA	UA748FM	0	AMU3F7748312
T05-8/1M	T	E-	E+	V-M	T*	R	V+	N									TBA222	UA741HM	0	AMU5B7741312
T05-8/1M	T	E-	E+	V-M	T*	R	V+	N									TBA221	UA741HC	0	AMU5B7741393
T05-8/1M	FT	E-	E+	V-	T*	R	V+	F*									TBC0748	UA748HM	0	AMU5B7748312
T05-8/1M	FT	E-	E+	V-	T*	R	V+	F*									TBB0748	UA748HC	0	AMU5B7748393
DIL-14/1C	E+	N	A2	A*2	V-	N	R	R*	N	V+	A1	A*1	N	E-			SN52733J	UA733DM	0	AMU6A7733312
DIL-14/1C	E+	N	A2	A*2	V-	N	R	R*	N	V+	A1	A*1	N	E-			SN72733J	UA733DC	0	AMU6A7733393
DIL-14/1C	N	N	T	E-	E+	V-	N	N	T*	R	V+	N	N	N			LM741D	UA741DM	0	AMU6A7741312
DIL-14/1C	N	N	T	E-	E+	V-	N	N	T*	R	V+	N	N	N			TBA221A	UA741DC	0	AMU6A7741393
DIL-14/1C	N	N	FT	E-	E+	V-	N	N	T*	R	V+	F*	N	N			SN52748JA	UA748DM	0	AMU6A7748312
DIL-14/1C	N	N	FT	E-	E+	V-	N	N	T*	R	V+	F*	N	N			SN72748J	UA748DC	0	AMU6A7748393
DIL-14/1C	E-1	E+1	T1	V-	T2	E+2	E-2	T*2	V+2	R2	N	R1	V+	T*1			SFC2747KM	UA747DM	0	AMU6W7747312
DIL-14/1C	E-1	E+1	T1	V-	T2	E+2	E-2	T*2	V+2	R2	N	R1	V+1	T*1			TBB0747A	UA747DC	0	AMU6W7747393
DIM-7/5P	E+	N	V+	G	V-	R	N											B100		
DIM-5/4P	E+	E-	V+	V-	R													C118A	0	C118A
DIM-5/4P	E+	E-	V+	V-	R												C118A		0	C118B
DIM-5/4P	E+	E-	V+	V-	R													C118B	0	C118C
T08-12/1M	E+	E-	N	V+	T	V-	GM	R	T*	N	N								0	C218A
T08-12/1M	E+	E-	N	V+	T	V-	GM	R	T*	N	N							C218A	0	C218B
T08-12/1M	E+	E-	N	V+	T	V-	GM	R	T*	N	N							C218B	0	C218C
T08-12/1M	E+	E-	N	V+	N	V-	GM	R	N	N	N								0	C228A
T08-12/1M	E+	E-	N	V+	N	V-	GM	R	N	N	N							C228A	0	C228B
T08-12/1M	E+	E-	N	V+	N	V-	GM	R	N	N	N							C228B	0	C228C
T08-12/1M	E+	E-	N	V+	N	V-	GM	R	Q	Q*	N							M238A	0	C238A
T08-12/1M	E+	E-	N	V+	N	V-	GM	R	Q	Q*	N							C238A	0	C238B
T08-12/1M	E+	E-	N	V+	N	V-	GM	R	Q	Q*	N							C238B	0	C238C

TYPE NUMBER	M F R	A P P	C M P	GBP MIN	SLEW RATE MIN	V_S^+ MAX	V_S^- MAX	T_{op} MAX	A_{VOL} MIN	V_{IO} MAX	I_B MAX	I_{IO} MAX	P_{TOT} MAX	I_{OUT} MIN	V_{OUT} MIN	V_{ICM} MAX	$V_{ID\pm}$ MAX	dV_{IO}/dT MAX	P_O MAX	I_O MAX	CM RR MIN	PS RR MIN	R_{IN} MIN
C238C	BLU	FET	INT	10MHZ	50V/uS	+22V	-22V	125C	.	2MV	10pA	.	.	20MA	10V	15V	30V	50uV/C	.	12MA	86dB	70dB	10G
C438A	BLU	XSR	INT	10MHZ	50V/uS	+22V	-22V	125C	90dB	1MV	5pA	.	.	20MA	10V	15V	30V	10uV/C	.	12MA	86dB	70dB	10G
C438B	BLU	XSR	INT	10MHZ	50V/uS	+22V	-22V	125C	90dB	1MV	5pA	.	.	20MA	10V	15V	30V	25uV/C	.	12MA	86dB	70dB	10G
C438C	BLU	XSR	INT	10MHZ	50V/uS	+22V	-22V	125C	90dB	2MV	10pA	.	.	20MA	10V	15V	30V	50uV/C	.	12MA	86dB	70dB	10G
CA101AG	RCU	GPU	EXT	.	.	+22V	-22V	125C	94dB	2MV	75NA	10NA	500MWF	5MA	12V	15V	30V	15uV/C	.	3MA	80dB	80dB	1.5M
CA101AS	RCU	GPU	EXT	.	.	+22V	-22V	125C	94dB	2MV	75NA	10NA	500MWF	5MA	12V	15V	30V	15uV/C	.	3MA	80dB	80dB	1.5M
CA101AT	RCU	GPU	EXT	.	.	+22V	-22V	125C	94dB	2MV	75NA	10NA	500MWF	5MA	12V	15V	30V	15uV/C	.	3MA	80dB	80dB	1.5M
CA101G	RCU	GPU	EXT	.	.	+22V	-22V	125C	94dB	5MV	500NA	200NA	500MWF	5MA	12V	15V	30V	15uV/C	.	3MA	70dB	70dB	300K
CA101S	RCU	GPU	EXT	.	.	+22V	-22V	125C	94dB	5MV	500NA	200NA	500MWF	5MA	12V	15V	30V	15uV/C	.	3MA	70dB	70dB	300K
CA101T	RCU	GPU	EXT	.	.	+22V	-22V	125C	94dB	5MV	500NA	200NA	500MWF	5MA	12V	15V	30V	10uV/C	.	3MA	70dB	70dB	300K
CA107G	RCU	GPK	INT	.	.	+22V	-22V	125C	94dB	2MV	75NA	10NA	500MWF	5MA	12V	15V	30V	15uV/C	.	3MA	80dB	80dB	1.5M
CA107S	RCU	GPK	INT	.	.	+22V	-22V	125C	94dB	2MV	75NA	10NA	500MWF	5MA	12V	15V	30V	15uV/C	.	3MA	80dB	80dB	1.5M
CA107T	RCU	GPK	INT	.	.	+22V	-22V	125C	94dB	2MV	75NA	10NA	500MWF	5MA	12V	15V	30V	15uV/C	.	3MA	80dB	80dB	1.5M
CA108AS	RCU	SBA	EXT	.	.	+20V	-20V	125C	98dB	0.5MV	2NA	0.2NA	500MWF	1MA	13V	15V	1V	5uV/C	.	.6MA	96dB	96dB	30M
CA108AT	RCU	SBA	EXT	.	.	+20V	-20V	125C	98dB	0.5MV	2NA	0.2NA	500MWF	1MA	13V	15V	1V	5uV/C	.	.6MA	96dB	96dB	30M
CA108S	RCU	SBA	EXT	.	.	+20V	-20V	125C	96dB	2MV	2NA	0.2NA	500MWF	1MA	13V	15V	1V	15uV/C	.	.6MA	85dB	80dB	30M
CA108T	RCU	SBA	EXT	.	.	+20V	-20V	125C	96dB	2MV	2NA	0.2NA	500MWF	1MA	13V	15V	1V	15uV/C	.	.6MA	85dB	80dB	30M
CA111G	RCU	CPR	EXT	.	.	+18V	-18V	125C	100dB	3MV	100NA	10NA	500MWF	8MA	.	15V	30V		.	6MA	.	.	.
CA111S	RCU	CPR	EXT	.	.	+18V	-18V	125C	100dB	3MV	100NA	10NA	500MWF	8MA	.	15V	30V		.	6MA	.	.	.
CA111T	RCU	CPR	EXT	.	.	+18V	-18V	125C	100dB	3MV	100NA	10NA	500MWF	.	.	15V	30V		.	6MA	.	.	.
CA124E	RCU	QGK	INT	.	.	+16V	-16V	125C	94dB	5MV	150NA	30NA	900MWF	.	.	16V	16V	35uV/C	.	2MA	70dB	65dB	.
CA124G	RCU	QGK	INT	.	.	+16V	-16V	125C	94dB	5MV	150NA	30NA	900MWF	.	.	16V	16V	35uV/C	.	2MA	70dB	65dB	.
CA139AE	RCU	QCP	EXT	.	.	+18V	-18V	125C	94dB	2MV	100NA	25NA	900MWF	.	.	18V	18V		.	2MA	.	.	.
CA139AG	RCU	QCP	EXT	.	.	+18V	-18V	125C	94dB	2MV	100NA	25NA	900MWF	.	.	18V	18V		.	2MA	.	.	.
CA139E	RCU	QCP	EXT	.	.	+18V	-18V	125C	88dB	5MV	100NA	25NA	900MWF	.	.	18V	18V		.	2MA	.	.	.
CA139G	RCU	QCP	EXT	.	.	+18V	-18V	125C	88dB	5MV	100NA	25NA	900MWF	.	.	18V	18V		.	2MA	.	.	.
CA201AG	RCU	GPU	EXT	.	.	+22V	-22V	85C	94dB	2MV	75NA	10NA	500MWF	5MA	12V	15V	30V	15uV/C	.	3MA	80dB	80dB	500K
CA201AS	RCU	GPU	EXT	.	.	+22V	-22V	85C	94dB	2MV	75NA	10NA	500MWF	5MA	12V	15V	30V	15uV/C	.	3MA	80dB	80dB	500K
CA201AT	RCU	GPU	EXT	.	.	+22V	-22V	85C	94dB	2MV	75NA	10NA	500MWF	5MA	12V	15V	30V	15uV/C	.	3MA	80dB	80dB	500K
CA201G	RCU	GPU	EXT	.	.	+22V	-22V	85C	86dB	7.5MV	1.5uA	0.5uA	500MWF	5MA	12V	15V	30V	30uV/C	.	3MA	65dB	70dB	100K
CA201S	RCU	GPU	EXT	.	.	+22V	-22V	85C	86dB	7.5MV	1.5uA	0.5uA	500MWF	5MA	12V	15V	30V	30uV/C	.	3MA	65dB	70dB	100K
CA201T	RCU	GPU	EXT	.	.	+22V	-22V	85C	86dB	7.5MV	1.5uA	0.5uA	500MWF	5MA	12V	15V	30V	30uV/C	.	3MA	65dB	70dB	100K
CA207G	RCU	GPK	INT	.	.	+22V	-22V	85C	94dB	2MV	75NA	10NA	500MWF	5MA	12V	15V	30V	15uV/C	.	3MA	80dB	80dB	1.5M
CA207S	RCU	GPK	INT	.	.	+22V	-22V	85C	94dB	2MV	75NA	10NA	500MWF	5MA	12V	15V	30V	15uV/C	.	3MA	80dB	80dB	1.5M
CA207T	RCU	GPK	INT	.	.	+22V	-22V	85C	94dB	2MV	75NA	10NA	500MWF	5MA	12V	15V	30V	15uV/C	.	3MA	80dB	80dB	1.5M
CA208AS	RCU	SBA	EXT	.	.	+20V	-20V	85C	98dB	0.5MV	2NA	0.2NA	500MWF	1MA	13V	15V	1V	5uV/C	.	.6MA	96dB	96dB	30M
CA208AT	RCU	SBA	EXT	.	.	+20V	-20V	85C	98dB	0.5MV	2NA	0.2NA	500MWF	1MA	13V	15V	1V	5uV/C	.	.6MA	96dB	96dB	30M
CA208S	RCU	SBA	EXT	.	.	+20V	-20V	85C	96dB	2MV	2NA	0.2NA	500MWF	1MA	13V	15V	1V	15uV/C	.	.6MA	85dB	80dB	30M
CA208T	RCU	SBA	EXT	.	.	+20V	-20V	85C	96dB	2MV	2NA	0.2NA	500MWF	1MA	13V	15V	1V	15uV/C	.	.6MA	85dB	80dB	30M
CA211G	RCU	CPR	EXT	.	.	+18V	-18V	85C	100dB	3MV	100NA	10NA	500MWF	.	.	15V	30V		.	6MA	.	.	.
CA211S	RCU	CPR	EXT	.	.	+18V	-18V	85C	100dB	3MV	100NA	10NA	500MWF	.	.	15V	30V		.	6MA	.	.	.
CA211T	RCU	CPR	EXT	.	.	+18V	-18V	85C	100dB	3MV	100NA	10NA	500MWF	.	.	15V	30V		.	6MA	.	.	.
CA224E	RCU	QGK	INT	.	.	+16V	-16V	85C	94dB	5MV	150NA	30NA	900MWF	.	.	16V	16V	35uV/C	.	2MA	70dB	65dB	.
CA224G	RCU	QGK	INT	.	.	+16V	-16V	85C	94dB	5MV	150NA	30NA	900MWF	.	.	16V	16V	35uV/C	.	2MA	70dB	65dB	.
CA239AE	RCU	QCP	EXT	.	.	+18V	-18V	85C	94dB	2MV	250NA	50NA	900MWF	6MA	.	18V	18V		.	2MA	.	.	.
CA239AG	RCU	QCP	EXT	.	.	+18V	-18V	85C	94dB	2MV	250NA	50NA	900MWF	6MA	.	18V	18V		.	2MA	.	.	.
CA239E	RCU	QCP	EXT	.	.	+18V	-18V	85C	88dB	5MV	250NA	50NA	900MWF	6MA	.	18V	18V		.	2MA	.	.	.
CA239G	RCU	QCP	EXT	.	.	+18V	-18V	85C	88dB	5MV	250NA	50NA	900MWF	6MA	.	18V	18V		.	2MA	.	.	.
CA301AE	RCU	GPU	EXT	.	.	+18V	-18V	70C	88dB	7.5MV	250NA	50NA	500MWF	5MA	12V	15V	30V	30uV/C	.	3MA	70dB	70dB	500K
CA301AG	RCU	GPU	EXT	.	.	+18V	-18V	70C	88dB	7.5MV	250NA	50NA	500MWF	5MA	12V	15V	30V	30uV/C	.	3MA	70dB	70dB	500K
CA301AGH	RCU	GPU	EXT	.	.	+18V	-18V	70C	88dB	7.5MV	250NA	50NA		5MA	12V	15V	30V	30uV/C	.	3MA	70dB	70dB	500K
CA301AH	RCU	GPU	EXT	.	.	+18V	-18V	70C	88dB	7.5MV	250NA	50NA		5MA	12V	15V	30V	30uV/C	.	3MA	70dB	70dB	500K
CA301AS	RCU	GPU	EXT	.	.	+18V	-18V	70C	88dB	7.5MV	250NA	50NA	500MWF	5MA	12V	15V	30V	30uV/C	.	3MA	70dB	70dB	500K
CA301AT	RCU	GPU	EXT	.	.	+18V	-18V	70C	88dB	7.5MV	250NA	50NA	500MWF	5MA	12V	15V	30V	30uV/C	.	3MA	70dB	70dB	500K
CA307E	RCU	GPK	INT	.	.	+18V	-18V	70C	84dB	7.5MV	250NA	50NA	500MWF	5MA	12V	15V	30V	30uV/C	.	.	70dB	70dB	0.5M
CA307G	RCU	GPK	INT	.	.	+18V	-18V	70C	84dB	7.5MV	250NA	50NA	500MWF	5MA	12V	15V	30V	30uV/C	.	.	70dB	70dB	0.5M
CA307GH	RCU	GPK	INT	.	.	+18V	-18V	70C	84dB	7.5MV	250NA	50NA	500MWF	5MA	12V	15V	30V	30uV/C	.	.	70dB	70dB	0.5M
CA307H	RCU	GPK	INT	.	.	+18V	-18V	70C	84dB	7.5MV	250NA	50NA	500MWF	5MA	12V	15V	30V	30uV/C	.	.	70dB	70dB	0.5M
CA307S	RCU	GPK	INT	.	.	+18V	-18V	70C	84dB	7.5MV	250NA	50NA	500MWF	5MA	12V	15V	30V	30uV/C	.	.	70dB	70dB	0.5M
CA307T	RCU	GPK	INT	.	.	+18V	-18V	70C	84dB	7.5MV	250NA	50NA	500MWF	5MA	12V	15V	30V	30uV/C	.	.	70dB	70dB	0.5M

LEFT HAND PAGE

APP = application (codes at APP.E.)

CMRR = common mode rejection ratio

CMP = compensation (frequency)

dV_{IO}/dT = input offset voltage temperature drift

GBP = gain bandwidth product

I_B = input bias current

I_{IO} = input bias offset current

I_Q = quiescent supply current

MFR = manufacturer (codes at App.C.)

P_D = quiescent power consumer

PSRR = power supply rejection ratio

V_{ICM} = common mode input voltage rating

V_{IDF} = differential input voltage rating

V_{IO} = input offset voltage

V_S = dc supply voltage

RIGHT HAND PAGE

Lead out coding summary (details at APP.G.) for different cases (APP.F.)

A = gain adjust

B = bias adjust

C = case

E– = inverting input

E+ = non-inverting input

F,F* = input frequency compensation

G = ground

J = high level input

K = output, open collector

L = output, open emitter

M = metal case

N = not connected

Q = special terminal

R,R' = outputs

S = strobe

T,T* = offset balance

V+ = +ve dc supply

V– = –ve dc supply

W = guard ring

X = blank position, no lead

++ = +ve supplementary dc supply

–– = –ve supplementary dc supply

◊,◊* = output frequency compensation

CASE (APP F)	LD1	LD2	LD3	LD4	LD5	LD6	LD7	LD8	LD9	LD10	LD11	LD12	LD13	LD14	LD15	LD16	EUROPE SUBSTITUTE	USA SUBSTITUTE	I S	TYPE NUMBER
DIM-14/1M	N	N	N	E-	E+	V-	N	M	N	R	V+	N	N	N					0	C438A
DIM-14/1M	N	N	N	E-	E+	V-	N	M	N	R	V+	N	N	N				C438A	0	C438B
DIM-14/1M	N	N	N	E-	E+	V-	N	M	N	R	V+	N	N	N				C438B	0	C438C
DIL-8/1P	FT	E-	E+	V-	T*	R	V+	F*										LM101AV	0	CA101AG
TO5-8/3M	FT	E-	E+	V-M	T*	R	V+	F*											0	CA101AS
TO5-8/1M	FT	E-	E+	V-M	T*	R	V+	F*									UA101AH	LM101AH*	0	CA101AT
DIL-8/1P	FT	E-	E+	V-	T*	R	V+	F*									SN52101AJP	LM101AN8	0	CA101G
TO5-8/3M	FT	E-	E+	V-	T*	R	V+	F*										CA101AS	0	CA101S
TO5-8/1M	FT	E-	E+	V-M	T*	R	V+	F*									SFC2101A	LM101H	0	CA101T
DIL-8/1P	N	E-	E+	V-	N	R	V+	N									SN52107JP	LM107J	0	CA107G
TO5-8/3M	N	E-	E+	V-M	N	R	V+	N											0	CA107S
TO5-8/1M	N	E-	E+	V-M	N	R	V+	N									SFC2107M	LM107H	0	CA107T
TO5-8/3M	F	E-	E+	V-M	N	R	V+	F*											0	CA108AS
TO5-8/1M	F	E-	E+	V-M	N	R	V+	F*									SFC2108A	LM108AH	0	CA108AT
TO5-8/3M	F	E-	E+	V-M	N	R	V+	F*											0	CA108S
TO5-8/1M	F	E-	E+	V-M	N	R	V+	F*									SFC2108M	LM108H	0	CA108T
DIL-8/1P	G	E+	E-	V-	T	T*S	R	V+									SN52111JP	UA111R	0	CA111G
TO5-8/3M	G	E+	E-	V-	T	T*S	R	V+											0	CA111S
TO5-8/1M	G	E+	E-	V-	T	T*S	R	V+									SFC2111M	LM111H	0	CA111T
DIL-14/1P	R1	E-1	E+1	V+	E+2	E-2	R2	R3	E-3	E+3	G	E+4	E-4	R4			MLM124L	LM124D	0	CA124E
DIL-14/1P	R1	E-1	E+1	V+	E+2	E-2	R2	R3	E-3	E+3	G	E+4	E-4	R4			MLM124L	LM124D	0	CA124G
DIL-14/1P	R2	R1	V+	E-1	E+1	E-2	E+2	E+3	E-3	E-4	E+4	G	R4	R3			MLM139AL	LM139AD	0	CA139AE
DIL-14/1P	R2	R1	V+	E-1	E+1	E-2	E+2	E+3	E-3	E-4	E+4	G	R4	R3			MLM139AL	LM139AD	0	CA139AG
DIL-14/1P	R2	R1	V+	E-1	E+1	E-2	E+2	E+3	E-3	E-4	E+4	G	R4	R3			MLM139L	LM139D	0	CA139E
DIL-14/1P	R2	R1	V+	E-1	E+1	E-2	E+2	E+3	E-3	E-4	E+4	G	R4	R3			MLM139L	LM139D	0	CA139G
DIL-8/1P	FT	E-	E+	V-	T*	R	V+	F*											0	CA201AG
TO5-8/3M	FT	E-	E+	V-M	T*	R	V+	F*											0	CA201AS
TO5-8/1M	FT	E-	E+	V-M	T*	R	V+	F*									UA201AH	LM201AH	0	CA201AT
DIL-8/1P	FT	E-	E+	V-	T*	R	V+	F*									SFC301ADC	LM201J	0	CA201G
TO5-8/3M	FT	E-	E+	V-M	T*	R	V+	F*										CA201AS	0	CA201S
TO5-8/1M	FT	E-	E+	V-M	T*	R	V+	F*									UA201H	LM201H	0	CA201T
DIL-8/1P	N	E-	E+	V-	N	R	V+	N										LM207J	0	CA207G
TO5-8/3M	N	E-	E+	V-M	N	R	V+	N											0	CA207S
TO5-8/1M	N	E-	E+	V-M	N	R	V+	N									SFC2207	LM207H	0	CA207T
TO5-8/3M	F	E-	E+	V-M	N	R	V+	F*											0	CA208AS
TO5-8/1M	F	E-	E+	V-M	N	R	V+	F*									SFC2208A	LM208AH	0	CA208AT
TO5-8/3M	F	E-	E+	V-M	N	R	V+	F*										CA208AS	0	CA208S
TO5-8/1M	F	E-	E+	V-M	N	R	V+	F*									SFC2208	LM208H	0	CA208T
DIL-8/1P	G	E+	E-	V-	T	T*S	R	V+										SN52111JP	0	CA211G
TO5-8/3M	G	E+	E-	V-	T	T*S	R	V+											0	CA211S
TO5-8/1M	G	E+	E-	V-	T	T*S	R	V+									SFC2211	LM211H	0	CA211T
DIL-14/1P	R1	E-1	E+1	V+	E+2	E-2	R2	R3	E-3	E+3	G	E+4	E-4	R4			SG224J	LM224D	0	CA224E
DIL-14/1P	R1	E-1	E+1	V+	E+2	E-2	R2	R3	E-3	E+3	G	E+4	E-4	R4			SG224J	LM224D	0	CA224G
DIL-14/1P	R2	R1	V+	E-1	E+1	E-2	E+2	E+3	E-3	E-4	E+4	G	R4	R3			MLM239AL	LM239AD	0	CA239AE
DIL-14/1P	R2	R1	V+	E-1	E+1	E-2	E+2	E+3	E-3	E-4	E+4	G	R4	R3			MLM239AL	LM239AD	0	CA239AG
DIL-14/1P	R2	R1	V+	E-1	E+1	E-2	E+2	E+3	E-3	E-4	E+4	G	R4	R3			MLM239L	LM239D	0	CA239E
DIL-14/1P	R2	R1	V+	E-1	E+1	E-2	E+2	E+3	E-3	E-4	E+4	G	R4	R3			MLM239L	LM239D	0	CA239G
DIL-8/1P	FT	E-	E+	V-	T*	R	V+	F*									SFC2301ADC	LM301AJ	0	CA301AE
DIL-8/1P	FT	E-	E+	V-	T*	R	V+	F*									SFC2301ADC	LM301AJ	0	CA301AG
CHP																			0	CA301AGH
CHP																			0	CA301AH
TO5-8/3M	FT	E-	E+	V-M	T*	R	V+	F*											0	CA301AS
TO5-8/1M	FT	E-	E+	V-M	T*	R	V+	F*									UA301AH	LM301AH	0	CA301AT
DIL-8/1P	N	E-	E+	V-	N	R	V+	N									UA307T	LM307J	0	CA307E
DIL-8/1P	N	E-	E+	V-	N	R	V+	N									UA307T	LM307J	0	CA307G
CHP																			0	CA307GH
CHP																			0	CA307H
TO5-8/3M	N	E-	E+	V-M	N	R	V+	N											0	CA307S
TO5-8/1M	N	E-	E+	V-M	N	R	V+	N									SFC2307	LM307H	0	CA307T
TO5-8/3M	F	E-	E+	V-M	N	R	V+	F*											0	CA308AS

TYPE NUMBER	M F R	A P P	C M P	GBP MIN	SLEW RATE MIN	V_S^+ MAX	V_S^- MAX	T_{op} MAX	A_{VOL} MIN	V_{IO} MAX	I_B MAX	I_{IO} MAX	P_{TOT} MAX	I_{OUT} MIN	V_{OUT} MIN	V_{ICM} MAX	V_{IDF} MAX	dV_{IO}/dT MAX	P_Q MAX	I_O MAX	CM RR MIN	PS RR MIN	R_{IN} MIN
CA308AS	RCU	SBA	EXT	.	.	+18V	-18V	70C	98dB	0.5MV	7NA	1NA	500MWF	1MA	13V	15V	1V	5uV/C	.	.6MA	96dB	96dB	10M
CA308AT	RCU	SBA	EXT	.	.	+18V	-18V	70C	98dB	0.5MV	7NA	1NA	500MWF	1MA	13V	15V	1V	5uV/C	.	.6MA	96dB	96dB	10M
CA308H	RCU	SBA	EXT	.	.	+18V	-18V	70C	88dB	7.5MV	7NA	1NA	.	1MA	13V	15V	1V	30uV/C	.	.6MA	80dB	80dB	10M
CA308S	RCU	SBA	EXT	.	.	+18V	-18V	70C	88dB	7.5MV	7NA	1NA	500MWF	1MA	13V	15V	1V	30uV/C	.	.6MA	80dB	80dB	10M
CA308T	RCU	SBA	EXT	.	.	+18V	-18V	70C	88dB	7.5MV	7NA	1NA	500MWF	1MA	13V	15V	1V	30uV/C	.	.6MA	80dB	80dB	10M
CA311E	RCU	CPR	EXT	.	.	+18V	-18V	70C	100dB	7.5MV	250NA	50NA	500MWF	.	.	15V	30V	.	.	8MA	.	.	.
CA311G	RCU	CPR	EXT	.	.	+18V	-18V	70C	100dB	7.5MV	250NA	50NA	500MWF	.	.	15V	30V	.	.	8MA	.	.	.
CA311H	RCU	CPR	EXT	.	.	+18V	-18V	70C	100dB	7.5MV	250NA	50NA	.	.	.	15V	30V	.	.	8MA	.	.	.
CA311S	RCU	CPR	EXT	.	.	+18V	-18V	70C	100dB	7.5MV	250NA	50NA	500MWF	.	.	15V	30V	.	.	8MA	.	.	.
CA311T	RCU	CPR	EXT	.	.	+18V	-18V	70C	100dB	7.5MV	250NA	50NA	500MWF	.	.	15V	30V	.	.	8MA	.	.	.
CA324E	RCU	QGK	INT	.	.	+16V	-16V	70C	88dB	7MV	250NA	50NA	900MWF	.	.	16V	16V	35uV/C	.	2MA	65dB	65dB	.
CA324G	RCU	QGK	INT	.	.	+16V	-16V	70C	88dB	7MV	250NA	50NA	900MWF	.	.	16V	16V	35uV/C	.	2MA	65dB	65dB	.
CA324GH	RCU	QGK	INT	.	.	+16V	-16V	70C	88dB	7MV	250NA	50NA	.	.	.	16V	16V	35uV/C	.	2MA	65dB	65dB	.
CA324H	RCU	QGK	INT	.	.	+16V	-16V	70C	88dB	7MV	250NA	50NA	.	.	.	16V	16V	35uV/C	.	2MA	65dB	65dB	.
CA324HG	RCU	QGK	INT	.	.	+16V	-16V	70C	88dB	7MV	250NA	50NA	.	.	.	16V	16V	35uV/C	.	2MA	65dB	65dB	.
CA339AE	RCU	QCP	EXT	.	.	+18V	-18V	70C	94dB	2MV	250NA	50NA	900MWF	6MA	.	18V	18V	.	.	2MA	.	.	.
CA339AG	RCU	QCP	EXT	.	.	+18V	-18V	70C	94dB	2MV	250NA	50NA	900MWF	6MA	.	18V	18V	.	.	2MA	.	.	.
CA339E	RCU	QCP	EXT	.	.	+18V	-18V	70C	88dB	5MV	250NA	50NA	900MWF	6MA	.	18V	18V	.	.	2MA	.	.	.
CA339G	RCU	QCP	EXT	.	.	+18V	-18V	70C	88dB	5MV	250NA	50NA	900MWF	6MA	.	18V	18V	.	.	2MA	.	.	.
CA339H	RCU	QCP	EXT	.	.	+18V	-18V	70C	88dB	5MV	250NA	50NA	.	6MA	.	18V	18V	.	.	2MA	.	.	.
CA339HG	RCU	QCP	EXT	.	.	+18V	-18V	70C	88dB	5MV	250NA	50NA	.	6MA	.	18V	18V	.	.	2MA	.	.	.
CA741CE	RCU	GPK	INT	.	0.2V/uS	+18V	-18V	70C	86dB	6MV	500NA	200NA	310MWF	5MA	12V	15V	30V	.	.	3MA	70dB	76dB	300K
CA741CG	RCU	GPK	INT	.	0.2V/uS	+18V	-18V	70C	86dB	6MV	500NA	200NA	310MWF	5MA	12V	15V	30V	.	.	3MA	70dB	76dB	300K
CA741CGH	RCU	GPK	INT	.	0.2V/uS	+18V	-18V	70C	86dB	6MV	500NA	200NA	.	5MA	12V	15V	30V	.	.	3MA	70dB	76dB	300K
CA741CH	RCU	GPK	INT	.	0.2V/uS	+18V	-18V	70C	86dB	6MV	500NA	200NA	.	5MA	12V	15V	30V	.	.	3MA	70dB	76dB	300K
CA741CHG	RCU	GPK	INT	.	0.2V/uS	+18V	-18V	70C	86dB	6MV	500NA	200NA	.	5MA	12V	15V	30V	.	.	3MA	70dB	76dB	300K
CA741CS	RCU	GPK	INT	.	0.2V/uS	+18V	-18V	70C	86dB	6MV	500NA	200NA	500MWF	5MA	12V	15V	30V	.	.	3MA	70dB	76dB	300K
CA741CT	RCU	GPK	INT	.	0.2V/uS	+18V	-18V	70C	86dB	6MV	500NA	200NA	500MWF	5MA	12V	15V	30V	.	.	3MA	70dB	76dB	300K
CA741E	RCU	GPK	INT	.	0.2V/uS	+22V	-22V	125C	94dB	5MV	500NA	200NA	500MWF	7MA	12V	15V	30V	.	85MW	3MA	70dB	76dB	300K
CA741G	RCU	GPK	INT	.	0.2V/uS	+22V	-22V	125C	94dB	5MV	500NA	200NA	500MWF	7MA	12V	15V	30V	.	85MW	3MA	70dB	76dB	300K
CA741L	RCU	GPK	INT	.	0.2V/uS	+22V	-22V	125C	94dB	5MV	500NA	200NA	.	7MA	12V	15V	30V	.	85MW	3MA	70dB	76dB	300K
CA741S	RCU	GPK	INT	.	0.2V/uS	+22V	-22V	125C	94dB	5MV	500NA	200NA	500MWF	7MA	12V	15V	30V	.	85MW	3MA	70dB	76dB	300K
CA741T	RCU	GPK	INT	.	0.2V/uS	+22V	-22V	125C	94dB	5MV	500NA	200NA	500MWF	7MA	12V	15V	30V	.	85MW	3MA	70dB	76dB	300K
CA747CE	RCU	DGK	INT	.	0.2V/uS	+18V	-18V	70C	88dB	6MV	500NA	200NA	800MW	5MA	12V	15V	30V	.	85MW	3MA	70dB	76dB	300K
CA747CG	RCU	DGK	INT	.	0.2V/uS	+18V	-18V	70C	88dB	6MV	500NA	200NA	800MW	5MA	12V	15V	30V	.	85MW	3MA	70dB	76dB	300K
CA747CGH	RCU	DGK	INT	.	0.2V/uS	+18V	-18V	70C	88dB	6MV	500NA	200NA	.	5MA	12V	15V	30V	.	85MW	3MA	70dB	76dB	300K
CA747CH	RCU	DGK	INT	.	0.2V/uS	+18V	-18V	70C	88dB	6MV	500NA	200NA	.	5MA	12V	15V	30V	.	85MW	3MA	70dB	76dB	300K
CA747CHG	RCU	DGK	INT	.	0.2V/uS	+18V	-18V	70C	88dB	6MV	500NA	200NA	.	5MA	12V	15V	30V	.	85MW	3MA	70dB	76dB	300K
CA747CT	RCU	DGK	INT	.	0.2V/uS	+18V	-18V	70C	88dB	6MV	500NA	200NA	800MWF	5MA	12V	15V	30V	.	85MW	3MA	70dB	76dB	300K
CA747E	RCU	DGK	INT	.	0.2V/uS	+22V	-22V	125C	94dB	5MV	500NA	200NA	800MW	5MA	12V	15V	30V	.	85MW	3MA	70dB	76dB	300K
CA747G	RCU	DGK	INT	.	0.2V/uS	+22V	-22V	125C	94dB	5MV	500NA	200NA	800MWF	5MA	12V	15V	30V	.	85MW	3MA	70dB	76dB	300K
CA747T	RCU	DGK	INT	.	0.2V/uS	+22V	-22V	125C	94dB	5MV	500NA	200NA	500MWF	5MA	12V	15V	30V	.	85MW	3MA	70dB	76dB	300K
CA748CCH	RCU	GPU	EXT	.	0.2V/uS	+18V	-18V	70C	86dB	6MV	500NA	200NA	.	5MA	12V	15V	30V	.	85MW	3MA	70dB	76dB	300K
CA748CE	RCU	GPU	EXT	.	0.2V/uS	+18V	-18V	70C	86dB	6MV	500NA	200NA	500MWF	5MA	12V	15V	30V	.	85MW	3MA	70dB	76dB	300K
CA748CG	RCU	GPU	EXT	.	0.2V/uS	+18V	-18V	70C	86dB	6MV	500NA	200NA	.	5MA	12V	15V	30V	.	85MW	3MA	70dB	76dB	300K
CA748CGH	RCU	GPU	EXT	.	0.2V/uS	+18V	-18V	70C	86dB	6MV	500NA	200NA	.	5MA	12V	15V	30V	.	85MW	3MA	70dB	76dB	300K
CA748CHG	RCU	GPU	EXT	.	0.2V/uS	+18V	-18V	70C	86dB	6MV	500NA	200NA	.	5MA	12V	15V	30V	.	85MW	3MA	70dB	76dB	300K
CA748CS	RCU	GPU	EXT	.	0.2V/uS	+18V	-18V	70C	86dB	6MV	500NA	200NA	.	5MA	12V	15V	30V	.	85MW	3MA	70dB	76dB	300K
CA748CT	RCU	GPU	EXT	.	0.2V/uS	+18V	-18V	70C	86dB	6MV	500NA	200NA	500MWF	5MA	12V	15V	30V	.	85MW	3MA	70dB	76dB	300K
CA748E	RCU	GPU	EXT	.	.25V/uS	+22V	-22V	125C	94dB	5MV	500NA	200NA	500MWF	5MA	10V	15V	30V	.	85MW	3MA	70dB	76dB	300K
CA748G	RCU	GPU	EXT	.	.25V/uS	+22V	-22V	125C	94dB	5MV	500NA	200NA	500MWF	5MA	10V	15V	30V	.	85MW	3MA	70dB	76dB	300K
CA748S	RCU	GPU	EXT	.	0.2V/uS	+22V	-22V	125C	94dB	5MV	500NA	200NA	500MWF	5MA	12V	15V	30V	.	85MW	3MA	70dB	76dB	300K
CA748T	RCU	GPU	EXT	.	0.2V/uS	+22V	-22V	125C	94dB	5MV	500NA	200NA	500MWF	5MA	12V	15V	30V	.	85MW	3MA	70dB	76dB	300K
CA1458E	RCU	DGK	INT	.5MHZ	0.3V/uS	+18V	-18V	75C	86dB	6MV	0.5uA	0.2uA	680MWF	5MA	12V	15V	30V	50uV/C	170MW	6MA	70dB	76dB	300K
CA1458G	RCU	DGK	INT	.5MHZ	0.3V/uS	+18V	-18V	75C	86dB	6MV	0.5uA	0.2uA	680MWF	5MA	12V	15V	30V	50uV/C	170MW	6MA	70dB	76dB	300K
CA1458HG	RCU	DGK	INT	.5MHZ	0.3V/uS	+18V	-18V	75C	86dB	6MV	0.5uA	0.2uA	.	5MA	12V	15V	30V	50uV/C	170MW	6MA	70dB	76dB	300K
CA1458S	RCU	DGK	INT	.5MHZ	0.3V/uS	+18V	-18V	75C	86dB	6MV	0.5uA	0.2uA	680MWF	5MA	12V	15V	30V	50uV/C	170MW	6MA	70dB	76dB	300K
CA1458T	RCU	DGK	INT	.5MHZ	0.3V/uS	+18V	-18V	75C	86dB	6MV	0.5uA	0.2uA	680MWF	5MA	12V	15V	30V	50uV/C	170MW	6MA	70dB	76dB	300K
CA1558E	RCU	DGK	INT	.5MHZ	0.3V/uS	+22V	-22V	125C	94dB	5MV	0.5uA	0.2uA	680MWF	5MA	12V	15V	30V	50uV/C	150MW	5MA	70dB	76dB	300K
CA1558G	RCU	DGK	INT	.5MHZ	0.3V/uS	+22V	-22V	125C	94dB	5MV	0.5uA	0.2uA	680MWF	5MA	12V	15V	30V	50uV/C	150MW	5MA	70dB	76dB	300K

detailed explanations of
umn heading notations, see
p. A.

so for ready references the
ore important abbreviations
ed in the column headings are
ted below:

FT HAND PAGE

- PP = application (codes at APP.E.)
- MRR = common mode rejection ratio
- MP = compensation (frequency)
- $/dT$ = input offset voltage temperature drift
- BP = gain bandwidth product
- = input bias current
- = input bias offset current
- = quiescent supply current
- MFR = manufacturer (codes at App.C.)
- = quiescent power consumer
- SRR = power supply rejection ratio
- ICM = common mode input voltage rating
- IOF = differential input voltage rating
- IO = input offset voltage
- S = dc supply voltage

RIGHT HAND PAGE

Lead out coding summary
details at APP.G.) for different
cases (APP.F.)

- A = gain adjust
- B = bias adjust
- C = case
- E— = inverting input
- E+ = non-inverting input
- F,F* = input frequency compensation
- G = ground
- J = high level input
- K = output, open collector
- L = output, open emitter
- M = metal case
- N = not connected
- Q = special terminal
- R,R* = outputs
- S = strobe
- T,T* = offset balance
- V+ = +ve dc supply
- V— = —ve dc supply
- W = guard ring
- X = blank position, no lead
- + + = +ve supplementary dc supply
- — — = —ve supplementary dc supply
- ◊,◊* = output frequency compensation

CASE (APP F)	LD 1	LD 2	LD 3	LD 4	LD 5	LD 6	LD 7	LD 8	LD 9	LD 10	LD 11	LD 12	LD 13	LD 14	LD 15	LD 16	EUROPE SUBSTITUTE	USA SUBSTITUTE	IS	TYPE NUMBER
TO5-8/1M	F	E-	E+	V-M	N	R	V+	F*									SFC2308A	LM308AH	O	CA308AT
CHP																			O	CA308H
TO5-8/3M	F	E-	E+	V-M	N	R	V+	F*										CA308AS	O	CA308S
TO5-8/1M	F	E-	E+	V-M	N	R	V+	F*									SFC2308	LM308H	O	CA308T
DIL-8/1P	G	E+	E-	V-	T	T*S	R	V+									UA311R	LM311N	O	CA311E
DIL-8/1P	G	E+	E-	V-	T	T*S	R	V+									SFC2311DC	LM311N	O	CA311G
CHP																			O	CA311H
																			O	CA311S
TO5-8/1M	G	E+	E-	V-	T	T*S	R	V+									SFC2311	LM311H	O	CA311T
DIL-14/1P	R1	E-1	E+1	V+	E+2	E-2	R2	R3	E-3	E+3	G	E+4	E-4	R4			MLM324J	LM324N	O	CA324E
DIL-14/1P	R1	E-1	E+1	V+	E+2	E-2	R2	R3	E-3	E+3	G	E+4	E-4	R4			MLM324J	LM324N	O	CA324G
CHP																			O	CA324GH
CHP																			O	CA324H
CHP																			O	CA324HG
DIL-14/1P	R2	R1	V+	E-1	E+1	E-2	E+2	E+3	E-3	E-4	E+4	G	R4	R3			MLM339AL	LM339AD	O	CA339AE
DIL-14/1P	R2	R1	V+	E-1	E+1	E-2	E+2	E+3	E-3	E-4	E+4	G	R4	R3			MLM339AL	LM339AD	O	CA339AG
DIL-14/1P	R2	R1	V+	E-1	E+1	E-2	E+2	E+3	E-3	E-4	E+4	G	R4	R3			MLM339L	LM339D	O	CA339E
DIL-14/1P	R2	R1	V+	E-1	E+1	E-2	E+2	E+3	E-3	E-4	E+4	G	R4	R3			MLM339L	LM339D	O	CA339G
CHP																			O	CA339H
CHP																			O	CA339HG
DIL-8/1P	T	E-	E+	V-	T*	R	V+	N									TBA221B	UA741TC	O	CA741CE
DIL-8/1P	T	E-	E+	V-	T*	R	V+	N									TBA221B	UA741TC	O	CA741CG
CHP																			O	CA741CGH
CHP																			O	CA741CH
CHP																			O	CA741CHG
TO5-8/3M	T	E-	E+	V-M	T*	R	V+	N											O	CA741CS
TO5-8/1M	T	E-	E+	V-M	T*	R	V+	N									TBA221	UA741HC	O	CA741CT
DIL-8/1P	T	E-	E+	V-	T*	R	V+	N											O	CA741E
DIL-8/1P	T	E-	E+	V-	T*	R	V+	N											O	CA741G
BML																			O	CA741L
TO5-8/3M	T	E-	E+	V-M	T*	R	V+	N									TBA222	UA741HM	O	CA741S
TO5-8/1M	T	E-	E+	V-M	T*	R	V+	N											O	CA741T
DIL-14/1P	E-1	E+1	T1	V-	T2	E+2	E-2	T*2	V+2	R2	N	R1	V+1	T*1			TBB0747A	UA747DC	O	CA747CE
DIL-14/1P	E-1	E+1	T1	V-	T2	E+2	E-2	T*2	V+2	R2	N	R1	V+1	T*1			TBB0747A	UA747DC	O	CA747CG
CHP																			O	CA747CGH
CHP																			O	CA747CH
CHP																			O	CA747CHG
TO5-10/1M	R1	V+1	E-1	E+1	V-	E+2	E-2	V+2	R2	N							TBB0747	UA747HC	O	CA747CT
DIL-14/1P	E-1	E+1	T1	V-	T2	E+2	E-2	T*2	V+2	R2	N	R1	V+	T*1			SFC2747KM	UA747DM	O	CA747E
DIL-14/1P	E-1	E+1	T1	V-	T2	E+2	E-2	T*2	V+2	R2	N	R1	V+	T*1			SFC2747KM	UA747DM	O	CA747G
TO5-10/1M	R1	V+1	E-1	E+1	V-	E+2	E-2	V+2	R2	N							SFC2747M	UA747HM	O	CA747T
CHP																			O	CA748CCH
DIL-8/1P	FT	E-	E+	V-	T*	R	V+	F*									TBB0748B	UA748TC	O	CA748CE
DIL-8/1P	FT	E-	E+	V-	T*	R	V+	F*									TBB0748B	UA748TC	O	CA748CG
CHP																		CA748CGH	O	CA748CGH
CHP																			O	CA748CH
CHP																			O	CA748CHG
CHP																			O	CA748CS
TO5-8/3M	FT	E-	E+	V-	T*	R	V+	F*									TBB0748	UA748HC	O	CA748CT
TO5-8/1M	FT	E-	E+	V-	T*	R	V+	F*									SN52748JP	LM748J	O	CA748E
DIL-8/1P	FT	E-	E+	V-	T*	R	V+	F*									SN52748JP	LM748J	O	CA748G
DIL-8/1C	FT	E-	E+	V-	T*	R	V+	F*											O	CA748S
TO5-8/3M	FT	E-	E+	V-	T*	R	V+	F*									TBC0748	UA748HM	O	CA748T
DIL-8/1P	R1	E-1	E+1	V-	E+2	E-2	R2	V+									TBB1458B	MC1458U	O	CA1458E
DIL-8/1P	R1	E-1	E+1	V-	E+2	E-2	R2	V+									TBB1458B	MC1458U	O	CA1458G
CHP																			O	CA1458HG
TO5-8/3M	R1	E-1	E+1	V-	E+2	E-2	R2	V+									TBB1458	MC1458G	O	CA1458S
TO5-8/1M	R1	E-1	E+1	V-	E+2	E-2	R2	V+									TBB1458	MC1458G	O	CA1458T
DIL-8/1P	R1	E-1	E+1	V-	E+2	E-2	R2	V+									LM1558J	MC1558U	O	CA1558E
DIL-8/1P	R1	E-1	E+1	V-	E+2	E-2	R2	V+									LM1558J	MC1558U	O	CA1558G
TO5-8/3M	R1	E-1	E+1	V-	E+2	E-2	R2	V+											O	CA1558S

TYPE NUMBER	MFR	APP	CMP	GBP MIN	SLEW RATE MIN	V_S^+ MAX	V_S^- MAX	T_{op} MAX	A_{VOL} MIN	V_{IO} MAX	I_B MAX	I_{IO} MAX	P_{TOT} MAX	I_{OUT} MIN	V_{OUT} MIN	V_{ICM} MAX	V_{IDF} MAX	dV_{IO}/dT MAX	P_Q MAX	I_Q MAX	CMRR MIN	PSRR MIN	R_{IN} MIN
CA1558S	RCU	DGK	INT	.5MHZ	0.3V/uS	+22V	-22V	125C	94dB	5MV	0.5uA	0.2uA	680MWF	5MA	12V	15V	30V	50uV/C	150MW	5MA	70dB	76dB	300K
CA1558T	RCU	DGK	INT	.5MHZ	0.3V/uS	+22V	-22V	125C	94dB	5MV	0.5uA	0.2uA	680MWF	5MA	12V	15V	30V	50uV/C	150MW	5MA	70dB	76dB	300K
CA3000H	RCU	BDO	EXT			+8V	-8V	125C	28dB	5MV	36uA	10uA			3V	2V	4V		60MW		70dB		70K
CA3000	RCU	BDO	EXT			+8V	-8V	125C	28dB	5MV	36uA	10uA	450MWF		3V	2V	4V		60MW		70dB		70K
CA3001	RCU	BDO	EXT	.1GHZ		+10V	-10V	125C	16dB	7MV	36uA	10uA	450MWF	40MA	2V	2.5V	4V				70dB		
CA3001H	RCU	BDO	EXT	.1GHZ		+10V	-10V	125C	16dB	.7MV	36uA	10uA		40MA	2V	2.5V	4V				70dB		
CA3005	RCU	BDO	EXT			+12V	-12V	125C	16dB	5MV	40uA	10uA	300MWF			2.5V	3.5V		50MW		90dB		
CA3005H	RCU	BDO	EXT			+12V	-12V	125C	16dB	5MV	40uA	10uA				2.5V	3.5V		50MW		90dB		
CA3006	RCU	BDO	EXT			+12V	-12V	125C	16dB	1MV	40uA	10uA	300MWF			2.5V	3.5V		50MW		90dB		
CA3008	RCU	GPU	EXT			+10V	-10V	125C	57dB	5MV	12uA	5uA	300MWF	10MA	2V	0.5V	5V		60MW		70dB	60dB	10K
CA3008A	RCU	GPU	EXT		1V/uS	+10V	-10V	125C	57dB	2MV	4uA	1.5uA	600MWF		2V	0.5V	5V		80MW		70dB	60dB	15K
CA3010	RCU	GPU	EXT			+10V	-10V	125C	57dB	5MV	12uA	5uA	300MWF		2V	0.5V	5V		60MW		70dB	60dB	10K
CA3010A	RCU	GPU	EXT		1V/uS	+10V	-10V	125C	57dB	2MV	4uA	1.5uA	600MWF		2V	0.5V	5V		80MW		70dB	60dB	15K
CA3015	RCU	GPU	EXT			+20V	-20V	125C	66dB	5MV	24uA	5uA	600MWF	10MA	6V	0.5V	5V		250MW		80dB	66dB	5K
CA3015A	RCU	GPU	EXT		2V/uS	+20V	-20V	125C	66dB	2MV	6uA	1.6uA	600MWF		6V	0.5V	5V		250MW		80dB	60dB	7.5K
CA3015H	RCU	GPU	EXT			+20V	-20V	125C	66dB	5MV	24uA	5uA		10MA	6V	0.5V	5V		250MW		80dB	66dB	5K
CA3015L	RCU	GPU	EXT			+20V	-20V	125C	66dB	5MV	24uA	5uA		10MA	6V	0.5V	5V		250MW		80dB	66dB	5K
CA3016	RCU	GPU	EXT			+20V	-20V	125C	66dB	5MV	25uA	5uA	600MWF	10MA	6V	0.5V	5V		250MW		80dB	66dB	5K
CA3016A	RCU	GPU	EXT		2V/uS	+20V	-20V	125C	66dB	2MV	6uA	1.6uA	600MWF		6V	0.5V	5V		250MW		80dB	60dB	7.5K
CA3026	RCU	DBD	EXT	.2GHZ		+10V	-10V	125C	22dB	5MV	24uA	2uA	600MWF					5uV/C			90dB		
CA3026H	RCU	DBD	EXT	.2GHZ		+10V	-10V	125C	22dB	5MV	24uA	2uA						5uV/C			90dB		
CA3029	RCU	GPU	EXT			+10V	-10V	85C	57dB	5MV	12uA	5uA	300MWF	10MA	2V	0.5V	5V		60MW		70dB	60dB	10K
CA3029A	RCU	GPU	EXT		1V/uS	+10V	-10V	80C	57dB	2MV	4uA	1.5uA	300MWF		2V	0.5V	5V		80MW		70dB	60dB	15K
CA3030	RCU	GPU	EXT			+20V	-20V	85C	66dB	5MV	24uA	5uA	600MWF	10MA	6V	0.5V	5V		250MW		80dB	66dB	5K
CA3030A	RCU	GPU	EXT		2V/uS	+20V	-20V	80C	66dB	2MV	6uA	1.6uA	300MWF		6V	0.5V	5V		250MW		80dB	60dB	7.5K
CA3031/702A	OBS	WBA	EXT		0.5V/uS	+14V	-7V	125C	68dB	2MV	5uA	0.5uA	300MWF	.3MA	5V	1.5V	5V	10uV/C	120MW	7MA	80dB	74dB	16K
CA3032/702C	OBS	WBA	EXT		0.5V/uS	+14V	-7V	70C	60dB	10MV	15uA	5uA	300MWF	.3MA	5V	1.5V	5V	20uV/C	125MW	7MA	65dB	70dB	6K
CA3033	RCU	HCO	EXT	50MHZ	1V/uS	+26V	-26V	125C	84dB	5MV	350NA	35NA	1.2WF	17MA	9V	3.5V	10V	30uV/C	180MW		84dB	66dB	
CA3033A	RCU	HCO	EXT	75MHZ	1V/uS	+26V	-26V	125C	87dB	5MV	180NA	25NA	1.2WF	38MA	11V	4.7V	10V	30uV/C	300MW		93dB	66dB	
CA3033H	RCU	HCO	EXT	50MHZ	1V/uS	+26V	-26V	125C	84dB	5MV	350NA	35NA		17MA	9V	3.5V	10V	30uV/C	180MW		84dB	66dB	
CA3037	RCU	GPU	EXT			+10V	-10V	125C	57dB	5MV	12uA	5uA	300MWF	10MA	2V	0.5V	5V		60MW		70dB	60dB	10K
CA3037A	RCU	GPU	EXT		1V/uS	+10V	-10V	125C	57dB	2MV	4uA	1.5uA	600MWF		2V	0.5V	5V		80MW		70dB	60dB	15K
CA3038	RCU	GPU	EXT			+20V	-20V	125C	66dB	5MV	24uA	5uA	600MWF	10MA	6V	0.5V	1V		250MW		80dB	66dB	5K
CA3038A	RCU	GPU	EXT		2V/uS	+20V	-20V	125C	66dB	2MV	6uA	1.6uA	600MWF		6V	0.5V	5V		250MW		80dB	60dB	7.5K
CA3047	RCU	HCO	EXT	50MHZ	1V/uS	+26V	-26V	70C	84dB	5MV	350NA	35NA	750MWF	17MA	9V	3.5V	10V	30uV/C	180MW		84dB	66dB	
CA3047A	RCU	HCO	EXT	75MHZ	1V/uS	+26V	-26V	70C	87dB	5MV	180NA	25NA	750MWF	38MA	11V	4.7V	10V	30uV/C	300MW		93dB	66dB	
CA3049H	RCU	DBD	EXT	.5MHZ		+10V	-10V	125C	18dB	5MV	33uA	3uA						10uV/C			90dB		
CA3049L	RCU	DBD	EXT	.5MHZ		+10V	-10V	125C	18dB	5MV	33uA	3uA						10uV/C			90dB		
CA3049T	RCU	DDD	EXT	.5MHZ		+10V	-10V	125C	18dB	5MV	33uA	3uA	600MWF					10uV/C			90dB		
CA3054	RCU	DBD	EXT	.2GHZ		+10V	-10V	125C	22dB	5MV	24uA	2uA	750MWF					5uV/C			90dB		
CA3054H	RCU	DBD	EXT	.2GHZ		+10V	-10V	125C	22dB	5MV	24uA	2uA						5uV/C			90dB		
CA3054L	RCU	DBD	EXT	.2GHZ		+10V	-10V	125C	22dB	5MV	24uA	2uA						5uV/dT			90dB		
CA3060AD	RCU	TOT	EXT		2V/uS	+18V	-18V	125C		5MV	5uA	1uA	490MWF	.2MA	12V	18V	5V		36MW	2MA	70dB	78dB	10K
CA3060BD	RCU	TOT	EXT		2V/uS	+18V	-18V	125C		5MV	5uA	1uA	490MWF	2MA	12V	18V	5V		36MW	2MA	70dB	78dB	10K
CA3060D	RCU	TOT	EXT		2V/uS	+7V	-7V	125C		5MV	5uA	1uA	490MWF	.2MA	4.5V	7V	5V		15MW	2MA	70dB	78dB	10K
CA3060E	RCU	TOT	EXT		2V/uS	+18V	-18V	85C		5MV	5uA	1uA	490MWF	.2MA	4.5V	18V	5V		15MW	2MA	70dB	78dB	10K
CA3060H	RCU	TOT	EXT		2V/uS	+18V	-18V	125C		5MV	5uA	1uA		.2MA	4.5V	18V	5V		15MW	2MA	70dB	78dB	10K
CA3078AS	RCU	PRA	EXT		.01V/uS	+18V	-18V	125C	92dB	3.5MV	12NA	2.5NA	50MWF	.5MA	5.1V	18V	6V	30uV/C		.2MA	80dB	76dB	0.3M
CA3078AT	RCU	PRA	EXT		.01V/uS	+18V	-18V	125C	92dB	3.5MV	12NA	2.5NA	50MWF	.5MA	5.1V	18V	6V	30uV/C		.2MA	80dB	76dB	0.3M
CA3078H	RCU	PRA	EXT		.01V/uS	+7V	-7V	125C	88dB	4.5MV	170NA	32NA		.5MA	5.1V	7V	6V	30uV/C		.2MA	80dB	76dB	.15M
CA3078S	RCU	PRA	EXT		.01V/uS	+7V	-7V	70C	88dB	4.5MV	170NA	32NA	500MWF	.5MA	5.1V	7V	6V	30uV/C		.2MA	80dB	76dB	.15M
CA3078T	RCU	PRA	EXT		.01V/uS	+7V	-7V	70C	88dB	4.5MV	170NA	32NA	500MWF	.5MA	5.1V	7V	6V	30uV/C		.2MA	80dB	76dB	.15M
CA3080	RCU	OTA	INT		25V/uS	+18V	-18V	70C		5MV	5uA	0.6uA	125MWF	.3MA	12V	18V	5V		36MW	2MA	80dB	76dB	10K
CA3080A	RCU	OTA	INT		25V/uS	+18V	-18V	125C		2MV	5uA	0.6uA	125MWF	.3MA	12V	18V	5V		36MW	2MA	80dB	76dB	10K
CA3080AS	RCU	OTA	INT		25V/uS	+18V	-18V	125C		2MV	5uA	0.6uA	125MWF	.3MA	12V	18V	5V		36MW	2MA	80dB	76dB	10K
CA3080E	RCU	OTA	INT		25V/uS	+18V	-18V	70C		5MV	5uA	0.6uA	125MWF	.3MA	12V	18V	5V		36MW	2MA	80dB	76dB	10K
CA3080H	RCU	OTA	INT		25V/uS	+18V	-18V	125C		5MV	5uA	0.6uA		.3MA	12V	18V	5V		36MW	2MA	80dB	76dB	10K
CA3080S	RCU	OTA	INT		25V/uS	+18V	-18V	70C		5MV	5uA	0.6uA	125MWF	.3MA	12V	18V	5V		36MW	2MA	80dB	76dB	10K
CA3094AF	RCU	PRA	EXT	15MHZ	0.2V/uS	+18V	-18V	85C	86dB	5MV	0.5uA	0.2uA	500MWF	50MA	11V	18V	5V	20uV/C	12MW		70dB	76dB	50K
CA3094AS	RCU	PRA	EXT	15MHZ	0.2V/uS	+18V	-18V	125C	86dB	5MV	0.5uA	0.2uA	630MWF	50MA	11V	18V	5V	20uV/C	12MW		70dB	76dB	50K

68

For detailed explanations of column heading notations, see App. A.

Also for ready references the more important abbreviations used in the column headings are listed below:

LEFT HAND PAGE

APP = application (codes at APP.E.)
CMRR = common mode rejection ratio
CMP = compensation (frequency)
dV_{in}/dT = input offset voltage temperature drift
GBP = gain bandwidth product
I_d = input bias current
I_{io} = input bias offset current
I_q = quiescent supply current
MFR = manufacturer (codes at App.C.)
P_q = quiescent power consumer
PSRR = power supply rejection ratio
V_{icm} = common mode input voltage rating
V_{ni} = differential input voltage rating
V_{io} = input offset voltage
V_s = dc supply voltage

RIGHT HAND PAGE

Lead out coding summary (details at APP.G.) for different cases (APP F.)

A = gain adjust
B = bias adjust
C = case
E− = inverting input
E+ = non inverting input
F,F* = input frequency compensation
G = ground
J = high level input
K = output, open collector
L = output, open emitter
M = metal case
N = not connected
Q = special terminal
R,R* = outputs
S = strobe
T,T* = offset balance
V+ = +ve dc supply
V− = −ve dc supply
W = guard ring
X = blank position, no lead
++ = +ve supplementary dc supply
−− = −ve supplementary dc supply
∮,∮* = output frequency compensation

CASE (APP F)	LD 1	LD 2	LD 3	LD 4	LD 5	LD 6	LD 7	LD 8	LD 9	LD 10	LD 11	LD 12	LD 13	LD 14	LD 15	LD 16	EUROPE SUBSTITUTE	USA SUBSTITUTE	IS	TYPE NUMBER
TO5-8/1M	R1	E-1	E+1	V-	E+2	E-2	R2	V+									TBC1458	MC1558G	O	CA1558T
CHP																			O	CA3000H
TO5-10/1M	E+	B	V-	B2	B1	E-	N	R*	V+	R									O	CA3000
TO5-12/1M	E-	B	V-M	B2	B1	E+	N	K*	V+	G	K	N							O	CA3001
CHP																			O	CA3001H
TO5-12/1M	E-	Q1	B	B1	B2	Q2	N	V-	E+	K*	K	B*						CA3006	O	CA3005
CHP																			O	CA3005H
TO5-12/1M	E-	Q1	B	B1	B2	Q2	N	V-	E+	K*	K	B*							O	CA3006
FLP-14/3C	F1	G	E-	E+	N	V-	N	Q	F2	F*2	ø	L	V+	F*1				CA3008A	O	CA3008
FLP-14/3C	F1	G	E-	E+	N	V-	N.	Q	F2	F*2	ø		V+	F*1					O	CA3008A
TO5-12/1M	G	E+	E-	V-	Q	F2	F*2	ø	L	V+	F*1	F1						CA3010	O	CA3010
TO5-12/1M	G	E+	E-	V-	Q	F2	F*2	ø	L	V+	F*1	F1							O	CA3010A
TO5-12/1M	G	E+	E-	V-	Q	F2	F*2	ø	L	V+	F*1	F1						CA3015A	O	CA3015
TO5-12/1M	G	E+	E-	V-	C	F2	F*2	ø	L	V+	F*1	F1							O	CA3015A
CHP																			O	CA3015H
BML																			O	CA3015L
FLP-14/3C	F1	G	E-	E+	N	V-	N	Q	F2	F*2	ø	L	V+	F*1				CA3016A	O	CA3016
FLP-14/3C	F1	G	E-	E+	N	V-	N	Q	F2	F*2	ø	L	V+	F*1					O	CA3016A
TO5-12/1M	E-1	B1	V-1	E+2	K*2	K2	E-2	B2	V-2	E+1	K*1	K1							O	CA3026
CHP																			O	CA3026H
DIL-14/1P	F1	G	E-	E+	N	V-	N	Q	F2	F*2	ø	L	V+	F*1				CA3029A	O	CA3029
DIL-14/1P	F1	G	E-	E+	N	V-	N	Q	F2	F*2	ø	L	V+	F*1					O	CA3029A
DIL-14/1P	F1	G	E-	E+	N	V-	N	Q	F2	F*2	ø	L	V+	F*1				CA3030A	O	CA3030
DIL-14/1P	F1	G	E-	E+	N	V-	N	Q	F2	F*2	ø	L	V+	F*1					O	CA3030A
TO5-8/1M	G	E-	E+	V-M	F	ø	R	V+									UA702AHM	SN52702AL	O	CA3031/702A
TO5-8/1M	G	E-	E+	V-M	F	ø	R	V+									UA702HC	SN72702L	O	CA3032/702C
DIL-14/1C	ø*	G	R	ø	T	V+	T*	F	J+	E+	E-	J-	F*	V-				CA3033A	O	CA3033
DIL-14/1C	ø*	G	R	ø	T	V+	T*	F	J+	E+	E-	J-	F*	V-					O	CA3033A
CHP																			O	CA3033H
DIL-14/1C	F1	G	E-	E+	N	V-	N	Q	F2	F*2	ø	L	V+	F*1				CA3037A	O	CA3037
DIL-14/1C	F1	G	E-	E+	N	V-	N	Q	F2	F*2	ø	L	V+	F*1					O	CA3037A
DIL-14/1C	F+	G	E-	E+	N	V-	N	Q	F2	F*2	ø	L	V+	F*1				CA3038A	O	CA3038
DIL-14/1C	F1	G	E-	E+	N	V-	N	Q	F2	F*2	ø	L	V+	F*1					O	CA3038A
DIL-14/1P	ø*	G	R	ø	T	V+	T*	F	J+	E+	E-	J-	F*	V-				CA3047A	O	CA3047
DIL-14/1P	ø*	G	R	ø	T	V+	T*	F	J+	E+	E-	J-	F*	V-					O	CA3047A
CHP																			O	CA3049H
BML																			O	CA3049L
TO5-12/1M	E-1	B1	V-1	E+2	K*2	K2	E-2	B2	V-M	E+1	K*1	K1							O	CA3049T
DIL-14/1P	K1	E-1	B1	V-1	Q	E+2	K*2	K2	E-2	N	B2	V-2	E+1	K*1					O	CA3054
CHP																			O	CA3054H
BML																			O	CA3054L
DIL-16/1C	Q	Q*	V+	E-3	E+3	B3	K3	V-	K2	B2	E+2	E-2	E-1	E+1	B1	K1		CA3060BD	O	CA3060D
DIL-16/1C	Q	Q*	V+	E-3	E+3	B3	K3	V-	K2	B2	E+2	E-2	E-1	E+1	B1	K1			O	CA3060BD
DIL-16/1C	Q	Q*	V+	E-3	E+3	B3	K3	V-	K2	B2	E+2	E-2	E-1	E+1	B1	K1		CA3060AD	O	CA3060D
DIL-16/1P	Q	Q*	V+	E-3	E+3	B3	K3	V-	K2	B2	E+2	E-2	E-1	E+1	B1	K1		CA3060AD	O	CA3060E
CHP																			O	CA3060H
TO5-8/3M	ø	E-	E+	V-	B	R	V+	ø*											O	CA3078AS
TO5-8/1M	ø	E-	E+	V-	B	R	V+	ø*											O	CA3078AT
CHP																			O	CA3078H
TO5-8/3M	ø	E-	E+	V-	B	R	V+	ø*										CA3078AS	O	CA3078S
TO5-8/1M	ø	E-	E+	V-	B	R	V+	ø*										CA3078AT	O	CA3078T
TO5-8/1M	N	E-	E+	V-M	B	R	V+	N										CA3080A	O	CA3080
TO5-8/1M	N	E-	E+	V-M	B	R	V+	N											O	CA3080A
TO5-8/3M	N	E-	E+	V-M	B	R	V+	N											O	CA3080AS
DIL-8/1P	N	E-	E+	V-	B	R	V+	N											O	CA3080E
CHP																			O	CA3080H
TO5-8/3M	N	E-	E+	V-M	B	R	V+	N										CA3080AS	O	CA3080S
DIL-8/1P	FS	E-	E+	V-	B	L	V+	K											O	CA3094AE
TO5-8/3M	FS	E-	E+	V-M	B	L	V+	K										CA3094BS	O	CA3094AS
TO5-8/1M	FS	E-	E+	V-M	B	L	V+	K										CA3094BT	O	CA3094AT

TYPE NUMBER	MFR	APP	CMP	GBP MIN	SLEW RATE MIN	V_S^+ MAX	V_S^- MAX	T_{OP} MAX	A_{VOL} MIN	V_{IO} MAX	I_B MAX	I_{IO} MAX	P_{TOT} MAX	I_{OUT} MIN	V_{OUT} MIN	V_{ICM} MAX	V_{IDF} MAX	dV_{IO}/dT MAX	P_O MAX	I_O MAX	CM RR MIN	PS RR MIN	R_{IN} MIN
CA3094AT	RCU	PRA	EXT	15MHZ	0.2V/uS	+18V	-18V	125C	86dB	5MV	0.5uA	0.2uA	630MWF	50MA	11V	18V	5V	20uV/C	12MW	.	70dB	76dB	5C
CA3094BS	RCU	PRA	EXT	15MHZ	0.2V/uS	+22V	-22V	125C	86dB	5MV	0.5uA	0.2uA	630MWF	50MA	11V	22V	5V	20uV/C	12MW	.	70dB	76dB	5C
CA3094BT	RCU	PRA	EXT	15MHZ	0.2V/uS	+22V	-22V	125C	86dB	5MV	0.5uA	0.2uA	630MWF	50MA	11V	22V	5V	20uV/C	12MW	.	70dB	76dB	5C
CA3094E	RCU	PRA	EXT	15MHZ	0.2V/uS	+12V	-12V	85C	86dB	5MV	0.5uA	0.2uA	500MWF	50MA	11V	12V	5V	20uV/C	12MW	.	70dB	76dB	5C
CA3094H	RCU	PRA	EXT	15MHZ	0.2V/uS	+12V	-12V	125C	86dB	5MV	0.5uA	0.2uA		50MA	11V	12V	5V	20uV/C	12MW	.	70dB	76dB	5C
CA3094S	RCU	PRA	EXT	15MHZ	0.2V/uS	+12V	-12V	125C	86dB	5MV	0.5uA	0.2uA	630MWF	50MA	11V	12V	5V	20uV/C	12MW	.	70dB	76dB	5C
CA3094T	RCU	PRA	EXT	15MHZ	0.2V/uS	+12V	-12V	125C	86dB	5MV	0.5uA	0.2uA	630MWF	50MA	11V	12V	5V	20uV/C	12MW	.	70dB	76dB	5C
CA1458GH	RCU	DGK	INT	.5MHZ	0.3V/uS	+18V	-18V	75C	86dB	6MV	0.5uA	0.2uA		5MA	12V	15V	30V	50uV/C	170MW	.	70dB	76dB	300
CA1458	RCU	DGK	INT	.5MHZ	0.3V/uS	+18V	-18V	125C	86dB	6MV	0.5uA	0.2uA		5MA	12V	15V	30V	50uV/C	170MW	.	70dB	76dB	300
CA3100H	RCU	WBA	EXT	10MHZ	8V/uS	+18V	-18V	125C	56dB	5MV	2uA	0.4uA		15MA	9V	15V	12V	.	.	11MA	76dB	60dB	10
CA3100T	RCU	WBA	EXT	10MHZ	8V/uS	+18V	-18V	125C	56dB	5MV	2uA	0.4uA	630MWF	15MA	9V	15V	12V	.	.	11MA	76dB	60dB	10
CA3100S	RCU	WBA	EXT	10MHZ	8V/uS	+18V	-18V	125C	56dB	5MV	2uA	0.4uA	630MWF	15MA	9V	15V	12V	.	.	11MA	76dB	60dB	10
CA3102E	RCU	DBD	EXT	.5MHZ	.	+10V	-10V	125C	18dB	5MV	33uA	3uA	750MWF	.	.	.	.	10uV/C	.	.	90dB	.	.
CA3102H	RCU	DBD	EXT	.5MHZ	.	+10V	-10V	125C	18dB	5MV	33uA	3uA		.	.	.	.	10uV/C	.	.	90dB	.	.
CA3130AS	RCU	FET	EXT	5MHZ	10V/uS	+8V	-8V	125C	94dB	5MV	40pA	20pA	630MWF	12MA	12V	10V	.	30uV/C	.	15MA	70dB	76dB	0.5
CA3130AT	RCU	FET	EXT	5MHZ	10V/uS	+8V	-8V	125C	94dB	5MV	40pA	20pA	630MWF	12MA	12V	10V	.	30uV/C	.	15MA	70dB	76dB	0.5
CA3130BS	RCU	FET	EXT	5MHZ	10V/uS	+8V	-8V	125C	100dB	2MV	20pA	10pA	630MWF	12MA	12V	10V	.	15uV/C	.	15MA	86dB	80dB	0.5
CA3130BT	RCU	FET	EXT	5MHZ	10V/uS	+8V	-8V	125C	100dB	2MV	20pA	10pA	630MWF	12MA	12V	10V	.	15uV/C	.	15MA	86dB	80dB	0.5
CA3130S	RCU	FET	EXT	5MHZ	10V/uS	+8V	-8V	125C	94dB	15MV	50pA	30pA	630MWF	12MA	12V	10V	.	30uV/C	.	15MA	70dB	76dB	0.5
CA3130T	RCU	FET	EXT	5MHZ	10V/uS	+8V	-8V	125C	94dB	15MV	50pA	30pA	630MWF	12MA	12V	10V	.	30uV/C	.	15MA	70dB	76dB	0.5
CA3140AS	RCU	FET	INT	3M7HZ	7V/uS	+18V	-18V	125C	86dB	5MV	40pA	20pA	630MWF	18MA	12V	18V	8V	.	180MW	6MA	70dB	76dB	0.5
CA3140AT	RCU	FET	INT	3M7HZ	7V/uS	+18V	-18V	125C	86dB	5MV	40pA	20pA	630MWF	18MA	12V	18V	8V	.	180MW	6MA	70dB	76dB	0.5
CA3140BS	RCU	FET	INT	3M7HZ	7V/uS	+22V	-22V	125C	94dB	2MV	30pA	10pA	630MWF	18MA	12V	22V	8V	.	180MW	6MA	86dB	80dB	0.5
CA3140BT	RCU	FET	INT	3M7HZ	7V/uS	+22V	-22V	125C	94dB	2MV	30pA	10pA	630MWF	18MA	12V	22V	8V	.	180MW	6MA	86dB	80dB	0.5
CA3140H	RCU	FET	INT	3M7HZ	7V/uS	+18V	-18V	125C	86dB	15MV	50pA	30pA		18MA	12V	18V	8V	.	180MW	6MA	70dB	76dB	0.5
CA3140S	RCU	FET	INT	3M7HZ	7V/uS	+18V	-18V	125C	86dB	15MV	50pA	30pA	630MWF	18MA	12V	18V	8V	.	180MW	6MA	70dB	76dB	0.5
CA3140T	RCU	FET	INT	3M7HZ	7V/uS	+18V	-18V	125C	86dB	15MV	50pA	30pA	630MWF	18MA	12V	18V	8V	.	180MW	6MA	70dB	76dB	0.5
CA3401E	RCU	QCD	INT	2MHZ	.15V/uS	+18V	-18V	125C	60dB	.	300NA	.	625MWF	5MA	5V	.	.	.	.	14MA	50dB	100	
CA3401G	RCU	QCD	INT	2MHZ	.15V/uS	+18V	-18V	125C	60dB	.	300NA	.	625MWF	5MA	5V	.	.	.	.	14MA	50dB	100	
CA3401H	RCU	QCD	INT	2MHZ	.15V/uS	+18V	-18V	125C	60dB	.	300NA	.		5MA	5V	.	.	.	.	14MA	50dB	100	
CA3401HG	RCU	QGU	INT	2MHZ	.15V/uS	+18V	-18V	125C	60dB	.	300NA	.		5MA	5V	.	.	.	.	14MA	50dB	100	
CA6078AS	RCU	PRA	EXT	.	.	+18V	-18V	125C	92dB	3.5MV	12NA	2.5NA	250MWF	5MA	13V	18V	6V	.	.	25UA	80dB		
CA6078AT	RCU	PRA	EXT	.	.	+18V	-18V	125C	92dB	3.5MV	12NA	2.5NA	250MWF	5MA	13V	18V	6V	.	.	25UA	80dB		
CA6741T	RCU	LNA	INT	.	.	+22V	-22V	125C	94dB	5MV	500NA	200NA	500MWF	5MA	12V	15V	30V	.	.	3MA	70dB		
CA6741S	RCU	LNA	INT	.	.	+22V	-22V	125C	94dB	5MV	500NA	200NA	500MWF	5MA	12V	15V	30V	.	.	3MA	70dB		
CN515T	FEG	GPU	EXT	.	.	+18V	-18V	125C	88dB	5MV	500NA	200NA	300MWF	.	12V	10V	5V	20uV/C	165MW	.	80dB	76dB	150
F418A	BLU	LBC	INT	.5MHZ	0.5V/uS	+18V	-18V	125C	88dB	1MV	1pA	.		5MA	10V	15V	30V	10uV/C	.	4MA	70dB	70dB	
F418B	BLU	LBC	INT	.5MHZ	0.5V/uS	+18V	-18V	125C	88dB	1MV	1pA	.		5MA	10V	15V	30V	25uV/C	.	4MA	70dB	70dB	
F418C	BLU	LBC	INT	.5MHZ	0.5V/uS	+18V	-18V	125C	88dB	2MV	1pA	.		5MA	10V	15V	30V	50uV/C	.	4MA	70dB	70dB	
HA1-2111	HAU	CPR	EXT	.	.	+18V	-18V	125C	100dB	3MV	100NA	10NA	500MWF	.	15V	30V	.	.	.	6MA			
HA1-2211	HAU	CPR	EXT	.	.	+18V	-18V	85C	100dB	3MV	100NA	10NA	500MWF	.	15V	30V	.	.	.	6MA			
HA1-2311	HAU	CPR	EXT	.	.	+18V	-18V	75C	100dB	7.5MV	100NA	10NA	500MWF	.	15V	30V	.	.	.	8MA			
HA1-2400	HAU	PRA	EXT	4MHZ	5V/uS	+22V	-22V	125C	94dB	4MV	200NA	50NA	300MWF	5MA	10V	10V	22V	.	.	6MA	80dB	80dB	1
HA1-2404	HAU	PRA	EXT	4MHZ	5V/uS	+22V	-22V	85C	94dB	4MV	200NA	50NA	300MWF	5MA	10V	10V	22V	.	.	6MA	80dB	80dB	1
HA1-2405	HAU	PRA	EXT	4MHZ	5V/uS	+22V	-22V	75C	94dB	9MV	250NA	50NA	300MWF	5MA	10V	10V	22V	.	.	6MA	74dB	74dB	1
HA1-2620	HAU	WBA	EXT	30MHZ	5V/uS	+22V	-22V	125C	100dB	4MV	15NA	15NA	300MWF	15MA	10V	11V	12V	.	.	4MA	80dB	80dB	6
HA1-2622	HAU	WBA	EXT	30MHZ	4V/uS	+22V	-22V	125C	98dB	4MV	25NA	25NA	300MWF	10MA	10V	11V	12V	.	.	4MA	74dB	74dB	4
HA1-2625	HAU	WBA	EXT	30MHZ	4V/uS	+22V	-22V	75C	98dB	5MV	25NA	25NA	300MWF	10MA	10V	11V	12V	.	.	4MA	74dB	74dB	4
HA1-2650	HAU	DHS	INT	.	2V/uS	+20V	-20V	125C	88dB	3MV	100NA	30NA	300MWF	5MA	13V	15V	30V	25uV/C	.	3MA	80dB	80dB	
HA1-2655	HAU	DHS	INT	.	2V/uS	+20V	-20V	75C	86dB	5MV	200NA	60NA	300MWF	5MA	13V	15V	30V	25uV/C	.	4MA	74dB	74dB	
HA1-2700	HAU	LOP	INT	.3MHZ	10V/uS	+22V	-22V	125C	106dB	3MV	20NA	10NA	300MWF	6MA	12V	11V	18V	.	.	.2UA	86dB	86dB	
HA1-2704	HAU	LOP	INT	.3MHZ	10V/uS	+22V	-22V	85C	106dB	3MV	20NA	10NA	300MWF	6MA	12V	11V	18V	.	.	.2UA	86dB	86dB	
HA1-2705	HAU	LOP	INT	.3MHZ	10V/uS	+22V	-22V	75C	106dB	5MV	40NA	15NA	300MWF	6MA	12V	11V	18V	.	.	.2UA	80dB	80dB	
HA1-2730	HAU	DPR	INT	.	0.3V/uS	+18V	-18V	125C	92dB	3MV	20NA	10NA	500MWF	2MA	12V	15V	30V	.	.	.3MA	80dB	76dB	
HA1-2735	HAU	DPR	INT	.	0.3V/uS	+18V	-18V	75C	88dB	5MV	30NA	10NA	500MWF	2MA	12V	15V	30V	.	.	.3MA	80dB	76dB	
HA2-909	HAU	GPK	INT	2MHZ	1.2V/uS	+25V	-25V	125C	88dB	5MV	300NA	150NA	300MWF	6MA	12V	12V	7V	.	.	3MA	80dB	80dB	20
HA2-911	HAU	GPK	INT	2MHZ	1.2V/uS	+25V	-25V	75C	86dB	6MV	500NA	300NA	300MWF	5MA	11V	12V	7V	.	.	3MA	74dB	74dB	10
HA2-2000	HAU	DVF	INT	.	30V/uS	+17V	-17V	125C	0dB	20MV	20pA	20pA	300MWF	.	10V	10V	17V	.	.	2MA	80dB	80dB	
HA2-2000A	HAU	DVF	INT	.	30V/uS	+17V	-17V	125C	0dB	10MV	20pA	20pA	300MWF	.	10V	10V	17V	.	.	2MA	80dB	80dB	
HA2-2005	HAU	DVF	INT	.	30V/uS	+17V	-17V	75C	0dB	50MV	20pA	20pA	300MWF	.	10V	10V	17V	.	.	2MA	70dB	70dB	

For detailed explanations of column heading notations, see App. A.
Also for ready references the more important abbreviations used in the column headings are listed below.

LEFT HAND PAGE

APP = application (codes at APP.E.)

CMRR = common mode rejection ratio

CMP = compensation (frequency)

dV_{io}/dT = input offset voltage temperature drift

GBP = gain bandwidth product

I_B = input bias current

I_{io} = input bias offset current

I_Q = quiescent supply current

MFR = manufacturer (codes at App.C.)

P_Q = quiescent power consumer

PSRR = power supply rejection ratio

V_{cm} = common mode input voltage rating

V_{IDF} = differential input voltage rating

V_{io} = input offset voltage

V_S = dc supply voltage

RIGHT HAND PAGE

Lead out coding summary (details at APP.G.) for different cases (APP.F.)

A = gain adjust

B = bias adjust

C = case

E— = inverting input

E+ = non-inverting input

F,F* = input frequency compensation

G = ground

J = high level input

K = output, open collector

L = output, open emitter

M = metal case

N = not connected

Q = special terminal

R,R* = outputs

S = strobe

T,T* = offset balance

V+ = +ve dc supply

V— = —ve dc supply

W = guard ring

X = blank position, no lead

++ = +ve supplementary dc supply

—— = —ve supplementary dc supply

ø,ø* = output frequency compensation

CASE APP.F	LD1	LD2	LD3	LD4	LD5	LD6	LD7	LD8	LD9	LD10	LD11	LD12	LD13	LD14	LD15	LD16	EUROPE SUBSTITUTE	USA SUBSTITUTE	I/S	TYPE NUMBER
TO5-8/3M	FS	E-	E+	V-M	B	L	V+	K											0	CA3094BS
TO5-8/1M	FS	E-	E+	V-M	B	L	V+	K											0	CA3094BT
DIL-8/1P	FS	E-	E+	V-	B	L	V+	K										CA3094AE	0	CA3094E
CHP																			0	CA3094H
TO5-8/3M	FS	E-	E+	V-M	B	L	V+	K										CA3094AS	0	CA3094S
TO5-8/1M	FS	E-	E+	V-M	B	L	V+	K										CA3094AT	0	CA3094T
CHP																			0	CA1458GH
CHP																			0	CA1458H
CHP																			0	CA3100H
TO5-8/1M	TF	E-	E+	V-	T*	R	V+	ø											0	CA3100T
TO5-8/3M	TF	E-	E+	V-	T*	R	V+	ø											0	CA3100S
DIL-14/1P	E-1	B1	V-1	E+1	Q	K*2	K	E-2	V-2	B2	E+2	Q	K*1	K1					0	CA3102E
CHP																			0	CA3102H
TO5-8/3M	TF	E-	E+.	V-M	T*	R	V+	SF*										CA3130BS	0	CA3130AS
TO5-8/1M	TF	E-	E+	V-M	T*	R	V+	SF*										CA3130BT	0	CA3130AT
TO5-8/3M	TF	E-	E+	V-M	T*	R	V+	SF*											0	CA3130BS
TO5-8/1M	TF	E-	E+	V-M	T*	R	V+	SF*											0	CA3130BT
TO5-8/3M	TF	E-	E+	V-M	T*	R	V+	SF*										CA3130AS	0	CA3130S
TO5-8/1M	TF	E-	E+	V-M	T*	R	V+	SF*										CA3130AT	0	CA3130T
TO5-8/3M	T	E-	E+	V-M	T*	R	V+	S										CA3140BS	0	CA3140AS
TO5-8/1M	T	E-	E+	V-M	T*	R	V+	S										CA3140BT	0	CA3140AT
TO5-8/1M	T	E-	E+	V-M	T*	R	V+	S											0	CA3140BS
TO5-8/1M	T	E-	E+	V-M	T*	R	V+	S											0	CA3140BT
CHP																			0	CA3140H
TO5-8/3M	T	E-	E+	V-M	T*	R	V+	S										CA3140AS	0	CA3140S
TO5-8/1M	T	E-	E+	V-M	T*	R	V+	S										C3140AT	0	CA3140T
DIL-14/1P	E+2	E+1	E-1	R1	R2	E-2	G	E-3	R3	R4	E-4	E+4	E+3	V+			LM3301N	MC3301P	0	CA3401E
DIL-14/1P	E+2	E+1	E-1	R1	R2	E-2	G	E-3	R3	R4	E-4	E+4	E+3	V+			LM3301N	MC3301P	0	CA3401G
CHP																			0	CA3401H
CHP																			0	CA3401HG
TO5-8/3M	F	E-	E+	V-	B	R	V+	F*											0	CA6078AS
TO5-8/1M	F	E-	E+	V-	B	R	V+	F*											0	CA6078AT
TO5-8/1M	T	E-	E+	V-	T*	R	V+	N											0	CA6741T
TO5-8/3M	T	E-	E+	V-	T*	R	V+	N											0	CA6741S
TO5-8/1M	F	E-	E+	V-M	ø	ø*R	V+	F*									TAA522	UA709A-T05	0	CN515T
DIM-14/1M	N	N	T	E-	E+	V-	N	N	T*	R	V+	N	N	M					0	F418A
DIM-14/1M	N	N	T	E-	E+	V-	N	N	T*	R	V+	N	N	M				F418A	0	F418B
DIM-14/1M	N	N	T	E-	E+	V-	N	N	T*	R	V+	N	N	M				F418B	0	F418C
DIL-14/1M	N	G	E+	E-	N	V-	T	T*S	R	N	V+	N	N	N			SN52111J	LM111D	0	HA1-2111
DIL-14/1M	N	G	E+	E-	N	V-	T	T*S	R	N	V+	N	N	N				LM211D	0	HA1-2211
DIL-14/1M	N	G	E+	E-	N	V-	T	T*S	R	N	V+	N	N	N			SFC2311EC	LM311D	0	HA1-2311
DIL-16/1P	E+1	E-1	E+2	E-2	E-3	E+3	E-4	E+4	V-	R	V+	F	G	S	Q	Q*			0	HA1-2400
DIL-16/1P	E+1	E-1	E+2	E-2	E-3	E+3	E-4	E+4	V-	R	V+	F	G	S	Q	Q*		HA1-2400	0	HA1-2404
DIL-16/1P	E+1	E-1	E+2	E-2	E-3	E+3	E-4	E+4	V-	R	V+	F	G	S	Q	Q*		HA1-2404	0	HA1-2405
DIL-14/1C	N	N	T	E-	E+	V-	N	N	T*	R	V+	N	N	ø					0	HA1-2620
DIL-14/1C	N	N	T	E-	E+	V-	N	N	T*	R	V+	N	N	ø				HA1-2620	0	HA1-2622
DIL-14/1C	N	N	T	E-	E+	V-	N	N	T*	R	V+	N	N	ø				HA1-2622	0	HA1-2625
DIL-14/1M	N	R1	T1	T*1	E-1	E+1	V-	E+2	E-2	T2	T*2	R2	N	V+					0	HA1-2650
DIL-14/1M	N	R1	T1	T*1	E-1	E+1	V-	E+2	E-2	T2	T*2	R2	N	V+				HA1-2650	0	HA1-2655
DIL-14/1M	N	T	W	E-	E+	W	V-	N	N	R	V+	T*	N	N					0	HA1-2700
DIL-14/1M	N	T	W	E-	E+	W	V-	N	N	R	V+	T*	N	N				HA1-2700	0	HA1-2704
DIL-14/1M	N	T	W	E-	E+	W	V-	N	N	R	V+	T*	N	N				HA1-2704	0	HA1-2705
DIL-14/1M	B1	R1	T1	T*1	E-1	E+1	V-M	E+2	E-2	T2	T*2	R2	B2	V+					0	HA1-2730
DIL-14/1M	B1	R1	T1	T*1	E-1	E+1	V-M	E+2	E-2	T2	T*2	R2	B2	V+				HA1-2730	0	HA1-2735
TO5-8/1M	M	E-	E+	V-	N	R	V+	ø											0	HA2-909
TO5-8/1M	M	E-	E+	V-	N	R	V+	ø										HA2-909	0	HA2-911
TO5-8/1M	T	E+2	E+1	V-	T*	V+	R1	R2										HA2-2000A	0	HA2-2000
TO5-8/1M	T	E+2	E+1	V-	T*	V+	R1	R2											0	HA2-2000A
TO5-8/1M	T	E+2	E+1	V-	T*	V+	R1	R2										HA2-2005A	0	HA2-2005
TO5-8/1M	T	E+2	E+1	V-	T*	V+	R1	R2											0	HA2-2005A

TYPE NUMBER	M F R	A P P	C M P	GBP MIN	SLEW RATE MIN	Vs+ MAX	Vs- MAX	Top MAX	Avol MIN	Vio MAX	Ib MAX	Iio MAX	Ptot MAX	Iout MIN	Vout MIN	Vicm MAX	Vidf MAX	dVic/dT MAX	Pq MAX	Io MAX	CM RR MIN	PS RR MIN	Rin MIN
HA2-2005A	HAU	DVF	INT	30V/uS		+17V	-17V	75C	0dB	10MV	20pA	20pA	300MWF		10V	10V	17V			2MA	70dB	70dB	0.1T
HA2-2050	HAU	XSR	EXT	1MHZ	50V/uS	+17V	-17V	125C	78dB	25MV	20pA	20pA	300MWF	10MA	10V	10V	15V			8MA	74JB	74dB	0.1T
HA2-2050A	HAU	XSR	EXT	1MHZ	50V/uS	+17V	-17V	125C	78dB	14MV	20pA	20pA	300MW	10MA	10V	10V	15V			8MA	74dB	74dB	0.1T
HA2-2055	HAU	XSR	EXT	1MHZ	50V/uS	+17V	-17V	75C	78dB	60MV	20pA	20pA	300MWF	10MA	10V	10V	15V			8MA	70dB	70dB	0.1T
HA2-2055A	HAU	XSR	EXT	1MHZ	50V/uS	+17V	-17V	75C	78dB	14MV	20pA	20pA	300MWF	10MA	10V	10V	15V			8MA	70dB	70dB	0.1T
HA2-2060	HAU	WBA	EXT	30MHZ	10V/uS	+17V	-17V	125C	98dB	25MV	20pA	20pA	300MWF	10MA	10V	10V	12V			6MA	74dB	74dB	0.1T
HA2-2060A	HAU	WBA	EXT	30MHZ	10V/uS	+17V	-17V	125C	98dB	12MV	20pA	20pA	300MWF	10MA	10V	10V	12V			6MA	74dB	74dB	0.1T
HA2-2065	HAU	WBA	EXT	30MHZ	10V/uS	+17V	-17V	75C	98dB	65MV	20pA	20pA	300MWF	10MA	10V	10V	12V			6MA	70dB	70dB	0.1T
HA2-2065A	HAU	WBA	EXT	30MHZ	10V/uS	+17V	-17V	75C	98dB	12MV	20pA	20pA	300MWF	10MA	10V	10V	12V			6MA	70dB	70dB	0.1T
HA2-2111	HAU	CPR	EXT			+18V	-18V	125C	100dB	3MV	100NA	10NA	500MWF			15V	30V			6MA			
HA2-2211	HAU	CPR	EXT			+18V	-18V	85C	100dB	3MV	100NA	10NA	500MWF			15V	30V			6MA			
HA2-2311	HAU	CPR	EXT			+18V	-18V	75C	100dB	7.5MV	100NA	10NA	500MWF			15V	30V			8MA			
HA2-2500	HAU	XSR	INT	30MHZ	25V/uS	+20V	-20V	125C	86dB	5MV	200NA	25NA	300MWF	10MA	10V	10V	15V	5uV/C		6MA	80dB	80dB	25M
HA2-2500-2	ING	XSR	INT	30MHZ	25V/uS	+20V	-20V	125C	86dB	5MV	200NA	25NA	300MWF	10MA	10V	15V	15V	100uV/C		6MA	80dB	80dB	25M
HA2-2502	HAU	XSR	INT	30MHZ	20V/uS	+20V	-20V	125C	84dB	8MV	250NA	50NA	300MWF	10MA	10V	10V	15V	5uV/C		6MA	74dB	74dB	20M
HA2-2502-2	ING	XSR	INT	30MHZ	20V/uS	+20V	-20V	125C	84dB	8MV	250NA	50NA	300MWF	10MA	10V	15V	15V	100uV/C		6MA	74dB	74dB	20M
HA2-2505	HAU	XSR	INT	30MHZ	20V/uS	+20V	-20V	75C	84dB	8MV	200NA	25NA	300MWF	10MA	10V	10V	15V	5uV/C		6MA	74dB	74dB	20M
HA2-2505-5	ING	XSR	INT	30MHZ	20V/uS	+20V	-20V	75C	84dB	8MV	200NA	25NA	300MWF	10MA	10V	15V	15V	100uV/C		6MA	74dB	74dB	20M
HA2-2510	HAU	XSR	INT	40MHZ	50V/uS	+20V	-20V	125C	80dB	8MV	200NA	25NA	300MWF	10MA	10V	10V	15V			6MA	80dB	80dB	50M
HA2-2512	HAU	XSR	INT	40MHZ	50V/uS	+20V	-20V	125C	78dB	10MV	250NA	50NA	300MWF	10MA	10V	10V	15V			6MA	74dB	74dB	40M
HA2-2515	HAU	XSR	INT	40MHZ	50V/uS	+20V	-20V	75C	78dB	10MV	250NA	50NA	300MWF	10MA	10V	10V	15V			6MA	74dB	74dB	40M
HA2-2520	HAU	XSR	EXT	20MHZ	100V/uS	+20V	-20V	125C	80dB	8MV	200NA	25NA	300MWF	10MA	10V	10V	15V	60uV/C		6MA	80dB	80dB	50M
HA2-2520-2	ING	XSR	EXT	20MHZ	100V/uS	+20V	-20V	125C	80dB	8MV	200NA	25NA	300MWF	10MA	10V	15V	15V	100uV/C		6MA	80dB	80dB	50M
HA2-2522	HAU	XSR	EXT	20MHZ	80V/uS	+20V	-20V	125C	78dB	10MV	250NA	50NA	300MWF	10MA	10V	10V	15V	75uV/C		6MA	74dB	74dB	40M
HA2-2522-2	ING	XSR	EXT	20MHZ	80V/uS	+20V	-20V	125C	78dB	10MV	250NA	50NA	300MWF	10MA	10V	15V	15V	100uV/C		6MA	74dB	74dB	40M
HA2-2525	HAU	XSR	EXT	20MHZ	80V/uS	+20V	-20V	75C	78dB	10MV	250NA	50NA	300MWF	10MA	10V	10V	15V	90uV/C		6MA	74dB	74dB	40M
HA2-2525-5	ING	XSR	EXT	20MHZ	80V/uS	+20V	-20V	75C	78dB	10MV	250NA	50NA	300MWF	10MA	10V	15V	15V	150uV/C		6MA	74dB	74dB	40M
HA2-2530	HAU	XSR	EXT	90MHZ	280V/uS	+20V	-20V	125C	100dB	3MV	100NA	20NA	550MWF	25MA	10V			20uV/C		6MA	86dB	86dB	1M
HA2-2535	HAU	XSR	EXT	90MHZ	250V/uS	+20V	-20V	75C	100dB	5MV	200NA	20NA	550MWF	25MA	10V			20uV/C		6MA	80dB	80dB	1M
HA2-2600	HAU	HIR	INT	4MHZ	4V/uS	+22V	-22V	125C	100dB	4MV	10NA	10NA	300MWF	15MA	10V	11V	12V	25uV/C		4MA	80dB	80dB	100M
HA2-2602	HAU	HIR	INT	4MHZ	4V/uS	+22V	-22V	125C	98dB	5MV	25NA	25NA	300MWF	10MA	10V	11V	12V	25uV/C		4MA	74dB	74dB	40M
HA2-2605	HAU	HIR	INT	4MHZ	4V/uS	+22V	-22V	75C	98dB	5MV	25NA	25NA	300MWF	10MA	10V	11V	12V	25uV/C		4MA	74dB	74dB	40M
HA2-2620	HAU	WBA	EXT	30MHZ	5V/uS	+22V	-22V	125C	100dB	4MV	15NA	15NA	300MWF	15MA	10V	11V	12V			4MA	80dB	80dB	65M
HA2-2622	HAU	WBA	EXT	30MHZ	4V/uS	+22V	-22V	125C	98dB	4MV	25NA	25NA	300MWF	10MA	10V	11V	12V			4MA	74dB	74dB	40M
HA2-2625	HAU	WBA	EXT	30MHZ	4V/uS	+22V	-22V	75C	98dB	5MV	25NA	25NA	300MWF	10MA	10V	11V	12V			4MA	74dB	74dB	40M
HA2-2630	HAU	VFA	INT		200V/uS	+20V	-20V	125C	-2dB	200MV	150uA			1WF	0.4A	10V	20V			20MA		66dB	1M
HA2-2635	HAU	VFA	INT		200V/uS	+20V	-20V	75C	-2dB	200MV	150uA			1WF	0.4A	10V	20V			23MA		66dB	1M
HA2-2640	HAU	HVO	INT	2MHZ	2V/uS	+50V	-50V	125C	100dB	4MV	25NA	12NA	680MWF	12MA	35V	35V	37V	50uV/C		4MA	80dB	80dB	50M
HA2-2645	HAU	HVO	INT	2MHZ	2V/uS	+50V	-50V	75C	100dB	6MV	30NA	30NA	680MWF	10MA	35V	35V	37V	50uV/C		5MA	74dB	74dB	40M
HA2-2650	HAU	DHS	INT		2V/uS	+20V	-20V	125C	88dB	3MV	100NA	30NA	300MWF	5MA	13V	15V	30V	25uV/C		3MA	80dB	80dB	5M
HA2-2655	HAU	DHS	INT		2V/uS	+20V	-20V	75C	86dB	5MV	200NA	60NA	300MWF	5MA	13V	15V	30V	25uV/C		4MA	74dB	74dB	5M
HA2-2700	HAU	LQP	INT	.3MHZ	10V/uS	+22V	-22V	125C	106dB	3MV	20NA	10NA	300MWF	6MA	12V	11V	18V			.2UA	86dB	86dB	
HA2-2704	HAU	LQP	INT	.3MHZ	10V/uS	+22V	-22V	85C	106dB	3MV	20NA	10NA	300MWF	6MA	12V	11V	18V			.2UA	86dB	86dB	
HA2-2705	HAU	LQP	INT	.3MHZ	10V/uS	+22V	-22V	75C	106dB	5MV	20NA	10NA	300MWF	6MA	12V	11V	18V			.2UA	86dB	86dB	
HA2-2720	HAU	PRA	INT		0.3V/uS	+22V	-22V	125C	92dB	3MV	20NA	10NA	300MWF	2MA	12V	15V	30V			.3MA	80dB	80dB	2M
HA2-2725	HAU	PRA	INT		0.3V/uS	+22V	-22V	75C	88dB	5MV	30NA	10NA	300MWF	2MA	12V	15V	30V			.3MA	74dB	74dB	2M
HA2-2900	HAU	CHP	EXT	1MHZ	1V/uS	+21V	-21V	125C	120dB	60uV	1NA	0.5NA	300MWF	10MA	10V	10V	15V	0.6uV/C		5MA	120dB	120dB	10M
HA2-2904	HAU	CHP	EXT	1MHZ	1V/uS	+21V	-21V	85C	140dB	50uV	1NA	0.5NA	300MWF	10MA	10V	10V	15V	0.4uV/C		5MA	130dB	130dB	10M
HA2-2905	HAU	CHP	EXT	1MHZ	1V/uS	+21V	-21V	75C	120dB	80uV	1NA	0.5NA	300MWF	7MA	10V	10V	15V	0.6uV/C		5MA	120dB	120dB	10M
HA9-909	HAU	GPK	INT	2MHZ	1.2V/uS	+25V	-25V	125C	88dB	5MV	300NA	150NA	300MWF	6MA	12V	12V	7V			3MA	80dB	80dB	200K
HA9-911	HAU	GPK	INT	2MHZ	1.2V/uS	+25V	-25V	75C	86dB	6MV	500NA	300NA	300MWF	5MA	11V	12V	7V			3MA	74dB	74dB	100K
HA9-2500	HAU	XSR	INT	30MHZ	25V/uS	+20V	-20V	125C	86dB	5MV	200NA	25NA	300MWF	10MA	10V	10V	15V	5uV/C		6MA	80dB	80dB	25M
HA9-2500-2	ING	XSR	INT	30MHZ	25V/uS	+20V	-20V	125C	86dB	5MV	200NA	25NA	300MWF	10MA	10V	15V	15V	100uV/C		6MA	80dB	80dB	25M
HA9-2502	HAU	XSR	INT	30MHZ	20V/uS	+20V	-20V	125C	84dB	8MV	250NA	50NA	300MWF	10MA	10V	10V	15V	5uV/C		6MA	80dB	80dB	20M
HA9-2502-2	ING	XSR	INT	30MHZ	20V/uS	+20V	-20V	125C	84dB	8MV	250NA	50NA	300MWF	10MA	10V	15V	15V	100uV/C		6MA	74dB	74dB	20M
HA9-2505	HAU	XSR	INT	30MHZ	20V/uS	+20V	-20V	75C	84dB	8MV	250NA	50NA	300MWF	10MA	10V	10V	15V	5uV/C		6MA	74dB	74dB	20M
HA9-2505-5	ING	XSR	INT	30MHZ	20V/uS	+20V	-20V	75C	84dB	8MV	250NA	50NA	300MWF	10MA	10V	15V	15V	100uV/C		6MA	74dB	74dB	20M
HA9-2510	HAU	XSR	INT	40MHZ	50V/uS	+20V	-20V	125C	80dB	8MV	200NA	25NA	300MWF	10MA	10V	10V	15V			6MA	80dB	80dB	50M
HA9-2512	HAU	XSR	INT	40MHZ	50V/uS	+20V	-20V	125C	78dB	10MV	250NA	50NA	300MWF	10MA	10V	10V	15V			6MA	74dB	74dB	40M
HA9-2515	HAU	XSR	INT	40MHZ	50V/uS	+20V	-20V	75C	78dB	10MV	250NA	50NA	300MWF	10MA	10V	10V	15V			6MA	74dB	74dB	40M

Also for ready references the more important abbreviations used in the column headings are listed below:

LEFT HAND PAGE

- APP = application (codes at APP.E.)
- CMRR = common mode rejection ratio
- CMP = compensation (frequency)
- dV_{in}/dT = input offset voltage temperature drift
- GBP = gain bandwidth product
- I_b = input bias current
- I_{io} = input bias offset current
- I_Q = quiescent supply current
- MFR = manufacturer (codes at App.C.)
- P_Q = quiescent power consumer
- PSRR = power supply rejection ratio
- V_{icm} = common mode input voltage rating
- V_{idf} = differential input voltage rating
- V_{io} = input offset voltage
- V_s = dc supply voltage

RIGHT HAND PAGE

Lead out coding summary (details at APP.G.) for different cases (APP.F.)

- A = gain adjust
- B = bias adjust
- C = case
- E− = inverting input
- E+ = non-inverting input
- F,F* = input frequency compensation
- G = ground
- J = high level input
- K = output, open collector
- L = output, open emitter
- M = metal case
- N = not connected
- Q = special terminal
- R,R* = outputs
- S = strobe
- T,T* = offset balance
- V+ = +ve dc supply
- V− = −ve dc supply
- W = guard ring
- X = blank position, no lead
- ++ = +ve supplementary dc supply
- −− = −ve supplementary dc supply
- ø,ø* = output frequency compensation

CASE (APP F)	LD1	LD2	LD3	LD4	LD5	LD6	LD7	LD8	LD9	LD10	LD11	LD12	LD13	LD14	LD15	LD16	EUROPE SUBSTITUTE	USA SUBSTITUTE	IS	TYPE NUMBER
TO5-8/1M	T	E-	E+	V-	T*	R	V+	ø										HA2-2050A	0	HA2-2050
TO5-8/1M	T	E-	E+	V-	T*	R	V+	ø											0	HA2-2050A
TO5-8/1M	T	E-	E+	V-	T*	R	V+	ø										HA2-2055	0	HA2-2055
TO5-8/1M	T	E-	E+	V-	T*	R	V+	ø										HA2-2050A	0	HA2-2055A
TO5-8/1M	T	E-	E+	V-	T*	R	V+	ø										HA2-2060A	0	HA2-2060
TO5-8/1M	T	E-	E+	V-	T*	R	V+..ø													0	HA2-2060A
TO5-8/1M	T	E-	E+	V-	T*	R	V+	ø										HA2-2065A	0	HA2-2065
TO5-8/1M	T	E-	E+	V-	T*	R	V+	ø											0	HA2-2065A
TO5-8/1M	G	E+	E-	V-	T	T*S	R	V+									SFC2111M	LM111H	0	HA2-2111
TO5-8/1M	G	E+	E-	V-	T	T*S	R	V+									SFC2211	LM211H	0	HA2-2211
TO5-8/1M	G	E+	E-	V-	T	T*S	R	V+									SFC2311	LM311H	0	HA2-2311
TO5-8/1M	T	E-	E+	V-	T*	R	V+	ø										HA2-2500-2	0	HA2-2500
TO5-8/1M	T	E-	E+	V-	T*	R	V+	ø										HA2-2500	0	HA2-2500-2
TO5-8/1M	T	E-	E+	V-	T*	R	V+	ø										HA2-2502-2	0	HA2-2502
TO5-8/1M	T	E-	E+	V-	T*	R	V+	ø										HA2-2502	0	HA2-2502-2
TO5-8/1M	T	E-	E+	V-	T*	R	V+	ø										HA2-2505-5	0	HA2-2505
TO5-8/1M	T	E-	E+	V-	T*	R	V+	ø										HA2-2505	0	HA2-2505-5
TO5-8/1M	T	E-	E+	V-	T*	R	V+	ø										HA2-2510	0	HA2-2510
TO5-8/1M	T	E-	E+	V-	T*	R	V+	ø										HA2-2510	0	HA2-2512
TO5-8/1M	T	E-	E+	V-	T*	R	V+	ø										HA2-2512	0	HA2-2515
TO5-8/1M	T	E-	E+	V-	T*	R	V+	ø										HA2-2520-2	0	HA2-2520
TO5-8/1M	T	E-	E+	V-	T*	R	V+	ø										HA2-2520	0	HA2-2520-2
TO5-8/1M	T	E-	E+	V-	T*	R	V+	ø										HA2-2522-2	0	HA2-2522
TO5-8/1M	T	E-	E+	V-	T*	R	V+	ø										HA2-2522	0	HA2-2522-2
TO5-8/1M	T	E-	E+	V-	T*	R	V+	ø										HA2-2525-2	0	HA2-2525
TO5-8/1M	T	E-	E+	V-	T*	R	V+	ø										HA2-2525	0	HA2-2525-2
TO5-8/1M	N	E-	E+	V-	ø	Rø*	V+	M											0	HA2-2530
TO5-8/1M	N	E-	E+	V-	ø	Rø*	V+	M										HA2-2530	0	HA2-2535
TO5-8/1M	T	E-	E+	V-	T*	R	V+	ø											0	HA2-2600
TO5-8/1M	T	E-	E+	V-	T*	R	V+	ø										HA2-2600	0	HA2-2602
TO5-8/1M	T	E-	E+	V-	T*	R	V+	ø										HA2-2602	0	HA2-2605
TO5-8/1M	T	E-	E+	V-	T*	R	V+	ø											0	HA2-2620
TO5-8/1M	T	E-	E+	V-	T*	R	V+	ø										HA2-2620	0	HA2-2622
TO5-8/1M	T	E-	E+	V-	T*	R	V+	ø										HA2-2622	0	HA2-2625
TO8-12/2M	Q	V-	E+	N	N	N	V+	Q*	R	N	N	N							0	HA2-2630
TO8-12/2M	Q	V-	E+	N	N	N	V+	Q*	R	N	N	N						HA2-2630	0	HA2-2635
TO5-8/1M	T	E-	E+	V-	T*	R	V+	N											0	HA2-2640
TO5-8/1M	T	E-	E+	V-	T*	R	V+	N										HA2-2640	0	HA2-2645
TO5-8/1M	R1	E-1	E+1	V-	E+2	E-2	R2	V+											0	HA2-2650
TO5-8/1M	R1	E-1	E+1	V-	E+2	E-2	R2	V+										HA2-2650		HA2-2655
TO5-8/1M	T	E-	E+	V-M	N	R	V+	T*											0	HA2-2700
TO5-8/1M	T	E-	E+	V-M	N	R	V+	T*										HA2-2700	0	HA2-2704
TO5-8/1M	T	E-	E+	V-M	N	R	V+	T*										HA2-2704	0	HA2-2705
TO5-8/1M	T	E-	E+	V-	T*	R	V+	B									HA2-2720	HA776HM	0	HA2-2720
TO5-8/1M	T	E-	E+	V-	T*	R	V+	B										HA776HC	0	HA2-2725
TO5-8/1M	Q1	E-	E+	V-	Q2	R	V+	Q3											0	HA2-2900
TO5-8/1M	Q1	E-	E+	V-	Q2	R	V+	Q3										HA2-2900	0	HA2-2904
TO5-8/1M	Q1	E-	E+	V-	Q2	R	V+	Q3										HA2-2904	0	HA2-2905
FLP-14/3G	ø	N	N	E-	E+	V-	T	B	N	R	V+	N	N	C					0	HA9-909
FLP-14/3G	ø		N	E-	E+	V-	T	B	N	R	V+	N	N	C				HA9-909	0	HA9-911
FLP-14/3G	N	ø	T	E-	E+	N	N	N*	M	V-	T*	R	V+	N				HA9-2500-2	0	HA9-2500
FLP-14/3G	N	ø	T	E-	E+	N	N	N	C	V-	T*	R	V+	N				HA9-2500	0	HA9-2500-2
FLP-14/3G	N	ø	T	E-	E+	N	N	N	M	V-	T*	R	V+	N				HA9-2502-2	0	HA9-2502
FLP-14/3G	N	ø	T	E-	E+	N	N	N	C	V-	T*	R	V+	N				HA9-2502	0	HA9-2502-2
FLP-14/3G	N	ø	T	E-	E+	N	N	N	M	V-	T*	R	V+	N				HA9-2505-5	0	HA9-2505
FLP-14/3G	N	ø	T	E-	E+	N	N	N	C	V-	T*	R	V+	N				HA9-2505	0	HA9-2505-5
FLP-14/3G	N	ø	T	E-	E+	N	N	N	C	V-	T*	R	V+	N					0	HA9-2510
FLP-14/3G	N	ø	T	E-	E+	N	N	N	C	V-	T*	R	V+	N				HA9-2510	0	HA9-2512
FLP-14/3G	N	ø	T	E-	E+	N	N	N	C	V-	T*	R	V+	N				HA9-2512	0	HA9-2515
FLP-14/3G	N	ø	T	E-	E+	N	N	N	C	V-	T*	R	V+	N				HA9-2520-2	0	HA9-2520

TYPE NUMBER	MFR	APP	CMP	GBP MIN	SLEW RATE MIN	Vs+ MAX	Vs− MAX	Top MAX	Avol MIN	Vio MAX	Ib MAX	Iio MAX	Ptot MAX	Iout MIN	Vout MIN	Vicm MAX	Vidf MAX	dVio/dT MAX	Pq MAX	Iq MAX	CMRR MIN	PSRR MIN	Rin MIN
HA9-2520	HAU	XSR	EXT	20MHZ	100V/uS	+20V	−20V	125C	80dB	8MV	200NA	25NA	300MWF	10MA	10V	10V	15V	60uV/C	.	6MA	80dB	80dB	50M
HA9-2520-2	ING	XSR	EXT	20MHZ	100V/uS	+20V	−20V	125C	80dB	8MV	200NA	25NA	300MWF	10MA	10V	15V	15V	100uV/C	.	6MA	80dB	80dB	50M
HA9-2522	HAU	XSR	EXT	20MHZ	80V/uS	+20V	−20V	125C	78dB	10MV	250NA	50NA	300MWF	10MA	10V	10V	15V	75uV/C	.	6MA	74dB	74dB	40M
HA9-2522-2	ING	XSR	EXT	20MHZ	80V/uS	+20V	−20V	125C	78dB	10MV	250NA	50NA	300MWF	10MA	10V	15V	15V	100uV/C	.	6MA	74dB	74dB	40M
HA9-2525	HAU	XSR	EXT	20MHZ	80V/uS	+20V	−20V	75C	78dB	10MV	250NA	50NA	300MWF	10MA	10V	10V	15V	90uV/C	.	6MA	74dB	74dB	40M
HA9-2525-5	ING	XSR	EXT	20MHZ	80V/uS	+20V	−20V	75C	78dB	10MV	250NA	50NA	300MWF	10MA	10V	15V	15V	150uV/C	.	6MA	74dB	74dB	40M
HA9-2600	HAU	HIR	INT	4MHZ	4V/uS	+22V	−22V	125C	100dB	4MV	10NA	10NA	300MWF	15MA	10V	11V	12V	25uV/C	.	4MA	80dB	80dB	100M
HA9-2602	HAU	HIR	INT	4MHZ	4V/uS	+22V	−22V	125C	98dB	5MV	25NA	25NA	300MWF	10MA	10V	11V	12V	25uV/C	.	4MA	74dB	74dB	40M
HA9-2605	HAU	HIR	INT	4MHZ	4V/uS	+22V	−22V	75C	98dB	5MV	25NA	25NA	300MWF	10MA	10V	11V	12V	25uV/C	.	4MA	74dB	74dB	40M
HA-4741-2	HAU	QGK	INT	1MHZ	0.5V/uS	+20V	−20V	125C	94dB	3MV	200NA	30NA	880MWF	5MA	12V	15V	30V	20uV/C	.	5MA	80dB	80dB	2M
HA-4741-5	HAU	QGK	INT	1MHZ	0.5V/uS	+20V	−20V	75C	88dB	5MV	300NA	50NA	880MWF	5MA	12V	15V	30V	20uV/C	.	7MA	80dB	80dB	2M
ICH8500	ING	LBC	INT	.	0.1V/uS	+18V	−18V	85C	86dB	50MV	0.1pA	.	500MWF	1MA	11V	10V	0.5V	.	.	.	60dB	.	.
ICH8500A	ING	LBC	INT	.	0.1V/uS	+18V	−18V	85C	86dB	50MV	.01pA	.	500MWF	1MA	11V	10V	0.5V	85uV/C	.	.	60dB	.	.
ICH8500ATV	ING	LBC	INT	.	0.1V/uS	+18V	−18V	85C	86dB	50MV	.01pA	.	500MWF	1MA	11V	10V	0.5V	85uV/C	.	.	60dB	.	.
ICH8500TV	ING	LBC	INT	.	0.1V/uS	+18V	−18V	85C	86dB	50MV	0.1pA	.	500MWF	1MA	11V	10V	0.5V	.	.	.	60dB	.	.
ICL101A-LNDD	ING	LNA	EXT	.2MHZ	.15V/uS	+20V	−20V	125C	88dB	3MV	100NA	20NA	500MWF	5MA	12V	15V	30V	15uV/C	.	3MA	80dB	80dB	.
ICL101A-LNFB	ING	LNA	EXT	.2MHZ	.15V/uS	+20V	−20V	125C	88dB	3MV	100NA	20NA	500MWF	5MA	12V	15V	30V	15uV/C	.	3MA	80dB	80dB	.
ICL101A-LNTY	ING	LNA	EXT	.2MHZ	.15V/uS	+20V	−20V	125C	88dB	3MV	100NA	20NA	500MWF	5MA	12V	15V	30V	15uV/C	.	3MA	80dB	80dB	.
ICL108-LN-TY	ING	LNA	EXT	.1MHZ	0.1V/uS	+20V	−20V	125C	88dB	3MV	3NA	0.4NA	500MWF	1MA	13V	15V	1V	15uV/C	.	1MA	85dB	80dB	.
ICL301A-LNPA	ING	LNA	EXT	.2MHZ	.15V/uS	+15V	−15V	70C	84dB	10M	300NA	70NA	500MWF	5MA	12V	15V	30V	30uV/C	.	3MA	70dB	70dB	.
ICL301A-LNTY	ING	LNA	EXT	.2MHZ	.15V/uS	+15V	−15V	70C	84dB	10MV	300NA	70NA	500MWF	5MA	12V	15V	30V	30uV/C	.	3MA	70dB	70dB	.
ICL308-LN-TY	ING	LNA	EXT	.1MHZ	C.1V/uS	+20V	−20V	70C	84dB	10MV	10NA	1.5NA	500MWF	1MA	13V	15V	1V	30uV/C	.	1MA	80dB	80dB	.
ICL741CHSPA	ING	GPK	INT	.	0.7V/uS	+18V	−18V	70C	94dB	5MV	500NA	200NA	500MWF	5MA	12V	15V	30V	.	.	3MA	70dB	77dB	300K
ICL741CHSTY	ING	GPK	INT	.	0.7V/uS	+18V	−18V	70C	88dB	6MV	500NA	200NA	500MWF	5MA	12V	15V	30V	.	.	3MA	70dB	77dB	300K
ICL741C-LNPA	ING	LNA	INT	.2MHZ	.15V/uS	+15V	−15V	70C	84dB	7.5MV	0.8uA	0.3uA	310MWF	5MA	12V	15V	30V	50uV/C	.	3MA	70dB	76dB	.
ICL741C-LNTY	ING	LNA	INT	.2MHZ	.15V/uS	+15V	−15V	70C	84dB	7.5MV	0.8uA	0.3uA	500MWF	5MA	12V	15V	30V	50uV/C	.	3MA	80dB	80dB	.
ICL741CTY	ING	GPK	INT	.	0.2V/uS	+18V	−18V	70C	86dB	6MV	500NA	200NA	500MWF	5MA	12V	15V	30V	.	.	3MA	70dB	76dB	300K
ICL741-LN-DD	ING	LNA	INT	.2MHZ	.15V/uS	+20V	−20V	125C	88dB	6MV	1.5uA	500NA	670MWF	5MA	12V	15V	30V	30uV/C	.	3MA	70dB	76dB	.
ICL741-LN-FB	ING	LNA	INT	.2MHZ	.15V/uS	+20V	−20V	125C	88dB	6MV	1.5uA	500NA	570MWF	5MA	12V	15V	30V	30UV/C	.	3MA	70dB	76dB	.
ICL741-LN-TY	ING	LNA	INT	.2MHZ	.15V/uS	+20V	−20V	125C	88dB	6MV	1.5uA	500NA	500MWF	5MA	12V	15V	30V	30uV/C	.	3MA	70dB	76dB	.
ICL741MHSDD	ING	GPK	INT	.	0.7V/uS	+18V	−18V	125C	94dB	5MV	500MA	200NA	500MWF	5MA	12V	15V	30V	.	.	3MA	70dB	77dB	300K
ICL741MHSFD	ING	GPK	INT	.	0.7V/uS	+18V	−18V	125C	94dB	5MV	500NA	200NA	500MWF	5MA	12V	15V	30V	.	.	3MA	70dB	77dB	300K
ICL741MHSTY	ING	GPK	INT	.	0.7V/uS	+18V	−18V	125C	94dB	5MV	500NA	200NA	500MWF	5MA	12V	15V	30V	.	.	3MA	70dB	77dB	300K
ICL741TY	ING	GPK	INT	.	0.2V/uS	+22V	−22V	125C	94dB	5MV	500NA	200NA	500MWF	7MA	12V	15V	30V	.	85MW	3MA	70dB	76dB	300K
ICL748CTY	ING	GPU	EXT	.	0.2V/uS	+18V	−18V	70C	86dB	6MV	500NA	200NA	500MWF	5MA	12V	15V	30V	.	85MW	3MA	70dB	76dB	300K
ICL748TY	ING	GPU	EXT	.	0.2V/uS	+22V	−22V	125C	94dB	5MV	500NA	200NA	500MWF	5MA	12V	15V	30V	.	85MW	3MA	70dB	76dB	300K
ICL8001CTZ	ING	CPR	EXT	.	.	+18V	−18V	70C	84dB	5MV	250NA	50NA	500MWF	.	3V	18V	15V	30uV/C	60MW	.	70dB	70dB	3M
ICL8001MTZ	ING	CPR	EXT	.	.	+18V	−18V	125C	84dB	3MV	100NA	20NA	500MWF	.	3V	18V	15V	20uV/C	60MW	.	70dB	70dB	3M
ICL8007ACTV	ING	FET	INT	.3MHZ	2.5V/uS	+18V	−18V	70C	86dB	30MV	1pA	0.5pA	500MWF	5MA	12V	15V	30V	50uV/C	180MW	6MA	86dB	74dB	0.5T
ICL8007AMTV	ING	FET	INT	.3MHZ	2.5V/uS	+18V	−18V	125C	86dB	30MV	1pA	0.5pA	500MWF	5MA	12V	15V	30V	50uV/C	156MW	5MA	86dB	74dB	0.5T
ICL8007C-1	ING	FET	INT	.3MHZ	3V/uS	+18V	−18V	70C	94dB	2MV	10pA	5pA	500MWF	5MA	12V	15V	30V	5uV/C	180MW	6MA	70dB	70dB	0.1T
ICL8007C-2	ING	FET	INT	.3MHZ	3V/uS	+18V	−18V	70C	94dB	2MV	10pA	5pA	500MWF	5MA	12V	15V	30V	15uV/C	180MW	6MA	70dB	70dB	0.1T
ICL8007C-3	ING	FET	INT	.3MHZ	3V/uS	+18V	−18V	70C	94dB	4MV	20pA	5pA	500MWF	5MA	12V	15V	30V	30uV/C	180MW	6MA	70dB	70dB	0.1T
ICL8007C-4	ING	FET	INT	.3MHZ	3V/uS	+18V	−18V	70C	94dB	10MV	10pA	5pA	500MWF	5MA	12V	15V	30V	10uV/C	180MW	6MA	70dB	70dB	0.1T
ICL8007C-5	ING	FET	INT	.3MHZ	3V/uS	+18V	−18V	70C	94dB	10MV	10pA	5pA	500MWF	5MA	12V	15V	30V	15uV/C	180MW	6MA	70dB	70dB	0.1T
ICL8007CTV	ING	FET	INT	.3MHZ	2V/uS	+18V	−18V	70C	86dB	50MV	50pA	5pA	500MWF	5MA	12V	15V	30V	75uV/C	180MW	6MA	70dB	64dB	0.5T
ICL8007M-2	ING	FET	INT	.3MHZ	3V/uS	+18V	−18V	125C	94dB	2MV	10pA	5pA	500MWF	5MA	12V	15V	30V	15uV/C	156MW	6MA	70dB	70dB	100G
ICL8007M-5	ING	FET	INT	.3MHZ	3V/uS	+18V	−18V	125C	94dB	10MV	10pA	5pA	500MWF	5MA	12V	15V	30V	15uV/C	156MW	6MA	70dB	70dB	100G
ICL8007MTV	ING	FET	INT	.3MHZ	2V/uS	+18V	−18V	125C	94dB	20MV	20pA	2pA	500MWF	5MA	12V	15V	30V	75uV/C	156MW	6MA	70dB	70dB	0.5T
ICL8008CPA	ING	GPK	INT	.	0.1V/uS	+15V	−15V	70C	86dB	6MV	25NA	20NA	500MWF	5MA	12V	15V	30V	75uV/C	85MW	3MA	70dB	76dB	5M
ICL8008CTY	ING	GPK	INT	.	0.1V/uS	+15V	−15V	70C	86dB	6MV	25NA	20NA	500MWF	5MA	12V	15V	30V	75uV/C	85MW	3MA	70dB	76dB	5M
ICL8008MTY	ING	GPK	INT	.	0.1V/uS	+15V	−15V	125C	86dB	5MV	10NA	5NA	500MWF	5MA	12V	15V	30V	35uV/C	85MW	3MA	70dB	76dB	5M
ICL8021CTA	ING	PRA	INT	.1MHZ	.03V/uS	+18V	−18V	70C	94dB	6MV	30NA	10NA	300MWF	1MA	12V	15V	15V	25uV/C	0.6MW	.	70dB	76dB	3M
ICL8021MTA	ING	PRA	INT	.1MHZ	.03V/uS	+18V	−18V	125C	94dB	3MV	20NA	7.5NA	300MWF	1MA	12V	15V	15V	25uV/C	.48MW	.	70dB	76dB	3M
ICL8022CDD	ING	DPR	INT	.1MHZ	.03V/uS	+18V	−18V	70C	94dB	6MV	30NA	10NA	300MWF	1MA	12V	15V	15V	25uV/C	0.6MW	.	70dB	76dB	3M
ICL8022MDD	ING	DPR	INT	.1MHZ	.03V/uS	+18V	−18V	125C	94dB	3MV	20NA	7.5NA	300MWF	1MA	12V	15V	15V	25uV/C	.48MW	.	70dB	76dB	3M
ICL8023CDE	ING	TPR	INT	.1MHZ	.03V/uS	+18V	−18V	70C	94dB	6MV	30NA	10NA	300MWF	1MA	12V	15V	15V	25uV/C	0.6MW	.	70dB	76dB	3M
ICL8023MDE	ING	TPR	INT	.1MHZ	.03V/uS	+18V	−18V	125C	94dB	3MV	20NA	7.5NA	300MWF	1MA	12V	15V	15V	25uV/C	.48MW	.	70dB	76dB	3M
ICL8043CDE	ING	DFE	INT	.3MHZ	2V/uS	+18V	−18V	70C	86dB	50MV	50pA	5pA	500MWF	5MA	12V	15V	30V	75uV/C	204MW	7MA	70dB	64dB	0.5T
ICL8043CPE	ING	DFE	INT	.3MHZ	2V/uS	+18V	−18V	70C	86dB	50MV	50pA	5pA	500MWF	5MA	12V	15V	30V	75uV/C	204MW	7MA	70dB	64dB	0.5T

Also for ready references the more important abbreviations used in the column headings are listed below:

LEFT HAND PAGE

APP = application (codes at APP.E.)

CMRR = common mode rejection ratio

CMP = compensation (frequency)

dV_{IO}/dT = input offset voltage temperature drift

GBP = gain bandwidth product

I_B = input bias current

I_{IO} = input bias offset current

I_Q = quiescent supply current

MFR = manufacturer (codes at App.C.)

P_Q = quiescent power consumer

PSRR = power supply rejection ratio

V_{ICM} = common mode input voltage rating

V_{IDF} = differential input voltage rating

V_{IO} = input offset voltage

V_S = dc supply voltage

RIGHT HAND PAGE

Lead out coding summary (details at APP.G.) for different cases (APP.F.)

A = gain adjust

B = bias adjust

C = case

E— = inverting input

E+ = non-inverting input

F,F* = input frequency compensation

G = ground

J = high level input

K = output, open collector

L = output, open emitter

M = metal case

N = not connected

Q = special terminal

R,R* = outputs

S = strobe

T,T* = offset balance

V+ = +ve dc supply

V— = —ve dc supply

W = guard ring

X = blank position, no lead

++ = +ve supplementary dc supply

—— = —ve supplementary dc supply

ø,ø* = output frequency compensation

CASE (APP F)	LD 1	LD 2	LD 3	LD 4	LD 5	LD 6	LD 7	LD 8	LD 9	LD 10	LD 11	LD 12	LD 13	LD 14	LD 15	LD 16	EUROPE SUBSTITUTE	USA SUBSTITUTE	I S	TYPE NUMBER
FLP-14/3G	N	ø	T	E-	E+	N	N	N	C	V-	T*	R	V+	N					HA9-2520	0	HA9-2520-2
FLP-14/3G	N	ø	T	E-	E+	N	N	N	C	V-	T*	R	V+	N					HA9-2522-2	0	HA9-2522
FLP-14/3G	N	ø	T	E-	E+	N	N	N	C	V-	T*	R	V+	N					HA9-2522-2	0	HA9-2522-2
FLP-14/3G	N	ø	T	E-	E+	N	N	N	C	V-	T*	R	V+	N					HA9-2525-5	0	HA9-2525
FLP-14/3G	N	ø	T	E-	E+	N	N	N.	C	V-	T*	R	V+	N					HA9-2525	0	HA9-2525-5
FLP-10/3G	N	T	E-	E+	V-	T*	R	V+	ø	N										0	HA9-2600
FLP-10/3G	N	T	E-	E+	V-	T*	R	V+	ø	N									HA9-2600	0	HA9-2602
FLP-10/3G	N	T	E-	E+	V-	T*	R	V+	ø	N									HA9-2602	0	HA9-2605
DIL-14/1C	R1	E-1	E+1	V+	E+2	E-2	R2	R3	E-3	E+3	V-	E+4	E-4	R4			MC4741L	LM148D	0	HA4741-1
DIL-14/1C	R1	E-1	E+1	V+	E+2	E-2	R2	R3	E-3	E+3	V-	E+4	E-4	R4			MC4741P	LM248D	0	HA4741-5
TO5-8/1M	T	E-	E+	V-	T*	R	V+	MW										ICH8500TV	0	ICH8500
TO5-8/1M	T	E-	E+	V-	T*	R	V+	MW										ICH8500ATV	0	ICH8500A
TO5-8/1M	T	E-	E+	V-	T*	R	V+	MW											0	ICH8500ATV
TO5-8/1M	T	E-	E+	V-	T*	R	V+	MW										ICH8500ATV	0	ICH8500TV
DIL-14/1C	N	N	TF	E-	E+	V-	N	N	T*	R	V+	F*	N	N					101ALN-DIL	0	ICL101A-LNDD
FLP-10/3C	N	TF	E-	E+	V-	T*	R	V+	F*	N									101ALN-FLP	0	ICL101A-LNFB
TO5-8/1M	TF	E-	E+	V-	T*	R	V+	F*											101ALN-TO5	0	ICL101A-LNTY
TO5-8/1M	F	E-	E+	V-	N	R	V+	F*											108-LN-TO5	0	ICL108-LN-TY
DIL-8/1P	T	E-	E+	V-	T*	R	V+	F*											301ALNDIL8	0	ICL301A-LNPA
TO5-8/1M	TF	E-	E+	V-	T*	R	V+	F*											301ALN-TO5	0	ICL301A-LNTY
TO5-8/1M	F	E-	E+	V-		R	V+	F*											308-LN-TO5	0	ICL308-LN-TY
DIL-8/1P	T	E-	E+	V-	T*	R	V+	N											741CHSDIL8	0	ICL741CHSPA
TO5-8/1M	T	E-	E+	V-M	T*	R	V+	N											741CHS-TO5	0	ICL741CHSTY
DIL-8/1P	T	E-	E+	V-	T*	R	V+	N											741CLNDIL8	0	ICL741C-LNPA
TO5-8/1M	T	E-	E+	V-	T*	R	V+	N											741CLN-TO5	0	ICL741C-LNTY
TO5-8/1M	T	E-	E+	V-M	T*	R	V+	N									TBA221	UA741HC		0	ICL741CTY
DIL-14/1C	N	N	T	E-	E+	V-	N	N	T	R	V+	N	N	N					741-LN-DIL	.	ICL741-LN-DD
FLP-10/3C	N	T	E-	E+	V-	T*	R	V+	N	N									741-LN-FLP	0	ICL741-LN-FB
TO5-8/1M	T	E-	E+	V-M	T*	R	V+	N											741-LN-TO5	0	ICL741-LN-TY
DIL-14/1C	N	N	T	E-	E+	V-	N	N	T*	R	V+	N	N	N					741MHS-DIL	0	ICL741MHSDD
FLP-10/3C	N	T	E-	E+	V-	T*	R	V+	N	N									741MHS-FLP	0	ICL741MHSFD
TO5-8/1M	T	E-	E+	V-M	T*	R	V+	N											741MHS-TO5	0	ICL741MHSTY
TO5-8/1M	T	E-	E+	V-M	T*	R	V+	N									TBA222	UA741HM		0	ICL741TY
TO5-8/1M	FT	E-	E+	V-	T*	R	V+	F*									TBB0748	UA748HC		0	ICL748CTY
TO5-8/1M	FT	E-	E+	V-	T*	R	V+	F*									TBC0748	UA748HM		0	ICL748TY
TO5-10/1M	E-	N	T	T*	V-M	G	R	V+	++	E+									ICL8001MTZ	0	ICL8001CTZ
TO5-10/1M	E-	N	T	T*	V-M	G	R	V+	++	E+										0	ICL8001MTZ
TO5-8/1M	T	E-	E+	V-	T*	R	V+	MW											8007AMTV	0	ICL8007ACTV
TO5-8/1M	T	E-	E+	V-	T*	R	V+	MW												0	ICL8007AMTV
TO5-8/1M	N	E-	E+	V-M	N	R	V+	N											ICL8007M-2	0	ICL8007C-1
TO5-8/1M	N	E-	E+	V-M	N	R	V+	N											ICL8007M-2	0	ICL8007C-2
TO5-8/1M	N	E-	E+	V-M	N	R	V+	N											ICL8007C-2	0	ICL8007C-3
TO5-8/1M	T	E-	E+	V-M	T*	R	V+	N											ICL8007M-5	0	ICL8007C-4
TO5-8/1M	T	E-	E+	V-M	T*	R	V+	N											ICL8007M-5	0	ICL8007C-5
TO5-8/1M	T	E-	E+	V-	T*	R	V+	N											ICL8007MTV	0	ICL8007CTV
TO5-8/1M	N	E-	E+	V-M	N	R	V+	N												0	ICL8007M-2
TO5-8/1M	T	E-	E+	V-M	T*	R	V+	N												0	ICL8007M-5
TO5-8/1M	T	E-	E+	V-	T*	R	V+	N												0	ICL8007MTV
DIL-8/1P	T	E-	E+	V-	T*	R	V+										N5556V	RC1556NB		0	ICL8008CPA
TO5-8/1M	T	E-	E+	V-	T*	R	V+	N									N5556T	MC1456G		0	ICL8008CTY
TO5-8/1M	T	E-	E+	V-	T*	R	V+	N									S5556T	MC1556G		0	ICL8008MTY
TO5-8/1M	T	E-	E+	V-	T*	R	V+	B									SG4250T	LM4250CH		0	ICL8021CTA
TO5-8/1M	T	E-	E+	V-	T*	R	V+	B									SG4250CT	LM4250CH		0	ICL8021MTA
DIL-14/1C	E+1	V-	T1	R1	R2	B2	T2	E-2	E+2	T*2	V+	B1	T*1	E-1					ICL8022MDD	0	ICL8022CDD
DIL-14/1C	E+1	V-	T1	R1	R2	B2	T2	E-2	E+2	T*2	V+	B1	T*1	E-1						0	ICL8022MDD
DIL-16/1C	N	E-1	E+1	R2	V23	B3	E-3	E+3	V-	R3	B2	E-2	E+2	R1	V+1	B1			ICL8023MDE	0	ICL8023CDE
DIL-16/1C	N	E-1	E+1	R2	V23	B3	E-3	E+3	V-	R3	B2	E-2	E+2	R1	V+1	B1				0	ICL8023MDE
DIL-16/1C	E+1	E-1	N	T1	T*1	V-	R1	N	N	R2	V+	T2	T*2	N	E-2	E+2			ICL8043MDE	0	ICL8043CDE
DIL-16/1P	E+1	E-1	N	T1	T*1	V-	R1	N	N	R2	V+	T2	T*2	N	E-2	E+2				0	ICL8043CPE
DIL-16/1C	E+1	E-1	N	T1	T*1	V-	R1	N	N	R2	V+	T2	T*2	N	E-2	E+2				0	ICL8043MDE

TYPE NUMBER	MFR	APP	CMP	GBP MIN	SLEW RATE MIN	V_S^+ MAX	V_S^- MAX	T_{op} MAX	A_{VOL} MIN	V_{IO} MAX	I_B MAX	I_{IO} MAX	P_{TOT} MAX	I_{OUT} MIN	V_{OUT} MIN	V_{ICM} MAX	V_{IDF} MAX	dV_{IO}/dT MAX	P_Q MAX	I_Q MAX	CMRR MIN	PSRR MIN	R_{IN} MIN
ICL8043MDE	ING	DFE	INT	.3MHZ	2V/uS	+18V	-18V	125C	94dB	20MV	20pA	2pA	500MWF	5MA	12V	15V	30V	75uV/C	180MW	6MA	70dB	70dB	0.5T
JSFC2301A	THF	GPU	EXT	.	.	+18V	-18V	70C	88dB	7.5MV	250NA	50NA	500MWH	5MA	10V	15V	30V	20uV/C	.	3MA	70dB	50dB	500K
JSFC2307	THF	GPK	INT	.	.	+18V	-18V	70C	88dB	7.5MV	250NA	50NA	500MWF	5MA	10V	15V	30V	20uV/C	.	3MA	70dB	50dB	500K
JSFC2709C	THF	GPU	EXT	.	.	+18V	-18V	70C	84dB	7.5MV	1.5uA	0.5uA	300MWH	5MA	10V	10V	5V	15uV/C	.	5MA	65dB	74dB	50K
JSFC2710C	THF	CPR	EXT	.	.	+14V	-7V	70C	60dB	5MV	25uA	5uA	300MWH	5MA	1V	7V	5V	20uV/C	.	9MA	70dB	.	.
JSFC2711C	THF	DCP	EXT	.	.	+14V	-7V	70C	57dB	5MV	100uA	15uA	300MWH	5MA	1V	7V	5V	20uV/C	.	9MA	.	.	.
JSFC2741C	THF	GPU	INT	.	.	+18V	-18V	70C	86dB	6MV	500NA	200NA	500MWH	5MA	10V	15V	30V	.	.	3MA	7CdB	76dB	300K
JSFC2861	THF	GPU	EXT	.	.	+10V	-10V	70C	80dB	11MV	1.5uA	.33uA	500MWH	4MA	9V	10V	2V	.	.	2MA	80dB	.	50K
L115T1	SGG	XSR	EXT	.	10V/uS	+18V	-18V	70C	80dB	7.5MV	1.5uA	.25uA	500MWF	5MA	15V	15V	15V	.	300MW	10MA	74dB	68dB	300K
L141B1	SGG	GPK	INT	.	0.2V/uS	+18V	-18V	70C	86dB	6MV	500NA	200NA	500MWF	5MA	12V	15V	30V	.	85MW	3MA	70dB	76dB	300K
L141T1	SGG	GPK	INT	.	0.2V/uS	+18V	-18V	70C	86dB	6MV	500NA	200NA	500MWF	5MA	12V	15V	30V	.	85MW	3MA	70dB	76dB	300K
L141T2	SGG	GPK	INT	.	0.3V/uS	+22V	-22V	125C	94dB	5MV	500NA	200NA	500MWF	6MA	13V	15V	30V	.	85MW	3MA	70dB	76dB	300K
L144AL	SLG	TPR	INT	.1MHZ	0.1V/uS	+18V	-18V	125C	70dB	5MV	200NA	50NA	250MWF	.	10V	18V	30V	.	.	.4MA	80dB	80dB	.
L144AP	SLG	TPR	INT	.1MHZ	0.1V/uS	+18V	-18V	125C	70dB	5MV	200NA	50NA	400MWF	.	10V	18V	30V	.	.	.4MA	80dB	80dB	.
L144BL	SLG	TPR	INT	.1MHZ	0.1V/uS	+18V	-18V	85C	70dB	5MV	200NA	50NA	250MWF	.	10V	15V	30V	.	.	.4MA	80dB	80dB	.
L144BP	SLG	TPR	INT	.1MHZ	0.1V/uS	+18V	-18V	85C	70dB	5MV	200NA	50NA	400MWF	.	10V	15V	30V	.	.	.4MA	80dB	80dB	.
L144CJ	SLG	TPR	INT	.1MHZ	0.1V/uS	+18V	-18V	70C	60dB	10V	250NA	70NA	155MWF	.	10V	15V	30V	.	.	.4MA	70dB	80dB	.
L147B1	SGG	DGK	INT	.	0.2V/uS	+18V	-18V	70C	88dB	6MV	500NA	200NA	500MWF	5MA	12V	15V	30V	.	85MW	3MA	70dB	76dB	300K
L148T1	SGG	GPU	EXT	.	.25V/uS	+18V	-18V	70C	86dB	6MV	500NA	200NA	500MWF	5MA	10V	15V	30V	.	85MW	3MA	70dB	76dB	300K
L148T2	SGG	GPU	EXT	.	.25V/uS	+22V	-22V	125C	94dB	5MV	500NA	200NA	500MWF	5MA	10V	15V	30V	.	85MW	3MA	70dB	76dB	300K
L161AL	SLG	QCP	EXT	.	.	+18V	-18V	125C	86dB	2.5MV	15NA	2NA	200MWF	.1MA	2.5V	18V	36V	.	.	.3MA	75dB	65dB	.
L161AP	SLG	QCP	EXT	.	.	+18V	-18V	125C	86dB	2.5MV	15NA	2NA	225MWF	.	2.5V	18V	36V	.	.	.3MA	75dB	65dB	.
L161BL	SLG	QCP	EXT	.	.	+18V	-18V	85C	86dB	2.5MV	15NA	2NA	200MWF	.	2.5V	18V	36V	.	.	.3MA	75dB	65dB	.
L161BP	SLG	QCP	EXT	.	.	+18V	-18V	85C	86dB	2.5MV	15NA	2NA	225MWF	.	2.5V	18V	36V	.	.	.3MA	75dB	65dB	.
L161CJ	SLG	QCP	EXT	.	.	+18V	-18V	70C	83dB	5MV	20NA	3NA	120MWF	.	2.5V	18V	36V	.	.	.3MA	75dB	65dB	.
LD101	ADU	GPU	EXT	.	.	+22V	-22V	125C	94dB	5MV	500NA	200NA	.	5MA	12V	15V	30V	15uV/C	.	3MA	70dB	70dB	300K
LD101A	ADU	GPU	EXT	.	.	+22V	-22V	125C	94dB	2MV	75NA	10NA	.	5MA	12V	15V	30V	15uV/C	.	3MA	80dB	80dB	1.5M
LD102	ADU	VFA	INT	.	.	+18V	-18V	125C	0dB	5MV	10NA	.	.	1MA	10V	.	.	30uV/C	.	6MA	.	60dB	10G
LD106	ADU	CPR	EXT	.	.	+15V	-15V	125C	84dB	2MV	20uA	3uA	.	50MA	2.5V	.	.	10uV/C	163MW	.	.	.	.
LD107	ADU	GPK	INT	.	.	+22V	-22V	125C	94dB	2MV	75NA	10NA	.	5MA	12V	15V	30V	15uV/C	.	3MA	80dB	80dB	1.5M
LD108	ADU	SBA	EXT	.	.	+20V	-20V	125C	96dB	2MV	2NA	0.2NA	.	1MA	13V	15V	1V	15uV/C	.	.6MA	85dB	80dB	30M
LD108A	ADU	SBA	EXT	.	.	+20V	-20V	125C	98dB	0.5MV	2NA	0.2NA	.	.	13V	15V	1V	5uV/C	.	.6MA	96dB	96dB	30M
LD110	ADU	VFA	INT	.	15V/uS	+18V	-18V	125C	0dB	4MV	3NA	.	.	.	10V	15V	15V	50uV/C	.	6MA	.	70dB	10G
LD111	ADU	CPR	EXT	.	.	+18V	-18V	125C	100dB	3MV	100NA	10NA	.	.	.	15V	30V	.	.	6MA	.	.	.
LD112	ADU	SBA	INT	.	.	+20V	-20V	125C	94dB	2MV	2NA	0.2NA	.	1MA	13V	14V	14V	15uV/C	.	.6MA	85dB	80dB	30M
LD118	ADU	XSR	INT	.	50V/uS	+20V	-20V	125C	94dB	4MV	250NA	50NA	.	6MA	12V	15V	1V	.	.	8MA	80dB	70dB	1M
LD119	ADU	DCP	INT	.	.	+18V	-18V	125C	80dB	4MV	500NA	75NA	.	.	.	15V	5V	.	.	12MA	.	.	.
LD124	ADU	QGK	INT	.	.	+16V	-16V	125C	94dB	5MV	150NA	30NA	.	.	.	16V	16V	35uV/C	.	2MA	70dB	65dB	.
LD124A	ADU	QGK	INT	.	.	+16V	-16V	125C	94dB	2MV	50NA	10NA	.	.	.	16V	16V	20uV/C	.	2MA	70dB	65dB	.
LD139	ADU	QCP	EXT	.	.	+18V	-18V	125C	88dB	5MV	100NA	25NA	.	.	.	18V	18V	.	.	2MA	.	.	.
LD139A	ADU	QCP	EXT	.	.	+18V	-18V	125C	94dB	2MV	100NA	25NA	.	.	.	18V	18V	.	.	2MA	.	.	.
LD148	ADU	QGK	INT	.3MHZ	0.2V/uS	+22V	-22V	125C	94dB	5MV	100NA	25NA	.	5MA	12V	22V	44V	.	.	1MA	70dB	77dB	800K
LD149	ADU	QGK	INT	1MHZ	0.5V/uS	+22V	-22V	125C	94dB	5MV	100NA	25NA	.	5MA	12V	22V	44V	.	.	1MA	70dB	77dB	800K
LD155	ADU	FET	INT	.5MHZ	2V/uS	+22V	-22V	125C	94dB	5MV	100pA	20pA	.	5MA	12V	20V	40V	20uV/C	.	4MA	85dB	85dB	0.1T
LD155A	ADU	FET	INT	.5MHZ	3V/uS	+22V	-22V	125C	94dB	2MV	50pA	10pA	.	5MA	12V	20V	40V	5uV/C	.	4MA	85dB	85dB	0.1T
LD156	ADU	HSR	INT	1MHZ	7.5V/uS	+22V	-22V	125C	94dB	5MV	100pA	20pA	.	5MA	12V	20V	40V	20uV/C	.	7MA	85dB	85dB	0.1T
LD156A	ADU	HSR	INT	4MHZ	10V/uS	+22V	-22V	125C	94dB	2MV	50pA	10pA	.	5MA	12V	20V	40V	5uV/C	.	4MA	85dB	85dB	0.1T
LD157	ADU	XSR	INT	4MHZ	6V/uS	+22V	-22V	125C	94dB	5MV	100pA	50pA	.	5MA	12V	20V	40V	20uV/C	.	7MA	85dB	85dB	0.1T
LD157A	ADU	XSR	INT	15MHZ	8V/uS	+22V	-22V	125C	94dB	2MV	50pA	10pA	.	5MA	12V	20V	40V	5uV/C	.	7MA	85dB	85dB	0.1T
LD216	ADU	LBC	INT	.	.	+20V	-20V	85C	86dB	10MV	150pA	50pA	.	1MA	13V	15V	14V	.	.	.8MA	80dB	80dB	300M
LD216A	ADU	LBC	INT	.	.	+20V	-20V	85C	92dB	3MV	50pA	15pA	.	1MA	13V	15V	14V	.	.	.6MA	80dB	80dB	2G
LD301	ADU	GPU	EXT	.	.	+18V	-18V	70C	83dB	10MV	2uA	.75uA	.	.	12V	15V	30V	30uV/C	.	3MA	65dB	70dB	100K
LD301A	ADU	GPU	EXT	.	.	+18V	-18V	70C	88dB	7.5MV	250NA	50NA	.	5MA	12V	15V	30V	30uV/C	.	3MA	70dB	70dB	500K
LD302	ADU	VFA	INT	.	.	+18V	-18V	70C	0dB	15MV	30NA	.	.	1MA	10V	.	.	90uV/C	.	6MA	.	60dB	10G
LD306	ADU	CPR	EXT	.	.	+15V	-15V	70C	84dB	5MV	25uA	5uA	.	50MA	2.5V	.	.	20uV/C	163MW	.	.	.	.
LD307	ADU	GPK	INT	.	.	+18V	-18V	70C	84dB	7.5MV	250NA	50NA	.	5MA	12V	15V	30V	30uV/C	.	.	70dB	70dB	0.5M
LD308	ADU	SBA	EXT	.	.	+18V	-18V	70C	88dB	7.5MV	7NA	1NA	.	1MA	13V	15V	1V	30uV/C	.	.6MA	80dB	80dB	10M
LD308A	ADU	SBA	EXT	.	.	+18V	-18V	70C	98dB	0.5MV	7NA	1NA	.	1MA	13V	15V	1V	5uV/C	.	.6MA	96dB	96dB	10M
LD310	ADU	VFA	INT	.	15V/uS	+18V	-18V	70C	0dB	7.5MV	7NA	.	.	1MA	10V	15V	15V	50uV/C	.	6MA	.	70dB	10G
LD311	ADU	CPR	EXT	.	.	+18V	-18V	70C	100dB	7.5MV	250NA	50NA	.	.	.	15V	30V	.	.	8MA	.	.	.

For detailed explanations of column heading notations, see App. A.
Also for ready references the more important abbreviations used in the column headings are listed below:

LEFT HAND PAGE

APP = application (codes at APP.E.)
CMRR = common mode rejection ratio
CMP = compensation (frequency)
dV_{io}/dT = input offset voltage temperature drift
GBP = gain bandwidth product
I_B = input bias current
I_{io} = input bias offset current
I_c = quiescent supply current
MFR = manufacturer (codes at App.C.)
P_0 = quiescent power consumer
PSRR = power supply rejection ratio
V_{icm} = common mode input voltage rating
V_{idr} = differential input voltage rating
V_{io} = input offset voltage
V_s = dc supply voltage

RIGHT HAND PAGE

Lead out coding summary (details at APP.G.) for different cases (APP.F.)

A = gain adjust
B = bias adjust
C = case
E— = inverting input
E+ = non-inverting input
F,F* = input frequency compensation
G = ground
J = high level input
K = output, open collector
L = output, open emitter
M = metal case
N = not connected
Q = special terminal
R,R* = outputs
S = strobe
T,T* = offset balance
V· = +ve dc supply
V = —ve dc supply
W = guard ring
X = blank position, no lead
++ = +ve supplementary dc supply
—— = —ve supplementary dc supply
ø,ø* = output frequency compensation

CASE (APP F)	LD 1	LD 2	LD 3	LD 4	LD 5	LD 6	LD 7	LD 8	LD 9	LD 10	LD 11	LD 12	LD 13	LD 14	LD 15	LD 16	EUROPE SUBSTITUTE	USA SUBSTITUTE	IS	TYPE NUMBER
CHP																			O	JSFC2301A
CHP																			O	JSF2307
CHP																			O	JSFC2709C
CHP																			O	JSF2710C
CHP																			O	JSF2711C
CHP																			O	JSFC2741C
CHP																			O	JSF.C2861
TO5-10/1M	F	Q	E-	E+	V-	R	ø	V+	ø*	F*								UA715HC	O	L115T1
DIL-14/1P	N	N	T	E-	E+	V-	N	N	T*	R	V+	N	N	N			TBA221A	UA741DC	O	L141B1
TO5-8/1M	T	E-	E+	V-M	T*	R	V+	N									TBA221	UA741HC	O	L141T1
TO5-8/1M	T	E-	E+	V-M	T*	R	V+	N									TBA222	UA741HM	O	L141T2
FLP-14/3G	B	R1	E-2	E+2	E+3	E-3	N	N	R3	V-	R2	E+1	E-1	V+					O	L144AL
DIL-14/1C	B	R1	E-2	E+2	E+3	E-3	N	N	R3	V-	R2	E+1	E-1	V+					O	L144AP
FLP-14/3G	B	R1	E-2	E+2	E+3	E-3	N	N	R3	V-	R2	E+1	E-1	V+				L144AL	O	L144BL
DIL-14/1C	B	R1	E-2	E+2	E+3	E-3	N	N	R3	V-	R2	E+1	E-1	V+				L144AP	O	L144BP
DIL-14/1P	B	R1	E-2	E+2	E+3	E-3	N	N	R3	V-	R2	E+1	E-1	V+				L144BP	O	L144CJ
DIL-14/1P	E-1	E+1	T1	V-	T2	E+2	E-2	T*2	V+	R2	N	R1	V+	T*1			TBB0747A	UA747DC	O	L147B1
TO5-8/1M	FT	E-	E+	V-M	T*	R	V+	F*									TBB0748	UA748HC	O	L148T1
TO5-8/1M	FT	E-	E+	V-M	T*	R	V+	F*									TBC0748	UA748HM	O	L148T2
FLP-16/1G	E+1	E-1	E+2	E-2	E-3	E+3	E-4	E+4	V-	R4	R3	R2	R1	N	B	V+			O	L161AL
DIL-14/1C	E+1	E-1	E+2	E-2	E-3	E+3	E-4	E+4	V-	R4	R3	R2	R1	N	B	V+			O	L161AP
FLP-16/1G	E+1	E-1	E+2	E-2	E-3	E+3	E-4	E+4	V-	R4	R3	R2	R1	N	B	V+		L161AL	O	L161BL
DIL-14/1C	E+1	E-1	E+2	E-2	E-3	E+3	E-4	E+4	V-	R4	R3	R2	R1	N	B	V+		L161AP	O	L161BP
DIL-14/1P	E+1	E-1	E+2	E-2	E-3	E+3	E-4	E+4	V-	R4	R3	R2	R1	N	B	V+		L161BP	O	L161CJ
CHP																	AM101-DICE	AMLD101	O	LD101
CHP																		AMLD101A	O	LD101A
CHP																		AMLD102	O	LD102
CHP																		AM106-DICE	O	LD106
CHP																	AM107-DICE	AMLD107	O	LD107
CHP																		AMLD108	O	LD108
CHP																	AMLD108A	AM108ADICE	O	LD108A
CHP																	AM110-DICE	AMLD110	O	LD110
CHP																	AM111-DICE	AMLD111	O	LD111
CHP																	AM112-DICE	AMLD112	O	LD112
CHP																	AM118-DICE	AMLD118	O	LD118
CHP																	AM119-DICE	AMLD119	O	LD119
CHP																	AM124-DICE	AMLD124	O	LD124
CHP																	AM124ADICE	AMLD124A	O	LD124A
CHP																	AM139-DICE	AMLD139	O	LD139
CHP																		AMLD139A	O	LD139A
CHP																		AMLD148	O	LD148
CHP																	AM149-DICE	AMLD149	O	LD149
CHP																		AMLD155	O	LD155
CHP																		AMLD155A	O	LD155A
CHP																		AMLD156	O	LD156
CHP																		AMLD156A	O	LD156A
CHP																		AMLD157	O	LD157
CHP																		AMLD157A	O	LD157A
CHP																		AMLD216	O	LD216
CHP																		AMLD216A	O	LD216A
CHP																		AMLD301	O	LD301
CHP																		AMLD301A	O	LD301A
CHP																		AMLD302	O	LD302
CHP																		AMLD306	O	LD306
CHP																			O	LD307
CHP																			O	LD308
CHP																			O	LD308A
CHP																			O	LD310
CHP																			O	LD311
CHP																			O	LD312

TYPE NUMBER	M F R	A P P	C M P	GBP MIN	SLEW RATE MIN	V_S^+ MAX	V_S^- MAX	T_{op} MAX	A_{VOL} MIN	V_{IO} MAX	I_B MAX	I_{IO} MAX	P_{TOT} MAX	I_{OUT} MIN	V_{OUT} MIN	V_{ICM} MAX	V_{IDF} MAX	dV_{IO}/dT MAX	P_O MAX	I_O MAX	CM RR MIN	PS RR MIN	R_{IN} MIN
LDJ12	ADU	SBA	INT	.	.	+18V	-18V	70C	88dB	7.5MV	7NA	1NA	.	1MA	13V	14V	14V	15uV/C	.	.8MA	80dB	80dB	10M
LD316	ADU	LBC	INT	.	.	+20V	-20V	70C	86dB	10MV	150pA	50pA	.	1MA	13V	15V	14V	.	.	.8MA	80dB	80dB	300M
LD316A	ADU	LBC	INT	.	.	+20V	-20V	70C	92dB	-3MV	50pA	15pA	.	1MA	13V	15V	14V	.	.	.6MA	80dB	80dB	2G
LD318	ADU	XSR	INT	.	50V/uS	+20V	-20V	70C	88dB	10MV	500NA	200NA	.	6MA	12V	15V	1V	.	.	8MA	70dB	65dB	500K
LD319	ADU	DCP	INT	.	.	+18V	-18V	70C	78dB	8MV	1uA	0.2uA	.	.	.	15V	5V	.	.	12MA	.	.	.
LD324	ADU	QGK	INT	.	.	+16V	-16V	70C	88dB	7MV	250NA	50NA	.	.	.	16V	16V	35uV/C	.	2MA	65dB	65dB	.
LD324A	ADU	QGK	INT	.	.	+16V	-16V	70C	88dB	3MV	100NA	30NA	.	.	.	16V	16V	30uV/C	.	2MA	65dB	65dB	.
LD339	ADU	QCP	EXT	.	.	+18V	-18V	70C	88dB	5MV	250NA	50NA	.	6MA	.	18V	18V	.	.	2MA	.	.	.
LD339A	ADU	QCP	EXT	.	.	+18V	-18V	70C	94dB	2MV	250NA	50NA	.	6MA	.	18V	18V	.	.	2MA	.	.	.
LD348	ADU	QGK	INT	.3MHZ	0.2V/uS	+18V	-18V	70C	88dB	6MV	200NA	50NA	.	5MA	12V	18V	36V	.	.	1MA	70dB	77dB	800K
LD349	ADU	QGK	INT	1MHZ	0.5V/uS	+18V	-18V	70C	88dB	6MV	200NA	50NA	.	5MA	12V	18V	36V	.	.	1MA	70dB	77dB	800K
LD355	ADU	FET	INT	.5MHZ	2V/uS	+18V	-18V	70C	88dB	10MV	100pA	20pA	.	5MA	12V	16V	30V	20uV/C	.	4MA	80dB	80dB	0.1T
LD355A	ADU	FET	INT	.5MHZ	3V/uS	+18V	-18V	70C	94dB	2MV	50pA	10pA	.	5MA	12V	16V	30V	5uV/C	.	4MA	85dB	85dB	0.1T
LD356	ADU	HSR	INT	1MHZ	7.5V/uS	+18V	-18V	70C	88dB	10MV	200pA	50pA	500MWF	5MA	12V	16V	30V	20uV/C	.	10MA	80dB	80dB	0.1T
LD356A	ADU	HSR	INT	4MHZ	10V/uS	+18V	-18V	70C	94dB	2MV	50pA	10pA	.	5MA	12V	16V	30V	5uV/C	.	10MA	85dB	85dB	0.1T
LD357	ADU	XSR	INT	4MHZ	6V/uS	+18V	-18V	70C	88dB	10MV	200pA	50pA	.	5MA	12V	16V	30V	20uV/C	.	10MA	80dB	80dB	0.1T
LD357A	ADU	XSR	INT	15MHZ	8V/uS	+18V	-18V	70C	94dB	2MV	50pA	10pA	.	5MA	12V	16V	30V	5uV/C	.	10MA	85dB	85dB	0.1T
LD592	ADU	BDO	EXT	20MHZ	.	+8V	-8V	125C	50dB	5MV	20uA	3uA	.	2MA	1.5V	6V	5V	.	.	24MA	60dB	50dB	2K
LD592C	ADU	BDO	EXT	20MHZ	.	+8V	-8V	70C	48dB	6MV	30uA	5uA	.	2MA	1.5V	6V	5V	.	.	24MA	60dB	50dB	2K
LD1458	ADU	DGK	INT	.5MHZ	0.3V/uS	+18V	-18V	75C	86dB	6MV	0.5uA	0.2uA	.	5MA	12V	15V	30V	50uV/C	170MW	6MA	70dB	76dB	300K
LD1558	ADU	DGK	INT	.5MHZ	0.3V/uS	+22V	-22V	125C	94dB	5MV	0.5uA	0.2uA	.	5MA	12V	15V	30V	50uV/C	150MW	5MA	70dB	76dB	300K
LF111D	NAU	CPR	EXT	.	.	+18V	-18V	125C	100dB	4MV	50pA	25pA	500MWF	8MA	.	15V	30V	.	.	6MA	.	.	.
LF111F	NAU	CPR	EXT	.	.	+18V	-18V	125C	100dB	4MV	50pA	25pA	500MWF	8MA	.	15V	30V	.	.	6MA	.	.	.
LF111H	NAU	CPR	EXT	.	.	+18V	-18V	125C	100dB	4MV	50pA	25pA	500MWF	8MA	.	15V	30V	.	.	6MA	.	.	.
LF152D	NAU	PIA	EXT	70KHZ	0.3V/uS	+22V	-22V	125C	60dB	15MV	20pA	10pA	900MWF	5MA	10V	22V	44V	30uV/C	.	2MA	.	66dB	1T
LF155AH	NAU	FET	INT	.5MHZ	3V/uS	+22V	-22V	125C	94dB	2MV	50pA	10pA	670MWF	5MA	12V	20V	40V	5uV/C	.	4MA	85dB	85dB	0.1T
LF155H	NAU	FET	INT	.5MHZ	2V/uS	+22V	-22V	125C	94dB	5MV	100pA	20pA	670MWF	5MA	12V	20V	40V	20uV/C	.	4MA	85dB	85dB	0.1T
LF155T	MUG	FET	INT	.5MHZ	2V/uS	+22V	-22V	125C	94dB	5MV	100pA	20pA	670MWF	5MA	12V	20V	40V	20uV/C	.	4MA	85dB	85dB	0.1T
LF156AH	NAU	HSR	INT	4MHZ	10V/uS	+22V	-22V	125C	94dB	2MV	50pA	10pA	670MWF	5MA	12V	20V	40V	5uV/C	.	4MA	85dB	85dB	0.1T
LF156H	NAU	HSR	INT	1MHZ	7.5V/uS	+22V	-22V	125C	94dB	5MV	100pA	20pA	670MWF	5MA	12V	20V	40V	20uV/C	.	7MA	85dB	85dB	0.1T
LF156T	MUG	HSR	INT	1MHZ	7.5V/uS	+22V	-22V	125C	94dB	5MV	100pA	20pA	670MWF	5MA	12V	20V	40V	20uV/C	.	7MA	85dB	85dB	0.1T
LF157AH	NAU	XSR	INT	15MHZ	8V/uS	+22V	-22V	125C	94dB	2MV	50pA	10pA	670MWF	5MA	12V	20V	40V	5uV/C	.	7MA	85dB	85dB	0.1T
LF157H	NAU	XSR	INT	4MHZ	6V/uS	+22V	-22V	125C	94dB	5MV	100pA	50pA	670MWF	5MA	12V	20V	40V	20uV/C	.	7MA	85dB	85dB	0.1T
LF157T	MUG	XSR	INT	4MHZ	6V/uS	+22V	-22V	125C	94dB	5MV	100pA	50pA	670MW	5MA	12V	20V	40V	20uV/C	.	7MA	85dB	85dB	0.1T
LF211D	NAU	CPR	EXT	.	.	+18V	-18V	85C	100dB	4MV	50pA	25pA	500MWF	8MA	.	15V	30V	.	.	6MA	.	.	.
LF211F	NAU	CPR	EXT	.	.	+18V	-18V	85C	100dB	4MV	50pA	25pA	500MWF	8MA	.	15V	30V	.	.	6MA	.	.	.
LF211H	NAU	CPR	EXT	.	.	+18V	-18V	85C	100dB	4MV	50pA	25pA	500MWF	8MA	.	15V	30V	.	.	6MA	.	.	.
LF252D	NAU	PIA	EXT	70KHZ	0.3V/uS	+18V	-18V	85C	60dB	30MV	40pA	20pA	900MWF	5MA	10V	18V	36V	50uV/C	.	2MA	.	60dB	1T
LF255H	NAU	FET	INT	.5MHZ	2V/uS	+22V	-22V	85C	94dB	5MV	100pA	20pA	570MWF	5MA	12V	20V	40V	20uV/C	.	4MA	85dB	85dB	0.1T
LF255T	MUG	FET	INT	.5MHZ	2V/uS	+22V	-22V	85C	94dB	5MV	100pA	20pA	570MWF	5MA	12V	20V	40V	20uV/C	.	4MA	85dB	85dB	0.1T
LF256H	NAU	HSR	INT	1MHZ	7.5V/uS	+22V	-22V	85C	94dB	5MV	100pA	20pA	570MWF	5MA	12V	20V	40V	20uV/C	.	7MA	85dB	85dB	0.1T
LF256T	MUG	HSR	INT	1MHZ	7.5V/uS	+22V	-22V	85C	94dB	5MV	100pA	20pA	570MWF	5MA	12V	20V	40V	20uV/C	.	7MA	85dB	85dB	0.1T
LF257H	NAU	XSR	INT	4MHZ	6V/uS	+22V	-22V	85C	94dB	5MV	100pA	20pA	570MWF	5MA	12V	20V	40V	20uV/C	.	7MA	85dB	85dB	0.1T
LF257T	MUG	XSR	INT	4MHZ	6V/uS	+22V	-22V	85C	94dB	5MV	100pA	20pA	570MWF	5MA	12V	20V	40V	20uV/C	.	7MA	85dB	85dB	0.1T
LF311D	NAU	CPR	EXT	.	.	+18V	-18V	70C	100dB	10MV	150pA	75pA	500MWF	8MA	.	15V	30V	.	.	8MA	.	.	.
LF311F	NAU	CPR	EXT	.	.	+18V	-18V	70C	100dB	10MV	150pA	75pA	500MWF	8MA	.	15V	30V	.	.	8MA	.	.	.
LF311H	NAU	CPR	EXT	.	.	+18V	-18V	70C	100dB	10MV	150pA	75pA	500MWF	8MA	.	15V	30V	.	.	8MA	.	.	.
LF352D	NAU	PIA	EXT	70KHZ	0.3V/uS	+18V	-18V	70C	60dB	30MV	40pA	20pA	900MWF	5MA	10V	18V	36V	50uV/C	.	2MA	.	60dB	1T
LF355AH	NAU	FET	INT	.5MHZ	3V/uS	+18V	-18V	70C	94dB	2MV	50pA	10pA	500MWF	5MA	12V	16V	30V	5uV/C	.	4MA	85dB	85dB	0.1T
LF355H	NAU	FET	INT	.5MHZ	2V/uS	+18V	-18V	70C	88dB	10MV	100pA	20pA	500MWF	5MA	12V	16V	30V	20uV/C	.	4MA	80dB	80dB	0.1T
LF355N	NAU	FET	INT	.5MHZ	2V/uS	+18V	-18V	70C	88dB	10MV	100pA	20pA	500MWF	5MA	12V	16V	30V	20uV/C	.	4MA	80dB	80dB	0.1T
LF355T	MUG	FET	INT	.5MHZ	2V/uS	+18V	-18V	70C	88dB	10MV	100pA	20pA	500MWF	5MA	12V	16V	30V	20uV/C	.	4MA	80dB	80dB	0.1T
LF356AH	NAU	HSR	INT	4MHZ	10V/uS	+18V	-18V	70C	94dB	2MV	50pA	10pA	500MWF	5MA	12V	16V	30V	5uV/C	.	10MA	85dB	85dB	0.1T
LF356H	NAU	HSR	INT	1MHZ	7.5V/uS	+18V	-18V	70C	88dB	10MV	200pA	50pA	500MWF	5MA	12V	16V	30V	20uV/C	.	10MA	80dB	80dB	0.1T
LF356N	NAU	HSR	INT	1MHZ	7.5V/uS	+18V	-18V	70C	88dB	10MV	200pA	50pA	500MWF	5MA	12V	16V	30V	20uV/C	.	10MA	80dB	80dB	0.1T
LF356T	MUG	HSR	INT	1MHZ	7.5V/uS	+18V	-18V	70C	88dB	10MV	200pA	50pA	500MWF	5MA	12V	16V	30V	20uV/C	.	10MA	80dB	80dB	0.1T
LF357AH	NAU	XSR	INT	15MHZ	8V/uS	+18V	-18V	70C	94dB	2MV	50pA	10pA	500MWF	5MA	12V	16V	30V	5uV/C	.	10MA	85dB	85dB	0.1T
LF357H	NAU	XSR	INT	4MHZ	6V/uS	+18V	-18V	70C	88dB	10MV	200pA	50pA	500MWF	5MA	12V	16V	30V	20uV/C	.	10MA	80dB	80dB	0.1T
LF357N	NAU	XSR	INT	4MHZ	6V/uS	+18V	-18V	70C	88dB	10MV	200pA	50pA	500MWF	5MA	12V	16V	30V	20uV/C	.	10MA	80dB	80dB	0.1T
LF357T	MUG	XSR	INT	4MHZ	6V/uS	+18V	-18V	70C	88dB	10MV	200pA	50pA	500MWF	5MA	12V	16V	30V	20uV/C	.	10MA	80dB	80dB	0.1T

For detailed explanations of column heading notations, see App. A.

Also for ready references the more important abbreviations used in the column headings are listed below:

LEFT HAND PAGE
APP = application (codes at APP.E.)
CMRR = common mode rejection ratio
CMP = compensation (frequency)
dV_{io}/dT = input offset voltage temperature drift
GBP = gain bandwidth product
I_B = input bias current
I_{io} = input bias offset current
I_Q = quiescent supply current
MFR = manufacturer (codes at App.C.)
P_U = quiescent power consumer
PSRR = power supply rejection ratio
V_{ICM} = common mode input voltage rating
V_{IDR} = differential input voltage rating
V_{io} = input offset voltage
V_S = dc supply voltage

RIGHT HAND PAGE
Lead out coding summary (details at APP.G.) for different cases (APP.F.)
A = gain adjust
B = bias adjust
C = case
E− = inverting input
E+ = non-inverting input
F,F* = input frequency compensation
G = ground
J = high level input
K = output, open collector
L = output, open emitter
M = metal case
N = not connected
Q = special terminal
R,R* = outputs
S = strobe
T,T* = offset balance
V+ = +ve dc supply
V− = −ve dc supply
W = guard ring
X = blank position, no lead
++ = +ve supplementary dc supply
−− = −ve supplementary dc supply
ϕ,ϕ^* = output frequency compensation

CASE (APP F)	LD 1	LD 2	LD 3	LD 4	LD 5	LD 6	LD 7	LD 8	LD 9	LD 10	LD 11	LD 12	LD 13	LD 14	LD 15	LD 16	EUROPE SUBSTITUTE	USA SUBSTITUTE	ISS	TYPE NUMBER
CHP																			0	LD316
CHP																			0	LD316A
CHP																			0	LD318
CHP																			0	LD319
CHP																			0	LD324
CHP																			0	LD324A
CHP																			0	LD339
CHP																			0	LD339A
CHP																			0	LD348
CHP																			0	LD349
CHP																		AMLD355	0	LD355
CHP																		AMLD355A	0	LD355A
CHP																		AMLD356	0	LD356
CHP																		AMLD356A	0	LD356A
CHP																		AMLD357	0	LD357
CHP																		AMLD357A	0	LD357A
CHP																			0	LD592
CHP																			0	LD592C
CHP																			0	LD1458
CHP																			0	LD1558
DIL-14/1M	N	G	E+	E−	N	V−	T	T*S	R	N	V+	N	N	N				UAF111D	0	LF111D
FLP-10/3M	G	E+	E−	N	V−	T	T*S	N	R	V+									0	LF111F
T05-8/1M	G	E+	E−	V−	T	T*S	R	V+										UAF111H	0	LF111H
DIL-16/1M	B	RT	E+	A1	G	A2	F	R	V+	V−	F*	A*2	Q	A*1	E−	T*			0	LF152D
T05-8/1M	T	E−	E+	V−	T*	R	V+	N										UAF155AHM	0	LF155AH
T05-8/1M	T	E−	E+	V−	T*	R	V+	N										UAF155HM	0	LF155H
T05-8/1M	T	E−	E+	V−	T*	R	V+	N									UAF155HM	LF155H	0	LF155T
T05-8/1M	T	E−	E+	V−	T*	R	V+	N										UAF156AHM	0	LF156AH
T05-8/1M	T	E−	E+	V−	T*	R	V+	N										UAF156HM	0	LF156H
T05-8/1M	T	E−	E+	V−	T*	R	V+	N									UAF156HM	LF156H	0	LF156T
T05-8/1M	T	E−	E+	V−	T*	R	V+	N										UAF157AHM	0	LF157AH
T05-8/1M	T	E−	E+	V−	T*	R	V+	N										UAF157HM	0	LF157H
T05-8/1M	T	E−	E+	V−	T*	R	V+	N									UAF157HM	LF157H	0	LF157T
DIL-14/1M	N	G	E+	E−	N	V−	T	T*S	R	N	V+	N	N	N				LF111D	0	LF211D
FLP-10/3M	G	E+	E−	N	V−	T	T*S	N	R	V+								LF111F	0	LF211F
T05-8/1M	G	E+	E−	V−	T	T*S	R	V+									UAF111D	LF111H	0	LF211H
DIL-16/1M	B	RT	E+	A1	G	A2	F	R	V+	V−	F*	A*2	Q	A*1	E−	T*		LF152D	0	LF252D
T05-8/1M	T	E−	E+	V−	T*	R	V+	N										AMLF255H	0	LF255H
T05-8/1M	T	E−	E+	V−	T*	R	V+	N										LF255H	0	LF255T
T05-8/1M	T	E−	E+	V−	T*	R	V+	N										AMLF256H	0	LF256H
T05-8/1M	T	E−	E+	V−	T*	R	V+	N										LF256H	0	LF256T
T05-8/1M	T	E−	E+	V−	T*	R	V+	N										AMLF257H	0	LF257H
T05-8/1M	T	E−	E+	V−	T*	R	V+	N										LF257H	0	LF257T
DIL-14/1M	N	G	E+	E−	N	V−	T	T*S	R	N	V+	N	N	N				UAF311D	0	LF311D
FLP-10/3M	G	E+	E−	N	V−	T	T*S	N	R	V+								LF211F	0	LF311F
T05-8/1M	G	E+	E−	V−	T	T*S	R	V+										UAF311H	0	LF311H
DIL-16/1M	B	RT	E+	A1	G	A2	F	R	V+	V−	F*	A*2	Q	A*1	E−	T*		LF252D	0	LF352D
T05-8/1M	T	E−	E+	V−	T*	R	V+	N										UAF355AHC	0	LF355AH
T05-8/1M	T	E−	E+	V−	T*	R	V+	N										UAF355HC	0	LF355H
DIL-8/1P	T	E−	E+	V−	T*	R													0	LF355N
T05-8/1M	T	E−	E+	V−	T*	R	V+	N									UAF355HC	LF355H	0	LF355T
T05-8/1M	T	E−	E+	V−	T*	R	V+	N										UA356AHC	0	LF356AH
T05-8/1M	T	E−	E+	V−	T*	R	V+	N										UA356HC	0	LF356H
DIL-8/1P	T	E−	E+	V−	T*	R												AMLF356N	0	LF356N
T05-8/1M	T	E−	E+	V−	T*	R	V+	N									UAF356HC	LF356H	0	LF356T
T05-8/1M	T	E−	E+	V−	T*	R	V+	N										UAF357AHC	0	LF357AH
T05-8/1M	T	E−	E+	V−	T*	R	V+	N										UAF357HC	0	LF357H
DIL-8/1P	T	E−	E+	V−	T*	R												AMLF357N	0	LF357N
T05-8/1M	T	E−	E+	V−	T*	R	V+	N									UAF357HC	LF357H	0	LF357T
T05-8/1M	T	E−	E+	V−M	T*	R	V+	N										UA740HM	0	LF13741H

TYPE NUMBER	M F R	A P P	C M P	GBP MIN	SLEW RATE MIN	V_S^+ MAX	V_S^- MAX	T_{op} MAX	A_{VOL} MIN	V_{IO} MAX	I_B MAX	I_{IO} MAX	P_{TOT} MAX	I_{OUT} MIN	V_{OUT} MIN	V_{ICM} MAX	V_{IDF} MAX	dV_{IO}/dT MAX	P_Q MAX	I_Q MAX	CM RR MIN	PS RR MIN	R_{IN} MIN
LF13741H	NAU	FET	T	.3MHZ	0.2V/uS	+18V	-18V	70C	88dB	15MV	200pA	50pA	500MWF		12V	16V	30V	30uV/C		4MA	70dB	77dB	0.1T
LFD111	ADU	CPR	EXT	.	.	+18V	-18V	125C	100dB	4MV	50pA	25pA	.	8MA	.	15V	30V	.	.	6MA			
LFD311	ADU	CPR	EXT	.	.	+18V	-18V	70C	100dB	10MV	150pA	75pA	.	8MA	.	15V	30V	.	.	8MA			
LH0001ACD	NAU	XLP	EXT	.	.	+20V	-20V	85C	88dB	5MV	200NA	60NA	400MWF	5MA	10V	20V	7V	15uV/C	.	.2MA	70dB	70dB	
LH0001ACF	NAU	XLP	EXT	.	.	+20V	-20V	85C	88dB	5MV	200NA	60NA	400MWF	5MA	10V	20V	7V	15uV/C	.	.2MA	70dB	70dB	
LH0001ACH	NAU	XLP	EXT	.	.	+20V	-20V	85C	88dB	5MV	200NA	60NA	400MWF	5MA	10V	20V	7V	15uV/C	.	.2MA	70dB	70dB	
LH0001AD	NAU	XLP	EXT	.	.	+20V	-20V	125C	88dB	2.5MV	100NA	20NA	400MWF	5MA	10V	20V	7V	15uV/C	.	.2MA	70dB	70dB	
LH0001AF	NAU	XLP	EXT	.	.	+20V	-20V	125C	88dB	2.5MV	100NA	20NA	400MWF	5MA	10V	20V	7V	15uV/C	.	.2MA	70dB	70dB	
LH0001AH	NAU	XLP	EXT	.	.	+20V	-20V	125C	88dB	2.5MV	100NA	20NA	400MWF	5MA	10V	20V	7V	15uV/C	.	.2MA	70dB	70dB	
LH0001H	NAU	XLP	EXT	.	.	+20V	-20V	125C	88dB	1MV	100NA	20NA	400MWF	5MA	10V	20V	7V	.	.	.1MA	70dB	70dB	500K
LH0003CH	NAU	WBA	EXT	.	.	+20V	-20V	85C	86dB	3MV	2uA	0.2uA	500MWF	0.1A	10V	20V	7V	20uV/C	.	3MA	70dB	70dB	25K
LH0003H	NAU	WBA	EXT	.	.	+20V	-20V	125C	86dB	3MV	2uA	0.2uA	500MWF	0.1A	10V	20V	7V	20uV/C	.	.2MA	70dB	70dB	25K
LH0004CH	NAU	HVO	EXT	.	.	+45V	-45V	85C	90dB	1.5MV	120NA	45NA	400MWF	6MA	30V	45V	7V	20uV/C	.	.2MA	70dB	70dB	.
LH0004H	NAU	HVO	EXT	.	.	+45V	-45V	125C	90dB	1MV	100NA	20NA	400MWF	6MA	30V	45V	7V	20uV/C	.	.2MA	70dB	70dB	.
LH0005AH	NAU	WBA	EXT	.	.	+20V	-20V	125C	72dB	3MV	25NA	5NA	400MWF	50MA	6V	20V	15V	50uV/C	.	5MA	60dB	60dB	1M
LH0005CH	NAU	WBA	EXT	.	.	+20V	-20V	85C	66dB	10MV	100NA	25NA	400MWF	50MA	6V	20V	15V	.	.	5MA	50dB	50dB	0.5M
LH0005H	NAU	WBA	EXT	.	.	+20V	-20V	125C	66dB	10MV	50NA	20NA	400MWF	50MA	6V	20V	15V	100uV/C	.	5MA	55dB	55dB	1M
LH0020CG	NAU	PIA	EXT	.	.	+22V	-22V	85C	94dB	6MV	500NA	200NA	1.5WF	45MA	14V	15V	30V	.	.	6MA	90dB	90dB	300K
LH0020G	NAU	PIA	EXT	.	.	+22V	-22V	125C	100dB	2.5MV	250NA	50NA	1.5WF	45MA	14V	15V	30V		.	5MA	90dB	90dB	600K
LH0021CK	NAU	HCO	INT	.	1V/uS	+18V	-18V	125C	100dB	6MV	500NA	200NA	5WF	1A	13V	15V	30V	.	120MW	4MA	70dB	70dB	300K
LH0021K	NAU	HCO	INT	.	1.5V/uS	+18V	-18V	125C	100dB	3MV	300NA	100NA	5WF	1.1A	13V	15V	30V	.	105MW	4MA	70dB	80dB	300K
LH0022CD	NAU	FET	INT	.3MHZ	1V/uS	+22V	-22V	85C	97dB	6MV	25pA	5pA	500MWF	10MA	10V	15V	30V	15uV/C	85MW	3MA	70dB	70dB	0.1T
LH0C22CF	NAU	FET	INT	.3MHZ	1V/uS	+22V	-22V	85C	97dB	6MV	25pA	5pA	500MWF	10MA	10V	15V	30V	15uV/C	85MW	3MA	70dB	70dB	0.1T
LH0022CH	NAU	FET	INT	.3MHZ	1V/uS	+22V	-22V	85C	97dB	6MV	25pA	5pA	500MWF	10MA	10V	15V	30V	15uV/C	85MW	3MA	70dB	70dB	0.1T
LH0022D	NAU	FET	INT	.3MHZ	1.5V/uS	+22V	-22V	125C	100dB	4MV	10pA	2pA	500MWF	10MA	10V	15V	30V	10uV/C	75MW	3MA	80dB	80dB	0.1T
LH0022F	NAU	FET	INT	.3MHZ	1.5V/uS	+22V	-22V	125C	100dB	4MV	10pA	2pA	500MWF	10MA	10V	15V	30V	10uV/C	75MW	3MA	80dB	80dB	0.1T
LH0022H	NAU	FET	INT	.3MHZ	1.5V/uS	+22V	-22V	125C	100dB	4MV	10pA	2pA	500MWF	10MA	10V	15V	30V	10uV/C	75MW	3MA	80dB	80dB	0.1T
LH0024CH	NAU	XSR	EXT	.	250V/uS	+18V	-18V	85C	70dB	8MV	40uA	15uA	600MWF	10MA	10V	18V	5V	125uV/C	.	14MA	50dB	50dB	.
LH0024H	NAU	XSR	EXT	.	400V/uS	+18V	-18V	125C	72dB	4MV	30uA	5uA	600MWF	10MA	12V	18V	5V	100uV/C	.	14MA	50dB	50dB	.
LH0032CG	NAU	XSR	EXT	.	350V/uS	+18V	-18V	85C	60dB	15MV	200pA	50pA	1.5WF	10MA	10V	18V	30V	125uV/C	.	22MA	50dB	50dB	.
LH0032G	NAU	XSR	EXT	.	350V/uS	+18V	-18V	125C	60dB	5MV	100pA	25pA	1.5WF	10MA	10V	18V	30V	125uV/C	.	20MA	50dB	50dB	.
LH0033CG	NAU	VFA	INT	50MHZ	1KV/uS	+20V	-20V	85C	0dB	20MV	.15NA		1.5WF	90MA	12V	20V	.	200uV/C	720MW	24MA			10G
LH0033CJ	NAU	VFA	INT	50MHZ	1KV/uS	+20V	-20V	85C	0dB	20MV	.15NA		1.5WF	90MA	12V	20V	.	200uV/C	720MW	24MA			10G
LH0033G	NAU	VFA	INT	50MHZ	1KV/uS	+20V	-20V	125C	0dB	10MV	0.1NA		1.5WF	90MA	12V	20V	.	200uV/C	660MW	22MA			10G
LH0033J	NAU	VFA	INT	50MHZ	1KV/uS	+20V	-20V	125C	0dB	10MV	0.1NA		1.5WF	90MA	12V	20V	.	200uV/C	660MW	22MA			10G
LH0041CG	NAU	HCO	INT	.	1V/uS	+18V	-18V	85C	100dB	6MV	500NA	200NA	1.5WF	.13A	13V	15V	30V	.	120MW	4MA	70dB	70dB	300K
LH0041CJ	NAU	HCO	INT	.	1V/uS	+18V	-18V	85C	100dB	6MV	500NA	200NA	1.5WF	.13A	13V	15V	30V	.	120MW	4MA	70dB	70dB	300K
LH0041G	NAU	HCO	INT	.	1.5V/uS	+18V	-18V	85C	100dB	3MV	300NA	100NA	1.5WF	.13A	13V	15V	30V	.	105MW	4MA	70dB	80dB	300K
LH0042CD	NAU	FET	INT	.3MHZ	1V/uS	+22V	-22V	85C	88dB	20MV	50pA	10pA	500MWF	10MA	10V	15V	30V	50uV/C	120MW	4MA	70dB	70dB	0.1T
LH0042CF	NAU	FET	INT	.3MHZ	1V/uS	+22V	-22V	85C	88dB	20MV	50pA	10pA	500MWF	10MA	10V	15V	30V	50uV/C	120MW	4MA	70dB	70dB	0.1T
LH0042CH	NAU	FET	INT	.3MHZ	1V/uS	+22V	-22V	85C	88dB	20MV	50pA	10pA	500MWF	10MA	10V	15V	30V	50uV/C	120MW	4MA	70dB	70dB	0.1T
LH0042D	NAU	FET	INT	.3MHZ	1.5V/uS	+22V	-22V	125C	94dB	20MV	25pA	5pA	500MWF	10MA	10V	15V	30V	25uV/C	105MW	4MA	70dB	70dB	0.1T
LH0042F	NAU	FET	INT	.3MHZ	1.5V/uS	+22V	-22V	125C	94dB	20MV	25pA	5pA	500MWF	10MA	10V	15V	30V	25uV/C	105MW	4MA	70dB	70dB	0.1T
LH0042H	NAU	FET	INT	.3MHZ	1.5V/uS	+22V	-22V	125C	94dB	20MV	25pA	5pA	500MWF	10MA	10V	15V	30V	25uV/C	105MW	4MA	70dB	70dB	0.1T
LH0044ACH	FAU	PIA	INT	.2MHZ	.01V/uS	+20V	-20V	125C	120dB	25uV	15NA	2.5NA	600MWF	1MA	13V	15V	15V	0.5uV/C	90MW	3MA	120dB	120dB	5M
LH0044AH	FAU	PIA	INT	.2MHZ	.01V/uS	+20V	-20V	125C	120dB	25uV	15NA	2.5NA	600MWF	1MA	13V	15V	15V	0.5uV/C	90MW	3MA	120dB	120dB	.
LH0044BH	FAU	PIA	INT	.2MHZ	.01V/uS	+20V	-20V	85C	114dB	50uV	30NA	5NA	600MWF	1MA	12V	15V	15V	0.5uV/C	120MW	4MA	114dB	114dB	2.5M
LH0044CH	FAU	PIA	INT	.2MHZ	.01V/uS	+20V	-20V	85C	114dB	100uV	30NA	5NA	600MWF	1MA	12V	15V	15V	1uV/C	120MW	4MA	114dB	114dB	2.5M
LH0044H	FAU	PIA	INT	.2MHZ	.01V/uS	+20V	-20V	85C	114dB	50uV	30NA	5NA	600MWF	1MA	12V	15V	15V	1uV/C	120MW	4MA	114dB	114dB	2.5M
LH0052CD	NAU	FET	INT	.3MHZ	1V/uS	+22V	-22V	125C	97dB	1MV	5pA	0.2pA	500MWF	10MA	10V	15V	30V	10uV/C	120MW	4MA	76dB	76dB	0.1T
LH0052CH	NAU	FET	INT	.3MHZ	1V/uS	+22V	-22V	85C	97dB	1MV	5pA	0.2pA	500MWF	10MA	10V	15V	30V	10uV/C	120MW	4MA	76dB	76dB	0.1T
LH0052D	NAU	FET	INT	.3MHZ	1.5V/uS	+22V	-22V	125C	100dB	0.5MV	1pA	0.1pA	500MWF	10MA	10V	15V	30V	5uV/C	105MW	4MA	80dB	80dB	0.1T
LH0052H	NAU	FET	INT	.3MHZ	1.5V/uS	+22V	-22V	125C	100dB	0.5MV	1pA	0.1pA	500MWF	10MA	10V	15V	30V	5uV/C	105MW	4MA	80dB	80dB	0.1T
LH0061CK	NAU	HCO	INT	2MHZ	25V/uS	+18V	-18V	85C	88dB	10MV	500NA	200NA	5WF	0.5A	10V	15V	1V	25uV/C	450MW	15MA	60dB	50dB	300K
LH0061K	NAU	HCO	INT	2MHZ	25V/uS	+18V	-18V	125C	94dB	4MV	300NA	100NA	5WF	0.5A	10V	15V	1V	25uV/C	300MW	10MA	70dB	70dB	300K
LH0062CD	NAU	FET	INT	3MHZ	50V/uS	+20V	-20V	85C	88dB	15MV	65pA	5pA	500MWF	10MA	12V	15V	30V	35uV/C	360MW	12MA	70dB	70dB	0.1T
LH0062CH	NAU	FET	INT	3MHZ	50V/uS	+20V	-20V	85C	88dB	15MV	65pA	5pA	500MWF	10MA	12V	15V	30V	35uV/C	360MW	12MA	70dB	70dB	0.1T
LH0062D	NAU	FET	INT	3MHZ	50V/uS	+20V	-20V	125C	96dB	5MV	10pA	2pA	500MWF	10MA	12V	15V	30V	25uV/C	240MW	8MA	80dB	80dB	0.1T
LH0062H	NAU	FET	INT	3MHZ	50V/uS	+20V	-20V	125C	96dB	5MV	10pA	2pA	500MWF	10MA	12V	15V	30V	25uV/C	240MW	8MA	80dB	80dB	
LH0063CK	NAU	VFA	INT	99MHZ	2KV/uS	+20V	-20V	85C	0dB	50MV	.2NA		5WF	0.2A	10V	20V		1V/C	2.40W	80MA			10G

For detailed explanations of column heading notations, see App. A.

Also for ready references the more important abbreviations used in the column headings are listed below:

LEFT HAND PAGE

APP = application (codes at APP.E.)

CMRR = common mode rejection ratio

CMP = compensation (frequency)

dV_{io}/dT = input offset voltage temperature drift

GBP = gain bandwidth product

I_B = input bias current

I_{io} = input bias offset current

I_Q = quiescent supply current

MFR = manufacturer (codes at App.C.)

P_Q = quiescent power consumer

PSRR = power supply rejection ratio

V_{ICM} = common mode input voltage rating

V_{iDF} = differential input voltage rating

V_{io} = input offset voltage

V_S = dc supply voltage

RIGHT HAND PAGE

Lead out coding summary (details at APP.G.) for different cases (APP.F.)

A = gain adjust
B = bias adjust
C = case
E— = inverting input
E+ = non-inverting input
F,F* = input frequency compensation
G = ground
J = high level input
K = output, open collector
L = output, open emitter
M = metal case
N = not connected
Q = special terminal
R,R* = outputs
S = strobe
T,T* = offset balance
V+ = +ve dc supply
V— = —ve dc supply
W = guard ring
X = blank position, no lead
+ + = +ve supplementary dc supply
— — = —ve supplementary dc supply
ø,ø* = output frequency compensation

CASE (APP F)	LD 1	LD 2	LD 3	LD 4	LD 5	LD 6	LD 7	LD 8	LD 9	LD 10	LD 11	LD 12	LD 13	LD 14	LD 15	LD 16	EUROPE SUBSTITUTE	USA SUBSTITUTE	I/S	TYPE NUMBER
CHP																			0	LFD111
CHP																		AMLFD311	0	LFD311
DIL-14/1M	E-	F	V-	N	øS	N	R	N	V+	N	ø*	N	N	E+				LH0001AD	0	LH0001ACD
FLP-10/3G	E+	ø*	E-	F	øS	V-	R	R	V+	N	N							LH0001AF	0	LH0001ACF
TO5-10/1M	ø*	E+	V-	E-	F	N	N	R	V+	øS								LH0001AH	0	LH0001ACH
DIL-14/1M	E-	F	V-	N	øS	N	R	N	V+	N	ø*	N	N	E+				LH0001AD	0	LH0001AD
FLP-10/3G	E+	ø*	E-	F	øS	V-	R	R	V+	N	N								0	LH0001AF
TO5-10/1M	ø*	E+	V-	E-	F	N	N	R	V+	øS									0	LH0001AH
TO5-10/1M	F	E+	V-	E-	ø	B	B*	R	V+	F*									0	LH0001H
TO5-10/1M	F	E+	V-	E-	F*	N	G	R	V+	øS								LH0003H	0	LH0003CH
TO5-10/1M	F	E+	V-	E-	F*	N	G	R	V+	øS									0	LH0003H
TO5-10/1M	F	E+	V-	E-	F*	B	B*	R	V+	øS								LH0004H	0	LH0004CH
TO5-10/1M	F	E+	V-	E-	F*	B	B*	R	V+	øS									0	LH0004H
TO5-10/1M	E+	V+	V-	F	N	G	R	V-	ø	F*									0	LH0005AH
TO5-10/1M	E+	V+	V-	F	N	G	R	V-	ø	F*								LH0005H	0	LH0005CH
TO5-10/1M	E+	V+	V-	F	N	G	R	V-	ø	F*								LH0005AH	0	LH0005H
TO8-12/1M	N	ø	F	F*	E+	E-	T	T*	G	V-	R	V+						LH0020G	0	LH0020CG
TO8-12/1M	N	ø	F	F*	E+	E-	T	T*	G	V-	R	V+							0	LH0020G
TO3-8/2M	Q	V+	G	F	E-	E+	V-	Q*										LH0021K	0	LH0021CK
TO3-8/1M	Q	V+	G	F	E-	E+	V-	Q*											0	LH0021K
DIL-14/1M	N	N	E-	N	V-	T	N	R	V+	T*	N	E+	N	N				LH0022D	0	LH0022CD
FLP-10/3G	E+	N	N	V-	T	V+	T*	Q	N	E-								LH0022F	0	LH0022CF
TO5-8/1M	T	E-	E+	V-	T*	R	V+	N										LH0022H	0	LH0022CH
DIL-14/1M	N	N	E-	N	V-	T	N	R	V+	T*	N	E+	N	N					0	LH0022D
FLP-10/3G	E+	N	V-	T	R	V+	T*	Q	N	E-									0	LH0022F
TO5-8/1M	T	E-	E+	V-	T*	R	V+	N											0	LH0022H
TO5-8/1M	ø	E-	E+	V-	TF*	R	V+	T*F										LH0024H	0	LH0024CH
TO5-8/1M	ø	E-	E+	V-	TF*	R	V+	T*F											0	LH0024H
TO8-12/1M	N	ø	F*T	T*F	E-	E+	N	N	N	V-	R	V+						LH0032G	0	LH0032CG
TO8-12/1M	N	ø	F*T	T*F	E-	E+	N	N	N	V-	R	V+							0	LH0032G
TO8-12/1M	V*+	N	N	N	E+	T	T*	N	V*-	V-	R	V+						LH0033G	0	LH0033CG
DIM-8/3C	E+	T	T*	R	V-	V*-	V*+	V+										LH0033J	0	LH0033CJ
TO8-12/1M	V*+	N	N	N	E+	T	T*	N	V*-	V-	R	V+							0	LH0033G
DIM-8/3C	E+	T	T*	R	V-	V*-	V*+	V+											0	LH0033J
TO8-12/1M	Q	F	G	N	E-	E+	T	T*	Q*	V-	R	V+						LH0041G	0	LH0041CG
DIM-8/3C	V-	N	G	V+	F	R	E-	E+											0	LH0041CJ
TO8-12/1M	Q	F	G	N	E-	E+	T	T*	Q*	V-	R	V+							0	LH0041G
DIL-14/1M	N	N	E-	N	V-	T	N	R	V+	T*	N	E+	N	N				LH0042D	0	LH0042CD
FLP-10/3G	E+	N	V-	T	R	V+	T*	Q	N	E-								LH0042F	0	LH0042CF
TO5-8/1M	T	E-	E+	V-	T*	R	V+	N										LH0042H	0	LH0042CH
DIL-14/1M	N	N	E-	N	V-	T	N	R	V+	T*	N	E+	N	N					0	LH0042D
FLP-10/3G	E+	N	V-	T	R	V+	T*	Q	N	E-									0	LH0042F
TO5-8/1M	T	E-	E+	V-	T*	R	V+	N											0	LH0042H
TO5-8/1M	F*	E-	E+	V-	N	R	V+	F										LH0044AH	0	LH0044ACH
TO5-8/1M	F*	E-	E+	V-	N	R	V+	F											0	LH0044AH
TO5-8/1M	F*	E-	E+	V-	N	R	V+	F										LH0044AH	0	LH0044BH
TO5-8/1M	F*	E-	E+	V-	N	R	V+	F										LH0044H	0	LH0044CH
TO5-8/1M	F*	E-	E+	V-	N	R	V+	F										LH0044AH	0	LH0044H
DIL-14/1M	N	N	E-	N	V-	T	N	R	V+	T*	N	E+	N	N				LH0052D	0	LH0052CD
TO5-8/1M	T	E-	E+	V-	T*	R	V+	N										LH0052H	0	LH0052CH
DIL-14/1M	N	N	E-	N	V-	T	N	R	V+	T*	N	E+	N	N					0	LH0052D
TO5-8/1M	T	E-	E+	V-	T*	R	V+	N											0	LH0052H
TO3-8/2M	Q	V+	G	F	E-	E+	V-	Q*										LH0061K	0	LH0061CK
TO3-8/2M	Q	V+	G	F	E-	E+	V-	Q*											0	LH0061K
DIL-14/1M	N	N	E-	N	V-	T*F	ø	R	V+	F*T	N	E+	N	N				LH0062D	0	LH0062CD
TO5-8/1M	T*F	E-	E+	V-	F*T	R	V+	ø										LH0062H	0	LH0062CH
DIL-14/1M	N	N	E-	N	V-	T*F	ø	R	V+	F*T	N	E+	N	N					0	LH0062D
TO5-8/1M	T*F	E-	E+	V-	F*T	R	V+	ø											0	LH0062H
TO3-8/2M	V+	V*+	R	E+	T	T*	V-	V*-										LH0063K	0	LH0063CK
TO3-8/2M	V+	V*+	R	E+	T	T*	V-	V*-											0	LH0063K

TYPE NUMBER	MFR	APP	CMP	GBP MIN	SLEW RATE MIN	Vs+ MAX	Vs- MAX	Top MAX	Avol MIN	Vio MAX	Ib MAX	Iio MAX	Ptot MAX	Iout MIN	Vout MIN	Vicm MAX	Vidf MAX	dVio/dT MAX	PQ MAX	Iq MAX	CMRR MIN	PSRR MIN	Rin MIN
LH0063K	NAU	VFA	INT	99MHZ	2KV/uS	+20V	-20V	125C	0dB	25MV	2NA		5WF	0.2A	10V	20V		1V/C	2.25W	75MA			10G
LH101F	NAU	GPK	INT			+22V	-22V	125C	88dB	5MV	500NA	200NA	500MWF	5MA	12V	15V	30V	15uV/C		3MA	70dB	70dB	300K
LH101H	NAU	GPK	INT			+22V	-22V	125C	88dB	5MV	500NA	200NA	500MWF	5MA	12V	15V	30V	15uV/C		3MA	70dB	70dB	300K
LH201F	NAU	GPK	INT			+22V	-22V	70C	86dB	7.5MV	1.5uA	0.5uA	250MW	5MA	12V	15V	30V	30uV/C		3MA	65dB	70dB	150K
LH201H	NAU	GPK	INT			+22V	-22V	70C	86dB	7.5MV	1.5uA	0.5uA	250MW	5MA	12V	15V	30V	30uV/C		3MA	65dB	70dB	150K
LH740ACH	NAU	FET	INT	.3MHZ	2V/uS	+22V	-22V	85C	94dB	20MV	500pA	150pA	500MWF	5MA	12V	15V	5V	25uV/C		4MA	80dB	80dB	0.1T
LH740AH	NAU	FET	INT	.3MHZ	2V/uS	+22V	-22V	125C	94dB	15MV	200pA	100pA	500MWF	5MA	12V	15V	5V	25uV/C		4MA	80dB	80dB	0.1T
LH2101AD	NAU	DGU	EXT			+22V	-22V	125C	94dB	2MV	75NA	10NA	500MWF	5MA	12V	15V	30V	15uV/C		3MA	80dB	80dB	1.5M
LH2101AF	MUG	DGU	EXT			+22V	-22V	125C	94dB	2MV	75NA	10NA	500MWF	5MA	12V	15V	30V	15uV/C		3MA	80dB	80dB	1.5M
LH2101AF	NAU	DGU	EXT			+22V	-22V	125C	94dB	2MV	75NA	10NA	500MWF	5MA	12V	15V	30V	15uV/C		3MA	80dB	80dB	1.5M
LH2108AD	NAU	DSB	EXT			+20V	-20V	125C	94dB	0.5MV	2NA	0.2NA	500MWF	1MA	13V	15V	1V	5uV/C		.6MA	96dB	96dB	30M
LH2108AF	MUG	DSB	EXT			+20V	-20V	125C	94dB	0.5MV	2NA	0.2NA	500MWF	1MA	13V	15V	1V	5uV/C		.6MA	96dB	96dB	30M
LH2108D	NAU	DSB	EXT			+20V	-20V	125C	94dB	2MV	2NA	0.2NA	500MWF	1MA	13V	15V	1V	15uV/C		.4MA	85dB	80dB	30M
LH2108F	MUG	DSB	EXT			+20V	-20V	125C	94dB	2MV	2NA	0.2NA	500MWF	1MA	13V	15V	1V	15uV/C		.4MA	85dB	80dB	30M
LH2108F	NAU	DSB	EXT			+20V	-20V	125C	94dB	2MV	2NA	0.2NA	500MWF	1MA	13V	15V	1V	15uV/C		.4MA	85dB	80dB	30M
LH2110D	NAU	DVF	INT		15V/uS	+18V	-18V	125C	0dB	4MV	3NA		500MWF	1MA	10V	15V	15V	50uV/C		6MA		70dB	10G
LH2110F	NAU	DVF	INT		15V/uS	+18V	-18V	125C	0dB	4MV	3NA		500MWF	1MA	10V	15V	15V	50uV/C		6MA		70dB	10G
LH2111D	NAU	DCP	EXT			+18V	-18V	125C	100dB	3MV	100NA	10NA	500MWF	8MA		15V	30V			6MA			
LH2111F	NAU	DCP	EXT			+18V	-18V	125C	100dB	3MV	100NA	10NA	500MWF	8MA		15V	30V			6MA			
LH2201AD	NAU	DGU	EXT			+22V	-22V	85C	94dB	2MV	75NA	10NA	500MWF	5MA	12V	15V	30V	15uV/C		4MA	80dB	80dB	1.5M
LH2201AF	MUG	DGU	EXT			+22V	-22V	85C	94dB	2MV	75NA	10NA	500MWF	5MA	12V	15V	30V	15uV/C		4MA	80dB	80dB	1.5M
LH2201AF	NAU	DGU	EXT			+22V	-22V	85C	94dB	2MV	75NA	10NA	500MWF	5MA	12V	15V	30V	15uV/C		4MA	80dB	80dB	1.5M
LH2208AD	NAU	DSB	EXT			+20V	-20V	85C	94dB	0.5MV	2NA	0.2NA	500MWF	1MA	13V	15V	1V	5uV/C		.6MA	96dB	96dB	30M
LH2208AF	MUG	DSB	EXT			+20V	-20V	85C	94dB	0.5MV	2NA	0.2NA	500MWF	1MA	13V	15V	1V	5uV/C		.5MA	96dB	96dB	30M
LH2208D	NAU	DSB	EXT			+20V	-20V	85C	94dB	2MV	2NA	0.2NA	500MWF	1MA	13V	15V	1V	5uV/C		.4MA	85dB	80dB	30M
LH2208F	MUG	DSB	EXT			+20V	-20V	85C	94dB	2MV	2NA	0.2NA	500MWF	1MA	13V	15V	1V	15uV/C		.4MA	85dB	80dB	30M
LH2208F	NAU	DSB	EXT			+20V	-20V	85C	94dB	2MV	2NA	0.2NA	500MWF	1MA	13V	15V	1V	15uV/C		.4MA	85dB	80dB	30M
LH2210D	NAU	DVF	INT		15V/uS	+18V	-18V	85C	0dB	4MV	3NA		500MWF	1MA	10V	15V	15V	50uV/C		6MA		70dB	10G
LH2210F	NAU	DVF	INT		15V/uS	+18V	-18V	85C	0dB	4MV	3NA		500MWF	1MA	10V	15V	15V	50uV/C		6MA		70dB	10G
LH2211D	NAG	DCP	EXT			+18V	-18V	85C	100dB	3MV	100NA	10NA	500MWF	8MA		15V	30V			6MA			
LH2211F	NAG	DCP	EXT			+18V	-18V	85C	100dB	3MV	100NA	10NA	500MWF	8MA		15V	30V			6MA			
LH2301AD	NAU	DGU	EXT			+22V	-22V	70C	88dB	7.5MV	250NA	50NA	500MWF	5MA	12V	15V	30V	30uV/C		3MA	70dB	70dB	0.5M
LH2301AF	MUG	DGU	EXT			+22V	-22V	70C	88dB	7.5MV	250NA	50NA	500MWF	5MA	12V	15V	30V	30uV/C		3MA	70dB	70dB	0.5M
LH2301AF	NAU	DGU	EXT			+22V	-22V	70C	88dB	7.5MV	250NA	50NA	500MWF	5MA	12V	15V	30V	30uV/C		3MA	70dB	70dB	0.5M
LH2308AD	NAU	DSB	EXT			+20V	-20V	70C	88dB	0.5MV	7NA	1NA	500MWF	1MA	13V	15V	1V	5uV/C		.8MA	96dB	96dB	10M
LH2308AF	MUG	DSB	EXT			+20V	-20V	70C	88dB	0.5MV	7NA	1NA	500MWF	1MA	13V	15V	1V	5uV/C		.8MA	96dB	96dB	10M
LH2308D	NAU	DSB	EXT			+20V	-20V	70C	88dB	7.5MV	7NA	1NA	500MWF	1MA	13V	15V	1V	30uV/C			80dB	80dB	10M
LH2308F	MUG	DSB	EXT			+20V	-20V	70C	88dB	7.5MV	7NA	1NA	500MWF	1MA	13V	15V	1V	30uV/C			80dB	80dB	10M
LH2308F	NAU	DSB	EXT			+20V	-20V	70C	88dB	7.5MV	7NA	1NA	500MWF	1MA	13V	15V	1V	30uV/C			80dB	80dB	10M
LH2310D	NAU	DVF	INT		15V/uS	+18V	-18V	70C	0dB	7.5MV	7NA		500MWF	1MA	10V	15V	15V	50uV/C		6MA		70dB	10G
LH2310F	NAU	DVF	INT		15V/uS	+18V	-18V	70C	0dB	7.5MV	7NA		500MWF	1MA	10V	15V	15V	50uV/C		6MA		70dB	10G
LH2311D	NAU	DCP	EXT			+18V	-18V	70C	100dB	7.5MV	250NA	50NA	500MWF	8MA		15V	30V			8MA			
LH2311F	NAU	DCP	EXT			+18V	-18V	70C	100dB	7.5MV	250NA	50NA	500MWF	8MA		15V	30V			8MA			
LH24250CD	NAU	DPR	INT			+18V	-18V	70C	95dB	6MV	30NA	10NA	500MWF	5MA	10V	15V	15V				70dB	76dB	3M
LH24250CF	NAU	DPR	INT			+18V	-18V	70C	95dB	6MV	30NA	10NA	500MWF	5MA	10V	15V	15V				70dB	76dB	3M
LH24250D	NAU	DPR	INT			+18V	-18V	125C	100dB	3MV	15NA	5NA	500MWF	5MA	10V	15V	15V				70dB	76dB	3M
LH24250F	NAU	DPR	INT			+18V	-18V	125C	100dB	3MV	15NA	5NA	500MWF	5MA	10V	15V	15V				70dB	76dB	3M
LM101AD	NAU	GPU	EXT			+22V	-22V	125C	94dB	2MV	75NA	10NA	500MWF	5MA	12V	15V	30V	15uV/C		3MA	80dB	80dB	1.5M
LM101AF	NAU	GPU	EXT			+22V	-22V	125C	94dB	2MV	75NA	10NA	500MWF	5MA	12V	15V	30V	15uV/C		3MA	80dB	80dB	1.5M
LM101AF	SJU	GPU	EXT			+22V	-22V	125C	94dB	2MV	75NA	10NA	500MWF	5MA	12V	15V	30V	15uV/C			80dB	80dB	1.5M
LM101AH	NAU	GPU	EXT			+22V	-22V	125C	94dB	2MV	75NA	10NA	500MWF	5MA	12V	15V	30V	15uV/C		3MA	80dB	80dB	1.5M
LM101AJ	NAU	GPU	EXT			+22V	-22V	125C	94dB	2MV	75NA	10NA	500MWF	5MA	12V	15V	30V	15uV/C		3MA	80dB	80dB	1.5M
LM101AJ-14	NAU	GPU	EXT			+22V	-22V	125C	94dB	2MV	75NA	10NA	500MWF	5MA	12V	15V	30V	15uV/C		3MA	80dB	80dB	1.5M
LM101AN(8)	MUG	GPU	EXT			+22V	-22V	125C	94dB	2MV	75NA	10NA	500MWF	5MA	12V	15V	30V	15uV/C		3MA	80dB	80dB	1.5M
LM101AT	SJU	GPU	EXT			+22V	-22V	125C	94dB	2MV	75NA	10NA	500MWF	5MA	12V	15V	30V	15uV/C			80dB	80dB	1.5M
LM101AV	MUG	GPU	EXT			+22V	-22V	125C	94dB	2MV	75NA	10NA	500MWF	5MA	12V	15V	30V	15uV/C		3MA	80dB	80dB	1.5M
LM101D	SJU	GPU	EXT			+22V	-22V	125C	94dB	5MV	1.5uA	0.5uA	500MWF	5MA	12V	15V	30V	15uV/C		3MA	70dB	70dB	300K
LM101F	MUG	GPU	EXT			+22V	-22V	125C	94dB	5MV	1.5uA	0.5uA	500MWF	5MA	12V	15V	30V	15uV/C		3MA	70dB	70dB	300K
LM101F	NAU	GPU	EXT			+22V	-22V	125C	94dB	5MV	1.5uA	0.5uA	500MWF	5MA	12V	15V	30V	15uV/C		.3MA	70dB	70dB	300K
LM101H	NAU	GPU	EXT			+22V	-22V	125C	94dB	5MV	500NA	200NA	500MWF	5MA	12V	15V	30V	15uV/C		3MA	70dB	70dB	300K

For detailed explanations of column heading notations, see App. A.

Also for ready references the more important abbreviations used in the column headings are listed below:

LEFT HAND PAGE

- APP = application (codes at APP.E.)
- CMRR = common mode rejection ratio
- CMP = compensation (frequency)
- dV_{IO}/dT = input offset voltage temperature drift
- GBP = gain bandwidth product
- I_B = input bias current
- I_{IO} = input bias offset current
- i_Q = quiescent supply current
- MFR = manufacturer (codes at App.C.)
- P_Q = quiescent power consumer
- PSRR = power supply rejection ratio
- V_{ICM} = common mode input voltage rating
- V_{IDF} = differential input voltage rating
- V_{IO} = input offset voltage
- V_S = dc supply voltage

RIGHT HAND PAGE

Lead out coding summary (details at APP.G.) for different cases (APP.F.)

- A = gain adjust
- B = bias adjust
- C = case
- E— = inverting input
- E+ = non-inverting input
- F,F* = input frequency compensation
- G = ground
- J = high level input
- K = output, open collector
- L = output, open emitter
- M = metal case
- N = not connected
- Q = special terminal
- R,R* = outputs
- S = strobe
- T,T* = offset balance
- V+ = +ve dc supply
- V— = —ve dc supply
- W = guard ring
- X = blank position, no lead
- + + = +ve supplementary dc supply
- — — = —ve supplementary dc supply
- ø,ø* = output frequency compensation

CASE (APP F)	LD 1	LD 2	LD 3	LD 4	LD 5	LD 6	LD 7	LD 8	LD 9	LD 10	LD 11	LD 12	LD 13	LD 14	LD 15	LD 16	EUROPE SUBSTITUTE	USA SUBSTITUTE	IS	TYPE NUMBER
FLP-10/3M	N	N	E-	E+	V-	N	R	V+	N	N							SFC2107PM	LM107F	0	LH101F
T05-8/1M	N	E-	E+	V-	N	R	V+	N									SFC2107M	LM107H	0	LH101H
FLP-10/3M	N	N	E-	E+	V-	N	P	V+	N	N							SFC2207PT	LM207F	0	LH201F
T05-8/1M	N	E-	E+	V-	N	R	V+	N									SFC2207	LM207H	0	LH201H
T05-8/1M	T	E-	E+	V-	T*	R	V+	N										NE536	0	LH740ACH
T05-8/1M	T	E-	E+	V-	T*	R	V+	N										SU536	0	LH740AH
DIL-16/1M	V+1	ø*1	Tø1	E-1	E+1	V-	T*2	R2	V+2	ø*2	Tø2	E-2	E+2	T*1	N	R1	AM1501DM	AMLH2101AD	0	LH2101AD
DIL-16/1C	V+1	ø*1	Tø1	E-1	E+1	V-	T*2	R2	V+2	ø*2	Tø2	E-2	E+2	T*1	N	R1		LH2101AD	0	LH2101AF
FLP-16/1M	V+1	ø*1	Tø1	E-1	E+1	V-	T*2	R2	V+2	ø*2	Tø2	E-2	E+2	T*1	N	R1	AM1501FM	AMLH2101AF	0	LH2101AF
DIL-16/1M	V+1	ø1	F1	E-1	E+1	V-	N	R2	V+2	ø2	F2	E-2	E+2	N	N	R1		LH2108AF	0	LH2108AD
DIL-16/1C	V+1	ø1	F1	E-1	E+1	V-	N	R2	V+2	ø2	F2	E-2	E+2	N	N	R1		LH2108AD	0	LH2108AF
DIL-16/1M	V+1	ø1	F1	E-1	E+1	V-	N	R2	V+2	ø2	F2	E-2	E+2	N	N	R1		LH2108AD	0	LH2108D
DIL-16/1C	V+1	ø1	F1	E-1	E+1	V-	N	R2	V+2	ø2	F2	E-2	E+2	N	N	R1		LH2108AF	0	LH2108F
FLP-16/3G	V+1	ø1	F1	E-1	E+1	V-	N	R2	V+2	ø2	F2	E-2	E+2	N	N	R1			0	LH2108F
DIL-16/1M	V+1	T1	T*1	E+1	L2	V-	N	R2	V+2	T2	T*2	E+2	L1	N	N	R1			0	LH2110D
FLP-16/3M	V+1	T1	T*1	E+1	L2	V-	N	R2	V+2	T2	T*2	E+2	L1	N	N	R1			0	LH2110F
DIL-16/1M	V+1	L1	E+1	E-1	V-	T*2	TS2	K2	V+2	L2	E+2	E-2	T*1	TS1	K1	N	AM1500DM	AMLH2111D	0	LH2111D
FLP-16/3M	V+1	L1	E+1	E-1	V-	T*2	TS2	K2	V+2	L2	E+2	E-2	T*1	TS1	K1	N	AM1500FM	AMLH2111F	0	LH2111F
DIL-16/1M	V+1	ø*1	Tø1	E-1	E+1	V-	T*2	R2	V+2	ø*2	Tø2	E-2	E+2	T*1	N	R1	AM1501DL	AMLH2201AD	0	LH2201AD
DIL-16/1C	V+1	ø*1	Tø1	E-1	E+1	V-	T*2	R2	V+2	ø*2	Tø2	E-2	E+2	T*1	N	R1		LH2201AD	0	LH2201AF
FLP-16/3M	V+1	ø*1	Tø1	E-1	E+1	V-	T*2	R2	V+2	ø*2	Tø2	E-2	E+2	T*1	N	R1	AM1501FL	LH2101AF	0	LH2201AF
DIL-16/1M	V+1	ø1	F1	E-1	E+1	V-	N	R2	V+2	ø2	F2	E-2	E+2	N	N	R1		LH2208AF	0	LH2208AD
DIL-16/1C	V+1	ø1	F1	E-1	E+1	V-	N	R2	V+2	ø2	F2	E-	E+2	N	N	R1		LH2208AD	0	LH2208AF
DIL-16/1M	V+1	ø1	F1	E-1	E+1	V-	N	R2	V+2	ø2	F2	E-2	E+2	N	N	R1		LH2208F	0	LH2208D
DIL-16/1C	V+1	ø1	F1	E-1	E+1	V-	N	R2	V+2	ø2	F2	E-2	E+2	N	N	R1		LH2208D	0	LH2208F
FLP-16/3G	V+1	ø1	F1	E-1	E+1	V-	N	R2	V+2	ø2	F2	E-2	E+2	N	N	R1		LH2108F	0	LH2208F
DIL-16/1M	V+1	T1	T*1	E+1	L2	V-	N	R2	V+2	T2	T*2	E+2	L1	N	N	R1		LH2110D	0	LH2210D
FLP-16/3M	V+1	T1	T*1	E+1	L2	V-	N	R2	V+2	T2	T*2	E+2	L1	N	N	R1		LH2110F	0	LH2210F
DIL-16/1M	V+1	L1	E+1	E-1	V-	T*2	TS2	K2	V+2	L2	E+2	E-2	T*1	TS1	K1	N	AM1500DL	AMLH2111D	0	LH2211D
FLP-16/3M	V+1	L1	E+1	E-1	V-	T*2	TS2	K2	V+2	L2	E+2	E-2	T*1	TS1	K1	N	AM1500FL	AMLH2211F	0	LH2211F
DIL-16/1M	V+1	ø*1	Tø1	E-1	E+1	V-	T*2	R2	V+2	ø*2	Tø2	E-2	E+2	T*1	N	R1	AM1501DC	LH2201AD	0	LH2301AD
DIL-16/1C	V+	ø*1	Tø1	E-1	E+1	V-	T*2	R2	V+2	ø*2	Tø2	E-2	E+2	T*1	N	R1		LH2301AD	0	LH2301AF
FLP-16/3M	V+1	ø*1	Tø1	E-1	E+1	V-	T*2	R2	V+2	ø*2	Tø2	E-2	E+2	T*1	N	R1	AM1501FC	AMLH2301AF	0	LH2301AF
DIL-16/1M	V+1	ø1	F1	E-1	E+1	V-	N	R2	V+2	ø2	F2	E-2	E+2	N	N	R1		LH2308AF	0	LH2308AD
DIL-16/1C	V+1	ø1	F1	E-1	E+1	V-	N	R2	V+2	ø2	F2	E-2	E+2	N	N	R1		LH2308AD	0	LH2308AF
DIL-16/1M	V+1	ø1	F1	E-1	E+1	V-	N	R2	V+2	ø2	F2	E-2	E+2	N	N	R1		LH2308F	0	LH2308D
DIL-16/1C	V+1	ø1	F1	E-1	E+1	V-	N	R2	V+2	ø2	F2	E-2	E+2	N	N	R1		LH2308D	0	LH2308F
FLP-16/3G	V+1	ø1	F1	E-1	E+1	V-	N	R2	V+2	ø2	F2	E-2	E+2	N	N	R1		LH2208F	0	LH2308F
DIL-16/1M	V+1	T1	T*1	E+1	L2	V-	N	R2	V+2	T2	T*2	E+2	L1	N	N	R1		LH2210D	0	LH2310D
FLP-16/3M	V+1	T1	T*1	E+1	L2	V-	N	R2	V+2	T2	T*2	E+2	L1	N	N	R1		LH2210F	0	LH2310F
DIL-16/1M	V+1	L1	E+1	E-1	V-	T*2	TS2	K2	V+2	L2	E+2	E-2	T*1	TS1	K1	N	AM1500DC	AMLH2311D	0	LH2311D
FLP-16/3M	V+1	L1	E+1	E-1	V-	T*2	TS2	K2	V+2	L2	E+2	E-2	T*1	TS1	K1	N	AM1500FC	AMLH2311F	0	LH2311F
DIL-16/1M	V+1	B1	T1	E-1	E+1	V-	T*2	R2	V+2	B2	T2	E-2	E+3	T*1	N	R1		LH24250D	0	LH24250CD
FLP-16/3M	V+1	B1	T1	E-1	E+1	V-	T*2	R2	V+2	B2	T2	E-2	E+3	T*1	N	R1		LH24250F	0	LH24250CF
DIL-16/1C	V+1	B1	T1	E-1	E+1	V-	T*2	R2	V+2	B2	T2	E-2	E+3	T*1	N	R1			0	LH24250D
FLP-16/3M	V+1	B1	T1	E-1	E+1	V-	T*2	R2	V+2	B2	T2	E-2	E+3	T*1	N	R1			0	LH24250F
DIL-14/1M	N	N	FT	E-	E+	V-	N	N	T*	R	V+	F*	N	N				UA101AD	0	LM101AD
FLP-10/3M	N	FT	E-	E+	V-	T*	R	V+	F*	N							SFC2101APM	UA101AF	0	LM101AF
DIL-14/1C	N	N	FT	E-	E+	V-	N	N	T*	R	V+	F*	N	N			UA101AD	LM101AJ14	0	LM101AF
T05-8/1M	FT	E-	E+	V-M	T*	R	V+	F*									SFC2101A	UA101AH	0	LM101AH
DIL-14/1M	N	N	FT	E-	E+	V-	N	N	T*	R	V+	F*	N	N				UA101AD	0	LM101AJ
DIL-14/1M	N	N	FT	E-	E+	V-	N	N	T*	R	V+	F*	N	N				UA101AD	0	LM101AJ-14
DIL-8/1P	FT	E-	E+	V-	T*	R	V+	F*									SN52101AJP	LM101AV	0	LM101AN(8)
T05-8/1M	FT	E-	E+	V-M	T*	R	V+	F*									SFC2101A	LM101AH	0	LM101AT
DIL-8/1P	FT	E-	E+	V-	T*	R	V+	F*									SN52101AJP		0	LM101AV
DIL-14/1C	N	N	FT	E-	E+	V-	N	N	T*	R	V+	F*	N	N			UA101AD	LM101AJ14	0	LM101D
DIL-14/1C	N	N	FT	E-	E+	V-	N	N	T*	R	V+	F*	N	N			UA101AD	LM101J14	0	LM101F
FLP-10/3M	N	FT	E-	E+	V-	T*	R	V+	F*	N							SFC2101APM	UA101AF	0	LM101F
T05-8/1M	FT	E-	E+	V-M	T*	R	V+	F*									SFC2101A	UA101H	0	LM101H
DIL-14/1M	N	N	FT	E-	E+	V-	N	N	T*	R	V+	F*	N	N				UA101D	0	LM101J14

TYPE NUMBER	M F R	A P P	C M P	GBP MIN	SLEW RATE MIN	Vs+ MAX	Vs− MAX	Top MAX	Avol MIN	Vio MAX	Ib MAX	Iio MAX	Ptot MAX	Iout MIN	Vout MIN	Vicm MAX	Vidf MAX	dVio/dT MAX	Po MAX	Io MAX	CM RR MIN	PS RR MIN	Rin MIN
LM101J14	NAU	GPU	EXT	.	.	+22V	−22V	125C	94dB	5MV	1.5uA	0.5uA	500MWF	5MA	12V	15V	30V	15uV/C	.	3MA	70dB	70dB	300K
LM101N−14	SJU	GPU	EXT	.	.	+22V	−22V	125C	94dB	5MV	1.5uA	0.5uA	500MWF	5MA	12V	15V	30V	15uV/C	.	3MA	70dB	70dB	300K
LM101Q	SJU	GPU	EXT	.	.	+22V	−22V	125C	94dB	5MV	1.5uA	0.5uA	500MWF	5MA	12V	15V	30V	15uV/C	.	3MA	70dB	70dB	300K
LM101T	MUG	GPU	EXT	.	.	+22V	−22V	125C	94dB	5MV	1.5uA	0.5uA	500MWF	5MA	12V	15V	30V	15uV/C	.	3MA	70dB	70dB	300K
LM102D	ADU	VFA	INT	.	.	+18V	−18V	125C	0dB	5MV	10NA	.	500MWF	1MA	10V	.	.	30uV/C	.	6MA	.	60dB	10G
LM102F	ING	VFA	INT	.	.	+18V	−18V	125C	0dB	5MV	10NA	.	500MWF	1MA	10V	.	.	30uV/C	.	6MA	.	60dB	10G
LM102H	NAU	VFA	INT	.	.	+18V	−18V	125C	0dB	5MV	10NA	.	500MWF	1MA	10V	.	.	30uV/C	.	6MA	.	60dB	10G
LM106F	NAU	CPR	EXT	.	.	+15V	−15V	125C	84dB	2MV	20uA	3uA	600MWF	50MA	2.5V	.	.	10uV/C	163MW	.	.	.	.
LM106H	NAU	CPR	EXT	.	.	+15V	−15V	125C	84dB	2MV	20uA	3uA	600MWF	50MA	2.5V	.	.	10uV/C	163MW	.	.	.	.
LM107D	NAU	GPK	INT	.	.	+22V	−22V	125C	94dB	2MV	75NA	10NA	500MWF	5MA	12V	15V	30V	15uV/C	.	3MA	80dB	80dB	1.5M
LM107F	MUG	GPK	INT	.	.	+22V	−22V	125C	94dB	2MV	75NA	10NA	500MWF	5MA	12V	15V	30V	15uV/C	.	3MA	80dB	80dB	1.5M
LM107F	NAU	GPK	INT	.	.	+22V	−22V	125C	94dB	2MV	75NA	10NA	500MWF	5MA	12V	15V	30V	15uV/C	.	3MA	80dB	80dB	1.5M
LM107H	NAU	GPK	INT	.	.	+22V	−22V	125C	94dB	2MV	75NA	10NA	500MWF	5MA	12V	15V	30V	15uV/C	.	3MA	80dB	80dB	1.5M
LM107J	NAU	GPK	INT	.	.	+22V	−22V	125C	94dB	2MV	75NA	10NA	500MWF	5MA	12V	15V	30V	15uV/C	.	3MA	80dB	80dB	1.5M
LM107J−14	NAU	GPK	INT	.	.	+22V	−22V	125C	94dB	2MV	75NA	10NA	500MWF	5MA	12V	15V	30V	15uV/C	.	3MA	80dB	80dB	1.5M
LM107N	SJU	GPK	INT	.	.	+22V	−22V	125C	94dB	2MV	75NA	10NA	500MWF	5MA	12V	15V	30V	15uV/C	.	3MA	80dB	80dB	1.5M
LM107T	MUG	GPK	INT	.	.	+22V	−22V	125C	94dB	2MV	75NA	10NA	500MWF	5MA	12V	15V	30V	15uV/C	.	3MA	80dB	80dB	1.5M
LM108AD	NAU	SBA	EXT	.	.	+20V	−20V	125C	98dB	0.5MV	2NA	0.2NA	500MWF	1MA	13V	15V	1V	5uV/C	.	.6MA	96dB	96dB	30M
LM108AF	MUG	SBA	EXT	.	.	+20V	−20V	125C	98dB	0.5MV	2NA	0.2NA	500MWF	1MA	13V	15V	1V	5uV/C	.	.6MA	96dB	96dB	30M
LM108AF	NAU	SBA	EXT	.	.	+20V	−20V	125C	98dB	0.5MV	2NA	0.2NA	500MWF	1MA	13V	15V	1V	5uV/C	.	.6MA	96dB	96dB	30M
LM108AH	NAU	SBA	EXT	.	.	+20V	−20V	125C	98dB	0.5MV	2NA	0.2NA	500MWF	1MA	13V	15V	1V	5uV/C	.	.6MA	96dB	96dB	30M
LM108AJ	NAU	SBA	EXT	.	.	+20V	−20V	125C	98dB	0.5MV	2NA	0.2NA	500MWF	1MA	13V	15V	1V	5uV/C	.	.6MA	96dB	96dB	30M
LM108AT	SJU	SBA	EXT	.	.	+20V	−20V	125C	98dB	0.5MV	2NA	0.2NA	500MWF	1MA	13V	15V	1V	5uV/C	.	.6MA	96dB	96dB	30M
LM108D	NAU	SBA	EXT	.	.	+20V	−20V	125C	96dB	2MV	2NA	0.2NA	500MWF	1MA	13V	15V	1V	15uV/C	.	.6MA	85dB	80dB	30M
LM108F	NAU	SBA	EXT	.	.	+20V	−20V	125C	96dB	2MV	2NA	0.2NA	500MWF	1MA	13V	15V	1V	15uV/C	.	.6MA	85dB	80dB	30M
LM108F	SJU	SBA	EXT	.	.	+20V	−20V	125C	96dB	2MV	2NA	0.2NA	500MWF	1MA	13V	15V	1V	15uV/C	.	.6MA	85dB	80dB	30M
LM108H	NAU	SBA	EXT	.	.	+20V	−20V	125C	96dB	2MV	2NA	0.2NA	500MWF	1MA	13V	15V	1V	15uV/C	.	.6MA	85dB	80dB	30M
LM108J	NAU	SBA	EXT	.	.	+20V	−20V	125C	96dB	2MV	2NA	0.2NA	500MWF	1MA	13V	15V	1V	15uV/C	.	.6MA	85dB	80dB	.
LM108T	SJU	SBA	EXT	.	.	+20V	−20V	125C	96dB	2MV	2NA	0.2NA	500MWF	1MA	13V	15V	1V	15uV/C	.	.6MA	85dB	80dB	30M
LM110D	NAU	VFA	INT	.	15V/uS	+18V	−18V	125C	0dB	4MV	3NA	.	500MWF	1MA	10V	15V	15V	50uV/C	.	6MA	.	70dB	10G
LM110F	NAU	VFA	INT	.	15V/uS	+18V	−18V	125C	0dB	4MV	3NA	.	500MWF	1MA	10V	15V	15V	50uV/C	.	6MA	.	70dB	10G
LM110H	NAU	VFA	INT	.	15V/uS	+18V	−18V	125C	0dB	4MV	3NA	.	500MWF	1MA	10V	15V	15V	50uV/C	.	6MA	.	70dB	10G
LM110J	NAU	VFA	INT	.	15V/uS	+18V	−18V	125C	0dB	4MV	3NA	.	500MWF	1MA	10V	15V	15V	50uV/C	.	6MA	.	70dB	10G
LM111D	NAU	CPR	EXT	.	.	+18V	−18V	125C	100dB	3MV	100NA	10NA	500MWF	.	.	15V	30V	.	.	6MA	.	.	.
LM111F	MUG	CPR	EXT	.	.	+18V	−18V	125C	100dB	3MV	100NA	10NA	500MWF	.	.	15V	30V	.	.	6MA	.	.	.
LM111F	NAU	CPR	EXT	.	.	+18V	−18V	125C	100dB	3MV	100NA	10NA	500MWF	.	.	15V	30V	.	.	6MA	.	.	.
LM111H	NAU	CPR	EXT	.	.	+18V	−18V	125C	100dB	3MV	100NA	10NA	500MWF	.	.	15V	30V	.	.	6MA	.	.	.
LM111J	NAU	CPR	EXT	.	.	+18V	−18V	125C	100dB	3MV	100NA	10NA	500MWF	.	.	15V	30V	.	.	6MA	.	.	.
LM111T	SJU	CPR	EXT	.	.	+18V	−18V	125C	100dB	3MV	100NA	10NA	500MWF	.	.	15V	30V	.	.	6MA	.	.	.
LM112D	NAU	SBA	INT	.	.	+20V	−20V	125C	94dB	2MV	2NA	0.2NA	500MWF	1MA	13V	14V	14V	15uV/C	.	.6MA	85dB	80dB	30M
LM112F	NAU	SBA	INT	.	.	+20V	−20V	125C	94dB	2MV	2NA	0.2NA	500MWF	1MA	13V	14V	14V	15uV/C	.	.6MA	85dB	80dB	30M
LM112H	NAU	SBA	INT	.	.	+20V	−20V	125C	94dB	2MV	2NA	0.2NA	500MWF	1MA	13V	14V	14V	15uV/C	.	.6MA	85dB	80dB	30M
LM118D	NAU	XSR	INT	.	50V/uS	+20V	−20V	125C	94dB	4MV	250NA	50NA	500MWF	6MA	12V	15V	1V	.	.	8MA	80dB	70dB	1M
LM118F	NAU	XSR	INT	.	50V/uS	+20V	−20V	125C	94dB	4MV	250NA	50NA	500MWF	6MA	12V	15V	1V	.	.	8MA	80dB	70dB	1M
LM118H	NAU	XSR	INT	.	50V/uS	+20V	−20V	125C	94dB	4MV	250NA	50NA	500MWF	6MA	12V	15V	1V	.	.	8MA	80dB	70dB	1M
LM119D	NAU	DCP	INT	.	.	+18V	−18V	125C	80dB	4MV	500NA	75NA	500MWF	.	.	15V	5V	.	.	12MA	.	.	.
LM119F	MUG	DCP	INT	.	.	+18V	−18V	125C	80dB	4MV	500NA	75NA	500MWF	.	.	15V	5V	.	.	12MA	.	.	.
LM119F	NAU	DCP	INT	.	.	+18V	−18V	125C	80dB	4MV	500NA	75NA	500MWF	.	.	15V	5V	.	.	12MA	.	.	.
LM119H	NAU	DCP	INT	.	.	+18V	−18V	125C	80dB	4MV	500NA	75NA	500MWF	.	.	15V	5V	.	.	12MA	.	.	.
LM119J	NAU	DCP	INT	.	.	+18V	−18V	125C	80dB	4MV	500NA	75NA	500MWF	.	.	15V	5V	.	.	12MA	.	.	.
LM119K	MUG	DCP	INT	.	.	+18V	−18V	125C	80dB	4MV	500NA	75NA	500MWF	.	.	15V	5V	.	.	12MA	.	.	.
LM121AD	NAU	PIA	EXT	.	.	+20V	−20V	125C	24dB	0.4MV	10NA	0.5NA	500MWF	.	.	15V	15V	0.2uV/C	.	2MA	126dB	120dB	4M
LM121AF	NAU	PIA	EXT	.	.	+20V	−20V	125C	24dB	0.4MV	10NA	0.5NA	500MWF	.	.	15V	15V	0.2uV/C	.	2MA	126dB	120dB	4M
LM121AH	NAU	PIA	EXT	.	.	+20V	−20V	125C	24dB	0.4MV	10NA	0.5NA	500MWF	.	.	15V	15V	0.2uV/C	.	2MA	126dB	120dB	4M
LM121D	NAU	PIA	EXT	.	.	+20V	−20V	125C	24dB	0.7MV	10NA	1NA	500MWF	.	.	15V	15V	1uV/C	.	2MA	120dB	120dB	4M
LM121F	NAU	PIA	EXT	.	.	+20V	−20V	125C	24dB	0.7MV	10NA	1NA	500MWF	.	.	15V	15V	1uV/C	.	2MA	120dB	120dB	4M
LM121H	NAU	PIA	EXT	.	.	+20V	−20V	125C	24dB	0.7MV	10NA	1NA	500MWF	.	.	15V	15V	1uV/C	.	2MA	120dB	120dB	4M
LM124AD	NAU	QGK	INT	.	.	+16V	−16V	125C	94dB	2MV	50NA	10NA	900MWF	.	.	16V	16V	20uV/C	.	2MA	70dB	65dB	.
LM124AF	NAU	QGK	INT	.	.	+16V	−16V	125C	94dB	2MV	50NA	10NA	800MWF	.	.	16V	16V	20uV/C	.	2MA	70dB	65dB	.
LM124AJ	NAU	QGK	INT	.	.	+16V	−16V	125C	94dB	2MV	50NA	10NA	900MWF	.	.	16V	16V	20uV/C	.	2MA	70dB	65dB	.

For detailed explanations of column heading notations, see App. A.

Also for ready references the more important abbreviations used in the column headings are listed below:

LEFT HAND PAGE

APP = application (codes at APP.E.)
CMRR = common mode rejection ratio
CMP = compensation (frequency)
dV_{io}/dT = input offset voltage temperature drift
GBP = gain bandwidth product
I_B = input bias current
I_{io} = input bias offset current
I_0 = quiescent supply current
MFR = manufacturer (codes at App.C.)
P_0 = quiescent power consumer
PSRR = power supply rejection ratio
V_{ICM} = common mode input voltage rating
V_{IDF} = differential input voltage rating
V_{io} = input offset voltage
V_S = dc supply voltage

RIGHT HAND PAGE

Lead out coding summary (details at APP.G.) for different cases (APP.F.)

A = gain adjust
B = bias adjust
C = case
E— = inverting input
E+ = non-inverting input
F,F* = input frequency compensation
G = ground
J = high level input
K = output, open collector
L = output, open emitter
M = metal case
N = not connected
Q = special terminal
R,R* = outputs
S = strobe
T,T* = offset balance
V+ = +ve dc supply
V— = —ve dc supply
W = guard ring
X = blank position, no lead
++ = +ve supplementary dc supply
—— = —ve supplementary dc supply
ø,ø* = output frequency compensation

CASE (APP.F)	LD1	LD2	LD3	LD4	LD5	LD6	LD7	LD8	LD9	LD10	LD11	LD12	LD13	LD14	LD15	LD16	EUROPE SUBSTITUTE	USA SUBSTITUTE	IS	TYPE NUMBER
DIL-14/1P	N	N	FT	E-	E+	V-	N	N	T*	R	V+	F*	N	N			UA101AD	LM101AJ14	0	LM101N-14
DIL-10/1P	N	FT	E-	E+	V-	T*	R	N	V+	F*	N								0	LM101Q
T05-8/1M	FT	E-	E+	V-M	T*	R	V+	F*									SFC2101A	LM101H	0	LM101T
DIL-14/1C	N	N	T	N	E+	V-	N	N	L	R	V+	T*	N	N					0	LM102D
FLP-10/3G	N	T	I.	E+	V-	L	R	V+	T*	N							AMLM102F	102(FLP)	0	LM102F
T05-8/1M	T	N	E+	V-	L	R	V+	T*										UA102M	0	LM102H
FLP-14/3G	N	G	E+	E-	N	V-	S1	S2	R	N	V+	N	N	N				SN52106FA	0	LM106F
T05-8/1M	G	E+	E-	V-M	S1	S2	R	V+										SN52106L	0	LM106H
DIL-14/1M	N	N	N	E-	E+	V-	N	N	N	R	V+	N	N	N				SN52107JA	0	LM107D
DIL-14/1C	N	N	N	E-	E+	V-	N	N	N	R	V+	N	N	N			SN52107JA	LM107D	0	LM107F
FLP-10/3G	N	N	E-	E+	V-	N	R	V+	N								SFC2107PM		0	LM107F
T05-8/1M	N	E-	E+	V-M	N	R	V+	N									SFC2107M	UA107H	0	LM107H
DIL-8/1C	N	E-	E+	V-	N	R	V+	N										SN52107JP	0	LM107J
DIL-14/1C	N	N	N	E-	E+	V-	N	N	N	R	V+	N	N	N				SN52107JA	0	LM107J-14
DIL-8/1P	N	E-	E+	V-	N	R	V+	N									SN52107JP	LM107J	0	LM107N
T05-8/1M	N	E-	E+	V-M	N	R	V+	N									UA107H	LM107H	0	LM107T
DIL-14/1G	N	F	N	E-	E+	N	V-	N	N	R	V+	F*	N	N			SN52108AJA	UA108AD	0	LM108AD
DIL-14/1C	N	F	N	E-	E+	N	V-	N	N	R	V+	F*	N	N			UA108AD	LM108AJ	0	LM108AF
FLP-10/3G	N	N	E-	E+	N	V-	R	V+	F*	F								UA108AF	0	LM108AJ
T05-8/1M	F	E-	E+	V-M	N	R	V+	F*									SFC2108A	UA108AH	0	LM108AH
DIL-14/1C	N	F	N	E-	E+	N	V-	N	N	N	R	V+	F*	N	N		SN52108AJA	UA108AD	0	LM108AJ
T05-8/1M	F	E-	E+	V-M	N	R	V+	F*									SFC2108A	LM108AH	0	LM108AT
DIL-14/1M	N	F	N	E-	E+	V-	N	N	R	V+	F*	N	N				SN52108JA	UA108D	0	LM108D
FLP-10/3G	N	N	E-	E+	N	V-	R	V+	F*	F							SFC2108PM	UA108F	0	LM108F
DIL-14/1C	N	F	N	E-	E+	N	V-	N	N	N	R	V+	F*	N	N		UA108D	LM108D	0	LM108F
T05-8/1M	F	E-	E+	V-M	N	R	V+	F*									SFC2108M	UA108H	0	LM108H
DIL-14/1C	N	F	N	E-	E+	N	V-	N	N	N	R	V+	F*	N	N		SN52108JA	UA108D	0	LM108J
T05-8/1M	F	E-	E+	V-M	N	R	V+	F*									SFC2108M	UA108H	0	LM108T
DIL-14/1M	N	N	T	N	E+	V-	N	N	L	R	V+	T*	N	N				SN52110JA	0	LM110D
FLP-10/3G	N	T	N	E+	V-	L	R	V+	T*	N							MLM110F		0	LM110F
T05-8/1M	T	N	E+	V-	L	R	V+	T*									SFC2110M	UA110M	0	LM110H
DIL-14/1C	N	N	T	N	E+	V-	N	N	L	R	V+	T*	N	N				SN52110JA	0	LM110J
DIL-14/1M	N	G	E+	E-	N	V-	T	T*S	R	N	V+	N	N	N				SN52111J	0	LM111D
DIL-14/1C	N	G	E+	E-	N	V-	T	T*S	R	N	V+	N	N	N			LM111D	LM111J	0	LM111F
FLP-10/3G	G	E+	E-	N	V-	T	T*S	N	R	V+								SN52111FA	0	LM111F
T05-8/1M	G	E+	E-	V-	T	T*S	R	V+									SFC2111M	SG111T	0	LM111H
DIL-14/1C	N	G	E+	E-	N	V-	T	T*S	R	N	V+	N	N	N				LM111D	0	LM111J
T05-8/1M	G	E+	E-	V-	T	T*S	R	V+									SFC2111M	LM111H	0	LM111T
DIL-14/1M	N	T	W	E-	E+	W*	N	N	F	R	V+	T*	N	N					0	LM112D
FLP-10/3G	N	W	E-	E+	W*	V-	R	V+	T	T*								MLM112F	0	LM112F
T05-8/1M	T	E-	E+	V-	F	R	V+	T*											0	LM112H
DIL-14/1M	N	N	T*F	E-	E+	V-	N	N	F*T	R	V+	ø	N	N				SN52118JA	0	LM118D
FLP-10/3G	N	T*F	E-	E+	V-	F*T	R	V+	ø	N								SN52118FA	0	LM118F
T05-8/1M	T*F	E-	E+	V-	F*T	R	V+	ø									TDC0118CM	SN52118L	0	LM118H
DIL-14/1M	N	N	G1	E+1	E-1	V-	R2	G2	E+2	E-2	V+	R1	N	N			TDC0119DC	MLM119D	0	LM119D
DIL-14/1C	N	N	G1	E+1	E-1	V-	R2	G2	E+2	E-2	V+	R1	N	N			TDC0119DC	LM119J	0	LM119F
FLP-10/3G	R1	G1	E+1	E-1	V-	R2	G2	E+2	E-2	V+								LM119F	0	LM119F
T05-10/1M	R1	G1	E+1	E-1	V-	R2	G2	E+2	E-2	V+							TDC0119CM	MLM119H	0	LM119H
DIL-14/1C	N	G1	E+1	E-1	V-	R2	G2	E+2	E-2	V+	R1	N	N				TDC0119DC	LM119D	0	LM119J
T05-10/1M	R1	G1	E+1	E-1	V-	R2	G2	E+2	E-2	V+							TDC0119CM	LM119H	0	LM119K
DIL-14/1M	N	R	W	E-	E+	W*	V-	N	T	T*	V+	R*	N	N					0	LM121AD
FLP-10/3G	R	W	E-	E+	W*	V-	T	T*	V+	R*									0	LM121AF
T05-8/1M	R	E-	E+	V-	T	T*	V+	R*											0	LM121AH
DIL-14/1M	N	R	W	E-	E+	W*	V-	N	T	T*	V+	R*	N	N				LM121AD	0	LM121D
FLP-10/3G	R	W	E-	E+	W*	V-	T	T*	V+	R*								LM121AF	0	LM121F
T05-8/1M	R	E-	E+	V-	T	T*	V+	R*										LM121AH	0	LM121H
DIL-14/1M	R1	E-1	E+1	V+	E+2	E-2	R2	R3	E-3	E+3	G	E+4	E-4	R4					0	LM124AD
FLP-14/3G	R1	E-1	E+1	V+	E+2	E-2	R2	R3	E-3	E+3	G	E+4	E-4	R4					0	LM124AF
DIL-14/1C	R1	E-1	E+1	V+	E+2	E-2	R2	R3	E-3	E+3	G	E+4	E-4	R4					0	LM124AJ
DIL-14/1M	R1	E-1	E+1	V+	E+2	E-2	R2	R3	E-3	E+3	G	E+4	E-4	R4				MLM124L	0	LM124D

TYPE NUMBER	M F R	A P P	C M P	GBP MIN	SLEW RATE MIN	V_S^+ MAX	V_S^- MAX	T_{OP} MAX	A_{VOL} MIN	V_{IO} MAX	I_{IB} MAX	I_{IC} MAX	P_{TOT} MAX	I_{OUT} MIN	V_{OUT} MIN	V_{ICM} MAX	V_{IDF} MAX	dV_{IO}/dT MAX	P_Q MAX	I_Q MAX	CM RR MIN	PS RR MIN	R_{IN} MIN
LM124D	NAU	QGK	INT	.	.	+16V	-16V	125C	94dB	5MV	150NA	30NA	900MWF			16V	16V	35uV/C		2MA	70dB	65dB	.
LM124DDD	ING	QGK	INT			+16V	-16V	125C	94dB	5MV	150NA	30NA	900MWF			16V	16V	35uV/C		2MA	70dB	65dB	
LM124DICE	ING	QGK	INT			+16V	-16V	125C	94dB	5MV	150NA	30NA				16V	16V	35uV/C		2MA	70dB	65dB	
LM124F	MUG	QGK	INT	.	.	+16V	-16V	125C	94dB	5MV	150NA	30NA	900MWF			16V	16V	35uV/C		2MA	70dB	65dB	.
LM124F	NAU	QGK	INT			+16V	-16V	125C	94dB	5MV	150NA	30NA	800MWF			16V	16V	35uV/C		2MA	70dB	65dB	
LM124J	NAU	QGK	INT			+16V	-16V	125C	94dB	5MV	150NA	30NA	900MWF			16V	16V	35uV/C		2MA	70dB	65dB	
LM139A	SJU	QCP	EXT			+18V	-18V	125C	94dB	5MV	100NA	25NA	900MWF			18V	18V	.		2MA			
LM139AD	NAU	QCP	EXT			+18V	-18V	125C	94dB	2MV	100NA	25NA	900MWF			18V	18V	.		2MA			
LM139ADDD	ING	QCP	EXT			+18V	-18V	125C	94dB	2MV	100NA	25NA	900MWF			18V	18V			2MA			
LM139AF	MUG	QCP	EXT			+18V	-18V	125C	94dB	2MV	100NA	25NA	900MWF			18V	18V			2MA			
LM139AF	NAU	QCP	EXT			+18V	-18V	125C	94dB	2MV	100NA	25NA	800MWF			18V	18V			2MA			
LM139AJ	NAU	QCP	EXT			+18V	-18V	125C	94dB	2MV	100NA	25NA	900MWF			18V	18V			2MA			
LM139D	NAU	QCP	EXT			+18V	-18V	88C	88dB	5MV	100NA	25NA	900MWF			18V	18V			2MA			
LM139DDD	ING	QCP	EXT			+18V	-18V	125C	88dB	5MV	100NA	25NA	900MWF			18V	18V			2MA			
LM139DICE	ING	QCP	EXT			+18V	-18V	125C	88dB	5MV	100NA	25NA				18V	18V			2MA			
LM139F	MUG	QCP	EXT			+18V	-18V	125C	88dB	5MV	100NA	25NA	900MWF			18V	18V			2MA			
LM139F	NAU	QCP	EXT			+18V	-18V	125C	88dB	5MV	100NA	25NA	800MWF			18V	18V			2MA			
LM139H	TDG	QCP	EXT			+18V	-18V	125C	88dB	5MV	100NA	25NA	900MWF			18V	18V			2MA			
LM139J	NAU	QCP	EXT			+18V	-18V	125C	88dB	5MV	100NA	25NA	900MWF			18V	18V			2MA			
LM139L	TDG	QCP	EXT			+18V	-18V	125C	88dB	5MV	100NA	25NA	900MWF			18V	18V			2MA			
LM143D	NAU	HVO	INT	.3MHZ	1V/uS	+40V	-40V	125C	100dB	5MV	30NA	2NA	680MWF	4MA	22V	40V	80V			4MA	80dB	80dB	.
LM143F	NAU	HVO	INT	.3MHZ	1V/uS	+40V	-40V	125C	100dB	5MV	30NA	2NA	680MWF	4MA	22V	40V	80V			4MA	80dB	80dB	
LM143H	NAU	HVO	INT	.3MHZ	1V/uS	+40V	-40V	125C	100dB	5MV	30NA	2NA	680MWF	4MA	22V	40V	80V			4MA	80dB	80dB	
LM144D	NAU	HVO	EXT	.3MHZ	1V/uS	+40V	-40V	125C	100dB	5MV	20NA	3NA	680MWF	4MA	22V	40V	80V			4MA	80dB	80dB	
LM144F	NAU	HVO	EXT	.3MHZ	1V/uS	+40V	-40V	125C	100dB	5MV	20NA	3NA	680MWF	4MA	22V	40V	80V			4MA	80dB	80dB	
LM144H	NAU	HVO	EXT	.3MHZ	1V/uS	+40V	-40V	125C	100dB	5MV	20NA	3NA	680MWF	4MA	22V	40V	80V			4MA	80dB	80dB	
LM148D	NAU	QGK	INT	.3MHZ	0.2V/uS	+22V	-22V	125C	94dB	5MV	100NA	25NA	900MWF	5MA	12V	22V	44V			1MA	70dB	77dB	800K
LM148F	NAU	QGK	INT	.3MHZ	0.2V/uS	+22V	-22V	125C	94dB	5MV	100NA	25NA	670MWF	5MA	12V	22V	44V			1MA	70dB	77dB	800K
LM149D	NAU	QGK	INT	1MHZ	0.5V/uS	+22V	-22V	125C	94dB	5MV	100NA	25NA	900MWF	5MA	12V	22V	44V			1MA	70dB	77dB	800K
LM149F	NAU	QGK	INT	1MHZ	0.5V/uS	+22V	-22V	125C	94dB	5MV	100NA	25NA	670MWF	5MA	12V	22V	44V			1MA	70dB	77dB	800K
LM158AH	NAU	DGK	INT	.	.	+16V	-16V	125C	94dB	2MV	50NA	10NA	500MWF	10MA		16V	32V	15uV/C		3MA	70dB	65dB	.
LM158H	NAU	DGK	INT	.	.	+16V	-16V	125C	94dB	5MV	150NA	30NA	500MWF	10MA		16V	32V	30uV/C		3MA	70dB	65dB	
LM158T	MUG	DGK	INT	.	.	+16V	-16V	125C	94dB	5MV	150NA	30NA	500MWF	10MA		16V	32V	30uV/C		3MA	70dB	65dB	
LM160D	NAU	CPR	EXT			+8V	-8V	125C		5MV	20uA	3uA				4V	5V	40uV/C		32MA			5K
LM160F	NAU	CPR	EXT	.	.	+8V	-8V	125C		5MV	20uA	3uA				4V	5V	40uV/C		32MA			5K
LM160H	NAU	CPR	EXT			+8V	-8V	125C		5MV	20uA	3uA				4V	5V	40uV/C		32MA			5K
LM160J-14	NAU	CPR	EXT			+8V	-8V	125C		5MV	20uA	3uA				4V	5V	40uV/C		32MA			5K
LM161D	NAU	CPP	EXT			+16V	-16V	125C	60dB	3MV	20uA	3uA	600MWF	18MA		6V	5V	.		18MA			8K
LM161F	NAU	CPR	EXT			+16V	-16V	125C	60dB	3MV	20uA	3uA	600MWF	18MA		6V	5V			18MA			8K
LM161H	NAU	CPR	EXT			+16V	-16V	125C	60dB	3MV	20uA	3uA	500MWF	18MA		6V	5V	.		18MA			8K
LM161J	NAU	CPR	EXT			+16V	-16V	125C	60dB	3MV	20uA	3uA	600MWF	18MA		6V	5V	.		18MA			8K
LM193AH	NAU	DCP	EXT			+18V	-18V	125C	94dB	2MV	100NA	25NA	900MWF	6MA		18V	36V	.		3MA			
LM193H	NAU	DCP	EXT			+18V	-18V	125C	94dB	5MV	100NA	25NA	900MWF	6MA		18V	36V	.		3MA			
LM193T	MUG	DCP	EXT			+18V	-18V	125C	94dB	5MV	100NA	25NA	900MWF	6MA		18V	36V	.		3MA			
LM201AA	MUG	GPU	EXT	.	.	+22V	-22V	85C	94dB	2MV	75NA	10NA	500MWF	5MA	12V	15V	30V	15uV/C		3MA	80dB	80dB	500K
LM201AD	NAU	GPU	EXT	.	.	+22V	-22V	85C	94dB	2MV	75NA	10NA	500MWF	5MA	12V	15V	30V	15uV/C		3MA	80dB	80dB	500K
LM201AF	SJU	GPU	EXT			+22V	-22V	85C	94dB	2MV	75NA	10NA	500MWF	5MA	12V	15V	30V	15uV/C		.	80dB	80dB	1.5M
LM201AF	NAU	GPU	EXT			+22V	-22V	85C	94dB	2MV	75NA	10NA	500MWF	5MAA	12V	15V	30V	15uV/C		3MA	80dB	80dB	500K
LM201AH	NAU	GPU	EXT			+22V	-22V	85C	94dB	2MV	75NA	10NA	500MWF	5MA	12V	15V	30V	15uV/C		3MA	80dB	80dB	500K
LM201AJ	TDG	GPU	EXT			+22V	-22V	85C	94dB	2MV	75NA	10NA	500MWF	5MA	12V	15V	30V	15uV/C		3MA	80dB	80dB	500K
LM201AJ-14	NAU	GPU	EXT			+22V	-22V	85C	94dB	2MV	75NA	10NA	500MWF	5MA	12V	15V	30V	15uV/C		3MA	80dB	80dB	500K
LM201AN(8)	MUG	GPU	EXT			+22V	-22V	85C	94dB	2MV	75NA	10NA	500MWF	5MA	12V	15V	30V	15uV/C		.	80dB	80dB	1.5M
LM201AN(14)	MUG	GPU	EXT			+22V	-22V	85C	94dB	2MV	75NA	10NA	500MWF	5MA	12V	15V	30V	15uV/C		3MA	80dB	80dB	500K
LM201AT	SJU	GPU	EXT			+22V	-22V	85C	94dB	2MV	75NA	10NA	500MWF	5MA	12V	15V	30V	15uV/C		.	80dB	80dB	1.5M
LM201AV	SJU	GPU	EXT			+22V	-22V	85C	94dB	2MV	75NA	10NA	500MWF	5MA	12V	15V	30V	15uV/C		.	80dB	80dB	1.5M
LM201D	SJU	GPU	EXT			+22V	-22V	85C	86dB	7.5MV	1.5uA	0.5uA	500MWF	5MA	12V	15V	30V	30uV/C		3MA	65dB	70dB	100K
LM201F	NAU	GPU	EXT			+22V	-22V	85C	86dB	7.5MV	1.5uA	0.5uA	500MWF	5MA	12V	15V	30V	30uV/C		3MA	65dB	70dB	100K
LM201H	NAU	GPU	EXT			+22V	-22V	85C	86dB	7.5MV	1.5uA	0.5uA	500MWF	5MA	12V	15V	30V	30uV/C		.3MA	65dB	70dB	100K
LM201J	NAU	GPU	EXT			+22V	-22V	85C	86dB	7.5MV	1.5uA	0.5uA	500MWF	5MA	12V	15V	30V	30uV/C		3MA	65dB	70dB	100K
LM201J-14	NAU	GPU	EXT			+22V	-22V	85C	86dB	7.5MV	1.5uA	0.5uA	500MWF	5MA	12V	15V	30V	30uV/C		3MA	65dB	70dB	100K

For detailed explanations of column heading notations, see App. A.

Also for ready references the more important abbreviations used in the column headings are listed below:

LEFT HAND PAGE

APP = application (codes at APP.E.)

CMRR = common mode rejection ratio

CMP = compensation (frequency)

dV_{io}/dT = input offset voltage temperature drift

GBP = gain bandwidth product

I_g = input bias current

I_{io} = input bias offset current

I_o = quiescent supply current

MFR = manufacturer (codes at App.C.)

P_o = quiescent power consumer

PSRR = power supply rejection ratio

V_{icm} = common mode input voltage rating

V_{idf} = differential input voltage rating

V_{io} = input offset voltage

V_s = dc supply voltage

RIGHT HAND PAGE

Lead out coding summary (details at APP.G.) for different cases APP.F.)

A = gain adjust

B = bias adjust

C = case

E− = inverting input

E+ = non-inverting input

F,F* = input frequency compensation

G = ground

J = high level input

K = output, open collector

: = output, open emitter

M = metal case

N = not connected

Q = special terminal

R,R* = outputs

S = strobe

T,T* = offset balance

V+ = +ve dc supply

V− = −ve dc supply

W = guard ring

X = blank position, no lead

++ = +ve supplementary dc supply

−− = −ve supplementary dc supply

6,6* = output frequency compensation

CASE (APP F)	LD1	LD2	LD3	LD4	LD5	LD6	LD7	LD8	LD9	LD10	LD11	LD12	LD13	LD14	LD15	LD16	EUROPE SUBSTITUTE	USA SUBSTITUTE	S	TYPE NUMBER
DIL-14/1C	R1	E-1	E+1	V+	E+2	E-2	R2	R3	E-3	E+3	G		E+4	E-4	R4		MLM124L	LM124D	0	LM124DDD
CHP																			0	LM124DICE
DIL-14/1C	R1	E-1	E+1	V+	E+2	E-2	R2	R3	E-3	E+3	G		E+4	E-4	R4		LM124AJ	MLM124L	0	LM124F
FLP-14/3G	R1	E-1	E+1	V+	E+2	E-2	R2	R3	E-3	E+3	G		E+4	E-4	R4			AMLM124F	0	LM124F
DIL-14/1C	R1	E-1	E+1	V+	E+2	E-2	R2	R3	E-3	E+3	G		E+4	E-4	R4			MLM124L	0	LM124J
DIL-14/1P	R2	R1	V+	E-1	E+1	E-2	E+2	E+3	E-3	E-4	E+4	G	R4	R3				LM139D	0	LM139A
DIL-14/1M	R2	R1	V+	E-1	E+1	E-2	E+2	E+3	E-3	E-4	E+4	G	R4	R3				LM139AD	0	LM139AD
DIL-14/1C	R2	R1	V+	E-1	E+1	E-2	E+2	E+3	E-3	E-4	E+4	G	R4	R3			MLM139AL	LM139AD	0	LM139ADDD
DIL-14/1C	R2	R1	V+	E-1	E+1	E-2	E+2	E+3	E-3	E-4	E+4	G	R4	R3			MLM139AL	LM1391AD	0	LM139AF
FLP-14/3G	R2	R1	V+	E-1	E+1	E-2	E+2	E+3	E-3	E-4	E+4	G	R4	R3					0	LM139AF
DIL-14/1C	R2	R1	V+	E-1	E+1	E-2	E+2	E+3	E-3	E-4	E+4	G	R4	R3			MLM139AL		0	LM139AJ
DIL-14/1M	R2	R1	V+	E-1	E+1	E-2	E+2	E+3	E-3	E-4	E+4	G	R4	R3			MLM139L		0	LM139D
DIL-14/1C	R2	R1	V+	E-1	E+1	E-2	E+2	E+3	E-3	E-4	E+4	G	R4	R3			MLM139L	LM139D	0	LM139DDD
CHP																	LD139	AM139-DICE	0	LM139DICE
DIL-14/1C	R2	R1	V+	E-1	E+1	E-2	E+2	E+3	E-3	E-4	E+4	G	R4	R3			MLM139L	LM139D	0	LM139F
FLP-14/3G	R2	R1	V+	E-1	E+1	E-2	E+2	E+3	E-3	E-4	E+4	G	R4	R3				LM139AF	0	LM139F
FLP-14/3C	R2	R1	V+	E-1	E+1	E-2	E+2	E+3	E-3	E-4	E+4	G	R4	R3				LM139F	0	LM139H
DIL-14/1C	R2	R1	V+	E-1	E+1	E-2	E+2	E+3	E-3	E-4	E+4	G	R4	R3				MLM139L	0	LM139J
DIL-14/1C	R2	R1	V+	E-1	E+1	E-2	E+2	E+3	E-3	E-4	E+4	G	R4	R3			MLM139L	LM139J	0	LM139L
DIL-14/1M	N	N	T	E-	E+	V-	N	N	T*	R	V+	N	N						0	LM143D
FLP-10/3G	N	T	E-	E+	V-	T*	R	V+	N	N									0	LM143F
TO5-8/1M	T	E-	E+	V-	T*	R	V+	N										MC1536G	0	LM143H
DIL-14/1M	N	N	FT	E-	E+	V-	N	N		R	V+	F*	N	N					0	LM144D
FLP-10/3G	N	FT	E-	E+	V-	T*	R	V+	F*	N									0	LM144F
TO5-8/1M	FT	E-	E+	V-M	T*	R	V+	F*											0	LM144H
DIL-14/1M	R1	E-1	E+1	V+	E+2	E-2	R2	R3	E-3	E+3	V	E+4	E-4	R4				LM149D	0	LM148D
FLP-14/3G	R1	E-1	E+1	V+	E+2	E-2	R2	R3	E-3	E+3	V	E+4	E-4	R4				LM149F	0	LM148F
DIL-14/1M	R1	E-1	E+1	V+	E+2	E-2	R2	R3	E-3	E+3	V	E+4	E-4	R4					0	LM149D
FLP-14/3G	R1	E-1	E+1	V+	E+2	E-2	R2	R3	E-3	E+3	V	E+4	E-4	R4					0	LM149F
TO5-8/1M	R1	E-1	E+1	G	E+2	E-2	R2	V+											0	LM158AH
TO5-8/1M	R1	E-1	E+1	G	E+2	E-2	R2	V+										MLM158G	0	LM158H
TO5-8/1M	R1	E-1	E+1	G	E+2	E-2	R2	V+									MLM158G	LM158H	0	LM158T
DIL-14/1M	N	N	N	E-	E+	V-	N	N	G	R	R*	V+	N	N				UA760DM	0	LM160D
FLP-14/3G	N	N	N	E-	E+	V-	N		G	R	R*	V+	N	N				UA760HM	0	LM160F
TO5-8/1M	N	E-	E+	V-	G	R	R*	V+										UA760HM	0	LM160H
DIL-14/1C	N	N	N	E-	E+	V-	N	N	G	R	R*	V+	N	N				UA760DM	0	LM160J-14
DIL-14/1M	V+	N	E+	E-	N	V-	S2	R	G	R*	N	S1	++					LM161J	0	LM161D
FLP-14/3G	V+	N	E+	E-	N	V-	S2	R	G	R*	N	S1	++						0	LM161F
TO5-10/1M	E+	E-	V-	S2	R	G	R*	S1	++	V+									0	LM161H
DIL-14/1C	V+	N	E+	E-	N	V-	N	S2	R	G	R*	N	S1	++				LM161D	0	LM161J
TO5-8/1M	R1	E-1	E+1	G	E+2	E-2	R2	V+											0	LM193AH
TO5-8/1M	R1	E-1	E+1	G	E+2	E-2	R2	V+										LM193AH	0	LM193H
TO5-8/1M	R1	E-1	E+1	G	E+2	E-2	R2	V+										LM193H	0	LM193T
DIL-14/1P	N	N	FT	E-	E+	V-	N	N	T*	R	V+	F*	N	N			UA201AD	LM201AD	0	LM201AA
DIL-14/1M	N	N	FT	E-	E+	V-	N	N	T*	R	V+	F*	N	N			UA201AD	LM201AD	0	LM201AD
DIL-14/1C	N	N	FT	E-	E+	V-	N	N	T*	R	V+	F*	N	N			UA101AD	LM101AJ14	0	LM201AF
FLP-10/3M	N	FT	E-	E+	V-	T*	R	V+	F*	N	N						SFC2201APT	UA201AF	0	LM201AF
TO5-8/1M	FT	E-	E+	V-M	T*	R	V+	F*									SFC2101A	UA201AH	0	LM201AH
DIL-14/1C	N	N	FT	E-	E+	V-	N	N	T*	R	V+	F*	N	N			UA201AD		0	LM201AJ
DIL-14/1M	N	N	FT	E-	E+	V-	N	N	T*	R	V+	F*	N	N			UA201AD	LM201AD	0	LM201AJ-14
DIL-8/1P	FT	E-	E+	V-M	T*	R	V+	F*										LM201AV	0	LM201AN(8)
DIL-14/1P	N	N	FT	E-	E+	V-	N	N	T*	R	V+	F*	N	N			UA201AD	LM201AA	0	LM201AN(14)
TO5-8/1M	FT	E-	E+	V-M	T*	R	V+	F*									SFC2101A	LM201AH	0	LM201AT
DIL-8/1P	FT	E-	E+	V-M	T*	R	V+	F*										LM201AN(8)	0	LM201AV
DIL-14/1C	N	N	FT	E-	E+	V-	N	N	T*	R	V+	F*	N	N			LM201AD	LM201AJ14	0	LM201D
FLP-10/3M	N	FT	E-	E+	V-	T*	R	V+	F*	N	N						SFC2201APM	LM201AF	0	LM201F
TO5-8/1M	FT	E-	E+	V-M	T*	R	V+	F*									SFC2101A	UA201H	0	LM201H
DIL-8/1M	FT	E-	E+	V-	T*	R	V+	F*									SFC301ADC	UA301AT	0	LM201J
DIL-14/1M	N	N	FT	E-	E+	V-	N	N	T*	R	V+	F*	N	N				UA201D	0	LM201J-14
DIL-8/1P	FT	E-	E+	V-	T*	R	V+	F*										LM201J	0	LM201N

TYPE NUMBER	MFR	APP	CMP	GBP MIN	SLEW RATE MIN	V_S^+ MAX	V_S^- MAX	T_{OP} MAX	A_{VOL} MIN	V_{IO} MAX	I_B MAX	I_{IO} MAX	P_{TOT} MAX	I_{OUT} MIN	V_{OUT} MIN	V_{ICM} MAX	V_{IDF} MAX	dV_{IO}/dT MAX	P_Q MAX	I_Q MAX	CMRR MIN	PSRR MIN	R_{IN} MIN
LM201N	SJU	GPU	EXT	.	.	+22V	-22V	85C	86dB	7.5MV	1.5uA	0.5uA	500MWF	5MA	12V	15V	30V	30uV/C	.	3MA	65dB	70dB	100K
LM201N(8)	MUG	GPU	EXT	.	.	+22V	-22V	85C	86dB	7.5MV	1.5uA	0.5uA	500MWF	5MA	12V	15V	30V	30uV/C	.	3MA	65dB	70dB	100K
LM201N-14	SJU	GPU	EXT	.	.	+22V	-22V	85C	86dB	7.5MV	1.5uA	0.5uA	500MWF	5MA	12V	15V	30V	30uV/C	.	3MA	65dB	70dB	100K
LM201Q	SJU	GPU	EXT	.	.	+22V	-22V	85C	86dB	7.5MV	1.5uA	0.5uA	500MWF	5MA	12V	15V	30V	30uV/C	.	3MA	65dB	70dB	100K
LM201T	MUG	GPU	EXT	.	.	+22V	-22V	85C	86dB	7.5MV	1.5uA	0.5uA	500MWF	5MA	12V	15V	30V	30uV/C	.	3MA	65dB	70dB	100K
LM201V	MUG	GPU	EXT	.	.	+22V	-22V	85C	86dB	7.5MV	1.5uA	0.5uA	500MWF	5MA	12V	15V	30V	30uV/C	.	3MA	65dB	70dB	100K
LM202D	ADU	VFA	INT	.	.	+18V	-18V	85C	0dB	10MV	15NA	.	500MWF	1MA	10V	.	.	60uV/C	.	6MA	.	60dB	10G
LM202H	NAU	VFA	INT	.	.	+18V	-18V	85C	0dB	10MV	15NA	.	500MWF	1MA	10V	.	.	60uV/C	.	6MA	.	60dB	10G
LM206F	NAU	CPR	EXT	.	.	+15V	-15V	85C	84dB	2MV	20uA	3uA	600MWF	50MA	2.5V	.	.	10uV/C	163MW	.	.	.	.
LM206H	NAU	CPR	EXT	.	.	+15V	-15V	85C	84dB	2MV	20uA	3uA	600MWF	50MA	2.5V	.	.	10uV/C	163MW	.	.	.	.
LM207A	MUG	GPK	INT	.	.	+22V	-22V	85C	94dB	2MV	75NA	10NA	500MWF	5MA	12V	15V	30V	15uV/C	.	3MA	80dB	80dB	1.5M
LM207D	NAU	GPK	INT	.	.	+22V	-22V	85C	94dB	2MV	75NA	10NA	500MWF	5MA	12V	15V	30V	15uV/C	.	3MA	80dB	80dB	1.5M
LM207F	NAU	GPK	INT	.	.	+22V	-22V	85C	94dB	2MV	75NA	10NA	500MWF	5MA	12V	15V	30V	15uV/C	.	3MA	80dB	80dB	1.5M
LM207H	NAU	GPK	INT	.	.	+22V	-22V	85C	94dB	2MV	75NA	10NA	500MWF	5MA	12V	15V	30V	15uV/C	.	3MA	80dB	80dB	1.5M
LM207J	NAU	GPK	INT	.	.	+22V	-22V	85C	94dB	2MV	75NA	10NA	500MWF	5MA	12V	15V	30V	15uV/C	.	3MA	80dB	80dB	1.5M
LM207J-14	NAU	GPK	INT	.	.	+22V	-22V	85C	94dB	2MV	75NA	10NA	500MWF	5MA	12V	15V	30V	15uV/C	.	3MA	80dB	80dB	1.5M
LM207N	SJU	GPK	INT	.	.	+22V	-22V	85C	94dB	2MV	75NA	10NA	500MWF	5MA	12V	15V	30V	15uV/C	.	3MA	80dB	80dB	1.5M
LM207N(8)	MUG	GPK	INT	.	.	+22V	-22V	85C	94dB	2MV	75NA	10NA	500MWF	5MA	12V	15V	30V	15uV/C	.	3MA	80dB	80dB	1.5M
LM207N(14)	MUG	GPK	INT	.	.	+22V	-22V	85C	94dB	2MV	75NA	10NA	500MWF	5MA	12V	15V	30V	15uV/C	.	3MA	80dB	80dB	1.5M
LM207T	MUG	GPK	INT	.	.	+22V	-22V	85C	94dB	2MV	75NA	10NA	500MWF	5MA	12V	15V	30V	15uV/C	.	3MA	80dB	80dB	1.5M
LM207V	MUG	GPK	INT	.	.	+22V	-22V	85C	94dB	2MV	75NA	10NA	500MWF	5MA	12V	15V	30V	15uV/C	.	3MA	80dB	80dB	1.5M
LM208AD	NAU	SBA	EXT	.	.	+20V	-20V	85C	98dB	0.5MV	2NA	0.2NA	500MWF	1MA	13V	15V	1V	5uV/C	.	.6MA	96dB	96dB	30M
LM208AF	NAU	SBA	EXT	.	.	+20V	-20V	85C	98dB	0.5MV	2NA	0.2NA	500MWF	1MA	13V	15V	1V	5uV/C	.	.6MA	96dB	96dB	30M
LM208AF	SJU	SBA	EXT	.	.	+20V	-20V	85C	98dB	0.5MV	2NA	0.2NA	500MWF	1MA	13V	15V	1V	5uV/C	.	.6MA	96dB	96dB	30M
LM208AH	NAU	SBA	EXT	.	.	+20V	-20V	85C	98dB	0.5MV	2NA	0.2NA	500MWF	1MA	13V	15V	1V	5uV/C	.	.6MA	96dB	96dB	30M
LM208AJ	NAU	SBA	EXT	.	.	+20V	-20V	85C	98dB	0.5MV	2NA	0.2NA	500MWF	1MA	13V	15V	1V	5uV/C	.	.6MA	96dB	96dB	30M
LM208AT	MUG	SBA	EXT	.	.	+20V	-20V	85C	98dB	0.5MV	2NA	0.2NA	500MWF	1MA	13V	15V	1V	5uV/C	.	.6MA	96dB	96dB	30M
LM208D	NAU	SBA	EXT	.	.	+20V	-20V	85C	96dB	2MV	2NA	0.2NA	500MWF	1MA	13V	15V	1V	15uV/C	.	.6MA	85dB	80dB	30M
LM208F	NJU	SBA	EXT	.	.	+20V	-20V	85C	96dB	2MV	2NA	0.2NA	500MWF	1MA	13V	15V	1V	15uV/C	.	.6MA	85dB	80dB	30M
LM208F	SJU	SBA	EXT	.	.	+20V	-20V	85C	96dB	2MV	2NA	0.2NA	500MWF	1MA	13V	15V	1V	15uV/C	.	.6MA	85dB	80dB	30M
LM208H	NAU	SBA	EXT	.	.	+20V	-20V	85C	96dB	2MV	2NA	0.2NA	500MWF	1MA	13V	15V	1V	15uV/C	.	.6MA	85dB	80dB	30M
LM208J	NAU	SBA	EXT	.	.	+20V	-20V	85C	96dB	2MV	2NA	0.2NA	500MWF	1MA	13V	15V	1V	15uV/C	.	.6MA	85dB	80dB	30M
LM208T	SJU	SBA	EXT	.	.	+20V	-20V	85C	96dB	2MV	2NA	0.2NA	500MWF	1MA	13V	15V	1V	15uV/C	.	.6MA	85dB	80dB	30M
LM210D	NAU	VFA	INT	.	15V/uS	+18V	-18V	85C	0dB	4MV	3NA	.	500MWF	1MA	10V	15V	15V	50uV/C	.	6MA	.	70dB	10G
LM210F	NAU	VFA	INT	.	15V/uS	+18V	-18V	85C	0dB	4MV	3NA	.	500MWF	1MA	10V	15V	15V	50uV/C	.	6MA	.	70dB	10G
LM210H	NAU	VFA	INT	.	15V/uS	+18V	-18V	85C	0dB	4MV	3NA	.	500MWF	1MA	10V	15V	15V	50uV/C	.	6MA	.	70dB	10G
LM210J	NAU	VFA	INT	.	15V/uS	+18V	-18V	85C	0dB	4MV	3NA	.	500MWF	1MA	10V	15V	15V	50uV/C	.	6MA	.	70dB	10G
LM211A	MUG	CPR	EXT	.	.	+18V	-18V	85C	100dB	3MV	100NA	10NA	500MWF	.	.	15V	30V	.	.	6MA	.	.	.
LM211D	NAU	CPR	EXT	.	.	+18V	-18V	85C	100dB	3MV	100NA	10NA	500MWF	.	.	15V	30V	.	.	6MA	.	.	.
LM211F	MUG	CPR	EXT	.	.	+18V	-18V	85C	100dB	3MV	100NA	10NA	500MWF	.	.	15V	30V	.	.	6MA	.	.	.
LM211F	NAU	CPR	EXT	.	.	+18V	-18V	85C	100dB	3MV	100NA	10NA	500MWF	.	.	15V	30V	.	.	6MA	.	.	.
LM211H	NAU	CPR	EXT	.	.	+18V	-18V	85C	100dB	3MV	100NA	10NA	500MWF	.	.	15V	30V	.	.	6MA	.	.	.
LM211J	NAU	CPR	EXT	.	.	+18V	-18V	85C	100dB	3MV	100NA	10NA	500MWF	.	.	15V	30V	.	.	6MA	.	.	.
LM211N(8)	MUG	CPR	EXT	.	.	+18V	-18V	85C	100dB	3MV	100NA	10NA	500MWF	.	.	15V	30V	.	.	6MA	.	.	.
LM211N(14)	MUG	CPR	EXT	.	.	+18V	-18V	85C	100dB	3MV	100NA	10NA	500MWF	.	.	15V	30V	.	.	6MA	.	.	.
LM211T	SJU	CPR	EXT	.	.	+18V	-18V	85C	100dB	3MV	100NA	10NA	500MWF	.	.	15V	30V	.	.	6MA	.	.	.
LM211V	MUG	CPR	EXT	.	.	+18V	-18V	85C	100dB	3MV	100NA	10NA	500MWF	.	.	15V	30V	.	.	6MA	.	.	.
LM212D	NAU	SBA	INT	.	.	+20V	-20V	85C	94dB	2MV	2NA	0.2NA	500MWF	1MA	13V	14V	14V	15uV/C	.	.6MA	85dB	80dB	30M
LM212F	NAU	SBA	INT	.	.	+20V	-20V	85C	94dB	2MV	2NA	0.2NA	500MWF	1MA	13V	14V	14V	15uV/C	.	.6MA	85dB	80dB	30M
LM212H	NAU	SBA	INT	.	.	+20V	-20V	85C	94dB	2MV	2NA	0.2NA	500MWF	1MA	13V	14V	14V	15uV/C	.	.6MA	85dB	80dB	30M
LM216AD	NAU	LBC	INT	.	.	+20V	-20V	85C	92dB	3MV	50pA	15pA	500MWF	1MA	13V	15V	14V	.	.	.6MA	80dB	80dB	2G
LM216AF	NAU	LBC	INT	.	.	+20V	-20V	85C	92dB	3MV	50pA	15pA	500MWF	1MA	13V	15V	14V	.	.	.6MA	80dB	80dB	2G
LM216AH	NAU	LBC	INT	.	.	+20V	-20V	85C	92dB	3MV	50pA	15pA	500MWF	1MA	13V	15V	14V	.	.	.6MA	80dB	80dB	2G
LM216D	NAU	LBC	INT	.	.	+20V	-20V	85C	86dB	10MV	150pA	50pA	500MWF	1MA	13V	15V	14V	.	.	.6MA	80dB	80dB	2G
LM216F	NAU	LBC	INT	.	.	+20V	-20V	85C	86dB	10MV	150pA	50pA	500MWF	1MA	13V	15V	14V	.	.	.8MA	80dB	80dB	300M
LM216H	NAU	LBC	INT	.	.	+20V	-20V	85C	86dB	10MV	150pA	50pA	500MWF	1MA	13V	15V	14V	.	.	.8MA	80dB	80dB	300M
LM218D	NAU	XSR	INT	.	50V/uS	+20V	-20V	85C	94dB	4MV	250NA	50NA	500MWF	6MA	12V	15V	1V	.	.	.8MA	80dB	70dB	1M
LM218F	NAU	XSR	INT	.	50V/uS	+20V	-20V	85C	94dB	4MV	250NA	50NA	500MWF	6MA	12V	15V	1V	.	.	8MA	80dB	70dB	1M
LM218H	NAU	XSR	INT	.	50V/uS	+20V	-20V	85C	94dB	4MV	250NA	50NA	500MWF	6MA	12V	15V	1V	.	.	8MA	80dB	70dB	1M
LM219D	NAU	DCP	INT	.	.	+18V	-18V	85C	80dB	4MV	500NA	75NA	500MWF	.	.	15V	5V	.	.	12MA	.	.	.

For detailed explanations of column heading notations, see App. A.

Also for ready references the more important abbreviations used in the column headings are listed below:

LEFT HAND PAGE

APP = application (codes at APP. E.)

CMRR = common mode rejection ratio

CMP = compensation (frequency)

dV_{IO}/dT = input offset voltage temperature drift

GBP = gain bandwidth product

I_B = input bias current

I_{IO} = input bias offset current

I_Q = quiescent supply current

MFR = manufacturer (codes at App. C.)

P_Q = quiescent power consumer

PSRR = power supply rejection ratio

V_{ICM} = common mode input voltage rating

V_{IDR} = differential input voltage rating

V_{IO} = input offset voltage

V_S = dc supply voltage

RIGHT HAND PAGE

Lead out coding summary (details at APP G.) for different cases (APP.F.)

A = gain adjust

B = bias adjust

C = case

E-- = inverting input

E+ = non-inverting input

F,F* = input frequency compensation

G = ground

J = high level input

K = output, open collector

L = output, open emitter

M = metal case

N = not connected

Q = special terminal

R,R* = outputs

S = strobe

T,T* = offset balance

V+ = +ve dc supply

V = —ve dc supply

W = guard ring

X = blank position, no lead

+ + = +ve supplementary dc supply

— — = —ve supplementary dc supply

ϕ,ϕ^* = output frequency compensation

CASE (APP F)	LD 1	LD 2	LD 3	LD 4	LD 5	LD 6	LD 7	LD 8	LD 9	LD 10	LD 11	LD 12	LD 13	LD 14	LD 15	LD 16	EUROPE SUBSTITUTE	USA SUBSTITUTE	I S	TYPE NUMBER
DIL-8/1P	FT	E-	E+	V-	T*	R	V+	F*									LM201J	LM201V	0	LM201N(8)
DIL-14/1P	N	N	FT	E-	E+	V-	N	N	T*	R	V+	F*	N	N			UA201AD	LM201AJ14	0	LM201N-14
DIL-10/1P		FT	E-	E+	V-	T*	R	V+	F*	N								LM101Q	0	LM201Q
TO5-8/1M	FT	E-	E+	V-M	T*	R	V+	F*									SFC2201A	LM201H	0	LM201T
DIL-8/1P	FT	E-	E+	V-	T*	R	V+	F*										LM201J	0	LM201V
DIL-14/1C	N	N	T	N	E+	V-	N	N	L	R	V+	T*	N	N				LM102D	0	LM202D
TO5-8/1M	T	N	E+	V-	L	R	V+	T*										UA102M	0	LM202H
FLP-14/3G	N	G	E+	E-	N	V-	S1	S2	R	N	N	V-	N	N				SN72306FA	0	LM206F
TO5-8/1M	G	E+	E-	V-M	S1	S2	R	V+										SN52106L	0	LM206H
DIL-14/1P	N	N	N	E-	E+	V-	N	N	N	R	V+	N	N	N			SN52107JA	LM207D	0	LM207A
DIL-14/1M	N	N	N	E-	E+	V-	N	N	N	R	V+	N	N	N			SN52107JA		0	LM207D
FLP-10/3G	N	N	E-	E+	V-	N	R	V+	N	N							SFC2207PT		0	LM207F
TO5-8/1M	N	E-	E+	V-M	N	R	V+	N									SFC2207	UA207H	0	LM207H
DIL-8/1C	N	E-	E+	V-	N	R	V+	N										SN52107JP	0	LM207J
DIL-14/1C	N	N	N	E-	E+	V-	N	N	N	R	V+	N	N	N				SN52107JA	0	LM207J-14
DIL-8/1P	N	E-	E+	V-	N	R	V+	N									SN52107JP	LM207J	0	LM207N
DIL-8/1P	N	E-	E+	V-	N	R	V+	N									SN52107JP	LM207J	0	LM207N(8)
DIL-14/1P	N	N	N	E-	E+	V-	N	N	N	R	V+	N	N	N			SN52107JA	LM207D	0	LM207N(14)
TO5-8/1M	N	E-	E+	V-M	N	R	V+	N									UA207H	LM207H	0	LM207T
DIL-8/1P	N	E-	E+	V-	N	R	V+	N									SN52107JP	LM207J	0	LM207V
DIL-14/1G	N	F	N	E-	E+	N	V-	N	N	R	V+	F*	N	N			SN52108AJA	UA208AD	0	LM208AD
FLP-10/3G	N	N	E-	E+	N	V-	R	V+	F*	F								UA208AF	0	LM208AF
DIL-14/1C	N	F	N	E-	E+	N	V-	N	N	R	V+	F*	N	N			UA208AD	LM208AD	0	LM208AF
TO5-8/1M	F	E-	E+	V-M	N	R	V+	F*									SFC2208A	UA208AH	0	LM208AH
DIL-14/1C	N	F	N	E-	E+	N	V-	N	N	R	V+	F*	N	N			SN52108AJA	UA208AD	0	LM208AJ
TO5-8/1M	F	E-	E+	V-M	N	R	V+	F*									UA208AH	LM208AH	0	LM208AT
DIL-14/1M	N	F	N	E-	E+	N	V-	N	N	R	V+	F*	N	N			SN52108JA	UA208D	0	LM208D
FLP10/3G	N	N	E-	E+	N	V-	R	V+	F*	F							SFC2208PT	UA208F	0	LM208F
DIL-14/1C	N	F	N	E-	E+	N	V-	N	R	N	V+	F*	N	N			UA208D	LM208D	0	LM208F
TO5-8/1M	F	E-	E+	V-M	N	R	V+	F*									SFC2208	UA208H	0	LM208H
DIL-14/1C	N	F	N	E-	E+	N	V-	N	N	R	V+	F*	N	N			SN52108JA	UA208D	0	LM208J
TO5-8/1M	F	E-	E+	V-M	N	R	V+	F*									SFC2208	LM208H	0	LM208T
DIL-14/1M	N	N	T	N	E+	V-	N	N	L	R	V+	T*	N	N				SN52110JA	0	LM210D
FLP-10/3G	N	T	N	E+	V-	L	R	V+	T*	N								LM110F	0	LM210F
TO5-8/1M	T	N	E+	V-	L	R	V+	T*									SFC2210	UA110M	0	LM210H
DIL-14/1C	N	N	T	N	E+	V-	N	N	L	R	V+	T*	N	N				SN52110JA	0	LM210J
DIL-14/1P	N	G	E+	E-	N	V-	T	T*S	R	N	V+	N	N	N			SN52111J	LM211J	0	LM211A
DIL-14/1M	N	G	E+	E-	N	V-	T	T*S	R	N	V+	N	N	N				SN52111J	0	LM211D
DIL-14/1C	N	G	E+	E-	N	V-	T	T*S	R	N	V+	N	N	N			SN52111J	LM211D	0	LM211F
FLP-10/3G	G	E+	E-	N	V-	T	T*S	N	R	V+								SN52111FA	0	LM211F
TO5-8/1M	G	E+	E-	V-	T	T*S	R	V+									SFC2211	SG211T	0	LM211H
DIL-14/1C	N	G	E+	E-	N	V-	T	T*S	R	N	V+	N	N	N				SN52111J	0	LM211J
DIL-8/1P	G	E+	E-	V-	T	T*S	R	V+									LM211V	UA111R	0	LM211N(8)
DIL-14/1P	N	G	E+	E-	N	V-	T	T*S	R	N	V+	N	N	N			SN52111J	LM211J	0	LM211N(14)
TO5-8/1M	G	E+	E-	V-	T	T*S	R	V+									SFC2211	LM211H	0	LM211T
DIL-8/1P	G	E+	E-	V-	T	T*S	R	V+										UA111R	0	LM211V
DIL-14/1M	N	T	W	E-	E+	W*	V-	N	F	R	V+	T*	N	N				LM112D	0	LM212D
FLP-10/3G	N	W	E-	E+	W*	V-	R	V+	T	T*								LM112F	0	LM212F
TO5-8/1M	T	E-	E+	V-	F	R	V+	T*										LM112H	0	LM212H
DIL-14/1M	N	T	W	E-	E+	W*	V-	N	F	R	V+	T*	N	N					0	LM216AD
FLP-10/3G	N	W	E-	E+	W*	V-	R	V+	T	T*									0	LM216AF
TO5-8/1M	T	E-	E+	V-	F	R	V+	T*											0	LM216AH
DIL-14/1M	N	T	W	E-	E+	W*	V-	N	F	R	V+	T*	N	N				LM216AD	0	LM216D
FLP-10/3G	N	W	E-	E+	W*	V-	R	V+	T	T*								LM216AF	0	LM216F
TO5-8/1M	T	E-	E+	V-	F	R	V+	T*										LM216AH	0	LM216H
DIL-14/1M	N	N	T*F	E-	E+	V-	N	N	F*T	R	V+	ϕ	N	N				LM118D	0	LM218D
FLP-10/3G	N	T*F	E-	E+	V-	F*T	R	V+	ϕ	N								LM118F	0	LM218F
TO5-8/1M	T*F	E-	E+	V-	F*T	R	V+	ϕ									TDB0118CM	LM118H	0	LM218H
DIL-14/1M	N	N	G1	E+1	E-1	V-	R2	G2	E+2	E-2	V+	R1	N	N			TDE0119DP	LM219J	0	LM219D
DIL-14/1C	N	N	G1	E+1	E-1	V-	R2	G2	E+2	E-2	V+	R1	N	N			TDE0119DP	LM219J	0	LM219F

TYPE NUMBER	MFR	APP	CMP	GBP MIN	SLEW RATE MIN	Vs+ MAX	Vs- MAX	Top MAX	Avol MIN	Vio MAX	Ie MAX	Iio MAX	Ptot MAX	Iout MIN	Vout MIN	Vicm MAX	Vidr MAX	dVio/dT MAX	Po MAX	Io MAX	CMRR MIN	PSRR MIN	Rin MIN
LM219F	MUG	DCP	INT	.		+18V	-18V	85C	80dB	4MV	500NA	75NA	500MWF		.	15V	5V		.	12MA			
LM219F	NAU	DCP	INT	.		+18V	-18V	85C	80dB	4MV	500NA	75NA	500MWF		.	15V	5V		.	12MA			
LM219H	NAU	DCP	INT	.		+18V	-18V	85C	80dB	4MV	500NA	75NA	500MWF		.	15V	5V		.	12MA			
LM219J	NAU	DCP	INT	.		+18V	-18V	85C	80dB	4MV	500NA	75NA	500MWF		.	15V	5V		.	12MA			
LM219K	MUG	DCP	INT	.		+18V	-18V	85C	80dB	4MV	500NA	75NA	500MWF		.	15V	5V		.	12MA			
LM221AD	NAU	PIA	EXT	.		+20V	-20V	85C	24dB	0.4MV	10NA	0.5NA	500MWF		.	15V	15V	0.2uV/C		2MA	126dB	120dB	4M
LM221AF	NAU	PIA	EXT	.		+20V	-20V	85C	24dB	0.4MV	10NA	0.5NA	500MWF		.	15V	15V	0.2uV/C		2MA	126dB	120dB	4M
LM221AH	NAU	PIA	EXT	.		+20V	-20V	85C	24dB	0.4MV	10NA	0.5NA	500MWF		.	15V	15V	0.2uV/C		2MA	126dB	120dB	4M
LM221D	NAU	PIA	EXT	.		+20V	-20V	85C	24dB	0.7MV	10NA	1NA	500MWF		.	15V	15V	1uV/C		2MA	120dB	120dB	4M
LM221F	NAU	PIA	EXT	.		+20V	-20V	85C	24dB	0.7MV	10NA	1NA	500MWF		.	15V	15V	1uV/C		2MA	120dB	120dB	4M
LM221H	NAU	PIA	EXT	.		+20V	-20V	85C	24dB	0.7MV	10NA	1NA	500MWF		.	15V	15V	1uV/C		2MA	120dB	120dB	4M
LM224A	SJU	OGK	INT	.		+16V	-16V	85C	94dB	5MV	150NA	30NA	570MWF		.	16V	16V	35uV/C		2MA	70dB	65dB	.
LM224AD	NAU	OGK	INT	.		+16V	-16V	85C	94dB	3MV	80NA	15NA	900MWF		.	16V	16V	20uV/C		2MA	70dB	65dB	.
LM224AF	NAU	OGK	INT	.		+16V	-16V	85C	94dB	3MV	80NA	15NA	800MWF		.	16V	16V	20uV/C		2MA	70dB	65dB	.
LM224AJ	NAU	OGK	INT	.		+16V	-16V	85C	94dB	3MV	80NA	15NA	900MWF		.	16V	16V	20uV/C		2MA	70dB	65dB	.
LM224D	NAU	OGK	INT	.		+16V	-16V	85C	94dB	5MV	150NA	30NA	900MWF		.	16V	16V	35uV/C		2MA	70dB	65dB	.
LM224DD0	ING	OGK	INT	.		+16V	-16V	85C	94dB	5MV	150NA	30NA	900MWF		.	16V	16V	35uV/C		2MA	70dB	65dB	.
LM224F	MUG	OGK	INT	.		+16V	-16V	85C	94dB	5MV	150NA	30NA	900MWF		.	16V	16V	35uV/C		2MA	70dB	65dB	.
LM224F	NAU	OGK	INT	.		+16V	-16V	85C	94dB	5MV	150NA	30NA	800MWF		.	16V	16V	35uV/C		2MA	70dB	65dB	.
LM224J	NAU	OGK	INT	.		+16V	-16V	85C	94dB	5MV	150NA	30NA	900MWF		.	16V	16V	35uV/C		2MA	70dB	65dB	.
LM224N(14)	MUG	OGK	INT	.		+16V	-16V	85C	94dB	5MV	150NA	30NA	570MWF		.	16V	16V	35uV/C		2MA	70dB	65dB	.
LM239A	SJU	OCP	EXT	.		+18V	-18V	85C	88dB	5MV	250NA	50NA	900MWF	6MA	.	18V	18V			2MA	.	.	.
LM239AA	MUG	OCP	EXT	.		+18V	-18V	85C	94dB	2MV	250NA	50NA	900MWF	6MA	.	18V	18V			2MA	.	.	.
LM239AD	NAU	OCP	EXT	.		+18V	-18V	85C	94dB	2MV	250NA	50NA	900MWF	6MA	.	18V	18V			2MA	.	.	.
LM239ADDD	ING	OCP	EXT	.		+18V	-18V	85C	94dB	2MV	250NA	50NA	900MWF	6MA	.	18V	18V			2MA	.	.	.
LM239AF	MUG	OCP	EXT	.		+18V	-18V	85C	94dB	2MV	250NA	50NA	900MWF	6MA	.	18V	18V			2MA	.	.	.
LM239AF	NAU	OCP	EXT	.		+18V	-18V	85C	94dB	2MV	250NA	50NA	800MWF	6MA	.	18V	18V			2MA	.	.	.
LM239AJ	NAU	OCP	EXT	.		+18V	-18V	85C	94dB	2MV	250NA	50NA	900MWF	6MA	.	18V	18V			2MA	.	.	.
LM239AN(14)	MUG	OCP	EXT	.		+18V	-18V	85C	94dB	2MV	250NA	50NA	900MWF	6MA	.	18V	18V			2MA	.	.	.
LM239D	NAU	OCP	EXT	.		+18V	-18V	85C	88dB	5MV	250NA	50NA	900MWF	6MA	.	18V	18V			2MA	.	.	.
LM239DDD	ING	OCP	EXT	.		+18V	-18V	85C	88dB	5MV	250NA	50NA	900MWF	6MA	.	18V	18V			2MA	.	.	.
LM239F	SJU	OCP	EXT	.		+18V	-18V	85C	88dB	5MV	250NA	50NA	900MWF	6MA	.	18V	18V			2MA	.	.	.
LM239F	NAU	OCP	EXT	.		+18V	-18V	85C	88dB	5MV	250NA	50NA	800MWF	6MA	.	18V	18V			2MA	.	.	.
LM239J	NAU	OCP	EXT	.		+18V	-18V	85C	88dB	5MV	250NA	50NA	900MWF	6MA	.	18V	18V			2MA	.	.	.
LM239L	TDG	OCP	EXT	.		+18V	-18V	85C	88dB	5MV	250NA	50NA	900MWF	6MA	.	18V	18V			2MA	.	.	.
LM239N(14)	MUG	OCP	EXT	.		+18V	-18V	85C	88dB	5MV	250NA	50NA	900MWF	6MA	.	18V	18V			2MA	.	.	.
LM248D	NAU	OGK	INT	.3MHZ	0.2V/uS	+18V	-18V	85C	88dB	6MV	200NA	50NA	900MWF	5MA	12V	18V	36V			1MA	70dB	77dB	800K
LM248J	NAU	OGK	INT	.3MHZ	0.2V/uS	+18V	-18V	85C	88dB	6MV	200NA	50NA	900MWF	5MA	12V	18V	36V			1MA	70dB	77dB	800K
LM249D	NAU	OGK	INT	1MHZ	0.5V/uS	+18V	-18V	85C	88dB	6MV	200NA	50NA	900MWF	5MA	12V	18V	36V			1MA	70dB	77dB	800K
LM249J	NAU	OGK	INT	1MHZ	0.5V/uS	+18V	-18V	85C	88dB	6MV	200NA	50NA	900MWF	5MA	12V	18V	36V			1MA	70dB	77dB	800K
LM258AH	NAU	DGK	INT	.		+16V	-16V	85C	94dB	3MV	80NA	15NA	500MWF	10MA	.	16V	32V	15uV/C		3MA	70dB	65dB	.
LM258H	NAU	DGK	INT	.		+16V	-16V	85C	94dB	5MV	150NA	30NA	500MWF	10MA	.	16V	32V	30uV/C		3MA	70dB	65dB	.
LM258N(8)	MUG	DGK	INT	.		+16V	-16V	85C	94dB	5MV	150NA	30NA	500MWF	10MA	.	16V	32V	30uV/C		3MA	70dB	65dB	.
LM258T	MUG	DGK	INT	.		+16V	-16V	85C	94dB	5MV	150NA	30NA	500MWF	10MA	.	16V	32V	30uV/C		3MA	70dB	65dB	.
LM258V	MUG	DGK	INT	.		+16V	-16V	85C	94dB	5MV	150NA	30NA	500MWF	10MA	.	16V	32V	30uV/C		3MA	70dB	65dB	.
LM260D	NAU	CPR	EXT	.		+8V	-8V	85C		5MV	20uA	3uA		6MA	.	4V	5V	40uV/C		32MA	.	.	5K
LM260H	NAU	CPR	EXT	.		+8V	-8V	85C		5MV	20uA	3uA		6MA	.	4V	5V	40uV/C		32MA	.	.	5K
LM260J-14	NAU	CPR	EXT	.		+8V	-8V	85C		5MV	20uA	3uA		6MA	.	4V	5V	40uV/C		32MA	.	.	5K
LM261D	NAU	CPR	EXT	.		+16V	-16V	85C	60dB	3MV	20uA	3uA	600MWF	18MA	.	6V	5V			18MA	.	.	8K
LM261H	NAU	CPR	EXT	.		+16V	-16V	85C	60dB	3MV	20uA	3uA	600MWF	18MA	.	6V	5V			18MA	.	.	8K
LM261J	NAU	CPR	EXT	.		+16V	-16V	85C	60dB	3MV	20uA	3uA	600MWF	18MA	.	6V	5V			18MA	.	.	8K
LM293AH	NAU	DCP	EXT	.		+18V	-18V	85C	94dB	2MV	250NA	50NA	900MWF	6MA	.	18V	36V			3MA	.	.	.
LM293H	NAU	DCP	EXT	.		+18V	-18V	85C	94dB	5MV	250NA	50NA	900MWF	6MA	.	18V	36V			3MA	.	.	.
LM293N(8)	MUG	DCP	EXT	.		+18V	-18V	85C	94dB	5MV	250NA	50NA	570MWF	6MA	.	18V	36V			3MA	.	.	.
LM293T	MUG	DCP	EXT	.		+18V	-18V	85C	94dB	5MV	250NA	50NA	900MWF	6MA	.	18V	36V			3MA	.	.	.
LM293V	MUG	DCP	EXT	.		+18V	-18V	85C	94dB	5MV	250NA	50NA	570MWF	6MA	.	18V	36V			3MA	.	.	.
LM301AA	MUG	GPU	EXT	.		+18V	-18V	70C	88dB	7.5MV	250NA	50NA	500MWF	5MA	12V	15V	30V	30uV/C		3MA	70dB	70dB	500K
LM301AD	MUG	GPU	EXT	.		+18V	-18V	70C	88dB	7.5MV	250NA	50NA	500MWF	5MA	12V	15V	30V	30uV/C		3MA	70dB	70dB	500K
LM301AF	SJU	GPU	EXT	.		+18V	-18V	70C	88dB	7.5MV	250NA	50NA	500MWF	5MA	12V	15V	30V	30uV/C			70dB	70dB	500K
LM301AH	NAU	GPU	EXT	.		+18V	-18V	70C	88dB	7.5MV	250NA	50NA	500MWF	5MA	12V	15V	30V	30uV/C		3MA	70dB	70dB	500K

For detailed explanations of column heading notations, see App. A.
Also for ready references the more important abbreviations used in the column headings are listed below:

LEFT HAND PAGE

APP = application (codes at APP.E.)
CMRR = common mode rejection ratio
CMP = compensation (frequency)
dV_{io}/dT = input offset voltage temperature drift
GBP = gain bandwidth product
I_B = input bias current
I_{io} = input bias offset current
I_Q = quiescent supply current
MFR = manufacturer (codes at App.C.)
P_Q = quiescent power consumer
PSRR = power supply rejection ratio
V_{ICM} = common mode input voltage rating
V_{IOF} = differential input voltage rating
V_{IO} = input offset voltage
V_S = dc supply voltage

RIGHT HAND PAGE

Lead out coding summary details at APP.G.) for different cases (APP.F.)

A = gain adjust
B = bias adjust
C = case
E- = inverting input
E+ = non-inverting input
F* = input frequency compensation
G = ground
H = high level input
K = output, open collector
L = output, open emitter
M = metal case
N = not connected
S = special terminal
,R* = outputs
= strobe
,T* = offset balance
+ = +ve dc supply
- = -ve dc supply
W = guard ring
= blank position, no lead
+ + = +ve supplementary dc supply
- - = -ve supplementary dc supply
G* = output frequency compensation

CASE (APP F)	LD1	LD2	LD3	LD4	LD5	LD6	LD7	LD8	LD9	LD10	LD11	LD12	LD13	LD14	LD15	LD16	EUROPE SUBSTITUTE	USA SUBSTITUTE	IS	TYPE NUMBER
FLP-10/3G	R1	G1	E+1	E-1	V-	R2	G2	E+2	E-2	V+								LM119F	0	LM219F
T05-10/1M	R1	G1	E+1	E-1	V-	R2	G2	E+2	E-2	V+							TDE0119CM	LM119H	0	LM219H
DIL-14/1C	N	N	G1	E+1	E-1	V-	R2	G2	E+2	E-2	V+	R1	N	N			TDE0119DP	LM219D	0	LM219J
T05-10/1M	R1	G1	E+1	E-1	V-	R2	G2	E+2	E-2	V+							TDE0119CM	LM219H	0	LM219K
DIL-14/1M	N	R	W	E-	E+	W*	V-	N	T	T*	V+	R*	N	N				LM121AD	0	LM221AD
FLP-10/3G	R	W	E-	E+	W*	V-	T	T*	V+	R*								LM121AF	0	LM221AF
T05-8/1M	R	E-	E+	V-	T	T*	V+	R*										LM121AH	0	LM221AH
DIL-14/1M	N	R	W	E-	E+	W*	V-	N	T	T*	V+	R*	N	N				LM221AD	0	LM221D
FLP-10/3G	R	W	E-	E+	W*	V-	T	T*	V+	R*								LM221AF	0	LM221F
T05-8/1M	P	E-	E+	V-	T	T*	V+	R*										LM121H	0	LM221H
DIL-14/1P	R1	E-1	E+1	V+	E+2	E-2	R2	R3	E-3	E+3	G	E+4	E-4	R4			LM224D	LM224J	0	LM224A
DIL-14/1M	R1	E-1	E+1	V+	E+2	E-2	R2	R3	E-3	E+3	G	E+4	E-4	R4				LM124AD	0	LM224AD
FLP-14/3G	R1	E-1	E+1	V+	E+2	E-2	R2	R3	E-3	E+3	G	E+4	E-4	R4				LM124AF	0	LM224AF
DIL-14/1C	R1	E-1	E+1	V+	E+2	E-2	R2	R3	E-3	E+3	G	E+4	E-4	R4				LM124AD	0	LM224AJ
DIL-14/1M	R1	E-1	E+1	V+	E+2	E-2	R2	R3	E-3	E+3	G	E+4	E-4	R4				SG224J	0	LM224D
DIL-14/1M	R1	E-1	E+1	V+	E+2	E-2	R2	R3	E-3	E+3	G	E+4	E-4	R4			SG224J	LM224D	0	LM224DDD
DIL-14/1C	R1	E-1	E+1	V+	E+2	E-2	R2	R3	E-3	E+3	G	E+4	E-4	R4			LM224J	LM224D	0	LM224F
FLP-14/3G	R1	E-1	E+1	V+	E+2	E-2	R2	R3	E-3	E+3	G	E+4	E-4	R4				LM124F	0	LM224F
DIL-14/1C	R1	E-1	E+1	V+	E+2	E-2	R2	R3	E-3	E+3	G	E+4	E-4	R4				SG224J	0	LM224J
DIL-14/1P	R1	E-1	E+1	V+	E+2	E-2	R2	R3	E-3	E+3	G	E+4	E-4	R4			LM224D	LM224J	0	LM224N(14)
DIL-14/1P	R2	R1	V+	E-1	E+1	E-2	E+2	E+3	E-3	E-4	E+4	G	R4	R3			MLM239L	LM239D	0	LM239A
DIL-14/1P	R2	R1	V+	E-1	E+1	E-2	E+2	E+3	E-3	E-4	E+4	G	R4	R3			MLM239AL	LM239AJ	0	LM239AA
DIL-14/1M	R2	R1	V+	E-1	E+1	E-2	E+2	E+3	E-3	E-4	E+4	G	R4	R3				MLM239AL	0	LM239AD
DIL-14/1C	R2	R1	V+	E-1	E+1	E-2	E+2	E+3	E-3	E-4	E+4	G	R4	R3			MLM239AL	LM239AD	0	LM239ADDD
DIL-14/1C	R2	R1	V+	E-1	E+1	E-2	E+2	E+3	E-3	E-4	E+4	G	R4	R3			MLM239AL	LM239AD	0	LM239AF
FLP-14/3G	R2	R1	V+	E-1	E+1	E-2	E+2	E+3	E-3	E-4	E+4	G	R4	R3					0	LM239AF
DIL-14/1C	R2	R1	V+	E-1	E+1	E-2	E+2	E+3	E-3	E-4	E+4	G	R4	R3				MLM239AL	0	LM239AJ
DIL-14/1P	R2	R1	V+	E-1	E+1	E-2	E+2	E+3	E-3	E-4	E+4	G	R4	R3			MLM239AL	LM239AJ	0	LM239AN(14)
DIL-14/1M	R2	R1	V+	E-1	E+1	E-2	E+2	E+3	E-3	E-4	E+4	G	R4	R3				MLM239L	0	LM239D
DIL-14/1C	R2	R1	V+	E-1	E+1	E-2	E+2	E+3	E-3	E-4	E+4	G	R4	R3			MLM239L	LM239D	0	LM239DDD
DIL-14/1C	R2	R1	V+	E-1	E+1	E-2	E+2	E+3	E-3	E-4	E+4	G	R4	R3			MLM239L	LM239J	0	LM239F
FLP-14/3G	R2	R1	V+	E-1	E+1	E-2	E+2	E+3	E-3	E-4	E+4	G	R4	R3					0	LM239F
DIL-14/1C	R2	R1	V+	E-1	E+1	E-2	E+2	E+3	E-3	E-4	E+4	G	R4	R3				MLM239L	0	LM239J
DIL-14/1C	R2	R1	V+	E-1	E+1	E-2	E+2	E+3	E-3	E-4	E+4	G	R4	R3			MLM239L	LM239J	0	LM239L
DIL-14/1P	R2	R1	V+	E-1	E+1	E-2	E+2	E+3	E-3	E-4	E+4	G	R4	R3			MLM239L	LM239D	0	LM239N(14)
DIL-14/1M	R1	E-1	E+1	V+	E+2	E-2	R2	R3	E-3	E+3	V-	E+4	E-4	R4				LM249D	0	LM248D
DIL-14/1C	R1	E-1	E+1	V+	E+2	E-2	R2	R3	E-3	E+3	V-	E+4	E-4	R4				LM249J	0	LM248J
DIL-14/1M	R1	E-1	E+1	V+	E+2	E-2	R2	R3	E-3	E+3	V-	E+4	E-4	R4				LM248D	0	LM249D
DIL-14/1C	R1	E-1	E+1	V+	E+2	E-2	R2	R3	E-3	E+3	V-	E+4	E-4	R4				LM248D	0	LM249J
T05-8/1M	P1	E-1	E+1	G	E+2	E-2	R2	V+										LM158AH	0	LM258AH
T05-8/1M	R1	E-1	E+1	G	E+2	E-2	R2	V+										MLM158G	0	LM258H
DIL-8/1P	R1	E-1	E+1	G	E+2	E-2	R2	V+										LM258V	0	LM258N(8)
T05-8/1M	R1	E-1	E+1	G	E+2	E-2	R2	V+									MLM158G	LM258H	0	LM258T
DIL-8/1P	R1	E-1	E+1	G	E+2	E-2	R2	V+										LM258N8	0	LM258V
DIL-14/1M	N	N	N	E-	E+	V-	N	N	G	R	R*	V+	N	N				UA7600M	0	LM260D
T05-8/1M	N	E-	E+	V-	G	R	R*	V+										LM160H	0	LM260H
DIL-14/1C	N	N	N	E-	E+	V-	N	N	G	R	R*	V+	N	N				LM160D	0	LM260J-14
DIL-14/1M	V+	N	E+	E-	N	V-	N	S2	R	R*	N	S1	++					LM161D	0	LM261D
T05-10/1M	E+	E-	V-	S2	R	G	R*	S1	++	V+								LM161H	0	LM261H
DIL-14/1C	V+	N	E+	E-	N	V-	N	S2	R	R*	N	S1	++					LM161J	0	LM261J
T05-8/1M	R1	E-1	E+1	G	E+2	E-2	R2	V+										LM193AH	0	LM293AH
T05-8/1M	R1	E-1	E+1	G	E+2	E-2	R2	V+										LM193H	0	LM293H
DIL-8/1P	R1	E-1	E+1	G	E+2	E-2	R2	V+										LM2903N	0	LM293N(8)
T05-8/1M	R1	E-1	E+1	G	E+2	E-2	R2	V+										LM293H	0	LM293T
DIL-8/1P	R1	E-1	E+1	G	E+2	E-2	R2	V+										LM2903N	0	LM293V
DIL-14/1P	N	N	FT	E-	E+	V-	N	N	T*	R	V+	F*	N	N			UA301AD	LM301AJ14	0	LM301AA
MDL-8/2P	FT	E-	E+	V-	T*	R	V+	F*										TDA0301D	0	LM301AD
DIL-14/1C	N	N	FT	E-	E+	V-	N	N	T*	R	V+	F*	N	N			UA301AD	LM301AJ14	0	LM301AF
T05-8/1M	FT	E-	E+	V-M	T*	R	V+	F*									SFC2301AH	UA301AH	0	LM301AH
DIL-8/1C	FT	E-	E+	V-	T*	R	V+	F*									SFC2301ADC	UA301AT	0	LM301AJ

TYPE NUMBER	MFR	APP	CMP	GBP MIN	SLEW RATE MIN	V_S^+ MAX	V_S^- MAX	T_{op} MAX	A_{VOL} MIN	V_{IO} MAX	I_B MAX	I_{IO} MAX	P_{TOT} MAX	I_{OUT} MIN	V_{OUT} MIN	V_{ICM} MAX	V_{IDF} MAX	dV_{IO}/dT MAX	P_Q MAX	I_Q MAX	CM RR MIN	PS RR MIN	R_{IN} MIN
LM301AJ	NAU	GPU	EXT	.		+18V	-18V	70C	88dB	7.5MV	250NA	50NA	500MWF	5MA	12V	15V	30V	30uV/C	.	3MA	70dB	70dB	500K
LM301AJ	TDG	GPU	EXT	.		+18V	-18V	70C	88dB	7.5MV	250NA	50NA	500MWF	5MA	12V	15V	30V	30uV/C	.	3MA	70dB	70dB	500K
LM301AJ14	NAU	GPU	EXT	.		+18V	-18V	70C	88dB	7.5MV	250NA	50NA	500MWF	5MA	12V	15V	30V	30uV/C	.	3MA	70dB	70dB	500K
LM301AN	ING	GPU	EXT	.		+18V	-18V	70C	88dB	7.5MV	250NA	50NA	500MWF	5MA	12V	15V	30V	30uV/C	.	3MA	70dB	70dB	500K
LM301AN(8)	MUG	GPU	EXT			+18V	-18V	70C	88dB	7.5MV	250NA	50NA	500MWF	5MA	12V		30V	30uV/C			70dB	70dB	500K
LM301AN-8	TDG	GPU	EXT			+18V	-18V	70C	88dB	7.5MV	250NA	50NA	500MWF	5MA	12V	15V	30V	30uV/C		3MA	70dB	70dB	500K
LM301AT	SJU	GPU	EXT			+18V	-18V	70C	88dB	7.5MV	250NA	50NA	500MWF	5MA	12V	15V	30V	30uV/C			70dB	70dB	500K
LM301AV	SJU	GPU	EXT			+18V	-18V	70C	88dB	7.5MV	250NA	50NA	500MWF	·5MA	12V	15V	30V	30uV/C			70dB	70dB	500K
LM301D	ADU	GPU	EXT			+18V	-18V	70C	83dB	10MV	2uA	.75uA	500MWF	5MA	12V	15V	30V	30uV/C	.	3MA	65dB	70dB	100K
LM301F	ADU	GPU	EXT			+18V	-18V	70C	83dB	10MV	2uA	.75uA	500MWF		12V	15V	30V	30uV/C	.	3MA	65dB	70dB	100K
LM301H	ADU	GPU	EXT			+18V	-18V	70C	83dB	10MV	2uA	.75uA	500MWF	.	12V	15V	30V	30uV/C	.	3MA	65dB	70dB	100K
LM302D	ADU	VFA	INT			+18V	-18V	85C	0dB	15MV	30NA	.	500MWF	1MA	10V			90uV/C		6MA		60dB	10G
LM302H	NAU	VFA	INT			+18V	-18V	70C	0dB	15MV	30NA	.	500MWF	1MA	10V			90uV/C		6MA		60dB	10G
LM306H	NAU	CPR	EXT			+15V	-15V	70C	84dB	5MV	25uA	5uA	600MWF	50MA	2.5V			20uV/C	163MW				
LM307A	MUG	GPK	INT			+18V	-18V	70C	84dB	7.5MV	250NA	50NA	500MWF	5MA	12V	15V	30V	30uV/C	.		70dB	70dB	0.5M
LM307D	NAU	GPK	INT			+18V	-18V	70C	84dB	7.5MV	250NA	50NA	500MWF	5MA	12V	15V	30V	30uV/C	.		70dB	70dB	0.5M
LM307H	NAU	GPK	INT			+18V	-18V	70C	84dB	7.5MV	250NA	50NA	500MWF	5MA	12V	15V	30V	30uV/C	.		70dB	70dB	0.5M
LM307J	NAU	GPK	INT			+18V	-18V	70C	84dB	7.5MV	250NA	50NA	500MWF	5MA	12V	15V	30V	30uV/C	.		70dB	70dB	0.5M
LM307J-14	NAU	GPK	INT			+18V	-18V	70C	84dB	7.5MV	250NA	50NA	500MWF	5MA	12V	15V	30V	30uV/C	.		70dB	70dB	0.5M
LM307N	NAU	GPK	INT			+18V	-18V	70C	84dB	7.5MV	250NA	50NA	500MWF	5MA	12V	15V	30V	30uV/C	.		70dB	70dB	0.5M
LM307N(8)	MUG	GPK	INT			+18V	-18V	70C	84dB	7.5MV	250NA	50NA	500MWF	5MA	12V	15V	30V	30uV/C	.		70dB	70dB	0.5M
LM307N(14)	MUG	GPK	INT			+18V	-18V	70C	84dB	7.5MV	250NA	50NA	500MWF	5MA	12V	15V	30V	30uV/C	.		70dB	70dB	0.5M
LM307T	MUG	GPK	INT			+18V	-18V	70C	84dB	7.5MV	250NA	50NA	500MWF	5MA	12V	15V	30V	30uV/C	.		70dB	70dB	0.5M
LM307V	MUG	GPK	INT			+18V	-18V	70C	84dB	7.5MV	250NA	50NA	500MWF	5MA	12V	15V	30V	30uV/C	.		70dB	70dB	0.5M
LM308AD	NAU	SBA	EXT			+18V	-18V	70C	98dB	0.5MV	7NA	1NA	500MWF	1MA	13V	15V	1V	5uV/C	.	.6MA	96dB	96dB	10M
LM308AF	SJU	SBA	EXT			+18V	-18V	70C	98dB	0.5MV	7NA	1NA	500MWF	1MA	13V	15V	1V	5uV/C	.	.6MA	96dB	96dB	10M
LM308AH	NAU	SBA	EXT			+18V	-18V	70C	98dB	0.5MV	7NA	1NA	500MWF	1MA	13V	15V	1V	5uV/C	.	.6MA	96dB	96dB	10M
LM308AH-1	NAU	SBA	EXT			+18V	-18V	70C	98dB	0.5MV	7NA	1NA	500MWF	1MA	13V	15V	1V	1uV/C	.	.6MA	96dB	96dB	10M
LM308AH-2	NAU	SBA	EXT			+18V	-18V	70C	98dB	0.5MV	7NA	1NA	500MWF	1MA	13V	15V	1V	2uV/C	.	.6MA	96dB	96dB	10M
LM308AJ	NAU	SBA	EXT			+18V	-18V	70C	98dB	0.5MV	7NA	1NA	500MWF	1MA	13V	15V	1V	5uV/C	.	.6MA	96dB	96dB	10M
LM308AN	ADU	SBA	EXT			+18V	-18V	70C	98dB	0.5MV	7NA	1NA	500MWF	1MA	13V	15V	1V	5uV/C	.	.6MA	96dB	96dB	10M
LM308AT	MUG	SBA	EXT			+18V	-18V	70C	98dB	0.5MV	7NA	1NA	500MWF	1MA	13V	15V	1V	30uV/C	.	.6MA	80dB	80dB	10M
LM308D	NAU	SBA	EXT			+18V	-18V	70C	88dB	7.5MV	7NA	1NA	500MWF	1MA	13V	15V	1V	30uV/C	.	.6MA	80dB	80dB	10M
LM308D	MUG	SBA	EXT			+18V	-18V	70C	88dB	7.5MV	7NA	1NA		1MA	13V	15V	1V	30uV/Ċ	.	.6MA	80dB	80dB	10M
LM308H	NAU	SBA	EXT			+18V	-18V	70C	88dB	7.5MV	7NA	1NA	500MWF	1MA	13V	15V	1V	30uV/C	.	.6MA	80dB	80dB	10M
LM308J	NAU	SBA	EXT			+18V	-18V	70C	88dB	7.5MV	7NA	1NA	500MWF	1MA	13V	15V	1V	30uV/C	.	.6MA	80dB	80dB	10M
LM308N	NAU	SBA	EXT			+18V	-18V	70C	88dB	7.5MV	7NA	1NA	500MWF	1MA	13V	15V	1V	30uV/C	.	.6MA	80dB	80dB	10M
LM308N(8)	SJU	SBA	EXT			+18V	-18V	70C	88dB	7.5MV	7NA	1NA	500MWF	1MA	13V	15V	1V	30uV/C	.	.6MA	80dB	80dB	10M
LM308T	SJU	SBA	EXT			+18V	-18V	70C	88dB	7.5MV	7NA	1NA	500MWF	1MA	13V	15V	1V	30uV/C	.	.6MA	80dB	80dB	10M
LM308V	SJU	SBA	EXT			+18V	-18V	70C	88dB	7.5MV	7NA	1NA	500MWF	1MA	13V	15V	1V	30uV/C	.	.6MA	80dB	80dB	10M
LM310D	NAU	VFA	INT	15V/uS		+18V	-18V	70C	0dB	7.5MV	7NA	.	500MWF	1MA	10V	15V	15V	50uV/C	.	6MA		70dB	10G
LM310F	NAU	VFA	INT	15V/uS		+18V	-18V	70C	0dB	7.5MV	7NA	.	500MWF	1MA	10V	15V	15V	50uV/C	.	6MA		70dB	10G
LM310H	NAU	VFA	INT	15V/uS		+18V	-18V	70C	0dB	7.5MV	7NA	.	500MWF	1MA	10V	15V	15V	50uV/C	.	6MA		70dB	10G
LM310J	NAU	VFA	INT	15V/uS		+18V	-18V	70C	0dB	7.5MV	7NA	.	500MWF	1MA	10V	15V	15V	50uV/C	.	6MA		70dB	10G
LM310J-8	NAU	VFA	INT	15V/uS		+18V	-18V	70C	0dB	7.5MV	7NA	.	500MWF	1MA	10V	15V	15V	50uV/C	.	6MA		70dB	10G
LM310N	NAU	VFA	INT	15V/uS		+18V	-18V	70C	0dB	7.5MV	7NA	.	500MWF	1MA	10V	15V	15V	50uV/C	.	6MA		70dB	10G
LM311A	SJU	CPR	EXT			+18V	-18V	70C	100dB	7.5MV	250NA	50NA	500MWF	.	.	15V	30V		.	8MA			
LM311D	NAU	CPR	EXT		.	+18V	-18V	70C	100dB	7.5MV	250NA	50NA	500MWF			15V	30V		.	8MA			
LM311F	MUG	CPR	EXT		.	+18V	-18V	70C	100dB	7.5MV	250NA	50NA	500MWF			15V	30V			8MA			
LM311F	NAU	CPR	EXT			+18V	-18V	70C	100dB	7.5MV	250NA	50NA	500MWF			15V	30V			8MA			
LM311H	NAU	CPR	EXT			+18V	-18V	70C	100dB	7.5MV	250NA	50NA	500MWF			15V	30V		.	8MA			
LM311J	NAU	CPR	EXT			+18V	-18V	70C	100dB	7.5MV	250NA	50NA	500MWF			15V	30V		.	8MA			
LM311J-8	NAU	CPR	EXT			+18V	-18V	70C	100dB	7.5MV	250NA	50NA	500MWF			15V	30V		.	8MA			
LM311N	NAU	CPR	EXT			+18V	-18V	70C	100dB	7.5MV	250NA	50NA	500MWF			15V	30V		.	8MA			
LM311N(8)	MUG	CPR	EXT			+18V	-18V	70C	100dB	7.5MV	250NA	50NA	500MWF			15V	30V		.	8MA			
LM311N-8	TDG	CPR	EXT			+18V	-18V	70C	100dB	7.5MV	250NA	50NA	500MWF			15V	30V	.	.	8MA			
LM311N(14)	MUG	CPR	EXT			+18V	-18V	70C	100dB	7.5MV	250NA	50NA	500MWF			15V	30V	.	.	8MA			
LM311N-14	NAU	CPR	EXT			+18V	-18V	70C	100dB	7.5MV	250NA	50NA	500MWF			15V	30V	.	.	8MA			
LM311T	SJU	CPR	EXT			+18V	-18V	70C	100dB	7.5MV	250NA	50NA	500MWF			15V	30V		.	8MA			
LM311V	SJU	CPR	EXT		.	+18V	-18V	70C	100dB	7.5MV	250NA	50NA	500MWF		.	15V	30V		.	8MA			

For detailed explanations of column heading notations, see App. A.

Also for ready references the more important abbreviations used in the column headings are listed below:

LEFT HAND PAGE

- APP = application (codes at APP.E.)
- CMRR = common mode rejection ratio
- CMP = compensation (frequency)
- dV_{io}/dT = input offset voltage temperature drift
- GBP = gain bandwidth product
- I_B = input bias current
- I_{io} = input bias offset current
- I_0 = quiescent supply current
- MFR = manufacturer (codes at App.C.)
- P_0 = quiescent power consumer
- PSRR = power supply rejection ratio
- V_{icm} = common mode input voltage rating
- V_{idr} = differential input voltage rating
- V_{io} = input offset voltage
- V_s = dc supply voltage

RIGHT HAND PAGE

Lead out coding summary (details at APP.G.) for different cases (APP.F.)

- A = gain adjust
- B = bias adjust
- C = case
- E− = inverting input
- E+ = non-inverting input
- F,F* = input frequency compensation
- G = ground
- J = high level input
- K = output, open collector
- L = output, open emitter
- M = metal case
- N = not connected
- Q = special terminal
- R,R* = outputs
- S = strobe
- T,T* = offset balance
- V+ = +ve dc supply
- V− = −ve dc supply
- W = guard ring
- X = blank position, no lead
- ++ = +ve supplementary dc supply
- −− = −ve supplementary dc supply
- ∅,∅* = output frequency compensation

CASE (APP F)	LD 1	LD 2	LD 3	LD 4	LD 5	LD 6	LD 7	LD 8	LD 9	LD 10	LD 11	LD 12	LD 13	LD 14	LD 15	LD 16	EUROPE SUBSTITUTE	USA SUBSTITUTE	IS	TYPE NUMBER
DIL-14/1C	N	N	FT	E-	E+	V-	N	N	T*	R	V+	F*	N	N	.	.	UA301AD	LM301AJ14	O	LM301AJ
DIL-14/1P	N	N	FT	E-	E+	V-	N	N	T*	R	V+	F*	N	N	.	.	UA301AD	LM301AJ14	O	LM301AJ14
DIL-8/1P	FT	E-	E+	V-	T*	R	V+	F*	.	.	.	.	.	.	.	.	SFC2301ADC	UA301AT		LM301AN
DIL-8/1P	FT	E-	E+	V-	T*	R	V+	F*	.	.	.	.	.	.	.	.	SFC2301ADC	LM301AJ	O	LM301AN(8)
DIL-8/1P	FT	E-	E+	V-	T*	R	V+	F*	.	.	.	.	.	.	.	.	SFC2301ADC	LM301AJ	O	LM301AN-8
T05-8/1M	FT	E-	E+	V-M	T*	R	V+	F*	.	.	.	.	.	.	.	.	SFC2301A	LM301AH		LM301AT
DIL-8/1P	FT	E-	E+	V-	T*	R	V+	F*	.	.	.	.	.	.	.	.	SFC2301ADC	LM301AJ	O	LM301AV
DIL-14/1C	N	N	FT	E-	E+	V-	N	N	T*	R	V+	F*	N	N	.	.	UA301AD	LM301AJ14	O	LM301D
FLP-10/3C	N	FT	E-	E+	V-	T*	R	V+	F*	N	.	.	.	.	.	.	SFC2201APT	LM301AF		LM301F
T05-8/1M	FT	E-	E+	V-M	T*	R	V+	F*	.	.	.	.	.	.	.	.	SFC2301A	LM301AH	O	LM301H
DIL-14/1C	N	N	T	N	E+	V-	N	N	L	R	V+	T*	N	N	.	.		LM202D		LM302D
T05-8/1M	T	N	E+	V-	L	R	V+	T*	.	.	.	.	.	.	.	.		UA302C		LM302H
T05-8/1M	G.	E+	E-	V-M	S1	S2	R	V+	.	.	.	.	.	.	.	.		SN72306L		LM306H
DIL-14/1P	N	N	N	E-	E+	V-	N	N	N	R	V+	N	N	N	.	.	SN72307JA	LM307D		LM307A
DIL-14/1M	N	N	N	E-	E+	V-	N	N	N	R	V+	N	N	N	.	.		SN72307JA		LM307D
T05-8/1M	N	E-	E+	V-M	N	R	V+	N	.	.	.	.	.	.	.	.	SFC2307	UA307H	O	LM307H
DIL-8/1C	N	E-	E+	V-	N	R	V+	N	.	.	.	.	.	.	.	.	SFC2307DC	UA307T		LM307J
DIL-14/1C	N	N	N	E-	E+	V-	N	N	N	R	V+	N	N	N	.	.	SN72307JA	SN72307JA		LM307J-14
DIL-8/1P	N	E-	E+	V-	N	R	V+	N	.	.	.	.	.	.	.	.	SFC2307DC	SN72307JP		LM307N
DIL-8/1P	N	E-	E+	V-	N	R	V+	N	.	.	.	.	.	.	.	.	SFC2307DC	LM307V	O	LM307N(8)
DIL-14/1P	N	N	N	E-	E+	V-	N	N	N	R	V+	N	N	N	.	.	SN72307JA	LM307A		LM307N(14)
T05-8/1M	N	E-	E+	V-M	N	R	V+	N	.	.	.	.	.	.	.	.	UA307H	LM307H	O	LM307T
DIL-8/1P	N	E-	E+	V-	N	R	V+	N	.	.	.	.	.	.	.	.	SFC2307DC	LM307N		LM307V
DIL-14/1G	N	F	N	E-	E+	N	V-	N	N	R	V+	F*	N	N	.	.	SN72308AJA	UA308AD		LM308AD
DIL-14/1C	N	F	N	E-	E+	N	V-	N	N	R	V+	F*	N	N	.	.	UA308AD	LM308AD	O	LM308AF
T05-8/1M	F	E-	E+	V-M	N	R	V+	F*	.	.	.	.	.	.	.	.	SFC2308A	UA308AH		LM308AH
T05-8/1M	F	E-	E+	V-M	N	R	V+	F*	.	.	.	.	.	.	.	.			O	LM308AH-1
T05-8/1M	F	E-	E+	V-M	N	R	V+	F*	.	.	.	.	.	.	.	.			O	LM308AH-2
DIL-14/1C	N	F	N	E-	E+	V-	N	N	N	R	V+	F*	N	N	.	.	SN72308AJA	UA308AD		LM308AJ
DIL-8/1P	F	E-	E+	V-	N	R	V+	F*	.	.	.	.	.	.	.	.			O	LM308AN
T05-8/1M	F	E-	E+	V-M	N	R	V+	F*	.	.	.	.	.	.	.	.	UA308AH	LM308AH		LM308AT
DIL-14/1M	N	F	N	E-	E+	V-	N	N	N	R	V+	F*	N	N	.	.	SN72308JA	UA308D		LM308D
MDL-8/2P	F	E-	E+	V-	N	R	V+	F*	.	.	.	.	.	.	.	.			O	LM308D
T05-8/1M	F	E-	E+	V-M	N	R	V+	F*	.	.	.	.	.	.	.	.	SFC2308	UA308H		LM308H
DIL-14/1C	N	F	N	E-	E+	N	V-	N	N	R	V+	F*	N	N	.	.	SN72308AJA	UA308D		LM308J
DIL-8/1P	F	E-	E+	V-	N	R	V+	F*	.	.	.	.	.	.	.	.	SFC2308DC	SN72308JP		LM308N
DIL-8/1P	F	E-	E+	V-	N	R	V+	F*	.	.	.	.	.	.	.	.	SFC2308DC	LM308N.	O	LM308N(8)
T05-8/1M	F	E-	E+	V-	N	R	V+	F*	.	.	.	.	.	.	.	.	UA308H	LM308H	O	LM308T
DIL-8/1P	F	E-	E+	V-	N	R	V+	F*	.	.	.	.	.	.	.	.	SFC2308DC	LM308N		LM308V
DIL-14/1M	N	N	T	N	E+	V-	N	N	L	R	V+	T*	N	N	.	.	SFC2310EC	SN72310JA		LM310D
FLP-10/3G	N	T	N	E+	V-	L	R	V+	T*	N	.	.	.	.	.	.		LM210F		LM310F
T05-8/1M	T	N	E+	V-	L	R	V+	T*	.	.	.	.	.	.	.	.	SFC2310EC	UA310C		LM310H
DIL-14/1C	N	N	T	N	E+	V-	N	N	L	R	V+	T*	N	N	.	.	SFC2310EC	SN72310JA		LM310J
DIL-8/1C	T	N	E+	V-	L	R	V+	T*	.	.	.	.	.	.	.	.	SFC2310DC	SN72310JP		LM310J-8
DIL-8/1P	T	N	E+	V-	L	R	V+	T*	.	.	.	.	.	.	.	.	SFC2310DC	SN72310JP		LM310N
DIL-14/1P	N	G	E+	E-	N	V-	T	T*S	R	N	V+	N	N	N	.	.	SFC2311EC	LM311D		LM311A
DIL-14/1M	N	G	E+	E-	N	V-	T	T*S	R	N	V+	N	N	N	.	.	SFC2311EC	SN72311J		LM311D
DIL-14/1C	N	G	E+	E-	N	V-	T	T*S	R	E+	E-	N	N		.	.	SFC2311EC	LM311J		LM311D
FLP-10/3G	G	E+	E-	N	V-	T	T*S	N	R	V+	.	.	.	.	.	.		SN72311FA		LM311JF
T05-8/1M	G	E+	E-	V-	T	T*S	R	V+	.	.	.	.	.	.	.	.	SFC2311	UA311H		LM311H
DIL-14/1C	N	G	E+	E-	N	V-	T	T*S	R	N	V+	N	N	N	.	.	SFC2311EC	SN72311J		LM311J
DIL-8/1C	G	E+	E-	V-	T	T*S	R	V+	.	.	.	.	.	.	.	.	SFC2311DC	UA311R		LM311J-8
DIL-8/1P	G	E+	E-	V-	T	T*S	R	V+	.	.	.	.	.	.	.	.	SFC2311DC	UA311R		LM311N
DIL-8/1P	G	E+	E-	V-	T	T*S	R	V+	.	.	.	.	.	.	.	.	UA311R	LM311N		LM311N(8)
DIL-8/1P	G	E+	E-	V-	T	T*S	R	V+	.	.	.	.	.	.	.	.	SFC2311DC	LM311N		LM311N-8
DIL-14/1P	N	G	E+	E-	N	V-	T	T*S	R	N	V+	N	N	N	.	.	SFC2311EC	LM311D	O	LM311N(14)
DIL-14/1P	N	G	E+	E-	N	V-	T	T*S	R	N	V+	N	N	N	.	.	SFC2311EC	SN72311J	O	LM311N-14
T05-8/1M	G	E+	E-	V-	T	T*S	R	V+	.	.	.	.	.	.	.	.	UA311H	LM311H		LM311T
DIL-8/1P	G	E+	E-	V-	T	T*S	R	V+	.	.	.	.	.	.	.	.	UA311R	LM311N		LM311V
DIL-14/1M	N	T	W	E-	E+	W*	V-	N	F	R	V+	T*	N	N	.	.		LM212D	O	LM312D

TYPE NUMBER	M F R	A P P	C M P	GBP MIN	SLEW RATE MIN	V_S^+ MAX	V_S^- MAX	T_{op} MAX	A_{VOL} MIN	V_{IO} MAX	I_B MAX	I_{IO} MAX	P_{TOT} MAX	I_{OUT} MIN	V_{OUT} MIN	V_{ICM} MAX	V_{IDF} MAX	dV_{IO}/dT MAX	P_Q MAX	I_Q MAX	CM RR MIN	PS RR MIN	R_{IN} MIN
LM312D	NAU	SBA	INT	.	.	+18V	-18V	70C	88dB	7.5MV	7NA	1NA	500MWF	1MA	13V	14V	14V	15uV/C	.	.8MA	80dB	80dB	10M
LM312F	NAU	SBA	INT	.	.	+18V	-18V	70C	88dB	7.5MV	7NA	1NA	500MWF	1MA	13V	14V	14V	15uV/C	.	.8MA	80dB	80dB	10M
LM312H	NAU	SBA	INT	.	.	+18V	-18V	70C	88dB	7.5MV	7NA	1NA	500MWF	1MA	13V	14V	14V	15uV/C	.	.8MA	80dB	80dB	10M
LM316AD	NAU	LBC	INT	.	.	+20V	-20V	70C	92dB	3MV	50pA	15pA	500MWF	1MA	13V	15V	14V	.	.	.6MA	80dB	80dB	2G
LM316AF	NAU	LBC	INT	.	.	+20V	-20V	70C	92dB	3MV	50pA	15pA	500MWF	1MA	13V	15V	14V	.	.	.6MA	80dB	80dB	2G
LM316AH	NAU	LBC	INT	.	.	+20V	-20V	70C	92dB	3MV	50pA	15pA	500MWF	1MA	13V	15V	14V	.	.	.6MA	80dB	80dB	2G
LM316D	NAU	LBC	INT	.	.	+20V	-20V	70C	86dB	10MV	150pA	50pA	500MWF	1MA	13V	15V	14V	.	.	.8MA	80dB	80dB	300M
LM316F	NAU	LBC	INT	.	.	+20V	-20V	70C	86dB	10MV	150pA	50pA	500MWF	1MA	13V	15V	14V	.	.	.8MA	80dB	80dB	300M
LM316H	NAU	LBC	INT	.	.	+20V	-20V	70C	86dB	10MV	150pA	50pA	500MWF	1MA	13V	15V	14V	.	.	.8MA	80dB	80dB	300M
LM318D	NAU	XSR	INT	.	50V/uS	+20V	-20V	70C	88dB	10MV	500NA	200NA	500MWF	6MA	12V	15V	1V	.	.	8MA	70dB	65dB	500K
LM318F	ADU	XSR	INT	.	50V/uS	+20V	-20V	70C	88dB	10MV	500NA	200NA	500MWF	6MA	12V	15V	1V	.	.	8MA	70dB	65dB	500K
LM318H	NAU	XSR	INT	.	50V/uS	+20V	-20V	70C	88dB	10MV	500NA	200NA	500MWF	6MA	12V	15V	1V	.	.	8MA	70dB	65dB	500K
LM318N	NAU	XSR	INT	.	50V/uS	+20V	-20V	70C	88dB	10MV	500NA	200NA	500MWF	6MA	12V	15V	1V	.	.	8MA	70dB	65dB	500K
LM319A	MUG	DCP	INT	.	.	+18V	-18V	70C	78dB	8MV	1uA	0.2uA	500MWF	.	.	15V	5V	.	.	2MA	.	.	.
LM319D	NAU	DCP	INT	.	.	+18V	-18V	70C	78dB	8MV	1uA	0.2uA	500MWF	.	.	15V	5V	.	.	12MA	.	.	.
LM319F	MUG	DCP	INT	.	.	+18V	-18V	70C	78dB	8MV	1uA	0.2uA	500MWF	.	.	15V	5V	.	.	12MA	.	.	.
LM319F	NAU	DCP	INT	.	.	+18V	-18V	70C	78dB	8MV	1uA	0.2uA	500MWF	.	.	15V	5V	.	.	12MA	.	.	.
LM319H	NAU	DCP	INT	.	.	+18V	-18V	70C	78dB	8MV	1uA	0.2uA	500MWF	.	.	15V	5V	.	.	12MA	.	.	.
LM319J	NAU	DCP	INT	.	.	+18V	-18V	70C	78dB	8MV	1uA	0.2uA	500MWF	.	.	15V	5V	.	.	12MA	.	.	.
LM319K	MUG	DCP	INT	.	.	+18V	-18V	70C	78dB	8MV	1uA	0.2uA	500MWF	.	.	15V	5V	.	.	12MA	.	.	.
LM319N	NAU	DCP	INT	.	.	+18V	-18V	70C	78dB	8MV	1uA	0.2uA	500MWF	.	.	15V	5V	.	.	12MA	.	.	.
LM319N(14)	MUG	DCP	INT	.	.	+18V	-18V	70C	78dB	8MV	1uA	0.2uA	500MWF	.	.	15V	5V	.	.	2MA	.	.	.
LM321AD	NAU	PIA	EXT	.	.	+20V	-20V	70C	22dB	0.4MV	15NA	0.5NA	500MWF	.	.	15V	15V	0.2uV/C	.	3MA	126dB	120dB	2M
LM321AF	NAU	PIA	EXT	.	.	+20V	-20V	70C	22dB	0.4MV	15NA	0.5NA	500MWF	.	.	15V	15V	0.2uV/C	.	3MA	126dB	120dB	2M
LM321AH	NAU	PIA	EXT	.	.	+20V	-20V	70C	22dB	0.4MV	15NA	0.5NA	500MWF	.	.	15V	15V	0.2uV/C	.	3MA	126dB	120dB	2M
LM321D	NAU	PIA	EXT	.	.	+20V	-20V	70C	22dB	1.5MV	18NA	2NA	500MWF	.	.	15V	15V	1uV/C	.	3MA	114dB	114dB	2M
LM321F	NAU	PIA	EXT	.	.	+20V	-20V	70C	22dB	1.5MV	18NA	2NA	500MWF	.	.	15V	15V	1uV/C	.	3MA	114dB	114dB	2M
LM321H	NAU	PIA	EXT	.	.	+20V	-20V	70C	22dB	1.5MV	18NA	2NA	500MWF	.	.	15V	15V	1uV/C	.	3MA	114dB	114dB	2M
LM324A	SJU	QGK	INT	.	.	+16V	-16V	70C	83dB	7MV	250NA	50NA	570MWF	.	.	16V	16V	35uV/C	.	2MA	65dB	65dB	.
LM324AD	ADU	QGK	INT	.	.	+16V	-16V	70C	88dB	3MV	100NA	30NA	900MWF	.	.	16V	16V	30uV/C	.	2MA	65dB	65dB	.
LM324AJ	NAU	QGK	INT	.	.	+16V	-16V	70C	88dB	3MV	100NA	30NA	900MWF	.	.	16V	16V	30uV/C	.	2MA	65dB	65dB	.
LM324AN	NAU	QGK	INT	.	.	+16V	-16V	70C	88dB	3MV	100NA	30NA	900MWF	.	.	16V	16V	30uV/C	.	2MA	65dB	65dB	.
LM324D	ING	QGK	INT	.	.	+16V	-16V	70C	88dB	7MV	250NA	50NA	900MWF	.	.	16V	16V	35uV/C	.	2MA	65dB	65dB	.
LM324D	MUG	QGK	INT	.	.	+16V	-16V	70C	88dB	7MV	250NA	50NA	.	.	.	16V	16V	35uV/C	.	2MA	65dB	65dB	.
LM324DDD	ING	QGK	INT	.	.	+16V	-16V	70C	88dB	7MV	250NA	50NA	900MWF	.	.	16V	16V	35uV/C	.	2MA	65dB	65dB	.
LM324F	SJU	QGK	INT	.	.	+16V	-16V	70C	88dB	7MV	250NA	50NA	900MWF	.	.	16V	16V	35uV/C	.	2MA	65dB	65dB	.
LM324J	NAU	QGK	INT	.	.	+16V	-16V	70C	88dB	7MV	250NA	50NA	900MWF	.	.	16V	16V	35uV/C	.	2MA	65dB	65dB	.
LM324N	NAU	QGK	INT	.	.	+16V	-16V	70C	88dB	7MV	250NA	50NA	900MWF	.	.	16V	16V	35uV/C	.	2MA	65dB	65dB	.
LM324N(14)	SJU	QGK	INT	.	.	+16V	-16V	70C	88dB	7MV	250NA	50NA	570MWF	.	.	16V	16V	35uV/C	.	2MA	65dB	65dB	.
LM324NPD	ING	QGK	INT	.	.	+16V	-16V	70C	88dB	7MV	250NA	50NA	900MWF	.	.	16V	16V	35uV/C	.	.	.	.	.
LM339A	MUG	QCP	EXT	.	.	+18V	-18V	70C	88dB	5MV	250NA	50NA	900MWF	6MA	.	18V	18V	.	.	2MA	.	.	.
LM339AA	MUG	QCP	EXT	.	.	+18V	-18V	70C	94dB	2MV	250NA	50NA	900MWF	6MA	.	18V	18V	.	.	2MA	.	.	.
LM339AF	MUG	QCP	EXT	.	.	+18V	-16V	70C	94dB	2MV	250NA	50NA	900MWF	6MA	.	18V	18V	.	.	2MA	.	.	.
LM339AJ	NAU	QCP	EXT	.	.	+18V	-18V	70C	94dB	2MV	250NA	50NA	900MWF	6MA	.	18V	18V	.	.	2MA	.	.	.
LM339AN(14)	MUG	QCP	EXT	.	.	+18V	-18V	70C	94dB	2MV	250NA	50NA	900MWF	6MA	.	18V	18V	.	.	2MA	.	.	.
LM339F	MUG	QCP	EXT	.	.	+18V	-18V	70C	88dB	5MV	250NA	50NA	900MWF	6MA	.	18V	18V	.	.	2MA	.	.	.
LM339J	NAU	QCP	EXT	.	.	+18V	-18V	70C	88dB	5MV	250NA	50NA	900MWF	6MA	.	18V	18V	.	.	2MA	.	.	.
LM339L	TDG	QCP	EXT	.	.	+18V	-18V	70C	88dB	5MV	250NA	50NA	900MWF	6MA	.	18V	18V	.	.	2MA	.	.	.
LM339N(14)	MUG	QCP	EXT	.	.	+18V	-18V	70C	88dB	5MV	250NA	50NA	900MWF	6MA	.	18V	18V	.	.	2MA	.	.	.
LM339AD	NAU	QCP	EXT	.	.	+18V	-18V	70C	94dB	2MV	250NA	50NA	900MWF	6MA	.	18V	18V	.	.	2MA	.	.	.
LM339ADDD	ING	QCP	EXT	.	.	+18V	-18V	70C	94dB	2MV	250NA	50NA	900MWF	6MA	.	18V	18V	.	.	2MA	.	.	.
LM339AN	NAU	QCP	EXT	.	.	+18V	-18V	70C	94dB	2MV	250NA	50NA	900MWF	6MA	.	18V	18V	.	.	2MA	.	.	.
LM339ANPD	ING	QCP	EXT	.	.	+18V	-18V	70C	94dB	2MV	250NA	50NA	900MWF	6MA	.	18V	18V	.	.	2MA	.	.	.
LM339D	NAU	QCP	EXT	.	.	+18V	-18V	70C	88dB	5MV	250NA	50NA	900MWF	6MA	.	18V	18V	.	.	2MA	.	.	.
LM339DDD	ING	QCP	EXT	.	.	+18V	-18V	70C	88dB	5MV	250NA	50NA	900MWF	6MA	.	18V	18V	.	.	2MA	.	.	.
LM339N	NAU	QCP	EXT	.	.	+18V	-18V	70C	88dB	5MV	250NA	50NA	800MWF	.	.	18V	18V	.	.	2MA	.	.	.
LM339NPD	ING	QCP	EXT	.	.	+18V	-18V	70C	88dB	5MV	250NA	50NA	800MWF	.	.	18V	18V	.	.	2MA	.	.	.
LM343D	NAU	HVO	INT	.3MHZ	1V/uS	+34V	-34V	70C	97dB	8MV	40NA	10NA	680MWF	4MA	20V	34V	68V	.	.	5MA	70dB	74dB	.
LM343H	NAU	HVO	INT	.3MHZ	1V/uS	+34V	-34V	70C	97dB	8MV	40NA	10NA	680MWF	4MA	20V	34V	68V	.	.	5MA	70dB	74dB	.
LM344D	NAU	HVO	EXT	.3MHZ	1V/uS	+34V	-34V	70C	97dB	8MV	40NA	10NA	680MWF	4MA	20V	34V	68V	.	.	5MA	70dB	74dB	.

94

For detailed explanations of column heading notations, see App. A.
Also for ready references the more important abbreviations used in the column headings are listed below:

LEFT HAND PAGE
APP = application (codes at APP.E.)
CMRR = common mode rejection ratio
CMP = compensation (frequency)
dV_{IO}/dT = input offset voltage temperature drift
GBP = gain bandwidth product
I_B = input bias current
I_{IO} = input bias offset current
I_Q = quiescent supply current
MFR = manufacturer (codes at App.C.)
P_Q = quiescent power consumer
PSRR = power supply rejection ratio
V_{ICM} = common mode input voltage rating
V_{IDF} = differential input voltage rating
V_{IO} = input offset voltage
V_S = dc supply voltage

RIGHT HAND PAGE
Lead out coding summary (details at APP.G.) for different cases (APP.F
A = gain adjust
B = bias adjust
C = case
E− = inverting input
E+ = non-inverting input
F,F* = input frequency compensation
G = ground
J = high level input
K = output, open collector
L = output, open emitter
M = metal case
N = not connected
Q = special terminal
R,R* = outputs
S = strobe
T,T* = offset balance
V+ = +ve dc supply
V− = −ve dc supply
W = guard ring
X = blank position, no lead
++ = +ve supplementary dc supply
−− = −ve supplementary dc supply
∅,∅* = output frequency compensation

CASE (APP F)	LD 1	LD 2	LD 3	LD 4	LD 5	LD 6	LD 7	LD 8	LD 9	LD 10	LD 11	LD 12	LD 13	LD 14	LD 15	LD 16	EUROPE SUBSTITUTE	USA SUBSTITUTE	IS	TYPE NUMBER
FLP-10/3G	N	W	E−	E+	W*	V−	R	V+	T	T*	.	.	.	.				LM212F	O	LM312F
TO5-8/1M	T	E−	E+	V−	F	R	V+	T*	.	.								LM212H	O	LM312H
DIL-14/1M	N	T	W	E−	E+	W*	V−	N	F	R	V+	T*	N	N				LM216AD	O	LM316AD
FLP-10/3G	N	W	E−	E+	W*	V−	R	V+	T	T*	.	.						LM216AF	O	LM316AF
TO5-8/1M	T	E−	E+	V−	F	R	V+	T*										LM216AH	O	LM316AH
DIL-14/1M	N	T	W	E−	E+	W*	V−	N	F	R	V+	T*	N	N			MLM316D	LM216D	O	LM316D
FLP-10/3G	N	W	E−	E+	W*	V−	R	V+	T	T*								LM216F	O	LM316F
TO5-8/1M	T	E−	E+	V−	F	R	V+	T*										LM216H	O	LM316H
DIL-14/1M	N	N	T*F	E−	E+	V−	N	F*T	R	V+	∅	N	N					SN72318JA	O	LM318D
FLP-10/3C	N	T*F	E−	E+	V−	F*T	R	V+	∅	N								LM218F	O	LM318F
TO5-8/1M	T*F	E−	E+	V−	F*T	R	V+	∅									TDE0118CM	SN72318L	O	LM318H
DIL-8/1P	T*F	E−	E+	V−	F*T	R	V−	∅										SN72318JP	O	LM318N
DIL-14/1P	N	N	G1	E+1	E-1	V−	R2	G2	E+2	E-2	V+	R1	N	N			TDB01190P	LM319N	O	LM319A
DIL-14/1M	N	N	G1	E+1	E-1	V−	R2	G2	E+2	E-2	V+	R1	N	N			TDB0119CM	LM219D	O	LM319D
DIL-14/1C	N	N	G1	E+1	E-1	V−	R2	G2	E+2	E-2	V+	R1	N	N			TDB0119DP	LM319J	O	LM319F
FLP-10/3G	R1	G1	E+1	E-1	V−	R2	G2	E+2	E-2	V+								LM219F	O	LM319F
TO5-10/1M	R1	G1	E+1	E-1	V−	R2	G2	E+2	E-2	V+							TDB0119CM	LM219H	O	LM319H
DIL-14/1C	N	N	G1	E+1	E-1	V−	R2	G2	E+2	E-2	V+	R1	N	N			TDB0119DP	LM219J	O	LM319J
TO5-10/1M	R1	G1	E+1	E-1	V−	R2	G2	E+2	E-2	V+							TDB0119CM	LM319H	O	LM319K
DIL-14/1P	N	N	G1	E+1	E-1	V−	R2	G2	E+2	E-2	V+	R1	N	N			TDB0119DP	LM219N	O	LM319N
DIL-14/1P	N	N	G1	E+1	E-1	V−	R2	G2	E+2	E-2	V+	R1	N	N			TDB0119DP	LM319N	O	LM319N(14)
DIL-14/1M	N	R	W	E−	E+	W*	V−	N	T	T*	V+	R*	N	N				LM221AD	O	LM321AD
FLP-10/3G	R	W	E−	E+	W*	V−	T	T*	V+	R*								LM221AF	O	LM321AF
TO5-8/1M	R	E−	E+	V−	T	T*	V+	R*										LM221AH	O	LM321AH
DIL-14/1M	N	R	W	E−	E+	W*	V−	N	T	T*	V+	R*	N	N				LM321AD	O	LM321D
FLP-10/3G	R	W	E−	E+	W*	V−	T	T*	V+	R*								LM221AF	O	LM321F
TO5-8/1M	R	E−	E+	V−	T	T*	V+	R*										LM221H	O	LM321H
DIL-14/1P	R1	E-1	E+1	V+	E+2	E-2	R2	R3	E-3	E+3	G	E+4	E-4	R4			MLM324J	LM324N	O	LM324A
DIL-14/1M	R1	E-1	E+1	V+	E+2	E-2	R2	R3	E-3	E+3	G	E+4	E-4	R4				LM224AD	O	LM324AD
DIL-14/1C	R1	E-1	E+1	V+	E+2	E-2	R2	R3	E-3	E+3	G	E+4	E-4	R4				LM224AD	O	LM324AJ
DIL-14/1P	R1	E-1	E+1	V+	E+2	E-2	R2	R3	E-3	E+3	G	E+4	E-4	R4				LM224AD	O	LM324AN
DIL-14/1C	R1	E-1	E+1	V+	E+2	E-2	R2	R3	E-3	E+3	G	E+4	E-4	R4				MLM324L	O	LM324D
MDL-14/4P	R1	E-1	E+1	V+	E+2	E-2	R2	R3	E-3	E+3	G	E+4	E-4	R4				TDA0324D	O	LM324D
DIL-14/1C	R1	E-1	E+1	V+	E+2	E-2	R2	R3	E-3	E+3	G	E+4	E-4	R4			MLM324L	LM324N	O	LM324DDD
DIL-14/1C	R1	E-1	E+1	V+	E+2	E-2	R2	R3	E-3	E+3	G	E+4	E-4	R4			LM324N	LM324J	O	LM324F
DIL-14/1C	R1	E-1	E+1	V+	E+2	E-2	R2	R3	E-3	E+3	G	E+4	E-4	R4				MLM324L	O	LM324J
DIL-14/1M	R1	E-1	E+1	V+	E+2	E-2	R2	R3	E-3	E+3	G	E+4	E-4	R4				LM324N	O	LM324N
DIL-14/1P	R1	E-1	E+1	V+	E+2	E-2	R2	R3	E-3	E+3	G	E+4	E-4	R4			MLM324L	LM324N	O	LM324N(14)
DIL-14/1P	R1	E-1	E+1	V+	E+2	E-2	R2	R3	E-3	E+3	G	E+4	E-4	R4			MLM324L	LM324N	O	LM324NPD
DIL-14/1P	R2	R1	V+	E-1	E+1	E-2	E+2	E+3	E-3	E-4	E+4	G	R4	R3			MLM339L	LM339D	O	LM339A
DIL-14/1P	R2	R1	V+	E-1	E+1	E-2	E+2	E+3	E-3	E-4	E+4	G	R4	R3			MLM339AL	LM339AD	O	LM339AA
DIL-14/1C	R2	R1	V+	E-1	E+1	E-2	E+2	E+3	E-3	E-4	E+4	G	R4	R3			MLM339AL	LM339AJ	O	LM339AF
DIL-14/1C	R2	R1	V+	E-1	E+1	E-2	E+2	E+3	E-3	E-4	E+4	G	R4	R3			MLM339AL	LM339AJ	O	LM339AJ
DIL-14/1M	R2	R1	V+	E-1	E+1	E-2	E+2	E+3	E-3	E-4	E+4	G	R4	R3			MLM339AL	LM339AD	O	LM339AN(14)
DIL-14/1C	R2	R1	V+	E-1	E+1	E-2	E+2	E+3	E-3	E-4	E+4	G	R4	R3			MLM339L	LM339J	O	LM339F
DIL-14/1C	R2	R1	V+	E-1	E+1	E-2	E+2	E+3	E-3	E-4	E+4	G	R4	R3			MLM339L		O	LM339J
DIL-14/1C	R2	R1	V+	E-1	E+1	E-2	E+2	E+3	E-3	E-4	E+4	G	R4	R3			MLM339L	LM339J	O	LM339L
DIL-14/1M	R2	R1	V+	E-1	E+1	E-2	E+2	E+3	E-3	E-4	E+4	G	R4	R3			MLM339L	LM339D	O	LM339N(14)
DIL-14/1C	R2	R1	V+	E-1	E+1	E-2	E+2	E+3	E-3	E-4	E+4	G	R4	R3			MLM339AL	LM339AD	O	LM339AD
DIL-14/1C	R2	R1	V+	E-1	E+1	E-2	E+2	E+3	E-3	E-4	E+4	G	R4	R3			MLM339AL	LM339AD	O	LM339ADDD
DIL-14/1P	R2	R1	V+	E-1	E+1	E-2	E+2	E+3	E-3	E-4	E+4	G	R4	R3			MLM339AL		O	LM339AN
DIL-14/1P	R2	R1	V+	E-1	E+1	E-2	E+2	E+3	E-3	E-4	E+4	G	R4	R3			MLM339AL	LM339AN	O	LM339ANPD
DIL-14/1C	R2	R1	V+	E-1	E+1	E-2	E+2	E+3	E-3	E-4	E+4	G	R4	R3			MLM339L	LM339D	O	LM339D
DIL-14/1C	R2	R1	V+	E-1	E+1	E-2	E+2	E+3	E-3	E-4	E+4	G	R4	R3			MLM339D		O	LM339DDD
DIL-14/1P	R2	R1	V+	E-1	E+1	E-2	E+2	E+3	E-3	E-4	E+4	G	R4	R3			MLM339L		O	LM339N
DIL-14/1P	R2	R1	V+	E-1	E+1	E-2	E+2	E+3	E-3	E-4	E+4	G	R4	R3			MLM339L	LM339N	O	LM339NPD
DIL-14/1M	T	E−	E+	V−	N	N	N	T*	R	V+	N								O	LM343D
TO5-8/1M	T	E−	E+	V−	T*	R	V+	N										MC1436G	O	LM343H
DIL-14/1M	N	N	FT	E−	E+	V−	N	N	T*	R	V+	F*	N	N					O	LM344D
TO5-8/1M	FT	E−	E+	V−M	T*	R	V+	F*											O	LM344H

TYPE NUMBER	M F R	A P P	C M P	GBP MIN	SLEW RATE MIN	V_S^+ MAX	V_S^- MAX	T_{op} MAX	A_{VOL} MIN	V_{IO} MAX	I_B MAX	I_{IO} MAX	P_{TOT} MAX	I_{OUT} MIN	V_{OUT} MIN	V_{ICM} MAX	V_{IDF} MAX	dV_{IO}/dT MAX	P_Q MAX	I_O MAX	CM RR MIN	PS RR MIN	R_{IN} MIN
LM344H	NAU	HVO	EXT	.3MHZ	1V/uS	+34V	-34V	70C	97dB	8MV	40NA	10NA	680MWF	4MA	20V	34V	68V			5MA	70dB	74dB	.
LM348D	NAU	QGK	INT	.3MHZ	0.2V/uS	+18V	-18V	70C	88dB	6MV	200NA	50NA	900MWF	5MA	12V	18V	36V			1MA	70dB	77dB	800K
LM348J	NAU	QGK	INT	.3MHZ	0.2V/uS	+18V	-18V	70C	88dB	6MV	200NA	50NA	900MWF	5MA	12V	18V	36V			1MA	70dB	77dB	800K
LM348N	NAU	QGK	INT	.3MHZ	0.2V/uS	+18V	-18V	70C	88dB	6MV	200NA	50NA	900MWF	5MA	12V	18V	36V			1MA	70dB	77dB	800K
LM349D	NAU	QGK	INT	1MHZ	0.5V/uS	+18V	-18V	70C	88dB	6MV	200NA	50NA	900MWF	5MA	12V	18V	36V			1MA	70dB	77dB	800K
LM349J	NAU	QGK	INT	1MHZ	0.5V/uS	+18V	-18V	70C	88dB	6MV	200NA	50NA	900MWF	5MA	12V	18V	36V			1MA	70dB	77dB	800K
LM349N	NAU	QGK	INT	1MHZ	0.5V/uS	+18V	-18V	70C	88dB	6MV	200NA	50NA	900MWF	5MA	12V	18V	36V			1MA	70dB	77dB	800K
LM358AH	NAU	DGK	INT			+16V	-16V	70C	88dB	3MV	100NA	30NA	500MWF	10MA		16V	32V	20uV/C		3MA	65dB	65dB	
LM358AN	NAU	DGK	INT			+16V	-16V	70C	88dB	3MV	100NA	30NA	500MWF	10MA		16V	32V	20uV/C		3MA	65dB	65dB	
LM358D	MUG	DGK	INT			+16V	-16V	70C	88dB	7MV	250NA	50NA		10MA		16V	32V	30uV/C		3MA	65dB	65dB	
LM358H	NAU	DGK	INT			+16V	-16V	70C	88dB	7MV	250NA	50NA	500MWF	10MA		16V	32V	30uV/C		3MA	65dB	65dB	
LM358N	NAU	DGK	INT			+16V	-16V	70C	88dB	7MV	250NA	50NA	570MWF	10MA		16V	32V	30uV/C		3MA	65dB	65dB	
LM358N(8)	MUG	DGK	INT			+16V	-16V	70C	88dB	7MV	250NA	50NA	570MWF	10MA		16V	32V	30uV/C		3MA	65dB	65dB	
LM358T	MUG	DGK	INT			+16V	-16V	70C	88dB	7MV	250NA	50NA	500MWF	10MA		16V	32V	30uV/C		3MA	65dB	65dB	
LM358V	MUG	DGK	INT			+16V	-16V	70C	88dB	7MV	250NA	50NA	570MWF	10MA		16V	32V	30uV/C		3MA	65dB	65dB	
LM360D	NAU	CPR	EXT			+8V	-8V	70C		5MV	20uA	3uA		6MA		4V	5V	40uV/C		32MA			5K
LM360H	NAU	CPR	EXT			+8V	-8V	70C		5MV	20uA	3uA		6MA		4V	5V	40uV/C		32MA			5K
LM360J-14	NAU	CPR	EXT			+8V	-8V	70C		5MV	20uA	3uA		6MA		4V	5V	40uV/C		32MA			5K
LM360N-8	NAU	CPR	EXT			+8V	-8V	70C		5MV	20uA	3uA		6MA		4V	5V	40uV/C		32MA			5K
LM360N-14	NAU	CPR	EXT			+8V	-8V	70C		5MV	20uA	3uA		6MA		4V	5V	40uV/C		32MA			5K
LM361D	NAU	CPR	EXT			+16V	-16V	70C	60dB	5MV	30uA	5uA	600MWF	18MA		6V	5V			20MA			8K
LM361J	NAU	CPR	EXT			+16V	-16V	70C	60dB	5MV	30uA	5uA	600MWF	18MA		6V	5V			20MA			8K
LM361H	NAU	CPR	EXT			+16V	-16V	70C	60dB	5MV	30uA	5uA	600MWF	18MA		6V	5V			20MA			8K
LM361N	NAU	CPR	EXT			+16V	-16V	70C	60dB	5MV	30uA	5uA	600MWF	18MA		6V	5V			20MA			8K
LM381AN	NAU	DLN	INT	5MHZ		+20V	-20V	70C	94dB		1uA		800MWF	1MA	19V					20MA		100dB	40K
LM381N	NAU	DLN	INT	5MHZ		+20V	-20V	70C	94dB		1uA		800MWF	1MA	19V					20MA		100dB	40K
LM382N	NAU	DLN	INT	5MHZ		+20V	-20V	70C	90dB		2uA		800MWF	1MA						16MA		100dB	40K
LM387AN	NAU	DLN	INT	5MHZ		+20V	-20V	70C	94dB		3uA		660MWF	1MA						15MA		100dB	50K
LM387N	NAU	DLN	INT	5MHZ		+15V	-15V	70C	94dB		3uA		660MWF	1MA						15MA		100dB	50K
LM393AH	NAU	DCP	EXT			+18V	-18V	85C	94dB	2MV	250NA	50NA	900MWF	6MA		18V	36V			3MA			
LM393AN	NAU	DCP	EXT			+18V	-18V	85C	94dB	2MV	250NA	50NA	570MWF	6MA		18V	36V			3MA			
LM393H	NAU	DCP	EXT			+18V	-18V	85C	94dB	5MV	250NA	50NA	500MWF	6MA		18V	36V			3MA			
LM393N	NAU	DCP	EXT			+18V	-18V	85C	94dB	5MV	250NA	50NA	570MWF	6MA		18V	36V			3MA			
LM393N(8)	MUG	DCP	EXT			+18V	-18V	70C	94dB	5MV	250NA	50NA	570MWF	6MA		18V	36V			3MA			
LM393T	MUG	DCP	EXT			+18V	-18V	70C	94dB	5MV	250NA	50NA	900MWF	6MA		18V	36V			3MA			
LM393V	MUG	DCP	EXT			+18V	-18V	70C	94dB	5MV	250NA	50NA	570MWF	6MA		18V	36V			3MA			
LM709AJ	NAU	GPU	EXT	.3MHZ	.15V/uS	+18V	-18V	125C	88dB	2MV	200NA	50NA	670MWF	5MA	12V	10V	5V	10uV/C	108MW	4MA	80dB	80dB	350K
LM709CH	NAU	GPU	EXT	.3MHZ	.15V/uS	+18V	-18V	70C	84dB	7.5MV	1.5uA	0.5uA	500MWF	5MA	12V	10V	5V		200MW		65dB	74dB	50K
LM709CJ	NAU	GPU	EXT	.3MHZ	.15V/uS	+18V	-18V	70C	84dB	7.5MV	1.5uA	0.5uA	670MWF	5MA	12V	10V	5V		200MW		65dB	74dB	50K
LM709CN	NAU	GPU	EXT	.3MHZ	.15V/uS	+18V	-18V	70C	84dB	7.5MV	1.5uA	0.5uA	670MWF	5MA	12V	10V	5V		200MW		65dB	74dB	50K
LM709CN-8	NAU	GPU	EXT	.3MHZ	.15V/uS	+18V	-18V	70C	84dB	7.5MV	1.5uA	0.5uA	670MWF	5MA	12V	10V	5V		200MW		65dB	74dB	50K
LM709H	NAU	GPU	EXT	.3MHZ	.15V/uS	+18V	-18V	125C	88dB	5MV	500NA	200NA	500MWF	5MA	12V	10V	5V	15uV/C	165MW		70dB	76dB	150K
LM709J	NAU	GPU	EXT	.3MHZ	.15V/uS	+18V	-18V	125C	88dB	5MV	500NA	200NA	670MWF	5MA	12V	10V	5V	15uV/C	165MW		70dB	76dB	150K
LM710CH	NAU	CPR	EXT			+14V	-7V	70C	60dB	5MV	25uA	5uA	500MWF	1MA	1V	7V	5V	20uV/C	150MW	9MA	70dB		
LM710H	NAU	CPR	EXT			+14V	-7V	125C	61dB	2MV	20uA	3uA	500MWF	2MA	1V	7V	5V	10uV/C	150MW	9MA	80dB		
LM711CH	NAU	DCP	EXT			+14V	-7V	70C	57dB	5MV	100uA	25uA	500MWF			7V	5V	20uV/C	180MW		70dB		
LM711CN	NAU	DCP	EXT			+14V	-7V	70C	57dB	5MV	100uA	25uA	670MWF			7V	5V	20uV/C	180MW		70dB		
LM711H	NAU	DCP	EXT			+14V	-7V	125C	58dB	3.5MV	75uA	10uA	500MWF			7V	5V	20uV/C	180MW		70dB		
LM725AH	NAU	PIA	EXT			+22V	-22V	125C	120dB	0.5MV	75NA	5NA	500MWF	6MA	12V	22V	5V	2uV/C	120MW		120dB	106dB	500K
LM725AJ-14	NAU	PIA	EXT			+22V	-22V	125C	120dB	0.5MV	75NA	5NA	500MWF	6MA	12V	22V	5V	2uV/C	120MW		120dB	106dB	500K
LM725CH	NAU	PIA	EXT			+22V	-22V	70C	106dB	2.5MV	125NA	35NA	500MWF	5MA	12V	22V	5V	10uV/C	150MW		94dB	90dB	500K
LM725CJ-14	NAU	PIA	EXT			+22V	-22V	70C	106dB	2.5MV	125NA	35NA	500MWF	5MA	12V	22V	5V	10uV/C	150MW		94dB	90dB	500K
LM725CN	NAU	PIA	EXT			+22V	-22V	70C	106dB	2.5MV	125NA	35NA	500MWF	5MA	12V	22V	5V	10uV/C	150MW		94dB	90dB	500K
LM725D	NAU	PIA	EXT			+22V	-22V	125C	120dB	1MV	100NA	20NA	500MWF	5MA	12V	22V	5V	5uV/C	105MW		110dB	100dB	500K
LM725J-14	NAU	PIA	EXT			+22V	-22V	125C	120dB	1MV	100NA	20NA	500MWF	5MA	12V	22V	5V	5uV/C	105MW		110dB	100dB	500K
LM725H	NAU	PIA	EXT			+22V	-22V	125C	120dB	1MV	100NA	20NA	500MWF	5MA	12V	22V	5V	5uV/C	105MW		110dB	100dB	500K
LM733CD	NAU	BDO	EXT	20MHZ		+8V	-8V	70C	48dB	6MV	30uA	5uA	670MWF	2MA	3V	6V	5V			24MA	60dB	50dB	2K
LM733CH	NAU	BDO	EXT	20MHZ		+8V	-8V	70C	48dB	6MV	30uA	5uA	500MWF	2MA	3V	6V	5V			24MA	60dB	50dB	2K
LM733CJ	NAU	BDO	EXT	20MHZ		+8V	-8V	70C	48dB	6MV	30uA	5uA	670MWF	2MA	3V	6V	5V			24MA	60dB	50dB	2K
LM733CN	NAU	BDO	EXT	20MHZ		+8V	-8V	70C	48dB	6MV	30uA	5uA	670MWF	2MA	3V	6V	5V			24MA	60dB	50dB	2K

Also for ready references the more important abbreviations used in the column headings are listed below:

LEFT HAND PAGE

APP = application (codes at APP.E.)
CMRR = common mode rejection ratio
CMP = compensation (frequency)
dV_{IO}/dT = input offset voltage temperature drift
GBP = gain bandwidth product
I_B = input bias current
I_{IO} = input bias offset current
I_U = quiescent supply current
MFR = manufacturer (codes at App.C.)
P_U = quiescent power consumer
PSRR = power supply rejection ratio
V_{ICM} = common mode input voltage rating
V_{IDi} = differential input voltage rating
V_{IO} = input offset voltage
V_S = dc supply voltage

RIGHT HAND PAGE

Lead out coding summary (details at APP.G.) for different cases (APP.F.)

A = gain adjust
B = bias adjust
C = case
E— = inverting input
E+ = non-inverting input
F,F* = input frequency compensation
G = ground
J = high level input
K = output, open collector
L = output, open emitter
M = metal case
N = not connected
Q = special terminal
R,R* = outputs
S = strobe
T,T* = offset balance
V+ = +ve dc supply
V— = —ve dc supply
W = guard ring
X = blank position, no lead
++ = +ve supplementary dc supply
—— = —ve supplementary dc supply
ø,ø* = output frequency compensation

CASE (APP F)	LD 1	LD 2	LD 3	LD 4	LD 5	LD 6	LD 7	LD 8	LD 9	LD 10	LD 11	LD 12	LD 13	LD 14	LD 15	LD 16	EUROPE SUBSTITUTE	USA SUBSTITUTE	IS	TYPE NUMBER
DIL-14/1M	R1	E-1	E+1	V+	E+2	E-2	R2	R3	E-3	E+3	V-	E+4	E-4	R4				LM3490	0	LM348D
DIL-14/1C	R1	E-1	E+1	V+	E+2	E-2	R2	R3	E-3	E+3	V-	E+4	E-4	R4				LM3490	0	LM348J
DIL-14/1P	R1	E-1	E+1	V+	E+2	E-2	R2	R3	E-3	E+3	V-	E+4	E-4	R4				LM3490	0	LM348N
DIL-14/1M	R1	E-1	E+1	V+	E+2	E-2	R2	R3	E-3	E+3	V-	E+4	E-4	R4				LM348D	0	LM3490D
DIL-14/1C	R1	E-1	E+1	V+	E+2	E-2	R2	R3	E-3	E+3	V-	E+4	E-4	R4				LM348D	0	LM349J
DIL-14/1P	R1	E-1	E+1	V+	E+2	E-2	R2	R3	E-3	E+3	V-	E+4	E-4	R4				LM348D	0	LM349J
T05-8/1M	R1	E-1	E+1	G	E+2	E-2	R2	V+										LM258AH	0	LM358AH
DIL-8/1P	R1	E-1	E+1	G	E+2	E-2	R2	V+									MLM158U		0	LM358AN
MDL-8/2P	R1	E-1	E+1	G	E+2	E-2	R2	V+									TDA0358		0	LM358D
T05-8/1M	R1	E-1	E+1	G	E+2	E-2	R2	V+									MLM358G		0	LM358H
DIL-8/1P	R1	E-1	E+1	G	E+2	E-2	R2	V+									MLM358U		0	LM358N
DIL-8/1P	R1	E-1	E+1	G	E+2	E-2	R2	V+									MLM358U	LM358N	0	LM358N(8)
T05-8/1M	R1	E-1	E+1	G	E+2	E-2	R2	V+									MLM358G	LM358H	0	LM358T
DIL-8/1P	R1	E-1	E+1	G	E+2	E-2	R2	V+									MLM358U	LM358N	0	LM358V
DIL-14/1M	N	N	N	E-	E+	V-	N	N	G	R	R*	V+	N	N				UA7600C	0	LM360D
T05-8/1M	N	E-	E+	V-	G	R	R*	V+										UA760HC	0	LM360H
DIL-14/1C	N	N	N	E-	E+	V-	N	N	G	R	R*	V+	N	N				UA7600C	0	LM360J-14
DIL-8/1P	N	N	N	E-	E+	V-	G	R	R*	V+									0	LM360N-8
DIL-14/1P	N	N	N	E-	E+	V-	N	N	G	R	R*	V+	N	N				UA7600C	0	LM360N-14
DIL-14/1M	V+	N	E+	E-	N	V-	N	S2	R	G	R*	N	S1	++				LM261D	0	LM361D
DIL-14/1C	V+	N	E+	E-	N	V-	N	S2	R	G	R*	N	S1	++				LM261J	0	LM361J
T05-10/1M	E+	E-	V-	S2	R	G	R*	S1	++	V+								LM261H	0	LM361H
DIL-14/1P	V+	N	E+	E-	N	V-	N	S2	R	G	R*	N	S1	++				LM261D	0	LM361N
DIL-14/1P	E+1	E-1	Q1	G	F1	F*1	R1	R2	V+	F2	F*2	Q2	E-2	E+2					0	LM381AN
DIL-14/1P	E+1	E-1	Q1	G	F1	F*1	R1	R2	V+	F2	F*2	Q2	E-2	E+2				LM381AN	0	LM381N
DIL-14/1P	E+1	E-1	A1	G	A*1	A*1	R1	R2	A*2	A*2	V+	A2	E-2	E+2					0	LM382N
DIL-8/1P	E+1	E-1	G	R1	R2	V	E-2	E+2											0	LM387AN
DIL-8/1P	E+1	E-1	G	R1	R2	V	E-2	E+2										LM387AN	0	LM387N
DIL-8/1P	R1	E-1	E+1	G	E+2	E-2	R2	V+										LM293AH	0	LM393AH
DIL-8/1P	R1	E-1	E+1	G	E+2	E-2	R2	V+											0	LM393AN
T05-8/1M	R1	E-1	E+1	G	E+2	E-2	R2	V+										LM393AH	0	LM393H
DIL-8/1P	R1	E-1	E+1	G	E+2	E-2	R2	V+										LM393AN	0	LM393N
DIL-8/1P	R1	E-1	E+1	G	E+2	E-2	R2	V+									LM393V	LM393N	0	LM393N(8)
T05-8/1M	R1	E-1	E+1	G	E+2	E-2	R2	V+										LM393H	0	LM393T
DIL-8/1P	R1	E-1	E+1	G	E+2	E-2	R2	V+										LM393N	0	LM393V
DIL-14/1C	N	N	F	E-	E+	V-	N	N	ø	R	V+	F*	N	N				UA709ADM	0	LM709AJ
T05-8/1M	F	E-	E+	V-	ø	ø*R	V+	F*									TAA521	UA709HC	0	LM709CH
DIL-14/1C	N	N	F	E-	E+	V-	N	N	ø	R	V+	F*	N	N			TAA521A	UA709DC	0	LM709CJ
DIL-14/1P	N	N	F	E-	E+	V-	N	N	ø	R	V+	F*	N	N			TAA521A	UA709DC	0	LM709CN
DIL-8/1P	F	E-	E+	V-	ø	ø*R	V+	F*									MC1709U	UA709TC	0	LM709CN-8
T05-8/1M	F	E-	E+	V-	ø	ø*R	V+	F*									TAA522	UA709HM	0	LM709H
DIL-14/1C	N	N	F	E-	E+	V-	N	N	ø	R	V+	F*	N	N				UA709DM	0	LM709J
T05-8/1M	G	E+	E-	V-M	N	N	R	V+									SFC2710C	UA710HC	0	LM710CH
T05-8/1M	G	E+	E-	V-M	N	N	R	V+									SFC2710M	UA710HM	0	LM710H
T05-10/1M	G	S1	E-1	E+1	V-	E+2	E-2	S2	R	V+							SFC2711C	UA711HC	0	LM711CH
DIL-14/1P	N	E-1	E+1	N	E+2	E-2	N	N	S2	R	V+	N	S1	N			SFC2711EC	UA711DC	0	LM711CN
T05-10/1M	G	S1	E-1	E+1	V-	E+2	E-2	S2	R	V+							SFC2711M	UA711HM	0	LM711H
T05-8/1M	T	E-	E+	V-	ø	ø*R	V+	T*										UA725AHM	0	LM725AH
DIL-14/1C	N	N	T	E-	E+	V-	N	N	ø	ø*R	V+	T*	N	N					0	LM725AJ-14
T05-8/1M	T	E-	E+	V-	ø	ø*R	V+	T*										UA725HC	0	LM725CH
DIL-14/1C	N	N	T	E-	E+	V-	N	N	ø	ø*R	V+	T*	N	N			LM725J-14		0	LM725CJ-14
DIL-8/1P	T	E-	E+	V-	ø	ø*R	V+	T*											0	LM725CN
DIL-14/1M	N	N	T	E-	E+	V-	N	N	ø	ø*R	V+	T*	N	N				LM725AJ-14	0	LM725
DIL-14/1C	N	N	T	E-	E+	V-	N	N	ø	ø*R	V+	T*	N	N				LM725D	0	LM725J-14
T05-8/1M	T	E-	E+	V-	ø	ø*R	V+	T*										UA725HM	0	LM725H
DIL-14/1M	E+	N	A2	A*2	V-	N	R	R*	N	V+	A1	A*1	N	E-			SN72733J	UA733DC	0	LM733CD
T05-10/1M	E-	E+	A2	A*2	V-	R	R*	V+	A1	A*1							SN72733J	UA733HC	0	LM733CH
DIL-14/1C	E+	N	A2	A*2	V-	N	R	R*	N	V+	A1	A*1	N	E-			SN72733J	UA733DC	0	LM733CJ
DIL-14/1P	E+	N	A2	A*2	V-	N	R	R*	N	V+	A1	A*1	N	E-			SN72733J	UA733DC	0	LM733CN
DIL-14/1M	E+	N	A2	A*2	V-	N	R	R*	N	V+	A1	A*1	N	E-			SN52733J	UA733DM	0	LM733D

TYPE NUMBER	M F R	A P P	C M P	GBP MIN	SLEW RATE MIN	Vs+ MAX	Vs- MAX	Top MAX	Avol MIN	Vio MAX	Ib MAX	Iio MAX	Ptot MAX	Iout MIN	Vout MIN	Vicm MAX	Vidf MAX	dVio/dT MAX	Po MAX	Io MAX	CMRR MIN	PSRR MIN	Rin MIN
LM733D	NAU	BDO	EXT	20MHZ	.	+8V	-8V	125C	50dB	5MV	20uA	3uA	670MWF	2MA	3V	6V	5V	.	.	24MA	60dB	50dB	2K
LM733H	NAU	BDO	EXT	20MHZ	.	+8V	-8V	125C	50dB	5MV	20uA	3uA	500MWF	2MA	3V	6V	5V			24MA	60dB	50dB	2K
LM733J	NAU	BDO	EXT	20MHZ	.	+8V	-8V	125C	50dB	5MV	20uA	3uA	670MWF	2MA	3V	6V	5V			24MA	60dB	50dB	2K
LM741AD	NAU	GPK	INT	.4MHZ	0.3V/uS	+22V	-22V	125C	94dB	3MV	80NA	30NA	670MWF	7MA	16V	15V	30V	15uV/C	150MW		80dB	86dB	1M
LM741AF	NAU	GPK	INT	.4MHZ	0.3V/uS	+22V	-22V	125C	94dB	3MV	80NA	30NA	570MWF	7MA	16V	15V	30V	15uV/C	150MW		80dB	86dB	1M
LM741AH	NAU	GPK	INT	.4MHZ	0.3V/uS	+22V	-22V	125C	94dB	3MV	80NA	30NA	500MWF	7MA	16V	15V	30V	15uV/C	150MW		80dB	86dB	1M
LM741AJ-14	NAU	GPK	INT	.4MHZ	0.3V/uS	+22V	-22V	125C	94dB	3MV	80NA	30NA	670MWF	7MA	16V	15V	30V	15uV/C	150MW		80dB	86dB	1M
LM741CD	NAU	GPK	INT	.	0.2V/uS	+18V	-18V	70C	88dB	6MV	500NA	200NA	670MWF	5MA	12V	15V	30V	.	85MW	3MA	70dB	76dB	300K
LM741CH	NAU	GPK	INT	.	0.2V/uS	+18V	-18V	70C	88dB	6MV	500NA	200NA	670MWF	5MA	12V	15V	30V		85MW	3MA	70dB	76dB	300K
LM741CJ	NAU	GPK	INT	.	0.2V/uS	+18V	-18V	70C	88dB	6MV	500NA	200NA	310MWF	5MA	12V	15V	30V		85MW	3MA	70dB	76dB	300K
LM741CJ-14	NAU	GPK	INT	.	0.2V/uS	+18V	-18V	70C	88dB	6MV	500NA	200NA	670MWF	5MA	12V	15V	30V		85MW	3MA	70dB	76dB	300K
LM741CN	NAU	GPK	INT	.	0.2V/uS	+18V	-18V	70C	88dB	6MV	500NA	200NA	310MWF	5MA	12V	15V	30V		85MW	3MA	70dB	76dB	300K
LM741CN-14	NAU	GPK	INT	.	0.2V/uS	+18V	-18V	70C	88dB	6MV	500NA	200NA	670MWF	5MA	12V	15V	30V		85MW	3MA	70dB	76dB	300K
LM741D	NAU	GPK	INT	.	0.3V/uS	+22V	-22V	125C	94dB	5MV	500NA	200NA	670MWF	5MA	12V	15V	30V		85MW	3MA	70dB	76dB	300K
LM741ED	NAU	GPK	INT	.4MHZ	0.3V/uS	+22V	-22V	70C	94dB	6MV	80NA	30NA	670MWF	7MA	16V	15V	30V	15uV/C	150MW		80dB	86dB	1M
LM741EH	NAU	GPK	INT	.4MHZ	0.3V/uS	+22V	-22V	70C	94dB	6MV	80NA	30NA	500MWF	7MA	16V	15V	30V	15uV/C	150MW		80dB	80dB	1M
LM741EJ	NAU	GPK	INT	.4MHZ	0.3V/uS	+22V	-22V	70C	94dB	6MV	80NA	30NA	670MWF	7MA	16V	15V	30V	15uV/C	150MW		80dB	80dB	1M
LM741EJ-14	NAU	GPK	INT	.4MHZ	0.3V/uS	+22V	-22V	70C	94dB	6MV	80NA	30NA	670MWF	7MA	16V	15V	30V	15uV/C	150MW		80dB	86dB	1M
LM741EN	NAU	GPK	INT	.4MHZ	0.3V/uS	+22V	-22V	70C	94dB	6MV	80NA	30NA	310MWF	7MA	16V	15V	30V	15uV/C	150MW		80dB	86dB	1M
LM741F	NAU	GPK	INT	.	0.3V/uS	+22V	-22V	125C	94dB	5MV	500NA	200NA	570MWF	5MA	12V	15V	30V		85MW	3MA	70dB	76dB	300K
LM741H	NAU	GPK	INT	.	0.3V/uS	+22V	-22V	125C	94dB	5MV	500NA	200NA	500MWF	5MA	12V	15V	30V		85MW	3MA	70dB	76dB	300K
LM741J-14	NAU	GPK	INT	.	0.3V/uS	+22V	-22V	125C	94dB	5MV	500NA	200NA	670MWF	5MA	12V	15V	30V		85MW	3MA	70dB	76dB	300K
LM747AD	NAU	DGK	INT	.4MHZ	0.3V/uS	+22V	-22V	125C	94dB	3MV	80NA	30NA	670MWF	7MA	16V	15V	30V	15uV/C	150MW		80dB	86dB	1M
LM747AH	NAU	DGK	INT	.4MHZ	0.3V/uS	+22V	-22V	125C	94dB	3MV	80NA	30NA	500MWF	7MA	16V	15V	30V	15uV/C	150MW		80dB	86dB	1M
LM747AJ	NAU	DGK	INT	.4MHZ	0.3V/uS	+22V	-22V	125C	94dB	3MV	80NA	30NA	670MWF	7MA	16V	15V	30V	15uV/C	150MW		80dB	86dB	1M
LM747CD	NAU	DGK	INT	.	0.2V/uS	+18V	-18V	70C	88dB	6MV	500NA	200NA	670MWF	5MA	12V	15V	30V		85MW	3MA	70dB	76dB	300K
LM747CF	NAU	DGK	INT	.	0.2V/uS	+18V	-18V	70C	88dB	6MV	500NA	200NA	800MWF	5MA	12V	15V	30V		85MW	3MA	70dB	76dB	300K
LM747CH	NAU	DGK	INT	.	0.2V/uS	+18V	-18V	70C	88dB	6MV	500NA	200NA	500MWF	5MA	12V	15V	30V		85MW	3MA	70dB	76dB	300K
LM747CJ	NAU	DGK	INT	.	0.2V/uS	+18V	-18V	70C	88dB	6MV	500NA	200NA	670MWF	5MA	12V	15V	30V		85MW	3MA	70dB	76dB	300K
LM747CN	NAU	DGK	INT	.	0.2V/uS	+18V	-18V	70C	88dB	6MV	500NA	200NA	670MWF	5MA	12V	15V	30V		85MW	3MA	70dB	76dB	300K
LM747D	NAU	DGK	INT	.	0.2V/uS	+22V	-22V	125C	94dB	5MV	500NA	200NA	670MWF	5MA	12V	15V	30V		85MW	3MA	70dB	76dB	300K
LM747ED	NAU	DGK	INT	.4MHZ	0.3V/uS	+22V	-22V	70C	94dB	3MV	80NA	30NA	670MWF	7MA	16V	15V	30V	15uV/C	150MW		80dB	86dB	1M
LM747EH	NAU	DGK	INT	.4MHZ	0.3V/uS	+22V	-22V	70C	94dB	3MV	80NA	30NA	500MWF	7MA	16V	15V	30V	15uV/C	150MW		80dB	86dB	1M
LM747EJ	NAU	DGK	INT	.4MHZ	0.3V/uS	+22V	-22V	70C	94dB	3MV	80NA	30NA	670MWF	7MA	16V	15V	30V	15uV/C	150MW		80dB	86dB	1M
LM747EN	NAU	DGK	INT	.4MHZ	0.3V/uS	+22V	-22V	70C	94dB	3MV	80NA	30NA	670MWF	7MA	16V	15V	30V	15uV/C	150MW		80dB	86dB	1M
LM747F	NAU	DGK	INT	.	0.2V/uS	+22V	-22V	125C	94dB	5MV	500NA	200NA	800MWF	5MA	12V	15V	30V		85MW	3MA	70dB	76dB	300K
LM747H	NAU	DGK	INT	.	0.2V/uS	+22V	-22V	125C	94dB	5MV	500NA	200NA	500MWF	5MA	12V	15V	30V		85MW	3MA	70dB	76dB	300K
LM747-IAD	NAU	DGK	INT	.4MHZ	0.3V/uS	+22V	-22V	125C	94dB	3MV	80NA	30NA	670MWF	7MA	16V	15V	30V	15uV/C	150MW	2MA	80dB	86dB	1M
LM747-IAH	NAU	DGK	INT	.4MHZ	0.3V/uS	+22V	-22V	125C	94dB	3MV	80NA	30NA	500MWF	7MA	16V	15V	30V	15uV/C	150MW	2MA	80dB	86dB	1M
LM747-IAJ	NAU	DGK	INT	.4MHZ	0.3V/uS	+22V	-22V	125C	94dB	3MV	80NA	30NA	670MWF	7MA	16V	15V	30V	15uV/C	150MW	2MA	80dB	86dB	1M
LM747-ICD	NAU	DGK	INT	.	0.2V/uS	+18V	-18V	70C	88dB	6MV	500NA	200NA	670MWF	5MA	12V	15V	30V		85MW	2MA	70dB	76dB	300K
LM747-ICH	NAU	DGK	INT	.	0.2V/uS	+18V	-18V	70C	88dB	6MV	500NA	200NA	500MWF	5MA	12V	15V	30V		85MW	2MA	70dB	76dB	300K
LM747-ICJ	NAU	DGK	INT	.	0.2V/uS	+18V	-18V	70C	88dB	6MV	500NA	200NA	670MWF	5MA	12V	15V	30V		85MW	2MA	70dB	76dB	300K
LM747-ICN	NAU	DGK	INT	.	0.2V/uS	+18V	-18V	70C	88dB	6MV	500NA	200NA	670MWF	5MA	12V	15V	30V		85MW	2MA	70dB	76dB	300K
LM747-ID	NAU	DGK	INT	.	0.2V/uS	+22V	-22V	125C	94dB	5MV	500NA	200NA	670MWF	5MA	12V	15V	30V		85MW	2MA	70dB	76dB	300K
LM747-IED	NAU	DGK	INT	.4MHZ	0.3V/uS	+22V	-22V	70C	94dB	3MV	80NA	30NA	670MWF	7MA	16V	15V	30V	15uV/C	150MW	2MA	80dB	86dB	1M
LM747-IEH	NAU	DGK	INT	.4MHZ	0.3V/uS	+22V	-22V	70C	94dB	3MV	80NA	30NA	500MWF	7MA	16V	15V	30V	15uV/C	150MW	2MA	80dB	86dB	.
LM747-IEJ	NAU	DGK	INT	.4MHZ	0.3V/uS	+22V	-22V	70C	94dB	3MV	80NA	30NA	670MWF	7MA	16V	15V	30V	15uV/C	150MW	2MA	80dB	86dB	1M
LM747-IEN	NAU	DGK	INT	.4MHZ	0.3V/uS	+22V	-22V	70C	94dB	3MV	80NA	30NA	670MWF	7MA	16V	15V	30V	15uV/C	150MW	2MA	80dB	86dB	1M
LM747-IH	NAU	DGK	INT	.	0.2V/uS	+22V	-22V	125C	94dB	5MV	500NA	200NA	500MWF	5MA	12V	15V	30V		85MW	2MA	70dB	76dB	300K
LM747-IJ	NAU	DGK	INT	.	0.2V/uS	+22V	-22V	125C	94dB	5MV	500NA	200NA	670MWF	5MA	12V	15V	30V		85MW	2MA	70dB	76dB	300K
LM747J	NAU	DGK	INT	.	0.2V/uS	+22V	-22V	125C	94dB	5MV	500NA	200NA	670MWF	5MA	12V	15V	30V		85MW	3MA	70dB	76dB	300K
LM748CH	NAU	GPU	EXT	.	.25V/uS	+18V	-18V	70C	94dB	6MV	500NA	200NA	500MW	5MA	10V	15V	30V		85MW	3MA	70dB	76dB	300K
LM748CJ	NAU	GPU	EXT	.	.25V/uS	+18V	-18V	70C	94dB	6MV	500NA	200NA	310MWF	5MA	10V	15V	30V		85MW	3MA	70dB	76dB	300K
LM748J	NAU	GPU	EXT	.	.25V/uS	+22V	-22V	125C	94dB	5MV	500NA	200NA	310MWF	5MA	10V	15V	30V		85MW	3MA	70dB	76dB	300K
LM748CN	NAU	GPU	EXT	.	.25V/uS	+18V	-18V	70C	94dB	6MV	500NA	200NA	310MWF	5MA	10V	15V	30V		85MW	3MA	70dB	76dB	300K
LM748H	NAU	GPU	EXT	.	.25V/uS	+22V	-22V	125C	94dB	5MV	500NA	200NA	310MWF	5MA	10V	15V	30V		85MW	3MA	70dB	76dB	300K
LM1303N	NAU	DLN	EXT	.	.	+15V	-15V	75C	78dB	10MV	10uA	0.4uA	415MWF	7MA	7V				300MW	15MA	.	.	.
LM1414J	NAU	DCP	EXT	.	.	+14V	-7V	75C	60dB	5MV	25uA	5uA	600MWF	.5MA	2.5V	7V	5V	25uV/C		18MA	70dB	.	.
LM1414N	NAU	DCP	EXT	.	.	+14V	-7V	75C	60dB	5MV	25uA	5uA	600MWF	5MA	2.5V	7V	5V	25uV/C		18MA	70dB	.	.

For detailed explanations of column heading notations, see App. A.

Also for ready references the more important abbreviations used in the column headings are listed below

LEFT HAND PAGE

APP = application (codes at APP.E.)
CMRR = common mode rejection ratio
CMP = compensation (frequency)
dV_{io}/dT = input offset voltage temperature drift
GBP = gain bandwidth product
I_B = input bias current
I_{io} = input bias offset current
I_q = quiescent supply current
MFR = manufacturer (codes at App.C.)
P_Q = quiescent power consumer
PSRR = power supply rejection ratio
V_{icm} = common mode input voltage rating
V_{idf} = differential input voltage rating
V_{io} = input offset voltage
V_s = dc supply voltage

RIGHT HAND PAGE

Lead out coding summary (details at APP G.) for different cases (APP F)

A = gain adjust
B = bias adjust
C = case
E— = inverting input
E+ = non-inverting input
F,F* = input frequency compensation
G = ground
J = high level input
K = output, open collector
L = output, open emitter
M = metal case
N = not connected
Q = special terminal
R,R* = outputs
S = strobe
T,T* = offset balance
V+ = +ve dc supply
V- = -ve dc supply
W = guard ring
X = blank position, no lead
++ = +ve supplementary dc supply
-- = -ve supplementary dc supply
φ,φ* = output frequency compensation

CASE (APP F)	LD 1	LD 2	LD 3	LD 4	LD 5	LD 6	LD 7	LD 8	LD 9	LD 10	LD 11	LD 12	LD 13	LD 14	LD 15	LD 16	EUROPE SUBSTITUTE	USA SUBSTITUTE	ISS	TYPE NUMBER
T05-10/1M	E-	E+	A2	A*2	V-	R*	R	V+	A1	A*1							SN52733L	UA733HM	O	LM733H
DIL-14/1C	E+	N	A2	A*2	V-	N	R	R*	N	V+	A1	A*1	N	E-			SN52733J	UA733DM	O	LM733J
DIL-14/1M	N	N	T	E-	E+	V-	N	N	T*	R	V+	N	N	E-				UA741ADM	O	LM741AD
FLP-10/3G	N	T	E-	E+	V-	T*	R	V+	N	R	N							UA741AFM	O	LM741AF
T05-8/1M	T	E-	E+	V-M	T*	R	V+	N										UA741AHM	O	LM741AH
DIL-14/1C	N	N	T	E-	E+	V-	N	N	T*	R	V+	N	N	N				UA741ADM	O	LM741AJ-14
DIL-14/1M	N	N	T	E-	E+	V-	N	N	T*	R	V+	N	N	N			TBA221A	UA741DC	O	LM741CD
T05-8/1M	T	E-	E+	V-M	T*	R	V+	N									TBA221	UA741HC	O	LM741CH
DIL-8/1C	T	E-	E+	V-	T*	R	V+	N									TBA221B	UA741TC	O	LM741CJ
DIL-14/1C	N	N	T	E-	E+	V-	N	N	T*	R	V+	N					TBA221A	UA741DC	O	LM741CJ-14
DIL-8/1P	T	E-	E+	V-	T*	R	V+	N									TBA221B	UA741TC	O	LM741CN
DIL-14/1P	N	N	T	E-	E+	V-	N	N	T*	R	V+	N	N	N			TBA221A			LM741CN-14
DIL-14/1M	N	N	T	E-	E+	V-	N	N	T*	R	V+	N	N	N			SFC2741KM	UA741DM	O	LM741D
DIL-14/1M	N	N	T	E-	E+	V-	N	N	T*	R	V+	N	N	N				UA741EDC	O	LM741ED
T05-8/1M	T	E-	E+	V-M	T*	R	V+	N										UA741EHC	O	LM741EH
DIL-8/1C	T	E-	E+	V-	T*	R	V+	N											O	LM741EJ
DIL-14/1C	N	N	T	E-	E+	V-	N	N	T*	R	V+	N	N	N				UA741EDC	O	LM741EJ-14
DIL-8/1P	T	E-	E+	V-	T*	R	V+	N											O	LM741EN
FLP-10/3G	N	T	E-	E+	V-	T*	R	V+	N	N							SFC2741PM	UA741FM	O	LM741F
T05-8/1M	T	E-	E+	V-M	T*	R	V+	N									TBA222	UA741HM	O	LM741H
DIL-14/1C	N	N	T	E-	E+	V-	N	N	T*	R	V+	N	N	N			SFC2741KM	UA741DM	O	LM741J-14
DIL-14/1M	E-1	E+1	T1	V-	T2	E+2	E-2	T*2	V+	R2	N	R1	V+	T*1				UA747ADM	O	LM747AD
T05-10/1M	R1	V+	E-1	E+1	V-	E+2	E-2	V+	R2	N							TBC0747	UA747AHM	O	LM747AH
DIL-14/1C	E-1	E+1	T1	V-	T2	E+2	E-2	T*2	V+	R2	N	R1	V+	T*1				UA747ADM	O	LM747AJ
DIL-14/1M	E-1	E+1	T1	V-	T2	E+2	E-2	T*2	V+	R2	N	R1	V+	T*1			TBB0747A	UA747DC	O	LM747CD
FLP-14/3G	E-1	E+1	T1	V-	T2	E+2	E-2	T*2	V+	R2	N	R1	V+	T*1				SG747F	O	LM747CF
T05-10/1M	R1	V+	E-1	E+1	V-	E+2	E-2	V+	R2	N							TBB0747	UA747HC	O	LM747CH
DIL-14/1C	E-1	E+1	T1	V-	T2	E+2	E-2	T*2	V+	R2	N	R1	V+	T*1			TBB0747A	UA747DC	O	LM747CJ
DIL-14/1P	E-1	E+1	T1	V-	T2	E+2	E-2	T*2	V+	R2	N	R1	V+	T*1			TBB0747A	UA747DC	O	LM747CN
DIL-14/1M	E-1	E+1	T1	V-	T2	E+2	E-2	T*2	V+	R2	N	R1	V+	T*1			SFC2747KM	UA747DM	O	LM747D
DIL-14/1M	E-1	E+1	T1	V-	T2	E+2	E-2	T*2	V+	R2	N	R1	V+	T*1				UA747EDC	O	LM747ED
T05-10/1M	R1	V+	E-1	E+1	V-	E+2	E-2	V+	R2	N								UA747EHC	O	LM747EH
DIL-14/1C	E-1	E+1	T1	V-	T2	E+2	E-2	T*2	V+	R2	N	R1	V+	T*1				UA747EDC	O	LM747EJ
DIL-14/1P	E-1	E+1	T1	V-	T2	E+2	E-2	T*2	V+	R2	N	R1	V+	T*1					O	LM747EN
FLP-14/3G	E-1	E+1	T1	V-	T2	E+2	E-2	T*2	V+	R2	N	R1	V+	T*1					O	LM747F
T05-10/1M	R1	V+	E-1	E+1	V-	E+2	E-2	V+	R2	N							SFC2747M	UA747HM	O	LM747H
DIL-14/1M	E-1	E+1	T1	V-	T2	E+2	E-2	T*2	V+	R2	N	R1	V+1	T*1					O	LM747-IAD
T05-10/1M	R1	V+1	E-1	E+1	V-	E+2	E-2	V+2	R2	N									O	LM747-IAH
DIL-14/1C	E-1	E+1	T1	V-	T2	E+2	E-2	T*2	V+2	R2	N	R1	V+1	T*1					O	LM747-IAJ
DIL-14/1M	E-1	E+1	T1	V-	T2	E+2	E-2	T*2	V+2	R2	N	R1	V+1	T*1				UA747-IDC	O	LM747-ICD
T05-10/1M	R1	V+1	E-1	E+1	V-	E+2	E-2	V+2	R2	N								UA747-IHC	O	LM747-ICH
DIL-14/1C	E-1	E+1	T1	V-	T2	E+2	E-2	T*2	V+2	R2	N	R1	V+1	T*1				UA747-IDC	O	LM747-ICJ
DIL-14/1P	E-1	E+1	T1	V-	T2	E+2	E-2	T*2	V+2	R2	N	R1	V+1	T*1				UA747-IDC	O	LM747-ICN
DIL-14/1M	E-1	E+1	T1	V-	T2	E+2	E-2	T*2	V+2	R2	N	R1	V+1	T*1				UA747-IDM	O	LM747-ID
DIL-14/1M	E-1	E+1	T1	V-	T2	E+2	E-2	T*2	V+2	R2	N	R1	V+1	T*1					O	LM747-IED
T05-10/1M	R1	V+1	E-1	E+1	V-	E+2	E-2	V+2	R2	N									O	LM747-IEH
DIL-14/1C	E-1	E+1	T1	V-	T2	E+2	E-2	T*2	V+2	R2	N	R1	V+1	T*1					O	LM747-IEJ
DIL-14/1P	E-1	E+1	T1	V-	T2	E+2	E-2	T*2	V+2	R2	N	R1	V+1	T*1					O	LM747-IEN
T05-10/1M	R1	V+1	E-1	E+1	V-	E+2	E-2	V+2	R2	N								UA747-IHM	O	LM747-IH
DIL-14/1C	E-1	E+1	T1	V-	T2	E+2	E-2	T*2	V+2	R2	N	R1	V+1	T*1				UA747-IDM	O	LM747-IJ
DIL-14/1C	E-1	E+1	T1	V-	T2	E+2	E-2	T*2	V+	R2	N	R1	V+	T*1			SFC2747KM	UA747DM	O	LM747J
T05-8/1M	FT	E-	E+	V-M	T*	R	V+	F*									TBB0748	UA748HC	O	LM748CH
DIL-8/1C	FT	E-	E+	V-	T*	R	V+	F*									TBB0748B	UA748TC	O	LM748CJ
DIL-8/1C	FT	E-	E+	V-	T*	R	V+	F*										SN52748JP	O	LM748J
DIL-8/1P	FT	E-	E+	V-	T*	R	V+	F*									TBB0748B	UA748TC	O	LM748CN
T05-8/1M	FT	E-	E+	V-M	T*	R	V+	F*									TBC0748	UA748HM	O	LM748H
DIL-14/1P	R1	φ1	F1	F*1	E+1	E-1	V-	E-2	E+2	F*2	F2	φ2	R2	V+				MC1303P	O	LM1303N
DIL-14/1C	R1	S1	V+1	N	E+2	E-2	V-	R2	S2	V+2	G	E+1	E-1	V-				MC1414L	O	LM1414J
DIL-14/1P	R1	S1	V+	N	E+2	E-2	V-	R2	S2	V+2	G	E+1	E-1	V-				MC1414L	O	LM1414N
DIL-8/1C	R1	E-1	E+1	V-	E+2	E-2	R2	V+									TBB1458B	MC1458U	O	LM1458J

TYPE NUMBER	M F R	A P P	C M P	GBP MIN	SLEW RATE MIN	Vs+ MAX	Vs- MAX	T_OP MAX	Avol MIN	Vio MAX	Ib MAX	Iio MAX	Ptot MAX	Iout MIN	Vout MIN	Vicm MAX	Vidf MAX	dVio/dT MAX	Pq MAX	Io MAX	CMRR MIN	PSRR MIN	Rin MIN
LM1458J	NAU	DGK	INT			+18V	-18V	70C	86dB	6MV	500NA	200NA	500MWF	5MA	12V	15V	30V			6MA	70dB	77dB	300K
LM1458H	NAU	DGK	INT			+18V	-18V	70C	86dB	6MV	500NA	200NA	500MWF	5MA	12V	15V	30V			6MA	70dB	77dB	300K
LM1458N	NAU	DGK	INT			+18V	-18V	70C	86dB	6MV	500NA	200NA	400MWF	5MA	12V	15V	30V			6MA	70dB	77dB	300K
LM1458N-14	NAU	DGK	INT			+18V	-18V	70C	86dB	6MV	500NA	200NA	400MWF	5MA	12V	15V	30V			6MA	70dB	77dB	300K
LM1514J	NAU	DCP	EXT			+14V	-7V	125C	62dB	2MV	20uA	3uA	600MWF	5MA	2.5V	7V	5V	15uV/C		18MA	80dB		
LM1558H	NAU	DGK	INT			+22V	-22V	125C	94dB	5MV	500NA	200NA	500MWF	5MA	12V	15V	30V			5MA	70dB	77dB	300K
LM1558J	NAU	DGK	INT			+22V	-22V	125C	94dB	5MV	500NA	200NA	500MWF	5MA	12V	15V	30V			5MA	70dB	77dB	300K
LM1900D	NAU	QCD	INT	1MHZ	0.2V/uS	+36V		125C	66dB		100NA		900MWF	6MA	13V					12MA		50dB	500K
LM1900J	NAU	QCD	INT	1MHZ	0.2V/uS	+36V		125C	66dB		100NA		900MWF	6MA	13V					12MA		50dB	500K
LM2900D	NAU	QCD	INT	1MHZ	0.2V/uS	+32V		85C	62dB		200NA		900MWF	6MA	13V					10MA		50dB	500K
LM2900J	NAU	QCD	INT	1MHZ	0.2V/uS	+32V		85C	62dB		200NA		900MWF	6MA	13V					10MA		50dB	500K
LM2900N	NAU	QCD	INT	1MHZ	0.2V/uS	+32V		85C	62dB		200NA		900MWF	6MA	13V					10MA		50dB	500K
LM2901A	MUG	QCP	EXT			+18V	-18V	85C	88dB	7MV	250NA	50NA	900MWF	6MA		18V	18V			2MA			
LM2901F	MUG	QCP	EXT			+18V	-18V	85C	88dB	7MV	250NA	50NA	900MWF	6MA		18V	18V			2MA			
LM2901J	NAU	QCP	EXT			+18V	-18V	85C	88dB	7MV	250NA	50NA	900MWF	6MA		18V	18V			2MA			
LM2901L	TDG	QCP	EXT			+18V	-18V	85C	88dB	7MV	250NA	50NA	900MWF	6MA		18V	18V			2MA			
LM2901N	MUG	QCP	EXT			+18V	-18V	85C	88dB	7MV	250NA	50NA	900MWF	6MA		18V	18V			2MA			
LM2901N(14)	MUG	QCP	EXT			+18V	-18V	85C	88dB	7MV	250NA	50NA	900MWF	6MA		18V	18V			2MA			
LM2902A	MUG	QGK	INT	.3MHZ		+16V	-16V	70C	88dB	7MV	250NA	50NA	900MWF			16V	16V	35uV/C		2MA	50dB	50dB	
LM2902J	NAU	QGK	INT			+16V	-16V	70C	88dB	7MV	250NA	50NA	900MWF			16V	16V	35uV/C		2MA	50dB	50dB	
LM2902N	NAU	QGK	INT			+16V	-16V	70C	88dB	7MV	250NA	50NA	900MWF			16V	16V	35uV/C		2MA	50dB	50dB	
LM2902N(14)	MUG	QGK	INT	.3MHZ		+16V	-16V	70C	88dB	7MV	250NA	50NA	900MWF			16V	16V	35uV/C		2MA	50dB	50dB	
LM2903N	NAU	DCP	EXT			+18V	-18V	85C	88dB	7MV	250NA	50NA	570MWF	6MA		18V	36V						
LM2903N(8)	MUG	DCP	EXT			+18V	-18V	85C	88dB	7MV	250NA	50NA	570MWF	6MA		18V	36V						
LM2903V	MUG	DCP	EXT			+18V	-18V	85C	88dB	7MV	250NA	50NA	570MWF	6MA		18V	36V						
LM2904N	NAU	DGK	INT			+13V	-13V	85C	86dB	7MV	250NA	50NA	570MWF	10MA		16V	26V	30uV/C		3MA	50dB	50dB	
LM3301N	NAU	QCD	INT	1MHZ	0.2V/uS	+28V		85C	60dB		300NA		625MWF	6MA	13V					10MA		50dB	
LM3302J	NAU	QCP	EXT		20V/uS	+14V	-14V	85C	66dB	20MV	500NA	100NA	900MWF	2MA		9V	14V			2MA			
LM3302N	NAU	QCP	EXT		20V/uS	14V	14V	85C	66dB	20MV	500NA	100NA	900MWF	2MA		9V	14V			2MA			
LM3401N	NAU	QCD	INT	2MHZ	0.2V/uS	+18V		75C	60dB		300NA		625MWF	6MA	13V					10MA		50dB	
LM3900N	NAU	QCD	INT	1MHZ	0.2V/uS	+32V		70C	62dB		200NA		570MWF	6MA	13V					10MA		50dB	500K
LM4250CH	NAU	PRA	INT			+18V	-18V	70C	95dB	6MV	75NA	20NA	500MWF	1MA	12V	15V	30V		3MW	.1MA	70dB	74dB	
LM4250CJ	NAU	PRA	INT			+18V	-18V	70C	95dB	6MV	75NA	20NA	500MWF	1MA	12V	15V	30V		3MW	.1MA	70dB	74dB	
LM4250CN	NAU	PRA	INT			+18V	-18V	70C	95dB	6MV	75NA	20NA	500MWF	1MA	12V	15V	30V		3MW	.1MA	70dB	74dB	
LM4250F	NAU	PRA	INT			+18V	-18V	125C	100dB	5MV	50NA	10NA	500MWF	1MA	12V	15V	30V		2.7MW	90UA	70dB	76dB	
LM4250H	NAU	PRA	INT			+18V	-18V	125C	100dB	5MV	50NA	10NA	500MWF	1MA	12V	15V	30V		2.7MW	90UA	70dB	76dB	
LM4250J	NAU	PRA	INT			+18V	-18V	125C	100dB	5MV	50NA	10NA	500MWF	1MA	12V	15V	30V		2.7MW	90UA	70dB	76dB	
M238A	BLU	FET	INT	10MHZ	50V/uS	+22V	-22V	125C		1MV	5pA			20MA	10V	15V	30V	10uV/C		12MA	86dB	70dB	10G
M238B	BLU	FET	INT	10MHZ	50V/uS	+22V	-22V	125C		1MV	5pA			20MA	10V	15V	30V	25uV/C		12MA	86dB	70dB	10G
M238C	BLU	FET	INT	10MHZ	50V/uS	+22V	-22V	125C		2MV	10pA			20MA	10V	15V	30V	50uV/C		12MA	86dB	70dB	10G
MC1303P	MTU	DLN	EXT			+15V	-15V	75C	78dB	10MV	10uA	0.4uA	625MWF	.7MA	7V				400MW				
MC1410G	MTU	BDO	INT	.4GHZ		+8V	-8V	75C	36dB	33MV	100uA	30uA	680MWF		2V	6V	5V	10uV/C	220MW		20dB		3K
MC1414F	MTU	DCP	EXT			+14V	-7V	75C	60dB	5MV	25uA	5uA	500MWF	5MA	1V	7V	5V	25uV/C	150MW	9MA	70dB		
MC1414L	MTU	DCP	EXT			+14V	-7V	75C	60dB	5MV	25uA	5uA	625MWF	5MA	1V	7V	5V	25uV/C	150MW	9MA	70dB		
MC1414P	MTU	DCP	EXT			+14V	-7V	75C	60dB	5MV	25uA	5uA	625MWF	5MA	1V	7V	5V	25uV/C	150MW	9MA	70dB		
MC1420F	MTU	BDO	EXT	3MHZ	2V/uS	+8V	-8V	75C	64dB	15MV	4uA	0.2uA	500MWF		6V	0.5V	8V	10uV/C	240MW		60dB		300K
MC1420G	MTU	BDO	EXT	3MHZ	2V/uS	+8V	-8V	75C	64dB	15MV	4uA	0.2uA	680MWF		6V	0.5V	8V	10uV/C	240MW		60dB		300K
MC1430F	MTU	GPU	EXT		0.3V/uS	+8V	-8V	75C	69dB	10MV	15uA	4uA	500MWF	4MA	4V	2V	5V		150MW		65dB		5K
MC1430G	MTU	GPU	EXT		0.3V/uS	+8V	-8V	75C	69dB	10MV	15uA	4uA	680MWF	4MA	4V	2V	5V		150MW		65dB		5K
MC1430L	MTU	GPU	EXT		0.3V/uS	+8V	-8V	75C	69dB	10MV	15uA	4uA	1WF	4MA	4V	2V	5V		150MW		65dB		5K
MC1430P	MTU	GPU	EXT		0.3V/uS	+8V	-8V	75C	69dB	10MV	15uA	4uA	400MWF	4MA	4V	2V	5V		150MW		65dB		5K
MC1431F	MTU	GPU	EXT		0.3V/uS	+8V	-8V	75C	62dB	15MV	0.3uA	0.1uA	500MWF	4MA	4V	2V	5V		150MW		60dB		300K
MC1431G	MTU	GPU	EXT		0.3V/uS	+8V	-8V	75C	62dB	15MV	0.3uA	0.1uA	680MWF	4MA	4V	2V	5V		150MW		60dB		300K
MC1431L	MTU	GPU	EXT		0.3V/uS	+8V	-8V	75C	62dB	15MV	0.3uA	0.1uA	1WF	4MA	4V	2V	5V		150MW		60dB		300K
MC1431P	MTU	GPU	EXT		0.3V/uS	+8V	-8V	75C	62dB	15MV	0.3uA	0.1uA	400MWF	4MA	4V	2V	5V		150MW		60dB		300K
MC1433F	MTU	GPU	EXT		0.2V/uS	+18V	-18V	75C	90dB	7.5MV	2uA	0.5uA	500MWF	5MA	12V	8V	10V	30uV/C	170MW		80dB	76dB	300K
MC1433F	OBS	GPU	EXT		0.2V/uS	+18V	-18V	75C	90dB	7.5MV	2uA	0.5uA	500MWF	5MA	12V	8V	10V	30uV/C	170MW		80dB	76dB	300K
MC1433G	MTU	GPU	EXT		0.2V/uS	+18V	-18V	75C	90dB	7.5MV	2uA	0.5uA	680MWF	5MA	12V	8V	10V	30uV/C	170MW		80dB	76dB	300K
MC1433G	OBS	GPU	EXT		0.2V/uS	+18V	-18V	75C	90dB	7.5MV	2uA	0.5uA	680MWF	5MA	12V	8V	10V	30uV/C	170MW		80dB	76dB	300K
MC1433L	MTU	GPU	EXT		0.2V/uS	+18V	-18V	75C	90dB	7.5MV	2uA	0.5uA	1WF	5MA	12V	8V	10V	30uV/C	170MW		80dB	76dB	300K

For detailed explanations of column heading notations, see App. A.

Also for ready references the more important abbreviations used in the column headings are listed below:

LEFT HAND PAGE

APP = application (codes at APP.E.)
CMRR = common mode rejection ratio
CMP = compensation (frequency)
dV_{io}/dT = input offset voltage temperature drift
GBP = gain bandwidth product
I_B = input bias current
I_{io} = input bias offset current
I_q = quiescent supply current
MFR = manufacturer (codes at App.C.)
P = quiescent power consumer
PSRR = power supply rejection ratio
V_{ICM} = common mode input voltage rating
V_{idr} = differential input voltage rating
V_{io} = input offset voltage
V_S = dc supply voltage

RIGHT HAND PAGE

Lead out coding summary (details at APP.G.) for different cases (APP.F.)

A = gain adjust
B = bias adjust
C = case
E− = inverting input
E+ = non-inverting input
F,F* = input frequency compensation
G = ground
J = high level input
K = output, open collector
L = output, open emitter
M = metal case
N = not connected
Q = special terminal
R,R* = outputs
S = strobe
T,T* = offset balance
V+ = +ve dc supply
V− = −ve dc supply
W = guard ring
X = blank position, no lead
++ = +ve supplementary dc supply
−− = −ve supplementary dc supply
∅,∅* = output frequency compensation

CASE (APP F)	LD 1	LD 2	LD 3	LD 4	LD 5	LD 6	LD 7	LD 8	LD 9	LD 10	LD 11	LD 12	LD 13	LD 14	.D 15	LD 16	EUROPE SUBSTITUTE	USA SUBSTITUTE	IS	TYPE NUMBER
TO5-8/1M	R1	E-1	E+1	V-	E+2	E-2	R2	V+	.	.							TBB1458	MC1458G	Q	LM1458H
DIL-8/1P	R1	E-1	E+1	V-	E+2	E-2	R2	V+	.								TBB1458B	MC1458U	0	LM1458N
DIL-14/1P	N	R1	N	N	E-1	E+1	V-	E+2	E-2	N	N	R2	N	V+				MC1458L	0	LM1458N-14
DIL-14/1C	R1	S1	V+1	N	E+2	E-2	V-	R2	S2	V+2	G	E+1	E-1	V-				MC1514L	0	LM1514J
TO5-8/1M	R1	E-1	E+1	V-	E+2	E-2	R2	V+	.								TBC1458	MC1558G	0	LM1558H
DIL-8/1C	R1	E-1	E+1	V-	E+2	E-2	R2	V+	.									MC1558U	0	LM1558J
DIL-14/1M	E+1	E+2	E-2	R2	R1	E-1	G	E-3	R3	R4	E-4	E+4	E+3	V+					0	LM1900D
DIL-14/1C	E+1	E+2	E-2	R2	R1	E-1	G	E-3	R3	R4	E-4	E+4	E+3	V+					0	LM1900J
DIL-14/1M	E+1	E+2	E-2	R2	R1	E-1	G	E-3	R3	R4	E-4	E+4	E+3	V+				LM1900D	0	LM2900D
DIL-14/1C	E+1	E+2	E-2	R2	R1	E-1	G	E-3	R3	R4	E-4	E+4	E+3	V+				LM1900D	0	LM2900J
DIL-14/1P	E+1	E+2	E-2	R2	R1	E-1	G	E-3	R3	R4	E-4	E+4	E+3	V+				LM1900D	0	LM2900N
DIL-14/1P	R2	R1	V+	E-1	E+1	E-2	E+2	E+3	E-3	E-4	E+4	G	R4	R3			LM2901N	LM2901J	0	LM2901A
DIL-14/1C	R2	R1	V+	E-1	E+1	E-2	E+2	E+3	E-3	E-4	E+4	G	R4	R3				LM2901J	0	LM2901F
DIL-14/1P	R2	R1	V+	E-1	E+1	E-2	E+2	E+3	E-3	E-4	E+4	G	R4	R3				LM2901N	0	LM2901J
DIL-14/1C	R2	R1	V+	E-1	E+1	E-2	E+2	E+3	E-3	E-4	E+4	G	R4	R3				LM2901J	0	LM2901L
DIL-14/1P	R2	R1	V+	E-1	E+1	E-2	E+2	E+3	E-3	E-4	E+4	G	R4	R3				LM2901F	0	LM2901N
DIL-14/1P	R2	R1	V+	E-1	E+1	E-2	E+2	E+3	E-3	E-4	E+4	G	R4	R3			LM2901N	LM2901J	0	LM2901N(14)
DIL-14/1P	R1	E-1	E+1	V+	E+2	E-2	R2	R3	E-3	E+3	G	E+4	E-4	R4			MLM2902F	LM2902N	0	LM2902A
DIL-14/1C	R1	E-1	E+1	V+	E+2	E-2	R2	R3	E-3	E+3	G	E+4	E-4	R4			MLM2902P	LM2902J	0	LM2902J
DIL-14/1P	R1	E-1	E+1	V+	E+2	E-2	R2	R3	E-3	E+3	G	E+4	E-4	R4			MLM2902P	LM2902N	0	LM2902N
DIL-14/1P	R1	E-1	E+1	V+	E+2	E-2	R2	R3	E-3	E+3	G	E+4	E-4	R4			MLM2902P	LM2902N	0	LM2902N(14)
DIL-8/1P	R1	E-1	E+1	G	E+2	E-2	R2	V+										LM393AN	0	LM2903N
DIL-8/1P	R1	E-1	E+1	G	E+2	E-2	R2	V+									LM2903V	LM2903N	0	LM2903N(8)
DIL-8/1P	R1	E-1	E+1	G	E+2	E-2	R2	V+										LM2903N	0	LM2903V
DIL-8/1P	R1	E-1	E+1	G	E+2	E-2	R2	V+										LM358AN	0	LM2904N
DIL-14/1P	E+1	E+2	E-2	R2	R1	E-1	G	E-3	R3	R4	E-4	E+4	E+3	V+			UA3301P	MC3301P	0	LM3301N
DIL-14/1C	R2	R1	V+	E-1	E+1	E-2	E+2	E+3	E-3	E-4	E+4	G	R4	R3				MC3302L	0	LM3302J
DIL-14/1P	R2	R1	V+	E-1	E+1	E-2	E+2	E+3	E-3	E-4	E+4	G	R4	R3				MC3302L	0	LM3302N
DIL-14/1P	E+1	E+2	E-2	R2	R1	E-1	G	E-3	R3	R4	E-4	E+4	E+3	V+			UA3401P	MC3401P	0	LM3401N
DIL-14/1P	E+1	E+2	E-2	R2	R1	E-1	G	E-3	R3	R4	E-4	E+4	E+3	V+				LM2900D	0	LM3900N
TO5-8/1M	T	E-	E+	V-	T*	R	V+	B	.	.								SG4250CT	0	LM4250CH
DIL-8/1C	T	E-	E+	V-	T*	R	V+	B	.									LM4250J	0	LM4250CJ
DIL-8/1P	T	E-	E+	V-	T*	R	V+	B	.									LM4250J	0	LM4250CN
FLP-10/3G	N	T	E-	E+	V-	T*	R	V+	B	N	.	.						LM4250F	0	LM4250F
TO5-8/1M	T	E-	E+	V-	T*	R	V+	B	.								SG4250T		0	LM4250H
DIL-8/1C	T	E-	E+	V-	T*	R	V+	B	.										0	LM4250J
TO8-12/1M	E+	E-	N	V+	N	V-	GM	R	Q	Q*	N	N						M238A	0	M238A
TO8-12/1M	E+	E-	N	V+	N	V-	GM	R	Q	Q*	N	N						M238A	0	M238B
TO8-12/1M	E+	E-	N	V+	N	V-	GM	R	Q	Q*	N	N						M238B	0	M238C
DIL-14/1P	R1	∅1	F1	F*1	E+1	E-1	V-	E-2	E+2	F*2	F2	∅2	R2	V+			UA749C	LM1303N	0	MC1303P
TO5-8/1M	E+	V+	E-	M	R-	G	R+	V-	.									MC1510G	0	MC1410G
FLP-14/3C	R2	S2	V+2	N	E+1	E-1	V-	R1	S1	V+1	G	E+2	E-2	V-			MC1514J	LM1414J	0	MC1414F
DIL-14/1C	R2	S2	V+2	N	E+1	E-1	V-	R1	S1	V+1	G	E+2	E-2	V-				LM1414J	0	MC1414L
DIL-14/1P	R2	S2	V+2	N	E+1	E-1	V-	R1	S1	V+1	G	E+2	E-2	V-			LM1414J	MC1414L	0	MC1414P
FLP-10/3C	E1	E2	F2	F1	V-	R1	R2	F*1	F*2	V+	.	.						MC1520F	0	MC1420F
TO5-10/1M	F2	F1	V-	R1	R2	F*1	F*2	V+	E1	E2	.							MC1520G	0	MC1420G
FLP-10/3C	E+	E-	G	V-	R	V+	F	F*	∅	∅*	.							MC1530F	0	MC1430F
TO5-10/1M	E+	E-	G	V-	R	V+	F	F*	∅	∅*	.							MC1530G	0	MC1430G
DIL-14/1C	N	∅*	N	E+	N	E-	V-	G	N	N	R	V+	F	F*				MC1530L	0	MC1430L
DIL-14/1P	∅*	∅*	N	E+	N	E-	V-	G	N	N	R	V+	F	F*				MC1430L	0	MC1430P
FLP-10/3C	E+	E-	G	V-	R	V+	F	F*	∅	∅*	.							MC1531F	0	MC1431F
TO5-10/1M	E+	E-	G	V-	R	V+	F	F*	∅	∅*	.							MC1531G	0	MC1431G
DIL-14/1C	N	∅*	N	E+	N	E-	V-	G	N	N	R	V+	F	F*				MC1531L	0	MC1431L
DIL-14/1P	∅	∅*	N	E+	N	E-	V-	G	N	N	R	V+	F	F*			MC1531L	MC1431L	0	MC1431P
FLP-10/3G	E+	∅	V-	R	V+	V+	V+	F	F*	E-	.							MC1533F	0	MC1433F
FLP-10/3G	E+	∅	V-	R	V+	A	T	F	F*	E-	.								0	MC1433F
TO5-10/1M	E-	E+	∅	V-	R	V+	V+	V+	F	F*	.							MC1533G	0	MC1433G
TO5-10/1M	E-	E+	∅	V-	R	V+	A	T	F	F*	.								0	MC1433G
DIL-14/1C	N	F	F*	E-	E+	∅	V-	N	N	N	R	V+	V+	V+				MC1533L	0	MC1433L
DIL-14/1P	N	F	F*	E-	E+	∅	V-	N	N	N	R	V+	V+	V+				MC1433L	0	MC1433P

TYPE NUMBER	M F R	A P P	C M P	GBP MIN	SLEW RATE MIN	Vs⁺ MAX	Vs⁻ MAX	Top MAX	Avol MIN	Vio MAX	IB MAX	Iio MAX	Ptot MAX	Iout MIN	Vout MIN	Vicm MAX	Vidr MAX	dVio/dT MAX	Po MAX	Io MAX	CM RR MIN	PS RR MIN	Rin MIN
MC1433P	MTU	GPU	EXT	.	0.2V/uS	+18V	-18V	75C	90dB	7.5MV	2uA	0.5uA	400MWF	5MA	12V	8V	10V	30uV/C	170MW		80dB	76dB	300K
MC1433P	OBS	GPU	EXT	.	0.2V/uS	+18V	-18V	75C	90dB	7.5MV	2uA	0.51A	400MWF	5MA	12V	8V	10V	30uV/C	170MW		80dB	76dB	300K
MC1435F	MTU	DGU	EXT	.	0.2V/uS	+9V	-9V	75C	71dB	5MV	5uA	0.5uA	500MWF	.5MA	5V	9V	5V	15uV/C	180MW		60dB	70dB	10K
MC1435G	MTU	DGU	EXT	.	0.2V/uS	+9V	-9V	75C	71dB	5MV	5uA	0.5uA	680MWF	.5MA	5V	9V	5V	15uV/C	180MW		60dB	70dB	10K
MC1435L	MTU	DGU	EXT	.	0.2V/uS	+9V	-9V	75C	71dB	5MV	5uA	0.5uA	680MWF	.5MA	5V	9V	5V	15uV/C	180MW		60dB	70dB	10K
MC1435P	MTU	DGU	EXT	.	0.2V/uS	+9V	-9V	75C	71dB	5MV	5uA	0.5uA	400MWF	.5MA	5V	9V	.5V	15uV/C	180MW	.	60dB	70dB	10K
MC1436CG	MTU	HVO	INT	.3MHZ	0.5V/uS	+30V	-30V	75C	94dB	12MV	90NA	25NA	680MWF	1MA	20V	30V	60V	.	280MW	5MA	50dB	70dB	3M
MC1436G	MTU	HVO	INT	.3MHZ	0.5V/uS	+34V	-34V	75C	97dB	10MV	40NA	10NA	680MWF	1MA	20V	34V	68V	.	280MW	5MA	70dB	74dB	3M
MC1437L	MTU	DGU	EXT	.	0.1V/uS	+18V	-18V	75C	84dB	7.5MV	1.5uA	0.5uA	750MWF	5MA	12V	18V	5V	10uV/C	225MW		65dB	74dB	50K
MC1437P	MTU	DGU	EXT	.	0.1V/uS	+18V	-18V	75C	84dB	7.5MV	1.5uA	0.5uA	625MWF	5MA	12V	18V	5V	10uV/C	225MW	.	65dB	74dB	50K
MC1439G	MTU	GPU	EXT	.	0.5V/uS	+18V	-18V	75C	84dB	7.5MV	1uA	.15uA	680MWF	10MA	10V	18V	36V	15uV/C		7MA	80dB	74dB	50K
MC1439L	MTU	GPU	EXT	.	0.4V/uS	+18V	-18V	75C	84dB	7.5MV	1uA	.15uA	750MWF	10MA	10V	18V	36V	15uV/C		7MA	80dB	74dB	50K
MC1439P	MTU	GPU	EXT	.	0.4V/uS	+18V	-18V	75C	84dB	7.5MV	1uA	.15uA	625MWF	10MA	10V	18V	36V	15uV/C	.	7MA	80dB	74dB	50K
MC1439P1	MTU	GPU	EXT	.	0.4V/uS	+18V	-18V	75C	84dB	7.5MV	1uA	.15uA	625MWF	10MA	10V	18V	36V	15uV/C	.	7MA	80dB	74dB	50K
MC1439P2	MTU	GPU	EXT	.	0.4V/uS	+18V	-18V	75C	84dB	7.5MV	1uA	.15uA	625MWF	10MA	10V	18V	36V	15uV/C		7MA	80dB	74dB	50K
MC1456CG	MTU	SBA	INT	.5MHZ	1V/uS	+18V	-18V	75C	88dB	12MV	90NA	30NA	680MWF	5MA	10V	18V	18V	50uV/C	120MW	4MA	60dB	70dB	1M
MC1456CL	MTU	SBA	INT	.5MHZ	1V/uS	+18V	-18V	75C	88dB	12MV	90NA	30NA	680MWF	5MA	10V	18V	18V	50uV/C	120MW	4MA	60dB	70dB	1M
MC1456F	SJU	SBA	INT	.5MHZ	1V/uS	+18V	-18V	70C	96dB	10MV	30NA	10NA	500MWF	5MA	11V	18V	18V	40uV/C	90MW	3MA	70dB	74dB	1M
MC1456G	MTU	SBA	INT	.5MHZ	1V/uS	+18V	-18V	75C	96dB	10MV	30NA	10NA	680MWF	5MA	11V	18V	18V	40uV/C	90MW	3MA	70dB	74dB	1M
MC1456L	MTU	SBA	INT	.5MHZ	1V/uS	+18V	-18V	75C	96dB	10MV	30NA	10NA	680MWF	5MA	11V	18V	18V	40uV/C	90MW	3MA	70dB	74dB	1M
MC1456T	MUG	SBA	INT	.5MHZ	1V/uS	+18V	-18V	70C	96dB	10MV	30NA	10NA	500MWF	5MA	11V	18V	30V	50uV/C	90MW	3MA	70dB	74dB	1M
MC1456V	MUG	SBA	INT	.5MHZ	1V/uS	+18V	-18V	70C	96dB	10MV	30NA	10NA	680MWF	5MA	11V	18V	18V	40uV/C	90MW	3MA	70dB	74dB	1M
MC1458CG	MTU	DGK	INT	.5MHZ	0.3V/uS	+18V	-18V	75C	84dB	10MV	0.7uA	0.3uA	680MWF	4MA	11V	15V	30V	50uV/C	240MW	8MA	60dB	66dB	150K
MC1458CL	MTU	DGK	INT	.5MHZ	0.3V/uS	+18V	-18V	75C	84dB	10MV	0.7uA	0.3uA	750MWF	4MA	11V	15V	30V	50uV/C	240MW	8MA	60dB	66dB	150K
MC1458CNG	MTU	DLN	INT	.5MHZ	0.3V/uS	+18V	-18V	75C	84dB	10MV	0.7uA	0.3uA	680MWF	4MA	11V	15V	30V	50uV/C	240MW	8MA	60dB	66dB	150K
MC1458CNL	MTU	DLN	INT	.5MHZ	0.3V/uS	+18V	-18V	75C	84dB	10MV	0.7uA	0.3uA	750MWF	4MA	11V	15V	30V	50uV/C	240MW	8MA	60dB	66dB	150K
MC1458CNP1	MTU	DLN	INT	.5MHZ	0.3V/uS	+18V	-18V	75C	84dB	10MV	0.7uA	0.3uA	625MWF	4MA	11V	15V	30V	50uV/C	240MW	8MA	60dB	66dB	150K
MC1458CNP2	MTU	DLN	INT	.5MHZ	0.3V/uS	+18V	-18V	75C	84dB	10MV	0.7uA	0.3uA	625MWF	4MA	11V	15V	30V	50uV/C	240MW	8MA	60dB	66dB	150K
MC1458CP1	MTU	DGK	INT	.5MHZ	0.3V/uS	+18V	-18V	75C	84dB	10MV	0.7uA	0.3uA	625MWF	4MA	11V	15V	30V	50uV/C	240MW	8MA	60dB	66dB	150K
MC1458CP2	MTU	DGK	INT	.5MHZ	0.3V/uS	+18V	-18V	75C	84dB	10MV	0.7uA	0.3uA	625MWF	4MA	11V	15V	30V	50uV/C	240MW	8MA	60dB	66dB	150K
MC1458CU	MTU	DGK	INT	.5MHZ	0.3V/uS	+18V	-18V	75C	84dB	10MV	0.7uA	0.3uA	750MWF	4MA	11V	15V	30V	50uV/C	240MW	8MA	60dB	66dB	150K
MC1458D	MUG	DGK	INT	.5MHZ	0.3V/uS	+18V	-18V	75C	88dB	6MV	0.5uA	0.2uA		5MA	12V	15V	30V		125MW	6MA	70dB	76dB	300K
MC1458G	MTU	DGK	INT	.5MHZ	0.3V/uS	+18V	-18V	75C	86dB	6MV	0.5uA	0.2uA	680MWF	5MA	12V	15V	30V	50uV/C	170MW	6MA	70dB	76dB	300K
MC1458L	MTU	DGK	INT	.5MHZ	0.3V/uS	+18V	-18V	75C	86dB	6MV	0.5uA	0.2uA	750MWF	5MA	12V	15V	30V	50uV/C	170MW	6MA	70dB	76dB	300K
MC1458N(8)	MUG	DGK	INT	.5MHZ	0.3V/uS	+18V	-18V	75C	88dB	6MV	0.5uA	0.2uA	400MWF	5MA	12V	15V	30V		125MW	6MA	70dB	76dB	300K
MC1458NG	MTU	DLN	INT	.5MHZ	0.3V/uS	+18V	-18V	75C	86dB	6MV	0.5uA	0.2uA	680MWF	5MA	12V	15V	30V	50uV/C	170MW	6MA	70dB	76dB	300K
MC1458NL	MTU	DLN	INT	.5MHZ	0.3V/uS	+18V	-18V	75C	86dB	6MV	0.5uA	0.2uA	750MWF	5MA	12V	15V	30V	50uV/C	170MW	6MA	70dB	76dB	300K
MC1458NP	MTU	DLN	INT	.5MHZ	0.3V/uS	+18V	-18V	75C	86dB	6MV	0.5uA	0.2uA	625MWF	5MA	12V	15V	30V	50uV/C	170MW	6MA	70dB	76dB	300K
MC1458NP1	MTU	DLN	INT	.5MHZ	0.3V/uS	+18V	-18V	75C	86dB	6MV	0.5uA	0.2uA	625MWF	5MA	12V	15V	30V	50uV/C	170MW	6MA	70dB	76dB	300K
MC1458NP2	MTU	DLN	INT	.5MHZ	0.3V/uS	+18V	-18V	75C	86dB	6MV	0.5uA	0.2uA	625MWF	5MA	12V	15V	30V	50uV/C	170MW	6MA	70dB	76dB	300K
MC1458NU	MTU	DLN	INT	.5MHZ	0.3V/uS	+18V	-18V	75C	86dB	6MV	0.5uA	0.2uA	750MWF	5MA	12V	15V	30V	50uV/C	170MW	6MA	70dB	76dB	300K
MC1458P	MTU	DGK	INT	.5MHZ	0.3V/uS	+18V	-18V	75C	86dB	6MV	0.5uA	0.2uA	625MWF	5MA	12V	15V	30V	50uV/C	170MW	6MA	70dB	76dB	300K
MC1458P1	MTU	DGK	INT	.5MHZ	0.3V/uS	+18V	-18V	75C	86dB	6MV	0.5uA	0.2uA	625MWF	5MA	12V	15V	30V	50uV/C	170MW	6MA	70dB	76dB	300K
MC1458P2	MTU	DGK	INT	.5MHZ	0.3V/uS	+18V	-18V	75C	86dB	6MV	0.5uA	0.2uA	625MWF	5MA	12V	15V	30V	50uV/C	170MW	6MA	70dB	76dB	300K
MC1458SG	MTU	DHS	INT	.5MHZ	3V/uS	+18V	-18V	75C	86dB	6MV	0.5uA	0.2uA	680MWF	5MA	12V	15V	30V	50uV/C	170MW	6MA	70dB	76dB	300K
MC1458SL	MTU	DHS	INT	.5MHZ	3V/uS	+18V	-18V	75C	86dB	6MV	0.5uA	0.2uA	750MWF	5MA	12V	15V	30V	50uV/C	170MW	6MA	70dB	76dB	300K
MC1458SP	MTU	DHS	INT	.5MHZ	3V/uS	+18V	-18V	75C	86dB	6MV	0.5uA	0.2uA	625MWF	5MA	12V	15V	30V	50uV/C	170MW	6MA	70dB	76dB	300K
MC1458SP1	MTU	DHS	INT	.MHHZ	3V/uS	+18V	-18V	75C	86dB	6MV	0.5uA	0.2uA	625MWF	5MA	12V	15V	30V	50uV/C	170MW	6MA	70dB	76dB	300K
MC1458SU	MTU	DHS	INT	.5MHZ	3V/uS	+18V	-18V	75C	86dB	6MV	0.5uA	0.2uA	750MWF	5MA	12V	15V	30V	50uV/C	170MW	6MA	70dB	76dB	300K
MC1458T	SJU	DGK	INT	.5MHZ	0.3V/uS	+18V	-18V	75C	88dB	6MV	0.5uA	0.2uA	500MWF	5MA	12V	15V	30V		125MW	6MA	70dB	76dB	300K
MC1458U	MTU	DGK	INT	.5MHZ	0.3V/uS	+18V	-18V	75C	86dB	6MV	0.5uA	0.2uA	750MWF	5MA	12V	15V	30V	50uV/C	170MW	6MA	70dB	76dB	300K
MC1458V	SJU	DGK	INT	.5MHZ	0.3V/uS	+18V	-18V	75C	88dB	6MV	0.5uA	0.2uA	400MWF	5MA	12V	15V	30V	.	125MW	6MA	70dB	76dB	300K
MC1510G	MTU	BDO	INT	.4GHZ		+8V	-8V	125C	38dB	17MV	80uA	20uA	680MWF	.	2V	6V	5V	10uV/C	220MW	.	30dB	.	3K
MC1514L	MTU	DCP	EXT	.	.	+14V	-7V	125C	62dB	2MV	20uA	3uA	1WF	5MA	1V	7V	5V	15uV/C	150MW	9MA	80dB	.	
MC1514F	MTU	DCP	EXT	.	.	+14V	-7V	125C	62dB	2MV	20uA	3uA	500MWF	5MA	1V	7V	5V	15uV/C	150MW	9MA	80dB	.	
MC1520F	MTU	BDO	EXT	3MHZ	2V/uS	+8V	-8V	125C	66dB	10MV	2uA	0.1uA	500MWF		7V	0.5V	8V	10uV/C	240MW		75dB	66dB	500K
MC1520G	MTU	BDO	EXT	3MHZ	2V/uS	+8V	-8V	125C	66dB	10MV	2uA	0.1uA	680MWF		7V	0.5V	6V	10uV/C	240MW		75dB	66dB	500K
MC1530F	MTU	GPU	EXT	.	0.3V/uS	+9V	-9V	125C	73dB	5MV	10uA	2uA	500MWF	5MA	4.5V	2V	5V	10uV/C	150MW		70dB	74dB	10K
MC1530G	MTU	GPU	EXT	.	0.3V/uS	+9V	-9V	125C	73dB	5MV	10uA	2uA	680MWF	5MA	4.5V	2V	5V	10uV/C	150MW		70dB	74dB	10K
MC1530L	MTU	GPU	EXT	.	0.3V/uS	+9V	-9V	125C	73dB	5MV	10uA	2uA	1WF	5MA	4.5V	2V	5V	10uV/C	150MW		70dB	74dB	10K

For detailed explanations of column heading notations, see App. A.

Also for ready references the more important abbreviations used in the column headings are listed below:

LEFT HAND PAGE

APP = application (codes at APP.E.)
CMRR = common mode rejection ratio
CMP = compensation (frequency)
dV_{io}/dT = input offset voltage temperature drift
GBP = gain bandwidth product
I_B = input bias current
I_{io} = input bias offset current
I_Q = quiescent supply current
MFR = manufacturer (codes at App.C.)
P_Q = quiescent power consumer
PSRR = power supply rejection ratio
V_{icm} = common mode input voltage rating
V_{idf} = differential input voltage rating
V_{io} = input offset voltage
V_S = dc supply voltage

RIGHT HAND PAGE

Lead out coding summary (details at APP.G.) for different cases (APP.F.)

A = gain adjust
B = bias adjust
C = case
E− = inverting input
E+ = non-inverting input
F,F* = input frequency compensation
G = ground
J = high level input
K = output, open collector
L = output, open emitter
M = metal case
N = not connected
Q = special terminal
R,R* = outputs
S = strobe
T,T* = offset balance
V+ = +ve dc supply
V− = −ve dc supply
W = guard ring
X = blank position, no lead
++ = +ve supplementary dc supply
−− = −ve supplementary dc supply
ø,ø* = output frequency compensation

CASE (APP F)	LD1	LD2	LD3	LD4	LD5	LD6	LD7	LD8	LD9	LD10	LD11	LD12	LD13	LD14	LD15	LD16	EUROPE SUBSTITUTE	USA SUBSTITUTE	IS	TYPE NUMBER
DIL-14/1P	N	F	.*	E−	E+	ø	V−	N	N	N	R	V+	A	T					0	MC1433P
FLP-14/3C	R2	ø2	F2	F*2	E+2	E-2	V−	E-1	E+1	F1	F*1	ø1	R1	V+				MC1535F	0	MC1435F
T05/10/1M	V−	E-2	E+2	ø2	R2	V+	R1	ø1	E+1	E-1								MC1535G	0	MC1435G
DIL-14/1C	R2	ø2	F2	F*2	E+2	E-2	V−	E-1	E+1	F1	F*1	ø1	R1	V+			MC1535L	UA749DM	0	MC1435L
DIL-14/1P	R2	ø2	F2	F*2	E+2	E-2	V−	E-1	E+1	F1	F*1	ø1	R1	V+			TBA231	UA739DC	0	MC1435P
T05-8/1M	T	E−	E+	V+	T*	R	V+	N										LM343	0	MC1436CG
T05-8/1M	T	E−	E+	V+	T*	R	V+	N										LM343	0	MC1436G
DIL-14/1C	ø2	R2	F2	F*2	E-2	E+2	V−	E+1	E-1	F1	F*1	R1	ø1	V+			MC1537L	RC1437DC	0	MC1437L
DIL-14/1P	ø2	R2	F2	F*2	E-2	E+2	V−	E+1	E-1	F1	F*1	R1	ø1	V+			MC1537L	RC1437DC	0	MC1437P
T05-8/1M	F	E−	E+	V−	ø	R	V+	F*									TAA521	UA709HC	0	MC1439G
DIL-14/1C	N	N	F	E−	E+	V−	N	N	ø	R	V+	F*	N	N			TAA521A	UA709DC	0	MC1439L
DIL-14/1P	N	N	F	E−	E+	V−	N	N	ø	R	V+	F*	N	N			TAA521A	UA709DC	0	MC1439P
DIL-8/1P	F	E−	E+	V−	ø	R	V+	F*									LM709CN-8	UA709TC	0	MC1439P1
DIL-14/1P	N	N	F	E−	E+	V−	N	N	ø	R	V+	F*	N	N			TAA521A	UA709DC	0	MC1439P2
T05-8/1M	T	E−	E+	V−	T*	R	V+	N									LM343H	MC1456T	0	MC1456CG
DIL-14/1C	N	N	T	E−	E+	V−	N	N	T*	R	V+	N	N	N			LM343D	LM1456L	0	MC1456CL
DIL-14/1C	N	N	T	E−	E+	V−	N	N	T*	R	V+	N	N	N				MC1456L	0	MC1456F
T05-8/1M	T	E−	E+	V−	T*	R	V+	N									LM343H	MC1456T	0	MC1456G
DIL-14/1C	N	N	T	E−	E+	V−	N	N	T*	R	V+	N	N	N			LM343D	MC1456F	0	MC1456L
T05-8/1M	T	E−	E+	V−	T*	R	V+	N									MC1456G	N5556T	0	MC1456T
DIL-8/1P	T	E−	E+	V−	T*	R	V+	N									RC1556NB	N5556V	0	MC1456V
T05-8/1M	R1	E-1	E+1	V−	E+2	E-2	R2	V+									TBB1458	LM1458H	0	MC1458CG
DIL-14/1C	N	R1	N	N	E-1	E+1	V−	E+2	E-2	N	N	R2	N	V+			MC1458L	LM1458N14	0	MC1458CL
T05-8/1M	R1	E-1	E+1	V−	E+2	E-2	R2	V+										MC1558CG	0	MC1458CNG
DIL-14/1C	N	R1	N	N	E-1	E+1	V−	E+2	E-2	N	N	R2	N	V+				MC1458NL	0	MC1458CNL
DIL-8/1P	R1	E-1	E+1	V−	E+2	E-2	R2	V+										MC1458NP1	0	MC1458CNP1
																		MC1458NP2	0	MC1458CNP2
DIL-14/1P	N	R1	N	N	E-1	E+1	V−	E+2	E-2	N	N	R2	N	V+			TBB1458	LM1458J	0	MC1458CP1
DIL-8/1P	R1	E-1	E+1	V−	E+2	E-2	R2	V+									LM1458N14	MC1458L	0	MC1458CP2
DIL-8/1C	R1	E-1	E+1	V−	E+2	E-2	R2	V+									TBB1458B	LM1458J	0	MC1458CU
MDL-8/2P	R1	E-1	E+1	V−	E+2	E-2	R2	V+										TDA1458D	0	MC1458D
T05-8/1M	R1	E-1	E+1	V−	E+2	E-2	R2	V+									TBB1458	LM1458H	0	MC1458G
DIL-14/1C	N	R1	N	N	E-1	E+1	V−	E+2	E-2	N	N	R2	N	V+				LM1458N14	0	MC1458L
DIL-8/1P	R1	E-1	E+1	V−	E+2	E-2	R2	V+									TBB1458B	MC1458U	0	MC1458N(8)
T05-8/1M	R1	E-1	E+1	V−	E+2	E-2	R2	V+										MC1558NG	0	MC1458NG
DIL-14/1C	N	R1	N	N	E-1	E+1	V−	E+2	E-2	N	N	R2	N	V+				MC1558NL	0	MC1458NL
DIL-14/1P	N	R1	N	N	E-1	E+1	V−	E+2	E-2	N	N	R2	N	V+				MC1558NL	0	MC1458NP
DIL-8/1P	R1	E-1	E+1	V−	E+2	E-2	E+2	V+										MC1558NU	0	MC1458NP1
DIL-14/1P	N	R1	N	N	E-1	E+1	V−	E+2	E-2	N	N	R2	N	V+				MC1558NL	0	MC1458NP2
DIL-8/1C	R1	E-1	E+1	V−	E+2	E-2	R2	V+										MC1558NU	0	MC1458NU
DIL-14/1P	N	R1	N	N	E-1	E+1	V−	E+2	E-2	N	N	R2	N	V+				LM1458N14	0	MC1458P
DIL-8/1P	R1	E-1	E+1	V−	E+2	E-2	R2	V+									TBB1458B	LM1458J	0	MC1458P1
DIL-14/1P	N	R1	N	N	E-1	E+1	V−	E+2	E-2	N	N	R2	N	V+			MC1458L	LM1458N14	0	MC1458P2
T05-8/1M	R1	E-1	E+1	V−	E+2	E-2	R2	V+										MC1558SG	0	MC1458SG
DIL-14/1C	N	R1	N	N	E-1	E+1	V−	E+2	E-2	N	N	R2	N	V+				MC1558SL	0	MC1458SL
DIL-14/1P	N	R1	N	N	E-1	E+1	V−	E+2	E-2	N	N	R2	N	V+				MC1558SL	0	MC1458SP
DIL-8/1P	R1	E-1	E+1	V−	E+2	E-2	R2	V+										MC1558SP	0	MC1458SP1
DIL-8/1C	R1	E-1	E+1	V−	E+2	E-2	R2	V+										MC1558SU	0	MC1458SU
T05-8/1M	R1	E-1	E+1	V−	E+2	E-2	R2	V+									TBB1458	MC1458G	0	MC1458T
DIL-8/1C	R1	E-1	E+1	V−	E+2	E-2	R2	V+									TBB1458B	LM1458J	0	MC1458U
DIL-8/1P	R1	E-1	E+1	V−	E+2	E-2	R2	V+									TBB1458B	MC1458U	0	MC1458V
T05-8/1M	E+	V+	E−	M	R	G	R+	V−											0	MC1510G
DIL-14/1C	R2	S2	V+2	N	E+1	E-1	V−	R1	S1	V+1	G	E+2	E-2	V−				LM1514J	0	MC1514L
FLP-14/3C	R2	S2	V+2	N	E+1	E-1	V−	R1	S1	V+1	G	E+2	E-2	V−					0	MC1514F
FLP-10/3C	E1	E2	F2	F1	V−	R1	R2	F*1	F*2	V+									0	MC1520F
T05-10/1M	F2	F1	V−	R1	R2	F*1	F*2	V+	E1	E2									0	MC1520G
FLP-10/3C	E+	E−	G	V−	R	V+	F	F*	ø	ø*									0	MC1530F
T05-10/1M	E+	E−	G	V−	R	V+	F	F*	ø	ø*									0	MC1530G
DIL-14/1C	ø	ø*	N	E+	N	E−	V−	G	N	N	R	V+	F	F*					0	MC1530L
FLP-10/3C	E+	E−	G	V−	R	V+	F	F*	ø	ø*									0	MC1531

TYPE NUMBER	M F R	A P P	C M P	GBP MIN	SLEW RATE MIN	Vs⁺ MAX	Vs⁻ MAX	T_op MAX	A_VOL MIN	V_IO MAX	I_B MAX	I_IO MAX	P_TOT MAX	I_OUT MIN	V_OUT MIN	V_ICM MAX	V_IDF MAX	dV_IO/dT MAX	P_O MAX	I_O MAX	CM RR MIN	PS RR MIN	R_IN MIN
MC1531F	MTU	GPU	EXT	.	0.3V/uS	+9V	-9V	125C	68dB	10MV	150NA	25NA	500MWF	5MA	4.5V	2V	5V	30uV/C	150MW	.	65dB	74dB	1M
MC1531G	MTU	GPU	EXT	.	0.3V/uS	+9V	-9V	125C	68dB	10MV	150NA	25NA	680MWF	5MA	5MV	2V	5V	30uV/C	150MW	.	65dB	74dB	1M
MC1531L	MTU	GPU	EXT	.	0.3V/uS	+9V	-9V	125C	68dB	10MV	150NA	25NA	1WF	5MA	4.5V	2V	5V	30uV/C	150MW	.	65dB	74dB	1M
MC1533F	MTU	GPU	EXT	.	0.2V/uS	+20V	-20V	125C	92dB	5MV	1uA	.15uA	500MW	6MA	12V	20V	10V	20uV/C	170MW	.	90dB	76dB	500K
MC1533F	OBS	GPU	EXT	.	0.2V/uS	+20V	-20V	125C	92dB	5MV	1uA	.15uA	500MWF	6MA	12V	20V	10V	20uV/C	170MW	.	90dB	76dB	500K
MC1533G	MTU	GPU	EXT	.	0.2V/uS	+20V	-20V	125C	92dB	5MV	1uA	.15uA	680MWF	6MA	12V	20V	10V	20uV/C	170MW	.	90dB	76dB	500K
MC1533G	OBS	GPU	EXT	.	0.2V/uS	+20V	-20V	125C	92dB	5MV	1uA	.15uA	680MWF	6MA	12V	20V	10V	20uV/C	170MW	.	90dB	76dB	500K
MC1533L	MTU	GPU	EXT	.	0.2V/uS	+20V	-20V	125C	92dB	5MV	1uA	.15uA	1WF	6MA	12V	20V	10V	20uV/C	170MW	.	90dB	76dB	500K
MC1535F	MTU	DGU	EXT	.	0.2V/uS	+10V	-10V	125C	72dB	3MV	3uA	0.3uA	500MWF	.3MA	3V	10V	5V	15uV/C	150MW	.	70dB	70dB	10K
MC1535G	MTU	DGU	EXT	.	0.2V/uS	+10V	-10V	125C	72dB	3MV	3uA	0.3uA	680MW	.3MA	3V	10V	5V	15uV/C	150MW	.	70dB	70dB	10K
MC1535L	MTU	DGU	EXT	.	0.2V/uS	+10V	-10V	125C	72dB	3MV	3uA	0.3uA	680MWF	.3MA	3V	10V	5V	15uV/C	150MW	.	70dB	70dB	10K
MC1536G	MTU	HVO	INT	.3MHZ	0.5V/uS	+40V	-40V	125C	100dB	5MV	20NA	3NA	680MWF	1MA	30V	40V	80V	.	224MW	4MA	80dB	80dB	3M
MC1537L	MTU	DGU	EXT	.	0.1V/uS	+18V	-18V	125C	88dB	5MV	0.5uA	0.2uA	750MWF	5MA	12V	18V	5V	10uV/C	225MW	.	70dB	76dB	150K
MC1539G	MTU	GPU	EXT	.	0.4V/uS	+18V	-18V	125C	94dB	3MV	500NA	75NA	680MWF	10MA	10V	18V	36V	10uV/C	.	5MA	80dB	76dB	150K
MC1539L	MTU	GPU	EXT	.	0.4V/uS	+18V	-18V	125C	94dB	3MV	500NA	75NA	750MWF	10MA	10V	18V	36V	10uV/C	.	5MA	80dB	76dB	150K
MC1556F	MUG	SBA	INT	.5MHZ	1V/uS	+22V	-22V	125C	100dB	4MV	15NA	2NA	680MWF	6MA	12V	22V	22V	30uV/C	45MW	2MA	80dB	80dB	1.5M
MC1556G	MTU	SBA	INT	.5MHZ	1V/uS	+22V	-22V	125C	100dB	4MV	15NA	2NA	680MWF	6MA	12V	22V	22V	30uV/C	45MW	2MA	80dB	80dB	1.5M
MC1556L	MTU	SBA	INT	.5MHZ	1V/uS	+22V	-22V	125C	100dB	4MV	15NA	2NA	680MWF	6MA	12V	22V	22V	30uV/C	45MW	2MA	80dB	80dB	1.5M
MC1556N(8)	MUG	SBA	INT	.5MHZ	1V/uS	+22V	-22V	125C	100dB	4MV	15NA	2NA	680MWF	6MA	12V	22V	22V	30uV/C	45MW	2MA	80dB	80dB	1.5A
MC1556T	MUG	SBA	INT	.5MHZ	1V/uS	+22V	-22V	125C	100dB	4MV	15NA	2NA	500MWF	6MA	12V	22V	22V	30uV/C	45MW	2MA	80dB	80dB	1.5M
MC1556V	MUG	SBA	INT	.5MHZ	1V/uS	+22V	-22V	125C	100dB	4MV	15NA	2NA	680MWF	6MA	12V	22V	22V	30uV/C	45MW	2MA	80dB	80dB	1.5M
MC1558G	MTU	DGK	INT	.5MHZ	0.3V/uS	+22V	-22V	125C	94dB	5MV	0.5uA	0.2uA	680MW	5MA	12V	15V	30V	50uV/C	150MW	5MA	70dB	76dB	300K
MC1558L	MTU	DGK	INT	.5MHZ	0.3V/uS	+22V	-22V	125C	94dB	5MV	0.5uA	0.2uA	750MWF	5MA	12V	15V	30V	50uV/C	150MW	5MA	70dB	76dB	300K
MC1558N(8)	MUG	DGK	INT	.5MHZ	0.3V/uS	+22V	-22V	125C	94dB	5MV	0.5uA	0.2uA	400MWF	5MA	12V	15V	30V	.	150MW	5MA	70dB	76dB	300K
MC1558NG	MTU	DLN	INT	.5MHZ	0.3V/uS	+22V	-22V	125C	94dB	5MV	0.5uA	0.2uA	680MWF	5MA	12V	15V	30V	50uV/C	150MW	5MA	70dB	76dB	300K
MC1558NL	MTU	DLN	INT	.5MHZ	0.3V/uS	+22V	-22V	125C	94dB	5MV	0.5uA	0.2uA	750MWF	5MA	12V	15V	30V	50uV/C	150MW	5MA	70dB	76dB	300K
MC1558NU	MTU	DLN	INT	.5MHZ	0.3V/uS	+22V	-22V	125C	94dB	5MV	0.5uA	0.2uA	750MWF	5MA	12V	15V	30V	50uV/C	150MW	5MA	70dB	76dB	300K
MC1558SG	MTU	DHS	INT	.5MHZ	3V/uS	+22V	-22V	125C	94dB	5MV	0.5uA	0.2uA	680MWF	5MA	12V	15V	30V	50uV/C	150MW	5MA	70dB	76dB	300K
MC1558SL	MTU	DHS	INT	.5MHZ	3V/uS	+22V	-22V	125C	94dB	5MV	0.5uA	0.2uA	750MWF	5MA	12V	15V	30V	50uV/C	150MW	5MA	70dB	76dB	300K
MC1558SU	MTU	DHS	INT	.5MHZ	3V/uS	+22V	-22V	125C	94dB	5MV	0.5uA	0.2uA	750MWF	5MA	12V	15V	30V	50uV/C	150MW	5MA	70dB	76dB	300K
MC1558T	SJU	DGK	INT	.5MHZ	0.3V/uS	+22V	-22V	125C	94dB	5MV	0.5uA	0.2uA	500MWF	5MA	12V	15V	30V	.	150MW	5MA	70dB	76dB	300K
MC1558U	MTU	DGK	INT	.5MHZ	0.3V/uS	+22V	-22V	125C	94dB	5MV	0.5uA	0.2uA	750MWF	5MA	12V	15V	30V	50uV/C	150MW	5MA	70dB	76dB	300K
MC1558V	MUG	DGK	INT	.5MHZ	0.3V/uS	+22V	-22V	125C	94dB	5MV	0.5uA	0.2uA	400MWF	5MA	12V	15V	30V	.	150MW	5MA	70dB	76dB	300K
MC1709AF	MTU	GPU	EXT	.3MHZ	.15V/uS	+18V	-18V	125C	88dB	2MV	200NA	50NA	500MWF	5MA	12V	10V	5V	10uV/C	108MW	4MA	80dB	80dB	350K
MC1709AG	MTU	GPU	EXT	.3MHZ	.15V/uS	+18V	-18V	125C	88dB	2MV	200NA	50NA	680MWF	5MA	12V	10V	5V	10uV/C	108MW	4MA	80dB	80dB	350K
MC1709AL	MTU	GPU	EXT	.3MHZ	.15V/uS	+18V	-18V	125C	88dB	2MV	200NA	50NA	750MWF	5MA	12V	10V	5V	10uV/C	108MW	4MA	80dB	80dB	350K
MC1709CF	MTU	GPU	EXT	.3MHZ	.15V/uS	+18V	-18V	75C	82dB	7.5MV	1.5uA	0.5uA	500MWF	5MA	12V	10V	5V	.	200MW	.	65dB	74dB	50K
MC1709CG	MTU	GPU	EXT	.3MHZ	.15V/uS	+18V	-18V	75C	82dB	7.5MV	1.5uA	0.5uA	680MWF	5MA	12V	10V	5V	.	200MW	.	65dB	74dB	50K
MC1709CL	MTU	GPU	EXT	.3MHZ	.15V/uS	+18V	-18V	75C	82dB	7.5MV	1.5uA	0.5uA	750MWF	5MA	12V	10V	5V	.	200MW	.	65dB	74dB	50K
MC1709CP	MTU	GPU	EXT	.3MHZ	.15V/uS	+18V	-18V	75C	82dB	7.5MV	1.5uA	0.5uA	625MWF	5MA	12V	10V	5V	.	200MW	.	65dB	74dB	50K
MC1709CP1	MTU	GPU	EXT	.3MHZ	.15V/uS	+18V	-18V	75C	82dB	7.5MV	1.5uA	0.5uA	625MWF	5MA	12V	10V	5V	.	200MW	.	65dB	74dB	50K
MC1709CP2	MTU	GPU	EXT	.3MHZ	.15V/uS	+18V	-18V	75C	82dB	7.5MV	1.5uA	0.5uA	625MWF	5MA	12V	10V	5V	.	200MW	.	65dB	74dB	50K
MC1709CU	MTU	GPU	EXT	.3MHZ	.15V/uS	+18V	-18V	75C	82dB	7.5MV	1.5uA	0.5uA	750MWF	5MA	12V	10V	5V	.	200MW	.	65dB	74dB	50K
MC1709F	MTU	GPU	EXT	.3MHZ	.15V/uS	+18V	-18V	125C	88dB	5MV	500NA	200NA	500MWF	5MA	12V	10V	5V	15uV/C	165MW	.	70dB	76dB	150K
MC1709G	MTU	GPU	EXT	.3MHZ	.15V/uS	+18V	-18V	125C	88dB	5MV	500NA	200NA	680MWF	5MA	12V	10V	5V	15uV/C	165MW	.	70dB	76dB	150K
MC1709L	MTU	GPU	EXT	.3MHZ	.15V/uS	+18V	-18V	125C	88dB	5MV	500NA	200NA	750MWF	5MA	12V	10V	5V	15uV/C	165MW	.	70dB	76dB	150K
MC1709U	MTU	GPU	EXT	.3MHZ	.15V/uS	+18V	-18V	125C	88dB	5MV	500NA	200NA	750MWF	5MA	12V	10V	5V	15uV/C	165MW	.	70dB	76dB	150K
MC1710CF	MTU	CPR	EXT	.	.	+14V	-7V	75C	60dB	5MV	25uA	5uA	500MWF	5MA	1V	7V	5V	20uV/C	150MW	9MA	70dB	.	.
MC1710CG	MTU	CPR	EXT	.	.	+14V	-7V	75C	60dB	5MV	25uA	5uA	680MWF	5MA	1V	7V	5V	20uV/C	150MW	9MA	70dB	.	.
MC1710CL	MTU	CPR	EXT	.	.	+14V	-7V	75C	60dB	5MV	25uA	5uA	625MWF	5MA	1V	7V	5V	20uV/C	150MW	9MA	70dB	.	.
MC1710CP	MTU	CPR	EXT	.	.	+14V	-7V	75C	60dB	5MV	25uA	5uA	625MWF	5MA	1V	7V	5V	20uV/C	150MW	9MA	70dB	.	.
MC1710F	MTU	CPR	EXT	.	.	+14V	-7V	125C	62dB	2MV	20uA	3uA	500MWF	5MA	1V	7V	5V	10uV/C	150MW	9MA	80dB	.	.
MC1710G	MTU	CPR	EXT	.	.	+14V	-7V	125C	62dB	2MV	20uA	3uA	500MWF	5MA	1V	7V	5V	10uV/C	150MW	9MA	80dB	.	.
MC1710L	MTU	CPR	EXT	.	.	+14V	-7V	125C	62dB	2MV	20uA	3uA	625MWF	5MA	1V	7V	5V	10uV/C	150MW	9MA	80dB	.	.
MC1711CF	MTU	DCP	EXT	.	.	+14V	-7V	75C	57dB	5MV	100uA	25uA	500MWF	5MA	2.5V	7V	5V	20uV/C	200MW	.	.	.	.
MC1711CG	MTU	DCP	EXT	.	.	+14V	-7V	75C	57dB	5MV	100uA	25uA	680MWF	5MA	2.5V	7V	5V	20uV/C	200MW	.	.	.	.
MC1711CL	MTU	DCP	EXT	.	.	+14V	-7V	75C	57dB	5MV	100uA	25uA	625MWF	5MA	2.5V	7V	5V	20uV/C	200MW	.	.	.	.
MC1711CP	MTU	DCP	EXT	.	.	+14V	-7V	75C	57dB	5MV	100uA	25uA	625MWF	5MA	2.5V	7V	5V	20uV/C	200MW	.	.	.	.
MC1711F	MTU	DCP	EXT	.	.	+14V	-7V	125C	58dB	3.5MV	75uA	10uA	500MWF	5MA	2.5V	7V	5V	20uV/C	200MW	.	.	.	.
MC1711G	MTU	DCP	EXT	.	.	+14V	-7V	125C	58dB	3.5MV	75uA	10uA	680MW	5MA	2.5V	7V	5V	20uV/C	200MW	.	.	.	.

For detailed explanations of column heading notations, see App. A. Also for ready references the more important abbreviations used in the column headings are listed below:

LEFT HAND PAGE

APP = application (codes at APP.E.)
CMRR = common mode rejection ratio
CMP = compensation (frequency)
dV_{io}/dT = input offset voltage temperature drift
GBP = gain bandwidth product
I_B = input bias current
I_{io} = input bias offset current
I_0 = quiescent supply current
MFR = manufacturer (codes at App.C.)
P_Q = quiescent power consumer
PSRR = power supply rejection ratio
V_{icm} = common mode input voltage rating
V_{idf} = differential input voltage rating
V_{io} = input offset voltage
V_S = dc supply voltage

RIGHT HAND PAGE

Lead out coding summary (details at APP.G.) for different cases (APP.F.)
A = gain adjust
B = bias adjust
C = case
E- = inverting input
E+ = non-inverting input
F,F* = input frequency compensation
G = ground
J = high level input
K = output, open collector
L = output, open emitter
M = metal case
N = not connected
Q = special terminal
R,R* = outputs
S = strobe
T,T* = offset balance
V+ = +ve dc supply
V- = -ve dc supply
W = guard ring
X = blank position, no lead
++ = +ve supplementary dc supply
-- = -ve supplementary dc supply
ø,ø* = output frequency compensation

CASE (APP F)	LD1	LD2	LD3	LD4	LD5	LD6	LD7	LD8	LD9	LD10	LD11	LD12	LD13	LD14	LD15	LD16	EUROPE SUBSTITUTE	USA SUBSTITUTE	IS	TYPE NUMBER
TO5-10/1M	E+	E-	G	V-	R	V+	F	F*	ø	ø*									0	MC1531G
DIL-14/1C	ø	ø*	N	E+	N	E-	V-	G	N	N	R	V+	F	F*					0	MC1531L
FLP-10/3G	E+	ø	V-	R	V+	V+	V+	F	F*	E-									0	MC1533F
FLP-10/3G	E+	ø	V-	R	V+	A	T	F	F*	E-									0	MC1533F
TO5-10/1M	E-	E+	ø	V-	R	V+	V+	V+	F	F*									0	MC1533G
TO5-10/1M	E-	E+	ø	V-	R	V+	A	T	F	F*									0	MC1533G
DIL-14/1C	N	F	F*	E-	E+	ø	V-	N	N	N	R	V+	V+	V+					0	MC1533L
FLP-14/3C	R2	ø2	F2	F*2	E+2	E-2	V-	E-1	E+1	F1	F*1	ø1	R1	V+					0	MC1535F
TO5-10/1M	V-	E-2	E+2	ø2	R2	V+	R1	ø1	E+1	E-1									0	MC1535G
DIL-14/1C	R2	ø2	F2	F*2	E+2	E-2	V-	E-1	E+1	F1	F*1	ø1	R1	V+				UA749DM	0	MC1535L
TO5-8/1M	T	E-	E+	V+	T*	R	V+	N										LM143	0	MC1536G
DIL-14/1C	ø2	R2	F2	F*2	E-2	E+2	V-	E+1	E-1	F1	F*1	R1	ø1	V+				RM1537DC	0	MC1537L
TO5-8/1M	F	E-	E+	V-	ø	R	V+	F*									TAA522	UA709HM	0	MC1539G
DIL-14/1C	N	N	F	E-	E+	V-	N	N	ø	R	V+	F*	N	N			LM709J	UA709DM	0	MC1539L
DIL-14/1C	N	N	T	E-	E+	V-	N	N	T*	R	V+	N	N	N				MC1556L	0	MC1556F
TO5-8/1M	T	E-	E+	V-	T*	R	V+	N									LM143H	MC1556T	0	MC1556G
DIL-14/1C	N	N	T	E-	E+	V-	N	N	T*	R	V+	N	N	N			LM143D	MC1556F	0	MC1556L
DIL-8/1P	T	E-	E+	V-	T*	R	V+	N										S5556V	0	MC1556N(8)
TO5-8/1M	T	E-	E+	V-	T*	R	V+	N									S5556T	MC1556G	0	MC1556T
DIL-8/1P	T	E-	E+	V-	T*	R	V+	N										S5556V	0	MC1556V
TO5-8/1M	R1	E-1	E+1	V-	E+2	E-2	R2	V+									TBC1458	LM1558H	0	MC1558G
DIL-14/1C	N	R1	N	N	E-1	E+1	V-	E+2	E-2	N	N	R2	N	V+					0	MC1558L
DIL-8/1P	R1	E-1	E+1	V-	E+2	E-2	R2	V+									LM1558J	MC1558U	0	MC1558N(8)
TO5-8/1M	R1	E-1	E+1	V-	E+2	E-2	R2	V+											0	MC1558NG
DIL-14/1C	N	R1	N	N	E-1	E+1	V-	E+2	E-2	N	N	R2	N	V+					0	MC1558NL
DIL-8/1C	R1	E-1	E+1	V-	E+2	E-2	R2	V+											0	MC1558NU
TO5-8/1M	R1	E-1	E+1	V-	E+2	E-2	R2	V+											0	MC1558SG
DIL-14/1C	N	R1	N	N	E-1	E+1	V-	E+2	E-2	N	N	R2	N	V+					0	MC1558SL
DIL-8/1C	R1	E-1	E+1	V-	E+2	E-2	R2	V+											0	MC1558SU
TO5-8/1M	R1	E-1	E+1	V-	E+2	E-2	R2	V+									TBC1458	MC1558G	0	MC1558T
DIL-8/1C	R1	E-1	E+1	V-	E+2	E-2	R2	V+										LM1558J	0	MC1558U
DIL-8/1P	R1	E-1	E+1	V-	E+2	E-2	R2	V+									LM1558J	MC1558U	0	MC1558V
FLP-10/3C	N	F	E-	E+	V-	ø	R	V+	F*	N							SN52709AFA	UA709AFM	0	MC1709AF
TO5-8/1M	F	E-	E+	V-	ø	R	V+	F*										UA709AHM	0	MC1709AG
DIL-14/1C	N	N	F	E-	E+	V-	N	N	ø	R	V+	F*	N	N			LM709AJ	UA709ADM	0	MC1709AL
FLP-10/3C	N	F	E-	E+	V-	ø	R	V+	F*	N							SN52709AFA	UA709FM	0	MC1709CF
TO5-8/1M	F	E-	E+	V-	ø	R	V+	F*									TAA521	UA709HC	0	MC1709CG
DIL-14/1C	N	N	F	E-	E+	V-	N	N	ø	R	V+	F*	N	N			TAA521A	UA709DC	0	MC1709CL
DIL-14/1P	N	N	F	E-	E+	V-	N	N	ø	R	V+	F*	N	N			TAA521A	UA709DC	0	MC1709CP
DIL-8/1C	F	E-	E+	V-	ø	R	V+	F*									LM709CN8	UA709TC	0	MC1709CP1
DIL-14/1P	N	N	F	E-	E+	V-	N	N	ø	R	V+	F*	N	N			TAA521A	UA709DC	0	MC1709CP2
DIL-8/1C	F	E-	E+	V-	ø	R	V+	F*									LM709CN8	UA709TC	0	MC1709CU
FLP-10/3C	N	F	E-	E+	V-	ø	R	V+	F*	N								UA709FM	0	MC1709F
TO5-8/1M	F	E-	E+	V-	ø	R	V+	F*									TAA522	UA709HM	0	MC1709G
DIL-14/1C	N	N	F	E-	E+	V-	N	N	ø	R	V+	F*	N	N			LM709J	UA709DM	0	MC1709L
DIL-8/1C	F	E-	E+	V-	ø	R	V+	F*										SN52709AJP	0	MC1709U
FLP-10/3C	G	E+	E-	N	V-	R	N	V+	N	N							SFC2710PM	UA710FM	0	MC1710CF
TO5-8/1M	G	E+	E-	N	N	R	V+										SFC2710C	UA710HC	0	MC1710CG
DIL-14/1C	N	G	E+	E-	N	V-	N	N	R	N	V+	N	N	N			SFC2710EC	UA710DC	0	MC1710CL
DIL-14/1P	N	G	E+	E-	N	V-	N	N	R	N	V+	N	N	N			SFC2710EC	UA710DC	0	MC1710CP
FLP-10/3C	G	E+	E-	N	V-	R	N	V+	N	N							SFC2710PM	UA710FM	0	MC1710F
TO5-8/1M	G	E+	E-	V-	N	R	V+										SFC2710M	UA710HM	0	MC1710G
DIL-14/1C	N	G	E+	E-	N	V-	N	N	R	N	V+	N	N	N			SFC2710KM	UA710DM	0	MC1710L
FLP-10/1C	E-1	E+1	V-	E+2	E-2	N	N	S2	R	V+	S1						SFC2711PM	UA711FM	0	MC1711CF
TO5-10/1M	G	S1	E-1	E+1	V-	E+2	E-2	S2	R	V+							SFC2711C	UA711HC	0	MC1711CG
DIL-14/1C	N	E-1	E+1	V-	E+2	E-2	N	N	S2	R	V+	G	S1	N			SFC2711EC	UA711DC	0	MC1711CL
DIL-14/1P	N	E-1	E+1	V-	E+2	E-2	N	N	S2	R	V+	G	S1	N			SFC2711EC	UA711DC	0	MC1711CP
FLP-10/1C	E-1	E+1	V-	E+2	E-2	N	N	S2	R	V+	S1							UA711FM	0	MC1711F
TO5-10/1M	G	S1	E-1	E+1	V-	E+2	E-2	S2	R	V+							SFC2711M	UA711HM	0	MC1711G
DIL-14/1C	N	E-1	E+1	V-	E+2	E-2	N	N	S2	R	V+	G	S1	N			SFC2711KM	UA711DM	0	MC1711L

TYPE NUMBER	MFR	APP	CMP	GBP MIN	SLEW RATE MIN	V_S^+ MAX	V_S^- MAX	T_{op} MAX	A_{VOL} MIN	V_{IO} MAX	I_B MAX	I_{IO} MAX	P_{TOT} MAX	I_{OUT} MIN	V_{OUT} MIN	V_{ICM} MAX	V_{IDF} MAX	dV_{IO}/dT MAX	P_O MAX	I_O MAX	CMRR MIN	PSRR MIN	R_{IN} MIN
MC1711L	MTU	DCP	EXT	.	.	+14V	-7V	125C	58dB	3.5MV	75uA	10uA	625MWF	5MA	2.5V	7V	5V	20uV/C	200MW	.	.	.	.
MC1712CF	MTU	GPU	EXT	3MHZ	.	+13V	-8V	75C	66dB	5MV	7.5uA	2uA	500MWF	.3MA	5V	1.5V	5V	20uV/C	120MW	7MA	70dB	70dB	10K
MC1712CG	MTU	GPU	EXT	3MHZ	.	+13V	-8V	75C	66dB	5MV	7.5uA	2uA	680MWF	.3MA	5V	1.5V	5V	20uV/C	120MW	7MA	70dB	70dB	10K
MC1712CL	MTU	GPU	EXT	3MHZ	.	+13V	-8V	75C	66dB	5MV	7.5uA	2uA	625MWF	.3MA	5V	1.5V	5V	20uV/C	120MW	7MA	70dB	70dB	10K
MC1712CP	MTU	GPU	EXT	3MHZ	.	+13V	-8V	75C	66dB	5MV	7.5uA	2uA	400MWF	.3MA	5V	1.5V	5V	20uV/C	120MW	7MA	70dB	70dB	10K
MC1712F	MTU	GPU	EXT	3MHZ	.	+13V	-8V	125C	68dB	2MV	5uA	0.5uA	500MWF	.3MA	5V	1.5V	5V	10uV/C	120MW	7MA	80dB	74dB	16K
MC1712G	MTU	GPU	EXT	3MHZ	.	+13V	-8V	125C	68dB	2MV	5uA	0.5uA	680MWF	.3MA	5V	1.5V	5V	10uV/C	120MW	7MA	80dB	74dB	16K
MC1712L	MTU	GPU	EXT	3MHZ	.	+13V	-8V	125C	68dB	2MV	5uA	0.5uA	625MWF	.3MA	5V	1.5V	5V	10uV/C	120MW	7MA	80dB	74dB	16K
MC1733CG	MTU	BDO	INT	40MHZ	.	+8V	-8V	70C	40dB	6MV	30uA	5uA	500MWF	2MA	2V	6V	5V	.	.	24MA	60dB	50dB	2K
MC1733CL	MTU	BDO	INT	40MHZ	.	+8V	-8V	70C	40dB	6MV	30uA	5uA	500MWF	2MA	2V	6V	5V	.	.	24MA	60dB	50dB	2K
MC1733CP	MTU	BDO	INT	40MHZ	.	+8V	-8V	70C	40dB	6MV	30uA	5uA	500MWF	2MA	2V	6V	5V	.	.	24MA	60dB	50dB	2K
MC1733G	MTU	BDO	INT	40MHZ	.	+8V	-8V	125C	50dB	5MV	20uA	3uA	500MWF	2MA	2V	6V	5V	.	.	24MA	60dB	50dB	2K
MC1733L	MTU	BDO	INT	40MHZ	.	+8V	-8V	125C	50dB	5MV	20uA	3uA	500MWF	2MA	2V	6V	5V	.	.	24MA	60dB	50dB	2K
MC1741CF	MTU	GPK	INT	.	0.2V/uS	+18V	-18V	70C	86dB	6MV	500NA	200NA	500MWF	5MA	12V	15V	30V	.	85MW	3MA	70dB	76dB	300K
MC1741CG	MTU	GPK	INT	.	0.2V/uS	+18V	-18V	70C	86dB	6MV	500NA	200NA	680MWF	5MA	12V	15V	30V	.	85MW	3MA	70dB	76dB	300K
MC1741CL	MTU	GPK	INT	.	0.2V/uS	+18V	-18V	70C	86dB	6MV	500NA	200NA	750MWF	5MA	12V	15V	30V	.	85MW	3MA	70dB	76dB	300K
MC1741CP1	MTU	GPK	INT	.	0.2V/uS	+18V	-18V	70C	86dB	6MV	500NA	200NA	625MWF	5MA	12V	15V	30V	.	85MW	3MA	70dB	76dB	300K
MC1741CP2	MTU	GPK	INT	.	0.2V/uS	+18V	-18V	70C	86dB	6MV	500NA	200NA	625MWF	5MA	12V	15V	30V	.	85MW	3MA	70dB	76dB	300K
MC1741CU	MTU	GPK	INT	.	0.2V/uS	+18V	-18V	70C	86dB	6MV	500NA	200NA	750MWF	5MA	12V	15V	30V	.	85MW	3MA	70dB	76dB	300K
MC1741F	MTU	GPK	INT	.	0.3V/uS	+22V	-22V	125C	94dB	5MV	500NA	200NA	500MWF	5MA	12V	15V	30V	.	75MW	3MA	70dB	76dB	300K
MC1741G	MTU	GPK	INT	.	0.3V/uS	+22V	-22V	125C	94dB	5MV	500NA	200NA	680MWF	5MA	12V	15V	30V	.	75MW	3MA	70dB	76dB	300K
MC1741L	MTU	GPK	INT	.	0.3V/uS	+22V	-22V	125C	94dB	5MV	500NA	200NA	750MWF	5MA	12V	15V	30V	.	75MW	3MA	70dB	76dB	300K
MC1741NCF	MTU	GPK	INT	.	3V/uS	+18V	-18V	70C	86dB	6MV	500NA	200NA	500MWF	5MA	12V	15V	30V	.	85MW	3MA	70dB	76dB	300K
MC1741NCG	MTU	LNA	INT	.	0.2V/uS	+18V	-18V	70C	86dB	6MV	500NA	200NA	680MWF	5MA	12V	15V	30V	.	85MW	3MA	70dB	76dB	300K
MC1741NCL	MTU	LNA	INT	.	0.2V/uS	+18V	-18V	70C	86dB	6MV	500NA	200NA	750MWF	5MA	12V	15V	30V	.	85MW	3MA	70dB	76dB	300K
MC1741NCP	MTU	LNA	INT	.	0.2V/uS	+18V	-18V	70C	86dB	6MV	500NA	200NA	625MWF	5MA	12V	15V	30V	.	85MW	3MA	70dB	76dB	300K
MC1741NCP1	MTU	LNA	INT	.	0.2V/uS	+18V	-18V	70C	86dB	6MV	500NA	200NA	625MWF	5MA	12V	15V	30V	.	85MW	3MA	70dB	76dB	300K
MC1741NCU	MTU	LNA	INT	.	0.2V/uS	+18V	-18V	70C	86dB	6MV	500NA	200NA	750MWF	5MA	12V	15V	30V	.	85MW	3MA	70dB	76dB	300K
MC1741NF	MTU	LNA	INT	.	0.3V/uS	+22V	-22V	125C	94dB	5MV	500NA	200NA	500MWF	5MA	12V	15V	30V	.	75MW	3MA	70dB	76dB	300K
MC1741NG	MTU	LNA	INT	.	0.3V/uS	+22V	-22V	125C	94dB	5MV	500NA	200NA	680MWF	5MA	12V	15V	30V	.	75MW	3MA	70dB	76dB	300K
MC1741NL	MTU	LNA	INT	.	0.3V/uS	+22V	-22V	125C	94dB	5MV	500NA	200NA	750MWF	5MA	12V	15V	30V	.	75MW	3MA	70dB	76dB	300K
MC1741NU	MTU	LNA	INT	.	0.3V/uS	+22V	-22V	125C	94dB	5MV	500NA	200NA	625MWF	5MA	5MV	15V	30V	.	75MW	3MA	70dB	76dB	300K
MC1741SCG	MTU	HSR	INT	.	3V/uS	+18V	-18V	70C	86dB	6MV	500NA	200NA	680MWF	5MA	12V	15V	30V	.	85MW	3MA	70dB	76dB	300K
MC1741SCL	MTU	HSR	INT	.	3V/uS	+18V	-18V	70C	86dB	6MV	500NA	200NA	750MWF	5MA	12V	15V	30V	.	85MW	3MA	70dB	76dB	300K
MC1741SCP	MTU	HSR	INT	.	3V/uS	+18V	-18V	70C	86dB	6MV	500NA	200NA	625MWF	5MA	12V	15V	30V	.	85MW	3MA	70dB	76dB	300K
MC1741SCP1	MTU	HSR	INT	.	3V/uS	+18V	-18V	70C	86dB	6MV	500NA	200NA	625MWF	5MA	12V	15V	30V	.	85MW	3MA	70dB	76dB	300K
MC1741SCU	MTU	HSR	INT	.	3V/uS	+18V	-18V	70C	86dB	6MV	500NA	200NA	750MWF	5MA	12V	15V	30V	.	85MW	3MA	70dB	76dB	300K
MC1741SG	MTU	HSR	INT	.	3V/uS	+22V	-22V	125C	94dB	5MV	500NA	200NA	680MWF	5MA	12V	15V	30V	.	75MW	3MA	70dB	76dB	300K
MC1741SL	MTU	HSR	INT	.	3V/uS	+22V	-22V	125C	94dB	5MV	500NA	200NA	750MWF	5MA	12V	15V	30V	.	75MW	3MA	70dB	76dB	300K
MC1741SU	MTU	HSR	INT	.	3V/uS	+22V	-22V	125C	94dB	5MV	500NA	200NA	625MWF	5MA	12V	15V	30V	.	75MW	3MA	70dB	76dB	300K
MC1741U	MTU	GPK	INT	.	0.3V/uS	+22V	-22V	125C	94dB	5MV	500NA	200NA	625MWF	5MA	12V	15V	30V	.	75MW	3MA	70dB	76dB	300K
MC1747CF	MTU	DGK	INT	.	0.2V/uS	+18V	-18V	75C	88dB	6MV	500NA	200NA	750MWF	5MA	12V	15V	30V	.	85MW	3MA	70dB	76dB	300K
MC1747CG	MTU	DGU	INT	.	0.2V/uS	+18V	-18V	75C	88dB	6MV	500NA	200NA	500MWF	5MA	12V	15V	30V	.	85MW	3MA	70dB	76dB	300K
MC1747CL	MTU	DGU	INT	.	0.2V/uS	+18V	-18V	75C	88dB	6MV	500NA	200NA	670MWF	5MA	12V	15V	30V	.	85MW	3MA	70dB	76dB	300K
MC1747CP	MTU	DGU	INT	.	0.2V/uS	+18V	-18V	75C	88dB	6MV	500NA	200NA	670MWF	5MA	12V	15V	30V	.	85MW	3MA	70dB	76dB	300K
MC1747F	MTU	DGK	INT	.	0.2V/uS	+22V	-22V	125C	94dB	5MV	500NA	200NA	750MWF	5MA	12V	15V	30V	.	85MW	3MA	70dB	76dB	300K
MC1747G	MTU	DGK	INT	.	0.2V/uS	+22V	-22V	125C	94dB	5MV	500NA	200NA	500MWF	5MA	12V	15V	30V	.	85MW	3MA	70dB	76dB	300K
MC1747L	MTU	DGK	INT	.	0.2V/uS	+22V	-22V	125C	94dB	5MV	500NA	200NA	670MWF	5MA	12V	15V	30V	.	85MW	3MA	70dB	76dB	300K
MC1748CG	MTU	GPU	EXT	.	.25V/uS	+22V	-22V	75C	94dB	5MV	500NA	200NA	680MWF	5MA	12V	15V	30V	.	85MW	3MA	70dB	76dB	300K
MC1748CP1	MTU	GPU	EXT	.	.25V/uS	+18V	-18V	75C	86dB	6MV	500NA	200NA	300MWF	5MA	12V	15V	30V	.	85MW	3MA	70dB	76dB	300K
MC1748CU	MTU	GPU	EXT	.	.25V/uS	+18V	-18V	75C	86dB	6MV	500NA	200NA	300MWF	5MA	12V	15V	30V	.	85MW	3MA	70dB	76dB	300K
MC1748G	MTU	GPU	EXT	.	.25V/uS	+22V	-22V	125C	94dB	5MV	500NA	200NA	680MWF	5MA	12V	15V	30V	.	85MW	3MA	70dB	76dB	300K
MC1748U	MTU	GPU	EXT	.	.25V/uS	+18V	-18V	125C	86dB	6MV	500NA	200NA	310MWF	5MA	12V	15V	30V	.	85MW	3MA	70dB	76dB	300K
MC1776CG	MTU	PRA	INT	.	0.3V/uS	+18V	-18V	70C	94dB	6MV	50NA	25NA	680MWF	2MA	10V	15V	30V	.	6MW	.2UA	70dB	74dB	2M
MC1776G	MTU	PRA	INT	.	0.3V/uS	+18V	-18V	125C	100dB	5MV	50NA	15NA	680MWF	2MA	10V	15V	30V	.	6MW	.2UA	70dB	76dB	2M
MC1776L	MTU	PRA	INT	.	0.3V/uS	+18V	-18V	125C	100dB	5MV	50NA	15NA	670MWF	2MA	10V	15V	30V	.	6MW	.2UA	70dB	76dB	2M
MC3301P	MTU	QCD	INT	1M5HZ	0.2V/uS	+28V	.	85C	60dB	.	300NA	.	625MWF	6MA	13V	.	.	.	.	10MA	.	50dB	.
MC3302A	MUG	QCP	EXT	.	20V/uS	+14V	-14V	85C	66dB	20MV	500NA	100NA	900MWF	2MA	.	9V	28V	.	.	2MA	.	.	.
MC3302L	MTU	QCP	EXT	.	20V/uS	+14V	-14V	85C	66dB	20MV	500NA	100NA	900MWF	2MA	.	9V	28V	.	.	2MA	.	.	.
MC3302N(14)	MUG	QCP	EXT	.	20V/uS	+14V	-14V	85C	66dB	20MV	500NA	100NA	900MWF	2MA	.	9V	28V	.	.	2MA	.	.	.

For detailed explanations of column heading notations, see App. A.

Also for ready references the more important abbreviations used in the column headings are listed below:

LEFT HAND PAGE
APP = application (codes at APP.E.)
CMRR = common mode rejection ratio
CMP = compensation (frequency)
dV_{IO}/dT = input offset voltage temperature drift
GBP = gain bandwidth product
I_B = input bias current
I_{IO} = input bias offset current
I_Q = quiescent supply current
MFR = manufacturer (codes at App.C.)
P_Q = quiescent power consumer
PSRR = power supply rejection ratio
V_{ICM} = common mode input voltage rating
V_{IDF} = differential input voltage rating
V_{IO} = input offset voltage
V_S = dc supply voltage

RIGHT HAND PAGE
Lead out coding summary (details at APP.G.) for different cases (APP.F.)
A = gain adjust
B = bias adjust
C = case
E− = inverting input
E+ = non-inverting input
F,F* = input frequency compensation
G = ground
J = high level input
K = output, open collector
L = output, open emitter
M = metal case
N = not connected
Q = special terminal
R,R* = outputs
S = strobe
T,T* = offset balance
V+ = +ve dc supply
V− = −ve dc supply
W = guard ring
X = blank position, no lead
+ + = +ve supplementary dc supply
− − = −ve supplementary dc supply
φ.φ* = output frequency compensation

CASE (APP F)	LD 1	LD 2	LD 3	LD 4	LD 5	LD 6	LD 7	LD 8	LD 9	LD 10	LD 11	LD 12	LD 13	LD 14	LD 15	LD 16	EUROPE SUBSTITUTE	USA SUBSTITUTE	ISS	TYPE NUMBER
FLP-10/3C	N	G	E-	E+	V-	F	ø	R	N	V+	.	.	.	.			SN52702FA	UA702FM	0	MC1712CF
TO5-8/1M	G	E-	E+	V-M	F	ø	R	V+	.	.							SN72702L	UA702HC	0	MC1712CG
DIL-14/1C	N	N	G	E-	E+	V-	N	N	F	ø	R	N	V+	N			SN72702J	UA702DC	0	MC1712CL
DIL-14/1P	N	N	G	E-	E+	V-	N	N	F	ø	R	N	V+	N			SN72702J	UA702DC	0	MC1712CP
FLP-10/3C	N	G	E-	E+	V-	F	ø	R	N	V+							SN52702FA	UA702FM	0	MC1712F
TO5-8/1M	G	E-	E+	V-M	F	ø	R	V+									SN52702AL	UA702HM	0	MC1712G
DIL-14/1C	N	N	G	E-	E+	V-	N	N	F	ø	R	N	V+	N			SN52702J	UA702DM	0	MC1712L
TO5-10/1M	E-	E+	A2	A*2	V-	R	R*	V+	A1	A*1							SN72733L	UA733HC	0	MC1733CG
DIL-14/1C	E+	N	A2	A*2	V-	N	R	R*	N	V+	A1	A*1	N	E-			SN72733J	UA733DC	0	MC1733CL
DIL-14/1P	E+	N	A2	A*2	V-	N	R	R*	N	V+	A1	A*1	N	E-			SN72733J	UA733DC	0	MC1733CP
TO5-10/1M	E-	E+	A2	A*2	V-	R	R*	V+	A1	A*1							SN52733L	UA733HM	0	MC1733G
DIL-14/1C	E+	N	A2	A*2	V-	N	R	R*	N	V+	A1	A*1	N	E-			SN52733J	UA733DM	0	MC1733L
FLP-10/3C	N	T	E-	E+	V-	T*	R	V+	N	N							LM741F	UA741FM	0	MC1741CF
TO5-8/1M	T	E-	E+	V-M	T*	R	V+	N									TBA221	UA741HC	0	MC1741CG
DIL-14/1C	N	N	T	E-	E+	V-	N	N	T*	R	V+	N	N	N			TBA221A	UA741DC	0	MC1741CL
DIL-8/1P	T	E-	E+	V-	T*	V+	N	N									TBA221B	UA741TC	0	MC1741CP1
DIL-14/1P	N	N	T	E-	E+	V-	N	N	T*	R	V+	N	N	N			TBA221A	UA741DC	0	MC1741CP2
DIL-8/1C	T	E-	E+	V-	T*	V+	N	N									TBA221B	UA741TC	0	MC1741CU
FLP-10/3C	N	T	E-	E+	V-	T*	R	V+	N	N							LM741F	UA741FM	0	MC1741F
TO5-8/1M	T	E-	E+	V-M	T*	R	V+	N									TBA222	UA741HM	0	MC1741G
DIL-14/1C	N	N	T	E-	E+	V-	N	N	T*	R	V+	N	N	N			LM741D	UA741DM	0	MC1741L
FLP-10/3C	N	T	E-	E+	V-	T*	R	V+	N	N								741LNFB	0	MC1741NCF
TO5-8/1M	T	E-	E+	V-M	T*	R	V+	N										741CLNTY	0	MC1741NCG
DIL-14/1C	N	N	T	E-	E+	V-	N	N	T*	R	V+	N	N	N				741LNDD	0	MC1741NCL
DIL-14/1P	N	N	T	E-	E+	V-	N	N	T*	R	V+	N	N	N				741LNDD	0	MC1741NCP
DIL-8/1P	T	E-	E+	V-	T*	V+	N	N										741CLNPA	0	MC1741NCP1
DIL-8/1C	T	E-	E+	V-	T*	V+	N	N										741CLNPA	0	MC1741NCU
FLP-10/3C	N	T	E-	E+	V-	T*	R	V+	N	N								741LNFB	0	MC1741NF
TO5-8/1M	T	E-	E+	V-	T*	R	V+	N										741MLNTY	0	MC1741NG
DIL-14/1C	N	N	T	E-	E+	V-	N	N	T*	R	V+	N	N	N				741LNDD	0	MC1741NL
DIL-8/1C	T	E-	E+	V-	T*	V+	N	N										.	0	MC1741NU
TO5-8/1M	T	E-	E+	V-M	T*	R	V+	N										741CHS	0	MC1741SCG
DIL-14/1C	N	N	T	E-	E+	V-	N	N	T*	R	V+	N	N	N				741MHSDD	0	MC1741SCL
DIL-14/1P	N	N	T	E-	E+	V-	N	N	T*	R	V+	N	N	N				741MHSDD	0	MC1741SCP
DIL-8/1P	T	E-	E+	V-	T*	V+	N	N										741CHSPA	0	MC1741SCP1
DIL-8/1C	T	E-	E+	V-	T*	V+	N	N										741CHSPA	0	MC1741SCU
TO5-8/1M	T	E-	E+	V-M	T*	R	V+	N										741MHSTY	0	MC1741SG
DIL-14/1C	N	N	T	E-	E+	V-	N	N	T*	R	V+	N	N	N				741MHSDD	0	MC1741SL
DIL-8/1C	T	E-	E+	V-	T*	V+	N	N											0	MC1741SU
DIL-8/1C	T	E-	E+	V-	T*	V+	N	N											0	MC1741U
FLP-14/3C	E-1	E+1	T1	V-	T2	E+2	E-2	T*2	V+2	R2	N	R1	V+1	T*1				LM747CF	0	MC1747CF
TO5-10/1M	R1	V+1	E-1	E+1	V-	E+2	E-2	V+2	R2	N							TBB0747	UA747HC	0	MC1747CG
DIL-14/1C	E-1	E+1	T1	V-	T2	E+2	E-2	T*2	V+2	R2	N	R1	V+1	T*1			TBB0747A	UA747DC	0	MC1747CL
DIL-14/1P	E-1	E+1	T1	V-	T2	E+2	E-2	T*2	V+2	R2	N	R1	V+1	T*1			TBB0747A	UA747DC	0	MC1747CP
FLP-14/3C	E-1	E+1	T1	V-	T2	E+2	E-2	T*2	V+2	R2	N	R1	V+1	T*1				LM747F	0	MC1747F
TO5-10/1M	R1	V+1	E-1	E+1	V-	E+2	E-2	V+2	R2	N							SFC2747M	UA747HM	0	MC1747G
DIL-14/1C	E-1	E+1	T1	V-	T2	E+2	E-2	T*2	V+2	R2	N	R1	V+1	T*1			SFC2747KM	UA747DM	0	MC1747L
TO5-8/1M	FT	E-	E+	V-M	T*	R	V+	F*									TBB0748	UA748HC	0	MC1748CG
DIL-8/1P	FT	E-	E+	V-	T*	V+	F*										TBB0748B	UA748TC	0	MC1748CP1
DIL-8/1C	FT	E-	E+	V-	T*	V+	F*										TBB0748B	UA748TC	0	MC1748CU
TO5-8/1M	FT	E-	E+	V-M	T*	R	V+	F*									TBC0748	UA748HM	0	MC1748G
DIL-8/1C	FT	E-	E+	V-	T*	V+	F*										SN52748JP	LM748J	0	MC1748U
TO5-8/1M	T	E-	E+	V-	T*	R	V+	B										UA776HC	0	MC1776CG
TO5-8/1M	T	E-	E+	V-	T*	R	V+	B										UA776HM	0	MC1776G
DIL-14/1C	N	N	T	E-	E+	V-	N	N	T*	R	V+	B	N	N				UA776DM	0	MC1776L
DIL-14/1P	E+1	E+2	E-2	R2	R1	E-1	G	E-3	R3	R4	E-4	E+4	E+3	V+			UA3301P	LM3301N	0	MC3301P
DIL-14/1P	R2	R1	V+	E-1	E+1	E-2	E+2	E+3	E-3	E-4	E+4	G	R4	R3			MC3302L	LM3302J	0	MC3302A
DIL-14/1C	R2	R1	V+	E-1	E+1	E-2	E+2	E+3	E-3	E-4	E+4	G	R4	R3				LM3302J	0	MC3302L
DIL-14/1P	R2	R1	V+	E-1	E+1	E-2	E+2	E+3	E-3	E-4	E+4	G	R4	R3			MC3302L	LM3302J	0	MC3302N(14)
DIL-14/1P	R2	R1	V+	E-1	E+1	E-2	E+2	E+3	E-3	E-4	E+4	G	R4	R3				LM3302J	0	MC3302P

TYPE NUMBER	MFR	APP	CMP	GBP MIN	SLEW RATE MIN	Vs+ MAX	Vs- MAX	T_op MAX	A_vol MIN	V_io MAX	I_R MAX	I_io MAX	P_TOT MAX	I_out MIN	V_out MIN	V_icm MAX	V_idf MAX	dV_io/dT MAX	P_o MAX	I_o MAX	CMRR MIN	PSRR MIN	R_in MIN
MC3302P	MTU	QCP	EXT	.	20V/uS	+14V	-14V	85C	66dB	20MV	500NA	100NA	900MWF	2MA	.	9V	28V		.	2MA	.	.	.
MC3303L	MTU	QGK	INT	.3MHz	0.2V/uS	+18V	-18V	85C	86dB	8MV	500NA	75NA	670MWF	10MA	12V	18V	30V	30uV/C		7MA	70dB	76dB	300K
MC3303P	MTU	QGK	INT	.3MHz	0.2V/uS	+18V	-18V	85C	86dB	8MV	500NA	75NA	670MWF	10MA	12V	18V	30V	30uV/C		7MA	70dB	76dB	300K
MC3401L	MTU	QCD	INT	1M5HZ	0.2V/uS	+18V		75C	60dB	.	300NA		625MWF	6MA	13V	.	.	.		10MA	.	50dB	.
MC3401P	MTU	QCD	INT	1M5HZ	0.2V/uS	+18V		75C	60dB	.	300NA		625MWF	6MA	13V	.	.	.		10MA	.	50dB	.
MC3403L	MTU	QGK	INT	.3MHZ	0.2V/uS	+18V	-18V	70C	94dB	10MV	500NA	50NA	750MWF	5MA	12V	15V	30V	50uV/C		7MA	70dB	76dB	300K
MC3403P	MTU	QGK	INT	.3MHZ	0.2V/uS	+18V	-18V	70C	94dB	10MV	500NA	50NA	625MWF	5MA	12V	15V	30V	50uV/C		7MA	70dB	76dB	300K
MC3430L	MTU	QCP	EXT			+7V	-7V	70C	54dB	10MV	20uA	5uA	1WF			5V	6V				.	.	.
MC3430P	MTU	QCP	EXT			+7V	-7V	70C	54dB	10MV	20uA	5uA	1WF			5V	6V				.	.	.
MC3431L	MTU	QCP	EXT			+7V	-7V	70C	54dB	10MV	20uA	5uA	1WF			5V	6V				.	.	.
MC3431P	MTU	QCP	EXT			+7V	-7V	70C	54dB	10MV	20uA	5uA	1WF			5V	6V				.	.	.
MC3432L	MTU	QCP	EXT			+7V	-7V	70C	54dB	10MV	20uA	5uA	1WF			5V	6V				.	.	.
MC3432P	MTU	QCP	EXT			+7V	-7V	70C	54dB	10MV	20uA	5uA	1WF			5V	6V				.	.	.
MC3433L	MTU	QCP	EXT			+7V	-7V	70C	54dB	10MV	20uA	5uA	1WF			5V	6V				.	.	.
MC3433P	MTU	QCP	EXT			+7V	-7V	70C	54dB	10MV	20uA	5uA	1WF			5V	6V				.	.	.
MC3476G	MTU	PRA	INT	.	0.3V/uS	+18V	-18V	70C	94dB	6MV	50NA	25NA	680MWF	1MA	12V	18V	30V		6MW	.2MA	70dB	74dB	2M
MC3476P1	MTU	PRA	INT	.	0.3V/uS	+18V	-18V	70C	94dB	6MV	50NA	25NA	625MWF	1MA	12V	18V	30V		6MW	.2MA	70dB	74dB	2M
MC3503L	MTU	QGK	INT	.3MHZ	0.2V/uS	+18V	-18V	125C	88dB	5MV	500NA	50NA	750MWF	5MA	12V	15V	30V	40uV/C		4MA	70dB	76dB	300K
MC3503P	MTU	QGK	INT	.3MHZ	0.2V/uS	+18V	-18V	125C	88dB	5MV	500NA	50NA	625MWF	5MA	12V	15V	30V	40uV/C		4MA	70dB	86dB	300K
MC4558CG	MTU	DWB	INT	1MHZ	0.5V/uS	+18V	-18V	70C	86dB	6MV	500NA	200NA	500MWF	5MA	12V	15V	30V		170MW		70dB	76dB	300K
MC4558CP2	MTU	DWB	INT	1MHZ	0.5V/uS	+18V	-18V	70C	86dB	6MV	500NA	200NA	500MWF	5MA	12V	15V	30V		170MW		70dB	76dB	300K
MC4558CU	MTU	DWB	INT	1MHZ	0.5V/uS	+18V	-18V	70C	86dB	6MV	500NA	200NA	500MWF	5MA	12V	15V	30V		170MW		70dB	76dB	300K
MC4558G	MTU	DWB	INT	1MHZ	0.5V/uS	+22V	-22V	125C	94dB	5MV	500NA	200NA	500MWF	5MA	12V	15V	30V		170MW		70dB	76dB	300K
MC4558L	MTU	DWB	INT	1MHZ	0.5V/uS	+22V	-22V	125C	94dB	5MV	500NA	200NA	500MWF	5MA	12V	15V	30V		170MW		70dB	76dB	300K
MC4558U	MTU	DWB	INT	1MHZ	0.5V/uS	+22V	-22V	125C	86dB	5MV	500NA	200NA	500MWF	5MA	12V	15V	30V		170MW		70dB	76dB	300K
MC4741CL	MTU	QGK	INT	.3MHZ	0.2V/uS	+18V	-18V	70C	86dB	6MV	500NA	200NA	900MWF	5MA	12V	18V	36V	50uV/C		3MA	70dB	76dB	800K
MC4741CP	MTU	QGK	INT	.3MHZ	0.2V/uS	+18V	-18V	70C	86dB	6MV	500NA	200NA	900MWF	5MA	12V	18V	36V	50uV/C		3MA	70dB	76dB	800K
MC4741L	MTU	QGK	INT	.3MHZ	0.2V/uS	+22V	-22V	125C	94dB	5MV	500NA	200NA	900MWF	5MA	12V	22V	44V	50uV/C		3MA	70dB	76dB	800K
MC4741P	MTU	QGK	INT	.3MHZ	0.2V/uS	+22V	-22V	125C	94dB	5MV	500NA	200NA	900MWF	5MA	12V	22V	44V	50uV/C		3MA	70dB	76dB	800K
MCBC1709	MTU	GPU	EXT	.3MHZ	.15V/uS	+18V	-18V	125C	88dB	5MV	500NA	200NA		5MA	12V	10V	5V	15uV/C	165MW		70dB	76dB	150K
MCBC1710	MTU	CPR	EXT	.	.	+14V	-7V	125C	62dB	2MV	20uA	3uA		5MA	1V	7V	5V	10uV/C	150MW	9MA	80dB	.	.
MCBC1741	MTU	GPK	INT	.	0.3V/uS	+22V	-22V	125C	94dB	5MV	500NA	200NA		5MA	12V	15V	30V	.	75MW	3MA	70dB	76dB	300K
MCBC1748	MTU	GPU	EXT	.	.25V/uS	+22V	-22V	125C	94dB	5MV	500NA	200NA		5MA	12V	15V	30V		85MW	3MA	70dB	76dB	300K
MCC1410	MTU	BDO	INT	.4GHZ	.	+8V	-8V	75C	36dB	33MV	100uA	30uA		.	2V	6V	5V	10uV/C	220MW		20dB	.	3K
MCC1414	MTU	DCP	EXT	.	.	+14V	-7V	75C	60dB	5MV	25uA	5uA		5MA	1V	7V	5V	25uV/C	150MW	9MA	70dB	.	.
MCC1420	MTU	BDO	EXT	3MHZ	2V/uS	+8V	-8V	75C	64dB	15MV	4uA	0.2uA		.	6V	0.5V	8V	10uV/C	240MW		60dB	.	300K
MCC1430	MTU	GPU	EXT	.	0.3V/uS	+8V	-8V	75C	69dB	10MV	15uA	4uA		4MA	4V	2V	5V		150MW		65dB	.	5K
MCC1433	MTU	GPU	EXT	.	0.2V/uS	+18V	-18V	75C	90dB	7.5MV	2uA	0.5uA		5MA	12V	8V	10V	30uV/C	170MW		80dB	76dB	300K
MCC1435	MTU	DGU	EXT	.	0.2V/uS	+9V	-9V	75C	71dB	5MV	5uA	0.5uA		.5MA	5V	9V	5V	15uV/C	180MW		60dB	70dB	10K
MCC1436	MTU	HVO	INT	.3MHZ	0.5V/uS	+34V	-34V	75C	97dB	10MV	40NA	10NA		1MA	20V	34V	68V	.	280MW	5MA	70dB	74dB	3M
MCC1437	MTU	DGU	EXT	.	0.1V/uS	+18V	-18V	75C	84dB	7.5MV	1.5uA	0.5uA		5MA	12V	18V	5V	10uV/C	225MW		65dB	74dB	50K
MCC1439	MTU	GPU	EXT	.	0.5V/uS	+18V	-18V	75C	84dB	7.5MV	1uA	.15uA		10MA	10V	18V	36V	15uV/C		7MA	80dB	74dB	50K
MCC1456	MTJ	SBA	INT	.5MHZ	1V/uS	+18V	-18V	75C	96dB	10MV	30NA	10NA		5MA	11V	18V	18V	40uV/C	90MW	3MA	70dB	74dB	1M
MCC1458	MTU	DGK	INT	.5MHZ	0.5V/uS	+18V	-18V	75C	86dB	6MV	0.5uA	0.2uA		5MA	12V	15V	30V	50uV/C	170MW	6MA	70dB	76dB	300K
MCC1458S	MTU	DHS	INT	.5MHZ	3V/uS	+18V	-18V	75C	86dB	6MV	0.5uA	0.2uA		5MA	12V	15V	30V	50uV/C	170MW	6MA	70dB	76dB	300K
MCC1510	MTU	BDO	INT	0.4HZ	:	+8V	-8V	125C	38dB	17MV	80uA	20uA		.	2V	6V	5V	10uV/C	220MW		30dB	.	3K
MCC1514	MTU	DCP	EXT	.	.	+14V	-7V	125C	62dB	2MV	20uA	3uA		5MA	1V	7V	5V	10uV/C	150MW	9MA	80dB	.	.
MCC1520	MTU	BDO	EXT	3MHZ	2V/uS	+8V	-8V	125C	66dB	10MV	2uA	0.1uA		.	7V	0.5V	8V	10uV/C	240MW		75dB	66dB	500K
MCC1530	MTU	GPU	EXT	.	0.3V/uS	+9V	-9V	125C	73dB	5MV	10uA	2uA		5MA	4.5V	2V	5V	10uV/C	15MW		70dB	74dB	10K
MCC1533	MTU	GPU	EXT	.	0.2V/uS	+20V	-20V	125C	92dB	5MV	1uA	.15uA		6MA	12V	20V	10V	20uV/C	170MW		90dB	76dB	500K
MCC1535	MTU	DGU	EXT	.	0.2V/uS	+10V	-10V	125C	72dB	3MV	3uA	0.3uA		.3MA	3V	10V	5V	15uV/C	150MW		70dB	70dB	10K
MCC1536	MTU	HVO	INT	.3MHZ	0.5V/uS	+40V	-40V	125C	100dB	5M	20NA	3NA		1MA	30V	40V	80V	.	224MW	4MA	80dB	80dB	3M
MCC1537	MTU	DGU	EXT	.	0.1V/uS	+18V	-18V	125C	88dB	5MV	0.5uA	0.2uA		5MA	12V	18V	5V	10uV/C	225MW		70dB	76dB	150K
MCC1539	MTU	GPU	EXT	.	0.4V/uS	+18V	-18V	125C	94dB	3MV	500NA	75NA		10MA	10V	18V	36V	10uV/C		5MA	80dB	76dB	150K
MCC1556	MTU	SBA	INT	.5MHZ	1V/uS	+22V	-22V	125C	100dB	4MV	15NA	2NA		6MA	12V	22V	22V	30uV/C	45MW	2MA	80dB	80dB	1.5M
MCC1558	MTU	DGK	INT	.5MHZ	0.3V/uS	+22V	-22V	125C	94dB	5MV	0.5uA	0.2uA		5MA	12V	15V	30V	50uV/C	150MW	5MA	70dB	76dB	300K
MCC1558S	MTU	DHS	INT	.5MHZ	3V/uS	+22V	-22V	125C	94dB	5MV	0.5uA	0.2uA		5MA	12V	15V	30V	50uV/C	150MW	5MA	70dB	76dB	300K
MCC1709	MTU	GPU	EXT	.3MHZ	.15V/uS	+18V	-18V	125C	88dB	5MV	500NA	200NA		5MA	12V	10V	5V	15uV/C	165MW		70dB	76dB	150K
MCC1709A	MTU	GPU	EXT	.3MHZ	.15V/uS	+18V	-18V	125C	88dB	2MV	200NA	50NA		5MA	12V	10V	5V	10uV/C	108MW	4MA	80dB	80dB	350K
MCC1709C	MTU	GPU	EXT	.3MHZ	.15V/uS	+18V	-18V	75C	82dB	7.5MV	1.5uA	0.5uA		5MA	12V	10V	5V	.	200MW		65dB	74dB	50K

For detailed explanations of column heading notations, see App. A.

Also, for ready references the more important abbreviations used in the column headings are listed below:

LEFT HAND PAGE

APP = application (codes at APP.E.)

CMRR = common mode rejection ratio

CMP = compensation (frequency)

dV_{IO}/dT = input offset voltage temperature drift

GBP = gain bandwidth product

I_e = input bias current

I_{IO} = input bias offset current

I_0 = quiescent supply current

MFR = manufacturer (codes at App.C.)

P_0 = quiescent power consumer

PSRR = power supply rejection ratio

V_{ICM} = common mode input voltage rating

V_{IDR} = differential input voltage rating

V_{in} = input offset voltage

V_S = dc supply voltage

RIGHT HAND PAGE

Lead out coding summary (details at APP.G.) for different cases (APP.F.)

A = gain adjust
B = bias adjust
C = case
E— = inverting input
E+ = non-inverting input
F F* = input frequency compensation
G = ground
J = high level input
K = output, open collector
L = output, open emitter
M = metal case
N = not connected
Q = special terminal
R,R* = outputs
S = strobe
T,T* = offset balance
V+ = +ve dc supply
V = —ve dc supply
W = guard ring
X = blank position, no lead
+ + = +ve supplementary dc supply
— — = —ve supplementary dc supply
∅,∅* = output frequency compensation

CASE (APP F)	LD 1	LD 2	LD 3	LD 4	LD 5	LD 6	LD 7	LD 8	LD 9	LD 10	LD 11	LD 12	LD 13	LD 14	LD 15	LD 16	EUROPE SUBSTITUTE	USA SUBSTITUTE	IS	TYPE NUMBER
DIL-14/1C	R1	E-1	E+1	V+	E+2	E-2	R2	R3	E-3	E+3	G	E+4	E-4	R4	.	.	RV3403ADC	UA3303P	O	MC3303L
DIL-14/1C	R1	E-1	E+1	V+	E+2	E-2	R2	R3	E-3	E+3	G	E+4	E-4	R4	.	.	RV3403ADC	UA3303P	O	MC3303P
DIL-14/1C	E+1	E+2	E-2	R2	R1	E-1	G	E-3	R3	R4	E-4	E+4	E+3	V+	.	.	UA3401P	LM3401N	O	MC3401L
DIL-14/1P	E+1	E+2	E-2	R2	R1	E-1	G	E-3	R3	R4	E-4	E+4	E+3	V+	.	.	UA3401P	LM3401N	O	MC3401P
DIL-14/1C	R1	E-1	E+1	V+	E+2	E-2	R2	R3	E-3	E+3	G	E+4	E-4	R4	.	.		UA3403D	O	MC3403L
DIL-14/1P	R1	E-1	E+1	V+	E+2	E-2	R2	R3	E-3	E+3	G	E+4	E-4	R4	.	.		UA3403D	O	MC3403P
DIL-16/1C	E-1	E+1	R1	S	R3	E+3	E-3	G	E-4	E+4	R4	V-	R2	E+2	E-2	V+		MC3430P	O	MC3430L
DIL-16/1P	E-1	E+1	R1	S	R3	E+3	E-3	G	E-4	E+4	R4	V-	R2	E+2	E-2	V+		MC3430L	O	MC3430P
DIL-16/1C	E-1	E+1	R1	S	R3	E+3	E-3	G	E-4	E+4	R4	V-	R2	E+2	E-2	V+		MC3431P	O	MC3431L
DIL-16/1P	E-1	E+1	R1	S	R3	E+3	E-3	G	E-4	E+4	R4	V-	R2	E+2	E-2	V+		MC3431L	O	MC3431P
DIL-16/1C	E-1	E+1	K1	S	K3	E+3	E-3	G	E-4	E+4	K4	V-	K2	E+2	E-2	V+		MC3430L	O	MC3432L
DIL-16/1P	E-1	E+1	K1	S	K3	E+3	E-3	G	E-4	E+4	K4	V-	K2	E+2	E-2	V+		MC3430P	O	MC3432P
DIL-16/1C	E-1	E+1	K1	S	K3	E+3	E-3	G	E-4	E+4	K4	V-	K2	E+2	E-2	V+		MC3431L	O	MC3433L
DIL-16/1P	E-1	E+1	K1	S	K3	E+3	E-3	G	E-4	E+4	K4	V-	K2	E+2	E-2	V+		MC3431P	O	MC3433P
TO5-8/1M	T	E-	E+	V-	T*	R	V+	B	.	.	.	.	.	.	.	.	MC1776G	UA776HM	O	MC3476G
DIL-8/1P	T	E-	E+	V-	T*	R	V+	B	.	.	.	.	.	.	.	.		UA776TC	O	MC3476P1
DIL-14/1C	R1	E-1	E+1	V-	E+2	E-2	R2	R3	E-3	E+3	G	E+4	E-4	R4	.	.		UA3503D	O	MC3503L
DIL-14/1P	R1	E-1	E+1	V+	E+2	E-2	R2	R3	E-3	E+3	G	E+4	E-4	R4	.	.		UA3503D	O	MC3503P
TO5-8/1M	R1	E-1	E+1	V-	E+2	E-2	R2	V+	.	.	.	.	.	.	.	.		RC4558T	O	MC4558CG
DIL-14/1P	N	R1	N	N	E-1	E+1	V-	E+2	E-2	N	N	R2	N	V+	.	.		MC4558L	O	MC4558CP2
DIL-8/1C	R1	E-1	E+1	CU	E+2	E-2	R2	V+	.	.	.	.	.	.	.	.		RC4558NB	O	MC4558CU
TO5-8/1M	R1	E-1	E+1	V-	E+2	E-2	R2	V+	.	.	.	.	.	.	.	.		RM4558T	O	MC4558G
DIL-14/1C	N	R1	N	N	E-1	E+1	V-	E+2	E-2	N	N	R2	N	V+	.	.			O	MC4558L
DIL-8/1C	R1	E-1	E+1	V-	E+2	E-2	R2	V+	.	.	.	.	.	.	.	.		RV4558NB	O	MC4558U
DIL-14/1C	R1	E-1	E+1	V+	E+2	E-2	V	E-3	E+3	R3	V+	R4	E+4	E-4	.	.	LM348D	HA4741-5	O	MC4741CL
DIL-14/1P	R1	E-1	E+1	V+	E+2	E-2	V-	E-3	E+3	R3	V+	R4	E+4	E-4	.	.	LM348D	HA4741-5	O	MC4741CP
DIL-14/1C	R1	E-1	E+1	V+	E+2	E-2	V-	E-3	E+3	R3	V+	R4	E+4	E-4	.	.	LM148D	HA4741-2	O	MC4741L
DIL-14/1P	R1	E-1	E+1	V+	E+2	E-2	V-	E-3	E+3	R3	V+	R4	E+4	E-4	.	.	LM148D	HA4741-2	O	MC4741P
BML	.	.	.	.	.	.	.	.	.	.	.	.	.	.	.	.		.	O	MCBC1709
BML	.	.	.	.	.	.	.	.	.	.	.	.	.	.	.	.		.	O	MCBC1710
BML	.	.	.	.	.	.	.	.	.	.	.	.	.	.	.	.		.	O	MCBC1741
BML	.	.	.	.	.	.	.	.	.	.	.	.	.	.	.	.		.	O	MCBC1748
CHP	.	.	.	.	.	.	.	.	.	.	.	.	.	.	.	.		.	O	MCC1410
CHP	.	.	.	.	.	.	.	.	.	.	.	.	.	.	.	.		.	O	MCC1414
CHP	.	.	.	.	.	.	.	.	.	.	.	.	.	.	.	.		.	O	MCC1420
CHP	.	.	.	.	.	.	.	.	.	.	.	.	.	.	.	.		.	O	MCC1430
CHP	.	.	.	.	.	.	.	.	.	.	.	.	.	.	.	.		.	O	MCC1433
CHP	.	.	.	.	.	.	.	.	.	.	.	.	.	.	.	.		.	O	MCC1435
CHP	.	.	.	.	.	.	.	.	.	.	.	.	.	.	.	.		.	O	MCC1436
CHP	.	.	.	.	.	.	.	.	.	.	.	.	.	.	.	.		.	O	MCC1437
CHP	.	.	.	.	.	.	.	.	.	.	.	.	.	.	.	.		.	O	MCC1439
CHP	.	.	.	.	.	.	.	.	.	.	.	.	.	.	.	.		.	O	MCC1456
CHP	.	.	.	.	.	.	.	.	.	.	.	.	.	.	.	.		.	O	MCC1458
CHP	.	.	.	.	.	.	.	.	.	.	.	.	.	.	.	.		.	O	MCC1458S
CHP	.	.	.	.	.	.	.	.	.	.	.	.	.	.	.	.		.	O	MCC1510
CHP	.	.	.	.	.	.	.	.	.	.	.	.	.	.	.	.		.	O	MCC1514
CHP	.	.	.	.	.	.	.	.	.	.	.	.	.	.	.	.		.	O	MCC1520
CHP	.	.	.	.	.	.	.	.	.	.	.	.	.	.	.	.		.	O	MCC1530
CHP	.	.	.	.	.	.	.	.	.	.	.	.	.	.	.	.		.	O	MCC1533
CHP	.	.	.	.	.	.	.	.	.	.	.	.	.	.	.	.		.	O	MCC1535
CHP	.	.	.	.	.	.	.	.	.	.	.	.	.	.	.	.		.	O	MCC1536
CHP	.	.	.	.	.	.	.	.	.	.	.	.	.	.	.	.		.	O	MCC1537
CHP	.	.	.	.	.	.	.	.	.	.	.	.	.	.	.	.		.	O	MCC1539
CHP	.	.	.	.	.	.	.	.	.	.	.	.	.	.	.	.		.	O	MCC1556
CHP	.	.	.	.	.	.	.	.	.	.	.	.	.	.	.	.		.	O	MCC1558
CHP	.	.	.	.	.	.	.	.	.	.	.	.	.	.	.	.		.	O	MCC1558S
CHP	.	.	.	.	.	.	.	.	.	.	.	.	.	.	.	.		.	O	MCC1709
CHP	.	.	.	.	.	.	.	.	.	.	.	.	.	.	.	.		.	O	MCC1709A
CHP	.	.	.	.	.	.	.	.	.	.	.	.	.	.	.	.		.	O	MCC1709C
CFL	.	.	.	.	.	.	.	.	.	.	.	.	.	.	.	.		.	O	MCC1710

TYPE NUMBER	M F R	A P P	C M P	GBP MIN	SLEW RATE MIN	Vs+ MAX	Vs− MAX	T$_{OP}$ MAX	A$_{VOL}$ MIN	V$_{IO}$ MAX	I$_B$ MAX	I$_{IO}$ MAX	P$_{TOT}$ MAX	I$_{OUT}$ MIN	V$_{OUT}$ MIN	V$_{ICM}$ MAX	V$_{IDF}$ MAX	dV$_{IO}$/dT MAX	P$_Q$ MAX	I$_Q$ MAX	CM RR MIN	PS RR MIN	R$_{IN}$ MIN
MCC1710	MTU	CPR	EXT	.	.	+14V	−7V	125C	62dB	2MV	20uA	3uA	.	5MA	1V	7V	5V	10uV/C	150MW	9MA	80dB	.	.
MCC1710C	MTU	CPR	EXT	.	.	+14V	−7V	75C	60dB	5MV	25uA	5uA	.	5MA	1V	7V	5V	20uV/C	150MW	9MA	70dB	.	.
MCC1711	MTU	DCP	EXT	.	.	+14V	−7V	125C	58dB	3.5MV	75uA	10uA	.	5MA	2.5V	7V	5V	20uV/C	200MW	.	.	.	.
MCC1711C	MTU	DCP	EXT	.	.	+14V	−7V	75C	57dB	5MV	100uA	25uA	.	5MA	2.5V	7V	5V	20uV/C	200MW	.	.	.	.
MCC1712	MTU	GPU	EXT	3MHZ	.	+13V	−8V	125C	68dB	2MV	5uA	0.5uA	.	.3MA	5V	1.5V	5V	10uV/C	120MW	7MA	80dB	74dB	16K
MCC1712C	MTU	GPU	EXT	3MHZ	.	+13V	−8V	75C	66dB	5MV	7.5uA	2uA	.	.3MA	5V	1.5V	5V	20uV/C	120MW	7MA	70dB	70dB	10K
MCC1733	MTU	BDO	INT	40MHZ	.	+8V	−8V	125C	50dB	5MV	20uA	3uA	.	2MA	2V	6V	5V	.	.	24MA	60dB	50dB	2K
MCC1733C	MTU	BDO	INT	40MHZ	.	+8V	−8V	70C	40dB	6MV	30uA	5uA	.	2MA	2V	6V	5V	.	.	24MA	60dB	50dB	2K
MCC1741	MTU	GPK	INT	.	0.3V/uS	+22V	−22V	125C	94dB	5MV	500NA	200NA	.	5MA	12V	15V	30V	.	75MW	3MA	70dB	76dB	300K
MCC1741C	MTU	GPK	INT	.	0.2V/uS	+18V	−18V	70C	86dB	6MV	500NA	200NA	.	5MA	12V	15V	30V	.	85MW	3MA	70dB	76dB	300K
MCC1741S	MTU	HSR	INT	.	3V/uS	+22V	−22V	125C	94dB	5MV	500NA	200NA	.	5MA	12V	15V	30V	.	75MW	3MA	70dB	76dB	300K
MCC1741SC	MTU	HSR	INT	.	3V/uS	+18V	−18V	70C	86dB	6MV	500NA	200NA	.	5MA	12V	15V	30V	.	85MW	3MA	70dB	76dB	300K
MCC1747	MTU	DGK	INT	.	0.2V/uS	+22V	−22V	125C	94dB	5MV	500NA	200NA	.	5MA	12V	15V	30V	.	85MW	3MA	70dB	76dB	300K
MCC1747C	MTU	DGU	INT	.	0.2V/uS	+18V	−18V	75C	88dB	6MV	500NA	200NA	.	5MA	12V	15V	30V	.	85MW	3MA	70dB	76dB	300K
MCC1748	MTU	GPU	EXT	.	.25V/uS	+22V	−22V	125C	94dB	5MV	500NA	200NA	.	5MA	12V	15V	30V	.	85MW	3MA	70dB	76dB	300K
MCC1748C	MTU	GPU	EXT	.	.25V/uS	+22V	−22V	75C	94dB	5MV	500NA	200NA	.	5MA	12V	15V	30V	.	85MW	3MA	70d3	76dB	300K
MCC1776	MTU	PRA	INT	.	0.3V/uS	+18V	−18V	125C	100dB	5MV	50NA	15NA	.	2MA	10V	15V	30V	.	6MW	.2UA	70dB	76dB	2M
MCC1776C	MTU	PRA	INT	.	0.3V/uS	+18V	−18V	70C	94dB	6MV	50NA	25NA	.	2MA	10V	15V	30V	.	6MW	.2UA	70dB	74dB	2M
MCC3301	MTU	QCD	INT	1M5HZ	0.2V/uS	+28V	.	85C	60dB	.	300NA	.	.	6MA	13V	.	.	.	.	10MA	.	50dB	.
MCC3302	MTU	QCP	EXT	.	20V/uS	+14V	−14V	85C	66dB	20MV	500NA	100NA	.	2MA	.	9V	28V	.	.	2MA	.	.	.
MCC3303	MTU	QGK	INT	.3MHZ	0.2V/uS	+18V	−18V	85C	86dB	8MV	500NA	75NA	.	10MA	12V	18V	30V	30uV/C	.	7MA	70dB	76dB	300K
MCC3401	MTU	QCD	INT	1M5HZ	0.2V/uS	+18V	.	75C	60dB	.	300NA	.	.	6MA	13V	.	.	.	.	10MA	.	50dB	.
MCC3403	MTU	QGK	INT	.3MHZ	0.2V/uS	+18V	−18V	70C	94dB	10MV	500NA	50NA	.	5MA	12V	15V	30V	50uV/C	.	7MA	70dB	76dB	300K
MCC3430	MTU	QCP	EXT	.	.	+7V	−7V	70C	54dB	10MV	20uA	5uA	.	.	.	5V	6V	.	.	.	.	.	.
MCC3431	MTU	QCP	EXT	.	.	+7V	−7V	70C	54dB	10MV	20uA	5uA	.	.	.	5V	6V	.	.	.	.	.	.
MCC3432	MTU	QCP	EXT	.	.	+7V	−7V	70C	54dB	10MV	20uA	5uA	.	.	.	5V	6V	.	.	.	.	.	.
MCC3433	MTU	QCP	EXT	.	.	+7V	−7V	70C	54dB	10MV	20uA	5uA	.	.	.	5V	6V	.	.	.	.	.	.
MCC3476	MTU	PRA	INT	.	0.3V/uS	+18V	−18V	70C	94dB	6MV	50NA	25NA	.	1MA	12V	18V	30V	.	6MW	.2MA	70dB	74dB	2M
MCC3503	MTU	QGK	INT	.3MHZ	0.2V/uS	+18V	−18V	125C	88dB	5MV	500NA	50NA	.	5MA	12V	15V	30V	40uV/C	.	4MA	70dB	76dB	300K
MCC4741	MTU	QGK	INT	.3MHZ	0.2V/uS	+22V	−22V	125C	94dB	5MV	500NA	200NA	.	5MA	12V	22V	44V	50uV/C	.	3MA	70dB	76dB	800K
MCC4741C	MTU	QGK	INT	.3MHZ	0.2V/uS	+18V	−18V	70C	86dB	6MV	500NA	200NA	.	5MA	12V	15V	36V	50uV/C	.	3MA	70dB	76dB	800K
MCCF1458	MTU	DGK	INT	.5MHZ	0.3V/uS	+18V	−18V	75C	86dB	6MV	0.5uA	0.2uA	.	5MA	12V	15V	30V	50uV/C	170MW	6MA	70dB	76dB	300K
MCCF1558	MTU	DHS	INT	.5MHZ	0.3V/uS	+22V	−22V	125C	94dB	5MV	0.5uA	0.2uA	.	5MA	12V	15V	30V	50uV/C	150MW	5MA	70dB	76dB	300K
MCCF1709	MTU	GPU	EXT	.3MHZ	.15V/uS	+18V	−18V	125C	88dB	5MV	500NA	200NA	.	5MA	12V	10V	5V	15uV/C	165MW	.	70dB	76dB	150K
MCCF1709C	MTU	GPU	EXT	.3MHZ	.15V/uS	+18V	−18V	75C	82dB	7.5MV	1.5uA	0.5uA	.	5MA	12V	10V	5V	.	200MW	.	65dB	74dB	50K
MCCF1741	MTU	GPK	INT	.	0.3V/uS	+22V	−22V	125C	94dB	5MV	500NA	200NA	.	5MA	12V	15V	30V	.	75MW	3MA	70dB	76dB	300K
MCCF1741C	MTU	GPK	INT	.	0.2V/uS	+18V	−18V	70C	86dB	6MV	500NA	200NA	.	5MA	12V	15V	30V	.	85MW	3MA	70dB	76dB	300K
MCCF3303	MTU	QGK	INT	.3MHZ	0.2V/uS	+18V	−18V	85C	86dB	8MV	500NA	75NA	.	10MA	12V	18V	30V	30uV/C	.	7MA	70dB	76dB	300K
MCCF3403	MTU	QGK	INT	.3MHZ	0.2V/uS	+18V	−18V	70C	94dB	10MV	500NA	50NA	.	5MA	12V	15V	30V	50uV/C	.	7MA	70dB	76dB	300K
MCCF3503	MTU	QGK	INT	.3MHZ	0.2V/uS	+18V	−18V	125C	88dB	5MV	500NA	50NA	.	5MA	12V	15V	30V	40uV/C	.	4MA	70dB	76dB	300K
MCE7042	MTU	PIA	EXT	.	4V/uS	+18V	−18V	75C	94dB	10MV	300NA	200NA	400MWF	20MA	11V	10V	30V	.	150MW	.	70dB	70dB	.
MCE7042A	MTU	PIA	EXT	.	8V/uS	+22V	−22V	75C	100dB	5MV	150NA	100NA	400MWF	20MA	11V	10V	30V	.	150MW	.	80dB	74dB	.
MIC709-1	FAU	GPU	EXT	.3MHZ	.15V/uS	+18V	−18V	125C	88dB	5MV	500NA	200NA	500MWF	5MA	12V	10V	5V	15uV/C	165MW	.	70dB	76dB	150K
MIC709-1C	OBS	GPU	EXT	.	.	+18V	−18V	125C	88dB	5MV	500NA	200NA	300MWF	20MA	14V	10V	5V	10uV/C	165MW	.	70dB	76dB	150K
MIC709-5	OBS	GPU	EXT	.3MHZ	.15V/uS	+18V	−18V	70C	84dB	7.5MV	1.5uA	500NA	500MWF	5MA	12V	10V	5V	.	200MW	.	65dB	74dB	50K
MIC709-5C	OBS	GPU	EXT	.	.	+18V	−18V	70C	84dB	7.5MV	1.5uA	500NA	250MWF	20MA	14V	10V	5V	15uV/C	200MW	.	65dB	74dB	50K
MIC709AC	OBS	GPU	EXT	.3MHZ	.15V/uS	+18V	−18V	125C	88dB	2MV	200NA	50NA	500MWF	5MA	12V	10V	5V	10uV/C	108MW	.	80dB	80dB	350K
MIC709AD	OBS	GPU	EXT	.3MHZ	.15V/uS	+18V	−18V	125C	88dB	2MV	200NA	50NA	670MWF	5MA	12V	10V	5V	10uV/C	108MW	.	80dB	80dB	350K
MIC710-1B	OBS	CPR	EXT	.	.	+14V	−7V	125C	62dB	2MV	20uA	3uA	200MWF	10MA	2.5V	5V	7V	10uV/C	150MW	9MA	80dB	.	.
MIC710-1C	OBS	CPR	EXT	.	.	+14V	−7V	125C	62dB	2MV	20uA	3uA	300MWF	10MA	2.5V	5V	7V	10uV/C	150MW	9MA	80dB	.	.
MIC710-5B	OBS	CPR	EXT	.	.	+14V	−7V	70C	60dB	5MV	25uA	5uA	200MWF	10MA	2.5V	5V	7V	20uV/C	150MW	9MA	70dB	.	.
MIC710-5C	OBS	CPR	EXT	.	.	+14V	−7V	70C	60dB	5MV	25uA	5uA	300MWF	10MA	2.5V	5V	7V	20uV/C	150MW	9MA	70dB	.	.
MIC711-1B	OBS	DCP	EXT	.	.	+14V	−7V	125C	58dB	3.5MV	10uA	10uA	200MWF	50MA	2.5V	5V	5V	10uV/C	200MW	17MA	.	.	.
MIC711-1C	OBS	DCP	EXT	.	.	+14V	−7V	125C	58dB	3.5MV	10uA	10uA	300MWF	50MA	2.5V	5V	5V	10uV/C	200MW	17MA	.	.	.
MIC711-5B	OBS	DCP	EXT	.	.	+14V	−7V	70C	57dB	5MV	15uA	15uA	200MWF	50MA	2.5V	5V	5V	10uV/C	230MW	17MA	.	.	.
MIC711-5C	OBS	DCP	EXT	.	.	+14V	−7V	70C	57dB	5MV	15uA	15uA	300MWF	50MA	2.5V	5V	5V	10uV/C	230MW	17MA	.	.	.
MIC712-1B	OBS	WBA	EXT	3MHZ	.	+13V	−8V	125C	68dB	2MV	5uA	0.5uA	570MWF	.3MA	5V	5V	5V	10uV/C	120MW	7MA	80d3	74dB	16K
MIC712-1C	OBS	WBA	EXT	3MHZ	.	+13V	−8V	125C	68dB	2MV	5uA	0.5uA	500MWF	.3MA	5V	5V	5V	10uV/C	120MW	7MA	80dB	74dB	16K
MIC712-1D	OBS	WBA	EXT	3MHZ	.	+13V	−8V	125C	68dB	2MV	5uA	0.5uA	670MWF	.3MA	5V	5V	5V	10uV/C	120MW	7MA	80dB	74dB	16K
MIC712-5B	OBS	GPU	EXT	3MHZ	.	+13V	−8V	75C	66dB	5MV	7.5uA	2uA	500MWF	.3MA	5V	1.5V	5V	20uV/C	120MW	7MA	70dB	70dB	10K

CASE (APP.F)	LD 1	LD 2	LD 3	LD 4	LD 5	LD 6	LD 7	LD 8	LD 9	LD 10	LD 11	LD 12	LD 13	LD 14	LD 15	LD 16	EUROPE SUBSTITUTE	USA SUBSTITUTE	I/S	TYPE NUMBER
CHP																			O	MCC1710C
CHP																			O	MCC1711
CHP																			O	MCC1711C
CHP																			O	MCC1712
CHP																			O	MCC1712C
CHP																			O	MCC1733
CHP																			O	MCC1733C
CHP																			O	MCC1741
CHP																			O	MCC1741C
CHP																			O	MCC1741S
CHP																			O	MCC1741SC
CHP																			O	MCC1747
CHP																			O	MCC1747C
CHP																			O	MCC1748
CHP																			O	MCC1748C
CHP																			O	MCC1776
CHP																			O	MCC1776C
CHP																			O	MCC3301
CHP																			O	MCC3302
CHP																			O	MCC3303
CHP																			O	MCC3401
CHP																			O	MCC3403
CHP																			O	MCC3430
CHP																			O	MCC3431
CHP																			O	MCC3432
CHP																			O	MCC3433
CHP																			O	MCC3476
CHP																			O	MCC3503
CHP																			O	MCC4741
CHP																			O	MCC4741C
CFL																			O	MCCF1458
CFL																			O	MCCF1558
CFL																			O	MCCF1709
CFL																			O	MCCF1709C
CFL																			O	MCCF1741
CFL																			O	MCCF1741C
CFL																			O	MCCF3303
CFL																			O	MCCF3403
CFL																			O	MCCF3503
FLP-14/3C	N	N	N	E-	E+	V-	N	T	T*	R	V+	F	F*	N					O	MCE7042
FLP-14/3C	N	N	N	E-	E+	V-	N	T	T*	R	V+	F	F*	N					O	MCE7042A
T05-8/1M	F	E-	E+	V-	∅	∅*R	V+	F*									TAA522	UA709HM	O	MIC709-1
T05-8/1M	F	E-	E+	V-M	∅	∅*R	V+	F*									TAA522	UA709HM	O	MIC709-1C
T05-8/1M	F	E-	E+	V-	∅	∅*R	V+	F*									TAA521	UA709HC	O	MIC709-5
T05-8/1M	F	E-	E+	V-M	∅	∅*R	V+	F*									TAA521	UA709HC	O	MIC709-5C
T05-8/1M	F	E-	E+	V-	∅	∅*R	V+	F*									TAA522	UA709AHM	O	MIC709AC
DIL-14/1C	N	N	F	E-	E+	V-	N	N	∅	R	V+	F*	N	N			LM709AJ	UA709ADM	O	MIC709AD
FLP-10/3C	G	E+	E-	N	V-	R	N	V+	N	N							SFC2710PM	UA710FM	O	MIC710-1B
T05-8/1M	G	E+	E-	V-M	N	N	R	V+									SFC2710M	UA710HM	O	MIC710-1C
FLP-10/3C	G	E+	E-	N	V-	R	N	V+	N	N							SFC2710PM	UA710FM	O	MIC710-5B
T05-8/1M	G	E+	E-	V-M	N	N	R	V+									SFC2710C	UA710HC	O	MIC710-5C
FLP-10/3C	E-1	E+1	V-	E+2	E-2	S1	R	V+	G	S2							SFC2711PM	UA711FM	O	MIC711-1B
T05-10/1M	G	S1	E-1	E+1	V-M	E+2	E-2	S2	R	V+							SFC2711M	UA711HM	O	MIC711-1C
FLP-10/3C	E-1	E+1	V-	E+2	E-2	S1	R	V+	G	S2							SFC2711PM	UA711FM	D	MIC711-5B
T05-10/1M	G	S1	E-1	E+1	V-M	E+2	E-2	S2	R	V+							SFC2711C	UA711HC	D	MIC711-5C
FLP-10/3C	N	G	E-	E+	V-	F	∅	R	N	V+							SN52702AFA	UA702FM	D	MIC712-1B
T05-8/1M	G	E-	E+	V-M	F	∅	R	V+									SN52702AL	UA702HM	D	MIC712-1C
DIL-14/1C	N	N	G	E-	E+	V-	N	N	F	∅	R	N	V+	N			SN52702AJ	UA702DM	D	MIC712-1D
FLP-10/3C	N	G	E-	E+	V-	F	∅	R	N	V+							UA702FM	MC1712CF	D	MIC712-5B
T05-8/1M	G	E-	E+	V-M	F	∅	R	V+									SN72702L	UA702HC	D	MIC712-5C

TYPE NUMBER	M F R	A P P	C M P	GBP MIN	SLEW RATE MIN	V_S^+ MAX	V_S^- MAX	T_{op} MAX	A_{VOL} MIN	V_{IO} MAX	I_B MAX	I_{IO} MAX	P_{TOT} MAX	I_{OUT} MIN	V_{OUT} MIN	V_{ICM} MAX	V_{IDF} MAX	dV_{IO}/dT MAX	P_Q MAX	I_Q MAX	CM RR MIN	PS RR MIN	R_{IN} MIN
MIC712-5C	OBS	WBA	EXT	3MHZ	.	+13V	-8V	70C	66dB	5MV	7.5uA	2uA	500MWF	.3MA	5V	5V	5V	20uV/C	120MW	7MA	70dB	70dB	10K
MIC712-5D	OBS	WBA	EXT	3MHZ	.	+13V	-8V	70C	66dB	5MV	7.5uA	2uA	670MWF	.3MA	5V	5V	5V	20uV/C	120MW	7MA	70dB	70dB	10K
MIC730-1C	OBS	BDO	EXT	1MHZ	.	+15V	.	125C	40dB	2.5MV	7.5uA	1.5uA	500MWF	.	2V	4V	5V	.	156MW	13MA	70dB	.	5K
MIC730-5C	FAU	BDO	EXT	1MHZ	.	+15V	.	70C	40dB	5MV	16uA	3uA	500MWF	.	2V	4V	5V	.	156MW	13MA	60dB	.	2.5K
MIC741-1D	OBS	GPK	INT	4MHZ	0.3V/uS	+22V	-22V	125C	94dB	3MV	80NA	30NA	670MWF	5MA	12V	15V	30V	15uV/C	150MW	.	80dB	86dB	1M
MIC741-1C	OBS	GPK	INT	.	0.2V/uS	+22V	-22V	125C	94dB	5MV	500NA	200NA	500MWF	7MA	12V	15V	30V	.	85MW	3MA	70dB	76dB	300K
MIC741-5C	OBS	GPK	INT	.	0.2V/uS	+18V	-18V	70C	86dB	6MV	500NA	200NA	500MWF	5MA	12V	15V	30V	.	.	3MA	70dB	76dB	300K
MIC741-5U	FAU	GPK	INT	.	0.2V/uS	+18V	-18V	70C	86dB	6MV	500NA	200NA	670MWF	5MA	12V	15V	30V	.	85MW	3MA	70dB	76dB	300K
ML101AF	OBS	GPU	EXT	.	.	+22V	-22V	125C	94dB	2MV	75NA	10NA	500MWF	5MA	12V	15V	30V	15uV/C	.	3MA	80dB	80dB	1.5M
ML101AM	OBS	GPU	EXT	.	.	+22V	-22V	125C	94dB	2MV	75NA	10NA	500MWF	5MA	12V	15V	30V	15uV/C	.	3MA	80dB	80dB	1.5M
ML101AT	OBS	GPU	EXT	.	.	+22V	-22V	125C	94dB	2MV	75NA	10NA	500MWF	5MA	.	15V	30V	15uV/C	.	3MA	80dB	80dB	1.5M
ML101F	OBS	GPU	EXT	.	.	+22V	-22V	125C	94dB	5MV	1.5uA	0.5uA	500MWF	5MA	12V	15V	30V	15uV/C	.	3MA	70dB	70dB	300K
ML101M	OBS	GPU	EXT	.	.	+22V	-22V	125C	94dB	5MV	1.5uA	0.5uA	500MWF	5MA	12V	15V	30V	15uV/C	.	3MA	70dB	70dB	300K
ML101T	OBS	GPU	EXT	.	.	+22V	-22V	125C	94dB	5MV	500NA	200NA	500MWF	5MA	12V	15V	30V	15uV/C	.	3MA	70dB	70dB	300K
ML107F	OBS	GPK	INT	.	.	+22V	-22V	125C	94dB	2MV	75NA	10NA	500MWF	5MA	12V	15V	30V	15uV/C	.	3MA	80dB	80dB	1.5M
ML107M	OBS	GPK	INT	.	.	+22V	-22V	125C	94dB	2MV	75NA	10NA	500MWF	5MA	12V	15V	30V	15uV/C	.	3MA	80dB	80dB	1.5M
ML107T	OBS	GPK	INT	.	.	+22V	-22V	125C	94dB	2MV	75NA	10NA	500MWF	5MA	12V	15V	30V	15uV/C	.	3MA	80dB	80dB	1.5M
ML108AF	OBS	SBA	EXT	.	.	+20V	-20V	125C	98dB	0.5MV	2NA	0.2NA	500MWF	1MA	13V	15V	1V	5uV/C	.	.6MA	96dB	96dB	30M
ML108AM	OBS	SBA	EXT	.	.	+20V	-20V	125C	98dB	0.5MV	2NA	0.2NA	500MWF	1MA	13V	15V	1V	5uV/C	.	.6MA	96dB	96dB	30M
ML108AT	OBS	SBA	EXT	.	.	+20V	-20V	125C	98dB	0.5MV	2NA	0.2NA	500MWF	1MA	13V	15V	1V	5uV/C	.	.6MA	96dB	96dB	30M
ML108M	OBS	SBA	EXT	.	.	+20V	-20V	125C	96dB	2MV	2NA	0.2NA	500MWF	1MA	13V	15V	1V	15uV/C	.	.6MA	85dB	80dB	30M
ML108T	OBS	SBA	EXT	.	.	+20V	-20V	125C	96dB	2MV	2NA	0.2NA	500MWF	1MA	13V	15V	1V	15uV/C	.	.6MA	85dB	80dB	30M
ML111F	OBS	CPR	EXT	.	.	+18V	-18V	125C	100dB	3MV	100NA	10NA	500MWF	.	.	15V	30V	.	.	6MA	.	.	.
ML111M	OBS	CPR	EXT	.	.	+18V	-18V	125C	100dB	3MV	100NA	10NA	500MWF	.	.	15V	30V	.	.	6MA	.	.	.
ML111S	OBS	CPR	EXT	.	.	+18V	-18V	125C	100dB	3MV	100NA	10NA	500MWF	8MA	.	15V	30V	.	.	6MA	.	.	.
ML111T	OBS	CPR	EXT	.	.	+18V	-18V	125C	100dB	3MV	100NA	10NA	500MWF	.	.	15V	30V	.	.	6MA	.	.	.
ML118F	OBS	XSR	INT	.	50V/uS	+20V	-20V	125C	94dB	4MV	250NA	50NA	500MWF	6MA	12V	15V	1V	.	.	8MA	80dB	70dB	1M
ML118M	OBS	XSR	INT	.	50V/uS	+20V	-20V	125C	94dB	4MV	250NA	50NA	500MWF	6MA	12V	15V	1V	.	.	8MA	80dB	70dB	1M
ML118T	OBS	XSR	INT	.	50V/uS	+20V	-20V	125C	94dB	4MV	250NA	50NA	500MWF	6MA	12V	15V	1V	.	.	8MA	80dB	70dB	1M
ML201AF	OBS	GPU	EXT	.	.	+22V	-22V	85C	94dB	2MV	75NA	10NA	500MWF	5MA	12V	15V	30V	15uV/C	.	3MA	80dB	80dB	500K
ML201AM	OBS	GPU	EXT	.	.	+22V	-22V	85C	94dB	2MV	75NA	10NA	500MWF	5MA	12V	15V	30V	15uV/C	.	3MA	80dB	80dB	500K
ML201AT	OBS	GPU	EXT	.	.	+22V	-22V	85C	94dB	2MV	75NA	10NA	500MWF	5MA	12V	15V	30V	15uV/C	.	3MA	80dB	80dB	500K
ML201F	OBS	GPU	EXT	.	.	+22V	-22V	85C	94dB	7.5MV	1.5uA	0.5uA	500MWF	5MA	12V	15V	30V	30uV/C	.	3MA	65dB	70dB	100K
ML201M	OBS	GPU	EXT	.	.	+22V	-22V	85C	86dB	7.5MV	1.5uA	0.5uA	500MWF	5MA	12V	15V	30V	30uV/C	.	3MA	65dB	70dB	100K
ML201T	OBS	GPU	EXT	.	.	+22V	-22V	85C	86dB	7.5MV	1.5uA	0.5uA	500MWF	5MA	12V	15V	30V	30uV/C	.	3MA	65dB	70dB	100K
ML207F	OBS	GPK	INT	.	.	+22V	-22V	85C	94dB	2MV	75NA	10NA	500MWF	5MA	12V	15V	30V	15uV/C	.	3MA	80dB	80dB	1.5M
ML207M	OBS	GPK	INT	.	.	+22V	-22V	85C	94dB	2MV	75NA	10NA	500MWF	5MA	12V	15V	30V	15uV/C	.	3MA	80dB	80dB	1.5M
ML207T	OBS	GPK	INT	.	.	+22V	-22V	85C	94dB	2MV	75NA	10NA	500MWF	5MA	12V	15V	30V	15uV/C	.	3MA	80dB	80dB	1.5M
ML208AF	OBS	SBA	EXT	.	.	+20V	-20V	85C	98dB	0.5MV	2NA	0.2NA	500MWF	1MA	13V	15V	1V	5uV/C	.	.6MA	96dB	96dB	30M
ML208AM	OBS	SBA	EXT	.	.	+20V	-20V	85C	98dB	0.5MV	2NA	0.2NA	500MWF	1MA	13V	15V	1V	5uV/C	.	.6MA	96dB	96dB	30M
ML208AT	OBS	SBA	EXT	.	.	+20V	-20V	85C	98dB	0.5MV	2NA	0.2NA	500MWF	1MA	13V	15V	1V	5uV/C	.	.6MA	96dB	96dB	30M
ML208M	OBS	SBA	EXT	.	.	+20V	-20V	85C	96dB	2MV	2NA	0.2NA	500MWF	1MA	13V	15V	1V	15uV/C	.	.6MA	85dB	80dB	30M
ML208T	OBS	SBA	EXT	.	.	+20V	-20V	85C	96dB	2MV	2NA	0.2NA	500MWF	1MA	13V	15V	1V	15uV/C	.	.6MA	85dB	80dB	30M
ML211F	OBS	CPR	EXT	.	.	+18V	-18V	85C	100dB	3MV	100NA	10NA	500MWF	.	.	15V	30V	.	.	6MA	.	.	.
ML211M	OBS	CPR	EXT	.	.	+18V	-18V	85C	100dB	3MV	100NA	10NA	500MWF	.	.	15V	30V	.	.	6MA	.	.	.
ML211S	OBS	CPR	EXT	.	.	+18V	-18V	85C	100dB	3MV	100NA	10NA	500MWF	.	.	15V	30V	.	.	6MA	.	.	.
ML211T	OBS	CPR	EXT	.	.	+18V	-18V	85C	100dB	3MV	100NA	10NA	500MWF	.	.	15V	30V	.	.	6MA	.	.	.
ML218F	OBS	XSR	INT	.	50V/uS	+20V	-20V	85C	94dB	4MV	250NA	50NA	500MWF	6MA	12V	15V	1V	.	.	8MA	80dB	70dB	1M
ML218M	OBS	XSR	INT	.	50V/uS	+20V	-20V	85C	94dB	4MV	250NA	50NA	500MWF	6MA	12V	15V	1V	.	.	8MA	80dB	70dB	1M
ML218T	OBS	XSR	INT	.	50V/uS	+20V	-20V	85C	94dB	4MV	250NA	50NA	500MWF	6MA	12V	15V	1V	.	.	8MA	80dB	70dB	1M
ML301AP	OBS	GPU	EXT	.	.	+18V	-18V	70C	88dB	7.5MV	250NA	50NA	500MWF	5MA	12V	15V	30V	30uV/C	.	3MA	70dB	70dB	500K
ML301AS	OBS	GPU	EXT	.	.	+18V	-18V	70C	88dB	7.5MV	250NA	50NA	500MWF	5MA	12V	15V	30V	30uV/C	.	3MA	70dB	70dB	500K
ML301AT	OBS	GPU	EXT	.	.	+18V	-18V	70C	88dB	7.5MV	250NA	50NA	500MWF	5MA	12V	15V	30V	30uV/C	.	3MA	70dB	70dB	500K
ML301P	OBS	GPU	EXT	.	.	+18V	-18V	70C	83dB	10MV	2uA	.75uA	500MWF	.	12V	15V	30V	30uV/C	.	3MA	65dB	70dB	100K
ML301S	OBS	GPU	EXT	.	.	+18V	-18V	70C	83dB	10MV	2uA	.75uA	500MWF	.	12V	15V	30V	30uV/C	.	3MA	65dB	70dB	100K
ML301T	OBS	GPU	EXT	.	.	+18V	-18V	70C	83dB	10MV	2uA	.75uA	500MWF	.	12V	15V	30V	30uV/C	.	3MA	65dB	70dB	100K
ML307P	OBS	GPK	INT	.	.	+18V	-18V	70C	84dB	7.5MV	250NA	50NA	500MWF	5MA	12V	15V	30V	30uV/C	.	.	70dB	70dB	0.5M
ML307S	OBS	GPK	INT	.	.	+18V	-18V	70C	84dB	7.5MV	250NA	50NA	500MWF	5MA	12V	15V	30V	30uV/C	.	.	70dB	70dB	0.5M
ML307T	OBS	GPK	INT	.	.	+18V	-18V	70C	84dB	7.5MV	250NA	50NA	500MW	5MA	12V	15V	30V	30uV/C	.	.	70dB	70dB	0.5M
ML308AM	OBS	SBA	EXT	.	.	+18V	-18V	70C	98dB	0.5MV	7NA	1NA	500MWF	1MA	13V	15V	1V	5uV/C	.	.6MA	96dB	96dB	10M

112

ailed explanations of heading notations, see

r ready references the
mportant abbreviations
the column headings are
below:

LAND PAGE
- = application (codes at APP.E.)
- = common mode rejection ratio
- = compensation (frequency)
- = input offset voltage temperature drift
- = gain bandwidth product
- = input bias current
- = input bias offset current
- = quiescent supply current
- = manufacturer (codes at App.C.)
- = quiescent power consumer
- = power supply rejection ratio
- = common mode input voltage rating
- = differential input voltage rating
- = input offset voltage
- = dc supply voltage

HAND PAGE
out coding summary
ls at APP.G.) for different (APP.F.)
- = gain adjust
- = bias adjust
- = case
- = inverting input
- = non-inverting input
- = input frequency compensation
- = ground
- = high level input
- = output, open collector
- = output, open emitter
- = metal case
- = not connected
- = special terminal
- = outputs
- = strobe
- = offset balance
- = +ve dc supply
- = —ve dc supply
- = guard ring
- = blank position, no lead
- = +ve supplementary dc supply
- = —ve supplementary dc supply
- = output frequency compensation

CASE (APP F)	LD 1	LD 2	LD 3	LD 4	LD 5	LD 6	LD 7	LD 8	LD 9	LD 10	LD 11	LD 12	LD 13	LD 14	LD .15	LD 16	EUROPE SUBSTITUTE	USA SUBSTITUTE	IS	TYPE NUMBER
DIL-14/1C	N	N	G	E-	E+	V-	N	N	F	ø	R	N	V+	N	.	.	SN72702J	UA702DC	O	MIC712-5D
T05-8/1M	R*1	E-	E+	G	R1	R2	V+	R*2	.	.	.	.	.	.	.	.		UA730M	O	MIC730-1C
T05-8/1M	R*1	E-	E+	G	R1	R2	V+	R*2	.	.	.	.	.	.	.	.		UA730HC	O	MIC730-5C
DIL-14/1C	N	N	T	E-	E+	V-	N	N	T*	R	V+	N	N	N	.	.	LM741D	UA741DM	O	MIC741-1D
T05-8/1M	T	E-	E+	V-M	T*	R	V+	N	.	.	.	.	.	.	.	.	TBA222	UA741HM	O	MIC741-1C
T05-8/1M	T	E-	E+	V-M	T*	R	V+	N	.	.	.	.	.	.	.	.	TBA221	UA741HC	O	MIC741-5C
DIL-14/1C	N	N	T	E-	E+	V-	N	N	T*	R	V+	N	N	N	.	.	TBA221A	UA741DC	O	MIC741-5D
FLP-10/3C	N	FT	E-	E+	V-	T*	R	V+	F*	N	.	.	.	.	.	.	SFC2101APM	LM101AF	O	ML101AF
DIL-14/1C	N	N	FT	E-	E+	V-	N	N	T*	R	V+	F*	N	N	.	.	UA101AD	LM101AJ14	O	ML101AM
T05-8/1M	FT	E-	E+	V-M	T*	R	V+	F*	.	.	.	.	.	.	.	.	SFC2101A	LM101AH	O	ML101AT
FLP-10/3G	N	FT	E-	E+	V-	T*	R	V+	F*	N	.	.	.	.	.	.	SFC2101APM	LM101F	O	ML101F
DIL-14/1C	N	N	FT	E-	E+	V-	N	N	T*	R	V+	F*	N	N	.	.	UA101D	LM101J14	O	ML101M
T05-8/1M	FT	E-	E+	V-M	T*	R	V+	F*	.	.	.	.	.	.	.	.	SFC2101A	LM101H	O	ML101T
FLP-10/3C	N	N	E-	E+	V-	N	R	V+	N	N	.	.	.	.	.	.	SFC2107PM	LM107F	O	ML107F
DIL-14/1C	N	N	N	E-	E+	V-	N	N	N	R	V+	N	N	N	.	.	SN52107JA	LM107D	O	ML107M
T05-8/1M	N	E-	E+	V-M	N	R	V+	N	.	.	.	.	.	.	.	.	SFC2107M	LM107H	O	ML107T
FLP-10/3G	N	N	E-	E+	N	V-	R	V+	F*	F	.	.	.	.	.	.	UA108AF	LM108AF	O	ML108AF
DIL-14/1C	N	F	N	E-	E+	N	V-	N	N	R	V+	F*	N	N	.	.	UA108AD	LM108AD	O	ML108AM
T05-8/1M	F	E-	E+	V-M	N	R	V+	F*	.	.	.	.	.	.	.	.	SFC2108A	LM108AH	O	ML108AT
DIL-14/1C	N	F	N	E-	E+	N	V-	N	N	R	V+	F*	N	N	.	.	UA108D	LM108D	O	ML108M
T05-8/1M	F	E-	E+	V-M	N	R	V+	F*	.	.	.	.	.	.	.	.	SFC2108M	LM108H	O	ML108T
FLP-10/3G	G	E+	E-	N	V-	T	T*S	N	R	V+	.	.	.	.	.	.	SN52111FA	LM111F	O	ML111F
DIL-14/1C	N	G	E+	E-	N	V-	T	T*S	R	N	V+	N	N	N	.	.	SN52111J	LM111D	O	ML111M
DIL-8/1C	G	E+	E-	V-	T	T*S	R	V+	.	.	.	.	.	.	.	.	SFC2311DC	UA111R	O	ML111S
T05-8/1M	G	E+	E-	V-	T	T*S	R	V+	.	.	.	.	.	.	.	.	SFC2111M	LM111H	O	ML111T
FLP-10/3G	N	T*F	E-	E+	V-	F*T	R	V+	ø	N	.	.	.	.	.	.	AM118-FLP	LM118F	O	ML118F
DIL-14/1C	N	N	T*F	E-	E+	V-	N	N	F*T	R	V+	ø	N	N	.	.	SN52118JA	LM118D	O	ML118M
T05-8/1M	T*F	E-	E+	V-	F*T	R	V+	ø	.	.	.	.	.	.	.	.	TDC0118CM	LM118H	O	ML118T
FLP-10/3G	N	FT	E-	E+	V-	T*	R	V+	F*	N	.	.	.	.	.	.	SFC2201APT	LM201AF	O	ML201AF
DIL-14/1C	N	N	FT	E-	E+	V-	N	N	T*	R	V+	F*	N	N	.	.	UA201AD	LM201AJ14	O	ML201AM
T05-8/1M	FT	E-	E+	V-M	T*	R	V+	F*	.	.	.	.	.	.	.	.	SFC2101A	LM201AH	O	ML201AT
FLP-10/3G	N	FT	E-	E+	V-	T*	R	V+	F*	N	.	.	.	.	.	.	SFC2201APM	LM201F	O	ML201F
DIL-14/1C	N	N	FT	E-	E+	V-	N	N	T*	R	V+	F*	N	N	.	.	UA201D	LM201J14	O	ML201M
T05-8/1M	FT	E-	E+	V-M	T*	R	V+	F*	.	.	.	.	.	.	.	.	SFC2101A	LM201H	O	ML201T
FLP-10/3C	N	N	E-	E+	V-	N	R	V+	N	N	.	.	.	.	.	.	SFC2207PT	LM207F	O	ML207F
DIL-14/1C	N	N	N	E-	E+	V-	N	N	N	R	V+	N	N	N	.	.	SN52107JA	LM207D	O	ML207M
T05-8/1M	N	E-	E+	V-M	N	R	V+	N	.	.	.	.	.	.	.	.	SFC2207	LM207H	O	ML207T
FLP-10/3G	N	N	E-	E+	N	V-	R	V+	F*	F	.	.	.	.	.	.	UA208AF	LM208AF	O	ML208AF
DIL-14/1C	N	F	N	E-	E+	N	V-	N	N	R	V+	F*	N	N	.	.	UA208AD	LM208AD	O	ML208AM
T05-8/1M	F	E-	E+	V-M	N	R	V+	F*	.	.	.	.	.	.	.	.	SFC2208A	LM208AH	O	ML208AT
DIL-14/1C	N	F	N	E-	E+	N	V-	N	N	R	V+	F*	N	N	.	.	UA208D	LM208D	O	ML208M
T05-8/1M	F	E-	E+	V-M	N	R	V+	F*	.	.	.	.	.	.	.	.	SFC2208	LM208H	O	ML208T
FLP-10/3G	G	E+	E-	N	V-	T	T*S	N	R	V+	.	.	.	.	.	.	SN52111FA	LM211F	O	ML211F
DIL-14/1C	N	G	E+	E-	N	V-	T	T*S	R	N	V+	N	N	N	.	.	SN52111J	LM211D	O	ML211M
DIL-8/1C	G	E+	E-	V-	T	T*S	R	V+	.	.	.	.	.	.	.	.	SN52111JP	UA111R	O	ML211S
T05-8/1M	G	E+	E-	V-	T	T*S	R	V+	.	.	.	.	.	.	.	.	SFC2211	LM211H	O	ML211T
FLP-10/3G	N	T*F	E-	E+	V-	F*T	R	V+	ø	N	.	.	.	.	.	.		LM218F	O	ML218F
DIL-14/1C	N	N	T*F	E-	E+	V-	N	N	F*T	R	V+	ø	N	N	.	.		LM218D	O	ML218M
T05-8/1M	T*F	E-	E+	V-	F*T	R	V+	ø	.	.	.	.	.	.	.	.	TDB0118CM	LM218H	O	ML218T
DIL-14/1P	N	N	FT	E-	E+	V-	N	N	T*	R	V+	F*	N	N	.	.	UA301AD	LM301AJ14	O	ML301AP
DIL-8/1P	FT	E-	E+	V-	T*	R	V+	F*	.	.	.	.	.	.	.	.	SFC2301ADC	LM301AJ	O	ML301AS
T05-8/1M	FT	E-	E+	V-M	T*	R	V+	F*	.	.	.	.	.	.	.	.	SFC2301AH	LM301AH	O	ML301AT
DIL-14/1P	N	N	FT	E-	E+	V-	N	N	T*	R	V+	F*	N	N	.	.		LM301H	O	ML301P
DIL-8/1P	FT	E-	E+	V-	T*	R	V+	F*	.	.	.	.	.	.	.	.	SFC2301ADC	LM201J	O	ML301S
T05-8/1M	FT	E-	E+	V-M	T*	R	V+	F*	.	.	.	.	.	.	.	.	SFC2301A	LM301H	O	ML301T
DIL-14/1P	N	N	N	E-	E+	V-	N	N	N	R	V+	N	N	N	.	.	SN72307JA	LM307D	O	ML307P
DIL-8/1P	N	E-	E+	V-	N	R	V+	N	.	.	.	.	.	.	.	.	SFC2307DC	LM307N	O	ML307S
T05-8/1M	N	E-	E+	V-M	N	R	V+	N	.	.	.	.	.	.	.	.	SFC2307	LM307H	O	ML307T
DIL-14/1C	N	F	N	E-	E+	N	V-	N	N	R	V+	F*	N	N	.	.	SN72308AJA	LM308AD	O	ML308AM
T05-8/1M	F	E-	E+	V-M	N	R	V+	F*	.	.	.	.	.	.	.	.	SFC2308A	LM308A	O	ML308AT

TYPE NUMBER	MFR	APP	CMP	GBP MIN	SLEW RATE MIN	V_S^+ MAX	V_S^- MAX	T_{OP} MAX	A_{VOL} MIN	V_{IO} MAX	I_B MAX	I_{IO} MAX	P_{TOT} MAX	I_{OUT} MIN	V_{OUT} MIN	V_{ICM} MAX	V_{IDF} MAX	dV_{IO}/dT MAX	P_Q MAX	I_Q MAX	CMRR MIN	PSRR MIN	R_{IN} MIN
ML308AT	OBS	SBA	EXT	.	.	+18V	-18V	70C	98dB	0.5MV	7NA	1NA	500MWF	1MA	13V	15V	1V	5uV/C	.	.6MA	96dB	96dB	10M
ML308M	OBS	SBA	EXT	.	.	+18V	-18V	70C	88dB	7.5MV	7NA	1NA	500MWF	1MA	13V	15V	1V	30uV/C	.	.6MA	80dB	80dB	10M
ML308T	OBS	SBA	EXT	.	.	+18V	-18V	70C	88dB	7.5MV	7NA	1NA	500MWF	1MA	13V	15V	1V	30uV/C	.	.6MA	80dB	80dB	10M
ML311M	OBS	CPR	EXT	.	.	+18V	-18V	70C	100dB	7.5MV	250NA	50NA	500MWF	.	.	15V	30V	.	.	8MA	.	.	.
ML311P	OBS	CPR	EXT	.	.	+18V	-18V	70C	100dB	7.5MV	250NA	50NA	500MWF	.	.	15V	30V	.	.	8MA	.	.	.
ML311S	OBS	CPR	EXT	.	.	+18V	-18V	70C	100dB	7.5MV	250NA	50NA	500MWF	.	.	15V	30V	.	.	8MA	.	.	.
ML311T	OBS	CPR	EXT	.	.	+18V	-18V	70C	100dB	7.5MV	250NA	50NA	500MWF	.	.	15V	30V	.	.	8MA	.	.	.
ML318M	OBS	XSR	INT	.	50V/uS	+20V	-20V	70C	88dB	10MV	500NA	200NA	500MWF	6MA	12V	15V	1V	.	.	8MA	70dB	65dB	500K
ML318T	OBS	XSR	INT	.	50V/uS	+20V	-20V	70C	88dB	10MV	500NA	200NA	500MWF	6MA	12V	15V	1V	.	.	8MA	70dB	65dB	500K
ML709AF	OBS	GPU	EXT	.3MHZ	.15V/uS	+18V	-18V	125C	88dB	2MV	200NA	50NA	570MWF	5MA	12V	10V	5V	10uV/C	108MW	.	80dB	80dB	350K
ML709AM	OBS	GPU	EXT	.3MHZ	.15V/uS	+18V	-18V	125C	88dB	2MV	200NA	50NA	670MWF	5MA	12V	10V	5V	10uV/C	108MW	.	80dB	80dB	350K
ML709AT	OBS	GPU	EXT	.3MHZ	.15V/uS	+18V	-18V	125C	88dB	2MV	200NA	50NA	500MWF	5MA	12V	10V	5V	10uV/C	108MW	.	80dB	80dB	350K
ML709CM	OBS	GPU	EXT	.3MHZ	.15V/uS	+18V	-18V	70C	84dB	7.5MV	1.5uA	500NA	670MWF	5MA	12V	10V	5V	.	200MW	.	65dB	74dB	50K
ML709CP	OBS	GPU	EXT	.3MHZ	.15V/uS	+18V	-18V	70C	84dB	7.5MV	1.5uA	500NA	670MWF	5MA	12V	10V	5V	.	200MW	.	65dB	74dB	50K
ML709CT	OBS	GPU	EXT	.3MHZ	.15V/uS	+18V	-18V	70C	84dB	7.5MV	1.5uA	500NA	500MWF	5MA	12V	10V	5V	.	200MW	.	65dB	.74dB	50K
ML709F	OBS	GPU	EXT	.3MHZ	.15V/uS	+18V	-18V	125C	88dB	5MV	500NA	200NA	570MWF	5MA	12V	10V	5V	15uV/C	165MW	.	70dB	76dB	150K
ML709M	OBS	GPU	EXT	.3MHZ	.15V/uS	+18V	-18V	125C	88dB	5MV	500NA	200NA	670MWF	5MA	12V	10V	5V	15uV/C	165MW	.	70dB	76dB	150K
ML709T	OBS	GPU	EXT	.3MHZ	.15V/uS	+18V	-18V	125C	88dB	5MV	500NA	200NA	500MWF	5MA	12V	10V	5V	15uV/C	165MW	.	70dB	76dB	150K
ML741AF	OBS	GPK	INT	.4MHZ	0.3V/uS	+22V	-22V	125C	94dB	3MV	80NA	30NA	570MWF	10MA	16V	15V	30V	15uV/C	150MW	.	80dB	86dB	1M
ML741AM	OBS	GPK	INT	.4MHZ	0.3V/uS	+22V	-22V	125C	94dB	3MV	80NA	30NA	670MWF	10MA	16V	15V	30V	15uV/C	150MW	.	80dB	86dB	1M
ML741AT	OBS	GPK	INT	.4MHZ	0.3V/uS	+22V	-22V	125C	94dB	3MV	80NA	30NA	500MWF	10MA	16V	15V	30V	15uV/C	150MW	.	80dB	86dB	1M
ML741CM	OBS	GPK	INT	.	0.2V/uS	+18V	-18V	70C	86dB	6MV	500NA	200NA	670MWF	5MA	12V	15V	30V	.	85MW	3MA	70dB	76dB	300K
ML741CP	OBS	GPK	INT	.	0.2V/uS	+18V	-18V	70C	86dB	6MV	500NA	200NA	670MWF	5MA	12V	15V	30V	.	85MW	3MA	70dB	76dB	300K
ML741CS	OBS	GPK	INT	.	0.2V/uS	+18V	-18V	70C	86dB	6MV	500NA	200NA	310MWF	5MA	12V	15V	30V	.	.	3MA	70dB	76dB	300K
ML741CT	OBS	GPK	INT	.	0.2V/uS	+18V	-18V	70C	86dB	6MV	500NA	200NA	500MWF	5MA	12V	15V	30V	.	.	3MA	70dB	76dB	300K
ML741F	OBS	GPK	INT	.	0.2V/uS	+22V	-22V	125C	94dB	5MV	500NA	200NA	570MWF	5MA	12V	15V	30V	.	85MW	3MA	70dB	76dB	300K
ML741M	OBS	GPK	INT	.4MHZ	0.3V/uS	+22V	-22V	125C	94dB	3MV	80NA	30NA	670MWF	5MA	12V	15V	30V	15uV/C	150MW	.	80dB	86dB	1M
ML741T	OBS	GPK	INT	.	0.2V/uS	+22V	-22V	125C	94dB	5MV	500NA	200NA	500MWF	7MA	12V	15V	30V	.	85MW	3MA	70dB	76dB	300K
ML747CP	OBS	DGK	INT	.	0.2V/uS	+18V	-18V	70C	88dB	6MV	500NA	200NA	670MWF	85MA	12V	15V	30V	.	85MW	3MA	70dB	76dB	300K
ML747CT	OBS	DGK	INT	.	0.2V/uS	+18V	-18V	70C	88dB	6MV	500NA	200NA	500MWF	5MA	12V	15V	30V	.	85MW	3MA	70dB	76dB	300K
ML747F	OBS	DGK	INT	.	0.2V/uS	+22V	-22V	125C	94dB	5MV	500NA	200NA	800MWF	5MA	12V	15V	30V	.	85MW	3MA	70dB	76dB	300K
ML747M	OBS	DGK	INT	.	0.2V/uS	+22V	-22V	125C	94dB	5MV	500NA	200NA	670MWF	5MA	12V	15V	30V	.	85MW	3MA	70dB	76dB	300K
ML747T	OBS	DGK	INT	.	0.2V/uS	+22V	-22V	125C	94dB	5MV	500NA	200NA	500MWF	5MA	12V	15V	30V	.	85MW	3MA	70dB	76dB	300K
ML748CP	OBS	GPU	EXT	.	0.2V/uS	+22V	-22V	70C	86dB	6MV	500NA	200NA	670MWF	5MA	12V	15V	30V	.	85MW	3MA	70dB	76dB	300K
ML748CS	OBS	GPU	EXT	.	0.2V/uS	+18V	-18V	70C	86dB	6MV	500NA	200NA	310MWF	5MA	12V	15V	30V	.	85MW	3MA	70dB	76dB	300K
ML748CT	OBS	GPU	EXT	.	0.2V/uS	+18V	-18V	70C	86dB	6MV	500NA	200NA	500MWF	5MA	12V	15V	30V	.	85MW	3MA	70dB	76dB	300K
ML748F	OBS	GPU	EXT	.	0.2V/uS	+22V	-22V	125C	94dB	5MV	500NA	200NA	570MWF	5MA	12V	15V	30V	.	85MW	3MA	70dB	76dB	300K
ML748M	OBS	GPU	EXT	.	0.2V/uS	+22V	-22V	125C	94dB	5MV	500NA	200NA	670MWF	5MA	12V	15V	30V	.	85MW	3MA	70dB	76dB	300K
ML748S	OBS	GPU	EXT	.	.25V/uS	+22V	-22V	125C	94dB	5MV	500NA	200NA	310MWF	5MA	10V	15V	30V	.	85MW	3MA	70dB	76dB	300K
ML748T	OBS	GPU	EXT	.	0.2V/uS	+22V	-22V	125C	94dB	5MV	500NA	200NA	500MWF	5MA	12V	15V	30V	.	85MW	3MA	70dB	76dB	300K
ML1437P	OBS	DGU	INT	.	0.1V/uS	+18V	-18V	75C	84dB	7.5MV	1.5uA	0.5uA	750MWF	5MA	12V	18V	5V	10uV/C	225MW	.	65dB	74dB	50K
ML1436T	OBS	HVO	INT	3MHZ	0.5V/uS	+34V	-34V	75C	97dB	10MV	40NA	10NA	680MWF	1MA	20V	34V	68V	.	280MW	5MA	70dB	74dB	3M
ML1458S	OBS	DGK	INT	5MHZ	0.3V/uS	+18V	-18V	75C	86dB	6MV	0.5uA	0.2uA	750MWF	5MA	12V	15V	30V	50uV/C	170MW	6MA	70dB	76dB	300K
ML1458P	OBS	DGK	INT	.5MHZ	0.3V/uS	+18V	-18V	75C	86dB	6MV	0.5uA	0.2uA	750MWF	5MA	12V	15V	30V	50uV/C	170MW	6MA	70dB	76dB	300K
ML1458T	OBS	DGK	INT	.5MHZ	0.3V/uS	+18V	-18V	75C	86dB	6MV	0.5uA	0.2uA	680MWF	5MA	12V	15V	30V	50uV/C	170MW	6MA	70dB	76dB	300K
ML1536T	OBS	HVO	INT	.3MHZ	0.5V/uS	+40V	-40V	125C	100dB	5MV	20NA	3NA	680MWF	1MA	30V	40V	80V	.	224MW	4MA	80dB	80dB	3M
ML1537M	OBS	DGU	EXT	.	0.1V/uS	+18V	-18V	125C	88dB	5MV	0.5uA	0.2uA	750MWF	8MA	12V	18V	5V	10uV/C	225MW	.	70dB	76dB	150K
ML1558M	OBS	DGK	INT	.5MHZ	0.3V/uS	+22V	-22V	125C	94dB	5MV	0.5uA	0.2uA	750MWF	5MA	12V	15V	30V	50uV/C	150MW	5MA	70dB	76dB	300K
ML1558S	OBS	DGK	INT	.5MHZ	0.3V/uS	+22V	-22V	125C	94dB	5MV	0.5uA	0.2uA	750MWF	5MA	12V	15V	30V	50uV/C	150MW	5MA	70dB	76dB	300K
ML1558T	OBS	DGK	INT	.5MHZ	0.3V/uS	+22V	-22V	125C	94dB	5MV	0.5uA	0.2uA	680MWF	5MA	12V	15V	30V	50uV/C	150MW	5MA	70dB	76dB	300K
ML4250CS	OBS	PRA	INT	.	.	+18V	-18V	70C	95dB	6MV	75NA	20NA	500MWF	1MA	12V	15V	30V	.	3MW	.1MA	70dB	74dB	.
ML4250CT	OBS	PRA	INT	.	.	+18V	-18V	70C	95dB	6MV	75NA	20NA	500MWF	1MA	12V	15V	30V	.	3MW	.1MA	70dB	74dB	.
ML4250T	OBS	PRA	INT	.	.	+18V	-18V	125C	100dB	5MV	50NA	10NA	500MWF	1MA	12V	15V	30V	.	2.7MW	90UA	70dB	76dB	.
MLF111G	MTU	CPR	EXT	.	.	+18V	-18V	125C	100dB	4MV	50pA	25pA	500MWF	8MA	.	15V	30V	.	.	6MA	.	.	.
MLF111U	MTU	CPR	EXT	.	.	+18V	-18V	125C	100dB	4MV	50pA	25pA	500MWF	8MA	.	15V	30V	.	.	6MA	.	.	.
MLF155AG	MTU	FET	INT	.5MHZ	3V/uS	+22V	-22V	125C	94dB	2MV	50pA	10pA	670MWF	5MA	12V	20V	40V	5uV/C	.	4MA	85dB	85dB	0.1T
MLF155G	MTU	FET	INT	.5MHZ	2V/uS	+22V	-22V	125C	94dB	5MV	100pA	20pA	670MWF	5MA	12V	20V	40V	20uV/C	.	4MA	85dB	85dB	0.1T
MLF156AG	MTU	HSR	INT	4MHZ	10V/uS	+22V	-22V	125C	94dB	2MV	50pA	10pA	670MWF	5MA	12V	20V	40V	5uV/C	.	.4MA	85dB	85dB	0.1T
MLF156G	MTU	HSR	INT	1MHZ	7.5V/uS	+22V	-22V	125C	94dB	5MV	100pA	20pA	670MWF	5MA	12V	20V	40V	20uV/C	.	7MA	85dB	85dB	0.1T
MLF157AG	MTU	XSR	INT	15MHZ	8V/uS	+22V	-22V	125C	94dB	2MV	50pA	10pA	670MWF	5MA	12V	20V	40V	5uV/C	.	7MA	85dB	85dB	0.1T

For detailed explanations of column heading notations, see App. A.

Also for ready references the more important abbreviations used in the column headings are listed below:

LEFT HAND PAGE
APP = application (codes at APP.E.)
CMRR = common mode rejection ratio
CMP = compensation (frequency)
dV_{io}/dT = input offset voltage temperature drift
GBP = gain bandwidth product
I_b = input bias current
I_{io} = input bias offset current
I_q = quiescent supply current
MFR = manufacturer (codes at App.C.)
P_q = quiescent power consumer
PSRR = power supply rejection ratio
V_{icm} = common mode input voltage rating
V_{di} = differential input voltage rating
V_{io} = input offset voltage
V_s = dc supply voltage

RIGHT HAND PAGE
Lead out coding summary (details at APP.G.) for different cases (APP.F.)
A = gain adjust
B = bias adjust
C = case
E– = inverting input
E+ = non-inverting input
F,F* = input frequency compensation
G = ground
J = high level input
K = output, open collector
L = output, open emitter
M = metal case
N = not connected
Q = special terminal
R,R* = outputs
S = strobe
T,T* = offset balance
V+ = +ve dc supply
V– = –ve dc supply
W = guard ring
X = blank position, no lead
++ = +ve supplementary dc supply
--- = –ve supplementary dc supply
ø,ø* = output frequency compensation

CASE (APP F)	LD1	LD2	LD3	LD4	LD5	LD6	LD7	LD8	LD9	LD10	LD11	LD12	LD13	LD14	LD15	LD16	EUROPE SUBSTITUTE	USA SUBSTITUTE	ISS	TYPE NUMBER
DIL-14/1C	N	F	N	E-	E+	N	V-	N	N	R	V+	F*	N	N			SN72308JA	LM308D	O	ML308M
T05-8/1M	F	E-	E+	V-M	N	R	V+	F*									SFC2308	LM308H	O	ML308T
DIL-14/1C	N	G	E+	E-	N	V-	T	T*S	R								SFC2311EC	LM311D	O	ML311M
DIL-14/1P	N	G	E+	E-	N	V-	T	T*S	R								SFC2311EC	LM311D	O	ML311P
DIL-8/1P	G	E+	E-	V-	T	T*S	R	V+									SFC2311DC	LM311N	N	ML311S
T05-8/1M	G	E+	E-	V-	T	T*S	R	V+									SFC2311	LM311H	O	ML311T
DIL-14/1C	N	N	T*F	E-	E+	V-	N	N	F*T	R	V+	ø	N	N			SN72318JA	LM318D	O	ML318M
T05-8/1M	T*F	E-	E+	V-	F*T	R	V+	ø									TDE0118CM	LM318H	O	ML318T
FLP-10/3G	N	F	E-	E+	V-	ø	R	V+	F*	N							SN52709AFA	UA709AFM	O	ML709AF
DIL-14/1C	N	N	F	E-	E+	V-	N	N	ø	R	V+	F*	N	N			LM709AJ	UA709ADM	O	ML709AM
T05-8/1M	F	E-	E+	V-	ø	ø*R	V+	N									TAA522	UA709AHM	O	ML709AT
DIL-14/1C	N	N	N	E-	E+	V-	N	N		R	V+	F*	N	N			TAA521A	UA709DC	O	ML709CM
DIL-14/1P	N	N	F	E-	E+	V-	N	N	ø	R	V+	F*	N	N			TAA521A	UA709DC	O	ML709CP
T05-8/1M	F	E-	E+	V-	ø	ø*R	V+	F*									TAA521	UA709HC	O	ML709CT
FLP-10/3G	N	F	E-	E+	V	ø	R	V+	F*	N							SN52709AFA	UA709FM	O	ML709F
DIL-14/1C	N	N	F	E-	E+	V-	N	N	ø	R	V+	F*	N	N			SN52709AJ	UA709DM	O	ML709M
T05-8/1M	F	E-	E+	V-	ø	ø*R	V+	F*									TAA522	UA709HM	O	ML709T
FLP-10/3G	N	T	E-	E+	V-	T*	R	V+	N	N							SFC2741PM	UA741AFM	O	ML741AF
DIL-14/1C	N	N	T	E-	E+	V-	N	N	T*	R	V+	N	N	N			LM741AD	UA741ADM	O	ML741AM
T05-8/1M	T	E-	E+	V-M	T*	R	V+	N									TBA222	UA741AHM	O	ML741AT
DIL-14/1P	N	N	T	E-	E+	V-	N	N	T*	R	V+	N	N	N			TBA221A	UA741DC	O	ML741CM
DIL-14/1P	N	N	T	E-	E+	V-	N	N	T*	R	V+	N	N	N			TBA221A	UA741DC	O	ML741CP
DIL-8/1P	T	E-	E+	V-	T*	R	V+	N									TBA221B	UA741TC	O	ML741CS
T05-8/1M	T	E-	E+	V-M	T*	R	V+	N									TBA221	UA741HC	O	ML741CT
FLP-10/3G	N	T	E-	E+	V-	T*	R	V+	N								SFC2741PM	UA741FM	O	ML741F
DIL-14/1C	N	N	T	E-	E+	V-	N	N	T*	R	V+	N	N	N			LM741D	UA741DM	O	ML741M
T05-8/1M	T	E-	E+	V-M	T*	R	V+	N									TBA222	UA741HM	O	ML741T
DIL-14/1P	E-1	E+1	T1	V-	T2	E+2	E-2	T*2	V+2	R2	N	R1	V+1	T*1			TBB0747A	UA747DC	O	ML747CP
T05-10/1M	R1	V+1	E-1	E+1	V-	E+2	E-2	V+2	R2	N							TBB0747	UA747HC	O	ML747CT
FLP-14/3G	E-1	E+1	T1	V-	T2	E+2	E-2	T*2	V+	R2	N	R1	V+	T*1				LM747F	O	ML747F
DIL-14/1C	E-1	E+1	T1	V-	T2	E+2	E-2	T*2	V+2	R2	N	R1	V+1	T*1			SFC2747KM	UA747DM	O	ML747M
T05-10/1M	R1	V+1	E-1	E+1	V-	E+2	E-2	V+2	R2	N							TBC0747	UA747HM	O	ML747T
DIL-14/1P	N	N	FT	E-	E+	V-	N	N	T*	R	V+	F*	N	N			SN72748J	UA748DC	O	ML748CP
DIL-8/1P	FT	E-	E+	V-	T*	R	V+	F*									TBB0748	UA748TC	O	ML748CS
T05-8/1M	FT	E-	E+	V-	T*	R	V+	F*									TBB0748	UA748HC	O	ML748CT
FLP-10/3G	N	FT	E-	E+	V-	T*	R	V+	F*	N							SN52748FA	UA748FM	O	ML748F
DIL-14/1C	N	N	FT	E-	E+	V-	N	N	T*	R	V+	F*	N	N			SN52748JA	UA748DM	O	ML748M
DIL-8/1C	FT	E-	E+	V-	T*	R	V+	F*									SN52748JP	LM748J	O	ML748S
T05-8/1M	FT	E-	E+	V-	T*	R	V+	F*									TBC0748	UA748HM	O	ML748T
DIL-14/1P	ø2	R2	F2	F*2	E-2	E+2	V-	E+1	E-1	F1	F*1	R1	ø1	V+			RC1437DC	MC1437L	O	ML1437P
T05-8/1M	T	E-	E+	V+	T*	R	V+	N									LM343	MC1436G	O	ML1436T
DIL-8/1P	R1	E-1	E+1	V-	E+2	E-2	R2	V+									TBB1458B	MC1458U	O	ML1458S
DIL-14/1P	N	R1	N	N	E-1	E+1	V-	E+2	E-2	N	N	R2	N	V+			LM1458N14	MC1458L	O	ML1458P
T05-8/1M	R1	E-1	E+1	V-	E+2	E-2	R2	V+									TBB1458	MC1458G		ML1458T
T05-8/1M	T	E-	E+	V+	T*	R	V+	N									LM143	MC1536G	O	ML1536T
DIL-14/1C	ø2	R2	F2	F*2	E-2	E+2	V-	E+1	E-1	F1	F*1	R1	ø1	V+			RM1537DC	MC1537L	O	ML1537M
DIL-14/1C	N	R1	N	N	E-1	E+1	V-	E+2	E-2	N	N	R2	N	V+				MC1558L	O	ML1558M
DIL-8/1C	R1	E-1	E+1	V-	E+2	E-2	R2	V+									LM1558J	MC1558U	O	ML1558S
T05-8/1M	R1	E-1	E+1	V-	E+2	E-2	R2	V+									TBC1458	MC1558G	O	ML1558T
DIL-8/1P	T	E-	E+	V-	T*	R	V+	B										LM4250CJ	O	ML4250CS
T05-8/1M	T	E-	E+	V-	T*	R	V+	B									SG4250CT	LM4250CH	O	ML4250CT
T05-8/1M	T	E-	E+	V-	T*	R	V+	B									SG4250T	LM4250H	O	ML4250T
T05-8/1M	G	E+	E-	V-	T	T*S	R	V+									UAF111H	LF111H	O	MLF111G
DIL-8/1C	G	E+	E-	V-	T	T*S	R	V+											O	MLF111U
T05-8/1M	T	E-	E+	V-	T*	R	V+	N									UAF155AHM	LF155AH	O	MLF155AG
T05-8/1M	T	E-	E+	V-	T*	R	V+	N									UAF155HM	LF155H	O	MLF155G
T05-8/1M	T	E-	E+	V-	T*	R	V+	N									UAF156AHM	LF156AH	O	MLF156AG
T05-8/1M	T	E-	E+	V-	T*	R	V+	N									UAF156HM	LF156H	O	MLF156G
T05-8/1M	T	E-	E+	V-	T*	R	V+	N									UAF157AHM	LF157AH	O	MLF157AG
T05-8/1M	T	E-	E+	V-	T*	R	V+	N									UAF157HM	LF157H	O	MLF157G

TYPE NUMBER	MFR	APP	CMP	GBP MIN	SLEW RATE MIN	V_S^+ MAX	V_S^- MAX	T_{OP} MAX	A_{VOL} MIN	V_{IO} MAX	I_B MAX	I_{IO} MAX	P_{TOT} MAX	I_{OUT} MIN	V_{OUT} MIN	V_{ICM} MAX	V_{IDF} MAX	dV_{IO}/dT MAX	P_Q MAX	I_Q MAX	CMRR MIN	PSRR MIN	R_{IN} MIN
MLF157G	MTU	XSR	INT	4MHZ	6V/uS	+22V	-22V	125C	94dB	5MV	100pA	50pA	670MWF	5MA	12V	20V	40V	20uV/C	.	7MA	85dB	85dB	0.1T
MLF211G	MTU	CPR	EXT	.	.	+18V	-18V	85C	100dB	4MV	50pA	25pA	500MWF	8MA	.	15V	30V		.	6MA	.	.	.
MLF211P	MTU	CPR	EXT	.	.	+18V	-18V	85C	100dB	4MV	50pA	25pA	500MWF	8MA	.	15V	30V		.	6MA	.	.	.
MLF211U	MTU	CPR	EXT	.	.	+18V	-18V	85C	100dB	4MV	50pA	25pA	500MWF	8MA	.	15V	30V		.	6MA	.	.	.
MLF311G	MTU	CPR	EXT	.	.	+18V	-18V	70C	100dB	10MV	150pA	75pA	500MWF	8MA	.	15V	30V		.	8MA	.	.	.
MLF311P	MTU	CPR	EXT	.	.	+18V	-18V	70C	100dB	10MV	150pA	75pA	500MWF	8MA	.	15V	30V		.	8MA	.	.	.
MLF311U	MTU	CPR	EXT	.	.	+18V	-18V	70C	100dB	10MV	150pA	75pA	500MWF	8MA	.	15V	30V		.	8MA	.	.	.
MLF355AG	MTU	FET	INT	.5MHZ	3V/uS	+18V	-18V	70C	94dB	2MV	50pA	10pA	500MWF	5MA	12V	16V	30V	5uV/C	.	4MA	85dB	85dB	0.1T
MLF355G	MTU	FET	INT	.5MHZ	2V/uS	+18V	-18V	70C	88dB	10MV	100pA	20pA	500MWF	5MA	12V	16V	30V	20uV/C	.	4MA	80dB	80dB	0.1T
MLF356AG	MTU	HSR	INT	4MHZ	10V/uS	+18V	-18V	70C	94dB	2MV	50pA	10pA	500MWF	5MA	12V	16V	30V	5uV/C	.	10MA	85dB	85dB	0.1T
MLF356G	MTU	HSR	INT	1MHZ	7.5V/uS	+18V	-18V	70C	88dB	10MV	200pA	50pA	500MWF	5MA	12V	16V	30V	20uV/C	.	10MA	80dB	80dB	0.1T
MLF357AG	MTU	XSR	INT	15MHZ	8V/uS	+18V	-18V	70C	94dB	2MV	50pA	10pA	500MWF	5MA	12V	16V	30V	5uV/C	.	10MA	85dB	85dB	0.1T
MLF357G	MTU	XSR	INT	4MHZ	6V/uS	+18V	-18V	70C	88dB	10MV	200pA	50pA	500MWF	5MA	12V	16V	30V	20uV/C	.	10MA	80dB	80dB	0.1T
MLM101AG	MTU	GPU	EXT	.	.	+22V	-22V	125C	94dB	2MV	75NA	10NA	500MWF	5MA	12V	15V	30V	15uV/C	.	3MA	80dB	80dB	1.5M
MM101AU	MTU	GPU	EXT	.	.	+22V	-22V	125C	94dB	2MV	75NA	10NA	625MWF	5MA	12V	15V	30V	15uV/C	.	3MA	80dB	80dB	1.5M
MLM107G	MTU	GPK	INT	.	.	+22V	-22V	125C	94dB	2MV	75NA	10NA	500MWF	5MA	12V	15V	30V	15uV/C	.	3MA	80dB	80dB	1.5M
MLM107U	MTU	GPK	INT	.	.	+22V	-22V	125C	94dB	2MV	75NA	10NA	500MWF	5MA	12V	15V	30V	15uV/C	.	3MA	80dB	80dB	1.5M
MLM108AF	MTU	LBC	EXT	.	.	+20V	-20V	125C	98dB	0.5MV	2NA	0.2NA	500MWF	1MA	13V	15V	1V	5uV/C	.	.6MA	96dB	96dB	30M
MLM108AG	MTU	LBC	EXT	.	.	+20V	-20V	125C	98dB	0.5MV	2NA	0.2NA	500MWF	1MA	13V	15V	1V	5uV/C	.	.6MA	96dB	96dB	30M
MLM108AU	MTU	SBA	EXT	.	.	+20V	-20V	125C	98dB	0.5MV	2NA	0.2NA	500MWF	1MA	13V	15V	1V	5uV/C	.	.6MA	96dB	96dB	30M
MLM108F	MTU	LBC	EXT	.	.	+20V	-20V	125C	96dB	2MV	2NA	0.2NA	500MWF	1MA	13V	15V	1V	15uV/C	.	.6MA	85dB	80dB	30M
MLM108G	MTU	LBC	EXT	.	.	+20V	-20V	125C	96dB	2MV	2NA	0.2NA	500MWF	1MA	13V	15V	1V	15uV/C	.	.6MA	85dB	80dB	30M
MLM108C	MTU	LBC	EXT	.	.	+20V	-20V	125C	96dB	2MV	2NA	0.2NA	500MWF	1MA	13V	15V	1V	15uV/C	.	.6MA	85dB	80dB	30M
MLM110G	MTU	VFA	INT	.	15V/uS	+18V	-18V	125C	0dB	4MV	3NA	.	500MWF	1MA	10V	15V	15V	50uV/C	.	6MA	.	70dB	10G
MLM111F	MTU	CPR	EXT	.	.	+18V	-18V	125C	100dB	3MV	100NA	10NA	500MWF	.	.	15V	30V		.	6MA	.	.	.
MLM111G	MTU	CPR	EXT	.	.	+18V	-18V	125C	100dB	3MV	100NA	10NA	680MWF	.	.	15V	30V		.	6MA	.	.	.
MLM111L	MTU	CPR	EXT	.	.	+18V	-18V	125C	100dB	3MV	100NA	10NA	625MWF	.	.	15V	30V		.	6MA	.	.	.
MLM124L	MTU	QLQ	INT	.	.	+16V	-16V	125C	94dB	5MV	150NA	30NA	900MWF	.	.	16V	16V	35uV/C	.	2MA	70dB	65dB	.
MLM124P	MTU	QLQ	INT	.	.	+16V	-16V	125C	94dB	5MV	150NA	30NA	900MWF	.	.	16V	16V	35uV/C	.	2MA	70dB	65dB	.
MLM139AL	MTU	QCP	EXT	.	.	+18V	-18V	125C	94dB	2MV	100NA	25NA	900MWF	.	.	18V	18V		.	2MA	.	.	.
MLM139L	MTU	QCP	EXT	.	.	+18V	-18V	125C	88dB	5MV	100NA	25NA	900MWF	.	.	18V	18V		.	2MA	.	.	.
MLM158G	MTU	DLQ	INT	.	.	+16V	-16V	125C	94dB	5MV	150NA	30NA	500MWF	10MA	.	16V	32V	30uV/C	.	3MA	70dB	65dB	.
MLM158U	MTU	DLQ	INT	.	.	+16V	-16V	125C	94dB	5MV	150NA	30NA	500MWF	10MA	.	16V	32V	30uV/C	.	3MA	70dB	65dB	.
MLM201AG	MTU	GPU	EXT	.	.	+22V	-22V	85C	94dB	2MV	75NA	10NA	500MWF	5MA	12V	15V	30V	15uV/C	.	3MA	80dB	80dB	1.5M
MLM201AP1	MTU	GPU	EXT	.	.	+22V	-22V	85C	94dB	2MV	75NA	10NA	625MWF	5MA	12V	15V	30V	15uV/C	.	3MA	80dB	80dB	1.5M
MLM201AU	MTU	GPU	EXT	.	.	+22V	-22V	85C	94dB	2MV	75NA	10NA	625MWF	5MA	12V	15V	30V	15uV/C	.	3MA	80dB	80dB	1.5M
MLM207G	MTU	GPK	INT	.	.	+22V	-22V	85C	94dB	2MV	75NA	10NA	500MWF	5MA	12V	15V	30V	15uV/C	.	3MA	80dB	80dB	1.5M
MLM210G	MTU	VFA	INT	.	15V/uS	+18V	-18V	85C	0dB	4MV	3NA	.	500MWF	1MA	10V	15V	15V	50uV/C	.	6MA	.	70dB	10G
MLM211F	MTU	CPR	EXT	.	.	+18V	-18V	85C	100dB	3MV	100NA	10NA	500MWF	.	.	15V	30V		.	6MA	.	.	.
MLM211G	MTU	CPR	EXT	.	.	+18V	-18V	85C	100dB	3MV	100NA	10NA	680MWF	.	.	15V	30V		.	6MA	.	.	.
MLM211L	MTU	CPR	EXT	.	.	+18V	-18V	85C	100dB	3MV	100NA	10NA	625MWF	.	.	15V	30V		.	6MA	.	.	.
MLM224L	MTU	QGK	INT	.	.	+16V	-16V	85C	94dB	5MV	150NA	30NA	900MWF	.	.	16V	16V	35uV/C	.	2MA	70dB	65dB	.
MLM224P	MTU	QGK	INT	.	.	+16V	-16V	85C	94dB	5MV	150NA	30NA	570MWF	.	.	16V	16V	35uV/C	.	2MA	70dB	65dB	.
MLM239AL	MTU	QCP	EXT	.	.	+18V	-18V	85C	94dB	2MV	250NA	50NA	900MWF	6MA	.	18V	18V		.	2MA	.	.	.
MLM239AP	MTU	QCP	EXT	.	.	+18V	-18V	85C	94dB	2MV	250NA	50NA	900MWF	6MA	.	18V	18V		.	2MA	.	.	.
MLM239L	MTU	QCP	EXT	.	.	+18V	-18V	85C	88dB	5MV	250NA	50NA	900MWF	6MA	.	18V	18V		.	2MA	.	.	.
MLM239P	MTU	QCP	EXT	.	.	+18V	-18V	85C	88dB	5MV	250NA	50NA	900MWF	6MA	.	18V	18V		.	2MA	.	.	.
MLM301AG	MTU	GPU	EXT	.	.	+18V	-18V	75C	88dB	7.5MV	250NA	50NA	500MWF	5MA	12V	15V	30V	30uV/C	.	3MA	70dB	70dB	0.5M
MLM301AP1	MTU	GPU	EXT	.	.	+18V	-18V	75C	88dB	7.5MV	250NA	50NA	625MWF	5MA	12V	15V	30V	30uV/C	.	3MA	70dB	70dB	0.5M
MLM301AU	MTU	GPU	EXT	.	.	+18V	-18V	75C	88dB	7.5MV	250NA	50NA	625MWF	5MA	12V	15V	30V	30uV/C	.	3MA	70dB	70dB	0.5M
MLM307G	MTU	GPK	INT	.	.	+18V	-18V	70C	88dB	7.5MV	250NA	50NA	500MWF	5MA	12V	15V	30V	30uV/C	.	3MA	70dB	70dB	0.5M
MLM307P1	MTU	GPK	INT	.	.	+18V	-18V	70C	88dB	7.5MV	250NA	50NA	500MWF	5MA	12V	15V	30V	30uV/C	.	3MA	70dB	70dB	0.5M
MLM307U	MTU	GPK	INT	.	.	+18	-18	70C	88dB	7.5MV	250NA	50NA	500MWF	5MA	12V	15V	30V	30uV/C	.	3MA	70dB	70dB	0.5M
MLM308AF	MTU	LBC	EXT	.	.	+18V	-18V	70C	98dB	0.5MV	7NA	1NA	500MWF	1MA	13V	15V	1V	5uV/C	.	.6MA	96dB	96dB	10M
MLM308AG	MTU	LBC	EXT	.	.	+18V	-18V	70C	98dB	0.5MV	7NA	1NA	500MWF	1MA	13V	15V	1V	5uV/C	.	.6MA	96dB	96dB	10M
MLM308AP1	MTU	LBC	EXT	.	.	+18V	-18V	70C	98dB	0.5MV	7NA	1NA	500MWF	1MA	-13V	15V	1V	5uV/C	.	.6MA	96dB	96dB	10M
MLM308AU	MTU	LBC	EXT	.	.	+18V	-18V	70C	98dB	0.5MV	7NA	1NA	500MWF	1MA	13V	15V	1V	5uV/C	.	.6MA	96dB	96dB	10M
MLM308F	MTU	LBC	EXT	.	.	+18V	-18V	70C	88dB	7.5MV	7NA	1NA	500MWF	1MA	13V	15V	1V	30uV/C	.	.6MA	80dB	80dB	10M
MLM308G	MTU	LBC	EXT	.	.	+18V	-18V	70C	88dB	7.5MV	7NA	1NA	500MWF	1MA	13V	15V	1V	30uV/C	.	.6MA	80dB	80dB	10M
MLM308P1	MTU	LBC	EXT	.	.	+18V	-18V	70C	88dB	7.5MV	7NA	1NA	500MWF	1MA	13V	15V	1V	30uV/C	.	.6MA	80dB	80dB	10M

For detailed explanations of column heading notations, see App. A.

Also for ready references the more important abbreviations used in the column headings are listed below:

LEFT HAND PAGE

APP = application (codes at APP.E.)

CMRR = common mode rejection ratio

CMP = compensation (frequency)

dV_{io}/dT = input offset voltage temperature drift

GBP = gain bandwidth product

I_b = input bias current

I = input bias offset current

I_q = quiescent supply current

MFR = manufacturer (codes at App.C.)

P_q = quiescent power consumer

PSRR = power supply rejection ratio

V_{ICM} = common mode input voltage rating

V_{IDF} = differential input voltage rating

V_{io} = input offset voltage

V_S = dc supply voltage

RIGHT HAND PAGE

Lead out coding summary (details at APP.G.) for different cases (APP.F.)

A = gain adjust

B = bias adjust

C = case

E— = inverting input

E+ = non-inverting input

F,F* = input frequency compensation

G = ground

J = high level input

K = output, open collector

L = output, open emitter

M = metal case

N = not connected

Q = special terminal

R,R* = outputs

S = strobe

T,T* = offset balance

V+ = +ve dc supply

V— = —ve dc supply

W = guard ring

X = blank position, no lead

+ + = +ve supplementary dc supply

— — = —ve supplementary dc supply

Φ,Φ* = output frequency compensation

CASE (APP.F)	LD1	LD2	LD3	LD4	LD5	LD6	LD7	LD8	LD9	LD10	LD11	LD12	LD13	LD14	LD15	LD16	EUROPE SUBSTITUTE	USA SUBSTITUTE	IS	TYPE NUMBER
TO5-8/1M	G	E+	E-	V-	T	T*S	R	V+										LF211H	0	MLF211G
DIL-8/1P	G	E+	E-	V-	T	T*S	R	V+										MLF211U	0	MLF211P
DIL-8/1C	G	E+	E-	V-	T	T*S	R	V+										MLF111U	0	MLF211U
TO5-8/1M	G	E+	E-	V-	T	T*S	R	V+									UAF311H	LF311H	0	MLF311G
DIL-8/1P	G	E+	E-	V-	T	T*S	R	V+										MLF211U	0	MLF311P
DIL-8/1C	G	E+	E-	V-	T	T*S	R	V+										MLF211U	0	MLF311U
TO5-8/1M	T	E-	E+	V-	T*	R	V+	N									UAF355AHC	LF355AH	0	MLF355AG
TO5-8/1M	T	E-	E+	V-	T*	R	V+	N									UAF355HC	LF355H	0	MLF355G
TO5-8/1M	T	E-	E+	V-	T*	R	V+	N									UA356AHC	LF356AH	0	MLF356AG
TO5-8/1M	T	E-	E+	V-	T*	R	V+	N									UA356HC	LF356H	0	MLF356G
TO5-8/1M	T	E-	E+	V-	T*	R	V+	N									UAF357AHC	LF357AH	0	MLF357AG
TO5-8/1M	T	E-	E+	V-	T*	R	V+	N									UAF357HC	LF357H	0	MLF357G
TO5-8/1M	FT	E-	E+	V-M	T*	R	V+	F*									SFC2101A	LM101AH	0	MLM101AG
DIL-8/1C	FT	E-	E+	V-	T*	R	V+	F*										SN52101AJP	0	MLM101AU
TO5-8/1M	N	E-	E+	V-M	N	R	V+	N									SFC2107M	LM107H	0	MLM107G
DIL-8/1C	N	E-	E+	V-	N	R	V+	N									SN52107JP	LM107J	0	MLM107U
FLP-10/3C	N	N	E-	E+	N	V-	R	V+	F*	F							UA108AF	LM108AJ	0	MLM108AF
TO5-8/1M	F	E-	E+	V-	N	R	V+	F*									SFC2108A	LM108AH	0	MLM108AG
DIL-8/1C	F	E-	E+	V-	N	R	V+	F*									SFC2108A	LM108AH	0	MLM108AU
FLP-10/3C	N	N	E-	E+	N	V-	R	V+	F*	F							SFC2108PM	LM108F	0	MLM108F
TO5-8/1M	F	E-	E+	V-	N	R	V+	F*									SFC2108M	LM108H	0	MLM108G
DIL-8/1U	F	E-	E+	V-	N	R	V+	F*											0	MLM108U
TO5-8/1M	T	N	E+	V-	L	R	V+	T*									SFC2110M	LM110H	0	MLM110G
FLP-10/3C	G	E+	E-	N	V-	T	T*S	N	R	V+							SN52111FA	LM111F	0	MLM111F
TO5-8/1M	G	E+	E-	V-	T	T*S	R	V+									SFC2111M	LM111H	0	MLM111G
DIL-14/1C	N	G	E+	E-	N	V-	T	T*S	R	N	V+	N	N	N			SN52111J	LM111D	0	MLM111L
DIL-14/1C	R1	E-1	E+1	V+	E+2	E-2	R2	R3	E-3	E+3	G	E+4	E-4	R4				LM124D	0	MLM124L
DIL-14/1P	R1	E-1	E+1	V+	E+2	E-2	R2	R3	E-3	E+3	G	E+4	E-4	R4				LM124L	0	MLM124P
DIL-14/1C	R2	R1	V+	E-1	E+1	E-2	E+2	E+3	E-3	E-4	E+4	G	R4	R3				LM139AD	0	MLM139AL
DIL-14/1C	R2	R1	V+	E-1	E+1	E-2	E+2	E+3	E-3	E-4	E+4	G	R4	R3				LM139J	0	MLM139L
TO5-8/1M	R1	E-1	E+1	G	E+2	E-2	R2	V+										LM158H	0	MLM158G
DIL-8/1C	R1	E-1	E+1	G	E+2	E-2	R2	V+											0	MLM158U
TO5-8/1M	FT	E-	E+	V-M	T*	R	V+	F*									SFC2101A	LM201AH	0	MLM201AG
DIL-8/1P	FT	E-	E+	V-	T*	R	V+	F*										SN52101AJP	0	MLM201AP1
DIL-8/1C	FT	E-	E+	V-	T*	R	V+	F*										SN52101AJP	0	MLM201AU
TO5-8/1M	N	E-	E+	V-M	N	R	V+	N									SFC2207	LM207H	0	MLM207G
TO5-8/1M	T	N	E+	V-	L	R	V+	T*									SFC2210	LM210H	0	MLM210G
FLP-10/3C	G	E+	E-	N	V-	T	T*S	N	R	V+							SN52111FA	LM211F	0	MLM211F
TO5-8/1M	G	E+	E-	V-	T	T*S	R	V+									SFC2211	LM211H	0	MLM211G
DIL-14/1C	N	G	E+	E-	N	V-	T	T*S	R	N	V+	N	N	N			SN52111J	LM211D	0	MLM211L
DIL-14/1C	R1	E-1	E+1	V+	E+2	E-2	R2	R3	E-3	E+3	G	E+4	E-4	R4			SG224J	LM224J	0	MLM224L
DIL-14/1P	R1	E-1	E+1	V+	E+2	E-2	R2	R3	E-3	E+3	G	E+4	E-4	R4			SG224J	LM224D	0	MLM224P
DIL-14/1C	R2	R1	V+	E-1	E+1	E-2	E+2	E+3	E-3	E-4	E+4	G	R4	R3				LM239AD	0	MLM239AL
DIL-14/1P	R2	R1	V+	E-1	E+1	E-2	E+2	E+3	E-3	E-4	E+4	G	R4	R3				LM239AJ	0	MLM239AP
DIL-14/1C	R2	R1	V+	E-1	E+1	E-2	E+2	E+3	E-3	E-4	E+4	G	R4	R3				LM239J	0	MLM239L
DIL-14/1P	R2	R1	V+	E-1	E+1	E-2	E+2	E+3	E-3	E-4	E+4	G	R4	R3				LM239J	0	MLM239P
TO5-8/1M	FT	E-	E+	V-	T*	R	V+	F*									SFC2301A	LM301AH	0	MLM301AG
DIL-8/1P	FT	E-	E+	V-	T*	R	V+	F*									SFC2301ADC	LM301AJ	0	MLM301AP1
DIL-8/1C	FT	E-	E+	V-	T*	R	V+	F*									SFC2301ADC	LM301AJ	0	MLM301AU
TO5-8/1M	N	E-	E+	V-M	N	R	V+	N									SFC2307	LM307H	0	MLM307G
DIL-8/1P	N	E-	E+	V-	N	R	V+	N									SFC2307DC	LM307J	0	MLM307P1
DIL-8/1C	N	E-	E+	V-	N	R	V+	N									SFC2307DC	LM307J	0	MLM307U
FLP-10/3C	N	N	E-	E+	N	V-	R	V+	F*	F								LM208AF	0	MLM308AF
TO5-8/1M	F	E-	E+	V-M	N	R	V+	F*									SFC2308A	LM308AH	0	MLM308AG
DIL-8/1P	F	E-	E+	V-	N	R	V+	F*										MLM308AU	0	MLM308AP1
DIL-8/1C	F	E-	E+	V-	N	R	V+	F*										MLM308AP1	0	MLM308AU
FLP-10/3C	N	N	E-	E+	N	V-	R	V+	F*	F							SFC2208PT	LM208F	0	MLM308F
TO5-8/1M	F	E-	E+	V-M	N	R	V+	F*									SFC2308A	LM308A	0	MLM308AG
DIL-8/1P	F	E-	E+	V-	N	R	V+	F*									SFC2308DC	LM308N	0	MLM308P1
DIL-8/1C	F	E-	E+	V-	N	R	V+	F*									SFC2308DC	LM308N	0	MLM308U

TYPE NUMBER	M F R	A P P	C M P	GBP MIN	SLEW RATE MIN	V_S^+ MAX	V_S^- MAX	T_{op} MAX	A_{VOL} MIN	V_{IO} MAX	I_B MAX	I_{IO} MAX	P_{TOT} MAX	I_{OUT} MIN	V_{OUT} MIN	V_{ICM} MAX	V_{IDF} MAX	dV_{IO}/dT MAX	P_O MAX	I_O MAX	CM RR MIN	PS RR MIN	R_{IN} MIN
MLM308U	MTU	LBC	EXT	.		+18V	-18V	70C	88dB	7.5MV	7NA	1NA	500MWF	1MA	13V	15V	1V	30uV/C		.6MA	80dB	80dB	10M
MLM310G	MTU	VFA	INT	15V/uS		+18V	-18V	70C	0dB	7.5MV	7NA	.	500MWF	1MA	10V	15V	15V	50uV/C		6MA		70dB	10G
MLM311F	MTU	CPR	EXT	.		+18V	-18V	70C	100dB	7.5MV	250NA	50NA	500MWF			15V	30V			8MA		.	.
MLM311G	MTU	CPR	EXT	.		+18V	-18V	70C	100dB	7.5MV	250NA	50NA	680MWF			15V	30V			8MA			
MLM311L	MTU	CPR	EXT			+18V	-18V	70C	100dB	7.5MV	250NA	50NA	625MFW			15V	30V.			8MA			
MLM311P1	MTU	CPR	EXT	.		+18V	-18V	70C	100dB	7.5MV	250NA	50NA	625MWF			15V	30V			8MA			
MLM324L	MTU	QLQ	INT			+16V	-16V	70C	88dB	7MV	250NA	50NA	900MWF			16V	16V	35uV/C		2MA	65dB	65dB	
MLM324P	MTU	QLQ	INT			+16V	-16V	70C	88dB	7MV	250NA	50NA	900MWF			16V	16V	35uV/C		2MA	65dB	65dB	
MLM339AL	MTU	QCP	EXT			+18V	-18V	70C	94dB	2MV	250NA	50NA	900MWF	6MA		18V	18V			2MA			
MLM339AP	MTU	QCP	EXT			+18V	-18V	70C	94dB	2MV	250NA	50NA	900MWF	6MA		18V	18V			2MA			
MLM339L	MTU	QCP	EXT			+18V	-18V	70C	88dB	5MV	250NA	50NA	900MWF	6MA		18V	18V			2MA			
MLM339P	MTU	QCP	EXT			+18V	-18V	70C	88dB	5MV	250NA	50NA	900MWF	6MA		18V	18V			2MA			
MLM358G	MTU	DLQ	INT			+16V	-16V	70C	88dB	7MV	250NA	50NA	500MWF	10MA		16V	32V	30uV/C		3MA	65dB	65dB	
MLM358P1	MTU	DLQ	INT			+16V	-16V	70C	88dB	7MV	250NA	50NA	570MWF	10MA		16V	32V	30uV/C		3MA	65dB	65dB	
MLM358U	MTU	DLQ	INT			+16V	-16V	70C	88dB	7MV	250NA	50NA	570MWF	10MA		16V	32V	30uV/C		3MA	65dB	65dB	
MLM2901P	MTU	QCP	EXT			+18V	-18V	85C	88dB	7MV	250NA	50NA	900MWF	6MA		18V	18V			2MA			
MLM2902P	MTU	QGK	INT			+16V	-16V	70C	88dB	7MV	250NA	50NA	900MWF			16V	16V	35uV/C		2MA	50dB	50dB	
MLMC101A	MTU	GPU	EXT			+22V	-22V	125C	94dB	2MV	75NA	10NA		5MA	12V	15V	30V	15uV/C		3MA	80dB	80dB	1.5M
MLMC107	MTU	GPK	INT			+22V	-22V	125C	94dB	2MV	75NA	10NA		5MA	12V	15V	30V	15uV/C		3MA	80dB	80dB	1.5M
MLMC108	MTU	LBC	EXT			+20V	-20V	125C	96dB	2MV	2NA	0.2NA		1MA	13V	15V	1V	15uV/C		.6MA	85dB	80dB	30M
MLMC108A	MTU	LBC	EXT		.	+20V	-20V	125C	98dB	0.5MV	2NA	0.2NA		1MA	13V	15V	1V	5uV/C		.6MA	96dB	96dB	30M
MLMC110	MTU	VFA	INT	15V/uS		+18V	-18V	125C	0dB	4MV	3NA	.		1MA	10V	15V	15V	50uV/C		6MA		70dB	10G
MLMC111	MTU	CPR	EXT			+18V	-18V	125C	100dB	3MV	100NA	10NA				15V	30V			6MA	.	.	
MLMC124	MTU	QLQ	INT			+16V	-16V	125C	94dB	5MV	150NA	30NA				16V	16V	35uV/C		2MA	70dB	65dB	
MLMC139	MTU	QCP	EXT			+18V	-18V	125C	88dB	5MV	100NA	25NA				18V	18V			2MA			
MLMC139A	MTU	QCP	EXT			+18V	-18V	125C	94dB	2MV	100NA	25NA				18V	18V			2MA			
MLMC158	MFR	DLQ	INT			+16V	-16V	125C	94dB	5MV	150NA	30NA		10MA		16V	32V	30uV/C		3MA	70dB	65dB	
MLMC201A	MTU	GPU	EXT			+22V	-22V	85C	94dB	2MV	75NA	10NA		5MA	12V	15V	30V	15uV/C		3MA	80dB	80dB	1.5M
MLMC207	MTU	GPK	INT			+22V	-22V	85C	94dB	2MV	75NA	10NA		5MA	12V	15V	30V	15uV/C		3MA	80dB	80dB	1.5M
MLMC208	MTU	SBA	EXT			+20V	-20V	85C	96dB	2MV	2NA	0.2NA		1MA	13V	15V	1V	15uV/C		.6MA	85dB	80dB	30M
MLMC208A	MTU	SBA	EXT			+20V	-20V	85C	98dB	0.5MV	2NA	0.2NA		1MA	13V	15V	1V	5uV/C		.6MA	96dB	96dB	30M
MLMC210	MTU	VFA	INT	15V/uS		+18V	-18V	85C	0dB	4MV	3NA			1MA	10V	15V	15V	50uV/C		6MA		70dB	10G
MLMC211	MTU	CPR	EXT			+18V	-18V	85C	100dB	3MV	100NA	10NA				15V	30V			6MA	.	.	
MLMC224	MTU	QGK	INT			+16V	-16V	85C	94dB	5MV	150NA	30NA				16V	16V	35uV/C		2MA	70dB	65dB	
MLMC239	MTU	QCP	EXT			+18V	-18V	85C	88dB	5MV	250NA	50NA		6MA		18V	18V			2MA			
MLMC239A	MTU	QCP	EXT			+18V	-18V	85C	94dB	2MV	250NA	50NA		6MA		18V	18V			2MA			
MLMC258	MTU	DGK	INT			+16V	-16V	85C	94dB	5MV	150NA	30NA		10MA		16V	32V	30uV/C		3MA	70dB	65dB	
MLMC301A	MTU	GPU	EXT			+18V	-18V	75C	88dB	7.5MV	250NA	50NA		5MA	12V	15V	30V	30uV/C		3MA	70dB	70dB	0.5M
MLMC307	MTU	GPK	INT			+18V	-18V	70C	88dB	7.5MV	250NA	50NA		5MA	12V	15V	30V	30uV/C		3MA	70dB	70dB	0.5M
MLMC308	MTU	LBC	EXT			+18V	-18V	70C	88dB	7.5MV	7NA	1NA		1MA	13V	15V	1V	30uV/C		.6MA	80dB	80dB	10M
MLMC308A	MTU	LBC	EXT			+18V	-18V	70C	98dB	0.5MV	7NA	1NA		1MA	13V	15V	1V	5uV/C		.6MA	96dB	96dB	10M
MLMC310	MTU	VFA	INT	15V/uS		+18V	-18V	70C	0dB	7.5MV	7NA	.		1MA	10V	15V	15V	50uV/C		6MA		70dB	10G
MLMC311	MTU	CPR	EXT			+13V	-18V	70C	100dB	7.5MV	250NA	50NA				15V	30V			8MA			
MLMC324	MTU	QLQ	INT			+16V	-16V	70C	88dB	7MV	250NA	50NA				16V	16V	35uV/C		2MA	65dB	65dB	
MLMC339	MTU	QCP	EXT			+18V	-18V	70C	88dB	5MV	250NA	50NA		6MA		18V	18V			2MA			
MLMC339A	MTU	QCP	EXT			+18V	-18V	70C	94dB	2MV	250NA	50NA		6MA		18V	18V			2MA			
MLMC358	MTU	DLQ	INT			+16V	-16V	70C	88dB	7MV	250NA	50NA		10MA		16V	32V	30uV/C		3MA	65dB	65dB	
MLMC2902	MTU	QGK	INT			+16V	-16V	70C	88dB	7MV	250NA	50NA				16V	16V	35uV/C		2MA	50dB	50dB	
MONO-OP-01CJ	PRU	HSR	INT	5V/uS		+20V	-20V	70C	88dB	5MV	100NA	20NA	500MWF	6MA	12V	15V	30V	10uV/C	90MW		80dB	80dB	
MONO-OP-01CP	OBS	HSR	INT	5V/uS		+20V	-20V	70C	88dB	5MV	100NA	20NA	500MWF	6MA	12V	15V	30V	10uV/C	90MW		80dB	80dB	
MONO-OP-01CY	PRU	HSR	INT	5V/uS		+20V	-20V	70C	88dB	5MV	100NA	20NA	500MWF	6MA	12V	15V	30V	10uV/C	90MW		80dB	80dB	
MONO-OP-01EJ	PRU	HSR	INT	5V/uS		+22V	-22V	70C	94dB	2MV	50NA	2NA	500MWF	6MA	12V	15V	30V	8uV/C	90MW		80dB	80dB	
MONO-OP-01EP	OBS	HSR	INT	5V/uS		+22V	-22V	70C	94dB	2MV	50NA	2NA	500MWF	6MA	12V	15V	30V	8uV/C	90MW		80dB	80dB	
MONO-OP-01EY	PRU	HSR	INT	5V/uS		+22V	-22V	70C	94dB	2MV	50NA	2NA	500MWF	6MA	12V	15V	30V	8uV/C	90MW		80dB	80dB	
MONO-OP-01FJ	PRU	HSR	INT	5V/uS		+22V	-22V	125C	94dB	2MV	50NA	5NA	500MWF	6MA	12V	15V	30V	8uV/C	90MW		80dB	80dB	
MONO-OP-01FL	PRU	HSR	INT	5V/uS		+22V	-22V	125C	94dB	2MV	50NA	5NA	500MWF	6MA	12V	15V	30V	8uV/C	90MW		80dB	80dB	
MONO-OP-01FP	OBS	HSR	INT	5V/uS		+22V	-22V	125C	94dB	2MV	50NA	5NA	500MWF	6MA	12V	15V	30V	8uV/C	-90MW		80dB	80dB	
MONO-OP-01FY	PRU	HSR	INT	5V/uS		+22V	-22V	125C	94dB	2MV	50NA	5NA	500MWF	6MA	12V	15V	30V	8uV/C	90MW		80dB	80dB	
MONO-OP-01GJ	PRU	HSR	INT	5V/uS		+20V	-20V	125C	88dB	5MV	100NA	20NA	500MWF	6MA	12V	15V	30V	10uV/C	90MW		80dB	80dB	
MONO-OP-01GL	PRU	HSR	INT	5V/uS		+20V	-20V	125C	88dB	5MV	100NA	20NA	500MWF	6MA	12V	15V	30V	10uV/C	90MW		80dB	80dB	

For detailed explanations of column heading notations, see App. A.
Also for ready references the more important abbreviations used in the column headings are listed below:

LEFT HAND PAGE

APP = application (codes at APP.E.)
CMRR = common mode rejection ratio
CMP = compensation (frequency)
dV_{IO}/dT = input offset voltage temperature drift
GBP = gain bandwidth product
I_B = input bias current
I_{IO} = input bias offset current
I_Q = quiescent supply current
MFR = manufacturer (codes at App.C.)
P_G = quiescent power consumer
PSRR = power supply rejection ratio
V_{ICM} = common mode input voltage rating
V_{IOF} = differential input voltage rating
V_{IO} = input offset voltage
V_S = dc supply voltage

RIGHT HAND PAGE

Lead out coding summary (details at APP.G.) for different cases (APP.F.)

A = gain adjust
B = bias adjust
C = case
E− = inverting input
E+ = non-inverting input
F,F* = input frequency compensation
G = ground
J = high level input
K = output, open collector
L = output, open emitter
M = metal case
N = not connected
Q = special terminal
R,R* = outputs
S = strobe
T,T* = offset balance
V+ = +ve dc supply
V− = −ve dc supply
W = guard ring
X = blank position, no lead
++ = +ve supplementary dc supply
−− = −ve supplementary dc supply
ø,ø* = output frequency compensation

CASE (APP F)	LD1	LD2	LD3	LD4	LD5	LD6	LD7	LD8	LD9	LD10	LD11	LD12	LD13	LD14	LD15	LD16	EUROPE SUBSTITUTE	USA SUBSTITUTE	ISS	TYPE NUMBER
TO5-8/1M	T	N	E+	V−	L	R	V+	T*									SFC2310EC	LM310H	0	MLM310G
FLP-10/3C	G	E+	E−	N		V−	T	T*S	N	R	V+						SN72111FA	LM311F	0	MLM311F
TO5-8/1M	G	E+	E−	V−	T		T*S	R	V+								SFC2311	LM311H	0	MLM311G
DIL-14/1C	N	G	E+	E−	N	V−	T	T*S	R	N	V+	N	N	N			SFC2311EC	LM311D	0	MLM311L
DIL-8/1P	G	E+	E−	V−	T		T*S	R	V+								SFC2311DC	LM311N	0	MLM311P1
DIL-14/1C	R1	E−1	E+1	V+	E+2	E−2	R2	R3	E−3	E+3	G	E+4	E−4	R4				LM324J	0	MLM324L
DIL-14/1P	R1	E−1	E+1	V+	E+2	E−2	R2	R3	E−3	E+3	G	E+4	E−4	R4				LM324N	0	MLM324P
DIL-14/1C	R2	R1	V+	E−1	E+1	E−2	E+2	E+3	E−3	E−4	E+4	G	R4	R3				LM339AJ	0	MLM339AL
DIL-14/1P	R2	R1	V+	E−1	E+1	E−2	E+2	E+3	E−3	E−4	E+4	G	R4	R3				LM339AJ	0	MLM339AP
DIL-14/1C	R2	R1	V+	E−1	E+1	E−2	E+2	E+3	E−3	E−4	E+4	G	R4	R3				LM339J	0	MLM339L
DIL-14/1P	R2	R1	V+	E−1	E+1	E−2	E+2	E+3	E−3	E−4	E+4	G	R4	R3				LM339J	0	MLM339P
TO5-8/1M	R1	E−1	E+1	G	E+2	E−2	R2	V+										LM358H	0	MLM358G
DIL-8/1P	R1	E−1	E+1	G	E+2	E−2	R2	V+										LM358N	0	MLM358P1
DIL-8/1C	R1	E−1	E+1	G	E+2	E−2	R2	V+										LM358N	0	MLM358U
DIL-14/1P	R2	R1	V+	E−1	E+1	E−2	E+2	E+3	E−3	E−4	E+4	G	R4	R3				LM2901N	0	MLM2901P
DIL-14/1P	R1	E−1	E+1	V+	E+2	E−2	R2	R3	E−3	E+3	G	E+4	E−4	R4				LM2902N	0	MLM2902P
CHP																			0	MLMC101A
CHP																			0	MLMC107
CHP																			0	MLMC108
CHP																			0	MLMC108A
CHP																			0	MLMC110
CHP																			0	MLMC111
CHP																			0	MLMC124
CHP																			0	MLMC139
CHP																			0	MLMC139A
CHP																			0	MLMC158
CHP																			0	MLMC201A
CHP																			0	MLMC207
CHP																			0	MLMC208
CHP																			0	MLM208A
CHP																			0	MLMC210
CHP																			0	MLMC211
CHP																			0	MLMC224
CHP																			0	MLMC239
CHP																			0	MLMC239A
CHP																			0	MLMC258
CHP																			0	MLMC301A
CHP																			0	MLMC307
CHP																			0	MLMC308
CHP																			0	MLMC308A
CHP																			0	MLMC310
CHP																			0	MLMC311
CHP																			0	MLMC324
CHP																			0	MLMC339
CHP																			0	MLMC339A
CHP																			0	MLMC358
CHP																			0	MLMC2902
TO5-8/1M	T	E−	E+	V−M	T*	R	V+	N									OP-01GJ		0	MONO-OP-01CJ
DIL-14/1C	N	N	T	E−	E+	V−	N	N	T*	R	V+	N	N	N			OP-01CY		0	MONO-OP-01CP
DIL-14/1C	N	N	T	E−	E+	V−	N	N	T*	R	V+	N	N				OP-01GY		0	MONO-OP-01CY
TO5-8/1M	T	E−	E+	V−M	T*	R	V+	N									OP-01FJ		0	MONO-OP-01EJ
DIL-14/1C	N	N	T	E−	E+	V−	N	N	T*	R	V+	N	N	N			OP-01EY		0	MONO-OP-01EP
DIL-14/1C	N	N	T	E−	E+	V−	N	N	T*	R	V+	N	N				OP-01FY		0	MONO-OP-01EY
TO5-8/1M	T	E−	E+	V−M	T*	R	V+	N									OP-01FJ		0	MONO-OP-01FJ
FLP-10/3C	N		T	E−	E+	V−	T	R	V+	N	N						OP-01FL		0	MONO-OP-01FL
DIL-14/1C	N	N	T	E−	E+	V−	N	N	T*	R	V+	N	N	N			OP-01FY		0	MONO-OP-01FP
DIL-14/1C	N	N	T	E−	E+	V−	N	N	T*	R	V+	N	N				OP-01Y		0	MONO-OP-01FY
TO5-8/1M	T	E−	E+	V−M	T*	R	V+	N									OP-01GJ		0	MONO-OP-01G...
FLP-10/3C	N		T	E−	E+	V−	T	R	V+	N	N						OP-01GL		0	MONO-OP-01GL
DIL-14/1C	N	N	T	E−	E+	V−	N	N	T*	R	V+	N	N	N			OP-01GY		0	MONO-OP-01GF

TYPE NUMBER	MFR	APP	CMP	GBP MIN	SLEW RATE MIN	Vs+ MAX	Vs− MAX	Top MAX	Avol MIN	Vio MAX	Ib MAX	Iio MAX	Ptot MAX	Iout MIN	Vout MIN	Vicm MAX	Vidf MAX	dVio/dT MAX	Pq MAX	Iq MAX	CMRR MIN	PSRR MIN
MONO-OP-01GP	OBS	HSR	INT	.	5V/uS	+20V	−20V	125C	88dB	5MV	100NA	20NA	500MWF	6MA	12V	15V	30V	10uV/C	90MW	.	80dB	80dB
MONO-OP-01GY	PRU	HSR	INT	.	5V/uS	+20V	−20V	125C	88dB	5MV	100NA	20NA	500MWF	6MA	12V	15V	30V	10uV/C	90MW	.	80dB	80dB
MONO-OP-01HJ	PRU	HSR	INT	.	5V/uS	+22V	−22V	70C	94dB	0.7MV	30NA	2NA	500MWF	6MA	12V	15V	30V	5uV/C	60MW	.	90dB	90dB
MONO-OP-01HP	OBS	HSR	INT	.	5V/uS	+22V	−22V	70C	94dB	0.7MV	30NA	2NA	500MWF	6MA	12V	15V	30V	5uV/C	60MW	.	90dB	90dB
MONO-OP-01HY	PRU	HSR	INT	.	5V/uS	+22V	−22V	70C	94dB	0.7MV	30NA	2NA	500MWF	6MA	12V	15V	30V	5uV/C	60MW	.	90dB	90dB
MONO-OP-01J	PRU	HSR	INT	.	5V/uS	+22V	−22V	125C	94dB	0.7MV	30NA	2NA	500MWF	6MA	12V	15V	30V	5uV/C	60MW	.	90dB	90dB
MONO-OP-01L	PRU	HSR	INT	.	5V/uS	+22V	−22V	125C	94dB	0.7MV	30NA	2NA	500MWF	6MA	12V	15V	30V	5uV/C	60MW	.	90dB	90dB
MONO-OP-01P	OBS	HSR	INT	.	5V/uS	+22V	−22V	125C	94dB	0.7MV	30NA	2NA	500MWF	6MA	12V	15V	30V	5uV/C	60MW	.	90dB	90dB
MONO-OP-01Y	PRU	HSR	INT	.	5V/uS	+22V	−22V	125C	94dB	0.7MV	30NA	2NA	500MWF	6MA	12V	15V	30V	5uV/C	60MW	.	90dB	90dB.
MONO-OP-05AJ	PRU	PIA	INT	.3MHZ	0.1V/uS	+22V	−22V	125C	110dB	.15MV	2NA	2NA	500MWF	10MA	12V	22V	30V	0.9uV/C	120MW	.	114dB	100dB
MONO-OP-05AL	PRU	PIA	INT	.3MHZ	0.1V/uS	+22V	−22V	125C	110dB	.15MV	2NA	2NA	500MWF	10MA	12V	22V	30V	0.9uV/C	120MW	.	114dB	100dB
MONO-OP-05AY	PRU	PIA	INT	.3MHZ	0.1V/uS	+22V	−22V	125C	110dB	.15MV	2NA	2NA	500MWF	10MA	12V	22V	30V	0.9uV/C	120MW	.	114dB	100dB
MONO-OP-05CJ	PRU	PIA	INT	.3MHZ	0.1V/uS	+22V	−22V	70C	100dB	1.3MV	7NA	6NA	500MWF	10MA	12V	22V	30V	4.5uV/C	150MW	.	100dB	90dB
MONO-OP-05CY	PRU	PIA	INT	.3MHZ	0.1V/uS	+22V	−22V	70C	100dB	1.3MV	7NA	6NA	500MWF	10MA	12V	22V	30V	4.5uV/C	150MW	.	100dB	90dB
MONO-OP-05EJ	PRU	PIA	INT	.3MHZ	0.1V/uS	+22V	−22V	70C	106dB	0.5MV	4NA	3.8NA	500MWF	10MA	12V	22V	30V	2uV/C	120MW	.	110dB	94dB
MONO-OP-05EY	PRU	PIA	INT	.3MHZ	0.1V/uS	+22V	−22V	70C	106dB	0.5MV	4NA	3.8NA	500MWF	10MA	12V	22V	30V	2uV/C	120MW	.	110dB	94dB
MONO-OP-05J	PRU	PIA	INT	.3MHZ	0.1V/uS	+22V	−22V	125C	106dB	0.5MV	3NA	2.8NA	500MWF	10MA	12V	22V	30V	2uV/C	120MW	.	114dB	100dB
MONO-OP-05L	PRU	PIA	INT	.3MHZ	0.1V/uS	+22V	−22V	125C	106dB	0.5MV	3NA	2.8NA	500MWF	10MA	12V	22V	30V	2uV/C	120MW	.	114dB	100dB
MONO-OP-05Y	PRU	PIA	INT	.3MHZ	0.1V/uS	+22V	−22V	125C	106dB	0.5MV	3NA	2.8NA	500MWF	10MA	12V	22V	30V	2uV/C	120MW	.	114dB	100dB
MONO-OP-10AY	PRU	DPI	INT	.3MHZ	.06V/uS	+22V	−22V	125C	106dB	0.5MV	3NA	2.8NA	500MWF	10MA	12V	22V	30V	2uV/C	120MW	.	114dB	100dB
MONO-OP-10CY	PRU	DPI	INT	.3MHZ	.06V/uS	+22V	−22V	70C	102dB	1.3MV	7NA	6NA	500MWF	10MA	12V	22V	30V	4.5uV/C	150MW	.	100dB	90dB
MONO-OP-10EY	PRU	DPI	INT	.3MHZ	.06V/uS	+22V	−22V	70C	106dB	0.5MV	4NA	3.8NA	500MWF	10MA	12V	22V	30V	2uV/C	120MW	.	110dB	94dB
MONO-OP-10Y	PRU	DPI	INT	.3MHZ	.06V/uS	+22V	−22V	125C	106dB	0.5MV	3NA	2.8NA	500MWF	10MA	12V	22V	30V	2uV/C	120MW	.	114dB	100dB
N5556T	SJU	SBA	INT	.5MHZ	1V/uS	+18V	−18V	70C	96dB	10MV	30NA	10NA	500MWF	5MA	11V	18V	18V	40uV/C	90MW	3MA	70dB	74dB
N5556V	SJU	SBA	INT	.5MHZ	1V/uS	+18V	−18V	70C	96dB	10MV	30NA	10NA	680MWF	5MA	11V	18V	18V	40uV/C	90MW	3MA	70dB	74dB
N5558F	OBS	DGK	INT	.5MHZ	0.3V/uS	+18V	−18V	75C	86dB	6MV	0.5uA	0.2uA	750MWF	5MA	12V	15V	30V	50uV/C	170MW	6MA	70dB	76dB
N5558T	SJU	DGK	INT	.5MHZ	0.3V/uS	+18V	−18V	75C	88dB	6MV	0.5uA	0.2uA	500MWF	5MA	12V	15V	30V	.	125MW	6MA	70dB	76dB
N5558V	SJU	DGK	INT	.5MHZ	0.3V/uS	+18V	−18V	75C	88dB	6MV	0.5uA	0.2uA	400MWF	5MA	12V	15V	30V	.	125MW	6MA	70dB	76dB
N5709A	OBS	GPU	EXT	.3MHZ	.15V/uS	+18V	−18V	70C	84dB	7.5MV	1.5uA	500NA	670MWF	5MA	12V	10V	5V	.	200MW	.	65dB	74dB
N5709G	OBS	GPU	EXT	.3MHZ	.15V/uS	+18V	−18V	125C	88dB	5MV	500NA	200NA	500MWF	5MA	12V	10V	5V	15uV/C	165MW	.	70dB	76dB
N5709T	OBS	GPU	EXT	.3MHZ	.15V/uS	+18V	−18V	70C	84dB	7.5MV	1.5uA	500NA	500MWF	5MA	12V	10V	5V	.	200MW	.	65dB	74dB
N5709V	OBS	GPU	EXT	.3MHZ	.15V/uS	+18V	−18V	125C	88dB	5MV	500NA	200NA	750MWF	5MA	12V	10V	5V	15uV/C	165MW	.	70dB	76dB
N5710A	OBS	CPR	EXT	.	.	+14V	−6V	70C	60dB	5MV	25uA	5uA	670MWF	5MA	2.5V	7V	5V	20uV/C	150MW	.	70dB	.
N5710T	OBS	CPR	EXT	.	.	+14V	−6V	70C	60dB	5MV	25uA	5uA	500MWF	5MA	2.5V	7V	5V	20uV/C	150MW	.	70dB	.
N5711A	OBS	DCP	EXT	.	.	+14V	−7V	70C	57dB	5MV	100uA	15uA	670MWF	5MA	2.5V	7V	5V	20uV/C	230MW	.	.	.
N5711K	OBS	DCP	EXT	.	.	+14V	−7V	70C	57dB	5MV	100uA	15uA	500MWF	5MA	2.5V	7V	5V	20uV/C	230MW	.	.	.
N5733K	OBS	BDO	EXT	20MHZ	.	+8V	−8V	70C	48dB	6MV	30uA	5uA	500MWF	2MA	1.5V	6V	5V	.	.	24MA	60dB	50dB
N5741A	OBS	GPK	INT	.	0.2V/uS	+18V	−18V	70C	86dB	6MV	500NA	200NA	670MW	5MA	12V	15V	30V	.	85MW	3MA	70dB	76dB
N5741T	OBS	GPK	INT	.	0.2V/uS	+18V	−18V	70C	86dB	6MV	500NA	200NA	500MWF	5MA	12V	15V	30V	.	.	3MA	70dB	76dB
N5741V	OBS	GPK	INT	.	0.2V/uS	+18V	−18V	70C	86dB	6MV	500NA	200NA	310MWF	5MA	12V	15V	30V	.	.	3MA	70dB	76dB
N5747A	OBS	DGK	INT	.	0.2V/uS	+18V	−18V	70C	88dB	6MV	500NA	200NA	670MWF	5MA	12V	15V	30V	.	85MW	3MA	70dB	76dB
N5747F	OBS	DGK	INT	.	0.2V/uS	+18V	−18V	70C	88dB	6MV	500NA	200NA	670MWF	5MA	12V	15V	30V	.	85MW	3MA	70dB	76dB
N5748A	OBS	GPU	EXT	.	0.2V/uS	+22V	−22V	70C	86dB	6MV	500NA	200NA	670MWF	5MA	12V	15V	30V	.	85MW	3MA	70dB	76dB
N5748T	OBS	GPU	EXT	.	0.2V/uS	+18V	−18V	70C	86dB	6MV	500NA	200NA	500MWF	5MA	12V	15V	30V	.	85MW	3MA	70dB	76dB
N5748V	OBS	GPU	EXT	.	0.2V/uS	+18V	−18V	70C	86dB	6MV	500NA	200NA	310MWF	5MA	12V	15V	30V	.	85MW	3MA	70dB	76dB
NE515A	SJU	PIA	EXT	.	.	+6V	−6V	125C	68dB	3MV	31uA	.	.	30MA	5.3V	1.5V	5V	.	.	7MA	94dB	.
NE515G	SJU	PIA	EXT	.	.	+6V	−6V	75C	68dB	3MV	31uA	.	.	30MA	5.3V	1.5V	5V	.	.	7MA	94dB	.
NE515K	SJU	PIA	EXT	.	.	+6V	−6V	75C	68dB	3MV	31uA	.	.	30MA	5.3V	1.5V	5V	.	.	7MA	94dB	.
NE527A	MUG	CPR	EXT	.	.	+15V	−15V	70C	66dB	6MV	2uA	.75uA	600MWF	.	1V	6V	5V	.	.	.	.	.
NE527N(14)	MUG	CPR	EXT	.	.	+15V	−15V	70C	66dB	6MV	2uA	.75uA	600MWF	.	1V	6V	5V	.	.	.	.	.
NE527K	SJU	CPR	EXT	.	.	+15V	−15V	70C	66dB	6MV	2uA	0.5uA	600MWF	.	1V	6V	5V	.	.	.	.	.
NE529A	MUG	CPR	EXT	.	.	+15V	−15V	70C	66dB	6MV	20uA	5uA	600MWF	.	1V	6V	5V	.	.	.	.	.
NE529K	SJU	CPR	EXT	.	.	+15V	−15V	125C	66dB	6MV	20uA	5uA	500MWF	.	1V	6V	5V	.	.	.	.	.
NE529N(14)	MUG	CPR	EXT	.	.	+15V	−15V	70C	66dB	6MV	20uA	5uA	600MWF	.	1V	6V	5V	.	.	.	.	.
NE531N(8)	MUG	HSR	EXT	.5MHZ	20V/uS	+22V	−22V	70C	86dB	6MV	1.5uA	0.2NA	500MWF	.	15V	15V	.	.	300MW	10MA	70dB	76dB
NE531T	SJU	HSR	EXT	.5MHZ	15V/uS	+22V	−22V	70C	86dB	6MV	1.5uA	200NA	300MWF	.	15V	15V	.	.	300MW	10MA	70dB	76dB
NE531V	SJU	HSR	EXT	.5MHZ	20V/uS	+22V	−22V	70C	86dB	6MV	1.5uA	200NA	300MWF	.	15V	15V	.	.	300MW	10MA	70dB	76dB
NE532N(8)	SJU	DGK	INT	.3MHZ	.	+15V	−15V	70C	84dB	7.5MV	500NA	150NA	.	.	14V	14V	16V	20uV/C	.	2MA	60dB	80dB
NE532T	SJU	DGK	INT	.3MHZ	.	+15V	−15V	70C	84dB	7.5MV	500NA	150NA	.	.	14V	14V	16V	20uV/C	.	2MA	60dB	80dB
NE532V	SJU	DGK	INT	.3MHZ	.	+15V	−15V	70C	84dB	7.5MV	500NA	150NA	.	.	14V	14V	16V	20uV/C	.	2MA	60dB	80dB

For detailed explanations of column heading notations, see App. A.

Also for ready references the more important abbreviations used in the column headings are listed below:

LEFT HAND PAGE

- APP = application (codes at APP.E.)
- CMRR = common mode rejection ratio
- CMP = compensation (frequency)
- dV_{io}/dT = input offset voltage temperature drift
- GBP = gain bandwidth product
- I_B = input bias current
- I_{io} = input bias offset current
- I_Q = quiescent supply current
- MFR = manufacturer (codes at App.C.)
- P_Q = quiescent power consumer
- PSRR = power supply rejection ratio
- V_{ICM} = common mode input voltage rating
- V_{IDR} = differential input voltage rating
- V_{io} = input offset voltage
- V_S = dc supply voltage

RIGHT HAND PAGE

Lead out coding summary (details at APP.G.) for different cases (APP.F.)

- A = gain adjust
- B = bias adjust
- C = case
- E— = inverting input
- E+ = non-inverting input
- F,F* = input frequency compensation
- G = ground
- J = high level input
- K = output, open collector
- L = output, open emitter
- M = metal case
- N = not connected
- O = special terminal
- R,R* = outputs
- S = strobe
- T T* = offset balance
- V+ = +ve dc supply
- V = —ve dc supply
- W = guard ring
- X = blank position, no lead
- + + = +ve supplementary dc supply
- —— = —ve supplementary dc supply
- ø,ø* = output frequency compensation

CASE (APP F)	LD 1	LD 2	LD 3	LD 4	LD 5	LD 6	LD 7	LD 8	LD 9	LD 10	LD 11	LD 12	LD 13	LD 14	LD 15	LD 16	EUROPE SUBSTITUTE	USA SUBSTITUTE	I/S	TYPE NUMBER
DIL-14/1C	N	N	T	E-	E+	V-	N	N	T*	R	V+	N	N	N				OP-01GY	0	MONO-OP-01G
T05-8/1M	T	E-	E+	V-M	T*	R	V+	N										OP-01HJ	0	MONO-OP-01H
DIL-14/1C	N	N	T	E-	E+	V-	N	N	T*	R	V+	N	N	N				OP-01HY	0	MONO-OP-01H
DIL-14/1C	N	N	T	E-	E+	V-	N	N	T*	R	V+	N	N	N				OP-01HY	0	MONO-OP-01N
T05-8/1M	T	E-	E+	V-M	T*	R	V+	N										OP-01J		MONO-OP-01J
FLP-10/3C	N	T	E-	E+	V-	T	R	V+	N	N								OP-01L	0	MONO-OP-01L
DIL-14/1C	N	N	T	E-	E+	V-	N	N	T*	R	V+	N	N	N				OP-01Y	0	MONO-OP-01P
DIL-14/1C	N	N	T	E-	E+	V-	N	N	T*	R	V+	N	N	N				OP-01Y	0	MONO-OP-01Y
T05-8/1M	T	E-	E+	V-M	N	R	V+	T*										OP-05AJ	0	MONO-OP-05A
FLP-10/3C	N	T	E-	E+	V	N	R	V+	T*	N								OP-05AL		MONO-OP-05AL
DIL-14/1C	N	N	T	E-	E+	V-	N	N	N	R	V+	T*	N	N				OP-05AY	0	MONO-OP-05A
T05-8/1M	T	E-	E+	V-M	N	R	V+	T*										OP-05EY	0	MONO-OP-05C
DIL-14/1C	N	N	T	E-	E+	V-	N	N	N	R	V+	T*	N	N				OP-05CY	0	MONO-OP-05C
T05-8/1M	T	E-	E+	V-M	N	R	V+	T*										OP-05EJ	0	MONO-OP-05E
DIL-14/1C	N	N	T	E-	E+	V-	N	N	N	R	V+	T*	N	N				OP-05EY	0	MONO-OP-05E
T05-8/1M	T	E-	E+	V-M	N	R	V+	T*										OP-05J		MONO-OP-05J
FLP-10/3C	N	T	E-	E+	V-	N	R	V+	T*	N								OP-05L	0	MONO-OP-05L
DIL-14/1C	N	N	T	E-	E+	V-	N	N	N	R	V+	T*	N	N				OP-05Y	0	MONO-OP-05Y
DIL-14/1C	T1	T*1	E-1	E+1	V-2	R2	V+2	T2	T*2	E-2	E+2	V-1	R1	V+1				OP-10AY	0	MONO-OP-10A
DIL-14/1C	T1	T*1	E-1	E+1	V-2	R2	V+2	T2	T*2	E-2	E+2	V-1	R1	V+1				OP-10CY	0	MONO-OP-10C
DIL-14/1C	T1	T*1	E-1	E+1	V-2	R2	V+2	T2	T*2	E-2	E+2	V-1	R1	V+1				OP-10EY	0	MONO-OP-10E
DIL-14/1C	T1	T*1	E-1	E+1	V-2	R2	V+2	T2	T*2	E-2	E+2	V-1	R1	V+1				OP-10Y	0	MONO-OP-10Y
T05-8/1M	T	E-	E+	V-	T*	R	V+	N									MC1456G	MC1456T	0	N5556T
DIL-8/1P	T	E-	E+	V-	T*	R	V+	N									RC1556NB	MC1456V	0	N5556V
DIL-14/1C	N	R1	N	N	E-1	E+1	V-	E+2	E-2	N	N	R2	N	V+			LM1458N14	MC1458L	0	N5558F
T05-8/1M	R1	E-1	E+1	V-	E+2	E-2	R2	V+									TBB1458	MC1458G	0	N5558T
DIL-8/1P	R1	E-1	E+1	V-	E+2	E-2	R2	V+									TBB1458B	MC1458U	0	N5558V
DIL-14/1P	N	N	F	E-	E+	V-	N	N	ø	R	V+	F*	N	N			TAA521A	UA709DC	0	N5709A
FLP-10/3C	N	F	E-	E+	V-	ø	R	V+	F*	N							UA709FM	MC1709F	0	N5709G
T05-8/1M	F	E-	E+	V-	ø	ø*R	V+	F*									TAA521	UA709HC	0	N5709T
DIL-8/1P	F	E-	E+	V-	ø	R	V+	F*									SN52709AJP	MC1709U	0	N5709V
DIL-14/1P	N	G	E+	E-	N	V-	N	N	R	N	V+	N	N				SFC2710EC	UA7100C	0	N5710A
T05-8/1M	G	E+	E-	V-	N	N	R	V+									SFC2710C	UA710HC	0	N5710T
DIL-14/1P	N	E-1	E+1	V-	E+2	E-2	N	N	S2	R	V+	G	S1	N			SFC2711EC	UA711DC	0	N5711A
T05-10/1M	G	S1	E-1	E+1	V-	E+2	E-2	S2	R	V+							SFC2711C	UA711HC	0	N5711K
T05-10/1M	E-	E+	A2	A*2	V-	R	R*	V+	A1	A*1							LM733CH	UA733HC	0	N5733K
DIL-14/1P	N	N	T	E-	E+	V-	N	N	T*	R	V+	N	N	N			TBA221A	UA741DC	0	N5741A
T05-8/1M	T	E-	E+	V-M	T*	R	V+	N									TBA221	UA741HC	0	N5741T
DIL-8/1P	T	E-	E+	V-	T*	R	V+	N									TBA221B	UA741TC	0	N5741V
DIL-14/1P	E-1	E+1	T1	V-	T2	E+2	E-2	T*2	V+2	R2	N	R1	V+1	T*1			TBB0747A	UA747DC	0	N5747A
DIL-14/1C	E-1	E+1	T1	V-	T2	E+2	E-2	T*2	V+2	R2	N	R1	V+1	T*1			TBB0747A	UA747DC	0	N5747F
DIL-14/1P	N	N	FT	E-	E+	V-	N	N	T*	R	V+	F*	N	N			SN72748J	UA748DC	0	N5748A
T05-8/1M	FT	E-	E+	V-	T*	R	V+	F*									TBB0748	UA748HC	0	N5748T
DIL-8/1P	FT	E-	E+	V-	T*	R	V+	F*									TBB0748	UA748TC	0	N5748V
DIL-14/1P	E2	N	N	F	N	R2	V-	R1	N	N	F*	N	E1	V+				SE515A	0	NE515A
T05-10/1M	V-M	R1	N	F	E1	V+	E2	F*	N	R2								SE515Q	0	NE515G
T05-10/1M	V-M	R1	NM	F	E1	V+	E2	F*	NM	R2								SE515K	0	NE515K
DIL-14/1P	V+1	N	E	E*	N	V-1	N	S*	R*	G	R	N	S	V+2					0	NE527A
DIL-14/1P	V+1	N	E	E*	N	V-1	N	S*	R*	G	R	N	S	V+2				NE527A	0	NE527N(14)
T05-10/1M	E	E*	V-1	S*	R*	G	R	S	V+2	V+1								SE527K	0	NE527K
DIL-14/1P	V+1	N	E	E*	N	V-1	N	S*	R*	G	R	N	S	V+2					0	NE529A
T05-10/1M	E	E*	V-1	S*	R*	G	R	S	V+2	V+1								SE529K	0	NE529K
DIL-14/1P	V+	N	E	E*	N	V-1	N	S*	R*	G	R	N	S	V+2				NE529A	0	NE529N(14)
DIL-8/1P	T	E-	E+	V-	T*	R	V+	F										NE531V	0	NE531N(8)
T05-8/1M	T	E-	E+	V-	T*	R	V+	F									HA2-2505	SE531T	0	NE531T
DIL-8/1P	T	E-	E+	V-	T*	R	V+	F										SE531V	0	NE531V
DIL-8/1P	R1	E-1	E+1	V-	E+2	E-2	R2	V+										NE532V	0	NE532N(8)
T05-8/1M	R1	E-1	E+1	V-	E+2	E-2	R2	V+										SE532T	0	NE532T
DIL-8/1P	R1	E-1	E+1	V-	E+2	E-2	R2	V+										SA532V	0	NE532V
DIL-8/1P	T	E-	E+	V-	T*	R	V+	N										NE535V	0	NE535N(8)

TYPE NUMBER	MFR	APP	CMP	GBP MIN	SLEW RATE MIN	Vs+ MAX	Vs⁻ MAX	Topp* MAX	Avol MIN	Vio MAX	Ib MAX	Iio MAX	Ptot MAX	Iout MIN	Vout MIN	Vicm MAX	Vidf MAX	dVio/dT MAX	Pq MAX	Io MAX	CM RR MIN	PS RR MIN	Rin MIN
NE535N(8)	MUG	HSR	INT	.3MHZ	5V/uS	+18V	-18V	70C	84dB	6MV	200NA	80NA	500MWF	5MA	12V	13V	30V	10uV/C	.	3MA	80dB	100dB	.
NE535T	MVG	HSR	INT	.3MHZ	5V/uS	+18V	-18V	70C	84dB	6MV	200NA	80NA	500MWF	5MA	12V	13V	30V	10uV/C	.	3MA	80dB	100dB	.
NE535V	MUG	HSR	INT	.3MHZ	5V/uS	+18V	-18V	70C	84dB	6MV	200NA	80NA	500MWF	5MA	12V	13V	30V	10uV/C	.	3MA	80dB	100dB	.
NE536T	SJU	FET	INT	.5MHZ	3V/uS	+22V	-22V	70C	94dB	90MV	100pA	10pA	500MWF	17MA	10V	22V	30V	300uV/C	350MW	8MA	64dB	70dB	50G
NE540L	SJU	HCO	EXT		1G0V/uS	+22V	-22V	70C	.	10MV	5uA	1uA	1WF	80MA	.	.	.	.	.	20MA	70dB	60dB	10K
NE592A	SJU	BDO	EXT	20MHZ	.	+8V	-8V	70C	48dB	6MV	30uA	5uA	670MWF	2MA	1.5V	6V	5V			24MA	60dB	50dB	2K
NE592F	MUG	BDO	EXT	20MHZ	.	+8V	-8V	70C	48dB	6MV	30uA	5uA	670MWF	2MA	1.5V	6V	5V			24MA	60dB	50dB	2K
NE592K	SJU	BDO	EXT	20MHZ	.	+8V	-8V	70C	48dB	6MV	30uA	5uA	500MWF	2MA	1.5V	6V	5V			24MA	60dB	50dB	2K
NE592N(14)	MUG	BDO	EXT	20MHZ	.	+8V	-8V	70C	48dB	6MV	30uA	5uA	670MWF	2MA	1.5V	6V	5V			24MA	60dB	50dB	2K
NH0001CH	OBS	XLP	EXT	.	.	+20V	-20V	85C	88dB	1MV	100NA	20NA	400MWF	5MA	10V	20V	7V			.1MA	70dB	70dB	500K
NH0001H	OBS	XLP	EXT	.	.	+20V	-20V	125C	88dB	1MV	100NA	20NA	400MWF	5MA	10V	20V	7V			.1MA	70dB	70dB	500K
NH0003CH	OBS	WBA	EXT	.	.	+20V	-20V	85C	86dB	3MV	2uA	0.2uA	500MWF	0.1A	10V	20V	7V	20uV/C		3MA	70dB	70dB	25K
NH0003H	OBS	WBA	EXT	.	.	+20V	-20V	125C	86dB	3MV	2uA	0.2uA	500MWF	0.1A	10V	20V	7V	20uV/C		3MA	70dB	70dB	25K
NH0004CH	OBS	HVO	EXT	.	.	+45V	-45V	85C	90dB	1.5MV	120NA	45NA	400MWF	6MA	30V	45V	7V	20uV/C		.2MA	70dB	70dB	
NH0004H	OBS	HVO	EXT	.	.	+45V	-45V	125C	90dB	1MV	100NA	20NA	400MWF	6MA	30V	45V	7V	20uV/C		.2MA	70dB	70dB	
NH0005AH	OBS	WBA	EXT	.	.	+20V	-20V	125C	72dB	3MV	25NA	5NA	400MWF	50MA	6V	20V	15V	50uV/C		5MA	60dB	60dB	1M
NH0005CH	OBS	WBA	EXT	.	.	+20V	-20V	85C	66dB	10MV	100NA	25NA	400MWF	50MA	6V	20V	15V			5MA	50dB	50dB	0.5M
NH0005H	NAU	WBA	EXT	.	.	+20V	-20V	125C	66dB	10MV	50NA	20NA	400MWF	50MA	6V	20V	15V	100uV/C		5MA	55dB	55dB	1M
OP-01CJ	PRU	HSR	INT		5V/uS	+20V	-20V	70C	88dB	5MV	100NA	20NA	500MWF	6MA	12V	15V	30V	10uV/C	.90MW	.	80dB	80dB	.
OP-01CP	OBS	HSR	INT		5V/uS	+20V	-20V	70C	88dB	5MV	100NA	20NA	500MWF	6MA	12V	15V	30V	10uV/C	90MW	.	80dB	80dB	.
OP-01CY	PRU	HSR	INT		5V/uS	+20V	-20V	70C	88dB	5MV	100NA	20NA	500MWF	6MA	12V	15V	30V	10uV/C	90MW	.	80dB	80dB	.
OP-01EJ	PRU	HSR	INT		5V/uS	+22V	-22V	70C	94dB	2MV	50NA	2NA	500MWF	6MA	12V	15V	30V	8uV/C	90MW	.	80dB	80dB	.
OP-01EP	OBS	HSR	INT		5V/uS	+22V	-22V	70C	94dB	2MV	50NA	2NA	500MWF	6MA	12V	15V	30V	8uV/C	90MW	.	80dB	80dB	.
OP-01EY	PRU	HSR	INT		5V/uS	+22V	-22V	70C	94dB	2MV	50NA	2NA	500MWF	6MA	12V	15V	30V	8uV/C	90MW	.	80dB	80dB	.
OP-01FJ	PRU	HSR	INT		5V/uS	+22V	-22V	125C	94dB	2MV	50NA	5NA	500MWF	6MA	12V	15V	30V	8uV/C	90MW	.	80dB	80dB	.
OP-01FL	PRU	HSR	INT		5V/uS	+22V	-22V	125C	94dB	2MV	50NA	5NA	500MWF	6MA	12V	15V	30V	8uV/C	90MW	.	80dB	80dB	.
OP-01FP	OBS	HSR	INT		5V/uS	+22V	-22V	125C	94dB	2MV	50NA	5NA	500MWF	6MA	12V	15V	30V	8uV/C	90MW	.	80dB	80dB	.
OP-01FY	PRU	HSR	INT		5V/uS	+22V	-22V	125C	94dB	2MV	50NA	5NA	500MWF	6MA	12V	15V	30V	8uV/C	90MW	.	80dB	80dB	.
OP-01GJ	PRU	HSR	INT		5V/uS	+20V	-20V	125C	88dB	5MV	100NA	20NA	500MWF	6MA	12V	15V	30V	10uV/C	90MW	.	80dB	80dB	.
OP-01GL	PRU	HSR	INT		5V/uS	+20V	-20V	125C	88dB	5MV	100NA	20NA	500MWF	6MA	12V	15V	30V	10uV/C	90MW	.	80dB	80dB	.
OP-01GP	OBS	HSR	INT		5V/uS	+20V	-20V	125C	88dB	5MV	100NA	20NA	1WF	6MA	12V	15V	30V	10uV/C	90MW	.	80dB	80dB	.
OP-01GY	PRU	HSR	INT		5V/uS	+20V	-20V	125C	88dB	5MV	100NA	20NA	500MWF	6MA	12V	15V	30V	10uV/C	90MW	.	80dB	80dB	.
OP-01HJ	PRU	HSR	INT		5V/uS	+22V	-22V	70C	94dB	0.7MV	30NA	2NA	500MWF	6MA	12V	15V	30V	5uV/C	60MW	.	90dB	90dB	.
OP-01HP	OBS	HSR	INT		5V/uS	+22V	-22V	70C	94dB	0.7MV	30NA	2NA	500MWF	6MA	12V	15V	30V	5uV/C	60MW	.	90dB	90dB	.
OP-01HY	PRU	HSR	INT		5V/uS	+22V	-22V	70C	94dB	0.7MV	30NA	2NA	500MWF	6MA	12V	15V	30V	5uV/C	60MW	.	90dB	90dB	.
OP-01J	PRU	HSR	INT	.	5V/uS	+22V	-22V	125C	94dB	0.7MV	30NA	2NA	500MWF	6MA	12V	15V	30V	5uV/C	60MW	.	90dB	90dB	.
OP-01L	PRU	HSR	INT	.	5V/uS	+22V	-22V	125C	94dB	0.7MV	30NA	2NA	500MWF	6MA	12V	15V	30V	5uV/C	60MW	.	90dB	90dB	.
OP-01P	OBS	HSR	INT	.	5V/uS	+22V	-22V	125C	94dB	0.7MV	30NA	2NA	500MWF	6MA	12V	15V	30V	5uV/C	60MW	.	90dB	90dB	.
OP-01Y	PRU	HSR	INT	.	5V/uS	+22V	-22V	125C	94dB	0.7MV	30NA	2NA	500MWF	6MA	12V	15V	30V	5uV/C	60MW	.	90dB	90dB	.
OP-02AJ	PRU	LNA	INT	.8MHZ	.25V/uS	+22V	-22V	125C	100dB	0.5MV	30NA	2NA	500MWF	6MA	12V	22V	30V	8uV/C	60MW	.	90dB	90dB	3.8M
OP-02AY	PRU	LNA	INT	.8MHZ	.25V/uS	+22V	-22V	125C	100dB	0.5MV	30NA	2NA	500MWF	6MA	12V	22V	30V	8uV/C	60MW	.	90dB	90dB	3.8M
OP-02CJ	PRU	LNA	INT	.8MHZ	.25V/uS	+22V	-22V	70C	94dB	2MV	50NA	5NA	500MWF	6MA	12V	22V	30V	10uV/C	90MW	.	90dB	90dB	2.3M
OP-02CY	PRU	LNA	INT	.8MHZ	.25V/uS	+22V	-22V	70C	94dB	2MV	50NA	5NA	500MWF	6MA	12V	22V	30V	10uV/C	90MW	.	90dB	90dB	2.3M
OP-02EJ	PRU	LNA	INT	.8MHZ	.25V/uS	+22V	-22V	70C	100dB	0.5MV	30NA	2NA	500MWF	6MA	12V	22V	30V	8uV/C	60MW	.	90dB	90dB	3.8M
OP-02EY	PRU	LNA	INT	.8MHZ	.25V/uS	+22V	-22V	70C	100dB	0.5MV	30NA	2NA	500MWF	6MA	12V	22V	30V	8uV/C	60MW	.	90dB	90dB	3.8M
OP-02J	PRU	LNA	INT	.8MHZ	.25V/uS	+22V	-22V	125C	94dB	2MV	50NA	5NA	500MWF	6MA	12V	22V	30V	10uV/C	90MW	.	90dB	90dB	2.3M
OP-02Y	PRU	LNA	INT	.8MHZ	.25V/uS	+22V	-22V	125C	94dB	2MV	50NA	5NA	500MWF	6MA	12V	22V	30V	10uV/C	90MW	.	90dB	90dB	2.3M
OP-05AJ	PRU	PIA	INT	.3MHZ	0.1V/uS	+22V	-22V	125C	110dB	.15MV	2NA	2NA	500MWF	10MA	12V	22V	30V	0.9uV/C	120MW	.	114dB	100dB	30M
OP-05AL	PRU	PIA	INT	.3MHZ	0.1V/uS	+22V	-22V	125C	110dB	.15MV	2NA	2NA	500MWF	10MA	12V	22V	30V	0.9uV/C	120MW	.	114dB	100dB	30M
OP-05AY	PRU	PIA	INT	.3MHZ	0.1V/uS	+22V	-22V	125C	110dB	.15MV	2NA	2NA	500MWF	10MA	12V	22V	30V	0.9uV/C	120MW	.	114dB	100dB	30M
OP-05CJ	PRU	PIA	INT	.3MHZ	0.1V/uS	+22V	-22V	70C	100dB	1.3MV	7NA	6NA	500MWF	10MA	12V	22V	30V	4.5uV/C	150MW	.	100dB	90dB	8M
OP-05CY	PRU	PIA	INT	.3MHZ	0.1V/uS	+22V	-22V	70C	100dB	1.3MV	7NA	6NA	500MWF	10MA	12V	22V	30V	4.5uV/C	150MW	.	100dB	90dB	8M
OP-05EJ	PRU	PIA	INT	.3MHZ	0.1V/uS	+22V	-22V	70C	106dB	0.5MV	4NA	3.8NA	500MWF	10MA	12V	22V	30V	2uV/C	120MW	.	110dB	94dB	15M
OP-05EY	PRU	PIA	INT	.3MHZ	0.1V/uS	+22V	-22V	70C	106dB	0.5MV	4NA	3.8NA	500MWF	10MA	12V	22V	30V	2uV/C	120MW	.	110dB	94dB	15M
OP-05J	PRU	PIA	INT	.3MHZ	0.1V/uS	+22V	-22V	125C	106dB	0.5MV	3NA	2.8NA	500MWF	10MA	12V	22V	30V	2uV/C	120MW	.	114dB	100dB	20M
OP-05L	PRU	PIA	INT	.3MHZ	0.1V/uS	+22V	-22V	125C	106dB	0.5MV	3NA	2.8NA	500MWF	10MA	12V	22V	30V	2uV/C	120MW	.	114dB	100dB	20M
OP-05Y	PRU	PIA	INT	.3MHZ	0.1V/uS	+22V	-22V	125C	106dB	0.5MV	3NA	2.8NA	500MWF	10MA	12V	22V	30V	2uV/C	120MW	.	114dB	100dB	20M
OP-07AJ	PRU	LOV	INT	.3MHZ	.06V/uS	+22V	-22V	125C	110dB	25uV	2NA	2NA	500MW	10MA	12V	22V	30V	0.6uV/C	120MW	.	110dB	100dB	30M
OP-07AY	PRU	LOV	INT	.3MHZ	.06V/uS	+22V	-22V	125C	110dB	25uV	2NA	2NA	500MW	10MA	12V	22V	30V	0.6uV/C	120MW	.	110dB	100dB	30M
OP-07CJ	PRU	LOV	INT	.3MHZ	.06V/uS	+22V	-22V	70C	102dB	150uV	7NA	6NA	500MWF	10MA	12V	22V	30V	1.8uV/C	150MW	.	100dB	90dB	8M

For detailed explanations of
column heading notations, see
App. A.
Also for ready references the
more important abbreviations
used in the column headings are
listed below:

LEFT HAND PAGE

APP = application (codes at APP.E.)
CMRR = common mode rejection ratio
CMP = compensation (frequency)
dV_{io}/dT = input offset voltage temperature drift
GBP = gain bandwidth product
I_B = input bias current
I_{io} = input bias offset current
I_0 = quiescent supply current
MFR = manufacturer (codes at App.C.)
P_0 = quiescent power consumer
PSRR = power supply rejection ratio
V_{icm} = common mode input voltage rating
V_{idf} = differential input voltage rating
V_{io} = input offset voltage
V_s = dc supply voltage

RIGHT HAND PAGE

Lead out coding summary (details at APP G.) for different cases (APP F.)

A = gain adjust
B = bias adjust
C = case
E-- = inverting input
E+ = non-inverting input
F,F* = input frequency compensation
G = ground
J = high level input
K = output, open collector
L = output, open emitter
M = metal case
N = not connected
U = special terminal
R,R* = outputs
S = strobe
T,T* = offset balance
V+ = +ve dc supply
V- = -ve dc supply
W = guard ring
X = blank position, no lead
+ + = +ve supplementary dc supply
-- = -ve supplementary dc supply
6,6* = output frequency compensation

CASE (APP F)	LD 1	LD 2	LD 3	LD 4	LD 5	LD 6	LD 7	LD 8	LD 9	LD 10	LD 11	LD 12	LD 13	LD 14	LD 15	LD 16	EUROPE SUBSTITUTE	USA SUBSTITUTE	ISS	TYPE NUMBER
TO5-8/1M	T	E-	E+	V-	T*	R	V+	N	.	.	.	.	.	.	.	.		SE535T	0	SE535T
DIL-8/1P	T	E-	E+	V-	T*	R	V+	N	.	.	.	.	.	.	.	.		SE535V	0	NE535V
TO5-8/1M	T	E-	E+	V-	T*	R	V+	N	.	.	.	.	.	.	.	.	UA740HM	SU536T	0	NE536T
TO5-10/1M	B	E+	N	E-	B*	V-	L	Q	K	V+	.	.	.	.	.	.		SE540L	0	NE540L
DIL-14/1P	E+	N	A2	A*2	V-		N	R	R*	N	V+	A1	A*1	N	E-		SN72733J	UA733DC	0	NE592A
DIL-14/1C	E+	N	A2	A*2	V-		N	R	R*	N	V+	A1	A*1	N	E-		SN72733J	UA733DC	0	NE592F
TO5-10/1M	E-	E+	A2	A*2	V-	R	R*	V+	A1	A*1							SN72733L	UA733HC	0	NE592K
DIL-14/1P	E+	N	A2	A*2	V-	N	R	R*	N	V+	A1	A*1	N	E-			SN72733J	UA733DC	0	NE592N(14)
TO5-10/1M	F	E+	V-	E-	ø	B	B*	R	V+	F*	.	.	.	.				LH0001CH	0	NH0001CH
TO5-10/1M	F	E+	V-	E-	ø	B	B*	R	V+	F*	.	.	.	.				LH0001H	0	NH0001H
TO5-10/1M	F	E+	V-	E-	F*	N	G	R	V+	øS								LH0003CH	0	NH0003CH
TO5-10/1M	F	E+	V-	E-	F*	N	G	R	V+	øS								LH0003H	0	NH0003H
TO5-10/1M	F	E+	V-	E-	F*	B	B*	R	V+	øS								LH0004CH	0	NH0004CH
TO5-10/1M	F	E+	V-	E-	F*	B	B*	R	V+	øS								LH0004H	0	NH0004H
TO5-10/1M	E+	V+	V-	F	N	G	R	V-	ø	F*	.	.	.	.				LH0005AH	0	NH0005AH
TO5-10/1M	E+	V+	V-	F	N	G	R	V-	ø	F*	.	.	.	.				LH0005CH	0	NH0005CH
TO5-10/1M	E+	V+	V-	F	N	G	R	V-	ø	F*	.	.	.	.				LH0005H	0	NH0005H
TO5-8/1M	T	E-	E+	V-M	T*	R	V+	N										OP-01GJ	0	OP-01CJ
DIL-14/1C	N	N	T	E-	E+	V-	N	N	T*	R	V+	N	N	N				OP-01CY	0	OP-01CP
DIL-14/1C	N	N	T	E-	E+	V-	N	N	T*	R	V+	N	N	N				OP-01GY	0	OP-01CY
TO5-8/1M	T	E-	E+	V-M	T*	R	V+											OP-01FJ	0	OP-01EJ
DIL-14/1C	N	N	T	E-	E+	V-	N	N	T*	R	V+	N	N	N				OP-01EY	0	OP-01EP
DIL-14/1C	N	N	T	E-	E+	V-	N	N	T*	R	V+	N	N	N				OP-01FY	0	OP-01EY
TO5-8/1M	T	E-	E+	V-M	T*	R	V+	N										OP-01J	0	OP-01FJ
FLP-10/3G	N	T	E-	E+	V-	T	R	V+	N	N								OP-01L	0	OP-01FL
DIL-14/1C	N	N	T	E-	E+	V-	N	N	T*	R	V+	N	N	N				OP-01FY	0	OP-01FP
DIL-14/1C	N	N	T	E-	E+	V-	N	N	T*	R	V+	N	N	N				OP-01Y	0	OP-01FY
TO5-8/1M	T	E-	E+	V-M	T*	R	V+	N										OP-01FJ	0	OP-01GJ
FLP-10/3C	N	T	E-	E+	V-	T	R	V+	N	N								OP-01FL	0	OP-01GL
DIL-14/1C	N	N	T	E-	E+	V-	N	N	T*	R	V+	N	N	N				OP-01GY	0	OP-01GP
DIL-14/1C	N	N	T	E-	E+	V-	N	N	T*	R	V+	N	N	N				OP-01FY	0	OP-01GY
TO5-8/1M	T	E-	E+	V-M	T*	R	V+	N										OP-01J	0	OP-01HJ
DIL-14/1C	N	N	T	E-	E+	V-	N	N	T*	R	V+	N	N	N				OP-01HY	0	OP-01HP
DIL-14/1C	N	N	T	E-	E+	V-	N	N	T*	R	V+	N	N	N				OP-01Y	0	OP-01HY
TO5-8/1M	T	E-	E+	V-M	T*	R	V+	N										MONO-OP01J	0	OP-01J
FLP-10/3C	N	T	E-	E+	V-	T	R	V+	N	N								MONO-OP01L	0	OP-01L
DIL-14/1C	N	N	T	E-	E+	V-	N	N	T*	R	V+	N	N	N				OP-01Y	0	OP-01P
DIL-14/1C	N	N	T	E-	E+	V-	N	N	T*	R	V+	N	N	N				MONO-OP01Y	0	OP-01Y
TO5-8/1M	T	E-	E+	V-M	T*	R	V+	N											0	OP-02AJ
DIL-14/1C	N	N	T	E-	E+	V-	N	N	T*	R	V+	N	N	N					0	OP-02AY
TO5-8/1M	T	E-	E+	V-M	T*	R	V+	N										OP-02J	0	OP-02CJ
DIL-14/1C	N	N	T	E-	E+	V-	N	N	T*	R	V+	N	N	N				OP-02Y	0	OP-02CY
TO5-8/1M	T	E-	E+	V-M	T*	R	V+	N										OP-02AJ	0	OP-02EJ
DIL-14/1C	N	N	T	E-	E+	V-	N	N	T*	R	V+	N	N	N				OP-02AY	0	OP-02EY
TO5-8/1M	T	E-	E+	V-M	T*	R	V+	N										OP-02AJ	0	OP-02J
DIL-14/1C	N	N	T	E-	E+	V-	N	N	T*	R	V+	N	N	N				OP-02AY	0	OP-02Y
TO5-8/1M	T	E-	E+	V-M	N	R	V+	T*											0	OP-05AJ
FLP-10/3C	N	T	E-	E+	V-	N	R	V+	T*	N									0	OP-05AL
DIL-14/1C	N	N	T	E-	E+	V-	N	N	N	R	V+	T*	N	N					0	OP-05AY
TO5-8/1M	T	E-	E+	V-M	N	R	V+	T*										OP-05EJ	0	OP-05CJ
DIL-14/1C	N	N	T	E-	E+	V-	N	N	N	R	V+	T*	N	N				OP-05EY	0	OP-05CY
TO5-8/1M	T	E-	E+	V-M	N	R	V+	T*										OP-05J	0	OP-05EJ
DIL-14/1C	N	N	T	E-	E+	V-	N	N	N	R	V+	T*	N	N				OP-05Y	0	OP-05EY
TO5-8/1M	T	E-	E+	V-M	N	R	V+	T*										OP-05AJ	0	OP-05J
FLP-10/3C	N	T	E-	E+	V-	N	R	V+	T*	N								OP-05AL	0	OP-05L
DIL-14/1C	N	N	T	E-	E+	V-	N	N	N	R	V+	T*	N	N				OP-05AY	0	OP-05Y
TO5-8/1M	T	E-	E+	V-M	N	R	V+	T*											0	OP-07AJ
DIL-14/1C	N	N	T	E-	E+	V-	N	N	N	R	V+	T*	N	N					0	OP-07AY
TO5-8/1M	T	E-	E+	V-M	N	R	V+	T*										OP-07EJ	0	OP-07CJ
DIL-14/1C	N	N	T	E-	E+	V-	N	N	N	R	V+	T*	N	N				OP-07EY	0	OP-07CY

TYPE NUMBER	MFR	APP	CMP	GBP MIN	SLEW RATE MIN	Vs+ MAX	Vs- MAX	Tpp MAX	Avol MIN	Vio MAX	Ib MAX	Iio MAX	Ptot MAX	Iout MIN	Vout MIN	Vicm MAX	Vidf MAX	dVio/dT MAX	Po MAX	Io MAX	CMRR MIN	PSRR MIN
OP-07CY	PRU	LOV	INT	.3MHZ	.06V/uS	+22V	-22V	70C	102dB	150uV	7NA	6NA	500MWF	10MA	12V	22V	30V	1.8uV/C	150MW	.	100dB	90dB
OP-07EJ	PRU	LOV	INT	.3MHZ	.06V/uS	+22V	-22V	70C	106dB	75uV	4NA	3.8A	500MWF	10MA	12V	22V	30V	1.3uV/C	120MW	.	106dB	94dB
OP-07EY	PRU	LOV	INT	.3MHZ	.06V/uS	+22V	-22V	70C	106dB	75uV	4NA	3.8NA	500MWF	10MA	12V	22V	30V	1.3uV/C	120MW	.	106dB	94dB
OP-07J	PRU	LOV	INT	.3MHZ	.06V/uS	+22V	-22V	125C	106dB	75uV	3NA	2NA	500MW	10MA	12V	22V	30V	1.3uV/C	120MW	.	110dB	100dB
OP-07Y	PRU	LOV	INT	.3MHZ	.06V/uS	+22V	-22V	125C	106dB	75uV	3NA	2NA	500MW	10MA	12V	22V	30V	1.3uV/C	120MW	.	110dB	100dB
OP-10AY	PRU	DPI	INT	.3MHZ	.06V/uS	+22V	-22V	125C	106dB	0.5MV	3NA	2.8NA	500MWF	10MA	12V	22V	30V	2uV/C	120MW	.	114dB	100dB
OP-10CY	PRU	DPI	INT	.3MHZ	.06V/uS	+22V	-22V	70C	102dB	1.3MV	7NA	6NA	500MWF	10MA	12V	22V	30V	4.5uV/C	150MW	.	100dB	90dB
OP-10EY	PRU	DPI	INT	.3MHZ	.06V/uS	+22V	-22V	70C	106dB	0.5MV	4NA	3.8NA	500MWF	10MA	12V	22V	30V	2uV/C	120MW	.	110dB	94dB
OP-10Y	PRU	DPI	INT	.3MHZ	.06V/uS	+22V	-22V	125C	106dB	0.5MV	3NA	2.8NA	500MWF	10MA	12V	22V	30V	2uV/C	120MW	.	114dB	100dB
RC101ABL	RAU	GPU	EXT	.	.	+22V	-22V	70C	94dB	2MV	75NA	10NA	500MWF	5MA	12V	15V	30V	15uV/C	.	3MA	80dB	80dB
RC101AD	OBS	GPU	EXT	.	.	+18V	-18V	70C	88dB	7.5MV	250NA	50NA	500MWF	5MA	12V	15V	30V	30uV/C	.	3MA	70dB	70dB
RC101AQ	OBS	GPU	EXT	.	.	+22V	-22V	70C	94dB	2MV	75NA	10NA	500MWF	5MA	12V	15V	30V	15uV/C	.	3MA	80dB	80dB
RC101AT	OBS	GPU	EXT	.	.	+18V	-18V	70C	88dB	7.5MV	250NA	50NA	500MWF	5MA	12V	15V	30V	30uV/C	.	3MA	70dB	70dB
RC106BL	RAU	CPR	EXT	.	.	+15V	-15V	70C	84dB	5MV	25uA	5uA	600MWF	50MA	2.5V	.	.	20uV/C	163MW	.	.	.
RC107D	OBS	GPK	INT	.	.	+18V	-18V	70C	84dB	7.5MV	250NA	50NA	500MW	5MA	12V	15V	30V	30uV/C	.	.	70dB	70dB
RC107DN	OBS	GPK	INT	.	.	+18V	-18V	70C	84dB	7.5MV	250NA	50NA	500MWF	5MA	12V	15V	30V	30uV/C	.	.	70dB	70dB
RC107DP	OBS	GPK	INT	.	.	+18V	-18V	70C	84dB	7.5MV	250NA	50NA	500MWF	5MA	12V	15V	30V	30uV/C	.	.	70dB	70dB
RC107Q	OBS	GPK	INT	.	.	+18V	-18V	70C	84dB	7.5MV	250NA	50NA	500MW	5MA	12V	15V	30V	30uV/C	.	.	70dB	70dB
RC107T	OBS	GPK	INT	.	.	+18V	-18V	70C	84dB	7.5MV	250NA	50NA	500MWF	5MA	12V	15V	30V	30uV/C	.	.	70dB	70dB
RC108AD	OBS	SBA	EXT	.	.	+18V	-18V	70C	98dB	0.5MV	7NA	1NA	500MWF	1MA	13V	15V	1V	5uV/C	.	.6MA	96dB	96dB
RC108AT	OBS	SBA	EXT	.	.	+18V	-18V	70C	98dB	0.5MV	7NA	1NA	500MWF	1MA	13V	15V	1V	5uV/C	.	.6MA	96dB	96dB
RC108D	OBS	SBA	EXT	.	.	+18V	-18V	70C	88dB	7.5MV	7NA	1NA	500MWF	1MA	13V	15V	1V	30uV/C	.	.6MA	80dB	80dB
RC108Q	OBS	SBA	EXT	.	.	+20V	-20V	70C	96dB	2MV	2NA	0.2NA	500MWF	1MA	13V	15V	1V	15uV/C	.	.6MA	85dB	80dB
RC108T	OBS	SBA	EXT	.	.	+18V	-18V	70C	88dB	7.5MV	7NA	1NA	500MWF	1MA	13V	15V	1V	30uV/C	.	.6MA	80dB	80dB
RC702	OBS	WBA	EXT	3MHZ	.	+13V	-8V	70C	66dB	5MV	7.5uA	2MA	500MWF	.3MA	5V	5V	5V	20uV/C	120MW	7MA	70dB	70dB
RC702D	OBS	WBA	EXT	3MHZ	.	+13V	-8V	70C	66dB	5MV	7.5uA	2uA	670MWF	.3MA	5V	5V	5V	20uV/C	120MW	7MA	70dB	70dB
RC702DC	RAU	WBA	EXT	3MHZ	.	+13V	-8V	70C	66dB	5MV	7.5uA	2uA	670MWF	.3MA	5V	5V	5V	20uV/C	120MW	7MA	70dB	70dB
RC702DN	OBS	WBA	EXT	3MHZ	.	+13V	-8V	70C	66dB	5MV	7.5uA	2uA	670MWF	.3MA	5V	5V	5V	20uV/C	120MW	7MA	70dB	70dB
RC702DP	OBS	WBA	EXT	3MHZ	.	+13V	-8V	70C	66dB	5MV	7.5uA	2uA	670MWF	.3MA	5V	5V	5V	20uV/C	120MW	7MA	70dB	70dB
RC702Q	OBS	GPU	EXT	3MHZ	.	+13V	-8V	70C	66dB	5MV	7.5uA	2uA	500MWF	.3MA	5V	1.5V	5V	20uV/C	120MW	7MA	70dB	70dB
RC702T	RAU	WBA	EXT	3MHZ	.	+13V	-8V	70C	66dB	5MV	7.5uA	2uA	500MWF	.3MA	5V	5V	5V	20uV/C	120MW	7MA	70dB	70dB
RC709	OBS	GPU	EXT	.3MHZ	.15V/uS	+18V	-18V	70C	84dB	7.5MV	1.5uA	500NA	500MWF	5MA	12V	10V	5V	.	200MW	.	65dB	74dB
RC709	OBS	GPU	EXT	.3MHZ	.15V/uS	+18V	-18V	70C	84dB	7.5MV	1.5uA	500NA	670MWF	5MA	12V	10V	5V	.	200MW	.	65dB	74dB
RC709BL	RAU	GPU	EXT	.	.	+18V	-18V	70C	84dB	7.5MV	1.5uA	500NA	500MWF	10MA	12V	10V	5V	.	200MW	.	65dB	100dB
RC709DC	RAU	GPU	EXT	.3MHZ	.15V/uS	+18V	-18V	70C	84dB	7.5MV	1.5uA	500NA	670MWF	5MA	12V	10V	5V	.	200MW	.	65dB	74dB
RC709T	RAU	GPU	EXT	.3MHZ	.15V/uS	+18V	-18V	70C	84dB	7.5MV	1.5uA	500NA	500MWF	5MA	12V	10V	5V	.	200MW	.	65dB	74dB
RC710BL	RAU	CPR	EXT	.	.	+14V	-6V	70C	60dB	5MV	25uA	5uA	.	5MA	2.5V	7V	5V	20uV/C	150MW	.	70dB	.
RC710DC	RAU	CPR	EXT	.	.	+14V	-6V	70C	60dB	5MV	25uA	5uA	670MWF	5MA	2.5V	7V	5V	20uV/C	150MW	.	70dB	.
RC710DP	RAU	CPR	EXT	.	.	+14V	-6V	70C	60dB	5MV	25uA	5uA	670MWF	5MA	2.5V	7V	5V	20uV/C	150MW	.	70dB	.
RC710T	RAU	CPR	EXT	.	.	+14V	-6V	70C	60dB	5MV	25uA	5uA	500MWF	5MA	2.5V	7V	5V	20uV/C	150MW	.	70dB	.
RC711BL	RAU	DCP	EXT	.	.	+14V	-7V	70C	57dB	5MV	100uA	15uA	.	5MA	2.5V	7V	5V	20uV/C	230MW	.	.	.
RC711DC	RAU	DCP	EXT	.	.	+14V	-7V	70C	57dB	5MV	100uA	15uA	670MWF	5MA	2.5V	7V	5V	20uV/C	230MW	.	.	.
RC711T	RAU	DCP	EXT	.	.	+14V	-7V	70C	57dB	5MV	100uA	15uA	500MWF	5MA	2.5V	7V	5V	20uV/C	230MW	.	.	.
RC725T	RAU	PIA	EXT	.	.	+22V	-22V	70C	108dB	2.5MV	125NA	35NA	500MWF	5MA	12V	22V	5V	5uV/C	150MW	.	94dB	90dB
RC733T	RAU	BDO	EXT	20MHZ	.	+8V	-8V	70C	48dB	6MV	30uA	5uA	500MWF	2MA	1.5V	6V	5V	.	.	24MA	60dB	50dB
RC741D	RAU	GPK	INT	.	0.2V/uS	+18V	-18V	70C	86dB	6MV	500NA	200NA	670MWF	5MA	12V	15V	30V	.	85MW	3MA	70dB	76dB
RC741DB	RAU	GPK	INT	.	0.2V/uS	+18V	-18V	70C	86dB	6MV	500NA	200NA	670MWF	5MA	12V	15V	30V	.	85MW	3MA	70dB	76dB
RC741DN	RAU	GPK	INT	.	0.2V/uS	+18V	-18V	70C	86dB	6MV	500NA	200NA	310MWF	5MA	12V	15V	30V	.	.	3MA	70dB	76dB
RC741DP	RAU	GPK	INT	.	0.2V/uS	+18V	-18V	70C	86dB	6MV	500NA	200NA	670MWF	5MA	12V	15V	30V	.	85MW	3MA	70dB	76dB
RC741Q	RAU	GPK	INT	.	0.2V/uS	+18V	-18V	70C	86dB	6MV	500NA	200NA	500MWF	5MA	12V	15V	30V	.	85MW	3MA	70dB	76dB
RC741T	RAU	GPK	INT	.	0.2V/uS	+18V	-18V	70C	86dB	6MV	500NA	200NA	500MWF	5MA	12V	15V	30V	.	.	3MA	70dB	76dB
RC741TE	RAU	GPK	INT	.	0.2V/uS	+18V	-18V	70C	86dB	6MV	500NA	200NA	500MWF	5MA	12V	15V	30V	.	.	3MA	70dB	76dB
RC747DB	RAU	DGK	INT	.	0.2V/uS	+18V	-18V	70C	88dB	6MV	500NA	200NA	670MWF	5MA	12V	15V	30V	.	85MW	3MA	70dB	76dB
RC747DC	RAU	DGK	INT	.	0.2V/uS	+18V	-18V	70C	88dB	6MV	500NA	200NA	670MWF	5MA	12V	15V	30V	.	85MW	3MA	70dB	76dB
RC747DP	RAU	DGK	INT	.	0.2V/uS	+18V	-18V	70C	88dB	6MV	500NA	200NA	670MWF	5MA	12V	15V	30V	.	85MW	3MA	70dB	76dB
RC747T	RAU	DGK	INT	.	0.2V/uS	+18V	-18V	70C	88dB	6MV	500NA	200NA	500MWF	5MA	12V	15V	30V	.	85MW	3MA	70dB	76dB
RC748BL	RAU	GPU	EXT	.	0.2V/uS	+22V	-22V	70C	86dB	6MV	500NA	200NA	.	5MA	12V	15V	30V	.	85MW	3MA	70dB	76dB
RC748DP	RAU	GPU	EXT	.	0.2V/uS	+22V	-22V	70C	86dB	6MV	500NA	200NA	670MWF	5MA	12V	15V	30V	.	85MW	3MA	70dB	76dB
RC748NB	RAU	GPU	EXT	.	0.2V/uS	+18V	-18V	70C	86dB	6MV	500NA	200NA	310MWF	5MA	12V	15V	30V	.	85MW	3MA	70dB	76dB
RC748T	RAU	GPU	EXT	.	0.2V/uS	+18V	-18V	70C	86dB	6MV	500NA	200NA	500MWF	5MA	12V	15V	30V	.	85MW	3MA	70dB	76dB

etailed explanations of
n heading notations, see

or ready references the
important abbreviations
n the column headings are
below:

HAND PAGE
= application
  (codes at APP.E.)
R = common mode
  rejection ratio
= compensation
  (frequency)
T= input offset voltage
  temperature drift
= gain bandwidth
  product
= input bias current
= input bias offset
  current
= quiescent supply
  current
= manufacturer
  (codes at App.C.)
= quiescent power
  consumer
= power supply rejection
  ratio
= common mode input
  voltage rating
= differential input
  voltage rating
= input offset voltage
= dc supply voltage

HAND PAGE
out coding summary
s at APP.G.) for different
(APP.F.)
= gain adjust
= bias adjust
= case
= inverting input
= non-inverting input
= input frequency
  compensation
= ground
= high level input
= output, open collector
= output, open emitter
= metal case
= not connected
= special terminal
= outputs
= strobe
= offset balance
= +ve dc supply
= −ve dc supply
= guard ring
= blank position, no lead
= +ve supplementary dc
  supply
= −ve supplementary dc
  supply
= output frequency
  compensation

CASE (APP F)	LD 1	LD 2	LD 3	LD 4	LD 5	LD 6	LD 7	LD 8	LD 9	LD 10	LD 11	LD 12	LD 13	LD 14	LD 15	LD 16	EUROPE SUBSTITUTE	USA SUBSTITUTE	I S	TYPE NUMBER
T05-8/1M	T	E-	E+	V-M	N.	R	V+	T*	.	.	.	.	.	.				OP-07J	0	OP-07EJ
DIL-14/1C	N	N	T	E-	E+	V-	N	N	.	R	V+	T*	N	N				OP-07Y	0	OP-07EY
T05-8/1M	T	E-	E+	V-M	N	R	V+	T*	.	.	.	.	.	.				OP-07AJ	0	OP-07J
DIL-14/1C	N	N	T	E-	E+	V-	N	N	N	R	V+	T*	N	N				OP-07AY	0	OP-07Y
DIL-14/1C	T1	T*1	E-1	E+1	V-2	R2	V+2	T2	T*2	E-2	E+2	V-1	R1	V+1					0	OP-10AY
DIL-14/1C	T1	T*1	E-1	E+1	V-2	R2	V+2	T2	T*2	E-2	E+2	V-1	R1	V+1				OP-10EY	0	OP-10CY
DIL-14/1C	T1	T*1	E-1	E+1	V-2	R2	V+2	T2	T*2	E-2	E+2	V-1	R1	V+1				OP-10Y	0	OP-10EY
DIL-14/1C	T1	T*1	E-1	E+1	V-2	R2	V+2	T2	T*2	E-2	E+2	V-1	R1	V+1				OP-10AY	0	OP-10Y
BML	.	.	.	.	.	.	.	.											0	RC101ABL
DIL-14/1C	N	N	FT	E-	E+	V-	N	N	T*	R	V+	F*	N	N			UA301AD	LM301AJ14	0	RC101AD
FLP-10/3G	N	FT	E-	E+	V-	T*	R	V+	F*	N	.	.					SFC2201APM	LM201AF	0	RC101AQ
T05-8/1M	FT	E-	E+	V-M	T*	R	V+	F*	.	.							SFC2301A	LM301AH	0	RC101AT
BML																			0	RC106BL
DIL-14/1C	N	N	N	E-	E+	V-	N	N	N	R	V+	N	N	N			SN72307JA	LM307D	0	RC107D
DIL-8/1P	N	E-	E+	V-	N	R	V+	N									SFC2307DC	LM307J	0	RC107DN
DIL-14/1P	N	N	N	E-	E+	V-	N	N	N	R	V+	N	N	N			SN72307JA	LM307D	0	RC107DP
FLP-10/3C	N	N	E-	E+	V-	N	R	V+	N	N	.	.					SFC2207PT	LM207F	0	RC107Q
T05-8/1M	N	E-	E+	V-M	N	R	V+	N	.	.							SFC2307	LM307H	0	RC107T
DIL-14/1C	N	F	N	E-	E+	N	V-	N	N	R	V+	F*	N	N			UA308AD	LM308AD	0	RC108AD
T05-8/1M	F	E-.	E+	V-M	N	R	V+	F*	.	.							SFC2308A	LM308AH	0	RC108AT
DIL-14/1C	N	F	N	E-	E+	N	V-	N	N	R	V+	F*	N	N			UA308D	LM308D	0	RC108D
FLP-10/3G	N	N	E-	E+	N	V-	R	V+	F*	F	.	.					SFC2208PT	LM208F	0	RC108Q
T05-8/1M	F	E-	E+	V-M	N	R	V+	F*	.	.							SFC2308	LM308H	0	RC108T
T05-8/1M	G	E-	E+	V-M	F	ø	R	V+	.	.							SN72702L	UA702HC	0	RC702
DIL-14/1C	N	N	G	E-	E+	V-	N	N	F	ø	R	N	V+	N			SN72702J	UA702DC	0	RC702D
DIL-14/1C	N	N	G	E-	E+	V-	N	N	F	ø	R	N	V+	N			SN72702J	UA702DC	0	RC702D0
DIL-14/1P	N	N	G	E-	E+	V-	N	N	F	ø	R	N	V+	N			SN72702J	UA702DC	0	RC702DN
DIL-14/1P	N	N	G	E-	E+	V-	N	N	F	ø	R	N	V+	N			SN72702J	UA702DC	0	RC702DP
FLP-10/3G	N	G	E-	E+	V-	F	ø	R	N	V+								MC1712CF	0	RC702Q
T05-8/1M	G	E-	E+	V-M	F	ø	R	V+	.	.							MC1712CG	UA702HC	0	RC702T
T05-8/1M	F	E-	E+	V-	ø	ø*R	V+	F*	.	.							TAA521	UA709HC	0	RC709
DIL-14/1P	N	N	F	E-	E+	V-	N	N	ø	R	V+	F*	N	N			TAA521A	UA709DC	0	RC709
BML																			0	RC709
DIL-14/1C	N	N	F	E-	E+	V-	N	N	ø	R	V+	F*	N	N			TAA521	UA709DC	0	RC709DC
T05-8/1M	F	E-	E+	V-	ø	ø*R	V+	F*	.	.							TAA521	UA709HC	0	RC709T
BML																			0	RC710BL
DIL-14/1C	N	G	E+	E-	N	V-	N	N	R	N	V+	N	N	N			SFC2710EC	UA710DC	0	RC710DC
DIL-14/1P	N	G	E+	E-	N	V-	N	N	R	N	V+	N	N	N			SFC2710EC	UA710DC	0	RC710DP
T05-8/1M	G	E+	E-	V-	N	N	R	V+	.	.							SFC2710C	UA710HC	0	RC710T
BML																			0	RC711BL
DIL-14/1C	N	E-1	E+1	V-	E+2	E-2	N	N	S2	R	V+	G	S1	N			SFC2711EC	UA711DC	0	RC711DC
T05-10/1M	S1	E-1	E+1	V-	E+2	E-2	S2	R	V+			S1					SFC2711C	UA711HC	0	RC711T
T05-8/1M	T	E-	E+	V-	ø	ø*	V+	T*									LM725CH	UA725HC	0	RC725T
T05-10/1M	E-	E+	A2	A*2	V-	R	R*	V+	A1	A*1							SN72733L	UA733HC	0	RC733T
DIL-14/1C	N	N	T	E-	E+	V-	N	N	T*	R	V+	N	N	N			TBA221A	UA741DC	0	RC741D
DIL-14/1P	N	N	T	E-	E+	V-	N	N	T*	R	V+	N	N	N			TBA221A	UA741DC	0	RC741DB
DIL-8/1P	T	E-	E+	V-	T*	R	V+	N	.	.							TBA221B	UA741TC	0	RC741DN
DIL-14/1P	N	N	T	E-	E+	V-	N	N	T*	R	V+	N	N	N			TBA221A	UA741DC	0	RC741DP
FLP-10/3C	N	T	E-	E+	V-	T*	R	V+	N	N							LM741F	MC1741CF	0	RC741Q
T05-8/1M	T	E-	E+	V-M	T*	R	V+										TBA221	UA741HC	0	RC741T
T05-8/1M	T	E-	E+	V-M	T*	R	V+	.	.								TBA221	UA741HC	0	RC741TE
DIL-14/1P	E-1	E+1	T1	V-	T2	E+2	E-2	T*2	V+2	R2	N	R1	V+1	T*1			TBB0747A	UA747DC	0	RC747DB
DIL-14/1C	E-1	E+1	T1	V-	T2	E+2	E-2	T*2	V+2	R2	N	R1	V+1	T*1			TBB0747A	UA747DC	0	RC747DC
DIL-14/1P	E-1	E+1	T1	V-	T2	E+2	E-2	T*2	V+2	R2	N	R1	V+1	T*1			TBB0747A	UA747DC	0	RC747DP
T05-10/1M	R1	V+1	E-1	E+1	V-	E+2	E-2	V+2	R2	N							TBB0747	UA747HC	0	RC747T
BML																			0	RC748BL
DIL-14/1P	N	N	FT	E-	E+	V-	N	N	T*	R	V+	F*	N	N			SN72748J	UA748DC	0	RC748DP
DIL-8/1P	FT	E-	E+	V-	T*	R	V+	F*	.	.							TBB0748	UA748TC	0	RC748NB
T05-10/1M	FT	E-	E+	V-	T*	R	V+	F*	.	.							TBB0748	UA748HC	0	RC748T
DIL-14/1P	ø2	R2	F2	F*2	E-2	E+2	V-	E+1	E-1	F1	F*1	R1	ø1	V+				MC1437L	0	RC1437DB

TYPE NUMBER	MFR	APP	CMP	GBP MIN	SLEW RATE MIN	V_S^+ MAX	V_S^- MAX	T_{OP} MAX	A_{VOL} MIN	V_{IO} MAX	I_B MAX	I_{IO} MAX	P_{TOT} MAX	I_{OUT} MIN	V_{OUT} MIN	V_{ICM} MAX	V_{IDF} MAX	dV_{IO}/dT MAX	P_Q MAX	I_Q MAX	CM RR MIN	PS RR MIN	R_{IN} MIN
RC1437DB	RAU	DGU	EXT	.	0.1V/uS	+18V	-18V	75C	84dB	7.5MV	1.5uA	0.5uA	500MWF	5MA	12V	18V	5V	10uV/C	225MW	.	65dB	74dB	50K
RC1437DC	RAU	DGU	EXT	.	0.1V/uS	+18V	-18V	75C	84dB	7.5MV	1.5uA	0.5uA	500MWF	5MA	12V	18V	5V	10uV/C	225MW	.	65dB	74dB	50K
RC1414DC	RAU	DCP	EXT	.	.	+14V	-7V	75C	60dB	5MV	25uA	5uA	625MWF	5MA	1V	7V	5V	25uV/C	150MW	9MA	70dB	.	.
RC1458DN	RAU	DGK	INT	.5MHZ	0.3V/uS	+18V	-18V	70C	86dB	6MV	0.5uA	0.2uA	750MWF	5MA	12V	15V	30V	50uV/C	170MW	6MA	70dB	76dB	300K
RC1458NB	RAU	DGK	INT	.5MHZ	0.3V/uS	+18V	-18V	70C	86dB	6MV	0.5uA	0.2uA	750MWF	5MA	12V	15V	30V	50uV/C	170MW	6MA	70dB	76dB	300K
RC1458T	RAU	DGK	INT	.5MHZ	0.3V/uS	+18V	-18V	70C	86dB	6MV	0.5uA	0.2uA	680MWF	5MA	12V	15V	30V	50uV/C	170MW	6MA	70dB	76dB	300K
RC1556ANB	RAU	SBA	INT	.5MHZ	1V/uS	+18V	-18V	70C	96dB	5MV	30NA	10NA	500MWF	5MA	11V	18V	18V	40uV/C	90MW	3MA	70dB	74dB	1M
RC1556AT	RAU	SBA	INT	2MHZ	1V/uS	+18V	-18V	70C	96dB	5MV	30NA	10NA	500MWF	6MA	11V	18V	30V				70dB	74dB	2M
RC1556NB	RAU	SBA	INT	.5MHZ	1V/uS	+18V	-18V	70C	96dB	10MV	30NA	10NA	500MWF	5MA	11V	18V	18V	40uV/C	90MW	3MA	70dB	74dB	1M
RC1556T	RAU	SBA	INT	2MHZ	1V/uS	+18V	-18V	70C	96dB	10MV	30NA	10NA	500MWF	5MA	11V	18V	18V	40uV/C	90MW	3MA	70dB	74dB	1M
RC1709	RAU	GPU	EXT	.3MHZ	.15V/uS	+18V	-18V	70C	82dB	7.5MV	1.5uA	0.5uA	500MWF	5MA	12V	10V	5V		200MW	.	65dB	74dB	50K
RC1741BL	RAU	GPK	INT	.2V/uS	.	+18V	-18V	70C	86dB	6MV	500NA	200NA		5MA	12V	15V	30V			3MA	70dB	70dB	300K
RC3401DB	RAU	QCD	INT	1M5HZ	.	+18V	.	75C	60dB	.	300NA	.	625MWF	6MA	13V	.	.		.	10MA		50dB	.
RC3403ADB	RAU	QGK	INT	.3MHZ	0.2V/uS	+18V	-18V	70C	88dB	6MV	500NA	50NA	650MWF	6MA	15V	18V	30V	50uV/C	.	5MA	70dB	80dB	300K
RC3403ADC	RAU	QGK	INT	.3MHZ	0.5V/uS	+18V	-18V	70C	88dB	6MV	500NA	50NA	650MWF	6MA	13V	18V	36V		.	5MA	70dB	80dB	
RC4131NB	RAU	GPK	INT	2MHZ	1V/uS	+18V	-18V	70C	91dB	5MV	150NA	20NA	500MWF	7MA	16V	15V	30V	20uV/C	.	2MA	70dB	70dB	700K
RC4131T	RAU	GPK	INT	2MHZ	1V/uS	+18V	-18V	70C	91dB	5MV	150NA	20NA	500MWF	7MA	16V	15V	30V	20uV/C	.	2MA	70dB	70dB	700K
RC4136DB	RAU	QGK	INT	1MHZ	0.3V/uS	+18V	-18V	70C	86dB	6MV	500NA	200NA	800MWF	5MA	12V	15V	30V		340MW		70dB	76dB	300K
RC4136DC	RAU	QGK	INT	1MHZ	0.3V/uS	+18V	-18V	70C	86dB	6MV	500NA	200NA	800MWF	5MA	12V	15V	30V		340MW		70dB	76dB	300K
RC4137DB	RAU	QGK	INT	.3MHZ	0.5V/uS	+18V	-18V	70C	88dB	6MV	500NA	50NA	650MWF	6MA	13V	18V	36V		.	5MA	70dB	80dB	
RC4137DC	RAU	QGK	INT	.3MHZ	0.5V/uS	+18V	-18V	70C	88dB	6MV	500NA	50NA	650MWF	6MA	13V	18V	36V		.	5MA	70dB	80dB	
RC4531D	RAU	HSR	EXT	.	10V/uS	+18V	-18V	70C	86dB	6MV	1.5uA	200NA	500MWF	5MA	10V	15V	15V		300MW	10MA	70dB	76dB	300K
RC4531DN	RAU	HSR	EXT	.	10V/uS	+18V	-18V	70C	86dB	6MV	1.5uA	200NA	500MWF	5MA	10V	15V	15V		300MW	10MA	70dB	76dB	300K
RC4531T	RAU	HSR	EXT	.	10V/uS	+18V	-18V	70C	86dB	6MV	1.5uA	200NA	500MWF	5MA	10V	15V	15V		300MW	10MA	70dB	76dB	300K
RC4558NB	RAU	DWB	INT	1MHZ	0.5V/uS	+18V	-18V	70C	86dB	6MV	500NA	200NA	500MWF	5MA	12V	15V	30V		170MW		70dB	76dB	300K
RC4558T	RAU	DWB	INT	1MHZ	0.5V/uS	+18V	-18V	70C	86dB	6MV	500NA	200NA	500MWF	5MA	12V	15V	30V		170MW		70dB	76dB	300K
RC4739DB	RAU	DLN	INT	.	0.3V/uS	+18V	-18V	70C	86dB	6MV	500NA	200NA	500MWF	5MA	12V	15V	30V		170MW		70dB	76dB	300K
RM101ABL	RAU	GPU	EXT	.	.	+22V	-22V	125C	94dB	2MV	75NA	10NA	.	5MA	12V	15V	30V	15uV/C	.	3MA	80dB	80dB	1.5M
RM101AD	OBS	GPU	EXT	.	.	+22V	-22V	125C	94dB	2MV	75NA	10NA	500MWF	5MA	12V	15V	30V	15uV/C	.	3MA	80dB	80dB	1.5M
RMJ01AQ	OBS	GPU	EXT	.	.	+22V	-22V	125C	94dB	2MV	75NA	10NA	500MWF	5MA	12V	15V	30V	15uV/C	.	3MA	80dB	80dB	1.5M
RM101AT	OBS	GPU	EXT	.	.	+22V	-22V	125C	94dB	2MV	75NA	10NA	500MWF	5MA	12V	15V	30V	15uV/C	.	3MA	80dB	80dB	1.5M
RM101D	OBS	GPU	EXT	.	.	+22V	-22V	125C	94dB	5MV	1.5uA	0.5uA	500MW	5MA	12V	15V	30V	15uV/C	.	3MA	70dB	70dB	300K
RM101Q	OBS	GPU	EXT	.	.	+22V	-22V	125C	94dB	5MV	1.5uA	0.5uA	500MWF	5MA	12V	15V	30V	15uV/C	.	3MA	70dB	70dB	300K
RM101T	OBS	GPU	EXT	.	.	+22V	-22V	125C	94dB	5MV	1.5uA	0.5uA	500MWF	5MA	12V	15V	30V	15uV/C	.	3MA	70dB	70dB	300K
RM106BL	RAU	CPR	EXT	.	.	+15V	-15V	125C	84dB	2MV	20uA	3uA	.	50MA	2.5V	.	.	10uV/C	163MW	.	.	.	.
RM107D	OBS	GPK	INT	.	.	+22V	-22V	125C	94dB	2MV	75NA	10NA	500MWF	5MA	12V	15V	30V	15uV/C	.	3MA	80dB	80dB	1.5M
RM107Q	OBS	GPK	INT	.	.	+22V	-22V	125C	94dB	2MV	75NA	10NA	500MWF	5MA	12V	15V	30V	15uV/C	.	3MA	80dB	80dB	1.5M
RM107T	OBS	GPK	INT	.	.	+22V	-22V	125C	94dB	2MV	75NA	10NA	500MWF	5MA	12V	15V	30V	15uV/C	.	3MA	80dB	80dB	1.5M
RM108AD	OBS	SBA	EXT	.	.	+20V	-20V	125C	98dB	0.5MV	2NA	0.2NA	500MWF	1MA	13V	15V	1V	5uV/C	.	.6MA	96dB	96dB	30M
RM108AQ	OBS	SBA	EXT	.	.	+20V	-20V	125C	98dB	0.5MV	2NA	0.2NA	500MWF	1MA	13V	15V	1V	5uV/C	.	.6MA	96dB	96dB	30M
RM108AT	OBS	SBA	EXT	.	.	+20V	-20V	125C	98dB	0.5MV	2NA	0.2NA	500MWF	1MA	13V	15V	1V	5uV/C	.	.6MA	96dB	96dB	30M
RM108D	OBS	SBA	EXT	.	.	+20V	-20V	125C	96dB	2MV	2NA	0.2NA	500MWF	1MA	13V	15V	1V	15uV/C	.	.6MA	85dB	80dB	30M
RM108Q	OBS	SBA	EXT	.	.	+20V	-20V	125C	96dB	2MV	2NA	0.2NA	500MWF	1MA	13V	15V	1V	15uV/C	.	.6MA	85dB	80dB	30M
RM108T	OBS	SBA	EXT	.	.	+20V	-20V	125C	96dB	2MV	2NA	0.2NA	500MWF	1MA	13V	15V	1V	15uV/C	.	.6MA	85dB	80dB	30M
RM702AQ	OBS	WBA	EXT	.	0.5V/uS	+14V	-7V	125C	68dB	2MV	5uA	0.5uA	500MWF	.3MA	5V	1.5V	5V	10uV/C	120MW	7MA	80dB	74dB	16K
RM702AT	OBS	WBA	EXT	3MHZ	.	+13V	-8V	125C	68dB	2MV	5uA	0.5uA	670MWF	.3MA	5V	5V	5V	10uV/C	120MW	7MA	80dB	74dB	16K
RM702D	OBS	WBA	EXT	3MHZ	.	+13V	-8V	125C	68dB	2MV	5uA	0.5uA	670MWF	.3MA	5V	5V	5V	10uV/C	120MW	7MA	80dB	74dB	16K
RM702DC	RAU	WBA	EXT	.	0.5V/uS	+14V	-7V	125C	63dB	5MV	10uA	2uA	300MWF	3MA	5V	1.5V	5V	20uV/C	120MW	7MA	70dB	70dB	8K
RM702Q	RAU	WBA	EXT	3MHZ	.	+13V	-8V	125C	68dB	2MV	5uA	0.5uA	570MWF	.3MA	5V	5V	5V	10uV/C	120MW	7MA	80dB	74dB	16K
RM702T	RAU	WBA	EXT	3MHZ	.	+13V	-8V	125C	68dB	2MV	5uA	0.5uA	500MWF	.3MA	5V	5V	5V	10uV/C	120MW	7MA	80dB	74dB	16K
RM709ABL	RAU	GPU	EXT	.3MHZ	.15V/uS	+18V	-18V	125C	88dB	2MV	200NA	50NA	.	5MA	12V	10V	5V	10uV/C	108MW	.	80dB	80dB	350K
RM709ADC	RAU	GPU	EXT	.3MHZ	.15V/uS	+18V	-18V	125C	88dB	2MV	200NA	50NA	670MWF	5MA	12V	10V	5V	10uV/C	108MW	.	80dB	80dB	350K
RM709AQ	RAU	GPU	EXT	.3MHZ	.15V/uS	+18V	-18V	125C	88dB	2MV	200NA	50NA	500MWF	5MA	12V	10V	5V	10uV/C	108MW	.	80dB	80dB	350K
RM709AT	RAU	GPU	EXT	.3MHZ	.15V/uS	+18V	-18V	125C	88dB	2MV	200NA	50NA	500MWF	5MA	12V	10V	5V	10uV/C	108MW	.	80dB	80dB	350K
RM709BL	RAU	GPU	EXT	.3MHZ	.15V/uS	+18V	-18V	125C	88dB	5MV	500NA	200NA	.	5MA	12V	10V	5V	15uV/C	165MW		70dB	76dB	150K
RM709DC	FAU	GPU	EXT	.3MHZ	.15V/uS	+18V	-18V	125C	88dB	5MV	500NA	200NA	670MWF	5MA	12V	10V	5V	15uV/C	165MW		70dB	76dB	150K
RM709Q	FAU	GPU	EXT	.3MHZ	.15V/uS	+18V	-18V	125C	88dB	5MV	500NA	200NA	570MWF	5MA	12V	10V	5V	15uV/C	165MW		70dB	76dB	150K
RM709T	RAU	GPU	EXT	.3MHZ	.15V/uS	+18V	-18V	125C	88dB	5MV	500NA	200NA	500MWF	5MA	12V	10V	5V	15uV/C	165MW	.	70dB	76dB	150K
RM710ABL	RAU	CPR	EXT	.	.	+14V	-6V	125C	63dB	1MV	15uA	1uA	.	5MA	2.5V	7V	5V	5uV/C	150MW	9MA	90dB	.	.
RM710ADC	RAU	CPR	EXT	.	.	+14V	-6V	125C	63dB	1MV	15uA	1uA	670MWF	5MA	2.5V	7V	5V	5uV/C	150MW	9MA	90dB	.	.

LEFT HAND PAGE

APP = application (codes at APP.E.)
CMRR = common mode rejection ratio
CMP = compensation (frequency)
dV_{io}/dT = input offset voltage temperature drift
GBP = gain bandwidth product
I_b = input bias current
I_{io} = input bias offset current
I_i = quiescent supply current
MFR = manufacturer (codes at App.C.)
P = quiescent power consumer
PSRR = power supply rejection ratio
V_{icm} = common mode input voltage rating
V_{id} = differential input voltage rating
V_{io} = input offset voltage
V = dc supply voltage

RIGHT HAND PAGE

Lead out coding summary (details at APP.G.) for different cases (APP.F.)
A = gain adjust
B = bias adjust
C = case
E— = inverting input
E+ = non-inverting input
F F* = input frequency compensation
G = ground
J = high level input
K = output, open collector
L = output, open emitter
M = metal case
N = not connected
Q = special terminal
R,R* = outputs
S = strobe
T,T* = offset balance
V· = +ve dc supply
V— = —ve dc supply
W = guard ring
X = blank position, no lead
++ = +ve supplementary dc supply
—— = —ve supplementary dc supply
ø,ø* = output frequency compensation

CASE (APP F)	LD 1	LD 2	LD 3	LD 4	LD 5	LD 6	LD 7	LD 8	LD 9	LD 10	LD 11	LD 12	LD 13	LD 14	LD 15	LD 16	EUROPE SUBSTITUTE	USA SUBSTITUTE	I S	TYPE NUMBER
DIL-14/1C	ø2	R2	F2	F*2	E-2	E+2	V-	E+1	E-1	F1	F*1	R1	ø1	V+	.	.		MC1437L	O	RC1437DC
DIL-14/1C	R2	S2	V+2	N	E+1	E-1	V-	R1	S1	V+1	G	E+2	E-2	V-	.	.	LM1414J	MC1414L	O	RC1414DC
DIL-8/1P	R1	E-1	E+1	V-	E+2	E-2	R2	V+									TBB1458B	MC1458DN	O	RC1458DN
DIL-8/1P	R1	E-1	E+1	V-	E+2	E-2	R2	V+									TBB1458B	MC1458U	O	RC1458NB
T0S-8/1M	R1	E-1	E+1	V-	E+2	E-2	R2	V+									TBB1458	MC1458G	O	RC1458T
DIL-8/1P	T	E-	E+	V-	T*	R	V+	N										RC1556ANB	O	RC1556ANB
T0S-8/1M	T	E-	E+	V-	T*	R	V+	N										RC1556AT	O	RC1556AT
DIL-8/1P	T	E-	E+	V-	T*	R	V+	N										N5556V	O	RC1556NB
T0S-8/1M	T	E-	E+	V-	T*	R	V+	N									MC1456G	MC1456T	O	RC1556T
FLP-10/3G	N	F	E-	E+	V-	ø	R	V+	F*	N							SN52709AFA	MC1709CF	O	RC1709
BML																			O	RC1741BL
DIL-14/1P	E+1	E+2	E-2	R2	R1	E-1	G	E-3	R3	R4	E-4	E+4	E+3	V+			UA3401P	MC3401P	O	RC3401DB
DIL-14/1P	R1	E-1	E+1	V+	E+2	E-2	R2	R3	E-3	E+3	G	E+4	E-4	R4			UA3403D	MC3403L	O	RC3403ADB
DIL-14/1C	R1	E-1	E+1	V+	E+2	E-2	R2	R3	E-3	E+3	G	E+4	E-4	R4			RV3403ADC		O	RC3403ADC
DIL-8/1P	T	E-	E+	V-	T*	R	V+	N											O	RC4131NB
T0S-8/1M	T	E-	E+	V-M	T*	R	V+	N										RM4131T	O	RC4131T
DIL-14/1P	E-1	E+1	R1	R2	E+2	E-2	V-	E-3	E+3	R3	V+	R4	E+4	E-4			RV4136DB		O	DC4136DB
DIL-14/1P	E-1	E+1	R1	R2	E+2	E-2	V-	E-3	E+3	R3	V+	R4	E+4	E-4			RV4136DB		O	RC4136DC
DIL-14/1P	R1	E-1	E+1	V+	E+2	E-2	R2	R3	E-3	E+3	G	E+4	E-4	R4			RC3403ADB		O	RC4137DB
DIL-14/1C	R1	E-1	E+1	V+	E+2	E-2	R2	R3	E-3	E+3	G	E+4	E-4	R4			RC3403ADC		O	RC4137DC
DIL-14/1C	N	N	T	E-	E+	V-	N	N	T*	R	V+	ø0	N	N				RM4531D	O	RC4531D
DIL-8/1P	T	E-	E+	V-	T	R	V+	ø											O	RC4531DN
T0S-8/1M	T	E-	E+	V-	T*	R	V+	ø										RM4531T	O	RC4531T
DIL-8/1P	R1	E-1	E+1	V-	E+2	E-2	R2	V+										MC4558CU	O	RC4558NB
T0S-8/1M	R1	E-1	E+1	V-	E+2	E-2	R2	V+										MC4558CG	O	RC4558T
DIL-14/1P	R1	N	N	N	E+1	E-1	V-	E-2	E+2	N	N	N	R2	V+					O	RC4739DB
BML																			O	RM101ABL
DIL-14/1C	N	N	FT	E-	E+	V-	N	N	T*	R	V+	F*	N	N			UA101AD	LM101AJ14	O	RM101AD
FLP-10/3G	N	FT	E-	E+	V-	T*	R	V+	F*	N							SFC2101APM	LM101A⁻	O	RM101AQ
T0S-8/1M	FT	E-	E+	V-M	T*	R	V+	F*									SFC2101A	LM101AH	O	RM101AT
DIL-14/1C	N	N	F*	E-	E+	V-	N	N	T*	R	V+	F*	N	N			UA101AD	LM101J14	O	RM101D
FLP-10/3G	N	FT	E-	E+	V-	T*	R	V+	F*	N							SFC2101APM	LM101F	O	RM101Q
T0S-8/1M	FT	E-	E+	V-M	T*	R	V+	F*									SFC2101A	LM101H	O	RM101T
BML																			O	RM106BL
DIL-14/1C	N	N	N	E-	E+	V-	N	N	N		V+	N	N	N			SN52107JA	LM107D	O	RM107D
FLP-10/3G	N	N	E-	E+	V-	N	R	V+	N	N							SFC2107PM	LM107F	O	RM107Q
T0S-8/1M	N	E-	E+	V-M	N	R	V+	N									SFC2107M	LM107H	O	RM107T
DIL-14/1C	N	F	N	E-	E+	N	V-	N			V+	F*	N	N			UA108AD	LM108AD	O	RM108AD
FLP-10/3G	N	N	E-	E+	N	N	R	V+	F*	F								LM108AF	O	RM108AQ
T0S-8/1M	F	E-	E+	V-M	N	R	V+	F*									SFC2108A	LM108AH	O	RM108AT
DIL-14/1C	N	F	N	E-	E+	N	V-	N			V+	F*	N	N			UA108D	LM108D	O	RM108D
FLP-10/3G	N	N	E-	E+	N	N	R	V+	F*	F							SFC2108PM	LM108F	O	RM108Q
T0S-8/1M	F	E-	E+	V-M	N	R	V+	F*									SFC2108M	LM108H	O	RM108T
FLP-10/3G	N	G	E-	E+	V-M	F	ø	R	N	V+							SN52702AFA		O	RM702AQ
T0S-8/1M	G	E-	E+	V-M	F	ø	R	V+									SN52702AL	UA702HM	O	RM702AT
DIL-14/1C	N	N	G	E-	E+	V-	N	N	F	ø	R	N	V+	N			SN52702AJ	UA702DM	O	RM702D
DIL-14/1C	N	N	G	E-	E+	N	N	N	F	ø	R	N	V+	N			MC1712L	SN52702J	O	RM702DC
FLP-10/3G	N	G	E-	E+	V-	F	ø	R	V+								MC1712F	UA702FM	O	RM702Q
T0S-8/1M	G	E-	E+	V-M	F	ø	R	V+									MC1712G	UA702HM	O	RM702T
BML																			O	RM709ABL
DIL-14/1C	N	N	F	E-	E+	V-	N	N	ø	R	V+	F*	N	N			LM709AJ	UA709ADM	O	RM709ADC
FLP-10/3G	N	F	E-	E+	V-	ø	R	V+	F*	N							SN52709AFA	UA709AFM	O	RM709AQ
T0S-8/1M	F	E-	E+	V-	ø	ø*R	V+	F*									MC1709AG	UA709AHM	O	RM709AT
BML																			O	RM709BL
DIL-14/1C	N	N	F	E-	E+	V-	N	N	ø	R	V+	F*	N	N			LM709J	UA709DM	O	RM709DC
FLP-10/3G	N	F	E-	E+	V-	ø	R	V+	F*	N								UA709FM	O	RM709Q
T0S-8/1M	F	E-	E+	V-	ø	ø*R	V+	F*									TAA522	UA709HM	O	RM709T
BML																			O	RM710ABL
DIL-14/1C	N	G	E+	E-	N	V-	N	N	R	N	V+	N	N					RM710ADC	O	RM710ADC
FLP-10/3G	G	E+	E-	V-	R	N	V+	N										RM710AQ	O	RM710AT

TYPE NUMBER	MFR	APP	CMP	GBP MIN	SI FW RATE MIN	Vs+ MAX	Vs- MAX	Top MAX	Avol MIN	Vio MAX	Ib MAX	Iio MAX	Ptot MAX	Iout MIN	Vout MIN	Vicm MAX	Vidf MAX	dVio/dT MAX	Pq MAX	Iq MAX	CMRR MIN	PSRR MIN	Rin MIN
RM710AQ	RAU	CPR	EXT	.	.	+14V	-6V	125C	63dB	1MV	15uA	1uA	500MWF	5MA	2.5V	7V	5V	5uV/C	150MW	9MA	90dB	.	.
RM710AT	RAU	CPR	EXT	.	.	+14V	-6V	125C	63dB	1MV	15uA	1uA	500MWF	5MA	2.5V	7V	5V	5uV/C	150MW	9MA	90dB	.	.
RM710BL	RAU	CPR	EXT	.	.	+14V	-6V	125C	62dB	2MV	20uA	3uA	.	5MA	2.5V	7V	5V	10uV/C	150MW	.	80dB	.	.
RM710DC	RAU	CPR	EXT	.	.	+14V	-6V	125C	62dB	2MV	20uA	3uA	670MWF	5MA	2.5V	7V	5V	10uV/C	150MW	.	80dB	.	.
RM710Q	RAU	CPR	EXT	.	.	+14V	-6V	125C	62dB	2MV	20uA	3uA	570MWF	5MA	2.5V	7V	5V	10uV/C	150MW	.	80dB	.	.
RM710T	RAU	CPR	EXT	.	.	+14V	-7V	125C	62dB	2MV	20uA	3uA	500MWF	5MA	2.5V	7V	5V	10uV/C	150MW	.	80dB	.	.
RM711BL	RAU	DCP	EXT	.	.	+14V	-7V	125C	58dB	3.5MV	75uA	10uA	.	5MA	2.5V	7V	5V	20uV/C	200MW	.	.	.	.
RM711DC	RAU	DCP	EXT	.	.	+14V	-7V	125C	58dB	3.5MV	75uA	10uA	670MWF	5MA	2.5V	7V	5V	20uV/C	200MW	.	.	.	.
RM711Q	RAU	DCP	EXT	.	.	+14V	-7V	125C	58dB	3.5MV	75uA	10uA	570MWF	5MA	2.5V	7V	5V	20uV/C	200MW	.	.	.	.
RM711T	RAU	DCP	EXT	.	.	+14V	-7V	125C	58dB	3.5MV	75uA	10uA	500MWF	5MA	2.5V	7V	5V	20uV/C	200MW	.	.	.	.
RM725T	RAU	PIA	EXT	.	.	+22V	-22V	125C	120dB	1MV	100NA	20NA	500MWF	5MA	12V	22V	5V	5uV/C	150MW	.	110dB	100dB	500K
RM733T	RAU	BDO	.	20MHZ	.	+8V	-8V	125C	50dB	5MV	20uA	3uA	500MWF	2MA	1.5V	6V	5V	.	.	24MA	60dB	50dB	2K
RM741D	RAU	GPK	INT	.4MHZ	0.3V/uS	+22V	-22V	125C	94dB	3MV	80NA	30NA	670MWF	5MA	12V	15V	30V	15uV/C	150MW	.	80dB	86dB	1M
RM741Q	RAU	GPK	INT	.	0.2V/uS	+22V	-22V	125C	94dB	5MV	500NA	200NA	570MWF	5MA	12V	15V	30V	.	85MW	3MA	70dB	76dB	300K
RM741T	RAU	GPK	INT	.	0.2V/uS	+22V	-22V	125C	94dB	5MV	500NA	200NA	500MWF	7MA	12V	15V	30V	.	85MW	3MA	70dB	76dB	300K
RM741TE	RAU	GPK	INT	.	0.2V/uS	+22V	-22V	125C	94dB	5MV	500NA	200NA	500MWF	7MA	12V	15V	30V	.	85MW	3MA	70dB	76dB	300K
RM747DC	RAU	DGK	INT	.	0.2V/uS	+18V	-18V	70C	88dB	6MV	500NA	200NA	670MWF	5MA	12V	15V	30V	.	85MW	3MA	70dB	76dB	300K
RM747T	RAU	DGK	INT	.	0.2V/uS	+22V	-22V	125C	94dB	5MV	500NA	200NA	500MWF	5MA	12V	15V	30V	.	85MW	3MA	70dB	76dB	300K
RM748BL	RAU	GPU	EXT	.	0.2V/uS	+22V	-22V	125C	94dB	5MV	500NA	200NA	.	5MA	12V	15V	30V	.	85MW	3MA	70dB	76dB	300K
RM748T	RAU	GPU	EXT	.	0.2V/uS	+22V	-22V	125C	94dB	5MV	500NA	200NA	500MWF	5MA	12V	15V	30V	.	85MW	3MA	70dB	76dB	300K
RM1514DC	RAU	DCP	EXT	.	.	+14V	-7V	125C	62dB	2MV	20uA	3uA	1WF	1V	7V	5V	15uV/C	150MW	9MA	80dB	.	.
RM1537DC	RAU	DGU	EXT	.	0.1V/uS	+18V	-18V	125C	88dB	5MV	0.5uA	0.2uA	750MWF	5MA	12V	18V	5V	10uV/C	225MW	.	70dB	76dB	150K
RM1556AT	RAU	SBA	INT	2MHZ	1V/uS	+22V	-22V	125C	100dB	2MV	15NA	2NA	500MWF	6MA	12V	15V	30V	.	.	.	80dB	80dB	3M
RM1556T	RAU	SBA	INT	.5MHZ	1V/uS	+22V	-22V	125C	100dB	4MV	15NA	2NA	680MWF	6MA	12V	22V	22V	30uV/C	45MW	2MA	80dB	80dB	1.5M
RM1558T	RAU	DGK	INT	.5MHZ	0.3V/uS	+22V	-22V	125C	94dB	5MV	0.5uA	0.2uA	680MWF	5MA	12V	15V	30V	50uV/C	150MW	5MA	70dB	76dB	300K
RM1741BL	RAU	GPK	INT	.	0.2V/uS	+22V	-22V	125C	94dB	5MV	500NA	200NA	.	7MA	12V	15V	30V	.	85MW	3MA	70dB	76dB	300K
RM3503ADC	RAU	QGK	INT	.3MHZ	0.5V/uS	+18V	-18V	125C	94dB	4MV	400NA	50NA	650MWF	6MA	13V	18V	36V	.	.	4MA	70dB	86dB	.
RM44131T	RAU	GPK	INT	2MHZ	1V/uS	+22V	-22V	125C	94dB	2MV	50NA	10NA	500MWF	7MA	16V	15V	30V	15uV/C	.	2MA	80dB	80dB	2.2M
RM44136DC	RAU	QGK	INT	1MHZ	0.5V/uS	+22V	-22V	125C	94dB	5MV	500NA	200NA	800MWF	5MA	12V	15V	30V	.	340MW	.	70dB	76dB	300K
RM44531D	RAU	HSR	EXT	.	10V/uS	+22V	-22V	125C	94dB	5MV	500NA	200NA	500MWF	5MA	10V	15V	15V	.	210MW	7MA	70dB	76dB	300K
RM44531T	RAU	HSR	EXT	.	10V/uS	+22V	-22V	125C	94dB	5MV	500NA	200NA	500MWF	5MA	10V	15V	15V	.	210MW	7MA	70dB	76dB	300K
RM44558T	RAU	DWB	INT	1MHZ	0.5V/uS	+22V	-22V	125C	94dB	5MV	500NA	200NA	500MWF	5MA	12V	15V	30V	.	170MW	.	70dB	76dB	300K
RSN52709H	RAU	GPU	EXT	.	.	+18V	-18V	125C	88dB	5MV	500NA	200NA	300MWF	5MA	10V	10V	5V	15uV/C	165MW	6MA	70dB	76dB	150K
RSN52709Y	RAU	GPU	EXT	.	.	+18V	-18V	125C	88dB	5MV	500NA	200NA	.	5MA	12V	10V	5V	15uV/C	165MW	6MA	70dB	76dB	150K
RV3301DB	RAU	QCD	INT	1M5HZ	0.2V/uS	+28V	.	85C	60dB	.	300NA	.	625MWF	6MA	13V	.	.	.	.	10MA	.	50dB	.
RV3403ADB	RAU	QGK	INT	.3MHZ	0.5V/uS	+18V	-18V	85C	88dB	6MV	500NA	50NA	650MWF	6MA	13V	18V	36V	.	.	5MA	70dB	80dB	.
RV3403ADC	RAU	QGK	INT	.3MHZ	0.5V/uS	+18V	-18V	85C	88dB	6MV	500NA	50NA	650MWF	6MA	13V	18V	36V	.	.	5MA	70dB	80dB	.
RV4136DB	RAU	QGK	INT	1MHZ	0.3V/uS	+18V	-18V	85C	86dB	5MV	500NA	200NA	800MWF	5MA	12V	15V	30V	.	340MW	.	70dB	76dB	300K
RV4558NB	RAU	DWB	INT	1MHZ	0.5V/uS	+22V	-22V	85C	86dB	5MV	500NA	200NA	500MWF	5MA	12V	15V	30V	.	170MW	.	70dB	76dB	300K
S5556T	SJU	SBA	INT	.5MHZ	1V/uS	+22V	-22V	125C	100dB	4MV	15NA	2NA	500MWF	6MA	12V	22V	22V	30uV/C	45MW	2MA	80dB	80dB	1.5M
S5556V	SJU	SBA	INT	.5MHZ	1V/uS	+22V	-22V	125C	100dB	4MV	15NA	2NA	680MWF	6MA	12V	22V	22V	30uV/C	45MW	2MA	80dB	80dB	1.5M
S5558E	OBS	DGK	INT	.5MHZ	0.3V/uS	+22V	-22V	125C	94dB	5MV	0.5uA	0.2uA	750MWF	5MA	12V	15V	30V	50uV/C	150MW	5MA	70dB	76dB	300K
S5558T	SJU	DGK	INT	.5MHZ	0.3V/uS	+22V	-22V	125C	94dB	5MV	0.5uA	0.2uA	500MWF	5MA	12V	15V	30V	.	150MW	5MA	70dB	76dB	300K
S5709G	OBS	GPU	EXT	.3MHZ	.15V/uS	+18V	-18V	125C	88dB	5MV	500NA	200NA	570MWF	5MA	12V	10V	5V	15uV/C	165MW	.	70dB	76dB	150K
S5709T	OBS	GPU	EXT	.3MHZ	.15V/uS	+18V	-18V	125C	88dB	5MV	500NA	200NA	.	5MA	12V	10V	5V	15uV/C	165MW	.	70dB	76dB	150K
S5710T	OBS	CPR	EXT	.	.	+14V	-7V	125C	62dB	2MV	20uA	3uA	500MWF	5MA	2.5V	7V	5V	10uV/C	150MW	.	80dB	.	.
S5711K	OBS	DCP	EXT	.	.	+14V	-7V	125C	58dB	3.5MV	75uA	10uA	500MWF	5MA	2.5V	7V	5V	20uV/C	200MW	.	.	.	.
S5733K	OBS	BDO	EXT	20MHZ	.	+8V	-8V	125C	50dB	5MV	20uA	3uA	500MWF	2MA	1.5V	6V	5V	.	.	24MA	60dB	50dB	2K
S5741T	OBS	GPK	INT	.	0.2V/uS	+22V	-22V	125C	94dB	5MV	500NA	200NA	500MWF	7MA	12V	15V	30V	.	85MW	3MA	70dB	76dB	300K
SA532N(8)	MUG	DGK	INT	.3MHZ	.	+15V	-15V	85C	84dB	7.5MV	500NA	150NA	.	.	14V	14V	16V	20uV/C	.	2MA	60dB	80dB	.
SA532V	MUG	DGK	INT	.3MHZ	.	+15V	-15V	85C	84dB	7.5MV	500NA	150NA	.	.	14V	14V	16V	20uV/C	.	2MA	60dB	80dB	.
SA534N	MUG	QGK	INT	.	.	+16V	-16V	85C	94dB	5MV	150NA	30NA	570MWF	.	.	16V	16V	35uV/C	.	2MA	70dB	65dB	.
SA534N(14)	MUG	QGK	INT	.	.	+16V	-16V	85C	94dB	5MV	150NA	30NA	570MWF	.	.	16V	16V	35uV/C	.	2MA	70dB	65dB	.
SE515A	MUG	PIA	EXT	.	.	+6V	-6V	125C	71dB	2MV	24uA	.	.	30MA	5.6V	1.5V	5V	.	.	7MA	94dB	.	2K
SE515K	SJU	PIA	EXT	.	.	+6V	-6V	75C	71dB	2MV	24uA	.	.	30MA	5.6V	1.5V	5V	.	.	7MA	94dB	.	2K
SE515Q	SJU	PIA	EXT	.	.	+6V	-6V	75C	71dB	2MV	24uA	.	.	30MA	5.6V	1.5V	5V	.	.	7MA	94dB	.	.
SE527K	SJU	CPR	EXT	.	.	+15V	-15V	125C	66dB	4MV	2uA	.75uA	600MWF	.	1V	6V	5V	.	.	.	.	.	200K
SE529K	SJU	CPR	EXT	.	.	+15V	-15V	125C	66dB	4MV	12uA	3uA	600MWF	.	1V	6V	5V	.	.	.	.	.	3K
SE531T	SJU	HSR	EXT	.5MHZ	20V/uS	+22V	-22V	125C	94dB	5MV	500NA	200NA	300MWF	.	.	15V	15V	.	210MW	6MA	70dB	76dB	10M
SE531V	MUG	HSR	EXT	.5MHZ	20V/uS	+22V	-22V	125C	94dB	5MV	500NA	200NA	300MWF	.	.	15V	15V	.	210MW	7MA	70dB	76dB	10M

For detailed explanations of column heading notations, see App. A.
Also for ready references the more important abbreviations used in the column headings are listed below:

LEFT HAND PAGE

APP = application (codes at APP.E.)

CMRR = common mode rejection ratio

CMP = compensation (frequency)

dV_{IO}/dT = input offset voltage temperature drift

GBP = gain bandwidth product

I_B = input bias current

I_{IO} = input bias offset current

I_Q = quiescent supply current

MFR = manufacturer (codes at App.C.)

P_Q = quiescent power consumer

PSRR = power supply rejection ratio

V_{ICM} = common mode input voltage rating

V_{IDF} = differential input voltage rating

V_{IO} = input offset voltage

V_S = dc supply voltage

RIGHT HAND PAGE

Lead out coding summary (details at APP.G.) for different cases (APP.F.)

A = gain adjust

B = bias adjust

C = case

E− = inverting input

E+ = non-inverting input

F,F* = input frequency compensation

G = ground

J = high level input

K = output, open collector

L = output, open emitter

M = metal case

N = not connected

O = special terminal

R,R* = outputs

S = strobe

T,T* = offset balance

V+ = +ve dc supply

V− = −ve dc supply

W = guard ring

X = blank position, no lead

++ = +ve supplementary dc supply

−− = −ve supplementary dc supply

ø,ø* = output frequency compensation

CASE (APP F)	LD 1	LD 2	LD 3	LD 4	LD 5	LD 6	LD 7	LD 8	LD 9	LD 10	LD 11	LD 12	LD 13	LD 14	LD 15	LD 16	EUROPE SUBSTITUTE	USA SUBSTITUTE	I S	TYPE NUMBER
TO5-8/1M	G	E+	E-	V-	N	N	R	V+										RM710AT	0	RM710AT
BML																			0	RM710BL
DIL-14/1C	N	G	E+	E-	N	V-	N	N	R	N	V+	N	N	N			SFC2710KM	UA710DM	0	RM710DC
FLP-10/3G	G	E+	E-	N	V-	R	N	V+	N	N							SFC2710PM	UA710FM	0	RM710Q
TO5-8/1M	G	E+	E-	V-	N	N	R	V+									SFC2710M	UA710HM	0	RM710T
BML																			0	RM711BL
DIL-14/1C	N	E-1	E+1	V-	E+2	E-2	N	N	S2	R	V+	G	S1	N			SFC2711KM	UA711DM	0	RM711DC
FLP-10/3G	E-1	E+1	V-	E+2	E-2	S2	R	V+	G	S1							SFC2711PM	UA711FM	0	RM711Q
TO5-10/1M	G	S1	E-1	E+1	V-	E+2	E-2	S2	R	V+							SFC2711M	UA711HM	0	RM711T
TO5-8/1M	T	E-	E+	V-	ø	ø*	V+	T*									LM725H	UA725HM	0	RM725T
TO5-10/1M	E-	E+	A2	A*2	V-	R	R*	V+	A1	A*1							SN52733L	UA733HM	0	RM733T
DIL-14/1C	N	N	T	E-	E+	V-	N	N	T*	R	V+	N	N	N			LM741D	UA741DM	0	RM741D
FLP-10/3G	N	T	E-	E+	V-	T*	R	V+	N	N							LM741F	UA741FM	0	RM741Q
TO5-8/1M	T	E-	E+	V-M	T*	R	V+	N									TBA222	UA741HM	0	RM741T
TO5-8/1M	T	E-	E+	V-M	T*	R	V+	N									TBA222	UA741HM	0	RM741TE
DIL-14/1C	E-1	E+1	T1	V-	T2	E+2	E-2	T*2	V+2	R2	N	R1	V+1	T*1			TBB0747A	UA747DC	0	RM747DC
TO5-10/1M	R1	V+1	E-1	E+1	V-	E+2	E-2	V+2	R2	N							SFC2747M	UA747HM	0	RM747T
BML																			0	RM748BL
TO5-8/1M	FT	E-	E+	V-	T*	R	V+	F*									TBC0748	UA748HM	0	RM748T
DIL-14/1C	R2	S2	V+2	N	E+1	E-1	V-	R1	S1	V+1	G	E+2	E-2	V-			LM1514J	MC1514L	0	RM1514DC
DIL-14/1C	ø2	R2	F2	F*2	E-2	E+2	V-	E+1	E-1	F1	F*1	R1	ø1	V+				MC1537L	0	RM1537DC
TO5-8/1M	T	E-	E+	V-	T*	R	V+	N										RM1556AT	0	RM1556AT
TO5-8/1M	T	E-	E+	V-	T*	R	V+	N									S5556T	MC1556G	0	RM1556T
TO5-8/1M	R1	E-1	E+1	V-	E+2	E-2	R2	V+									TBC1458	MC1558G	0	RM1558T
BML																			0	RM1741BL
DIL-14/1C	R1	E-1	E+1	V+	E+2	E-2	R2	R3	E-3	E+3	G	E+4	E-4	R4					0	RM3503ADC
TO5-8/1M	T	E-	E+	V-M	T*	R	V+	N											0	RM4131T
DIL-14/1C	E-1	E+1	R1	R2	E+2	E-2	V-	E-3	E+3	R3	V+	R4	E+4	E-4					0	RM4136DC
DIL-14/1C	N	N	T	E-	E+	V-	N	N	T*	R	V+	ø	N	N					0	RM4531D
TO5-8/1M	T	E-	E+	V-	T*	R	V+	ø											0	RM4531T
TO5-8/1M	R1	E-1	E+1	V-	E+2	E-2	R2	V+										MC4558G	0	RM4558T
FLP-10/3G	N	F	E-	E+	V-	ø	R	V+	F*	N							UA709FM	SN52709FA	0	RSN52709H
CHP																			0	RSN52709Y
DIL-14/1P	E+1	E+2	E-2	R2	R1	E-1	G	E-3	R3	R4	E-4	E+4	E+3	V+			UA3301P	MC3301P	0	RV3301DB
DIL-14/1P	R1	E-1	E+1	V+	E+2	E-2	R2	R3	E-3	E+3	G	E+4	E-4	R4					0	RV3403ADB
DIL-14/1C	R1	E-1	E+1	V+	E+2	E-2	R2	R3	E-3	E+3	G	E+4	E-4	R4				RM3403ADC	0	RV3403ADC
DIL-14/1P	E-1	E+1	R1	R2	E+2	E-2	V-	E-3	E+3	R3	V+	R4	E+4	E-4				RM4136DC	0	RV4136DB
DIL-8/1P	R1	E-1	E+1	V-	E+2	E-2	R2	V+										MC4558U	0	RV4558NB
TO5-8/1M	T	E-	E+	V-	T*	R	V+	N									MC1556T	MC1556G	0	S5556T
DIL-8/1P	T	E-	E+	V-	T*	R	V+	N										MC1556V	0	S5556V
DIL-14/1C	N	R1	N	N	E-1	E+1	V-	E+2	E-2	N	N	R2	N	V+				MC1558L	0	S5558E
TO5-8/1M	R1	E-1	E+1	V-	E+2	E-2	R2	V+									TBC1458	MC1558G	0	S5558T
FLP-10/3C	N	F	E-	E+	V-	ø	R	V+	F*	N							SN52709SAFA	UA709FM	0	S5709G
TO5-8/1M	F	E-	E+	V-	ø	ø*R	V+	F*									TAA522	UA709HM	0	S5709T
TO5-8/1M	G	E+	E-	V-	N	N	R	V+									SFC2710M	UA710HM	0	S5710T
TO5-10/1M	G	S1	E-1	E+1	V-	E+2	E-2	S2	R	V+							SFC2711M	UA711HM	0	S5711K
TO5-10/1M	E-	E+	A2	A*2	V-	R	R*	V+	A1	A*1							LM733H	UA733HM	0	S5733K
TO5-8/1M	T	E-	E+	V-M	T*	R	V+	N									TBA222	UA741HM	0	S5741T
DIL-8/1P	R1	E-1	E+1	V-	E+2	E-2	R2	V+										SA532V	0	SA532N(8)
DIL-8/1P	R1	E-1	E+1	V-	E+2	E-2	R2	V+										SA532V	0	SA532V
DIL-14/1P	R1	E-1	E+1	V+	E+2	E-2	R2	R3	E-3	E+3	G	E+4	E-4	R4			LM224J	LM224D	0	SA534A
DIL-14/1P	R1	E-1	E+1	V+	E+2	E-2	R2	R3	E-3	E+3	G	E+4	E-4	R4			LM224J	LM224D	0	SA534N(14)
DIL-14/1P	E2	N	N	F	N	R2	V-	R1	N	N	F*	N	E1	V+					0	SE515A
TO5-10/1M	V-M	R1	NM	F	E1	V+	E2	F*	NM	R2									0	SE515K
TO5-10/1M	V-M	R1	N	F	E1	V+	E2	F*	N	R2									G	SE515O
TO5-10/1M	E	E*	V-1	S*	R*	G	R	S	V+2	V+1									0	SE527K
TO5-10/1M	E	E*	V-1	S*	R*	G	R	S	V+2	V+1									0	SE529K
TO5-8/1M	T	E-	E+	V-	T*	R	V+	F										HA-2505	0	SE531T
DIL-8/1P	T	E-	E+	V-	T*	R	V+	F											0	SE531V
TO5-8/1M	R1	E-1	E+1	V-	E+2	E-2	R2	V+											0	SE532T

TYPE NUMBER	M F R	A P P	C M P	GBP MIN	SLEW RATE MIN	V_S^+ MAX	V_S^- MAX	T_{op} MAX	A_{VOL} MIN	V_{IO} MAX	I_B MAX	I_{IO} MAX	P_{TOT} MAX	I_{OUT} MIN	V_{OUT} MIN	V_{ICM} MAX	V_{IDF} MAX	dV_{IO}/dT MAX	P_Q MAX	I_Q MAX	CM RR MIN	PS RR MIN	R_{IN} MIN
SE532T	SJU	DGK	INT	.3MHZ		+15V	-15V	125C	88dB	6MV	300NA	100NA	.	.	14V	14V	16V	20uV/C		2MA	60dB	80dB	
SE535N(8)	MUG	HSR	INT	.3MHZ	5V/uS	+22V	-22V	125C	88dB	3MV	100NA	20NA	500MWF	5MA	12V	13V	30V	1.5uV/C		3MA	80dB	100dB	
SE535T	MUG	HSR	INT	.3MHZ	5V/uS	+22V	-22V	125C	88dB	3MV	100NA	20NA	500MWF	5MA	12V	13V	30V	1.5uV/C		3MA	80dB	100dB	
SE535V	MUG	HSR	INT	.3MHZ	5V/uS	+22V	-22V	125C	88dB	3MV	100NA	20NA	500MWF	5MA	12V	13V	30V	1.5uV/C		3MA	80dB	100dB	
SE540L	SJU	HCO	EXT	.	100V/uS	+27V	-27V	125C	.	7MV	3uA	0.7uA	1WF	.12A						20MA	90dB	80dB	10K
SE592A	SJU	BDO	EXT	20MHZ		+8V	-8V	125C	50dB	5MV	20uA	3uA	670MWF	2MA	1.5V	6V	5V			24MA	60dB	50dB	2K
SE592F	MUG	BDO	EXT	20MHZ		+8V	-8V	125C	50dB	5MV	20uA	3uA	670MWF	2MA	1.5V	6V	5V			24MA	60dB	50dB	2K
SE592K	SJU	BDO	EXT	20MHZ		+8V	-8V	125C	50dB	5MV	20uA	3uA	500MWF	2MA	1.5V	6V	5V			24MA	60dB	50dB	2K
SE592N(14)	MUG	BDO	EXT	20MHZ		+8V	-8V	125C	50dB	5MV	20uA	3uA	670MWF	2MA	1.5V	6V	5V			24MA	60dB	50dB	2K
SF.C2101A	THF	GPU	EXT			+22V	-22V	125C	94dB	2MV	75NA	10NA	500MWF	5MA	12V	15V	30V	15uV/C		3MA	80dB	80dB	1.5M
SF.C2101APM	THF	GPU	EXT			+22V	-22V	125C	94dB	2MV	75NA	10NA	500MWF	2MA	12V	15V	30V	15uV/C		3MA	80dB	80dB	1.5M
SF.C2107M	THF	GPK	INT			+22V	-22V	125C	94dB	2MV	75NA	10NA	500MWF	2MA	12V	15V	30V	15uV/C		3MA	80dB	80dB	1.5M
SF.C2107PM	THF	GPK	INT			+22V	-22V	125C	94dB	2MV	75NA	10NA	500MWF	2MA	12V	15V	30V	15uV/C		3MA	80dB	80dB	1.5M
SF.C2108A	THF	SBA	EXT			+20V	-20V	125C	98dB	0.5MV	2NA	0.2NA	500MWF	1MA	13V	15V	1V	5uV/C		.6MA	96dB	96dB	30M
SF.C2108M	THF	SBA	EXT			+20V	-20V	125C	96dB	2MV	2NA	0.2NA	500MWF	1MA	13V	15V	1V	15uV/C		.6MA	85dB	80dB	30M
SF.C2108PM	THF	SBA	EXT			+20V	-20V	125C	96dB	2MV	2NA	0.2NA	500MWF	1MA	13V	15V	1V	15uV/C		.6MA	85dB	80dB	30M
SF.C2110M	THF	VFA	INT		15V/uS	+18V	-18V	125C	0dB	4MV	3NA	.	500MWF	1MA	10V	15V	15V	50uV/C		6MA	.	70dB	10G
SF.C2111M	THF	CPR	EXT			+18V	-18V	125C	100dB	3MV	100NA	10NA	500MWF		.	15V	30V			6MA	.	.	.
SF.C2118M	THF	XSR	INT		50V/uS	+20V	-20V	125C	94dB	4MV	250NA	50NA	500MWF	6MA	12V	15V	1V			8MA	80dB	70dB	1M
SF.C2201A	THG	GPU	EXT			+22V	-22V	85C	94dB	2MV	75NA	10NA	500MWF	5MA	12V	15V	30V	15uV/C		3MA	80dB	80dB	500K
SF.C2201APT	THG	GPU	EXT			+22V	-22V	85C	94dB	2MV	75NA	10NA	500MWF	5MA	12V	15V	30V	15uV/C		3MA	80dB	80dB	500K
SF.C2207	THF	GPK	INT			+22V	-22V	85C	94dB	2MV	75NA	10NA	500MWF	5MA	12V	15V	30V	15uV/C		3MA	80dB	80dB	1.5M
SF.C2207PT	THF	GPK	INT			+22V	-22V	85C	94dB	2MV	75NA	10NA	500MWF	5MA	12V	15V	30V	15uV/C		3MA	80dB	80dB	1.5M
SF.C2208	THF	SBA	EXT			+20V	-20V	85C	96dB	2MV	2NA	0.2NA	500MWF	1MA	13V	15V	1V	15uV/C		.6MA	85dB	80dB	30M
SF.C2208A	THF	SBA	EXT			+20V	-20V	85C	98dB	0.5MV	2NA	0.2NA	500MWF	1MA	13V	15V	1V	5uV/C		.6MA	96dB	96dB	30M
SF.C2208PT	THF	SBA	EXT			+20V	-20V	85C	96dB	2MV	2NA	0.2NA	500MWF	1MA	13V	15V	1V	15uV/C		.6MA	85dB	80dB	30M
SF.C2210	THF	VFA	INT		15V/uS	+18V	-18V	85C	0dB	4MV	3NA	.	500MWF	1MA	10V	15V	15V	50uV/C		6MA	.	70dB	10G
SF.C2211	THF	CPR	EXT			+18V	-18V	85C	100dB	3MV	100NA	10NA	500MWF		.	15V	30V			6MA	.	.	.
SF.C2218	THF	XSR	INT		50V/uS	+20V	-20V	85C	94dB	4MV	250NA	50NA	500MWF	6MA	12V	15V	1V			8MA	80dB	70dB	1M
SF.C2301A	THF	GPU	EXT			+18V	-18V	70C	88dB	7.5MV	250NA	50NA	500MWF	5MA	12V	15V	30V	30uV/C		3MA	70dB	70dB	500K
SF.C2301ADC	THG	GPU	EXT			+18V	-18V	70C	88dB	7.5MV	250NA	50NA	500MWF	5MA	12V	15V	30V	30uV/C		3MA	70dB	70dB	500K
SF.C2307	THF	GPK	INT			+18V	-18V	70C	84dB	7.5MV	250NA	50NA	500MWF	5MA	12V	15V	30V	30uV/C		.	70dB	70dB	0.5M
SF.C2307DC	THF	GPK	INT			+18V	-18V	70C	84dB	7.5MV	250NA	50NA	500MWF	5MA	12V	15V	30V	30uV/C		.	70dB	70dB	0.5M
SF.C2308	THF	SBA	EXT			+18V	-18V	70C	88dB	7.5MV	7NA	1NA	500MWF	1MA	13V	15V	1V	30uV/C		.6MA	80dB	80dB	10M
SF.C2308A	THF	SBA	EXT			+18V	-18V	70C	98dB	0.5MV	7NA	1NA	500MWF	1MA	13V	15V	1V	5uV/C		.6MA	96dB	96dB	10M
SF.C2308DC	THF	SBA	EXT			+18V	-18V	70C	88dB	7.5MV	7NA	1NA	500MWF	1MA	13V	15V	1V	30uV/C		.6MA	80dB	80dB	10M
SFC2310	THF	VFA	INT		15V/uS	+18V	-18V	70C	0dB	7.5MV	7NA	.	500MWF	1MA	10V	15V	15V	50uV/C		6MA	.	70dB	10G
SF.C2310DC	THF	VFA	INT		15V/uS	+18V	-18V	70C	0dB	7.5MV	7NA	.	500MWF	1MA	10V	15V	15V	50uV/C		6MA	.	70dB	10G
SF.C2310EC	THF	VFA	INT		15V/uS	+18V	-18V	70C	0dB	7.5MV	7NA	.	500MWF	1MA	10V	15V	15V	50uV/C		6MA	.	70dB	10G
SF.C2311	THF	CPR	EXT			+18V	-18V	70C	100dB	7.5MV	250NA	50NA	500MWF		.	15V	30V			8MA	.	.	.
SF.C2311DC	THF	CPR	EXT			+18V	-18V	70C	100dB	7.5MV	250NA	50NA	500MWF		.	15V	30V			8MA	.	.	.
SF.C2311EC	THF	CPR	EXT			+18V	-18V	70C	100dB	7.5MV	250NA	50NA	500MWF		.	15V	30V			8MA	.	.	.
SF.C2315DC	THF	GPU	EXT		3V/uS	+15V	-15V	70C	75dB	20MV	50NA	25NA	.	25MA	14V	13V	13V	50uV/C		3MA	60dB	74dB	1M
SF.C2318	THG	XSR	INT		50V/uS	+20V	-20V	70C	88dB	10MV	500NA	200NA	500MWF	6MA	12V	15V	1V			8MA	70dB	65dB	500K
SF.2318EC	THF	XSR	INT		50V/uS	+20V	-20V	70C	88dB	10MV	500NA	200NA	500MWF	6MA	12V	15V	1V			8MA	70dB	65dB	500K
SF.C2458C	THF	DGK	INT		0.3V/uS	+18V	-18V	70C	88dB	6MV	500NA	200NA	680MWF	5MA	12V	15V	30V			6MA	70dB	77dB	300K
SF.C2458DC	THF	DGK	INT			+18V	-18V	70C	86dB	6MV	500NA	200NA	500MWF	5MA	12V	15V	30V			6MA	70dB	77dB	300K
SF.C2458M	THF	DGK	INT	.5MHZ	0.3V/uS	+22V	-22V	125C	94dB	2MV	500NA	200NA	680MWF	5MA	12V	15V	30V		150MW	5MA	70dB	76dB	300K
SF.C2709A	THF	GPU	EXT	.3MHZ	.15V/uS	+18V	-18V	125C	88dB	2MV	200NA	50NA	300MWF	7MA	12V	10V	5V	10uV/C	108MW	4MA	80dB	80dB	350K
SF.C2709AE	THF	GPU	EXT	.3MHZ	.15V/uS	+18V	-18V	70C	88dB	3.5MV	500NA	300NA	300MWF	6MA	12V	10V	5V	10uV/C	200MW	6MA	70dB	76dB	120K
SF.C2709AP	THF	GPU	EXT	.3MHZ	.15V/uS	+18V	-18V	125C	88dB	2MV	200NA	50NA	250MWF	7MA	12V	10V	5V	10uV/C	108MW	4MA	80dB	80dB	350K
SF.C2709C	THF	GPU	EXT	.3MHZ	.15V/uS	+18V	-18V	70C	88dB	7.5MV	1.5uA	500NA	300MWF	7MA	12V	10V	5V		200MW	7MA	65dB	74dB	50K
SF.C2709DC	THF	GPU	EXT	.3MHZ	.15V/uS	+18V	-18V	70C	84dB	7.5MV	1.5uA	500NA	300MWF	7MA	12V	10V	5V		200MW	7MA	65dB	74dB	50K
SF.C2709EC	THF	GPU	EXT	.3MHZ	.15V/uS	+18V	-18V	70C	84dB	7.5MV	1.5uA	500NA	300MWF	7MA	12V	10V	5V		200MW	7MA	65dB	74dB	50K
SF.C2709ET	THF	GPU	EXT		.15V/uS	+18V	-18V	85C	88dB	5MV	750NA	300NA	300MWF	5MA	12V	10V	5V	20uV/C		7MA	65dB	74dB	70K
SF.C2709KM	THF	GPU	EXT	.3MHZ	.15V/uS	+18V	-18V	125C	88dB	5MV	500NA	200NA	300MWF	6MA	12V	10V	5V	15uV/C	165MW	6MA	70dB	76dB	150K
SF.C2709M	THF	GPU	EXT	.3MHZ	.15V/uS	+18V	-18V	125C	88dB	5MV	500NA	200NA	300MWF	6MA	12V	10V	5V	15uV/C	165MW	6MA	70dB	76dB	150K
SF.C2709PM	THF	GPU	EXT	.3MHZ	.15V/uS	+18V	-18V	125C	88dB	5MV	500NA	200NA	250MWF	6MA	12V	10V	5V	15uV/C	165MW	6MA	70dB	76dB	150K
SFC2709PT	THF	GPU	EXT		.15V/uS	+18V	-18V	85C	88dB	5MV	750NA	300NA	250MWF	5MA	12V	10V	5V	20uV/C		7MA	65dB	74dB	70K
SF.C2709T	THF	GPU	EXT		.15V/uS	+18V	-18V	85C	88dB	5MV	750NA	300NA	300MWF	5MA	12V	10V	5V	20uV/C		7MA	65dB	74dB	70K

For detailed explanations of column heading notations, see App. A.

Also for ready references the more important abbreviations used in the column headings are listed below:

LEFT HAND PAGE
APP = application (codes at APP.E.)
CMRR = common mode rejection ratio
CMP = compensation (frequency)
dV_{IO}/dT = input offset voltage temperature drift
GBP = gain bandwidth product
I_B = input bias current
I_{IO} = input bias offset current
I_Q = quiescent supply current
MFR = manufacturer (codes at App.C.)
P_Q = quiescent power consumer
PSRR = power supply rejection ratio
V_{ICM} = common mode input voltage rating
V_{IDR} = differential input voltage rating
V_{IO} = input offset voltage
V_S = dc supply voltage

RIGHT HAND PAGE
Lead out coding summary (details at APP.G.) for different cases (APP.F.)
A = gain adjust
B = bias adjust
C = case
E— = inverting input
E+ = non-inverting input
F,F* = input frequency compensation
G = ground
J = high level input
K = output, open collector
L = output, open emitter
M = metal case
N = not connected
Q = special terminal
R,R* = outputs
S = strobe
T,T* = offset balance
V+ = +ve dc supply
V— = —ve dc supply
W = guard ring
X = blank position, no lead
+ + = +ve supplementary dc supply
— — = —ve supplementary dc supply
ø,ø* = output frequency compensation

CASE (APP F)	LD1	LD2	LD3	LD4	LD5	LD6	LD7	LD8	LD9	LD10	LD11	LD12	LD13	LD14	LD15	LD16	EUROPE SUBSTITUTE	USA SUBSTITUTE	ISS	TYPE NUMBER
DIL-8/1P	T	E-	E+	V-	T*	R	V+	N										SE535V		SE535N(8)
T05-8/1M	T	E-	E+	V-	T*	R	V+	N											0	SE535T
DIL-8/1P	T	E-	E+	V-	T*	R	V+	N											0	SE535V
T05-10/1M	B	E+	N	E-	B*	V-	L	Q	K	V+									0	SE540L
DIL-14/1P	E+	N	A2	A*2	V-	N	R	R*	N	V+	A1	A*1	N	E-			SN52733J	UA733DM	0	SE592A
DIL-14/1C	E+	N	A2	A*2	V-	N	R	R*	N	V+	A1	A*1	N	E-			SN52733J	UA733DM	0	SE592F
T05-10/1M	E-	E+	A2	A*2	V-	R	R*	V+	A1	A*1							SN52733L	UA733HM	0	SE592K
DIL-14/1P	E+	N	A2	A*2	V-	N	R	R*	N	V+	A1	A*1	N	E-			SN52733J	UA733DM	0	SE592N(14)
T05-8/1M	FT	E-	E+	V-M	T*	R	V+	F*									UA2101A	LM101AH	0	SF.C2101A
FLP-10/3G	N	FT	E-	E+	V-	T*	R	V+	F*	N							UA101AF	LM101AF	0	SF.C2101APM
T05-8/1M	N	E-	E+	V-M	R	V+	N										UA107H	LM107H	0	SF.C2107
FLP-10/3G	N	N	E-	E+	V-	N	R	V+	N	N								LM107F	0	SF.C2107PM
T05-8/1M	F	E-	E+	V-M	N	R	V+	F*									UA108AH	LM108AH	0	SF.C2108A
T05-8/1M	F	E-	E+	V-M	N	R	V+	F*									UA108H	LM108	0	SF.C2108M
FLP-10/3G	N	N	E-	E+	N	V-	R	V+	F*	F							UA108F	LM108F	0	SF.C2108PM
T05-8/1M	T	N	E+	V-	L	R	V+	T*									UA110M	LM110H	0	SF.C2110M
T05-8/1M	G	E+	E-	V-	T	T*S	R	V+									SG111T	LM111H	0	SF.C2111M
T05-8/1M	T*F	E-	E+	V-	F*T	R	V+	ø									TDC0118CM	LM118H	0	SF.C2118M
T05-8/1M	FT	E-	E+	V-M	T*	R	V+	F*									UA201AH	LM201AH	0	SF.C2201A
FLP-10/3G	N	FT	E-	E+	V-	T*	R	V+	F*	N							UA201AF	LM201AF	0	SF.C2201APT
T05-8/1M	N	E-	E+	V-M	N	R	V+	N									UA207H	LM207H	0	SF.C2207
FLP-10/3G	N	N	E-	E+	V-	N	R	V+	N	N								LM207F	0	SF.C2207PT
T05-8/1M	F	E-	E+	V-M	N	R	V+	F*									UA208H	LM208H	0	SF.C2208
T05-8/1M	F	E-	E+	V-M	N	R	V+	F*									UA208AH	LM208AH	0	SF.C2208A
FLP-10/3G	N	N	E-	E+	N	V-	R	V+	F*	F							UA208F	LM208F	0	SF.C2208PT
T05-8/1M	T	N	E+	V-	L	R	V+	T*									UA110M	LM210H	0	SF.C2210
T05-8/1M	G	E+	E-	V-	T	T*S	R	V+									SG211T	LM211H	0	SF.2211
T05-8/1M	T*F	E-	E+	V-	F*T	R	V+	ø									TDB0118CM	LM218H	0	SF.C2218
T05-8/1M	FT	E-	E+	V-M	T*	R	V+	F*									UA301AH	LM301AH	0	SFC2301A
DIL-8/1G	FT	E-	E+	V-	T*	R	V+	F*									UA301AT	LM301AJ	0	SF.C2301ADC
T05-8/1M	N	E-	E+	V-M	N	R	V+	N									UA307H	LM307H	0	SF.C2307
DIL-8/1P	N	E-	E+	V-	N	R	V+	N									UA307T	LM307J	0	SF.C2307DC
T05-8/1M	F	E-	E+	V-M	N	R	V+	F*									UA308H	LM308H		SF.C2308
T05-8/1M	F	E-	E+	V-M	N	R	V+	F*									UA308AH	LM308AH	0	SF.C2308A
DIL-8/1P	F	E-	E+	V-	N	R	V+	F*									SN72308JP	LM308N	0	SF.C2308DC
T05-8/1M	T	N	E+	V-	L	R	V+	T*									UA310C	LM310H	0	SFC2310
DIL-8/1P	T	N	E+	V-	L	R	V+	T*									SN72310JP	LM310N	0	SF.C2310DC
DIL-14/1P	N	N	N	T	N	E+	V-	N	N	L	R	V+	T*	N	N		SN72310JA	LM310D	0	SF.C2310EC
T05-8/1M	G	E+	E-	V-	T	T*S	R	V+									UA311H	LM311H	0	SF.C2311
DIL-8/1P	G	E+	E-	V-	T	T*S	R	V+									UA311R	LM311N	0	SF.C2311DC
DIL-14/1P	N	G	E+	E-	N	V-	T	T*S	R	N	V+	N	N	N	N		SN72311J	LM311D	0	SF.C2311EC
DIL-6/1P	V+	E+	E-	V-	K	K*											TCA315A		0	SF.C2315DC
T05-8/1M	T*F	E-	E+	V-	F*T	R	V+	ø									TDE0118CM	LM318H	0	SF.C2318
DIL-14/1P	N	N	T*F	E-	E+	V-	N	N	F*T	R	V+	ø	N	N			SN72318JA	LM318D	0	SF.C2318EC
T05-8/1M	R1	E-1	E+1	V-	E+2	E-2	R2	V+									TBB1458	MC1458G	0	SF.C2458C
DIL-8/1P	R1	E-1	E+1	V-	E+2	E-2	R2	V+									TBB1458B	MC1458U	0	SF.C2458DC
T05-8/1M	R1	E-1	E+1	V-	E+2	E-2	R2	V+									TBC1458	MC1558G	0	SF.C2458M
T05-8/1M	F	E-	E+	V-	ø	ø*R	V+	F*									MC1709AG	UA709AHM	0	SF.C2709A
DIL-14/1P	N	N	F	E-	E+	V-	N	N	ø	R	V+	F*	N	N			LM709AJ	UA709ADM	0	SF.C2709AE
FLP-10/3G	N	F	E-	E+	V-	ø	R	V+	F*	N								UA709AFM	0	SF.C2709AP
T05-8/1M	F	E-	E+	V-	ø	ø*R	V+	F*									TAA521	UA709HC	0	SF.C2709C
DIL-8/1P	F	E-	E+	V-	ø	ø*R	V+	F*									LM709CN8	UA709TC	0	SF.C2709DC
DIL-14/1P	N	N	F	E-	E+	V-	N	N	ø	R	V+	F*	N	N			LM709AJ	UA709ADM	0	SF.C2709EC
DIL-14/1C	N	N	F	E-	E+	V-	N	N	ø	R	V+	F*	N	N			LM709AJ	UA709ADM	0	SF.C2709ET
DIL-14/1C	N	N	F	E-	E+	V-	N	N	ø	R	V+	F*	N	N			LM709J	UA709DM	0	SF.C2709KM
T05-8/1M	F	E-	E+	V-	ø	ø*R	V+	F*									TAA522	UA709HM	0	SF.C2709M
FLP-10/3G	N	F	E-	E+	V-	ø	R	V+	F*	N								UA709FM	0	SF.C2709PM
FLP-10/3G	N	F	E-	E+	V-	ø	R	V+	F*	N								UA709FM	0	SF.C2709PT
T05-8/1M	F	E-	E+	V-	ø	ø*R	V+	F*									TAA522	UA709HM	0	SF.C2709T
T05-8/1M	G	E+	E-	V-M	N	R	V+										LM710CH	UA710HC	0	SF.C2710C

TYPE NUMBER	M F R	A P P	C M P	GBP MIN	SLEW RATE MIN	Vs+ MAX	Vs- MAX	Top MAX	Avol MIN	Vio MAX	Ib MAX	Iio MAX	Ptot MAX.	Iout MIN	Vout MIN	Vicm MAX	Vidf MAX	dVio/dT MAX	Pq MAX	Io MAX	CMRR MIN	PSRR MIN	Rin MIN
SF.C2710C	TFH	CPR	EXT	.	.	+14V	-7V	70C	60dB	5MV	25uA	5uA	500MWF	5MA	1V	7V	5V	20uV/C	150MW	9MA	70dB	.	.
SF.C2710EC	TFH	CPR	EXT	.	.	+14V	-7V	70C	60dB	5MV	25uA	5uA	600MWF	5MA	1V	7V	5V	20uV/C	150MW	9MA	70dB	.	.
SF.C2710KM	TFH	CPR	EXT	.	.	+14V	-7V	125C	61dB	2MV	20uA	3uA	670MWF	5MA	1V	7V	5V	10uV/C	150MW	9MA	80dB	.	.
SF.C2710M	TFH	CPR	EXT	.	.	+14V	-7V	125C	61dB	2MV	20uA	3uA	500MWF	5MA	1V	7V	5V	10uV/C	150MW	9MA	80dB	.	.
SF.C2710PM	TFH	CPR	EXT	.	.	+14V	-7V	125C	61dB	2MV	20uA	3uA	570MWF	5MA	1V	7V	5V	10uV/C	150MW	9MA	80dB	.	.
SF.C2711C	THF	DCP	EXT	.	.	+14V	-7V	70C	57dB	5MV	100uA	15uA	500MWF	5MA	2.5V	7V	5V	20uV/C	230MW				
SF.C2711EC	THF	DCP	EXT	.	.	+14V	-7V	70C	57dB	5MV	100uA	15uA	670MWF	5MA	2.5V	7V	5V	20uV/C	230MW				
SF.C2711KM	THF	DCP	EXT	.	.	+14V	-7V	125C	58dB	3.5MV	75uA	10uA	670MWF	5MA	2.5V	7V	5V	20uV/C	200MW				
SF.C2711M	THF	DCP	EXT	.	.	+14V	-7V	125C	58dB	3.5MV	75uA	10uA	500MWF	5MA	2.5V	7V	5V	20uV/C	200MW				
SF.C2711PM	THF	DCP	EXT	.	.	+14V	-7V	125C	58dB	3.5MV	75uA	10uA	570MWF	5MA	2.5V	7V	5V	20uV/C	200MW				
SF.C2741C	THF	DGK	INT	.	0.2V/uS	+18V	-18V	70C	86dB	6MV	500NA	200NA	500MWF	5MA	12V	15V	30V			3MA	70dB	76dB	300K
SF.C2741DC	THF	GPK	INT	.	0.2V/uS	+18V	-18V	70C	86dB	6MV	500NA	200NA	500MWF	5MA	12V	15V	30V			3MA	70dB	76dB	300K
SF.C2741EC	THF	DGK	INT	.4MHZ	0.3V/uS	+22V	-22V	70C	94dB	3MV	80NA	30NA	500MW	10MA	16V	15V	30V	15uV/C	150MW		80dB	86dB	1M
SF.C2741KM	THF	DGK	INT	.4MHZ	0.3V/uS	+22V	-22V	125C	94dB	3MV	80NA	30NA	500MWF	5MA	12V	15V	30V	15uV/C	150MW		80dB	86dB	1M
SF.C2741M	THF	GPK	INT	.	0.2V/uS	+22V	-22V	125C	94dB	5MV	500NA	200NA	500MWF	5MA	12V	15V	30V		85MW	3MA	70dB	76dB	300K
SF.C2741PM	FAU	DGK	INT	.	0.2V/uS	+22V	-22V	125C	94dB	5MV	500NA	200NA	500MWF	5MA	12V	15V	30V		85MW	3MA	70dB	76dB	300K
SF.C2747C	THF	DGK	INT	.	0.2V/uS	+18V	-18V	70C	88dB	6MV	500NA	200NA	680MWF	5MA	12V	15V	30V		85MW	3MA	70dB	76dB	300K
SF.C2747EC	THF	DGK	INT	.4MHZ	0.3V/uS	+22V	-22V	70C	94dB	3MV	80NA	30NA	500MW	10MA	16V	15V	30V	15uV/C	150MW	3MA	80dB	86dB	1M
SF.C2747KM	THF	DGK	INT	.	0.2V/uS	+22V	-22V	125C	94dB	5MV	500NA	200NA	680MWF	5MA	12V	15V	30V		85MW	3MA	70dB	76dB	300K
SF.C2747M	THF	DGK	INT	.	0.2V/uS	+22V	-22V	125C	94dB	5MV	500NA	200NA	680MWF	5MA	12V	15V	30V		85MW	3MA	70dB	76dB	300K
SF.C2748C	THF	GPU	EXT	.	0.2V/uS	+18V	-18V	70C	86dB	6MV	500NA	200NA	500MWF	5MA	12V	15V	30V		85MW	3MA	70dB	76dB	300K
SF.C2748DC	THF	GPU	EXT	.	0.2V/uS	+18V	-18V	70C	86dB	6MV	500NA	200NA	500MWF	5MA	12V	15V	30V		85MW	3MA	70dB	76dB	300K
SF.C2748M	THF	GPU	EXT	.	0.2V/uS	+22V	-22V	125C	94dB	5MV	500NA	200NA	500MWF	5MA	12V	15V	30V		85MW	3MA	70dB	76dB	300K
SF.C2761C	THF	GPU	EXT	.	3V/uS	+18V	-18V	70C	81dB	6MV	1uA	0.3uA	500MWF	25MA	14V	18V	2V	35uV/C	190MW	3MA	65dB	74dB	50K
SF.C2761DC	THF	GPU	EXT	.	3V/uS	+18V	-18V	70C	81dB	6MV	1uA	0.3uA	500MWF	25MA	14V	18V	2V	15uV/C	190MW	3MA	65dB	74dB	50K
SF.C2761DT	THF	GPU	EXT	.	3V/uS	+18V	-18V	85C	81dB	6MV	1uA	0.3uA	500MWF	25MA	14V	18V	2V	15uV/C	190MW	3MA	65dB	74dB	50K
SF.C2761M	THF	GPU	EXT	.	3V/uS	+18V	-18V	125C	85dB	4MV	700NA	100NA	500MWF	25MA	14V	18V	2V	25uV/C	180MW	3MA	70dB	74dB	50K
SF.C2761PM	THF	GPU	EXT	.	3V/uS	+18V	-18V	125C	85dB	4MV	700NA	100NA	500MWF	25MA	14V	18V	2V	15uV/C	180MW	3MA	70dB	74dB	50K
SF.C2761T	THF	GPU	EXT	.	3V/uS	+18V	-18V	85C	81dB	6MV	1uA	0.3uA	500MWF	25MA	14V	18V	2V	35uV/C	190MW	3MA	65dB	74dB	50K
SF.C2776C	THF	PRA	INT	.	0.3V/uS	+18V	-18V	70C	94dB	6MV	50NA	25NA	500MWF	2MA	10V	15V	30V		6MW	.2MA	70dB	74dB	2M
SF.C2776DC	THF	PRA	INT	.	0.3V/uS	+18V	-18V	70C	94dB	6MV	50NA	25NA	310MWF	2MA	10V	15V	30V		6MW	.2MA	70dB	74dB	2M
SF.C2776EC	THF	PRA	INT	.	0.3V/uS	+18V	-18V	70C	106dB	5MV	50NA	15NA	670MWF	2MA	10V	15V	30V		6MW	.2MA	70dB	76dB	2M
SF.C2776KM	THF	PRA	INT	.	0.3V/uS	+18V	-18V	125C	106dB	5MV	50NA	15NA	670MWF	2MA	10V	15V	30V		6MW	.2MA	70dB	76dB	2M
SF.C2776M	THF	PRA	INT	.	0.3V/uS	+18V	-18V	125C	100dB	5MV	50NA	15NA	500MWF	2MA	10V	15V	30V		6MW	.2MA	70dB	76dB	2M
SF.2776PC	THF	PRA	INT	.	0.3V/uS	+18V	-18V	70C	94dB	6MV	50NA	25NA	250MWF	2MA	10V	15V	30V		6MW	.2MA	70dB	74dB	2M
SF.C2776PM	THF	PRA	INT	.	0.3V/uS	+18V	-18V	125C	106dB	5MV	50NA	15NA	250MWF	2MA	10V	15V	30V		6MW	.2MA	70dB	76dB	2M
SF.C2778C	THF	PRA	EXT	.	0.3V/uS	+18V	-18V	70C	94dB	6MV	50NA	25NA	500MWF	2MA	10V	15V	30V		5.7MW	.2UA	70dB	76dB	2M
SF.C2778DC	THF	PRA	EXT	.	0.3V/uS	+18V	-18V	70C	94dB	6MV	50NA	25NA	310MWF	2MA	10V	15V	30V		5.7MW	.2UA	70dB	76dB	2M
SF.C2778EC	THF	PRA	EXT	.	0.3V/uS	+18V	-18V	70C	94dB	6MV	50NA	25NA	670MWF	2MA	10V	15V	30V		5.7MW	.2UA	70dB	76dB	2M
SF.2778KM	THF	PRA	EXT	.	0.3V/uS	+18V	-18V	125C	100dB	5MV	50NA	15NA	670MWF	2MA	10V	15V	30V		5.4MW	.2UA	70dB	76dB	2M
SF.C2778M	THF	PRA	EXT	.	0.3V/uS	+18V	-18V	125C	100dB	5MV	50NA	15NA	500MWF	2MA	10V	15V	30V		5.4MW	.2UA	70dB	76dB	2M
SF.C2778PC	THF	PRA	EXT	.	0.3V/uS	+18V	-18V	70C	94dB	6MV	50NA	25NA	250MWF	2MA	10V	15V	30V		5.7MW	.2UA	70dB	76dB	2M
SF.C2778PM	THF	PRA	EXT	.	0.3V/uS	+18V	-18V	125C	100dB	5MV	50NA	15NA	250MWF	2MA	10V	15V	30V		5.4MW	.2UA	70dB	76dB	2M
SF.C2861C	THF	GPU	EXT	.	3V/uS	+10V	-10V	70C	75dB	10MV	1.0uA	0.3uA	500MWF	25MA	10V	.	10V	35uV/C		2MA	60dB	74dB	50K
SF.C2861DC	THF	GPU	EXT	.	3V/uS	+10V	-10V	70C	75dB	10MV	1.0uA	0.3uA	500MWF	25MA	10V	.	10V	35uV/C		2MA	60dB	74dB	50K
SF.C2861DT	THF	GPU	EXT	.	3V/uS	+10V	-10V	85C	75dB	10MV	1.0uA	0.3uA	500MWF	25MA	10V	.	10V	35uV/C		2MA	60dB	74dB	50K
SF.C2861M	THF	GPU	EXT	.	3V/uS	+10V	-10V	125C	85dB	4MV	0.7uA	0.1uA	500MWF	25MA	10V	.	10V	25uV/C		2MA	70dB	74dB	50K
SF.C2861PM	THF	GPU	EXT	.	3V/uS	+10V	-10V	125C	85dB	4MV	0.7uA	0.1uA	500MWF	25MA	10V	.	10V	25uV/C		2MA	70dB	74dB	50K
SF.C2861T	THF	GPU	EXT	.	3V/uS	+10V	-10V	85C	75dB	10MV	1.0uA	0.3uA	500MWF	25MA	10V	.	10V	35uV/C		2MA	60dB	74dB	50K
SG101AD	SKU	GPU	EXT	.	.	+22V	-22V	125C	94dB	2MV	75NA	10NA	500MWF	5MA	12V	15V	30V	15uV/C		3MA	80dB	80dB	1.5M
SG101AF	SKU	GPU	EXT	.	.	+22V	-22V	125C	94dB	2MV	75NA	10NA	500MWF	5MA	12V	15V	30V	15uV/C		3MA	80dB	80dB	1.5M
SG101AJ	SKU	GPU	EXT	.	.	+22V	-22V	125C	94dB	2MV	75NA	10NA	500MWF	5MA	12V	15V	30V	15uV/C		3MA	80dB	80dB	1.5M
SG101AT	SKU	GPU	EXT	.	.	+22V	-22V	125C	94dB	2MV	75NA	10NA	500MWF	5MA	12V	15V	30V	15uV/C		3MA	80dB	80dB	1.5M
SG101D	SKU	GPU	EXT	.	.	+22V	-22V	125C	94dB	5MV	1.5uA	0.5uA	500MWF	5MA	12V	15V	30V	15uV/C		3MA	70dB	70dB	300K
SG101F	SKU	GPU	EXT	.	.	+22V	-22V	125C	94dB	5MV	1.5uA	0.5uA	500MWF	5MA	12V	15V	30V	15uV/C		3MA	70dB	70dB	300K
SG101J	SKU	GPU	EXT	.	.	+22V	-22V	125C	94dB	5MV	1.5uA	0.5uA	500MWF	5MA	12V	15V	30V	15uV/C		3MA	70dB	70dB	300K
SG101T	SKU	GPU	EXT	.	.	+22V	-22V	125C	94dB	5MV	1.5uA	0.5uA	500MWF	5MA	12V	15V	30V	15uV/C		3MA	70dB	70dB	300K
SG102J	SKU	VFA	INT	.	15V/uS	+18V	-18V	125C	0dB	4MV	3NA	.	500MWF	1MA	10V	15V	15V	50uV/C		6MA	.	70dB	10G
SG102T	SKU	VFA	INT	.	.	+18V	-18V	125C	0dB	5MV	10NA	.	500MWF	1MA	10V	.	.	30uV/C		6MA	.	60dB	10G
SG107D	SKU	GPK	INT	.	.	+22V	-22V	125C	94dB	2MV	75NA	10NA	500MWF	5MA	12V	15V	30V	15uV/C		3MA	80dB	80dB	1.5M

For detailed explanations of column heading notations, see App. A.

Also for ready references the more important abbreviations used in the column headings are listed below:

LEFT HAND PAGE

APP = application (codes at APP.E.)
CMRR = common mode rejection ratio
CMP = compensation (frequency)
dV_{IO}/dT = input offset voltage temperature drift
GBP = gain bandwidth product
I_B = input bias current
I_{IO} = input bias offset current
I_0 = quiescent supply current
MFR = manufacturer (codes at App.C.)
P_0 = quiescent power consumer
PSRR = power supply rejection ratio
V_{ICM} = common mode input voltage rating
V_{IDF} = differential input voltage rating
V_{IO} = input offset voltage
V_S = dc supply voltage

RIGHT HAND PAGE

Lead out coding summary (details at APP.G.) for different cases (APP.F.)

A = gain adjust
B = bias adjust
C = case
E− = inverting input
E+ = non-inverting input
F,F* = input frequency compensation
G = ground
J = high level input
K = output, open collector
L = output, open emitter
M = metal case
N = not connected
Q = special terminal
R,R* = outputs
S = strobe
T,T* = offset balance
V+ = +ve dc supply
V− = −ve dc supply
W = guard ring
X = blank position, no lead
++ = +ve supplementary dc supply
−− = −ve supplementary dc supply
ø,ø* = output frequency compensation

CASE (APP F)	LD1	LD2	LD3	LD4	LD5	LD6	LD7	LD8	LD9	LD10	LD11	LD12	LD13	LD14	LD15	LD16	EUROPE SUBSTITUTE	USA SUBSTITUTE	I/S	TYPE NUMBER
DIL-14/1P	N	G	E+	E-	N	V-	N	N	R	N	V+	N	N	N				UA710DM	0	SF.C2710EC
DIL-14/1C	N	G	E+	E-	N	V-	N	N	R	R	N	V+	N	N	N			UA710DM	0	SF.C2710KM
T05-8/1M	G	E+	E-	V-M	N	N	R	V+									LM710H	UA710HM	0	SF.C2710M
FLP-10/3G	G	E+	E-	N	V-	R	N	V+	N	N								UA710FM	0	SF.C2710PM
T05-10/1M	G	S1	E-1	E+1	V-	E+2	E-2	S2	R	V+							LM711CH	UA711HC	0	SF.C2711C
DIL-14/1P	N	E-1	E+1	V-	E+2	E-2	N	N	S2	R	V+	G	S1	N			LM711CN	UA711DC	0	SF.C2711EC
DIL-14/1C•	N	E-1	E+1	V-	E+2	E-2	N	N	S2	R	V+	G	S1	N				UA711DM	0	SF.C2711KM
T05-10/1M	G	S1	E-1	E+1	V-	E+2	E-2	S2	R	V+							LM711H	UA711HM	0	SF.C2711M
FLP-10/3G	E-1	E+1	V-	E+2	E-2	S2	R	V+	G	S1								UA711FM	0	SF.C2711PM
T05-8/1M	T	E-	E+	V-M	T*	R	V+	N									TBA221	UA741HC	0	SF.C2741C
DIL-8/1P	T	E-	E+	V-	T*	R	V+	N									TBA221B	UA741TC	0	SF.C2741DC
DIL-14/1P	N	N	T	E-	E+	V-	N	N	T*	R	V+	N	N	N			LM741ED	UA741EDC	0	SF.C2741EC
DIL-14/1C	N	N	T	E-	E+	V-	N	N	T*	R	V+	N	N	N			LM741D	UA741DM	0	SF.C2741KM
T05-8/1M	T	E-	E+	V-M	T*	R	V+	N									TBA222	UA741HM	0	SFC2741M
FLP-10/3P	N	T	E-	E+	V-	T*	R	V+	N	N							LM741F	UA741FM	0	SF.C2741PM
T05-10/1M	R1	V+1	E-1	E+1	V-	E+2	E-2	V+2	R2	N							TBB0747	UA747HC	0	SF.C2747C
DIL-14/1P	E-1	E+1	T1	V-	T2	E+2	E-2	T*2	V+2	R2	N	R1	V+1	T*1			LM747ED	UA747EDC	0	SF.C2747EC
DIL-14/1C	E-1	E+1	T1	V-	T2	E+2	E-2	T*2	V+2	R2	N	R1	V+	T*1			LM747D	UA747DM	0	SF.C2747KM
T05-10/1M	R1	V+1	E-1	E+1	V-	E+2	E-2	R2	N								TBC0747	UA747HM	0	SF.C2747M
T05-8/1M	FT	E-	E+	V-	T*	R	V+	F*									TBB0748	UA748HC	0	SF.C2748C
DIL-8/1P	FT	E-	E+	V-	T*	R	V+	F*									TBB0748	UA748TC	0	SF.C2748DC
T05-8/1M	FT	E-	E+	V-	T*	R	V+	F*									TBC0748	UA748HM	0	SF.C2748M
T05-8/4M	X	V+	E+	E-	X	V-	K	ø									TAA761		0	SF.C2761C
DIL-6/1P	V+	E+	E-	V-	K	ø											TAA761A		0	SF.C2761DC
DIL-6/1P	V+	E+	E-	V-	K	ø											TAA765A		0	SF.C2761DT
T05-8/4M	X	V+	E+	E-	X	V-	K	ø									TAA762		0	SF.C2761M
FLP-10/3G	V+	N	E+	N	E-	V-	N	K	N	ø									0	SFC.2761PM
T05-8/4M	X	V+	E+	E-	X	V-	K	ø									TAA765		0	SF.C2761T
T05-8/1M	T	E-	E+	V-	T*	R	V+	B									MC1776CG	UA776HC	0	SF.C2776C
DIL-8/1P	T	E-	E+	V-	T*	R	V+	B										UA776TC	0	SF.C2776DC
DIL-14/1P	N	N	T	E-	E+	V-	N	N	T*	R	V+	B	N	N			MC1776L	UA776DM	0	SF.C2776EC
DIL-14/1P	N	N	T	E-	E+	V-	N	N	T*	R	V+	B	N	N			MC1776L	UA776DM	0	SF.C2776KM
T05-8/1M	T	E-	E+	V-	T*	R	V+	B									MC1776G	UA776HM	0	SF.C2776M
FLP-10/3G	N	T	E-	E+	V-	T*	R	V+	B	N									0	SF.C2776PC
FLP-10/3G	N	T	E-	E+	V-	T*	R	V+	B	N									0	SF.C2776PM
T05-8/1M	F	E-	E+	V-	TF*	R	V+	B											0	SFC2778C
DIL-8/1P	F	E-	E+	V-	TF*	R	V+	B											0	SF.C2778DC
DIL-14/1P	N	N	F	E-	E+	V-	N	N	TF*	R	V+	B	N	N					0	SF.C2778EC
DIL-14/1C	N	N	F	E-	E+	V-	N	N	TF*	R	V+	B	N	N					0	SF.C2778KM
T05-8/1M	F	E-	E+	V-	TF*	R	V+	B											0	SF.C2778M
FLP-10/3G	N	F	E-	E+	V-	TF*	R	V+	B	N									0	SFC2778PC
FLP-10/3G	N	F	E-	E+	V-	TF*	R	V+	B	N									0	SF.C2778PM
T05-8/1M	X	V+	E+	E-	X	V-	K	ø									TAA861		0	SF.C2861M
DIL-6/1P	V+	E+	E-	V-	K	ø											TAA861A		0	SF.C2861DC
DIL-6/1P	V+	E+	E-	V-	K	ø											TAA865A		0	SF.C2861DT
T05-8/1M	X	V+	E+	E-	X	V-	K	ø									TAA862		0	SF.C2861M
FLP-10/3G	V+	N	E+	N	E-	V-	N	K	N	ø									0	SF.C2861PM
T05-8/1M	X	V+	E+	E-	X	V-	K	ø									TAA865		0	SF.C2861T
DIL-14/1C	N	N	FT	E-	E+	V-	N	N	T*	R	V+	F*	N	N			UA101AD	LM101AJ14	0	SG101AD
FLP-10/3C	N	FT	E-	E+	V-	T*	R	V+	F*	N							SFC2101APM	LM101AF	0	SG101AF
DIL-14/1C	N	N	FT	E-	E+	V-	N	N	T*	R	V+	F*	N	N			UA101AD	LM101AJ14	0	SG101AJ
T05-8/1M	FT	E-	E+	V-M	T*	R	V+	F*									SFC2101A	LM101AH	0	SG101AT
DIL-14/1C	N	N	FT	E-	E+	V-	N	N	T*	R	V+	F*	N	N			UA101AD	LM101J14	0	SG101D
FLP-10/3C	N	FT	E-	E+	V-	T*	R	V+	F*	N							SFC2101APM	LM101F	0	SG101F
DIL-14/1C	N	N	FT	E-	E+	V-	N	N	T*	R	V+	F*	N	N			UA101AD	LM101J14	0	SG101J
T05-8/1M	FT	E-	E+	V-M	T*	R	V+	F*									SFC2101A	LM101H	0	SG101T
DIL-14/1C	N	N	T	N	E+	V-	N	N	L	R	V+	T*	N	N			SN52110JA	LM1100	0	SG102J
T05-8/1M	T	N	E+	V-	L	R	V+	T*									UA102M	LM102H	0	SG102T
DIL-14/1C	N	N	N	E-	E+	V-	N	N	N	R	V+	N	N	N			SN52107JA	LM107D	0	SG107D
FLP-10/3C	N	N	E-	E+	V-	N	R	V+	N	N							SFC2107PM	LM107F	0	SG107F

TYPE NUMBER	MFR	APP	CMP	GBP MIN	SLEW RATE MIN	V_S^+ MAX	V_S^- MAX	T_{op} MAX	A_{VOL} MIN	V_{IO} MAX	I_B MAX	I_{IO} MAX	P_{TOT} MAX	I_{OUT} MIN	V_{OUT} MIN	V_{ICM} MAX	V_{IDF} MAX	dV_{IO}/dT MAX	P_O MAX	I_O MAX	CM RR MIN	PS RR MIN	R_{IN} MIN
SG107F	SKU	GPK	INT	.	.	+22V	-22V	125C	94dB	2MV	75NA	10NA	500MWF	5MA	12V	15V	30V	15uV/C	.	3MA	80dB	80dB	1.5M
SG107J	SKU	GPK	INT	.	.	+22V	-22V	125C	94dB	2MV	75NA	10NA	500MWF	5MA	12V	15V	30V	15uV/C	.	3MA	80dB	80dB	1.5M
SG107T	SKU	GPK	INT	.	.	+22V	-22V	125C	94dB	2MV	75NA	10NA	500MWF	5MA	12V	15V	30V	15uV/C	.	3MA	80dB	80dB	1.5M
SG108AD	SKU	SBA	EXT	.	.	+20V	-20V	125C	98dB	0.5MV	2NA	0.2NA	500MWF	1MA	13V	15V	1V	5uV/C	.	.6MA	96dB	96dB	30M
SG108AF	SKU	SBA	EXT	.	.	+20V	-20V	125C	98dB	0.5MV	2NA	0.2NA	500MWF	1MA	13V	15V	1V	5uV/C	.	.6MA	96dB	96dB	30M
SG108AJ	SKU	SBA	EXT	.	.	+20V	-20V	125C	98dB	0.5MV	2NA	0.2NA	500MWF	1MA	13V	15V	1V	5uV/C	.	.6MA	96dB	96dB	30M
SG108AT	SKU	SBA	EXT	.	.	+20V	-20V	125C	98dB	0.5MV	2NA	0.2NA	500MWF	1MA	13V	15V	1V	5uV/C	.	.6MA	96dB	96dB	30M
SG108D	SKU	SBA	EXT	.	.	+20V	-20V	125C	96dB	2MV	2NA	0.2NA	500MWF	1MA	13V	15V	1V	15uV/C	.	.6MA	85dB	80dB	30M
SG108F	SKU	SBA	EXT	.	.	+20V	-20V	125C	96dB	2MV	2NA	0.2NA	500MWF	1MA	13V	15V	1V	15uV/C	.	.6MA	85dB	80dB	30M
SG108J	SKU	SBA	EXT	.	.	+20V	-20V	125C	96dB	2MV	2NA	0.2NA	500MWF	1MA	13V	15V	1V	15uV/C	.	.6MA	85dB	80dB	30M
SG108T	SKU	SBA	EXT	.	.	+20V	-20V	125C	96dB	2MV	2NA	0.2NA	500MWF	1MA	13V	15V	1V	15uV/C	.	.6MA	85dB	80dB	30M
SG110J	SKU	VFA	INT	15V/uS	.	+18V	-18V	125C	0dB	4MV	3NA	.	500MWF	1MA	10V	15V	15V	50uV/C	.	6MA	.	70dB	10G
SG110T	SKU	SBA	EXT	15V/uS	.	+18V	-18V	125C	0dB	4MV	3NA	.	500MWF	1MA	10V	15V	15V	50uV/C	.	6MA	.	70dB	10G
SG111D	SKU	CPR	EXT	.	.	+18V	-18V	125C	100dB	3MV	100NA	10NA	500MWF	.	.	15V	30V	.	.	6MA	.	.	.
SG111F	SKU	CPR	EXT	.	.	+18V	-18V	125C	100dB	3MV	100NA	10NA	500MWF	.	.	15V	30V	.	.	6MA	.	.	.
SG111J	SKU	CPR	EXT	.	.	+18V	-18V	125C	100dB	3MV	100NA	10NA	500MWF	.	.	15V	30V	.	.	6MA	.	.	.
SG111T	SKU	CPR	EXT	.	.	+18V	-18V	125C	100dB	3MV	100NA	10NA	500MWF	.	.	15V	30V	.	.	6MA	.	.	.
SG118F	SKU	XSR	INT	50V/uS	.	+20V	-20V	125C	94dB	4MV	250NA	50NA	500MWF	6MA	12V	15V	1V	.	.	8MA	80dB	70dB	1M
SG118J	SKU	XSR	INT	50V/uS	.	+20V	-20V	125C	94dB	4MV	250NA	50NA	500MWF	6MA	12V	15V	1V	.	.	8MA	80dB	70dB	1M
SG118T	SKU	XSR	INT	50V/uS	.	+20V	-20V	125C	94dB	4MV	250NA	50NA	500MWF	6MA	12V	15V	1V	.	.	8MA	80dB	70dB	1M
SG124F	SKU	QGK	INT	.	.	+16V	-16V	125C	94dB	5MV	150NA	30NA	800MWF	.	.	16V	16V	35uV/C	.	2MA	70dB	65dB	.
SG124J	SKU	QGK	INT	.	.	+16V	-16V	125C	94dB	5MV	150NA	30NA	900MWF	.	.	16V	16V	35uV/C	.	2MA	70dB	65dB	.
SG139AF	SKU	QCP	EXT	.	.	+18V	-18V	125C	94dB	2MV	100NA	25NA	800MWF	.	.	18V	18V	.	.	2MA	.	.	.
SG139AJ	SKU	QCP	EXT	.	.	+18V	-18V	125C	94dB	2MV	100NA	25NA	900MWF	.	.	18V	18V	.	.	2MA	.	.	.
SG139F	SKU	QCP	EXT	.	.	+18V	-18V	125C	88dB	5MV	100NA	25NA	800MW	.	.	18V	18V	.	.	2MA	.	.	.
SG139J	SKU	QCP	EXT	.	.	+18V	-18V	125C	88dB	5MV	100NA	25NA	900MWF	.	.	18V	18V	.	.	2MA	.	.	.
SG201AD	SKU	GPU	EXT	.	.	+22V	-22V	85C	94dB	2MV	75NA	10NA	500MWF	5MA	12V	15V	30V	15uV/C	.	3MA	80dB	80dB	500K
SG201AJ	SKU	GPU	EXT	.	.	+22V	-22V	85C	94dB	2MV	75NA	10NA	500MWF	5MA	12V	15V	30V	15uV/C	.	3MA	80dB	80dB	500K
SG201AM	SKU	GPU	EXT	.	.	+22V	-22V	85C	94dB	2MV	75NA	10NA	500MWF	5MA	12V	15V	30V	15uV/C	.	3MA	80dB	80dB	500K
SG201AN	SKU	GPU	EXT	.	.	+22V	-22V	85C	94dB	2MV	75NA	10NA	500MWF	5MA	12V	15V	30V	15uV/C	.	3MA	80dB	80dB	500K
SG201AT	SKU	GPU	EXT	.	.	+22V	-22V	85C	94dB	2MV	75NA	10NA	500MWF	5MA	12V	15V	30V	15uV/C	.	3MA	80dB	80dB	500K
SG201D	SKU	GPU	EXT	.	.	+22V	-22V	85C	86dB	7.5MV	1.5uA	0.5uA	500MWF	5MA	12V	15V	30V	30uV/C	.	3MA	65dB	70dB	100K
SG201J	SKU	GPU	EXT	.	.	+22V	-22V	85C	86dB	7.5MV	1.5uA	0.5uA	500MWF	5MA	12V	15V	30V	30uV/C	.	3MA	65dB	70dB	100K
SG201T	SKU	GPU	EXT	.	.	+22V	-22V	85C	86dB	7.5MV	1.5uA	0.5uA	500MWF	5MA	12V	15V	30V	30uV/C	.	3MA	65dB	70dB	100K
SG201M	SKU	GPU	EXT	.	.	+22V	-22V	85C	86dB	7.5MV	1.5uA	0.5uA	500MWF	5MA	12V	15V	30V	30uV/C	.	3MA	65dB	70dB	100K
SG201N	SKU	GPU	EXT	.	.	+22V	-22V	85C	86dB	7.5MV	1.5uA	0.5uA	500MWF	5MA	12V	15V	30V	30uV/C	.	3MA	65dB	70dB	100K
SG202J	SKU	VFA	INT	15V/uS	.	+18V	-18V	85C	0dB	4MV	3NA	.	500MWF	1MA	10V	15V	15V	50uV/C	.	6MA	.	70dB	10G
SG202N	SKU	VFA	INT	15V/uS	.	+18V	-18V	85C	0dB	4MV	3NA	.	500MWF	1MA	10V	15V	15V	50uV/C	.	6MA	.	70dB	10G
SG210J	SKU	VFA	INT	15V/uS	.	+18V	-18V	85C	0dB	4MV	3NA	.	500MWF	1MA	10V	15V	15V	50uV/C	.	6MA	.	70dB	10G
SG210N	SKU	VFA	INT	15V/uS	.	+18V	-18V	85C	0dB	4MV	3NA	.	500MWF	1MA	10V	15V	15V	50uV/C	.	6MA	.	70dB	10G
SG202M	SKU	VFA	INT	.	.	+18V	-18V	85C	0dB	10MV	15NA	.	500MWF	1MA	10V	.	.	60uV/C	.	6MA	.	60dB	10G
SG202T	SKU	VFA	INT	.	.	+18V	-18V	85C	0dB	10MV	15NA	.	500MWF	1MA	10V	.	.	60uV/C	.	6MA	.	60dB	10G
SG207J	SKU	GPK	INT	.	.	+22V	-22V	85C	94dB	2MV	75NA	10NA	500MWF	5MA	12V	15V	30V	15uV/C	.	3MA	80dB	80dB	1.5M
SG207M	SKU	GPK	INT	.	.	+22V	-22V	85C	94dB	2MV	75NA	10NA	500MWF	5MA	12V	15V	30V	15uV/C	.	3MA	80dB	80dB	1.5M
SG207N	SKU	GPK	INT	.	.	+22V	-22V	85C	94dB	2MV	75NA	10NA	500MWF	5MA	12V	15V	30V	15uV/C	.	3MA	80dB	80dB	1.5M
SG207T	SKU	GPK	INT	.	.	+22V	-22V	85C	94dB	2MV	75NA	10NA	500MWF	5MA	12V	15V	30V	15uV/C	.	3MA	80dB	80dB	1.5M
SG208AJ	SKU	SBA	EXT	.	.	+20V	-20V	85C	98dB	0.5MV	2NA	0.2NA	500MWF	1MA	13V	15V	1V	5uV/C	.	.6MA	96dB	96dB	30M
SG208AM	SKU	SBA	EXT	.	.	+20V	-20V	85C	98dB	0.5MV	2NA	0.2NA	500MWF	1MA	13V	15V	1V	5uV/C	.	.6MA	96dB	96dB	30M
SG208AT	SKU	SBA	EXT	.	.	+20V	-20V	85C	98dB	0.5MV	2NA	0.2NA	500MWF	1MA	13V	15V	1V	5uV/C	.	.6MA	96dB	96dB	30M
SG208J	SKU	SBA	EXT	.	.	+20V	-20V	85C	96dB	2MV	2NA	0.2NA	500MWF	1MA	13V	15V	1V	15uV/C	.	.6MA	85dB	80dB	30M
SG208T	SKU	SBA	EXT	.	.	+20V	-20V	85C	96dB	2MV	2NA	0.2NA	500MWF	1MA	13V	15V	1V	15uV/C	.	.6MA	85dB	80dB	30M
SG208M	SKU	SBA	EXT	.	.	+20V	-20V	85C	96dB	2MV	2NA	0.2NA	500MWF	1MA	13V	15V	1V	15uV/C	.	.6MA	85dB	80dB	30M
SG210M	SKU	VFA	INT	15V/uS	.	+18V	-18V	85C	0dB	4MV	3NA	.	500MWF	1MA	10V	15V	15V	50uV/C	.	6MA	.	70dB	10G
SG210T	SKU	VFA	INT	15V/uS	.	+18V	-18V	85C	0dB	4MV	3NA	.	500MWF	1MA	10V	15V	15V	50uV/C	.	6MA	.	70dB	10G
SG211J	SKU	CPR	EXT	.	.	+18V	-18V	85C	100dB	3MV	100NA	10NA	500MWF	.	.	15V	30V	.	.	6MA	.	.	.
SG211M	SKU	CPR	EXT	.	.	+18V	-18V	85C	100dB	3MV	100NA	10NA	500MWF	.	.	15V	30V	.	.	6MA	.	.	.
SG211N	SKU	CPR	EXT	.	.	+18V	-18V	85C	100dB	3MV	100NA	10NA	500MWF	.	.	15V	30V	.	.	6MA	.	.	.
SG211T	SKU	CPR	EXT	.	.	+18V	-18V	85C	100dB	3MV	100NA	10NA	500MWF	.	.	15V	30V	.	.	6MA	.	.	.
SG218J	SKU	XSR	INT	50V/uS	.	+20V	-20V	85C	94dB	4MV	250NA	50NA	500MWF	6MA	12V	15V	1V	.	.	8MA	80dB	70dB	1M
SG218M	SKU	XSR	INT	50V/uS	.	+20V	-20V	85C	94dB	4MV	250NA	50NA	500MWF	6MA	12V	15V	1V	.	.	8MA	80dB	70dB	1M

134

r detailed explanations of lumn heading notations, see op A.

so for ready references the ore important abbreviations sed in the column headings are sted below:

FT HAND PAGE

PP = application (codes at APP E.)

MRR = common mode rejection ratio

M? = compensation (frequency)

V_{io}/dT = input offset voltage temperature drift

BP = gain bandwidth product

= input bias current

= input bias offset current

= quiescent supply current

MFR = manufacturer (codes at App.C.)

P_u = quiescent power consumer

PSRR = power supply rejection ratio

V_{icm} = common mode input voltage rating

V_{idi} = differential input voltage rating

V_{io} = input offset voltage

V_s = dc supply voltage

RIGHT HAND PAGE

Lead out coding summary (details at APP.G.) for different cases (APP.F.)

A = gain adjust

B = bias adjust

C = case

E— = inverting input

E+ = non-inverting input

F,F* = input frequency compensation

G = ground

J = high level input

K = output, open collector

L = output, open emitter

M = metal case

N = not connected

Q = special terminal

R,R* = outputs

S = strobe

T,T* = offset balance

V+ = +ve dc supply

V− = −ve dc supply

W = guard ring

X = blank position, no lead

+ + = +ve supplementary dc supply

− − = −ve supplementary dc supply

φ,φ* = output frequency compensation

CASE (APP F)	LD 1	LD 2	LD 3	LD 4	LD 5	LD 6	LD 7	LD 8	LD 9	LD 10	LD 11	LD 12	LD 13	LD 14	LD 15	LD 16	EUROPE SUBSTITUTE	USA SUBSTITUTE	IS	TYPE NUMBER
DIL-14/1C	N	N	N	E-	E+	V-	N	N	N	R	V+	N	N	N			SN52107JA	LM107D	O	SG107J
TO5-8/1M	N	E-	E+	V-M	N	R	V+	N									SFC2107M	LM107H	O	SG107T
DIL-14/1C	N	F	N	E-	E+	N	V-	N	N	R	V+	F*	N	N			UA108AD	LM108AD	O	SG108AD
FLP-10/3C	N	N	E-	E+	N	V-	R	V+	F*	F								LM108AF	O	SG108AF
DIL-14/1C	N	F	N	E-	E+	N	V-	N	N	R	V+	F*	N	N			UA108AD	LM108AD	O	SG108AJ
TO5-8/1M	F	E-	E+	V-M	N	R	V+	F*									SFC2108A	LM108AH	O	SG108AT
DIL-14/1C	N	F	N	E-	E+	N	V-	N	N	R	V+	F*	N	N			UA108D	LM108D	O	SG108D
FLP-10/3C	N	N	E-	E+	N	V-	R	V+	F*	F							SFC2108PM	LM108F	O	SG108F
DIL-14/1C	N	F	N	E-	E+	N	V-	N	N	R	V+	F*	N	N			UA108D	LM108D	O	SG108J
TO5-8/1M	F	E-	E+	V-M	N	R	V+	F*									SFC2108M	LM108H	O	SG108T
DIL-14/1C	N	N	T	N	E+	V-	N	N	L	R	V+	T*	N	N			SN52110JA	LM110D	O	SG110J
TO5-8/1M	T	N	E+	V-	L	R	V+	T*									SFC2110M	LM110H	O	SG110T
DIL-14/1C	N	G	E+	E-	N	V-	T	T*S	R	N	V+	N	N	N			SN52111J	LM111D	O	SG111D
FLP-10/3C	G	E+	E-	N	V-	T	T*S	N	R	V+							SN52111FA	LM111F	O	SG111F
DIL-14/1C	N	G	E+	E-	N	V-	T	T*S	R	N	V+	N	N	N			SN52111J	LM111D	O	SG111J
TO5-8/1M	G	E+	E-	V-	T	T*S	R	V+									SFC2111M	LM111H	O	SG111T
FLP-10/3C	N	T*F	E-	E+	V-	F*T	R	V+	ø	N							SN52118FA	LM118F	O	SG118F
DIL-14/1C	N	N	T*F	E-	E+	V-	N	N	F*T	R	V+	ø					SN52118JA	LM118D	O	SG118J
TO5-8/1M	T*F	E-	E+	V-	F*T	R	V+	ø									TDC0118CM	LM118H	O	SG118T
FLP-14/3C	R1	E-1	E+1	V+	E+2	E-2	R2	R3	E-3	E+3	G	E+4	E-4	R4				LM124F	O	SG124F
DIL-14/1C	R1	E-1	E+1	V+	E+2	E-2	R2	R3	E-3	E+3	G	E+4	E-4	R4			MLM124L	LM124D	O	SG124J
FLP-14/3C	R2	R1	V+	E-1	E+1	E-2	E+2	E+3	E-3	E-4	E+4	G	R4	R3				LM139AF	O	SG139AF
DIL-14/1C	R2	R1	V+	E-1	E+1	E-2	E+2	E+3	E-3	E-4	E+4	G	R4	R3			MLM139AL	LM139AD	O	SG139AJ
FLP-14/3C	R2	R1	V+	E-1	E+1	E-2	E+2	E+3	E-3	E-4	E+4	G	R4	R3			LM139H	LM139F	O	SG139F
DIL-14/1C	R2	R1	V+	E-1	E+1	E-2	E+2	E+3	E-3	E-4	E+4	G	R4	R3			MLM139L	LM139D	O	SG139J
DIL-14/1C	N	N	FT	E-	E+	V-	N	N	T*	R	V+	F*	N	N			UA101AD	LM201AJ14	O	SG201AD
DIL-14/1C	N	N	FT	E-	E+	V-	N	N	T*	R	V+	F*	N	N			UA201AD	LM201AJ14	O	SG201AJ
DIL-8/1P	FT	E-	E+	V-M	T*	R	V+	F*											O	SG201AM
DIL-14/1P	N	N	FT	E-	E+	V-	N	N	T*	R	V+	F*	N	N			UA201AD	LM201AJ14	O	SG201AN
TO5-8/1M	FT	E-	E+	V-M	T*	R	V+	F*									SFC2101A	LM201AH	O	SG201AT
DIL-14/1C	N	N	FT	E-	E+	V-	N	N	T*	R	V+	F*	N	N				LM201J14	O	SG201D
DIL-14/1C	N	N	FT	E-	E+	V-	N	N	T*	R	V+	F*	N	N				LM201J14	O	SG201J
TO5-8/1M	FT	E-	E+	V-M	T*	R	V+	F*									SFC2201A	LM201H	O	SG201T
DIL-8/1P	FT	E-	E+	V-	T*	R	V+	F*										LM201J	O	SG201M
DIL-14/1P	N	N	FT	E-	E+	V-	N	N	T*	R	V+	F*	N	N				LM201J14	O	SG201N
DIL-14/1C	N	N	T	N	E+	V-	N	N	L	R	V+	T*	N	N			SN52110JA	LM210D	O	SG202J
DIL-14/1P	N	N	T	N	E+	V-	N	N	L	R	V+	T*	N	N			SN52110JA	LM210D	O	SG202N
DIL-14/1C	N	N	T	N	E+	V-	N	N	L	R	V+	T*	N	N			SN52110JA	LM210D	O	SG210J
DIL-14/1P	N	N	T	N	E+	V-	N	N	L	R	V+	T*	N	N			SN52110JA	LM210D	O	SG210N
DIL-8/1P	T	N	E+	V-	L	R	V+	T*										SN52110JP	O	SG202M
TO5-8/1M	T	N	E+	V-	L	R	V+	T*									UA102M	LM202H	O	SG202T
DIL-14/1C	N	N	N	E-	E+	V-	N	N	N	R	V+	N	N	N			SN52107JA	LM207D	O	SG207J
DIL-8/1P	N	E-	E+	V-	N	R	V+	N									SN52107JP	LM207J	O	SG207M
DIL-14/1P	N	N	N	E-	E+	V-	N	N	N	R	V+	N	N	N			SN52107JA	LM207D	O	SG207N
TO5-8/1M	N	E-	E+	V-M	N	R	V+	N									SFC2207	LM207H	O	SG207T
DIL-14/1C	N	F	N	E-	E+	N	V-	N	N	R	V+	F*	N	N			UA208AD	LM208AD	O	SG208AJ
DIL-8/1P	F	E-	E+	V-	N	R	V+	F*											O	SG208AM
TO5-8/1P	F	E-	E+	V-M	N	R	V+	F*									SFC2208A	LM208AH	O	SG208AT
DIL-14/1C	N	F	N	E-	E+	N	V-	N	N	R	V+	F*	N	N			UA208D	LM208D	O	SG208J
TO5-8/1M	F	E-	E+	V-M	N	R	V+	F*									SFC2208	LM208H	O	SG208T
DIL-8/1P	F	E-	E+	V-	N	R	V+	F*											O	SG208M
DIL-8/1P	T	N	E+	V-	L	R	V+	T*										SN52110JP	O	SG210M
TO5-8/1M	T	N	E+	V-	L	R	V+	T*									SFC2210	LM210H	O	SG210T
DIL-14/1C	N	G	E+	E-	N	V-	T	T*S	R	N	V+	N	N	N			SN52111J	LM211D	O	SG211J
DIL-8/1P	G	E+	E-	V-	T	T*S	R	V+											O	SG211M
DIL-14/1P	N	G	E+	E-	N	V-	T	T*S	R	N	V+	N	N	N			SN52111J	LM211D	O	SG211N
TO5-8/1M	G	E+	E-	V-	T	T*S	R	V+									SFC2211	LM211H	O	SG211T
DIL-14/1C	N	N	T*F	E-	E+	V-	N	N	F*T	R	V+	ø	N	N			SN52118JA	LM218D	O	SG218J
DIL-8/1P	T*F	E-	E+	V-	F*T	R	V+	ø										SN52118JP	O	SG218M
TO5-8/1M	T*F	E-	E+	V-	F*T	R	V+	ø									TDB0118CM	LM218H	O	SG218T

TYPE NUMBER	MFR	APP	CMP	GBP MIN	SLEW RATE MIN	V_S^+ MAX	V_S^- MAX	T_{op} MAX	A_{VOL} MIN	V_{IO} MAX	I_B MAX	I_{IO} MAX	P_{TOT} MAX	I_{OUT} MIN	V_{OUT} MIN	V_{ICM} MAX	V_{IDF} MAX	dV_{IO}/dT MAX	P_Q MAX	I_O MAX	CMRR MIN	PSRR MIN	R_{IN} MIN
SG218T	SKU	XSR	INT		50V/uS	+20V	-20V	85C	94dB	4MV	250NA	50NA	500MWF	6MA	12V	15V	1V			8MA	80dB	70dB	1
SG224J	SKU	QGK	INT			+16V	-16V	85C	94dB	5MV	150NA	30NA	900MWF			16V	16V	35uV/C		2MA	70dB	65dB	
SG224N	SKU	QGK	INT			+16V	-16V	85C	94dB	5MV	150NA	30NA	900MWF			16V	16V	35uV/C		2MA	70dB	65dB	
SG239AJ	SKU	QCP	EXT			+18V	-18V	85C	94dB	2MV	250NA	50NA	900MWF	6MA		18V	18V			2MA			
SG239AN	SKU	QCP	EXT			+18V	-18V	85C	94dB	2MV	250NA	50NA	900MWF	6MA		18V	18V			2MA			
SG239J	SKU	QCP	EXT			+18V	-18V	85C	88dB	5MV	250NA	50NA	900MWF	6MA		18V	18V			2MA			
SG239N	SKU	QCP	EXT			+18V	-18V	85C	88dB	5MV	250NA	50NA	900MWF	6MA		18V	18V			2MA			
SG301AD	SKU	GPU	EXT			+18V	-18V	70C	88dB	7.5MV	250NA	50NA	500MWF	5MA	12V	15V	30V	30uV/C		3MA	70dB	70dB	500K
SG301AF	SKU	GPU	EXT			+18V	-18V	70C	88dB	7.5MV	250NA	50NA	500MWF	5MA	12V	15V	30V	30uV/C			70dB	70dB	500K
SG301AJ	SKU	GPU	EXT			+18V	-18V	70C	88dB	7.5MV	250NA	50NA	500MWF	5MA	12V	15V	30V	30uV/C		3MA	70dB	70dB	500K
SG301AM	SKU	GPU	EXT			+18V	-18V	70C	88dB	7.5MV	250NA	50NA	500MWF	5MA	12V	15V	30V	30uV/C		3MA	70dB	70dB	500K
SG301AN	SKU	GPU	EXT			+18V	-18V	70C	88dB	7.5MV	250NA	50NA	500MWF	5MA	12V	15V	30V	30uV/C			70dB	70dB	500K
SG301AT	SKU	GPU	EXT			+18V	-18V	70C	88dB	7.5MV	250NA	50NA	500MWF	5MA	12V	15V	30V	30uV/C		3MA	70dB	70dB	500K
SG302J	SKU	VFA	INT			+18V	-18V	70C	0dB	15MV	30NA		500MWF	1MA	10V			90uV/C		6MA		60dB	10G
SG302M	SKU	VFA	INT		15V/uS	+18V	-18V	70C	0dB	7.5MV	7NA		500MWF	1MA	10V	15V	15V	50uV/C		6MA		70dB	10G
SG302N	SKU	VFA	INT			+18V	-18V	70C	0dB	15MV	30NA		500MWF	1MA	10V			90uV/C		6MA		60dB	10G
SG302T	SKU	VFA	INT			+18V	-18V	70C	0dB	15MV	30NA		500MWF	1MA	10V			90uV/C		6MA		60dB	10G
SG307D	SKU	GPK	INT			+18V	-18V	70C	84dB	7.5MV	250NA	50NA	500MWF	5MA	12V	15V	30V	30uV/C			70dB	70dB	0.5M
SG307F	SKU	GPK	INT			+18V	-18V	70C	84dB	7.5MV	250NA	50NA	500MWF	5MA	12V	15V	30V	30uV/C			70dB	70dB	0.5M
SG307J	SKU	GPK	INT			+18V	-18V	70C	84dB	7.5MV	250NA	50NA	500MWF	5MA	12V	15V	30V	30uV/C			70dB	70dB	0.5M
SG307M	SKU	GPK	INT			+18V	-18V	70C	84dB	7.5MV	250NA	50NA	500MWF	5MA	12V	15V	30V	30uV/C			70dB	70dB	0.5M
SG307N	SKU	GPK	INT			+18V	-18V	70C	84dB	7.5MV	250NA	50NA	500MWF	5MA	12V	15V	30V	30uV/C			70dB	70dB	0.5M
SG307T	SKU	GPK	INT			+18V	-18V	70C	84dB	7.5MV	250NA	50NA	500MWF	5MA	12V	15V	30V	30uV/C			70dB	70dB	0.5M
SG308AD	SKU	SBA	EXT			+18V	-18V	70C	98dB	0.5MV	7NA	1NA	500MWF	1MA	13V	15V	1V	5uV/C		.6MA	96dB	96dB	10M
SG308AF	SKU	SBA	EXT			+20V	-20V	70C	98dB	0.5MV	2NA	0.2NA	500MWF	1MA	13V	15V	1V	5uV/C		.6MA	96dB	96dB	30M
SG308AJ	SKU	SBA	EXT			+18V	-18V	70C	98dB	0.5MV	7NA	1NA	500MWF	1MA	13V	15V	1V	5uV/C		.6MA	96dB	96dB	10M
SG308AM	SKU	SBA	EXT			+18V	-18V	70C	98dB	0.5MV	7NA	1NA	500MWF	1MA	13V	15V	1V	5uV/C		.6MA	96dB	96dB	10M
SG308AN	SKU	SBA	EXT			+18V	-18V	70C	98dB	0.5MV	7NA	1NA	500MWF	1MA	13V	15V	1V	5uV/C		.6MA	96dB	96dB	10M
SG308AT	SKU	SBA	EXT			+18V	-18V	70C	98dB	0.5MV	7NA	1NA	500MWF	1MA	13V	15V	1V	5uV/C		.6MA	96dB	96dB	10M
SG308D	SKU	SBA	EXT			+18V	-18V	70C	88dB	7.5MV	7NA	1NA	500MWF	1MA	13V	15V	1V	30uV/C		.6MA	80dB	80dB	10M
SG308F	SKU	SBA	EXT			+20V	-20V	70C	96dB	2MV	2NA	0.2NA	500MWF	1MA	13V	15V	1V	15uV/C		.6MA	85dB	80dB	30M
SG308J	SKU	SBA	EXT			+18V	-18V	70C	88dB	7.5MV	7NA	1NA	500MWF	1MA	13V	15V	1V	30uV/C		.6MA	80dB	80dB	10M
SG308M	SKU	SBA	EXT			+18V	-18V	70C	88dB	7.5MV	7NA	1NA	500MWF	1MA	13V	15V	1V	30uV/C		.6MA	80dB	80dB	10M
SG308N	SKU	SBA	EXT			+18V	-18V	70C	88dB	7.5MV	7NA	1NA	500MWF	1MA	13V	15V	1V	30uV/C		.6MA	80dB	80dB	10M
SG308T	SKU	SBA	EXT			+18V	-18V	70C	88dB	7.5MV	7NA	1NA	500MWF	1MA	13V	15V	1V	30uV/C		.6MA	80dB	80dB	10M
SG310J	SKU	VFA	INT		15V/uS	+18V	-18V	70C	0dB	7.5MV	7NA		500MWF	1MA	10V	15V	15V	50uV/C		6MA		70dB	10G
SG310M	SKU	VFA	INT		15V/uS	+18V	-18V	70C	0dB	7.5MV	7NA		500MWF	1MA	10V	15V	15V	50uV/C		6MA		70dB	10G
SG310N	SKU	VFA	INT		15V/uS	+18V	-18V	70C	0dB	7.5MV	7NA		500MWF	1MA	10V	15V	15V	50uV/C		6MA		70dB	10G
SG310T	SKU	VFA	INT		15V/uS	+18V	-18V	70C	0dB	7.5MV	7NA		500MWF	1MA	10V	15V	15V	50uV/C		6MA		70dB	10G
SG311D	SKU	CPR	EXT			+18V		70C	100dB	7.5MV	250NA	50NA	500MWF			15V	30V			8MA			
SG311J	SKU	CPR	EXT			+18V	-18V	70C	100dB	7.5MV	250NA	50NA	500MWF			15V	30V			8MA			
SG311M	SKU	CPR	EXT			+18V	-18V	70C	100dB	7.5MV	250NA	50NA	500MWF			15V	30V			8MA			
SG311N	SKU	CPR	EXT			+18V	-18V	70C	100dB	7.5MV	250NA	50NA	500MWF			15V	30V			8MA			
SG311T	SKU	CPR	EXT			+18V	-18V	70C	100dB	7.5MV	250NA	50NA	500MWF			15V	30V			8MA			
SG318J	SKU	XSR	INT		50V/uS	+20V	-20V	70C	88dB	10MV	500NA	200NA	500MWF	6MA	12V	15V	1V			8MA	70dB	65dB	500K
SG318M	SKU	XSR	INT		50V/uS	+20V	-20V	70C	88dB	10MV	500NA	200NA	500MWF	6MA	12V	15V	1V			8MA	70dB	65dB	500K
SG318T	SKU	XSR	INT		50V/uS	+20V	-20V	70C	88dB	10MV	500NA	200NA	500MWF	6MA	12V	15V	1V			8MA	70dB	65dB	500K
SG324J	SKU	QGK	INT			+16V	-16V	70C	88dB	7MV	250NA	50NA	900MWF			16V	16V	35uV/C		2MA	65dB	65dB	
SG324N	SKU	QGK	INT			+16V	-16V	70C	88dB	7MV	250NA	50NA	900MWF			16V	16V	35uV/C		2MA	65dB	65dB	
SG339AJ	SKU	QCP	EXT			+18V	-18V	70C	94dB	2MV	250NA	50NA	900MWF	6MA		18V	18V			2MA			
SG339AN	SKU	QCP	EXT			+18V	-18V	70C	94dB	2MV	250NA	50NA	900MWF	6MA		18V	18V			2MA			
SG339J	SKU	QCP	EXT			+18V	-18V	70C	88dB	5MV	250NA	50NA	900MWF	6MA		18V	18V			2MA			
SG339N	SKU	QCP	EXT			+18V	-18V	70C	88dB	5MV	250NA	50NA	900MWF	6MA		18V	18V			2MA			
SG710CD	SKU	CPR	EXT			+14V	-6V	70C	60dB	5MV	25uA	5uA	670MWF	5MA	2.5V	7V	5V	20uV/C	150MW		70dB		
SG710CJ	SKU	CPR	EXT			+14V	-6V	70C	60dB	5MV	25uA	5uA	670MWF	5MA	2.5V	7V	5V	20uV/C	150MW		70dB		
SG710CN	SKU	CPR	EXT			+14V	-6V	70C	60dB	5MV	25uA	5uA	670MWF	5MA	2.5V	7V	5V	20uV/C	150MW		70dB		
SG710CT	SKU	CPR	EXT			+14V	-6V	70C	60dB	5MV	25uA	5uA	500MWF	5MA	2.5V	7V	5V	20uV/C	150MW		70dB		
SG710D	SKU	CPR	EXT			+14V	-6V	125C	62dB	2MV	20uA	3uA	670MWF	5MA	2.5V	7V	5V	10uV/C	150MW		80dB		
SG710J	SKU	CPR	EXT			+14V	-6V	125C	62dB	2MV	20uA	3uA	670MWF	5MA	2.5V	7V	5V	10uV/C	150MW		80dB		
SG710T	SKU	CPR	EXT			+14V	-7V	125C	62dB	2MV	20uA	3uA	500MWF	5MA	2.5V	7V	5V	10uV/C	150MW		80dB		

136

For detailed explanations of column heading notations, see App. A.

Also for ready references the more important abbreviations used in the column headings are listed below:

LEFT HAND PAGE

APP = application (codes at APP.E.)
CMRR = common mode rejection ratio
CMP = compensation (frequency)
dV_{IO}/dT = input offset voltage temperature drift
GBP = gain bandwidth product
I_B = input bias current
I_{IO} = input bias offset current
I_Q = quiescent supply current
MFR = manufacturer (codes at App.C.)
P_Q = quiescent power consumer
PSRR = power supply rejection ratio
V_{ICM} = common mode input voltage rating
V_{IDF} = differential input voltage rating
V_{IO} = input offset voltage
V_S = dc supply voltage

RIGHT HAND PAGE

Lead out coding summary (details at APP.G.) for different cases (APP.F.)

A = gain adjust
B = bias adjust
C = case
E— = inverting input
E+ = non-inverting input
F,F* = input frequency compensation
G = ground
J = high level input
K = output, open collector
L = output, open emitter
M = metal case
N = not connected
Q = special terminal
R,R* = outputs
S = strobe
T,T* = offset balance
V+ = +ve dc supply
V– = –ve dc supply
W = guard ring
X = blank position, no lead
++ = +ve supplementary dc supply
–– = –ve supplementary dc supply
ø,ø* = output frequency compensation

CASE (APP F)	LD 1	LD 2	LD 3	LD 4	LD 5	LD 6	LD 7	LD 8	LD 9	LD 10	LD 11	LD 12	LD 13	LD 14	LD 15	LD 16	EUROPE SUBSTITUTE	USA SUBSTITUTE	I S	TYPE NUMBER
DIL-14/1C	R1	E-1	E+1	V+	E+2	E-2	R2	R3	E-3	E+3	G	E+4	E-4	R4				LM224D	0	SG224J
DIL-14/1P	R1	E-1	E+1	V+	E+2	E-2	R2	R3	E-3	E+3	G	E+4	E-4	R4				LM224D	0	SG224N
DIL-14/1C	R2	R1	V+	E-1	E+1	E-2	E+2	E+3	E-3	E-4	E+4	G	R4	R3			MLM239AL	LM239AD	0	SG239AJ
DIL-14/1P	R2	R1	V+	E-1	E+1	E-2	E+2	E+3	E-3	E-4	E+4	G	R4	R3			MLM239AL	LM239AD	0	SG239AN
DIL-14/1C	R2	R1	V+	E-1	E+1	E-2	E+2	E+3	E-3	E-4	E+4	G	R4	R3			MLM239L	LM239D	0	SG239J
DIL-14/1P	R2	R1	V+	E-1	E+1	E-2	E+2	E+3	E-3	E-4	E+4	G	R4	R3			MLM239L	LM239D	0	SG239N
DIL-14/1C	N	N	FT	E-	E+	V-	N	N	T*	R	V+	F*	N	N			UA301AD	LM301AJ14	0	SG301AD
FLP-10/1C	N	FT	E-	E+	V-	T*	R	V+	F*	N								SFC2201APT	0	SG301AF
DIL-14/1C	N	N	FT	E-	E+	V-	N	N	T*	R	V+	F*	N	N			UA301AD	LM301AJ14	0	SG301AJ
DIL-8/1P	FT	E-	E+	V-	T*	R	V+	F*									SFC2301ADC	LM301AJ	0	SG301AM
DIL-14/1P	N	N	FT	E-	E+	V-	N	N	T*	R	V+	F*	N	N			UA301AD	LM301AJ14	0	SG301AN
TO5-8/1M	FT	E-	E+	V-M	T*	R	V+	F*	N								SFC2301A	LM301AH	0	SG301AT
DIL-14/1C	N	N	T	N	E+	V-	N	N	L	R	V+	T*	N	N			SFC2310EC	LM310D	0	SG302J
DIL-8/1P	T	N	E+	V-	L	R	V+	T*									SFC2310DC	LM310N	0	SG302M
DIL-14/1P	N	N	T	N	E+	V-	N	N	L	R	V+	T*	N	N			SFC2310EC	LM310D	0	SG302N
TO5-8/1M	T	N	E+	V-M	L	R	V+	T*									UA302C	LM302H	0	SG302T
DIL-14/1C	N	N	N	E-	E+	V-	N	N	N	R	V+	N	N	N			SN72307JA	LM307D	0	SG307D
FLP-10/3C	N	N	E-	E+	V-	N	R	V+	N	N							SFC2207PT	LM207F	0	SG307F
DIL-14/1C	N	N	N	E-	E+	V-	N	N	N	R	V+	N	N	N			SN72307JA	LM307D	0	SG307J
DIL-8/1P	N	E-	E+	V-	N	R	V+	N									SFC2307DC	LM307J	0	SG307M
DIL-14/1P	N	N	N	E-	E+	V-	N	N	N	R	V+	N	N	N			SN72307JA	LM307D	0	SG307N
TO5-8/1M	N	E-	E+	V-M	N	R	V+	N									SFC2307	LM307H	0	SG307T
DIL-14/1C	N	F	N	E-	E+	V-	N	N	N	R	V+	F*	N	N			UA308AD	LM308AD	0	SG308AD
FLP-10/3C	N	N	E-	E+	N	V-	R	V+	F*	F								LM208AF	0	SG308AF
DIL-14/1C	N	F	N	E-	E+	V-	N	N	N	R	V+	F*	N	N			UA308AD	LM308AD	0	SG308AJ
DIL-8/1P	F	E-	E+	V-	N	R	V+	F*									UA308AB	LM308AD	0	SG308AM
DIL-14/1P	N	F	N	E-	E+	V-	N	N	N	R	V+	F*	N	N			UA308AD	LM308AD	0	SG308AN
TO5-8/1M	F	E-	E+	V-M	N	R	V+	F*									SFC2308A	LM308AH	0	SG308AT
DIL-14/1C	N	F	N	E-	E+	V-	N	N	N	R	V+	F*	N	N			UA308D	LM308D	0	SG308D
FLP-10/3C	N	N	E-	E+	N	V-	R	V+	F*	F							SFC2208PT	LM208F	0	SG308F
DIL-14/1C	N	F	N	E-	E+	V-	N	N	N	R	V+	F*	N	N			UA308D	LM308D	0	SG308J
DIL-8/1P	F	E-	E+	V-	N	R	V+	F*									SFC2308DC	LM308N	0	SG308M
DIL-14/1P	N	F	N	E-	E+	V-	N	N	N	R	V+	F*	N	N			UA308D	LM308D	0	SG308N
TO5-8/1M	F	E-	E+	V-M	N	R	V+	F*									SFC2308	LM308H	0	SG308T
DIL-14/1C	N	N	T	N	E+	V-	N	N	L	R	V+	T*	N	N			SFC2310EC	LM310D	0	SG310J
DIL-8/1P	T	N	E+	V-	L	R	V+	T*									SFC2310DC	LM310N	0	SG310M
DIL-14/1P	N	N	T	N	E+	V-	N	N	L	R	V+	T*	N	N			SFC2310EC	LM310D	0	SG310N
TO5-8/1M	T	N	E+	V-M	L	R	V+	T*									SFC2310EC	LM310H	0	SG310T
DIL-14/1C	N	G	E+	E-	N	V-	T	T*S	R	V+	N	N	N				SFC2311EC	LM311D	0	SG311D
DIL-14/1C	N	G	E+	E-	N	V-	T	T*S	R	V+	N	N	N				SFC2311EC	LM311D	0	SG311J
DIL-8/1P	G	E+	E-	V-	T	T*S	R	V+									SFC2311DC	LM311N	0	SG311M
DIL-14/1P	N	G	E+	E-	N	V-	T	T*S	R	V+	N	N	N				SFC2311EC	LM311D	0	SG311N
TO5-8/1M	G	E+	E-	V-M	T	T*S	R	V+									SFC2311	LM311H	0	SG311T
DIL-14/1C	N	N	T*F	E-	E+	V-	N	N	F*T	R	V+	ø	N	N			SFC2318EC	LM318D	0	SG318J
DIL-8/1P	T*F	E-	E+	V-	F*T	R	V+	ø									SN72318P	LM318N	0	SG318M
TO5-8/1M	T*F	E-	E+	V-M	F*T	R	V+	ø									TDE0118CM	LM318H	0	SG318T
DIL-14/1C	R1	E-1	E+1	V+	E+2	E-2	R2	R3	E-3	E+3	G	E+4	E-4	R4			MLM324J	LM324N	0	SG324J
DIL-14/1P	R1	E-1	E+1	V+	E+2	E-2	R2	R3	E-3	E+3	G	E+4	E-4	R4			MLM324J	LM324N	0	SG324N
DIL-14/1C	R2	R1	V+	E-1	E+1	E-2	E+2	E+3	E-3	E-4	E+4	G	R4	R3			MLM339AL	LM339AJ	0	SG339AJ
DIL-14/1P	R2	R1	V+	E-1	E+1	E-2	E+2	E+3	E-3	E-4	E+4	G	R4	R3			MLM339AL	LM339AJ	0	SG339AN
DIL-14/1C	R2	R1	V+	E-1	E+1	E-2	E+2	E+3	E-3	E-4	E+4	G	R4	R3			MLM339L	LM339J	0	SG339J
DIL-14/1P	R2	R1	V+	E-1	E+1	E-2	E+2	E+3	E-3	E-4	E+4	G	R4	R3			MLM339L	LM339J	0	SG339N
DIL-14/1C	N	G	E+	E-	N	V-	N	N	R	N	V+	N	N				SFC2710EC	UA710DC	0	SG710CD
DIL-14/1C	N	G	E+	E-	N	V-	N	N	R	N	V+	N	N				SFC2710EC	UA710DC	0	SG710CJ
DIL-14/1P	N	G	E+	E-	N	V-	N	N	R	N	V+	N	N				SFC2710EC	UA710DC	0	SG710CN
TO5-8/1M	G	E+	E-	V-	N	N	R	V+									SFC2710C	UA710HC	0	SG710CT
DIL-14/1C	G	E+	E-	V-	N	N	R	V+									SFC2710KM	UA710DM	0	SG710D
DIL-14/1C	G	E+	E-	V-	N	N	R	V+									SFC2710KM	UA710DM	0	SG710J
TO5-8/1M	G	E+	E-	V-	N	N	R	V+									SFC2710M	UA710HM	0	SG710T
DIL-14/1C	N	E-1	E+1	V-	E+2	E-2	N	N	S2	R	V+	G	S1	N			SFC2711EC	UA711DC	0	SG711CD

TYPE NUMBER	MFR	APP	CMP	GBP MIN	SLEW RATE MIN	Vs+ MAX	Vs− MAX	T∞ MAX	AVOL MIN	VIO MAX	IB MAX	IIO MAX	PTOT MAX	IOUT MIN	VOUT MIN	VICM MAX	VIDI MAX	dVIO/dT MAX	PO MAX	IO MAX	CMRR MIN	PSRR MIN	PIN MIN
SG711CD	SKU	DCP	EXT	.		+14V	−7V	70C	57dB	5MV	100uA	15uA	670MWF	5MA	2.5V	7V	5V	20uV/C	230MW	.	.	.	.
SG711CF	SKU	DCP	EXT	.		+14V	−7V	70C	57dB	5MV	100uA	15uA	570MWF	5MA	2.5V	7V	5V	20uV/C	230MW	.	.	.	.
SG711CJ	SKU	DCP	EXT	.		+14V	−7V	70C	57dB	5MV	100uA	15uA	670MWF	5MA	2.5V	7V	5V	20uV/C	230MW	.	.	.	.
SG711CN	SKU	DCF	EXT	.		+14V	−7V	70C	57dB	5MV	100uA	15uA	670MWF	5MA	2.5V	7V	5V	20uV/C	230MW	.	.	.	.
SG711CT	SKU	DCP	EXT	.		+14V	−7V	70C	57dB	5MV	100uA	15uA	500MWF	5MA	2.5V	7V	5V	20uV/C	230MW	.	.	.	.
SG711D	SKU	DCP	EXT	.		+14V	−7V	125C	58dB	3.5MV	75uA	10uA	670MWF	5MA	2.5V	7V	5V	20uV/C	200MW	.	.	.	.
SG711F	SKU	DCP	EXT	.		+14V	−7V	125C	58dB	3.5MV	75uA	10uA	570MWF	5MA	2.5V	7V	5V	20uV/C	200MW	.	.	.	.
SG711J	SKU	DCP	EXT	.		+14V	−7V	125C	58dB	3.5MV	75uA	10uA	670MWF	5MA	2.5V	7V	5V	20uV/C	200MW	.	.	.	.
SG711T	SKU	DCP	EXT	.		+14V	−7V	125C	58dB	3.5MV	75uA	10uA	500MWF	5MA	2.5V	7V	5V	20uV/C	200MW	.	.	.	.
SG733CJ	SKU	BDO	EXT	20MHZ		+8V	−8V	70C	48dB	6MV	30uA	5uA	670MWF	2MA	1.5V	6V	5V		.	24MA	60dB	50dB	2K
SG733CN	SKU	BDO	EXT	20MHZ		+8V	−8V	70C	48dB	6MV	30uA	5uA	670MWF	2MA	1.5V	6V	5V			24MA	60dB	50dB	2K
SG733CT	SKU	BDO	EXT	20MHZ		+8V	−8V	70C	48dB	6MV	30uA	5uA	500MWF	2MA	1.5V	6V	5V			24MA	60dB	50dB	2K
SG733J	SKU	BDO	EXT	20MHZ		+8V	−8V	125C	50dB	5MV	20uA	3uA	670MWF	2MA	1.5V	6V	5V			24MA	60dB	50dB	2K
SG733T	SKU	BDO	EXT	20MHZ		+8V	−8V	125C	50dB	5MV	20uA	3uA	500MWF	2MA	1.5V	6V	5V			24MA	60dB	50dB	2K
SG741CD	SKU	GPK	INT	.	0.2V/uS	+18V	−18V	70C	86dB	6MV	500NA	200NA	670MWF	5MA	12V	15V	30V		85MW	3MA	70dB	76dB	300K
SG741CF	SKU	GPK	INT	.	0.2V/uS	+18V	−18V	70C	86dB	6MV	500NA	200NA	570MWF	5MA	12V	15V	30V		85MW	3MA	70dB	76dB	300K
SG741CJ	SKU	GPK	INT	.	0.2V/uS	+18V	−18V	70C	86dB	6MV	500NA	200NA	670MWF	5MA	12V	15V	30V		85MW	3MA	70dB	76dB	300K
SG741CM	SKU	GPK	INT	.	0.2V/uS	+18V	−18V	70C	86dB	5MV	500NA	200NA	310MWF	5MA	12V	15V	30V		.	3MA	70dB	76dB	300K
SG741CN	SKU	GPK	INT	.	0.2V/uS	+18V	−18V	70C	86dB	6MV	500NA	200NA	670MWF	5MA	12V	15V	30V		85MW	3MA	70dB	76dB	300K
SG741CT	SKU	GPK	INT	.	0.2V/uS	+18V	−13V	70C	86dB	6MV	500NA	200NA	500MWF	5MA	12V	15V	30V		.	3MA	70dB	76dB	300K
SG741D	SKU	GPK	INT	.4MHZ	0.3V/uS	+22V	−22V	125C	94dB	3MV	80NA	30NA	670MWF	5MA	12V	15V	30V	15uV/C	150MW	.	80dB	86dB	1M
SG741F	SKU	GPK	INT	.	0.2V/uS	+22V	−22V	125C	94dB	5MV	500NA	200NA	570MWF	5MA	12V	15V	30V		85MW	3MA	70dB	76dB	300K
SG741J	SKU	GPK	INT	.4MHZ	0.3V/uS	+22V	−22V	125C	94dB	3MV	80NA	30NA	670MWF	5MA	12V	15V	30V	15uV/C	150MW	.	80dB	86dB	1M
SG741SCJ	SKU	HSR	INT	.	5V/uS	+18V	−18V	70C	86dB	6MV	500NA	200NA	670MWF	5MA	12V	15V	30V		.	3MA	70dB	76dB	300K
SG741SCM	SKU	HSR	INT	.	5V/uS	+18V	−18V	70C	86dB	6MV	500NA	200NA	310MWF	5MA	12V	15V	30V		.	3MA	70dB	76dB	300K
SG741SCN	SKU	HSR	INT	.	5V/uS	+18V	−18V	70C	86dB	6MV	500NA	200NA	670MWF	5MA	12V	15V	30V		85MW	3MA	70dB	76dB	300K
SG741SCT	SKU	HSR	INT	.	5V/uS	+18V	−18V	70C	86dB	6MV	500NA	200NA	500MWF	5MA	12V	15V	30V		.	3MA	70dB	76dB	300K
SG741ST	SKU	HSR	INT	.	5V/uS	+22V	−22V	125C	94dB	5MV	500NA	200NA	500MWF	7MA	12V	15V	30V		85MW	3MA	70dB	76dB	300K
SG741T	SKU	GPK	INT	.	0.2V/uS	+22V	−22V	125C	94dB	5MV	500NA	200NA	500MWF	7MA	12V	15V	30V		85MW	3MA	70dB	76dB	300K
SG747CD	SKU	DGK	INT	.	0.2V/uS	+18V	−18V	70C	88dB	6MV	500NA	200NA	670MWF	5MA	12V	15V	30V		85MW	3MA	70dB	76dB	300K
SG747CJ	SKU	DGK	INT	.	0.2V/uS	+18V	−18V	70C	88dB	6MV	500NA	200NA	670MWF	5MA	12V	15V	30V		85MW	3MA	70dB	76dB	300K
SG747CN	SKU	DGK	INT	.	0.2V/uS	+18V	−18V	70C	88dB	6MV	500NA	200NA	670MWF	5MA	12V	15V	30V		85MW	3MA	70dB	76dB	300K
SG747CT	SKU	DGK	INT	.	0.2V/uS	+18V	−18V	70C	88dB	6MV	500NA	200NA	500MWF	5MA	12V	15V	30V		85MW	3MA	70dB	76dB	300K
SG747D	SKU	DGK	INT	.	0.2V/uS	+22V	−22V	125C	94dB	5MV	500NA	200NA	670MWF	5MA	12V	15V	30V		85MW	3MA	70dB	76dB	300K
SG747T	SKU	DGK	INT	.	0.2V/uS	+22V	−22V	125C	94dB	5MV	500NA	200NA	500MWF	5MA	12V	15V	30V		85MW	3MA	70dB	76dB	300K
SG748CD	SKU	GPU	EXT	.	0.2V/uS	+22V	−22V	70C	86dB	6MV	500NA	200NA	670MWF	5MA	12V	15V	30V		85MW	3MA	70dB	76dB	300K
SG748CJ	SKU	GPU	EXT	.	0.2V/uS	+22V	−22V	70C	86dB	6MV	500NA	200NA	670MWF	5MA	12V	15V	30V		85MW	3MA	70dB	76dB	300K
SG748CM	SKU	GPU	EXT	.	0.2V/uS	+18V	−18V	70C	86dB	6MV	500NA	200NA	310MWF	5MA	12V	15V	30V		85MW	3MA	70dB	76dB	300K
SG748CN	SKU	GPU	EXT	.	0.2V/uS	+22V	−22V	70C	86dB	6MV	500NA	200NA	670MWF	5MA	12V	15V	30V		85MW	3MA	70dB	76dB	300K
SG748CT	SKU	GPU	EXT	.	0.2V/uS	+18V	−18V	70C	86dB	6MV	500NA	200NA	500MWF	5MA	12V	15V	30V		85MW	3MA	70dB	76dB	300K
SG748D	SKU	GPU	EXT	.	0.2V/uS	+22V	−22V	125C	94dB	5MV	500NA	200NA	670MWF	5MA	12V	15V	30V		85MW	3MA	70dB	76dB	300K
SG748F	SKU	GPU	EXT	.	0.2V/uS	+22V	−22V	125C	94dB	5MV	500NA	200NA	570MWF	5MA	12V	15V	30V		85MW	3MA	70dB	76dB	300K
SG748J	SKU	GPU	EXT	.	0.2V/uS	+22V	−22V	125C	94dB	5MV	500NA	200NA	670MWF	5MA	12V	15V	30V		85MW	3MA	70dB	76dB	300K
SG748T	SKU	GPU	EXT	.	0.2V/uS	+22V	−22V	125C	94dB	5MV	500NA	200NA	500MWF	5MA	12V	15V	30V		85MW	3MA	70dB	76dB	300K
SG777CJ	SKU	PIA	EXT	.	0.2V/uS	+22V	−22V	70C	88dB	5MV	100NA	20NA	670MWF	5MA	12V	15V	30V	30uV/C	85MW	3MA	70dB	76dB	1M
SG777CM	SKU	PIA	EXT	.	0.2V/uS	+22V	−22V	70C	88dB	5MV	100NA	20NA	310MWF	5MA	12V	15V	30V	30uV/C	85MW	3MA	70dB	76dB	1M
SG777CN	SKU	PIA	EXT	.	0.2V/uS	+22V	−22V	70C	88dB	5MV	100NA	20NA	670MWF	5MA	12V	15V	30V	30uV/C	85MW	3MA	70dB	76dB	1M
SG777CT	SKU	PIA	EXT	.	0.2V/uS	+22V	−22V	70C	88dB	5MV	100NA	20NA	500MWF	5MA	12V	15V	30V	30uV/C	85MW	3MA	70dB	76dB	1M
SG1436CT	SKU	GPK	INT	.3MHZ	0.5V/uS	+30V	−30V	75C	94dB	12MV	90NA	25NA	680MWF	1MA	20V	30V	60V		280MW	5MA	50dB	70dB	3M
SG1436M	SKU	GPK	INT	.3MHZ	0.5V/uS	+34V	−34V	75C	97dB	10MV	40NA	10NA	680MWF	1MA	20V	34V	68V		280MW	5MA	70dB	74dB	3M
SG1436T	SKU	GPK	INT	.3MHZ	0.5V/uS	+34V	−34V	75C	97dB	10MV	40NA	10NA	680MWF	1MA	20V	34V	68V		280MW	5MA	70dB	74dB	3M
SG1456CT	SKU	GPK	INT	.5MHZ	1V/uS	+18V	−18V	75C	96dB	10MV	30NA	10NA	680MWF	5MA	11V	18V	18V	40uV/C	90MW	3MA	70dB	74dB	1M
SG1456T	SKU	GPK	INT	.5MHZ	1V/uS	+18V	−18V	75C	96dB	10MV	30NA	10NA	680MWF	5MA	11V	18V	18V	40uV/C	90MW	3MA	70dB	74dB	1M
SG1458CM	SKU	DGK	INT	.5MHZ	0.3V/uS	+18V	−18V	75C	84dB	10MV	0.7uA	0.3uA	625MWF	4MA	11V	15V	30V	50uV/C	240MW	8MA	60dB	66dB	150K
SG1458M	SKU	DGK	INT	.5MHZ	0.3V/uS	+18V	−18V	75C	86dB	6MV	0.5uA	0.2uA	625MWF	5MA	12V	15V	30V	50uV/C	170MW	6MA	70dB	76dB	300K
SG1458T	SKU	DGK	INT	.5MHZ	0.3V/uS	+18V	−18V	75C	86dB	6MV	0.5uA	0.2uA	680MWF	5MA	12V	15V	30V	50uV/C	170MW	6MA	70dB	76dB	300K
SG1536T	SKU	HVO	INT	.3MHZ	0.5V/uS	+40V	−40V	125C	100dB	5MV	20NA	3NA	680MWF	1MA	30V	40V	80V		224MW	4MA	80dB	80dB	3M
SG1556T	SKU	GFK	INT	.5MHZ	1V/uS	+22V	−22V	125C	100dB	4MV	15NA	2NA	680MWF	6MA	12V	22V	22V	30uV/C	45MW	2MA	80dB	80dB	1.5M
SG1558T	SKU	DGK	INT	.5MHZ	0.3V/uS	+22V	−22V	125C	94dB	5MV	0.5uA	0.2uA	680MWF	5MA	12V	15V	30V	50uV/C	150MW	5MA	70dB	76dB	300K

For detailed explanations of column heading notations, see App. A.

Also for ready references the more important abbreviations used in the column headings are listed below:

LEFT HAND PAGE

APP = application (codes at APP.E.)
CMRR = common mode rejection ratio
CMP = compensation (frequency)
dV_{IO}/dT = input offset voltage temperature drift
GBP = gain bandwidth product
I_B = input bias current
I_{IO} = input bias offset current
I_Q = quiescent supply current
MFR = manufacturer (codes at App C.)
P_Q = quiescent power consumer
PSRR = power supply rejection ratio
V_{ICM} = common mode input voltage rating
V_{IDF} = differential input voltage rating
V_{IO} = input offset voltage
V_S = dc supply voltage

RIGHT HAND PAGE

Lead out coding summary (details at APP.G.) for different cases (APP.F.)

A = gain adjust
B = bias adjust
C = case
E— = inverting input
E+ = non-inverting input
F,F* = input frequency compensation
G = ground
J = high level input
K = output, open collector
L = output, open emitter
M = metal case
N = not connected
Q = special terminal
R,R* = outputs
S = strobe
T,T* = offset balance
V+ = +ve dc supply
V— = —ve dc supply
W = guard ring
X = blank position, no lead
++ = +ve supplementary dc supply
—— = —ve supplementary dc supply
∅,∅* = output frequency compensation

CASE (APP F)	LD 1	LD 2	LD 3	LD 4	LD 5	LD 6	LD 7	LD 8	LD 9	LD 10	LD 11	LD 12	LD 13	LD 14	LD 15	LD 16	EUROPE SUBSTITUTE	USA SUBSTITUTE	I/S	TYPE NUMBER
FLP-10/3C	E-1	E+1	V-	E+2	E-2	S2	R	V+	G	S1	.	.	.	.			SFC2711PM	UA711FM	0	SG711CF
DIL-14/1C	N	E-1	E+1	V-	E+2	E-2	N	N	S2	R	V+	G	S1	N			SFC2711EC	UA711DC	0	SG711CJ
DIL-14/1P	N	E-1	E+1	V-	E+2	E-2	N	N	S2	R	V+	G	S1	N			SFC2711EC	UA711DC	0	SG711CN
TO5-10/1M	G	S1	E-1	E+1	V-	E+2	E-2	S2	R	V+	.	.					SFC2711C	UA711HC	0	SC711CT
DIL-14/1C	N	E-1	E+1	V-	E+2	E-2	N	N	S2	R	V+	G	S1	N			SFC2711KM	UA711DM	0	SG711D
FLP-10/3C	E-1	E+1	V-	E+2	E-2	S2	R	V+	G	S1	.	.					SFC2711PM	UA711FM	0	SG711F
DIL-14/1C	N	E-1	E+1	V-	E+2	E-2	N	N	S2	R	V+	G	S1	N			SFC2711KM	UA711DM	0	SG711J
TO5-10/1M	G	S1	E-1	E+1	V-	E+2	E-2	S2	R	V+	.						SFC2711M	UA711HM	0	SG711T
DIL-14/1C	E+	N	A2	A*2	V-	N	R	R*	N	V+	A1	A*1	N	E-			SN72733J	UA733DC	0	SG733CJ
DIL-14/1P	E+	N	A2	A*2	V-	N	R	R*	N	V+	A1	A*1	N	E-			SN72733J	UA733DC	0	SG733CN
TO5-10/1M	E-	E+	A2	A*2	V-	R	R*	V+	A1	A*1	.	.					SN72733L	UA733HC	0	SG733CT
DIL-14/1C	E+	N	A2	A*2	V-	N	R	R*	N	V+	A1	A*1	N	E-			SN52733J	UA733DM	0	SG733J
TO5-10/1M	E-	E+	A2	A*2	V-	R	R*	V+	A1	A*1	.						SN52733L	UA733HM	0	SG733T
DIL-14/1C	N	T	E-	E+	V-	N	R	N	T*	R	V+	N	N	N			TBA221A	UA741DC	0	SG741CD
FLP-10/3C	N	T	E-	E+	V-	T*	R	V+	N	N	.	.					LM741F	UA741FM	0	SG741CF
DIL-14/1C	N	N	T	E-	E+	V-	N	N	T*	R	V+	N	N	N			TBA221A	UA741DC	0	SG741CJ
DIL-8/1P	T	E-	E+	V-	T*	R	V+	N	.	.							TBA221B	UA741TC	0	SG741CM
DIL-14/1P	N	N	T	E-	E+	V-	N	N	T*	R	V+	N	N	N			TBA221A	UA741DC	0	SG741CN
TO5-8/1M	T	E-	E+	V-M	T*	R	V+	N	.								TBA221	UA741HC	C	SG741CT
DIL-14/1C	N	N	T	E-	E+	V-	N	N	T*	R	V+	N	N	N			LM741D	UA741DM	0	SG741D
FLP-10/3C	N	T	E-	E+	V-	T*	R	V+	N	N	.						LM741F	UA741FM	0	SG741F
DIL-14/1C	N	N	T	E-	E+	V-	N	N	T*	R	V+	N	N	N			LM741D	UA741DM	0	SG741J
DIL-14/1C	N	N	T	E-	E+	V-	N	N	T*	R	V+	N	N	N			.	741MHSDD	0	SG741SCJ
DIL-8/1P	T	E-	E+	V-	T*	R	V+	N	.	.							.	741CHSPA	0	SG741SCM
DIL-14/1P	N	N	T	E-	E+	V-	N	N	T*	R	V+	N	N	N			.	741MHSDD	0	SG741SCN
TO5-8/1M	T	E-	E+	V-M	T*	R	V+	N	.	.							.	741CHSTY	0	SG741SCT
TO5-8/1M	T	E-	E+	V-M	T*	R	V+	N	.	.							.	741MHSTY	0	SG741ST
TO5-8/1M	T	E-	E+	V-M	T*	R	V+	N	.	.							TBA222	UA741HM	0	SG741T
DIL-14/1C	E-1	E+1	T1	V-	T2	E+2	E-2	T*2	V+2	R2	R1	V+1	T*1	.			TBB0747A	UA747DC	0	SG747CD
DIL-14/1C	E-1	E+1	T1	V-	T2	E+2	E-2	T*2	V+2	R2	R1	V+1	T*1	.			TBB0747A	UA747DC	0	SG747CJ
DIL-14/1P	E-1	E+1	T1	V-	T2	E+2	E-2	T*2	V+2	R2	R1	V+1	T*1	.			TBB0747A	UA747DC	0	SG747CN
TO5-10/1M	R1	V+1	E-1	E+1	V-	E+2	E-2	V+2	R2	N	.	.					TBB0747	UA747HC	0	SG747CT
DIL-14/1C	E-1	E+1	T1	V-	T2	E+2	E-2	T*2	V+2	R2	R1	V+	T*1	.			SFC2747KM	UA747DM	0	SG747D
DIL-14/1C	E-1	E+1	T1	V-	T2	E+2	E-2	T*2	V+2	R2	R1	V+	T*1	.			SFC2747KM	UA747DM	0	SG747J
TO5-10/1M	R1	V+1	E-1	E+1	V-	E+2	E-2	V+2	R2	N	.	.					SFC2747M	UA747HM	0	SG747T
DIL-14/1C	N	N	FT	E-	E+	V-	N	N	T*	R	V+	F*	N	N			SN72748J	UA748DC	0	SG748CD
DIL-14/1C	N	N	FT	E-	E+	V-	N	N	T*	R	V+	F*	N	N			SN72748J	UA748DC	0	SG748CJ
DIL-8/1P	FT	E-	E+	V-	T*	R	V+	F*	.	.							TBB0748B	UA748TC	0	SG748CM
DIL-14/1P	N	N	FT	E-	E+	V-	N	N	T*	R	V+	F*	N	N			SN72748J	UA748DC	0	SG748CN
TO5-8/1M	FT	E-	E+	V-	T*	R	V+	F*	.	.							TBB0748	UA748HC	0	SG748CT
DIL-14/1C	N	N	FT	E-	E+	V-	N	N	T*	R	V+	F*	N	N			SN52748JA	UA748DM	0	SG748D
FLP-10/3C	N	FT	E-	E+	V-	T*	R	V+	F*	N	.	.					SN52748FA	UA748FM	0	SG748F
DIL-14/1C	N	N	FT	E-	E+	V-	N	N	T*	R	V+	F*	N	N			SN52748JA	UA748DM	0	SG748J
TO5-8/1M	FT	E-	E+	V-	T*	R	V+	F*	.	.							TBC0748	UA748HM	0	SG748T
DIL-14/1C	N	N	TF	E-	E+	V-	N	N	T*	R	V+	F*	N	N			.	UA777DC	0	SG777CJ
DIL-8/1P	TF	E-	E+	V-	T*	R	V+	F*	.	.							.	UA777TC	0	SG777CM
DIL-14/1P	N	N	TF	E-	E+	V-	N	N	T*	R	V+	F*	N	N			.	UA777DC	0	SG777CN
TO5-8/1M	TF	E-	E+	V-	T*	R	V+	F*	.	.							.	UA777HC	0	SG777T
TO5-8/1M	T	E-	E+	V+	T*	R	V+	N	.	.	?	.					.	MC1436CG	0	SG1436CT
DIL-8/1P	T	E-	E+	V-	T*	R	V+	N	.	.							.	.	0	SG1436M
TO5-8/1M	T	E-	E+	V+	T*	R	V+	N	.	.							.	MC1436G	0	SG1436T
TO5-8/1M	T	E-	E+	V-	T*	R	V+	N	.	.							.	MC1456CG	0	SG1456CT
TO5-8/1M	T	E-	E+	V-	T*	R	V+	N	.	.							.	MC1456G	0	SG1456T
DIL-8/1P	R1	E-1	E+1	V-	E+2	E-2	R2	V+	.	.							.	MC1458CP1	0	SG1458CM
DIL-8/1P	R1	E-1	E+1	V-	E+2	E-2	R2	V+	.	.							.	MC1458P1	0	SG1458M
TO5-8/1M	R1	E-1	E+1	V-	E+2	E-2	R2	V+	.	.							TBB1458	MC1458G	0	SG1458T
TO5-8/1M	T	E-	E+	V+	T*	R	V+	N	.	.							LM143	MC1536G	0	SG1536T
TO5-8/1M	T	E-	E+	V-	T*	R	V+	N	.	.							MC1556T	MC1556G	0	SG1556T
TO5-8/1M	R1	E-1	E+1	V-	E+2	E-2	R2	V+	.	.							LM1558H	MC1558G	0	SG1558T
DIL-14/1C	R2	R1	V+	E-1	E+1	E-2	E+2	E+3	E-3	E-4	E+4	G	R4	R3			LM3302J	MC3302L	0	SG3302J

TYPE NUMBER	MFR	APP	CMP	GBP MIN	SLEW RATE MIN	V_S^+ MAX	V_S^- MAX	T_{op} MAX	A_{VOL} MIN	V_{IO} MAX	I_B MAX	I_{IO} MAX	P_{TOT} MAX	I_{OUT} MIN	V_{OUT} MIN	V_{ICM} MAX	V_{IDF} MAX	dV_{IO}/dT MAX	P_Q MAX	I_Q MAX	CMRR MIN	PSRR MIN	R_{IN} MIN
SG3302J	SKU	QCP	EXT	.	20V/uS	+14V	-14V	85C	66dB	20MV	500NA	100NA	900MWF	2MA	.	9V	28V	.	.	2MA	.	.	.
SG3302N	SKU	QCP	EXT	.	20V/uS	+14V	-14V	85C	66dB	20MV	500NA	100NA	900MWF	2MA	.	9V	28V	.	.	2MA	.	.	.
SG4250CM	SKU	PRA	INT	.	.	+18V	-18V	70C	95dB	6MV	75NA	20NA	500MWF	1MA	12V	15V	30V	.	3MW	.1MA	70dB	74dB	.
SG4250CT	SKU	PRA	INT	.	.	+18V	-18V	70C	95dB	6MV	75NA	20NA	500MWF	1MA	12V	15V	30V	.	3MW	.1MA	70dB	74dB	.
SG4250T	SKU	PRA	INT	.	.	+18V	-18V	125C	100dB	5MV	50NA	10NA	500MWF	1MA	12V	15V	30V	.	2.7MW	90UA	70dB	76dB	.
SI-1050GS	SAJ	HPO	INT	0.1V/uS	.	+40V	-40V	80C	70dB	60MV	10uA	2uA	20WH	7A	30V	32V	6V	500uV/C	.	50MA	50dB	50dB	1K
SL701B	PLG	WBA	EXT	20MHZ	.	+14V	-14V	100C	66dB	5MV	1uA	0.3uA	600MWF	.	4V	.	1V	50uV/C	225MW	8MA	70dB	60dB	30K
SL701C	PLG	WBA	EXT	20MHZ	.	+14V	-14V	100C	66dB	20MV	3uA	1.8uA	600MWF	.	4V	.	1V	50uV/C	225MW	8MA	60dB	60dB	30K
SL702B	PLG	WBA	EXT	20MHZ	.	+14V	-14V	100C	66dB	5MV	1uA	0.3uA	600MWF	.	4V	.	1V	50uV/C	225MW	8MA	70dB	60dB	30K
SL702C	PLG	WBA	EXT	20MHZ	.	+14V	-14V	100C	66dB	20MV	3uA	1.8uA	600MWF	.	4V	.	1V	50uV/C	225MW	8MA	60dB	60dB	30K
SL717A	PLG	DCP	EXT	.	.	+14V	-7VV	70C	70dB	3MV	15uA	4uA	.	.	.	1V	5V	5V	.	35MA	70dB	.	.
SL717C	PLG	DCP	EXT	.	.	+14V	-7V	70C	66dB	15MV	30uA	15uA	.	.	.	5V	5V	.	.	35MA	70dB	.	.
SL751B/E	PLG	WBA	EXT	20MHZ	.	+14V	-14V	100C	66dB	5MV	1uA	0.3uA	600MWF	.	4V	.	1V	50uV/C	225MW	8MA	70dB	60dB	30K
SL751B/F	PLG	WBA	EXT	20MHZ	.	+14V	-14V	100C	66dB	5MV	1uA	0.3uA	600MWF	.	4V	.	1V	50uV/C	225MW	8MA	70dB	60dB	30K
SL751C/E	PLG	WBA	EXT	20MHZ	.	+14V	-14V	100C	66dB	20MV	3uA	1.8uA	600MWF	.	4V	.	1V	50uV/C	225MW	8MA	60dB	60dB	30K
SL751C/F	PLG	WBA	EXT	20MHZ	.	+14V	-14V	100C	66dB	20MV	3uA	1.8uA	600MWF	.	4V	.	1V	50uV/C	225MW	8MA	60dB	60dB	30K
SN52L022JP	TGU	DLP	INT	.3MHZ	0.2V/uS	+22V	-22V	125C	72dB	5MV	100NA	40NA	500MWF	1MA	10V	15V	30V	.	3MW	.2MA	60dB	76dB	.
SN52L022L	TGU	DLP	INT	.3MHZ	0.2V/uS	+22V	-22V	125C	72dB	5MV	100NA	40NA	500MWF	1MA	10V	15V	30V	.	3MW	.2MA	60dB	76dB	.
SN52L044JA	TGU	QLP	INT	.3MHZ	.15V/uS	+22V	-22V	125C	72dB	5MV	100NA	40NA	500MWF	1MA	10V	15V	30V	.	3MW	.1MA	60dB	76dB	.
SN5510FA	TGU	BDO	EXT	.1GHZ	.	+8V	-8V	100C	38dB	13MV	80uA	20uA	.	.	.	1V	5V	.	220MW	.	75dB	.	3K
SN5510JP	TGU	BDO	EXT	.1GHZ	.	+8V	-8V	100C	38dB	13MV	80uA	20uA	.	.	.	1V	5V	.	220MW	.	75dB	.	3K
SN5510L	TGU	BDO	EXT	.1GHZ	.	+8V	-8V	100C	38dB	13MV	80uA	20uA	.	.	.	1V	5V	.	220MW	.	75dB	.	3K
SN5511FA	TGU	BDO	EXT	50MHZ	.	+8V	-8V	125C	56dB	5MV	15uA	7uA	500MWF	2MA	1.2V	6V	6V	10uV/C	300MW	.	59dB	.	2K
SN5511L	TGU	BDO	EXT	50MHZ	.	+8V	-8V	125C	56dB	5MV	15uA	7uA	500MWF	2MA	1.2V	6V	6V	10uV/C	300MW	.	59dB	.	2K
SN5511N	TGU	BDO	EXT	50MHZ	.	+8V	-8V	125C	56dB	5MV	15uA	7uA	500MWF	2MA	1.2V	6V	6V	10uV/C	300MW	.	59dB	.	2K
SN5512L	TGU	BDO	EXT	.2GHZ	.	+8V	-8V	125C	48dB	5MV	80uA	3uA	500MWF	2MA	1.5V	6V	5V	.	.	25MA	74dB	50dB	2K
SN5512N	TGU	BDO	EXT	.2GHZ	.	+8V	-8V	125C	48dB	5MV	80uA	3uA	500MWF	2MA	1.5V	6V	5V	.	.	25MA	74dB	50dB	2K
SN5514JP	TGU	BDO	EXT	.2GHZ	.	+8V	-8V	125C	48dB	5MV	80uA	3uA	500MWF	2MA	1.5V	6V	5V	.	.	25MA	74dB	50dB	2K
SN5514L	TGU	BDO	EXT	.2GHZ	.	+8V	-8V	125C	48dB	5MV	80uA	3uA	500MWF	2MA	1.5V	6V	5V	.	.	25MA	74dB	50dB	2K
SN52101AFA	TGU	GPU	EXT	.	.	+22V	-22V	125C	94dB	2MV	75NA	10NA	500MWF	5MA	12V	15V	30V	15uV/C	.	3MA	80dB	80dB	1.5M
SN52101AJ	TGU	GPU	EXT	.	.	+22V	-22V	125C	94dB	2MV	75NA	10NA	500MWF	5MA	12V	15V	30V	15uV/C	.	3MA	80dB	80dB	1.5M
SN52101AJA	TGU	GPU	EXT	.	.	+22V	-22V	125C	94dB	2MV	75NA	10NA	500MWF	5MA	12V	15V	30V	15uV/C	.	3MA	80dB	80dB	1.5M
SN52101AJP	TGU	GPU	EXT	.	.	+22V	-22V	125C	94dB	2MV	75NA	10NA	500MWF	5MA	12V	15V	30V	15uV/C	.	3MA	80dB	80dB	1.5M
SN52101AL	TGU	GPU	EXT	.	.	+22V	-22V	125C	94dB	2MV	75NA	10NA	500MWF	5MA	12V	15V	30V	15uV/C	.	3MA	80dB	80dB	1.5M
SN52106FA	TGU	CPR	EXT	.	.	+15V	-15V	125C	84dB	3MV	45uA	7uA	600MWF	.	2.5V	.	.	10uV/C	163MW	.	.	.	.
SN52106J	TGU	CPR	EXT	.	.	+15V	-15V	125C	84dB	3MV	45uA	7uA	600MWF	.	2.5V	.	.	10uV/C	163MW	.	.	.	.
SN52106JP	TGU	CPR	EXT	.	.	+15V	-15V	125C	84dB	3MV	45uA	7uA	600MWF	.	2.5V	.	.	10uV/C	163MW	.	.	.	.
SN52106L	TGU	CPR	EXT	.	.	+15V	-15V	125C	84dB	3MV	45uA	7uA	600MWF	.	2.5V	.	.	10uV/C	163MW	.	.	.	.
SN52107FA	TGU	GPK	INT	.	.	+22V	-22V	125C	94dB	2MV	75NA	10NA	500MWF	5MA	12V	15V	30V	15uV/C	.	3MA	80dB	80dB	1.5M
SN52107JA	TGU	GPK	INT	.	.	+22V	-22V	125C	94dB	2MV	75NA	10NA	500MWF	5MA	12V	15V	30V	15uV/C	.	3MA	80dB	80dB	1.5M
SN52107JP	TGU	GPK	INT	.	.	+22V	-22V	125C	94dB	2MV	75NA	10NA	500MWF	5MA	12V	15V	30V	15uV/C	.	3MA	80dB	80dB	1.5M
SN52107L	TGU	GPK	INT	.	.	+22V	-22V	125C	94dB	2MV	75NA	10NA	500MWF	5MA	12V	15V	30V	15uV/C	.	3MA	80dB	80dB	1.5M
SN52108AFA	TGU	SBA	EXT	.	.	+20V	-20V	125C	98dB	0.5MV	2NA	0.2NA	500MWF	1MA	13V	15V	1V	5uV/C	.	.6MA	96dB	96dB	30M
SN52108AJA	TGU	SBA	EXT	.	.	+20V	-20V	125C	98dB	0.5MV	2NA	0.2NA	500MWF	1MA	13V	15V	1V	5uV/C	.	.6MA	96dB	96dB	30M
SN52108AJP	TGU	SBA	EXT	.	.	+20V	-20V	125C	98dB	0.5MV	2NA	0.2NA	500MWF	1MA	13V	15V	1V	5uV/C	.	.6MA	96dB	96dB	30M
SN52108AL	TGU	SBA	EXT	.	.	+20V	-20V	125C	98dB	0.5MV	2NA	0.2NA	500MWF	1MA	13V	15V	1V	5uV/C	.	.6MA	96dB	96dB	30M
SN52108FA	TGU	SBA	EXT	.	.	+20V	-20V	125C	94dB	2MV	2NA	0.2NA	500MWF	1MA	13V	15V	1V	15uV/C	.	.6MA	85dB	80dB	30M
SN52108JA	TGU	SBA	EXT	.	.	+20V	-20V	125C	94dB	2MV	2NA	0.2NA	500MWF	1MA	13V	15V	1V	15uV/C	.	.6MA	85dB	80dB	30M
SN52108JP	TGU	SBA	EXT	.	.	+20V	-20V	125C	94dB	2MV	2NA	0.2NA	500MWF	1MA	13V	15V	1V	15uV/C	.	.6MA	85dB	80dB	30M
SN52108L	TGU	SBA	EXT	.	.	+20V	-20V	125C	94dB	2MV	2NA	0.2NA	500MWF	1MA	13V	15V	1V	15uV/C	.	.6MA	85dB	80dB	30M
SN52110FA	TGU	VFA	INT	15V/uS	.	+18V	-18V	125C	0dB	4MV	3NA	.	500MWF	1MA	10V	15V	15V	50uV/C	.	6MA	.	70dB	10G
SN52110JA	TGU	VFA	INT	15V/uS	.	+18V	-18V	125C	0dB	4MV	3NA	.	500MWF	1MA	10V	15V	15V	50uV/C	.	6MA	.	70dB	10G
SN52110JP	TGU	VFA	INT	15V/uS	.	+18V	-18V	125C	0dB	4MV	3NA	.	500MWF	1MA	10V	15V	15V	50uV/C	.	6MA	.	70dB	10G
SN52110L	TGU	VFA	INT	15V/uS	.	+18V	-18V	125C	0dB	4MV	3NA	.	500MWF	1MA	10V	15V	15V	50uV/C	.	6MA	.	70dB	10G
SN52111FA	TGU	CPR	EXT	.	.	+18V	-18V	125C	100dB	3MV	100NA	10NA	500MWF	.	.	15V	30V	.	.	6MA	.	.	.
SN52111J	TGU	CPR	EXT	.	.	+18V	-18V	125C	100dB	3MV	100NA	10NA	500MWF	.	.	15V	30V	.	.	6MA	.	.	.
SN52111JP	TGU	CPR	EXT	.	.	+18V	-18V	125C	100dB	3MV	100NA	10NA	500MWF	.	.	15V	30V	.	.	6MA	.	.	.
SN52111L	TGU	CPR	EXT	.	.	+18V	-18V	125C	100dB	3MV	100NA	10NA	500MWF	.	.	15V	30V	.	.	6MA	.	.	.
SN52118FA	TGU	XSR	INT	50V/uS	.	+20V	-20V	125C	94dB	4MV	250NA	50NA	500MWF	6MA	12V	15V	1V	.	.	8MA	80dB	70dB	1M
SN52118JA	TGU	XSR	INT	50V/uS	.	+20V	-20V	125C	94dB	4MV	250NA	50NA	500MWF	6MA	12V	15V	1V	.	.	8MA	80dB	70dB	1M

For detailed explanations of column heading notations, see App. A.

Also for ready references the more important abbreviations used in the column headings are listed below:

LEFT HAND PAGE

APP = application (codes at APP.E.)

CMRR = common mode rejection ratio

CMP = compensation (frequency)

dV_{io}/dT = input offset voltage temperature drift

GBP = gain bandwidth product

I_B = input bias current

I_{io} = input bias offset current

I_o = quiescent supply current

MFR = manufacturer (codes at App.C.)

P_o = quiescent power consumer

PSRR = power supply rejection ratio

V_{icm} = common mode input voltage rating

V_{idf} = differential input voltage rating

V_{io} = input offset voltage

V_s = dc supply voltage

RIGHT HAND PAGE

Lead out coding summary (details at APP.G.) for different cases, (APP.F.)

A = gain adjust

B = bias adjust

C = case

E− = inverting input

E+ = non-inverting input

F,F* = input frequency compensation

G = ground

J = high level input

K = output, open collector

L = output, open emitter

M = metal case

N = not connected

Q = special terminal

R,R* = outputs

S = strobe

T,T* = offset balance

V+ = +ve dc supply

V− = −ve dc supply

W = guard ring

X = blank position, no lead

++ = +ve supplementary dc supply

−− = −ve supplementary dc supply

∮,∮* = output frequency compensation

CASE (APP F)	LD 1	LD 2	LD 3	LD 4	LD 5	LD 6	LD 7	LD 8	LD 9	LD 10	LD 11	LD 12	LD 13	LD 14	LD 15	LD 16	EUROPE SUBSTITUTE	USA SUBSTITUTE	IS	TYPE NUMBER
DIL-14/1P	R2	R1	V+	E-1	E+1	E-2	E+2	E+3	E-3	E-4	E+4	G	R4	R3			LM3302J	MC3302P	0	SG3302N
DIL-8/1P	T	E-	E+	V-	T*	R	V+	B									LM4250J	LM4250CN	0	SG4250CM
T05-8/1M	T	E-	E+	V-	T*	R	V+	B										LM4250CH	0	SG4250CT
T05-8/1M	T	E-	E+	V-	T*	R	V+	B										LM4250H	0	SG4250T
SIH-10/1M	F	F*	E+T	V-	V-	E-	R	N	V+	N									0	SI-1050GS
T05-8/1M	G	Q	V+	ø	E-	F	E+	V-											0	SL701B
T05-8/1M	G	Q	V+	ø	E-	F	E+	V-										SL701B	0	SL701C
T05-8/1M	G	L	V+	ø	E-	F	E+	V-											0	SL702B
T05-8/1M	G	L	V+	ø	E-	F	E+	V-										SL702B	0	SL702C
T05-10/2M	G	R2	E-2	E+2	V-	E+1	E-1	R1	V+	N									0	SL717A
T05-10/1M	G	R2	E-2	E+2	V-	E+1	E-1	R1	V+	N								SL717A	0	SL717C
DIL-14/1C	F	N	N	E+	V-	N	G	Q	L	N	V+	ø	N	E-					0	SL751B/E
FLP-14/3G	V-	G	N	Q	L	N	V+	ø	N	E-	N	F	N	E+					0	SL751B/F
DIL-14/1C	F	N	N	E+	V-	N	G	Q	L	N	V+	ø	N	E-				SL751B/E	0	SL751C/E
FLP-14/3G	V-	G	N	Q	L	N	V+	ø	N	E-	N	F	N	E+				SL751B/F	0	SL751C/F
DIL-8/1C	R1	E-1	E+1	V-	E+2	E-2	R2	V+											0	SN52L022JP
T05-8/1M	R1	E-1	E+1	V-	E+2	E-2	R2	V+											0	SN52L022L
DIL-16/1C	R1	E-1	E+1	V-	E+2	E-2	R2	V+	R3	E-3	E+3	V-	E+4	E-4	R4	V+*			0	SN52L044JA
FLP-10/3C	E	N	V+	N	E*	R*	G	N	R	V-									0	SN5510FA
DIL-8/1C	E	V+	E*	N	R*	G	R	V-											0	SN5510JP
T05-8/1M	E	V+	E*	M	R*	G	R	V-											0	SN5510L
FLP-10/3C	F	E	V+	E*	F*	R*	B	G	V-	R									0	SN5511FA
T05-10/1M	F	E	V+	E*	F*	R*	B	G	V-	R									0	SN5511L
DIL-14/1P	N	F	E	V+	E*	F*	N	R*	B	G	V-	R	N						0	SN5511N
T05-10/1M	T*	E	V+	E*	N	R*	G	R	V-M	T									0	SN5512L
DIL-14/1P	N	T*	E	V+	E*	N	N	R*	G	R	V-	T	N						0	SN5512N
DIL-8/1C	E	V+	E*	N	R*	G	R	V-											0	SN5514JP
T05-8/1M	E	V+	E*	M	R*	G	R	V-											0	SN5514L
FLP-14/3C	N	N	FT	E-	E+	V-	N	N	T*	R	V+	F*	N	N					0	SN52101AFA
DIL-14/1C	N	N	FT	E-	E+	V-	N	N	T*	R	V+	F*	N	N			UA101AD	LM101AJ14	0	SN52101AJ
DIL-14/1C	N	N	FT	E-	E+	V-	N	N	T*	R	V+	F*	N	N			UA101AD	LM101AJ14	0	SN52101AJA
DIL-8/1C	FT	E-	E+	V-	T*	R	V+	F*											0	SN52101AJP
T05-8/1M	FT	E-	E+	V-M	T*	R	V+	F*									SFC2101A	LM101AH	0	SN52101AL
FLP-14/3C	N	G	E+	E-	N	V-	S1	S2	R	N	V+	N	N	N				LM106F	0	SN52106FA
DIL-14/1C	N	G	E+	E-	N	V-	S1	S2	R	N	V+	N	N	N					0	SN52106J
DIL-8/1C	G	E+	E-	V-	S1	S2	R	V+											0	SN52106JP
T05-8/1M	G	E+	E-	V-M	S1	S2	R	V+										LM106H	0	SN52106L
FLP-14/3C	N	N	N	E-	E+	V-	N	N	N	N	V+	N	N	N					0	SN52107FA
DIL-14/1C	N	N	N	E-	E+	V-	N	N	N	R	V+	N	N					LM107D	0	SN52107JA
DIL-8/1C	N	E-	E+	V-	N	R	V+	N										LM107J	0	SN52107JP
T05-8/1M	N	E-	E+	V-M	N	R	V+	N									SFC2107M	LM107H	0	SN52107L
FLP-10/1C	N	N-	E-	E+	N	V-	R	V+	F*									LM108AF	0	SN52108AFA
DIL-14/1C	N	F	N	E-	E+	V-	N	N	F	R	V+	F*	N	N			UA108AD	LM108AD	0	SN52108AJA
DIL-8/1C	F	E-	E+	V-	N	R	V+	F*											0	SN52108AJP
T05-8/1M	F	E-	E+	V-M	N	R	V+	F*									SFC2108A	LM108AH	0	SN52108AL
FLP-10/1C	N	N	E-	E+	N	V-	R	V+	F	F*							SFC2108PM	LM108F	0	SN52108FA
DIL-14/1C	N	F	N	E-	E+	V-	N	N	R	V+	F*	N	N				UA108D	LM108D	0	SN52108JA
SIH-8/1C	F	E-	E+	V-	N	R	V+	F*											0	SN52108JP
T05-8/1M	F	E-	E+	V-M	N	R	V+	F*									SFC2108M	LM108H	0	SN52108L
FLP-14/3C	N	N	T	N	E+	V-	N	N	L	N	V+	T*	N	N					0	SN52110FA
DIL-14/1C	N	N	T	N	E+	V-	N	N	L	R	V+	T*	N	N				LM110D	0	SN52110JA
DIL-8/1C	T	N	E+	V-	L	R	V+	T*										SN52110P	0	SN52110JP
T05-8/1M	T	N	E+	V-	L	R	V+	T*									UA110M	LM110H	0	SN52110L
FLP-10/3C	G	E+	E-	N	V-	T	T*S	R	N									LM111F	0	SN52111FA
DIL-14/1C	N	G	E+	E-	N	V-	T	T*S	R	N	V+	N	N					LM111D	0	SN52111J
DIL-8/1C	G	E+	E-	V-	T	T*S	R	V+										UA111R	0	SN52111JP
T05-8/1M	G	E+	E-	V-	T	T*S	R	V+									SFC2111M	LM111H	0	SN52111L
FLP-10/3C	N	N*	E-	E+	V-	F*T	R	N										LM118F	0	SN52118FA
DIL-14/1C	N	N	T*F	E-	E+	V-	N	N	F*T	R	V+	ø	N	N				LM118D	0	SN52118JA
DIL-8/1C	T*F	E-	E+	V-	F*T	R	V+	ø											0	SN52118JP

TYPE NUMBER	MFR	APP	CMP	GBP MIN	SLEW RATE MIN	Vs+ MAX	Vs- MAX	Top MAX	Ávol MIN	V10 MAX	IB MAX	I10 MAX	PTOT MAX	IOUT MIN	VOUT MIN	V'ICM MAX	VIDF MAX	dV10/dT MAX	PQ MAX	IQ MAX	CM RR MIN	PS RR MIN	RIN MIN
SN52118JP	TGU	XSR	INT	.	50V/uS	+20V	-20V	125C	94dB	4MV	250NA	50NA	500MWF	6MA	12V	15V	1V		.	8MA	80dB	70dB	1M
SN52118L	TGU	XSR	INT	.	50V/uS	+20V	-20V	125C	94dB	4MV	250NA	50NA	500MWF	6MA	12V	15V	1V	.	.	8MA	80dB	70dB	1M
SN52505FA	TGU	DCP	EXT		30V/uS	+15V	-15V	125C		2MV	20uA	3uA	600MWF	.	12V	7V	5V	10uV/C		.			
SN52506J	TGU	DCP	EXT		30V/uS	+15V	-15V	125C		2MV	20uA	3uA	600MWF	.	12V	7V	5V	10uV/C		.			
SN52510FA	TGU	CPR	EXT		.	+14V	-7V	125C		2MV	15uA	3uA	300MWF	5MA	2.5V	7V	5V	10uV/C	150MW	9MA	80dB		
SN52510J	TGU	CPR	EXT		.	+14V	-7V	125C		2MV	15uA	3uA	300MWF	5MA	2.5V	7V	5V	10uV/C	150MW	9MA	80dB		
SN52510JP	TGU	CPR	EXT		.	+14V	-7V	125C		2MV	15uA	3uA	300MWF	5MA	2.5V	7V	5V	10uV/C	150MW	9MA	80dB		
SN52510L	TGU	CPR	EXT		.	+14V	-7V	125C		2MV	15uA	3uA	300MWF	5MA	2.5V	7V	5V	10uV/C	150MW	9MA	80dB		
SN52514J	TGU	DCP	EXT		.	+14V	-7V	125C		2MV	15uA	3uA	300MWF	5MA	2.5V	7V	5V	10uV/C	150MW	9MA	80dB		
SN52558JP	TGU	DGK	INT	.3MHZ	0.2V/uS	+22V	-22V	125C	94dB	5MV	500NA	200NA	500MWF	5MA	12V	15V	30V		85MW	3MA	70dB	76dB	300k
SN52558L	TGU	DGK	INT	.3MHZ	0.2V/uS	+22V	-22V	125C	94dB	5MV	500NA	200NA	500MWF	5MA	12V	15V	30V		85MW	3MA	70dB	76dB	300K
SN52660FA	TGU	GPU	EXT		.	+20V	-20V	125C	88dB	3MV	15NA	2NA	500MWF	1MA	13V	15V	1V	25uV/C	.	.8MA	80dB	80dB	4M
SN52660JA	TGU	GPU	EXT		.	+20V	-20V	125C	88dB	3MV	15NA	2NA	500MWF	1MA	13V	15V	1V	25uV/C	.	.8MA	80dB	80dB	4M
SN52660JP	TGU	GPU	EXT		.	+20V	-20V	125C	88dB	3MV	15NA	2NA	500MWF	1MA	13V	15V	1V	25uV/C	.	.8MA	80dB	80dB	4M
SN52660L	TGU	GPU	EXT		.	+20V	-20V	125C	88dB	3MV	15NA	2NA	500MWF	1MA	13V	15V	1V	25uV/C	.	.8MA	80dB	80dB	4M
SN52702AFA	TGU	WBA	EXT		0.5V/uS	+14V	-7V	125C	68dB	2MV	5uA	0.5uA	300MWF	.3MA	5V	1.5V	5V	10uV/C	120MW	7MA	80dB	74dB	16K
SN52702AJ	TGU	WBA	EXT		0.5V/uS	+14V	-7V	125C	68dB	2MV	5uA	0.5uA	300MWF	.3MA	5V	1.5V	5V	10uV/C	120MW	7MA	80dB	74dB	16K
SN52702AL	TGU	WBA	EXT		0.5V/uS	+14V	-7V	125C	68dB	2MV	5uA	0.5uA	300MWF	.3MA	5V	1.5V	5V	10uV/C	120MW	7MA	80dB	74dB	16K
SN52702FA	TGU	WBA	EXT		0.5V/uS	+14V	-7V	125C	63dB	5MV	10uA	2uA	300MWF	.3MA	5V	1.5V	5V	20uV/C	120MW	7MA	70dB	70dB	8K
SN52702J	TGG	WBA	EXT		0.5V/uS	+14V	-7V	125C	63dB	5MV	10uA	2uA	300MWF	.3MA	5V	1.5V	5V	20uV/C	120MW	7MA	70dB	70dB	8K
SN52702L	TGU	WBA	EXT		0.5V/uS	+14V	-7V	125C	63dB	3MV	10uA	2uA	300MWF	.3MA	5V	1.5V	5V	20uV/C	120MW	7MA	70dB	70dB	8K
SN52709AFA	TGU	GPU	EXT		.	+18V	-18V	125C	88dB	2MV	200NA	50NA	300MWF	5MA	12V	10V	5V	10uV/C	108MW	4MA	80dB	80dB	300K
SN52709AJ	TGU	GPU	EXT		.	+18V	-18V	125C	88dB	2MV	200NA	50NA	300MWF	5MA	12V	10V	5V	10uV/C	108MW	4MA	80dB	80dB	300K
SN52709AJA	TGU	GPU	EXT		.	+18V	-18V	125C	88dB	2MV	200NA	50NA	300MWF	5MA	12V	10V	5V	10uV/C	108MW	4MA	80dB	80dB	300K
SN52709AJP	TGU	GPU	EXT		.	+18V	-18V	125C	88dB	2MV	200NA	50NA	300MWF	5MA	12V	10V	5V	10uV/C	108MW	4MA	80dB	80dB	300K
SN52709AL	TGU	GPU	EXT		.	+18V	-18V	125C	88dB	2MV	200NA	50NA	300MWF	5MA	12V	10V	5V	10uV/C	108MW	4MA	80dB	80dB	300K
SN52709FA	TGU	GPU	INT		.	+18V	-18V	125C	88dB	5MV	500NA	200NA	300MWF	5MA	12V	10V	5V	15uV/C	165MW	6MA	70dB	76dB	150K
SN52709J	TGU	GPU	EXT		.	+18V	-18V	125C	88dB	5MV	500NA	200NA	300MWF	5MA	12V	10V	5V	15uV/C	165MW	6MA	70dB	76dB	150K
SN52709JA	TGU	GPU	EXT		.	+18V	-18V	125C	88dB	5MV	500NA	200NA	300MWF	5MA	12V	10V	5V	15uV/C	165MW	6MA	70dB	76dB	150K
SN52709JP	TGU	GPU	EXT		.	+18V	-18V	125C	88dB	5MV	500NA	200NA	300MWF	5MA	12V	10V	5V	15uV/C	165MW	6MA	70dB	76dB	150K
SN52709L	TGU	GPU	EXT		.	+18V	-18V	125C	88dB	5MV	500NA	200NA	300MWF	5MA	12V	10V	5V	15uV/C	165MW	6MA	70dB	76dB	150K
SN52710FA	TGU	CPR	EXT		.	+14V	-7V	125C	57dB	5MV	75uA	10uA	300MWF	5MA	1V	7V	5V	10uV/C	175MW	10MA	70dB		
SN52710J	TGU	CPR	EXT		.	+14V	-7V	125C	57dB	5MV	75uA	10uA	300MWF	5MA	1V	7V	5V	10uV/C	175MW	10MA	70dB		
SN52710JP	TGU	CPR	EXT		.	+14V	-7V	125C	57dB	5MV	75uA	10uA	300MWF	5MA	1V	7V	5V	10uV/C	175MW	10MA	70dB		
SN52710L	TGU	CPR	EXT		.	+14V	-7V	125C	57dB	5MV	75uA	10uA	300MWF	5MA	1V	7V	5V	10uV/C	175MW	10MA	70dB		
SN52711FA	TGU	DCP	EX		.	+14V	-7V	125C	58dB	3.5MV	75uA	10uA	300MWF	5MA	2.5V	7V	5V	20uV/C	200MW	.	70dB		
SN52711J	TGU	DCP	EXT		.	+14V	-7V	125C	58dB	3.5MV	75uA	10uA	300MWF	5MA	2.5V	7V	5V	20uV/C	200MW	.	70dB		
SN52711L	TGU	DCP	EXT		.	+14V	-7V	125C	58dB	3.5MV	75uA	10uA	300MWF	5MA	2.5V	7V	5V	20uV/C	200MW	.	70dB		
SN52733FA	TGU	BDO	EXT	20MHZ		+8V	-8V	125C	50dB	5MV	20uA	3uA	500MWF	2MA	2V	6V	5V		.	24MA	60dB	50dB	2K
SN52733J	TGU	BDO	EXT	20MHZ		+8V	-8V	125C	50dB	5MV	20uA	3uA	500MWF	2MA	2V	6V	5V		.	24MA	60dB	50dB	2K
SN52733L	TGU	BDO	EXT	20MHZ		+8V	-8V	125C	50dB	5MV	20uA	3uA	500MWF	2MA	2V	6V	5V		.	24MA	60dB	50dB	2K
SN52741FA	TGU	GPK	INT	.	0.2V/uS	+22V	-22V	125C	94dB	5MV	500NA	200NA	500MWF	5MA	12V	15V	30V		85MW	3MA	70dB	76dB	300K
SN52741J	TGU	GPK	INT	.	0.2V/uS	+22V	-22V	125C	94dB	5MV	500NA	200NA	500MWF	5MA	12V	15V	30V		85MW	3MA	70dB	76dB	300K
SN52741JA	TGU	GPK	INT	.	0.2V/uS	+22V	-22V	125C	94dB	5MV	500NA	200NA	500MWF	5MA	12V	15V	30V		85MW	3MA	70dB	76dB	300K
SN52741JP	TGU	GPK	INT	.	0.2V/uS	+22V	-22V	125C	94dB	5MV	500NA	200NA	500MWF	5MA	12V	15V	30V		85MW	3MA	70dB	76dB	300K
SN52741L	TGU	GPK	INT	.	0.2V/uS	+22V	-22V	125C	94dB	5MV	500NA	200NA	500MWF	5MA	12V	15V	30V		85MW	3MA	70dB	76dB	300K
SN52741N	TGU	GPK	INT	.	0.2V/uS	+22V	-22V	125C	94dB	5MV	500NA	200NA	500MWF	5MA	12V	15V	30V		85MW	3MA	70dB	76dB	300K
SN52741P	TGU	GPK	INT	.	0.2V/uS	+22V	-22V	125C	94dB	5MV	500NA	200NA	500MWF	5MA	12V	15V	30V		85MW	3MA	70dB	76dB	300K
SN52747FA	TGU	DGK	INT	.	0.2V/uS	+22V	-22V	125C	94dB	5MV	500NA	200NA	500MWF	5MA	12V	15V	30V		85MW	3MA	70dB	76dB	300K
SN52747J	TGU	DGK	INT	.	0.2V/uS	+22V	-22V	125C	94dB	5MV	500NA	200NA	500MWF	5MA	12V	15V	30V		85MW	3MA	70dB	76dB	300K
SN52747JA	TGU	DGK	INT	.	0.2V/uS	+22V	-22V	125C	94dB	5MV	500NA	200NA	500MWF	5MA	12V	15V	30V		85MW	3MA	70dB	76dB	300K
SN52747L	TGU	DGK	INT	.	0.2V/uS	+22V	-22V	125C	94dB	5MV	500NA	200NA	500MWF	5MA	12V	15V	30V		85MW	3MA	70dB	76dB	300K
SN52748FA	TGU	GPU	EXT	.	0.2V/uS	+22V	-22V	125C	94dB	5MV	500NA	200NA	500MW	5MA	10V	15V	30V		85MW	3MA	70dB	76dB	300K
SN52748J	TGU	GPU	EXT	.	0.2V/uS	+22V	-22V	125C	94dB	5MV	500NA	200NA	500MWF	5MA	10V	15V	30V		85MW	3MA	70dB	76dB	300K
SN52748JA	TGU	GPU	EXT	.	0.2V/uS	+22V	-22V	125C	94dB	5MV	500NA	200NA	500MWF	5MA	10V	15V	30V		85MW	3MA	70dB	76dB	300K
SN52748JP	TGU	GPU	EXT	.	0.2V/uS	+22V	-22V	125C	94dB	5MV	500NA	200NA	500MWF	5MA	10V	15V	30V		85MW	3MA	70dB	76dB	300K
SN52748L	TGU	GPU	EXT	.	0.2V/uS	+22V	-22V	125C	94dB	5MV	500NA	200NA	500MWF	5MA	10V	15V	30V		85MW	3MA	70dB	76dB	300K
SN52770FA	TGU	HSR	EXT	.5MHZ	1V/uS	+22V	-22V	125C	94dB	4MV	15NA	2NA	500MWF	6MA	12V	15V	30V		60MW	2MA	80dB	76dB	30M
SN52770JA	TGU	HSR	EXT	.5MHZ	1V/uS	+22V	-22V	125C	94dB	4MV	15NA	2NA	500MWF	6MA	12V	15V	30V		60MW	2MA	80dB	76dB	30M
SN52770JP	TGU	HSR	EXT	.5MHZ	1V/uS	+22V	-22V	125C	94dB	4MV	15NA	2NA	500MWF	6MA	12V	15V	30V		60MW	2MA	80dB	76dB	30M

Also for ready references the more important abbreviations used in the column headings are listed below:

LEFT HAND PAGE

APP = application (codes at APP.E.)

CMRR = common mode rejection ratio

CMP = compensation (frequency)

dV_{IO}/dT = input offset voltage temperature drift

GBP = gain bandwidth product

I_B = input bias current

I_{IO} = input bias offset current

I_Q = quiescent supply current

MFR = manufacturer (codes at App.C.)

P_Q = quiescent power consumer

PSRR = power supply rejection ratio

V_{ICM} = common mode input voltage rating

V_{IDF} = differential input voltage rating

V_{IO} = input offset voltage

V_S = dc supply voltage

RIGHT HAND PAGE

Lead out coding summary (details at APP.G.) for different cases (APP.F.)

A = gain adjust
B = bias adjust
C = case
E— = inverting input
E+ = non-inverting input
F,F* = input frequency compensation
G = ground
J = high level input
K = output, open collector
L = output, open emitter
M = metal case
N = not connected
Q = special terminal
R,R* = outputs
S = strobe
T,T* = offset balance
V+ = +ve dc supply
V- = -ve dc supply
W = guard ring
X = blank position, no lead
++ = +ve supplementary dc supply
—— = -ve supplementary dc supply
$\emptyset,\emptyset^*$ = output frequency compensation

CASE (APP F)	LD 1	LD 2	LD 3	LD 4	LD 5	LD 6	LD 7	LD 8	LD 9	LD 10	LD 11	LD 12	LD 13	LD 14	LD 15	LD 16	EUROPE SUBSTITUTE	USA SUBSTITUTE	IS	TYPE NUMBER
T05-8/1M	T*F	E-	E+	V-	F*T	R	V+	∅									TDC0118CM	LM118H	0	SN52118L
FLP-14/3C	S1	E-1	E+1	V-	E+2	E-2	S*2	N	R2	V+	R1	G	S*1						0	SN52506FA
DIL-14/1C	S1	E-1	E+1	V-	E+2	E-2	S*2	N	R2	V+	R1	G	S*1						0	SN52506J
FLP-10/3C	G	E+	E-	N	V-	R	S	V+	N	N									G	SN52510FA
DIL-14/1C	N	G	E+	E-	N	V-	N	N	R	S	V+	N	N	N					0	SN52510J
DIL-8/1C	G	E+	E-	V-	N	S	R	V+											0	SN52510JP
T05-8/1M	G	E+	E-	V-M	N	S	R	V+											0	SN52510L
DIL-14/1C	R1	S1	V+	N	E+2	E-2	V-	R2	S2	V+	G	E+1	E-1	V-					0	SN52514J
DIL-8/1C	R	E-1	E+1	V-	E+2	E-2	R2	V+									LM1558J	MC1558U	0	SN52558JP
T05-8/1M	R1	E-1	E+1	V-M	E+2	E-2	R2	V+									TBC1458	MC1558G	0	SN52558L
FLP-10/3C	N	N	E-	E+	N	V-	R	V+	F*	F							SFC2108PM	LM108F	0	SN52660FA
DIL-14/1C	N	F	N	E-	E+	N	N	N	N	R	V+	F*	N	N			UA108D	LM108D	0	SN52660JA
DIL-8/1C	F	E-	E+	V-	N	R	V+	F*											G	SN52660JP
T05-8/1M	F	E-	E+	V-M	N	R	V+	F*									SFC2108M	LM108H	0	SN52660L
FLP-10/3C	N	G	E-	E+	V-	F	∅	R	N	V+									0	SN52702AFA
DIL-14/1C	N	N	G	E-	E+	V-	N	N	F	∅	R	N	V+	N				UA702ADM	0	SN52702AJ
T05-8/1M	G	E-	E+	V-M	F	∅	R	V+										UA702AHM	0	SN52702AL
FLP-10/3C	N	G	E-	E+	V-	F	∅	R	N	V+							MC1712F	UA702FM	0	SN52702FA
DIL-14/1C	N	N	G	E-	E+	V-	N	N	F	∅	R	N	V+	N			MC1712L	UA702DM	0	SN52702J
T05-8/1M	G	E-	E+	V-M	F	∅	R	V+									MC1712G	UA702HM	0	SN52702L
FLP-10/3C	N	F	E-	E+	V-	∅	R	V+	F*	N								UA709AFM	0	SN52709AFA
DIL-14/1C	N	N	F	E-	E+	V-	N	N	∅	R	V+	F*	N	N			LM709AJ	UA709ADM	0	SN52709AJ
DIL-14/1C	N	N	F	E-	E+	V-	N	N	∅	R	V+	F*	N	N			LM709AJ	UA709ADM	0	SN52709AJA
DIL-8/1C	F	E-	E+	V-	∅	∅*R	V+	F*											0	SN52709AJP
T05-8/1M	F	E-	E+	V-M	∅	∅*R	V+	F*									MC1709AC	UA709AHM	0	SN52709AL
FLP-10/3C	N	F	E-	E+	V-	∅	R	V+	F*	N							SN52709AFA	UA709FM	0	SN52709FA
DIL-14/1C	N	N	F	E-	E+	V-	N	N	∅	R	V+	F*	N	N			LM709J	UA709DM	0	SN52709J
DIL-14/1C	N	N	F	E-	E+	V-	N	N	∅	R	V+	F*	N	N			LM709J*	UA709DM	0	SN52709JA
DIL-8/1C	F	E-	E+	V-	∅	∅*R	V+	F*									MC1709U	SN52709AJP	0	SN52709JP
T05-8/1M	F	E-	E+	V-M	∅	∅*R	V+	F*									TAA522	UA709L	0	SN52709L
FLP-10/3C	G	E+	E-	N	V-	R	N	V+	N	N							SFC2710PM	UA710FM	0	SN52710FA
DIL-14/1C	N	G	E+	E-	N	V-	N	N	R	N	V+	N	N	N			SFC2710KM	UA710DM	0	SN52710J
DIL-8/1C	G	E+	E-	V-	N	N	R	V+									SFC2710PM	UA710FM	0	SN52710JP
T05-8/1M	G	E+	E-	V-M	N	N	R	V+									SFC2710M	UA710HM	0	SN52710L
FLP-10/3C	E-1	E+1	V-	E+2	E-2	S2	R	V+	G	S1							SFC2711PM	UA711FM	0	SN52711FA
DIL-14/1C	N	E-1	E+1	V-	E+2	E-2	N	N	S2	R	V+	G	S1	N			SFC2711KM	UA711DM	0	SN52711J
T05-10/1M	G	S1	E-1	E+1	V-	E+2	E-2	S2	R	V+							LM711H	UA711HM	0	SN52711L
FLP-10/3C	E+	A2	A*2	V-	R	R*	V+	A1	A*1	E-								UA733FM	0	SN52733FA
DIL-14/1C	E+	N	A2	A*2	V-	N	R	R*	N	V+	A1	A*1	N	E+			LM733D	UA733DM	0	SN52733J
T05-10/1M	E-	E+	A2	A*2	V-	R	R*	V+	A1	A*1							LM733H	UA733HM	0	SN52733L
FLP-14/3C	N	N	T	E-	E+	V-	N	N	T*	R	V+	N	N	N					0	SN52741FA
DIL-14/1C	N	N	T	E-	E+	V-	N	N	T*	R	V+	N	N	N			LM741D	UA741DM	0	SN52741J
DIL-14/1C	N	N	T	E-	E+	V-	N	N	T*	R	V+	N	N	N			LM741D	UA741DM	0	SN52741JA
DIL-8/1C	T	E-	E+	V-	T*	R	V+	N										LM741EJ	0	SN52741JP
T05-8/1M	T	E-	E+	V-M	T*	R	V+	N									TBA222	UA741HM	0	SN52741L
DIL-14/1P	N	N	T	E-	E+	V-	N	N	T*	R	V+	N	N	N			LM741D	UA741DM	0	SN52741N
DIL-8/1P	T	E-	E+	V-	T*	R	V+	N										LM741EJ	0	SN52741P
FLP-14/3C	E-1	E+1	T1	V-	T2	E+2	E-2	T*2	V+	R2	N	R1	V+	T*1				LM747F	0	SN52747FA
DIL-14/1C	E-1	E+1	T1	V-	T2	E+2	E-2	T*2	V+	R2	N	R1	V+	T*1			SFC2747KM	UA747DM	0	SN52747J
DIL-14/1C	E-1	E+1	T1	V-	T2	E+2	E-2	T*2	V+	R2	N	R1	V+	T*1			SFC2747KM	UA747DM	0	SN52747JA
T05-10/1M	R1	V+1	E-1	E+1	V-	E+2	E-2	V+2	R2	N							SFC2747M	UA747HM	0	SN52747L
FLP-14/3C	N	N	FT	E-	E+	V-	N	N	T*	R	V+	F*	N	N					0	SN52748FA
DIL-14/1C	N	N	FT	E-	E+	V-	N	N	T*	R	V+	F*	N	N			SN52748JA	UA748DM	0	SN52748J
DIL-14/1C	N	N	FT	E-	E+	V-	N	N	T*	R	V+	F*	N	N			SN52748J	UA748DM	0	SN52748JA
DIL-8/1C	FT	E-	E+	V-	T*	R	V+	F*										LM748J	0	SN52748JP
T05-8/1M	FT	E-	E+	V-M	T*	R	V+	F*									TBC0748	UA748HM	0	SN52748L
FLP-14/3C	N	TF	N	E-	E+	V-	N	N	T*	R	V+	∅	N	N					0	SN52770FA
DIL-14/1C	N	TF	N	E-	E+	N	N	N	T*	R	V+	∅	N	N					0	SN52770JA
DIL-8/1C	TF	E-	E+	V-	T*	R	V+	∅											0	SN52770JP
T05-8/1M	TF	E-	E+	V-M	T*	R	V+	∅										HA2-2505	0	SN52770L

TYPE NUMBER	MFR	APP	CMP	GBP MIN	SLEW RATE MIN	Vs+ MAX	Vs− MAX	Top MAX	AVOL MIN	VIO MAX	IB MAX	IIO MAX	PTOT MAX	IOUT MIN	VOUT MIN	VICM MAX	VIDF MAX	dVIO/dT MAX	PO MAX	IO MAX	CMRR MIN	PSRR MIN	RIN MIN
SN52770L	TGU	HSR	EXT	.5MHZ	1V/uS	+22V	−22V	125C	94dB	4MV	15NA	2NA	500MWF	6MA	12V	15V	30V		60MW	2MA	80dB	76dB	30M
SN52771FA	TGU	HSR	INT	.5MHZ	1V/uS	+22V	−22V	125C	94dB	4MV	15NA	2NA	500MWF	6MA	12V	15V	30V		60MW	2MA	80dB	76dB	30M
SN52771JA	TGU	HSR	INT	.5MHZ	1V/uS	+22V	−22V	125C	94dB	4MV	15NA	2NA	500MWF	6MA	12V	15V	30V		60MW	2MA	80dB	76dB	30M
SN52771JP	TGU	HSR	INT	.5MHZ	1V/uS	+22V	−22V	125C	94dB	4MV	15NA	2NA	500MWF	6MA	12V	15V	30V		60MW	2MA	80dB	76dB	30M
SN52771L	TGU	HSR	INT	.5MHZ	1V/uS	+22V	−22V	125C	94dB	4MV	15NA	2NA	500MWF	6MA	12V	15V	30V		60MW	2MA	80dB	76dB	30M
SN52777FA	TGU	PIA	EXT		0.2V/uS	+22V	−22V	125C	94dB	2MV	25NA	3NA	500MWF	5MA	12V	15V	30V	15uV/C		4MA	80dB	80dB	2M
SN52777JA	TGU	PIA	EXT		0.2V/uS	+22V	−22V	125C	94dB	2MV	25NA	3NA	500MWF	5MA	12V	15V	30V	15uV/C		4MA	80dB	80dB	2M
SN52777JP	TGU	PIA	EXT		0.2V/uS	+22V	−22V	125C	94dB	2MV	25NA	3NA	500MWF	5MA	12V	15V	30V	15uV/C		4MA	80dB	.80dB	2M
SN52777L	TGU	PIA	EXT		0.2V/uS	+22V	−22V	125C	94dB	2MV	25NA	3NA	500MWF	5MA	12V	15V	30V	15uV/C		4MA	80dB	80dB	2M
SN52810FA	TGU	CPR	EXT			+14V	−7V	125C	82dB	2MV	15uA	3uA	300MWF	5MA	1V	7V	5V	10uV/C	150MW	9MA	80dB		
SN52810J	TGU	CPR	EXT			+14V	−7V	125C	82dB	2MV	15uA	3uA	300MWF	5MA	1V	7V	5V	10uV/C	150MW	9MA	80dB		
SN52810JP	TGU	CPR	INT			+14V	−7V	125C	82dB	2MV	15uA	3uA	300MWF	5MA	1V	7V	5V	10uV/C	150MW	9MA	80dB		
SN52810L	TGU	CPR	EXT			+14V	−7V	125C	82dB	2MV	15uA	3uA	300MWF	5MA	1V	7V	5V	10uV/C	150MW	9MA	80dB		
SN52811FA	TGU	DCP	EXT			+14V	−7V	125C	82dB	3.5MV	20uA	3uA	300MWF	5MA	1V	7V	5V	20uV/C	150MW	7MA	70dB		
SN52811J	TGU	DCP	EXT			+14V	−7V	125C	82dB	3.5MV	20uA	3uA	300MWF	5MA	1V	7V	5V	20uV/C	150MW	7MA	70dB		
SN52811L	TGU	DCP	EXT			+14V	−7V	125C	82dB	3.5MV	20uA	3uA	300MWF	5MA	1V	7V	5V	20uV/C	150MW	7MA	70dB		
SN52820J	TGU	DCP	EXT			+14V	−7V	125C	82dB	2MV	15uA	3uA	300MWF	5MA	1V	7V	5V	10uV/C	150MW	9MA	80dB		
SN72L022L	TGU	DLP	INT	.3MHZ	0.2V/uS	+18V	−18V	70C	60dB	5MV	250NA	80NA	500MWF	1MA	10V	15V	30V		4MW	.3MA	60dB	74dB	
SN72L022P	TGU	DLP	INT	.3MHZ	0.2V/uS	+18V	−18V	70C	60dB	5MV	250NA	80NA	500MWF	1MA	10V	15V	30V		4MW	.3MA	60dB	74dB	
SN72L044JA	TGU	QLP	INT	.3MHZ	.15V/uS	+18V	−18V	70C	60dB	5MV	250NA	80NA	500MWF	1MA	10V	15V	30V		4MW	.1MA	60dB	74dB	
SN72L044N	TGU	QLP	INT	.3MHZ	.15V/uS	+18V	−18V	70C	60dB	5MV	250NA	80NA	500MWF	1MA	10V	15V	30V		4MW	.1MA	60dB	74dB	
SN7510FA	TGU	BDO	EXT	.1GHZ		+8V	−8V	70C	36dB	30MV	100uA	30uA			1V	5V			220MW		70dB		3K
SN7510L	TGU	BDO	EXT	.1GHZ		+8V	−8V	70C	36dB	30MV	100uA	30uA			1V	5V			220MW		70dB		3K
SN7510P	TGU	BDO	EXT	.1GHZ		+8V	−8V	70C	36dB	30MV	100uA	30uA			1V	5V			220MW		70dB		3K
SN7511FA	TGU	BDO	EXT	50MHZ		+8V	−8V	70C	50dB	5MV	20uA	10uA	500MWF	1MA	.75V	6V	6V	10uV/C	300MW		52dB		2K
SN7511L	TGU	BDO	EXT	50MHZ		+8V	−8V	70C	50dB	5MV	20uA	10uA	500MWF	1MA	.75V	6V	6V	10uV/C	300MW		52dB		2K
SN7511N	TGU	BDO	EXT	50MHZ		+8V	−8V	70C	50dB	5MV	20uA	10uA	500MWF	1MA	.75V	6V	6V	10uV/C	300MW		52dB		2K
SN7512L	TGU	BDO	EXT	.2GHZ		+8V	−8V	70C	46dB	7MV	80uA	5uA	500MWF	2MA	1.5V	6V	5V			25MA	74dB	50dB	2K
SN7512N	TGU	BDO	EXT	.2GHZ		+8V	−8V	70C	46dB	7MV	80uA	5uA	500MWF	2MA	1.5V	6V	5V			25MA	74dB	50dB	2K
SN7514L	TGU	BDO	EXT	.2GHZ		+8V	−8V	70C	46dB	7MV	80uA	5uA	500MWF	2MA	1.5V	6V	5V			25MA	74dB	50dB	2K
SN7514P	TGU	BDO	EXT	.2GHZ		+8V	−8V	70C	46dB	7MV	80uA	3uA	500MWF	2MA	1.5V	6V	5V			25MA	74dB	50dB	2K
SN72301AFA	TGU	GPU	EXT			+18V	−18V	70C	88dB	7.5MV	250NA	50NA	500MWF	5MA	12V	15V	30V	30uV/C		3MA	70dB	70dB	500K
SN72301AJ	TGU	GPU	EXT			+18V	−18V	70C	88dB	7.5MV	250NA	50NA	500MWF	5MA	12V	15V	30V	30uV/C		3MA	70dB	70dB	500K
SN72301AL	TGU	GPU	EXT			+18V	−18V	70C	88dB	7.5MV	250NA	50NA	500MWF	5MA	12V	15V	30V	30uV/C		3MA	70dB	70dB	500K
SN72301AN	TGU	GPU	EXT			+18V	−18V	70C	88dB	7.5MV	250NA	50NA	500MWF	5MA	12V	15V	30V	30uV/C		3MA	70dB	70dB	500K
SN72301AP	TGU	GPU	EXT			+18V	−18V	70C	88dB	7.5MV	250NA	50NA	500MWF	5MA	12V	15V	30V	30uV/C		3MA	70dB	70dB	500K
SN72306FA	TGU	CPR	EXT			+15V	−15V	70C	84dB	5MV	25uA	5uA	600MWF	50MA	2.5V			20uV/C	163MW				
SN72306J	TGU	CPR	EXT			+15V	−15V	70C	84dB	5MV	25uA.	5uA	600MWF	50MA	2.5V			20uV/C	163MW				
SN72306L	TGU	CPR	EXT			+15V	−15V	70C	84dB	5MV	25uA	5uA	600MWF	50MA	2.5V			20uV/C	163MW				
SN72306N	TGU	CPR	EXT			+15V	−15V	70C	84dB	5MV	25uA	5uA	600MWF	50MA	2.5V			20uV/C	163MW				
SN72306P	TGU	CPR	EXT			+15V	−15V	70C	84dB	5MV	25uA	5uA	600MWF	50MA	2.5V			20uV/C	163MW				
SN72307FA	TGU	GPK	INT			+18V	−18V	70C	84dB	7.5MV	250NA	50NA	500MWF	5MA	12V	15V	30V	30uV/C			70dB	70dB	0.5M
SN72307JA	TGU	GPK	INT			+18V	−18V	70C	84dB	7.5MV	250NA	50NA	500MWF	5MA	12V	15V	30V	30uV/C			70dB	70dB	0.5M
SN72307L	TGU	GPK	INT			+18V	−18V	70C	84dB	7.5MV	250NA	50NA	500MWF	5MA	12V	15V	30V	30uV/C			70dB	70dB	0.5M
SN72307N	TGU	GPK	INT			+18V	−18V	70C	84dB	7.5MV	250NA	50NA	500MWF	5MA	12V	15V	30V	30uV/C			70dB	70dB	0.5M
SN72307P	TGU	GPK	INT			+18V	−18V	70C	84dB	7.5MV	250NA	50NA	500MWF	5MA	12V	15V	30V	30uV/C			70dB	70dB	0.5M
SN72308AFA	TGU	SBA	EXT			+18V	−18V	70C	98dB	0.5MV	7NA	1NA	500MWF	1MA	13V	15V	1V	5uV/C		.6MA	96dB	96dB	10M
SN72308AJA	TGU	SBA	EXT			+18V	−18V	70C	98dB	0.5MV	7NA	1NA	500MWF	1MA	13V	15V	1V	5uV/C		.6MA	96dB	96dB	10M
SN72308AL	TGU	SBA	EXT			+18V	−18V	70C	98dB	0.5MV	7NA	1NA	500MWF	1MA	13V	15V	1V	5uV/C		.6MA	96dB	96dB	10M
SN72308AN	TGU	SBA	EXT			+18V	−18V	70C	98dB	0.5MV	7NA	1NA	500MWF	1MA	13V	15V	1V	5uV/C		.6MA	96dB	96dB	10M
SN72308AP	TGU	SBA	EXT			+18V	−18V	70C	98dB	0.5MV	7NA	1NA	500MWF	1MA	13V	15V	1V	5uV/C		.6MA	96dB	96dB	10M
SN72308FA	TGU	SBA	EXT			+18V	−18V	70C	88dB	7.5MV	7NA	1NA	500MWF	1MA	13V	15V	1V	30uV/C		.6MA	80dB	80dB	10M
SN72308JA	TGU	SBA	EXT			+18V	−18V	70C	88dB	7.5MV	7NA	1NA	500MWF	1MA	13V	15V	1V	30uV/C		.6MA	80dB	80dB	10M
SN72308L	TGU	SBA	EXT			+18V	−18V	70C	88dB	7.5MV	7NA	1NA	500MWF	1MA	13V	15V	1V	30uV/C		.6MA	80dB	80dB	10M
SN72308N	TGU	SBA	EXT			+18V	−18V	70C	88dB	7.5MV	7NA	1NA	500MWF	1MA	13V	15V	1V	30uV/C		.6MA	80dB	80dB	10M
SN72308P	TGU	SBA	EXT			+18V	−18V	70C	88dB	7.5MV	7NA	1NA	500M...	1MA	13V	15V	1V	30uV/C		.6MA	80dB	80dB	10M
SN72310FA	TGU	VFA	INT		15V/uS	+18V	−18V	70C	0dB	7.5MV	7NA		500MWF	1MA	10V	15V	15V	50uV/C		6MA		70dB	10G
SN72310JA	TGU	VFA	INT		15V/uS	+18V	−18V	70C	0dB	7.5MV	7NA		500MWF	1MA	10V	15V	15V	50uV/C		6MA		70dB	10G
SN72310L	TGU	VFA	INT		15V/uS	+18V	−18V	70C	0dB	7.5MV	7NA		500MWF	1MA	10V	15V	15V	50uV/C		6MA		70dB	10G
SN72310N	TGU	VFA	INT		15V/uS	+18V	−18V	70C	0dB	7.5MV	7NA		500MWF	1MA	10V	15V	15V	50uV/C		6MA		70dB	10G

Also for ready references the more important abbreviations used in the column headings are listed below:

LEFT HAND PAGE

APP = application (codes at APP.E.)
CMRR = common mode rejection ratio
CMP. = compensation (frequency)
dV_{io}/dT = input offset voltage temperature drift
GBP = gain bandwidth product
I_B = input bias current
I_{io} = input bias offset current
I_c = quiescent supply current
MFR = manufacturer (codes at App C.)
P_o = quiescent power consumer
PSRR = power supply rejection ratio
V_{ICM} = common mode input voltage rating
V_{idf} = differential input voltage rating
V_{io} = input offset voltage
V_S = dc supply voltage

RIGHT HAND PAGE

Lead out coding summary (details at APP.G.) for different cases (APP.F.)

A = gain adjust
B = bias adjust
C = case
E— = inverting input
E+ = non-inverting input
F,F* = input frequency compensation
G = ground
J = high level input
K = output, open collector
L = output, open emitter
M = metal case
N = not connected
O = special terminal
R,R* = outputs
S = strobe
T,T* = offset balance
V+ = +ve dc supply
V— = —ve dc supply
W = guard ring
X = blank position, no lead
++ = +ve supplementary dc supply
—— = —ve supplementary dc supply
⌀,⌀* = output frequency compensation

CASE (APP F)	LD1	LD2	LD3	LD4	LD5	LD6	LD7	LD8	LD9	LD10	LD11	LD12	LD13	LD14	LD15	LD16	EUROPE SUBSTITUTE	USA SUBSTITUTE	ISS	TYPE NUMBER
FLP-14/3C	N	T	N	E-	E+	V-	N	T*	R	V+	N	N	N						0	SN52771FA
DIL-14/1C	N	T	N	E-	E+	N	V-	N	T*	R	V+	N	N	N					0	SN52771JA
DIL-8/1C	T	E-	E+	V-	T*	R	V+	N											0	SN52771JP
TO5-8/1M	T	E-	E+	V-M	T*	R	V+	N											0	SN52771L
FLP-10/3C	N	TF	E-	E+	V-	T*	R	V+	F*										0	SN52777FA
DIL-14/1C	N	N	TF	E-	E+	V-	N	N	T*	R	V+	F*	N	N					0	SN52777JA
DIL-8/1C	TF	E-	E+	V-	T*	R	V+	F*											0	SN52777JP
TO5-8/1M	TF	E-	E+	V-M	T*	R	V+	F*											0	SN52777L
FLP-10/3C	G	E+	E-	N	V-	R	N	V+	N	N									0	SN52810FA
DIL-14/1C	N	G	E+	E-	N	V-	N	N	R	N	V+	N	N	N					0	SN52810J
DIL-8/1C	G	E+	E-	V-	N	R	V+												0	SN52810JP
TO5-8/1M	G	E+	E-	V-M	N	R	V+												0	SN52810L
FLP-10/3C	E-1	E+1	V-	E+2	E-2	S2	R	V+	G	S1									0	SN52811FA
DIL-14/1C	N	E-1	E+1	V-	E+2	E-2	N	N	S2	R	V+	G	S1	N					0	SN52811J
TO5-10/1M	G	S1	E-1	E+1	V-	E+2	E-2	S2	R	V+									0	SN52811L
DIL-14/1C	R1	N	V+1	N	E+2	E-2	V-2	R2	N		V+2	G	E+1	E-1	V-1				0	SN52820J
TO5-8/1M	R1	E-1	E+1	V-	E+2	E-2	R2	V+											0	SN72L022L
DIL-8/1P	R1	E-1	E+1	V-	E+2	E-2	R2	V+											0	SN72L022P
DIL-16/1C	R1	E-1	E+1	V-	E+2	E-2	R2	N	R3	E-3	E+3	V-	E+4	E-4	R4	V+*			0	SN72L044JA
DIL-16/1P	R1	E-1	E+1	V-	E+2	E-2	R2	V+	R3	E-3	E+3	V-	E+4	E-4	R4	V+*			0	SN72L044N
FLP-10/3C	E	N	V+	N	E*	R*	G	N	R	V-							SN5510FA		0	SN7510FA
TO5-8/1M	E	V+	E*	M	R*	G	R	V-									SN5510L		0	SN7510L
DIL-8/1P	E.	V+	E*	N	R*	G	R	V-									SN5510JP		0	SN7510P
FLP-10/3C	F	E	V+	E*	F*	R*	B	G	V-	R							SN5511FA		0	SN7511FA
TO5-10/1M	F	E	V+	E*	F*	R*	B	G	V-	R							SN5511L		0	SN7511L
DIL-14/1P	N	F	E	V+	E*	F*	N	N	R*	B	G	V-	R	N			SN5511N		0	SN7511N
TO5-10/1M	T*	E	V+	E*	N	R*	G	R	V-M	T							SN5512L		0	SN7512L
DIL-14/1P	N	T*	E	V+	E*	N	N	N	R*	G	R	V-	T	N			SN5512N		0	SN7512N
TO5-8/1M	E+	V+	E*	M	R*	G	R	V-									SN5514L		0	SN7514L
DIL-8/1P	E	V+	E*	N	R*	G	R	V-									SN5514P		0	SN7514P
FLP-14/3C	N	N	FT	E-	E+	V-	N	N	T*	R	V+	F*	N	N					0	SN72301AFA
DIL-14/1C	N	N	FT	E-	E+	V-	N	N	T*	R	V+	F*	N	N			UA301AD	LM301AJ14	0	SN72301AJ
TO5-8/1M	FT	E-	E+	V-M	T*	R	V+	F*									SFC2301A	LM301AH	0	SN72301AL
DIL-14/1P	N	N	FT	E-	E+	V-	N	N	T*	R	V+	F*	N	N			UA301AD	LM301AJ14	0	SN72301AN
DIL-8/1P	FT	E-	E+	V-	T*	R	V+	F*									SFC2301ADC	LM301AJ	0	SN72301AP
FLP-14/3C	N	G	E+	E-	N	V-	S1	S2	R	N	V+	N	N					LM206F	0	SN72306FA
DIL-14/1C	N	G	E+	E-	N	V-	S1	S2	R	N	V+	N	N					SN52106J	0	SN72306J
TO5-8/1M	G	E+	E-	V-M	S1	S2	R	V+										LM306H	0	SN72306L
DIL-14/1P	N	G	E+	E-	N	V-	S1	S2	R	N	V+	N	N					SN52106N	0	SN72306N
DIL-8/1P	G	E+	E-	V-	S1	S2	R	V+										SN52106P	0	SN72306P
FLP-14/3C	N	N	N	E-	E+	V-	N	N	N	R	V+	N	N					SN52107FA	0	SN72307FA
DIL-14/1C	N	N	N	E-	E+	V-	N	N	N	R	V+	N	N	N			SN52107JA	LM307D	0	SN72307JA
TO5-8/1M	N	E-	E+	V-M	N	R	V+	N									SFC2307	LM307H	0	SN72307L
DIL-14/1P	N	N	N	E-	E+	V-	N	N	N	R	V+	N	N				SN52107JA	LM307D	0	SN72307N
DIL-8/1P	N	E-	E+	V-	N	R	V+	N									SFC2307DC	LM307J	0	SN72307P
FLP-10/1C	N	N	E-	E+	N	V-	R	V+	F	F*								SN52108AFA	0	SN72308AFA
DIL-14/1C	N	F	N	E-	E+	N	V-	N	N	R	V+	F*	N	N			UA308AD	LM308AD	0	SN72308AJA
TO5-8/1M	F	E-	E+	V-M	N	R	V+	F*									SFC2308A	LM308AH	0	SN72308AL
DIL-14/1P	N	F	N	E-	E+	N	V-	N	N	R	V+	F*	N	N			UA308AD	LM308AD	0	SN72308AN
DIL-8/1P	F	E-	E+	V-	N	R	V+	F*										SN52108AP	0	SN72308AP
FLP-10/1C	N	N	E-	E+	N	V-	R	V+	F	F*							SFC2208PT	LM208F	0	SN72308FA
DIL-14/1C	N	F	N	E-	E+	N	V-	N	N	R	V+	F*	N	N			UA308D	LM308D	0	SN72308JA
TO5-8/1M	F	E-	E+	V-M	N	R	V+	F*									SFC2308	LM308H	0	SN72308L
DIL-14/1P	N	F	N	E-	E+	N	V-	N	N	R	V+	F*	N	N			UA308D	LM308D	0	SN72308N
DIL-8/1P	F	E-	E+	V-	N	R	V+	F*									SFC2308DC	LM308N	0	SN72308P
FLP-14/3C	N	N	T	N	E+	V-	N	N	L	R	V+	T*	N	N				SN52110FA	0	SN72310FA
DIL-14/1C	N	N	T	N	E+	V-	N	N	N	L	R	V+	T*	N	N		SFC2310EC	LM3100	0	SN72310JA
TO5-8/1M	T	N	E+	V-	L	R	V+	T*									UA310C	LM310H	0	SN72310L
DIL-14/1P	N	N	T	N	E+	V-	N	N	N	L	R	V+	T*	N	N		SFC2310EC	LM3100	0	SN72310N
DIL-8/1P	T	N	E+	V-	L	R	V+	T*									SFC23100DC	LM310N	0	SN72310P

TYPE NUMBER	MFR	APP	CMP	GBP MIN	SLEW RATE MIN	V_S^+ MAX	V_S^- MAX	T_{MP} MAX	A_{VOL} MIN	V_{IO} MAX	I_B MAX	I_{IO} MAX	P_{TOT} MAX	I_{OUT} MIN	V_{OUT} MIN	V_{ICM} MAX	V_{IDF} MAX	dV_{IO}/dT MAX	P_O MAX	I_O MAX	CMRR MIN	PSRR MIN	R_{IN} MIN
SN72310P	TGU	VFA	INT		15V/uS	+18V	-18V	70C	0dB	7.5MV	7NA	.	500MWF	1MA	10V	15V	15V	50uV/C		6MA		70dB	10G
SN72311FA	TGU	CPR	EXT			+18V	-18V	70C	100dB	7.5MV	250NA	50NA	500MWF			15V	30V			8MA			
SN72311J	TGU	CPR	EXT			+18V	-18V	70C	100dB	7.5MV	250NA	50NA	500MWF			15V	30V			8MA			
SN72311L	TGU	CPR	EXT			+18V	-18V	70C	100dB	7.5MV	250NA	50NA	500MWF			15V	30V			8MA			
SN72311N	TGU	CPR	EXT			+18V	-18V	70C	100dB	7.5MV	250NA	50NA	500MWF			15V	30V			8MA			
SN72311P	TGU	CPR	EXT			+18V	-18V	70C	100dB	7.5MV	250NA	50NA	500MWF			15V	30V			8MA			
SN72318FA	TGU	XSR	INT		50V/uS	+20V	-20V	70C	88dB	10MV	500NA	200NA	500MWF	6MA	12V	15V	1V			8MA	70dB	65dB	500K
SN72318L	TGU	XSR	INT		50V/uS	+20V	-20V	70C	88dB	10MV	500NA	200NA	500MWF	6MA	12V	15V	1V			8MA	70dB	65dB	500K
SN72318N	TGU	XSR	INT		50V/uS	+20V	-20V	70C	88dB	10MV	500NA	200NA	500MWF	6MA	12V	15V	1V			8MA	70dB	65dB	500K
SN72318JA	TGU	XSR	INT		50V/uS	+20V	-20V	70C	88dB	10MV	500NA	200NA	500MWF	6MA	12V	15V	1V			8MA	70dB	65dB	500K
SN72318P	TGU	XSR	INT		50V/uS	+20V	-20V	70C	88dB	10MV	500NA	200NA	500MWF	6MA	12V	15V	1V			8MA	70dB	65dB	500K
SN72506FA	TGU	DCP	EXT		20V/uS	+15V	-15V	70C		5MV	25uA	5uA	600MWF		12V	7V	5V	20uV/C					
SN72506J	TGU	DCP	EXT		20V/uS	+15V	-15V	70C		5MV	25uA	5uA	600MWF		12V	7V	5V	20uV/C					
SN72506M	TGU	DCP	EXT		20V/uS	+15V	-15V	70C		5MV	25uA	5uA	600MWF		12V	7V	5V	20uV/C					
SN72510FA	TGU	CPR	EXT			+14V	-7V	70C		3.5MV	20uA	5uA	300MWF		2.5V	7V	5V	20uV/C	150MW	9MA	70dB		
SN72510J	TGU	CPR	EXT			+14V	-7V	70C		3.5MV	20uA	5uA	300MWF	5MA	2.5V	7V	5V	20uV/C	150MW	9MA	70dB		
SN72510L	TGU	CPR	EXT			+14V	-7V	70C		3.5MV	20uA	5uA	300MWF	5MA	2.5V	7V	5V	20uV/C	150MW	9MA	70dB		
SN72510N	TGU	CPR	EXT			+14V	-7V	70C		3.5MV	20uA	5uA	300MWF	5MA	2.5V	7V	5V	20uV/C	150MW	9MA	70dB		
SN72510P	TGU	CPR	EXT			+14V	-7V	70C		3.5MV	20uA	5uA	300MWF	5MA	2.5V	7V	5V	20uV/C	150MW	9MA	70dB		
SN72514J	TGU	DCP	EXT			+14V	-7V	70C		3.5MV	20uA	5uA	300MWF	5MA	2.5V	7V	5V	20uV/C	150MW	9MA	70dB		
SN72514N	TGU	DCP	EXT			+14V	-7V	70C		3.5MV	20uA	5uA	300MWF	5MA	2.5V	7V	5V	20uV/C	150MW	9MA	70dB		
SN72558L	TGU	DGK	INT	.3MHZ	0.2V/uS	+18V	-18V	70C	86dB	6MV	500NA	200NA	500MWF	5MA	12V	15V	30V		85MW	3MA	70dB	76dB	300K
SN72558P	TGU	DGK	INT	.3MHZ	0.2V/uS	+18V	-18V	70C	86dB	6MV	500NA	200NA	500MWF	5MA	12V	15V	30V		85MW	3MA	70dB	76dB	300K
SN72660FA	TGU	GPU	EXT			+18V	-18V	70C	88dB	4MV	15NA	2NA	500MWF	1MA	13V	15V	1V	30uV/C		.8MA	80dB	80dB	4M
SN72660JA	TGU	GPU	EXT			+18V	-18V	70C	88dB	4MV	15NA	2NA	500MWF	1MA	13V	15V	1V	30uV/C		.8MA	80dB	80dB	4M
SN72660L	TGU	GPU	EXT			+18V	-18V	70C	88dB	4MV	15NA	2NA	500MWF	1MA	13V	15V	1V	30uV/C		.8MA	80dB	80dB	4M
SN72660N	TGU	GPU	EXT			+18V	-18V	70C	88dB	4MV	15NA	2NA	500MWF	1MA	13V	15V	1V	30uV/C		.8MA	80dB	80dB	4M
SN72660P	TGU	GPU	EXT			+18V	-18V	70C	88dB	4MV	15NA	2NA	500MWF	1MA	13V	15V	1V	30uV/C		.8MA	80dB	80dB	4M
SN72702FA	TGU	WBA	EXT		0.5V/uS	+14V	-7V	70C	60dB	10MV	15uA	5uA	300MWF	.3MA	5V	1.5V	5V	20uV/C	125MW	7MA	65dB	70dB	6K
SN72702J	TGU	WBA	EXT		0.5V/uS	+14V	-7V	70C	60dB	10MV	15uA	5uA	300MWF	.3MA	5V	1.5V	5V	20uV/C	125MW	7MA	65dB	70dB	6K
SN72702L	TGU	WBA	EXT		0.5V/uS	+14V	-7V	70C	60dB	10MV	15uA	5uA	300MWF	.3MA	5V	1.5V	5V	20uV/C	125MW	7MA	65dB	70dB	6K
SN72702N	TGU	WBA	EXT		0.5V/uS	+14V	-7V	70C	60dB	10MV	15uA	5uA	300MWF	.3MA	5V	1.5V	5V	20uV/C	125MW	7MA	65dB	70dB	6K
SN72709DN	TGU	GPU	EXT			+18V	-18V	70C	84dB	7.5MV	1.5uA	0.5uA	300MWF	5MA	12V	10V	5V		200MW		65dB	74dB	50K
SN72709FA	TGU	GPU	EXT			+18V	-18V	70C	84dB	7.5MV	1.5uA	0.5uA	300MWF	5MA	12V	10V	5V		200MW		65dB	74dB	50K
SN72709J	TGU	GPU	EXT			+18V	-18V	70C	84dB	7.5MV	1.5uA	0.5uA	300MWF	5MA	12V	10V	5V		200MW		65dB	74dB	50K
SN72709L	TGU	GPU	EXT			+18V	-18V	70C	84dB	7.5MV	1.5uA	0.5uA	300MWF	5MA	12V	10V	5V		200MW		65dB	74dB	50K
SN72709N	TGU	GPU	EXT			+18V	-18V	70C	84dB	7.5MV	1.5uA	0.5uA	300MWF	5MA	12V	10V	5V		200MW		65dB	74dB	50K
SN72709P	TGU	GPU	EXT			+18V	-18V	70C	84dB	7.5MV	1.5uA	0.5uA	300MWF	5MA	12V	10V	5V		200MW		65dB	74dB	50K
SN72710FA	TGU	CPR	EXT			+14V	-7V	70C	57dB	7.5MV	100uA	15uA	300MWF	5MA	1V	7V	5V	20uV/C			65dB		
SN72710J	TGU	CPR	EXT			+14V	-7V	70C	57dB	7.5MV	100uA	15uA	300MWF	5MA	1V	7V	5V	20uV/C			65dB		
SN72710L	TGU	CPR	EXT			+14V	-7V	70C	57dB	7.5MV	100uA	15uA	300MWF	5MA	1V	7V	5V	20uV/C			65dB		
SN72710N	TGU	CPR	EXT			+14V	-7V	70C	57dB	7.5MV	100uA	15uA	300MWF	5MA	1V	7V	5V	20uV/C			65dB		
SN72710P	TGU	CPR	EXT			+14V	-7V	70C	57dB	7.5MV	100uA	15uA	300MWF	5MA	1V	7V	5V	20uV/C			65dB		
SN72711FA	TGU	DCP	EXT			+14V	-7V	70C	57dB	5MV	100uA	15uA	300MWF	5MA	2.5V	7V	5V	20uV/C	230MW		65dB		
SN72711J	TGU	DCP	EXT			+14V	-7V	70C	57dB	5MV	100uA	15uA	30.MWF	5MA	2.5V	7V	5V	20uV/C	230MW		65dB		
SN72711L	TGU	DCP	EXT			+14V	-7V	70C	57dB	5MV	100uA	15uA	300MWF	5MA	2.5V	7V	5V	20uV/C	230MW		65dB		
SN72711N	TGU	DCP	EXT			+14V	-7V	70C	57dB	5MV	100uA	15uA	300MWF	5MA	2.5V	7V	5V	20uV/C	230MW		65dB		
SN72720J	TGU	DCP	EXT			+14V	-7V	70C	57dB	7.5MV	100uA	10uA	300MWF	5MA	1V	7V	5V	20uV/C	150MW	10MA	65dB		
SN72720N	TGU	DCP	EXT			+14V	-7V	70C	57dB	7.5MV	100uA	10uA	300MWF	5MA	1V	7V	5V	20uV/C	150MW	10MA	65dB		
SN72733FA	TGU	BDO	EXT	20MHZ		+8V	-8V	70C	48dB	6MV	30uA	5uA	500MWF	2MA	2V	6V	5V			24MA	60dB	50dB	2K
SN72733J	TGU	BDO	EXT	20MHZ		+8V	-8V	70C	48dB	6MV	30uA	5uA	500MWF	2MA	2V	6V	5V			24MA	60dB	50dB	2K
SN72733L	TGU	BDO	EXT	20MHZ		+8V	-8V	70C	48dB	6MV	30uA	5uA	500MWF	2MA	2V	6V	5V			24MA	60dB	50dB	2K
SN72733N	TGU	BDO	EXT	20MHZ		+8V	-8V	70C	48dB	6MV	30uA	5uA	500MWF	2MA	2V	6V	5V			24MA	60dB	50dB	2K
SN72741DN	TGU	GPK	INT		0.2V/uS	+18V	-18V	70C	86dB	6MV	500NA	200NA	500MWF	5MA	12V	15V	30V		85MW	3MA	70dB	76dB	300K
SN72741FA	TGU	GPK	INT		0.2V/uS	+18V	-18V	70C	86dB	6MV	500NA	200NA	500MWF	5MA	12V	15V	30V		85MW	3MA	70dB	76dB	300K
SN72741J	TGU	GPK	INT		0.2V/uS	+18V	-18V	70C	86dB	6MV	500NA	200NA	500MWF	5MA	12V	15V	30V		85MW	3MA	70dB	76dB	300K
SN72741L	TGU	GPK	INT		0.2V/uS	+18V	-18V	70C	86dB	6MV	500NA	200NA	500MWF	5MA	12V	15V	30V		85MW	3MA	70dB	76dB	300K
SN72741N	TGU	GPK	INT		0.2V/uS	+18V	-18V	70C	86dB	6MV	500NA	200NA	500MWF	5MA	12V	15V	30V		85MW	3MA	70dB	76dB	300K
SN72741P	TGU	GPK	INT		0.2V/uS	+18V	-18V	70C	86dB	6MV	500NA	200NA	500MWF	5MA	12V	15V	30V		85MW	3MA	70dB	76dB	300K
SN72747FA	TGU	DGU	INT		0.2V/uS	+18V	-18V	70C	94dB	6MV	500NA	200NA	500MWF	5MA	12V	15V	30V		85MW	3MA	70dB	76dB	300K

or detailed explanations of column heading notations, see App. A.

Also for ready references the more important abbreviations used in the column headings are listed below:

LEFT HAND PAGE

APP = application (codes at APP.E.)

CMRR = common mode rejection ratio

CMP = compensation (frequency)

dV_{IO}/dT = input offset voltage temperature drift

GBP = gain bandwidth product

I_B = input bias current

I_{IO} = input bias offset current

I_Q = quiescent supply current

MFR = manufacturer (codes at App.C.)

P_U = quiescent power consumer

PSRR = power supply rejection ratio

V_{ICM} = common mode input voltage rating

V_{ins} = differential input voltage rating

V_{IO} = input offset voltage

V_c = dc supply voltage

RIGHT HAND PAGE

Lead out coding summary (details at APP.G.) for different cases (APP F.)

A = gain adjust

B = bias adjust

C = case

E- = inverting input

E+ = non-inverting input

F,F* = input frequency compensation

G = ground

J = high level input

K = output, open collector

L = output, open emitter

M = metal case

N = not connected

Q = special terminal

R,R* = outputs

S = strobe

T,T* = offset balance

V+ = +ve dc supply

V- = -ve dc supply

W = guard ring

X = blank position, no lead

++ = +ve supplementary dc supply

-- = -ve supplementary dc supply

∮,∮* = output frequency compensation

CASE (APP F)	LD1	LD2	LD3	LD4	LD5	LD6	LD7	LD8	LD9	LD10	LD11	LD12	LD13	LD14	LD15	LD16	EUROPE SUBSTITUTE	USA SUBSTITUTE	ISS	TYPE NUMBER
FLP-10/3C	G	E+	E-	N	V-	T	T*S	N	R	V+							SN52111FA	LM311F	0	SN72311FA
DIL-14/1C	N	G	E+	E-	N	V-	T	T*S	R	N	V+	N	N	N			SFC2311EC	LM311D	0	SN72311J
TO5-8/1M	G	E+	E-	V-	N	T	T*S	R	V+								SFC2311	LM311H	0	SN72311L
DIL-14/1P	N	G	E+	E-	N	V-	T	T*S	R	N	V+	N	N	N			SFC2311EC	LM311D	0	SN72311N
DIL-8/1P	G	E+	E-	V-	T	T*S	R	V+									SFC2311DC	UA311R	0	SN72311D
FLP-10/3C	N	T*	E-	E+	V-	F*T	R	V+	ø	N							SN52118FA	LM218F	0	SN72318FA
TO5-8/1M	T*F	E-	E+	V-	F*T	R	V+	ø									TDE0118CM	LM318H	0	SN72318L
DIL-14/1P	N	N	T*F	E-	E+	V-	N	N	F*T	R	V+	ø	N	N			SFC2318EC	LM318D	0	SN72318N
DIL-14/1C	N	N	T*F	E-	E+	V-	N	N	F*T	R	V+	ø	N	N			SFC2318EC	LM318D	0	SN72318JA
DIL-8/1P	T*F	E-	E+	V-	F*T	R	V+	ø										LM318N	0	SN72318P
FLP-14/3C	S1	E-1	E+1	V-	E+2	E-2	S2	S*2	N	R2	V+	R1	G	S*1				SN52506A	0	SN72506FA
DIL-14/1P	S1	E-1	E+1	V-	E+2	E-2	S2	S*2	N	R2	V+	R1	G	S*1				SN52506J	0	SN72506J
DIL-14/1P	S1	E-1	E+1	V-	E+2	E-2	S2	S*2	N	R2	V+	R1	G	S*1				SN52506N	0	SN72506N
FLP-10/3C	G	E+	E-	N	V-	R	S	V+	N	N								SN52510FA	0	SN72510FA
DIL-14/1C	N	G	E+	E-	N	V-	N	N	R	S	V+	N	N	N				SN52510L	0	SN72510J
TO5-8/1M	G	E+	E-	V-M	N	S	R	V+										SN52510L	0	SN72510L
DIL-14/1P	N	G	E+	E-	N	V-	N	N	R	S	V+	N	N	N				SN52510N	0	SN72510N
DIL-8/1P	G	E+	E-	V-	N	R	S	V+										SN52510P	0	SN72510P
DIL-14/1C	R1	S1	V+	N	E+2	E-2	V-	R2	S2	V+	G	E+1	E-1	V-					0	SN72514J
DIL-14/1P	R1	S	V+	N	E+2	E-2	V-	R2	S2	V+	G	E+1	E-1	V-					0	SN72514N
TO5-8/1M	R1	E-1	E+1	V-M	E+2	E-2	R2	V+									TBB1458	MC1458G	0	SN72558L
DIL-8/1P	R	E-1	E+1	V-	E+2	E-2	R2	V+									TBB1458B	MC1458U	0	SN72558P
FLP-10/3C	N	N	E-	E+	N	V-	N	V+	F*	F							SFC2208PT	LM208F	0	SN72660FA
DIL-14/1C	N	F	N	E-	E+	N	V-	N	N	R	V+	F*	N	N			UA308D	LM308D	0	SN72660JA
TO5-8/1M	F	E-	E+	V-M	N	R	V+	F*									SFC2308	LM308H	0	SN72660L
DIL-14/1P	N	F	N	E-	E+	N	V-	N	N	R	V+	F*	N	N			UA308D	LM308D	0	SN72660N
DIL-8/1P	F	E-	E+	V-	N	R	V+	F*										SN52660JP	0	SN72660P
FLP-10/3C	N	G	E-	E+	V-	F	ø	R	N	V+							MC1712CF	UA702FM	0	SN72702FA
DIL-14/1C	N	N	G	E-	E+	V-	N	N	F	ø	R	N	V+	N			MC1712CL	UA702DC	0	SN72702J
TO5-8/1M	G	E-	E+	V-M	F	ø	R	V+									MC1712CG	UA702HC	0	SN72702L
DIL-14/1P	N	N	G	E-	E+	V-	N	N	F	ø	R	N	V+	N			MC1712CL	UA702DC	0	SN72702N
DIL-14/1P	N	N	F	E-	E+	V-	N	N	ø	R	V+	F*	N	N			TAA521A	UA709DC	0	SN72709DN
FLP-10/3C	N	F	E-	E+	V-	ø	R	V+	F*	N							MC1709F	UA709FM	0	SN72709FA
DIL-14/1C	N	N	F	E-	E+	V-	N	N	ø	R	V+	F*	N	N			TAA521A	UA709DC	0	SN72709J
TO5-8/1M	F	E-	E+	V-M	ø	ø*R	V+	F*									TAA521	UA709HC	0	SN72709L
DIL-14/1P	N	N	F	E-	E+	V-	N	N	ø	R	V+	F*	N	N			TAA521A	UA709DC	0	SN72709N
DIL-8/1P	F	E-	E+	V-	ø	ø*R	V+	F*									LM709CN8	UA709TC	0	SN72709P
FLP-10/3C	G	E+	E-	N	V-	R	N	V+	N	N							SFC2710PM	UA710FM	0	SN72710FA
DIL-14/1C	N	G	E+	E-	N	V-	N	N	R	N	V+	N	N	N			SFC2710EC	UA710DC	0	SN72710J
TO5-8/1M	G	E+	E-	V-M	N	N	R	V+									SFC2710C	UA710HC	0	SN72710L
DIL-14/1P	N	G	E+	E-	N	V-	N	N	R	N	V+	N	N	N			SFC2710EC	UA710DC	0	SN72710N
DIL-8/1P	G	E+	E-																0	SN72710GP
FLP-10/3C	E-1	E+1	V-	E+2	E-2	S2	R	V+	G	S1							SFC2711PM	UA711FM	0	SN72711FA
DIL-14/1C	N	E-1	E+1	V-	E+2	E-2	N	N	S2	R	V+	G	S1	N			SFC2711EC	UA711DC	0	SN72711J
TO5-10/1M	G	S1	E-1	E+1	V-	E+2	E-2	S2	R	V+							SFC2711C	UA711HC	0	SN72711L
DIL-14/1P	N	E-1	E+1	V-	E+2	E-2	N	N	S2	R	V+	G	S1	N			SFC2711EC	UA711DC	0	SN72711N
DIL-14/1C	R1	N	V+1	N	E+2	E-2	V-2	R2	N	V+2	G	E+1	E-1	V-1					0	SN72720J
DIL-14/1P	R1	N	V+1	N	E+2	E-2	V-2	R2	N	V+2	G	E+1	E-1	V-1					0	SN72720N
FLP-10/3C	E+	A2	A*2	V-	R	R*	V+	A1	A*1	E-								UA733FM	0	SN72733FA
DIL-14/1C	E+	N	A2	A*2	V-	N	R	R*	N	V+	A1	A*1	N	E-				UA733DM	0	SN72733J
TO5-10/1M	E-	E+	A2	A*2	V-	R	R*	V+	A1	A*1							LM733CH	UA733HC	0	SN72733L
DIL-14/1P	E+	N	A2	A*2	V-	N	R	R*	N	V+	A1	A*1	N	E-				UA733DC	0	SN72733N
DIL-14/1P	N	N	T	E-	E+	V-	N	N	T*	R	V+	N	N	N			TBA221A	UA741DC	0	SN72741DN
FLP-14/3C	N	N	T	E-	E+	V-	N	N	T*	R	V+	N	N	N				SN52741FA	0	SN72741FA
DIL-14/1C	N	N	T	E-	E+	V-	N	N	T*	R	V+	N	N	N			TBA221A	UA741DC	0	SN72741J
TO5-8/1M	T	E-	E+	V-M	T*	R	V+										TBA221	UA741HC	0	SN72741L
DIL-14/1P	N	N	T	E-	E+	V-	N	N	T*	R	V+	N	N	N			TBA221A	UA741DC	0	SN72741N
DIL-8/1P	E-	E+	V-	T*	R	V+											TBA221B	UA741CJ	0	SN72741P
FLP-14/3C	E-1	E+1	T1	V-	T2	E+2	E-2	T*2	V+	R2	N	R1	V+	T*1				LM747CF	0	SN72747FA
DIL-14/1C	E-1	E+1	T1	V-	T2	E+2	E-2	T*2	V+	R2	N	R1	V+	T*1			TBB0747A	UA747DC	0	SN72747J

TYPE NUMBER	M F R	A P P	C M P	GBP MIN	SLEW RATE MIN	Vs+ MAX	Vs- MAX	Top MAX	Avol MIN	Vio MAX	Ib MAX	Iio MAX	Ptot MAX	Iout MIN	Vout MIN	Vicm MAX	Vidf MAX	dVio/dT MAX	Po MAX	Iu MAX	CM RR MIN	PS RR MIN	Rin MIN
SN72747J	TGU	DGU	INT	.	0.2V/uS	+18V	-18V	70C	94dB	6MV	500NA	200NA	500MWF	5MA	12V	15V	30V	.	85MW	3MA	70dB	76dB	300K
SN72747L	TGU	DGU	INT	.	0.2V/uS	+18V	-18V	70C	94dB	6MV	500NA	200NA	500MWF	5MA	12V	15V	30V	.	85MW	3MA	70dB	76dB	300K
SN72747N	TGU	DGU	INT	.	0.2V/uS	+18V	-18V	70C	94dB	6MV	500NA	200NA	500MWF	5MA	12V	15V	30V	.	85MW	3MA	70dB	76dB	300K
SN72748FA	TGU	GPU	EXT	.	0.2V/uS	+18V	-18V	70C	94dB	6MV	500NA	200NA	500MWF	5MA	10V	15V	30V	.	85MW	3MA	70dB	76dB	300K
SN72748J	TGU	GPU	EXT	.	0.2V/uS	+18V	-18V	70C	94dB	6MV	500NA	200NA	500MWF	5MA	10V	15V	30V	.	85MW	3MA	70dB	76dB	300K
SN72748L	TGU	GPU	EXT	.	0.2V/uS	+18V	-18V	70C	94dB	6MV	500NA	200NA	500MWF	5MA	10V	15V	30V	.	85MW	3MA	70dB	76dB	300K
SN72748N	TGU	GPU	EXT	.	0.2V/uS	+18V	-18V	70C	94dB	6MV	500NA	200NA	500MWF	5MA	10V	15V	30V	.	85MW	3MA	70dB	76dB	300K
SN72748P	TGU	GPU	EXT	.	0.2V/uS	+18V	-18V	70C	94dB	6MV	500NA	200NA	500MWF	5MA	10V	15V	30V	.	85MW	3MA	70dB	76dB	300K
SN72770FA	TGU	HSR	EXT	.5MHZ	1V/uS	+18V	-18V	70C	91dB	10MV	30NA	10NA	500MWF	6MA	12V	15V	30V	.	120MW	4MA	70dB	74dB	30M
SN72770JA	TGU	HSR	EXT	.5MHZ	1V/uS	+18V	-18V	70C	91dB	10MV	30NA	10NA	500MWF	6MA	12V	15V	30V	.	120MW	4MA	70dB	74dB	30M
SN72770L	TGU	HSR	EXT	.5MHZ	1V/uS	+18V	-18V	70C	91dB	10MV	30NA	10NA	500MWF	6MA	12V	15V	30V	.	120MW	4MA	70dB	74dB	30M
SN72770N	TGU	HSR	EXT	.5MHZ	1V/uS	+18V	-18V	70C	91dB	10MV	30NA	10NA	500MWF	6MA	12V	15V	30V	.	120MW	4MA	70dB	74dB	30M
SN72770P	TGU	HSR	EXT	.5MHZ	1V/uS	+18V	-18V	70C	91dB	10MV	30NA	10NA	500MWF	6MA	12V	15V	30V	.	120MW	4MA	70dB	74dB	30M
SN72771FA	TGU	HSR	INT	.5MHZ	1V/uS	+18V	-18V	70C	91dB	10MV	30NA	10NA	500MWF	6MA	12V	15V	30V	.	120MW	4MA	70dB	74dB	30M
SN72771JA	TGU	HSR	INT	.5MHZ	1V/uS	+18V	-18V	70C	91dB	10MV	30NA	10NA	500MWF	6MA	12V	15V	30V	.	120MW	4MA	70dB	74dB	30M
SN72771L	TGU	HSR	INT	.5MHZ	1V/uS	+18V	-18V	70C	91dB	10MV	30NA	10NA	500MWF	6MA	12V	15V	30V	.	120MW	4MA	70dB	74dB	30M
SN72771N	TGU	HSR	INT	.5MHZ	1V/uS	+18V	-18V	70C	91dB	10MV	30NA	10NA	500MWF	6MA	12V	15V	30V	.	120MW	4MA	70dB	74dB	30M
SN72771P	TGU	HSR	INT	.5MHZ	1V/uS	+18V	-18V	70C	91dB	10MV	30NA	10NA	500MWF	6MA	12V	15V	30V	.	120MW	4MA	70dB	74dB	30M
SN72777FA	TGU	PIA	EXT	.	0.2V/uS	+22V	-22V	70C	88dB	5MV	100NA	20NA	500MWF	5MA	12V	15V	30V	30uV/C	.	4MA	70dB	76dB	1M
SN72777JA	TGU	PIA	EXT	.	0.2V/uS	+22V	-22V	70C	88dB	5MV	100NA	20NA	500MWF	5MA	12V	15V	30V	30uV/C	.	4MA	70dB	76dB	1M
SN72777L	TGU	PIA	EXT	.	0.2V/uS	+22V	-22V	70C	88dB	5MV	100NA	20NA	500MWF	5MA	12V	15V	30V	30uV/C	.	4MA	70dB	76dB	1M
SN72777N	TGU	PIA	EXT	.	0.2V/uS	+22V	-22V	70C	88dB	5MV	100NA	20NA	500MWF	5MA	12V	15V	30V	30uV/C	.	4MA	70dB	76dB	1M
SN72777P	TGU	PIA	EXT	.	0.2V/uS	+22V	-22V	70C	88dB	5MV	100NA	20NA	500MWF	5MA	12V	15V	30V	30uV/C	.	4MA	70dB	76dB	1M
SN72810FA	TGU	CPR	EXT	.	.	+14V	-7V	70C	80dB	3.5MV	50uA	5uA	300MWF	5MA	1V	7V	5V	20uV/C	150MW	9MA	70dB	.	.
SN72810J	TGU	CPR	EXT	.	.	+14V	-7V	70C	80dB	3.5MV	50uA	5uA	300MWF	5MA	1V	7V	5V	20uV/C	150MW	9MA	70dB	.	.
SN72810L	TGU	CPR	EXT	.	.	+14V	-7V	70C	80dB	3.5MV	50uA	5uA	300MWF	5MA	1V	7V	5V	20uV/C	150MW	9MA	70dB	.	.
SN72810N	TGU	CPR	EXT	.	.	+14V	-7V	70C	80dB	3.5MV	50uA	5uA	300MWF	5MA	1V	7V	5V	20uV/C	150MW	9MA	70dB	.	.
SN72810P	TGU	CPR	EXT	.	.	+14V	-7V	70C	80dB	3.5MV	50uA	5uA	300MWF	5MA	1V	7V	5V	20uV/C	150MW	9MA	70dB	.	.
SN72811FA	TGU	DCP	EXT	.	.	+14V	-7V	70C	80dB	5MV	30uA	5uA	300MWF	5MA	1V	7V	5V	20uV/C	200MW	7MA	65dB	.	.
SN72811J	TGU	DCP	EXT	.	.	+14V	-7V	70C	80dB	5MV	30uA	5uA	300MWF	5MA	1V	7V	5V	20uV/C	200MW	7MA	65dB	.	.
SN72811L	TGU	DCP	EXT	.	.	+14V	-7V	70C	80dB	5MV	30uA	5uA	300MWF	5MA	1V	7V	5V	20uV/C	200MW	7MA	65dB	.	.
SN72811N	TGU	DCP	EXT	.	.	+14V	-7V	70C	80dB	5MV	30uA	5uA	300MWF	5MA	1V	7V	5V	20uV/C	200MW	7MA	65dB	.	.
SN72820J	TGU	DCP	EXT	.	.	+14V	-7V	70C	80dB	3.5MV	20uA	5uA	300MWF	5MA	1V	7V	5V	20uV/C	150MW	9MA	70dB	.	.
SN72820N	TGU	DCP	EXT	.	.	+14V	-7V	70C	80dB	3.5MV	20uA	5uA	300MWF	5MA	1V	7V	5V	20uV/C	150MW	9MA	70dB	.	.
SSS725AJ	PRU	PIA	EXT	.	.	+22V	-22V	125C	120dB	0.1MV	70NA	1NA	500MWF	11MA	12V	22V	30V	0.8uV/C	120MW	.	120dB	114dB	800K
SSS725AL	PRU	PIA	EXT	.	.	+22V	-22V	125C	120dB	0.1MV	70NA	1NA	500MWF	11MA	12V	22V	30V	0.8uV/C	120MW	.	120dB	114dB	800K
SSS725AP	OBS	PIA	EXT	.	.	+22V	-22V	125C	120dB	0.1MV	70NA	1NA	500MWF	11MA	12V	22V	30V	0.8uV/C	120MW	.	120dB	114dB	800K
SSS725AY	PRU	PIA	EXT	.	.	+22V	-22V	125C	120dB	0.1MV	70NA	1NA	500MWF	11MA	12V	22V	30V	0.8uV/C	120MW	.	120dB	114dB	800K
SSS725BJ	PRU	PIA	EXT	.	.	+22V	-22V	70C	120dB	.75MV	80NA	5NA	500MWF	11MA	12V	22V	30V	2.8uV/C	120MW	.	110dB	106dB	700K
SSS725BL	PRU	PIA	EXT	.	.	+22V	-22V	70C	120dB	.75MV	80NA	5NA	500MWF	11MA	12V	22V	30V	2.8uV/C	120MW	.	110dB	106dB	700K
SSS725BP	OBS	PIA	EXT	.	.	+22V	-22V	70C	120dB	.75MV	80NA	5NA	500MWF	11MA	12V	22V	30V	2.8uV/C	120MW	.	110dB	106dB	700K
SSS725BY	PRU	PIA	EXT	.	.	+22V	-22V	70C	120dB	.75MV	80NA	5NA	500MWF	11MA	12V	22V	30V	2.8uV/C	120MW	.	110dB	106dB	700K
SSS725CJ	PRU	PIA	EXT	.	.	+22V	-22V	70C	114dB	1.3MV	110NA	13NA	500MWF	11MA	12V	22V	30V	2.5uV/C	150MW	.	100dB	100dB	500K
SSS725CP	OBS	PIA	EXT	.	.	+22V	-22V	70C	114dB	1.3MV	110NA	13NA	500MWF	11MA	12V	22V	30V	2.5uV/C	150MW	.	100dB	100dB	500K
SSS725CY	PRU	PIA	EXT	.	.	+22V	-22V	70C	114dB	1.3MV	110NA	13NA	500MWF	11MA	12V	22V	30V	2.5uV/C	150MW	.	100dB	100dB	500K
SSS725EJ	PRU	PIA	EXT	.	.	+22V	-22V	70C	120dB	0.5MV	80NA	5NA	500MWF	11MA	12V	22V	30V	2uV/C	120MW	.	120dB	106dB	700K
SSS725EP	OBS	PIA	EXT	.	.	+22V	-22V	70C	120dB	0.5MV	80NA	5NA	500MWF	11MA	12V	22V	30V	2uV/C	120MW	.	120dB	106dB	700K
SSS725EY	PRU	PIA	EXT	.	.	+22V	-22V	70C	120dB	0.5MV	80NA	5NA	500MWF	11MA	12V	22V	30V	2uV/C	120MW	.	120dB	106dB	700K
SSS725J	PRU	PIA	EXT	.	.	+22V	-22V	125C	120dB	0.5MV	80NA	5NA	500MWF	11MA	12V	22V	30V	2uV/C	120MW	.	120dB	106dB	700K
SSS725L	PRU	PIA	EXT	.	.	+22V	-22V	125C	120dB	0.5MV	80NA	5NA	500MWF	11MA	12V	22V	30V	2uV/C	120MW	.	120dB	106dB	700K
SSS725P	OBS	PIA	EXT	.	.	+22V	-22V	125C	120dB	0.5MV	80NA	5NA	500MWF	11MA	12V	22V	30V	2uV/C	120MW	.	120dB	106dB	700K
SSS725Y	PRU	PIA	EXT	.	.	+22V	-22V	125C	120dB	0.5MV	80NA	5NA	500MWF	11MA	12V	22V	30V	2uV/C	120MW	.	120dB	106dB	700K
SSS741B.I	PRU	GPK	INT	.	.	+22V	-22V	125C	94dB	3MV	50NA	5NA	500MWF	5MA	12V	22V	30V	.	85MW	.	80dB	80dB	2M
SSS741BP	OBS	GPK	INT	.	.	+22V	-22V	125C	94dB	3MV	50NA	5NA	500MWF	5MA	12V	22V	30V	.	85MW	.	80dB	80dB	2M
SSS741BY	PRU	GPK	INT	.	.	+22V	-22V	125C	94dB	3MV	50NA	5NA	500MWF	5MA	12V	22V	30V	.	85MW	.	80dB	80dB	2M
SSS741CJ	PRU	GPK	INT	.	.	+18V	-18V	70C	86dB	6MV	100NA	25NA	500MWF	5MA	12V	18V	30V	.	85MW	.	70dB	76dB	1M
SSS741CP	OBS	GPK	INT	.	.	+18V	-18V	70C	86dB	6MV	100NA	25NA	500MWF	5MA	12V	18V	30V	.	85MW	.	70dB	76dB	1M
SSS741CY	PRU	GPK	INT	.	.	+18V	-18V	70C	86dB	6MV	100NA	25NA	500MWF	5MA	12V	18V	30V	.	85MW	.	70dB	76dB	1M
SSS741GJ	PRU	GPK	INT	.	.	+22V	-22V	125C	94dB	5MV	100NA	25NA	500MWF	5MA	12V	22V	30V	.	85MW	.	70dB	76dB	1M
SSS741GP	OBS	GPK	INT	.	.	+22V	-22V	125C	94dB	5MV	100NA	25NA	500MWF	5MA	12V	22V	30V	.	85MW	.	70dB	76dB	1M

For detailed explanations of column heading notations, see App. A.

Also for ready references the more important abbreviations used in the column headings are listed below:

LEFT HAND PAGE

APP = application (codes at APP.E.)
CMRR = common mode rejection ratio
CMP = compensation (frequency)
dV_{io}/dT = input offset voltage temperature drift
GBP = gain bandwidth product
I_B = input bias current
I_{io} = input bias offset current
I_Q = quiescent supply current
MFR = manufacturer (codes at App.C.)
P_Q = quiescent power consumer
PSRR = power supply rejection ratio
V_{ICM} = common mode input voltage rating
V_{IDR} = differential input voltage rating
V_{io} = input offset voltage
V_c = dc supply voltage

RIGHT HAND PAGE

Lead out coding summary (details at APP G.) for different cases (APP.F.):

A = gain adjust
B = bias adjust
C = case
E— = inverting input
E+ = non-inverting input
F,F* = input frequency compensation
G = ground
J = high level input
K = output, open collector
L = output, open emitter
M = metal case
N = not connected
Q = special terminal
R,R* = outputs
S = strobe
T,T* = offset balance
V+ = +ve dc supply
V— = —ve dc supply
W = guard ring
X = blank position, no lead
++ = +ve supplementary dc supply
—— = —ve supplementary dc supply
ø,ø* = output frequency compensation

CASE (APP F)	LD 1	LD 2	LD 3	LD 4	LD 5	LD 6	LD 7	LD 8	LD 9	LD 10	LD 11	LD 12	LD 13	LD 14	LD 15	LD 16	EUROPE SUBSTITUTE	USA SUBSTITUTE	IS	TYPE NUMBER
T05-10/1M	R1	V+1	E-1	E+1	V-		E+2	E-2	V+2	R2	N						TBB0747	UA747HC	0	SN72747L
DIL-14/1C	E-1	E+1	T1	V-	T2	E+2	E-2	T*2	V+	R2	N	R1	V+	T*1			TBB0747A	UA747DC	0	SN72747N
FLP-14/3C	N	N	FT	E-	E+	V-	N	N	T*	R	V+	F*	N	N				SN52748FA	0	SN72748FA
DIL-14/1C	N	N	FT	E-	E+	V-	N	N	T*	R	V+	F*	N	N				UA748DC	0	SN72748J
T05-8/1M	FT	E-	E+	V-M	T*	R	V+	F*									TBB0748	UA748HC	0	SN72748L
DIL-14/1P	N	N	FT	E-	E+	V-	N	N	T*	R	V+	F*	N	N			SN72748J	UA748DC	0	SN72748N
DIL-8/1P	FT	E-	E+	V-	T*	R	V+	F*									TBB0748B	UA748TC	0	SN72748P
FLP-14/3C	N	TF	N	E-	E+	N	V-	N	T*	R	V+	ø	N	N				SN52770FA	0	SN72770FA
DIL-14/1C	N	TF	N	E-	E+	N	V-	N	T*	R	V+	ø	N	N				SN52770JA	0	SN72770JA
T05-8/1M	TF	E-	E+	V-M	T*	R	V+	ø									SN52770L	HA2-2505	0	SN72770L
DIL-14/1P	N	TF	N	E-	E+	V-	N	N	T*	R	V+	ø	N	N				SN52770JA	0	SN72770N
DIL-8/1P	TF	E-	E+	V-	T*	R	V+	ø										SN52770JP	0	SN72770P
FLP-14/3C	N	T	N	E-	E+	N	V-	N	T*	R	V+	N	N	N				SN52771FA	0	SN72771FA
DIL-14/1C	N	T	N	E-	E+	N	V-	N	T*	R	V+	N	N	N				SN52771JA	0	SN72771JA
T05-8/1M	T	E-	E+	V-M	T*	R	V+	N										SN52771L	0	SN72771L
DIL-14/1P	N	T	N	E-	E+	N	V-	N	T*	R	V+	N	N	N				SN52771JA	0	SN72771N
DIL-8/1P	T	E-	E+	V-	T*	R	V+	N										SN52771JP	0	SN72771P
FLP-10/3C	N	TF	E-	E+	V-	T*	R	V+	F*									SN52777FA	0	SN72777FA
DIL-14/1C	N	N	TF	E-	E+	V-	N	N	T*	R	V+	F*	N	N			SN52777JA	UA777DC	0	SN72777JA
T05-8/1M	TF	E-	E+	V-M	T*	R	V+	F*										UA777HC	0	SN72777L
DIL-14/1P	N	N	TF	E-	E+	V-	N	N	T*	R	V+	F*	N	N				UA777DC	0	SN72777N
DIL-8/1P	TF	E-	E+	V-	T*	R	V+	F*									SN52777JP	UA777TC	0	SN72777P
FLP-10/3C	G	E+	E-	N	V-	R	N	V+	N	N								SN52810FA	0	SN72810FA
DIL-14/1C	N	G	E+	E-	N	V-	N	N	R	N	V+	N	N	N				SN52810J	0	SN72810J
T05-8/1M	G	E+	E-	V-M	N	N	R	V+										SN52810L	0	SN72810L
DIL-14/1P	N	G	E+	E-	N	V-	N	N	R	N	V+	N	N	N				SN52810J	0	SN72810N
DIL-8/1P	G	E+	E-	V-	N	N	R	V+										SN52810JP	0	SN72810P
FLP-10/3C	E-1	E+1	V-	E+2	E-2	S2	N	S1										SN52811FA	0	SN72811FA
DIL-14/1C	N	E-1	E+1	V-	E+2	E-2	N	N	S2	R	V+	G	S1	N				SN52811J	0	SN72811J
T05-10/1M	G	S1	E-1	E+1	V-		E+2	E-2	S2	R	V+							SN52811L	0	SN72811L
DIL-14/1P	N	E-1	E+1	V-	E+2	E-2	N	N	S2	R	V+	G	S1	N				SN52811J	0	SN72811N
DIL-14/1C	R1	N	N+1	V-	E+2	E-2	V-2	R2	N	V+2	G	E+1	E-1	V-1				SN52820J	0	SN72820J
DIL-14/1C	R1	N	V+1	N	E+2	E-2	V-2	R2	N	V+2	G	E+1	E-1	V-1				SN52820J	0	SN72820N
T05-8/1M	T	E-	E+	V-M	ø	R	V+	T*											0	SSS725AJ
FLP-10/3C	N	T	E-	E+	V-	ø	R	V+	T*	N									0	SSS725AL
DIL-14/1C	N	N	T	E-	E+	V-	N	N	ø	R	V+	T*	N	N				SSS725AY	0	SSS725AP
DIL-14/1C	N	N	T	E-	E+	V-	N	N	ø	R	V+	T*	N	N						SSS725AY
T05-8/1M	T	E-	E+	V-M	ø	R	V+	T*										UA725AHM	0	SSS725BJ
FLP-10/3C	N	T	E-	E+	V-	O	R	V+	T*	N								SSS725L	0	SSS725BL
DIL-14/1C	N	N	T	E-	E+	V-	N	N	ø	R	V+	T*	N	N				SSS725BY	0	SSS725BP
DIL-14/1C	N	N	T	E-	E+	V-	N	N	ø	R	V+	T*	N	N				LM725D	0	SSS725BY
T05-8/1M	T	E-	E+	V-M	ø	R	V+	T*										UA725HM	0	SSS725CJ
DIL-14/1C	N	N	T	E-	E+	V-	N	N	ø	R	V+	T*	N	N				SSS725CY	0	SSS725CP
DIL-14/1C	N	N	T	E-	E+	V-	N	N	ø	R	V+	T*	N	N				LM725J14	0	SSS725CY
T05-8/1M	T	E-	E+	V-M	ø	R	V+	T*										UA725EHC	0	SSS725EJ
DIL-14/1C	N	N	T	E-	E+	V-	N	N	ø	R	V+	T*	N	N				SSS725EY	0	SSS725EP
DIL-14/1C	N	N	T	E-	E+	V-	N	N	ø	R	V+	T*	N	N				LM725D	0	SSS725EY
T05-8/1M	T	E-	E+	V-M	ø	R	V+	N										UA725AHM	0	SSS725J
FLP-10/3C	N	T	E-	E+	V-	ø	R	V+	T*	N								UA725AHM	0	SSS725L
DIL-14/1C	N	N	T	E-	E+	V-	N	N	ø	R	V+	T*	N	N				SSS725Y	0	SSS725P
DIL-14/1C	N	N	T	E-	E+	V-	N	N	ø	R	V+	T*	N	N				LM725AJ14	0	SSS725Y
T05-8/1M	T	E-	E+	V-M	T*	R	V+	N									LM741AH	UA741AHM	0	SSS741BJ
DIL-14/1C	N	N	T	E-	E+	V-	N	N	T*	R	V+	N	N	N				SSS741BY	0	SSS741BP
DIL-14/1C	N	N	T	E-	E+	V-	N	N	T*	R	V+	N	N	N			LM741AD	UA741ADM	0	SSS741BY
T05-8/1M	T	E-	E+	V-M	T*	R	V+	N									LM741EH	UA741EHC	0	SSS741CJ
DIL-14/1C	N	N	T	E-	E+	V-	N	N	T*	R	V+	N	N	N				SSS741CY	0	SSS741CP
DIL-14/1C	N	N	T	E-	E+	V-	N	N	T*	R	V+	N	N	N			LM741ED	UA741EDC	0	SSS741CY
T05-8/1M	T	E-	E+	V-M	T*	R	V+	N									LM741AH	UA741AHM	0	SSS741GJ
DIL-14/1C	N	N	T	E-	E+	V-	N	N	T*	R	V+	N	N	N				SSS741GY	0	SSS741GP
DIL-14/1C	N	N	T	E-	E+	V-	N	N	T*	R	V+	N	N	N			LM741AD	UA741ADM	0	SSS741GY

TYPE NUMBER	M F R	A P P	C M P	GBP MIN	SLEW RATE MIN	V_S^+ MAX	V_S^- MAX	T_{OP} MAX	A_{VOL} MIN	V_{IO} MAX	I_B MAX	I_{IO} MAX	P_{TOT} MAX	I_{OUT} MIN	V_{OUT} MIN	V_{ICM} MAX	V_{IDF} MAX	dV_{IO}/dT MAX	P_Q MAX	I_O MAX	CM RR MIN	PS RR MIN	R_{IN} MIN
SSS741GY	PRU	GPK	INT	.	.	+22V	-22V	125C	94dB	5MV	100NA	25NA	500MWF	5MA	12V	22V	30V	.	85MW	.	70dB	76dB	1M
SSS741J	PRU	GPK	INT	.	.	+22V	-22V	125C	100dB	2MV	50NA	5NA	500MWF	5MA	12V	22V	30V	.	85MW	.	80dB	80dB	2M
SSS741P	OBS	GPK	INT	.	.	+22V	-22V	125C	100dB	2MV	50NA	5NA	500MWF	5MA	12V	22V	30V	.	85MW	.	80dB	80dB	2M
SSS741Y	PRU	GPK	INT	.	.	+22V	-22V	125C	100dB	2MV	50NA	5NA	500MWF	5MA	12V	22V	30V	.	85MW	.	80dB	80dB	2M
SSS747BK	PRU	DGK	INT	.	.	+22V	-22V	85C	94dB	3MV	50NA	5NA	500MWF	5MA	12V	22V	30V	.	85MW	.	80dB	80dB	2M
SSS747BM	PRU	DGK	INT	.	.	+22V	-22V	85C	94dB	3MV	50NA	5NA	500MWF	5MA	12V	22V	30V	.	85MW	.	80dB	80dB	2M
SSS747BP	OBS	DGK	INT	.	.	+22V	-22V	85C	94dB	3MV	50NA	5NA	500MWF	5MA	12V	22V	30V	.	80MW	.	80dB	80dB	2M
SSS747BY	PRU	DGK	INT	.	.	+22V	-22V	85C	94dB	3MV	50NA	5NA	500MWF	5MA	12V	22V	30V	.	85MW	.	80dB	80dB	2M
SSS747CK	PRU	DGK	INT	.	.	+22V	-22V	70C	94dB	5MV	100NA	25NA	500MWF	5MA	12V	22V	30V	.	85MW	.	70dB	76dB	1M
SSS747CP	OBS	DGK	INT	.	.	+22V	-22V	70C	94dB	5MV	100NA	25NA	500MWF	5MA	12V	22V	30V	.	85MW	.	70dB	76dB	1M
SSS747CY	PRU	DGK	INT	.	.	+22V	-22V	70C	94dB	5MV	100NA	25NA	500MWF	5MA	12V	22V	30V	.	85MW	.	70dB	76dB	1M
SSS747GK	PRU	DGK	INT	.	.	+22V	-22V	125C	94dB	5MV	100NA	25NA	500MWF	5MA	12V	22V	30V	.	85MW	.	70dB	76dB	1M
SSS747GM	PRU	DGK	INT	.	.	+22V	-22V	125C	94dB	5MV	100NA	25NA	500MWF	5MA	12V	22V	30V	.	85MW	.	70dB	76dB	1M
SSS747GP	OBS	DGK	INT	.	.	+22V	-22V	125C	94dB	5MV	100NA	25NA	500MWF	5MA	12V	22V	30V	.	85MW	.	70dB	76dB	1M
SSS747GY	PRU	DGK	INT	.	.	+22V	-22V	125C	94dB	5MV	100NA	25NA	500MWF	5MA	12V	22V	30V	.	85MW	.	70dB	76dB	1M
SSS747K	PRU	DGK	INT	.	.	+22V	-22V	125C	100dB	2MV	80NA	5NA	500MWF	5MA	12V	22V	30V	.	85MW	.	80dB	80dB	2M
SSS747M	PRU	DGK	INT	.	.	+22V	-22V	125C	100dB	2MV	80NA	5NA	500MWF	5MA	12V	22V	30V	.	85MW	.	80dB	80dB	2M
SSS747P	OBS	DGK	INT	.	.	+22V	-22V	125C	100dB	2MV	80NA	5NA	500MWF	5MA	12V	22V	30V	.	85MW	.	80dB	80dB	2M
SSS747Y	PRU	DGK	INT	.	.	+22V	-22V	125C	100dB	2MV	80NA	5NA	500MWF	5MA	12V	22V	30V	.	85MW	.	80dB	80dB	2M
SSS1458	PRU	DGK	INT	.	.	+22V	-22V	70C	94dB	5MV	100NA	25NA	500MWF	5MA	12V	22V	30V	.	85MW	.	70dB	76dB	1M
SSS1458J	PRU	DGK	INT	.	.	+22V	-22V	70C	94dB	5MV	100NA	25NA	500MWF	5MA	12V	22V	30V	.	85MW	.	70dB	76dB	1M
SSS1558	PRU	DGK	INT	.	.	+22V	-22V	125C	94dB	5MV	100NA	25NA	500MWF	5MA	12V	22V	30V	.	85MW	.	70dB	76dB	1M
SSS1558J	PRU	DGK	INT	.	.	+22V	-22V	125C	94dB	5MV	100NA	25NA	500MWF	5MA	12V	22V	30V	.	85MW	.	70dB	76dB	1M
SU536T	SJU	FET	INT	.5MHZ	3V/uS	+22V	-22V	85C	94dB	20MV	30pA	10pA	500MWF	17MA	10V	22V	30V	150uV/C	350MW	6MA	70dB	76dB	50G
TA7502AM	TOJ	LNA	EXT	.3MHZ	.15V/uS	+18V	-18V	125C	86dB	5MV	500NA	150NA	300MWF	5MA	12V	10V	6V	20uV/C	150MW	.	70dB	80dB	150K
TA7502BM	TOJ	GPU	EXT	.3MHZ	.15V/uS	+18V	-18V	75C	86dB	1MV	250NA	70NA	300MWF	5MA	12V	10V	6V	10uV/C	150MW	.	70dB	80dB	150K
TA7502CM	TOJ	GPU	EXT	.3MHZ	.15V/uS	+18V	-18V	75C	86dB	5MV	250NA	70NA	300MWF	5MA	12V	10V	6V	10uV/C	150MW	.	70dB	80dB	150K
TA7502M	TOJ	GPU	EXT	.3MHZ	.15V/uS	+18V	-18V	75C	86dB	5MV	500NA	150NA	300MWF	5MA	12V	10V	6V	20uV/C	150MW	.	70dB	80dB	150K
TA7502P	TOJ	GPU	EXT	.3MHZ	.15V/uS	+18V	-18V	75C	86dB	5MV	250NA	70NA	300MWF	5MA	12V	10V	6V	20uV/C	150MW	.	70dB	80dB	150K
TA7504M	TOJ	GPK	INT	.	0.2V/uS	+18V	-18V	75C	86dB	5MV	500NA	200NA	500MWF	5MA	12V	18V	30V	50uV/C	85MW	3MA	70dB	76dB	300K
TA7504P	TOJ	GPK	INT	.	0.2V/uS	+18V	-18V	75C	86dB	5MV	500NA	200NA	300MWF	5MA	12V	18V	30V	50uV/C	85MW	3MA	70dB	76dB	300K
TA7504S	TOJ	GPK	INT	.	0.2V/uS	+18V	-18V	75C	86dB	5MV	500NA	200NA	400MWF	5MA	12V	18V	30V	50uV/C	85MW	3MA	70dB	76dB	300K
TA7506M	TOJ	GPU	EXT	.	.	+18V	-18V	75C	88dB	5MV	250NA	50NA	500MWF	5MA	12V	18V	30V	30uV/C	.	3MA	70dB	70dB	500K
TA7506P	TOJ	GPU	EXT	.	.	+18V	-18V	75C	88dB	5MV	250NA	50NA	300MWF	5MA	12V	18V	30V	30uV/C	.	3MA	70dB	70dB	500K
TAA182	OBS	BDO	EXT	20MHZ	.	-12V	-12V	125C	58dB	20MV	800NA	250NA	600MWH	6MA	3.5V	6V	.	10uV/C	500MW	.	75dB	.	100K
TAA201	OBS	BDO	EXT	.	.	+25V	-24V	75C	32dB	10MV	1.2uA	30NA	200MWF	.	.	6V	.	20uV/C	33MW	3MA	70dB	.	75K
TAA202	OBS	BDO	EXT	.	.	+25V	-14V	125C	32dB	7MV	.	.	200MWF	.	6.5V	.	.	25uV/C	26MW	3MA	70dB	.	300K
TAA243	OBS	GPU	EXT	.	.	+14V	-7V	100C	59dB	15MV	15uA	5uA	200MWF	50UA	5V	1.5V	5V	.	125MW	.	65dB	.	6K
TAA241	OBS	WBA	EXT	3MHZ	.	+14V	-7V	70C	60dB	5MV	7.5uA	2uA	300MWF	.3MA	5V	1.5V	5V	20uV/C	120MW	7MA	70dB	70dB	10K
TAA242	OBS	WBA	EXT	3MHZ	.	+14V	-7V	125C	68dB	2MV	5uA	0.5uA	300MWF	.3MA	5V	1.5V	5V	10uV/C	120MW	7MA	80dB	74dB	16K
TAA445	OBS	GPU	EXT	.	.	+12V	-8V	125C	56dB	35MV	.	.	100MWF	1MA	5V	.	12V	.	.	.	18dB	.	3M
TAA495	AEW	GPU	EXT	.	.	+9V	-15V	125C	40dB	100MV	.	.	.	.	.	.	.	.	.	.	.	.	20K
TAA522	SIW	GPU	EXT	.3MHZ	.15V/uS	+18V	-18V	125C	88dB	5MV	0.5uA	0.2uA	200MWF	.	12V	10V	5V	.	200MW	.	65dB	74dB	150K
TAA522-709	SIW	GPU	EXT	.3MHZ	.15V/uS	+18V	-18V	125C	88dB	5MV	0.5uA	0.2uA	200MWF	.	12V	10V	5V	.	200MW	.	65dB	74dB	150K
TAA521A	SIW	GPU	EXT	.3MHZ	.15V/uS	+18V	-18V	70C	84dB	7.5MV	1.5uA	0.5uA	200MWF	5MA	12V	10V	5V	.	200MW	.	65dB	74dB	50K
TAA521A-709	SIW	GPU	EXT	.3MHZ	.15V/uS	+18V	-18V	70C	84dB	7.5MV	1.5uA	0.5uA	200MWF	5MA	12V	10V	5V	.	200MW	.	65dB	74dB	50K
TAA521	SIW	GPU	EXT	.3MHZ	.15V/uS	+18V	-18V	70C	84dB	7.5MV	1.5uA	0.5uA	200MWF	5MA	12V	10V	5V	.	200MW	.	65dB	74dB	150K
TAA521-709	SIW	GPU	EXT	.3MHZ	.15V/uS	+18V	-18V	70C	84dB	7.5MV	1.5uA	0.5uA	200MWF	5MA	12V	10V	5V	.	200MW	.	65dB	74dB	150K
TAA721	SIW	BDO	EXT	40MHZ	.	+8V	-8V	70C	37dB	.	100uA	30uA	180MWF	.4MA	2V	.	5V	.	230MW	25MA	75dB	.	3K
TAA722	SIW	BDO	EXT	40MHZ	.	+8V	-8V	125C	37dB	.	80uA	30uA	500MWF	.4MA	2V	.	5V	.	230MW	25MA	75dB	.	3K
TAA761	SIW	GPU	EXT	.	3V/uS	+18V	-18V	70C	81dB	6MV	1uA	0.3uA	500MWF	25MA	14V	12V	18V	35uV/C	.	3MA	65dB	74dB	50K
TAA761A	SIW	GPU	EXT	.	3V/uS	+18V	-18V	70C	81dB	6MV	1uA	0.3uA	500MWF	25MA	14V	12V	18V	35uV/C	.	3MA	65dB	74dB	50K
TAA761W	SIW	GPU	EXT	.	3V/uS	+18V	-18V	70C	81dB	6MV	1uA	0.3uA	500MWF	25MA	14V	12V	18V	35uV/C	.	3MA	65dB	74dB	50K
TAA762	SIW	GPU	EXT	.	3V/uS	+18V	-18V	125C	85dB	4MV	0.7uA	0.1uA	500MWF	25MA	14V	12V	18V	25uV/C	.	3MA	70dB	74dB	50K
TAA765	SIW	GPU	EXT	.	3V/uS	+18V	-18V	85C	81dB	6MV	1uA	0.3uA	500MWF	25MA	14V	12V	18V	35uV/C	.	3MA	65dB	74dB	50K
TAA765A	SIW	GPU	EXT	.	3V/uS	+18V	-18V	85C	81dB	6MV	1uA	0.3uA	500MWF	25MA	14V	12V	18V	35uV/C	.	3MA	65dB	74dB	50K
TAA765W	SIW	GPU	EXT	.	3V/uS	+18V	-18V	85C	81dB	6MV	1uA	0.3uA	500MWF	25MA	14V	12V	18V	35uV/C	.	3MA	65dB	74dB	50K
TAA861	SIW	GPU	EXT	.	3V/uS	+10V	-10V	70C	75dB	10MV	1uA	0.3uA	190MWF	25MA	10V	.	10V	35uV/C	.	2MA	60dB	74dB	50K
TAA861A	SIW	GPU	EXT	.	3V/uS	+10V	-10V	70C	75dB	10MV	1uA	0.3uA	190MWF	25MA	10V	.	10V	35uV/C	.	2MA	60dB	74dB	50K
TAA861W	SIW	GPU	EXT	.	3V/uS	+10V	-10V	70C	75dB	10MV	1uA	0.3uA	80MWF	25MA	10V	.	10V	35uV/C	.	2MA	60dB	74dB	50K

For detailed explanations of column heading notations, see App. A.

Also for ready references the more important abbreviations used in the column headings are listed below:

LEFT HAND PAGE

APP = application (codes at APP.E.)
CMRR = common mode rejection ratio
CMP = compensation (frequency)
dV_{io}/dT = input offset voltage temperature drift
GBP = gain bandwidth product
I_B = input bias current
I_{io} = input bias offset current
I_q = quiescent supply current
MFR = manufacturer (codes at App.C.)
P_q = quiescent power consumer
PSRR = power supply rejection ratio
V_{icm} = common mode input voltage rating
V_{idr} = differential input voltage rating
V_{io} = input offset voltage
V_S = dc supply voltage

RIGHT HAND PAGE

Lead out coding summary (details at APP.G.) for different cases (APP.F.)

A = gain adjust
B = bias adjust
C = case
E− = inverting input
E+ = non-inverting input
F,F* = input frequency compensation
G = ground
J = high level input
K = output, open collector
L = output, open emitter
M = metal case
N = not connected
Q = special terminal
R,R* = outputs
S = strobe
T,T* = offset balance
V+ = +ve dc supply
V− = −ve dc supply
W = guard ring
X = blank position, no lead
+ + = +ve supplementary dc supply
− − = −ve supplementary dc supply
ø,ø* = output frequency compensation

CASE (APP.F)	LD 1	LD 2	LD 3	LD 4	LD 5	LD 6	LD 7	LD 8	LD 9	LD 10	LD 11	LD 12	LD 13	LD 14	LD 15	LD 16	EUROPE SUBSTITUTE	USA SUBSTITUTE	IS	TYPE NUMBER
TO5-8/1M	T	E-	E+	V-M	T*	R	V+	N										RM4131T		SSS741J
DIL-14/1C	N	N	T	E-	E+	V-	N	N	T*	R	V+	N	N	N						SSS741P
DIL-14/1C	N	N	T	E-	E+	V-	N	N	T*	R	V+	N	N	N				UA741ADM		SSS741Y
TO5-10/1M	R1	V+	E-1	E+1	V-	E+2	E-2	V+2	R2	N							TBC0747	UA747AHM		SSS747BK
FLP-14/3G	E-1	E+1	T1	V-	T2	E+2	E-2	T*2	V+2	R2	N	R1	V+1	T*1				SSS747M	0	SSS747BM
DIL-14/1C	E-1	E+1	T1	V-	T2	E-2	E+2	T*2	V+2	R2	N	R1	V+1	T*1				SSS747BY	0	SSS747BP
DIL-14/1C	E-1	E+1	T1	V-	T2	E-2	E+2	T*2	V+2	R2	N	R1	V+1	T*1				UA747ADM	0	SSS747BY
TO5-10/1M	R1	V+	E-1	E+1	V-	E+2	E-2	V+2	R2	N							SFC2747M	UA741HM		SS747CK
DIL-14/1C	E-1	E+1	T1	V-	T2	E+2	E-2	T*2	V+2	R2	N	R1	V+1	T*1				SSS747CY	0	SSS747CP
DIL-14/1C	E-1	E+1	T1	V-	T2	E-2	E+2	T*2	V+2	R2	N	R1	V+1	T*1			LM747AD	UA747DM	0	SSS747CY
TO5-10/1M	R1	V+1	E-1	E+1	V-	E+2	E-2	V+2	R2	N							TBC0747	UA747AHM		SSS747GK
FLP-14/3G	E-1	E+1	T1	V-	T2	E+2	E-2	T*2	V+2	R2	N	R1	V+1	T*1			LM747F			SSS747GM
DIL-14/1C	E-1	E+1	T1	V-	T2	E+2	E-2	T*2	V+2	R2	N	R1	V+1	T*1				SSS747GY		SSS747GY
DIL-14/1C	E-1	E+1	T1	V-	T2	E+2	E-2	T*2	V+2	R2	N	R1	V+1	T*1			LM747AD	UA747ADM		SSS747GY
TO5-10/1M	R1	V+1	E-1	E+1	V-	E+2	E-2	V+2	R2	N									0	SSS747K
FLP-14/3G	E-1	E+1	T1	V-	T2	E-2	E+2	T*2	V+2	R2	N	R1	V+1	T*1					0	SSS747M
DIL-14/1C	E-1	E+1	T1	V-	T2	E-2	E+2	T*2	V+2	R2	N	R1	V+1	T*1				SSS747Y	0	SSS747P
DIL-14/1C	E-1	E+1	T1	V-	T2	E-2	E+2	T*2	V+2	R2	N	R1	V+1	T*1					0	SSS747Y
TO5-8/1M	R1	E-1	E+1	V-	E+2	E-2	R2	V+									TBC1458	MC1558G		SSS1458
TO5-8/1M	R1	E-1	E+1	V-	E+2	E-2	R2	V+									TBC1458	MC1558G		SSS1458J
TO5-8/1M	R1	E-1	E+1	V-	E+2	E-2	R2	V+										MC1558G		SSS1558
TO5-8/1M	R1	E-1	E+1	V-	E+2	E-2	R2	V+										MC1558G		SSS1558J
TO5-8/1M	T	E-	E+	V-	T*	R	V+	N										UA740HM	0	SU536T
TO5-8/1M	F	E-	E+	V-M	ø	ø*R	V+	F*									MC1709AG	UA709AHM	0	TA7502AM
TO5-8/1M	F	E-	E+	V-M	ø	ø*R	V+	F*									MC1709AG	UA709AHM		TA7502BM
TO5-8/1M	F	E-	E+	V-M	ø	ø*R	V+	F*									MC1709AG	UA709AHM	0	TA7502CM
TO5-8/1M	F	E-	E+	V-M	ø	ø*R	V+	F*									LM709H	UA709HM	0	TA7502M
DIL-8/1P	F	E-	E+	V-	ø	ø*R	V+	F*										SNS52709AJP	0	TA7502P
TO5-8/1M	T	E-	E+	V-M	T*	R	V+	N									TBA222	UA741HC	0	TA7504M
DIL-8/1P	T	E-	E+	V-	T*	R	V+	N									TBA221B	UA741TC	0	TA7504P
SIL-7/1P	T	E-	E+	V-	T*	R	V+	N												TA7504S
TO5-8/1M	FT	E-	E+	V-M	T*	R	V+	F*									SFC2301A	LM301AH	0	TA7506M
DIL-8/1P	FT	E-	E+	V-	T*	R	V+	F*									SFC2301ADC	LM301AJ	0	TA7506P
FLP-14/1G	V+	E-	F	F*	E+	V-	R	V+	R-	R+									0	TAA182
TO5-8/1M	E-	G	V-	E+	R*	N	V+	R											0	TAA201
FLP-14/1G	E-	G	N	N	V-	N	E+	R*	N	N	V+	N	N	R					0	TAA202
TO5-8/1M	G	E-	E+	V-	ø	ø*	R	V+											0	TAA243
TO5-8/1M	G	E-	E+	V-M	F	ø	R	V+									SNS52702L	UA702HM	0	TAA241
TO5-8/1M	G	E-	E+	V-M	F	ø	R	V+									SNS52702AL	UA702HM	0	TAA242
FLP-10/3G	V+	X	G	X	E+	V-	X	E-	X	K									0	TAA445
FLP-6/2P	V+	B	E+	V-	E-	K													0	TAA495
TO5-8/1M	F	E-	E+	V-M	ø	ø*R	V+	F*									LM709H	UA709HM	0	TAA522
TO5-8/1M	F	E-	E+	V-M	ø	ø*R	V+	F*									LM709H	UA709HM	0	TAA522-709
DIL-14/1P	N	N	F	E-	E+	V-	N	N	N	ø	R	V+	F*	N	N		LM709CJ	UA709DC	0	TAA521A
DIL-14/1P	N	N	F	E-	E+	V-	N	N	N	ø	R	V+	F*	N	N		LM709CJ	UA709DC	0	TAA521A-709
TO5-8/1M	F	E-	E+	V-M	ø	ø*R	V+	F*									LM709CH	UA709HC	0	TAA521
TO5-8/1M	F	E-	E+	V-M	ø	ø*R	V+	F*									LM709CH	UA709HC	0	TAA521-709
TO5-8/1M	E	V+	E*	N	R*	G	R	V-											0	TAA721
TO5-8/1M	E	V+	E*	N	R*	G	R	V-											0	TAA722
TO5-8/4M	X	V+	E+	E-	X	V-	K	ø									TAA762	SFC2761C	0	TAA761
DIL-6/1P	V+	E+	E-	V-	K	ø												SFC2761DC	0	TAA761A
FLP-6/2P	V+	E+	E-	V-	K	ø													0	TAA761W
TO5-8/4M	X	V+	E+	E-	X	V-	K	ø										SFC2761M	0	TAA762
TO5-8/4M	X	V+	E+	E-	X	V-	K	ø										SFC2761T	0	TAA765
DIL-6/1P	V+	E+	E-	V-	K	ø												SFC2761DT	0	TAA765A
FLP-6/2P	V+	E+	E-	V-	K	ø													0	TAA765W
TO5-8/4M	X	V+	E+	E-	X	V-	K	ø										TAA865	0	TAA861
DIL-6/1P	V+	E+	E-	V-	K	ø												TAA865A	0	TAA861A
FLP-6/2P	V+	E+	E-	V-	K	ø												TAA865W	0	TAA861W
TO5-8/4M	X	V+	E+	E-	X	V-	K	ø											0	TAA862

TYPE NUMBER	M F R	A P P	C M P	GBP MIN	SLEW RATE MIN	V_S^+ MAX	V_S^- MAX	T_{OC} MAX	A_{VOL} MIN	V_{IO} MAX	I_B MAX	I_{IO} MAX	P_{TOT} MAX	I_{OUT} MIN	V_{OUT} MIN	V_{ICM} MAX	V_{IDF} MAX	dV_{IO}/dT MAX	P_O MAX	$-I_O$ MAX	CM RR MIN	PS RR MIN	R_{IN} MIN
TAA862	SIW	GPU	EXT	.	3V/uS	+10V	-10V	125C	85dB	4MV	0.7uA	0.1uA	190MWF	25MA	10V	.	10V	25uV/C	.	2MA	70dB	74dB	50K
TAA862F	SIW	GPU	EXT	.	3V/uS	+10V	-10V	125C	85dB	4MV	0.7uA	0.1uA	190MWF	25MA	10V	.	10V	25uV/C	.	2MA	70dB	74dB	50K
TAA865	SIW	GPU	EXT	.	3V/uS	+10V	-10V	85C	75dB	10MV	1uA	0.3uA	190MWF	25MA	10V	.	10V	35uV/C	.	2MA	60dB	74dB	50K
TAA865A	SIW	GPU	EXT	.	3V/uS	+10V	-10V	85C	75dB	10MV	1uA	0.3uA	190MWF	25MA	10V	.	10V	35uV/C	.	2MA	60dB	74dB	50K
TAA865W	SIW	GPU	EXT	.	3V/uS	+10V	-10V	85C	75dB	10MV	1uA	0.3uA	80MWF	25MA	10V	.	10V	35uV/C	.	2MA	60dB	74dB	50K
TAA2761	SIW	DGK	INT	.	.	+15V	-15V	70C	80dB	6MV	1uA	0.3uA	240MWF	25MA	14V	12V	15V	40uV/C	.	2MA	65dB	80dB	100K
TAA2761A	SIW	DGK	INT	.	.	+15V	-15V	70C	80dB	6MV	1uA	0.3uA	320MWF	25MA	14V	12V	15V	40uV/C	.	2MA	65dB	80dB	100K
TAA2762	SIW	DGK	INT	.	.	+15V	-15V	125C	85dB	4MV	0.7uA	0.1uA	510MWF	25MA	14V	12V	15V	25uV/C	.	2MA	70dB	80dB	100K
TAA2765	SIW	DGK	INT	.	.	+15V	-15V	85C	80dB	6MV	1uA	0.3uA	310MWF	25MA	14V	12V	15V	40uV/C	.	2MA	65dB	80dB	100K
TAA2765A	SIW	DGK	INT	.	.	+15V	-15V	85C	80dB	6MV	1uA	0.3uA	320MWF	25MA	14V	12V	15V	40uV/C	.	2MA	65dB	80dB	100K
TAA4761A	SIW	OGK	INT	.	.	+15V	-15V	70C	80dB	6MV	1uA	0.3uA	300MWF	25MA	14V	.	15V	30uV/C	.	3MA	65dB	74dB	100K
TAA4765A	SIW	OGK	INT	.	.	+15V	-15V	85C	80dB	6MV	1uA	0.3uA	430MWF	25MA	14V	.	15V	30uV/C	.	3MA	65dB	74dB	100K
TBA221	SIW	GPK	INT	.	0.2V/uS	+18V	-18V	70C	86dB	6MV	500NA	200NA	237MWF	5MA	12V	15V	30V	.	85MW	3MA	70dB	76dB	300K
TBA221-741	SIW	GPK	INT	.	0.2V/uS	+18V	-18V	70C	86dB	6MV	500NA	200NA	237MWF	5MA	12V	15V	30V	.	85MW	3MA	70dB	76dB	300K
TBA221A	SIW	GPK	INT	.	0.2V/uS	+18V	-18V	70C	86dB	6MV	500NA	200NA	375MWF	5MA	12V	15V	30V	.	85MW	3MA	70dB	76dB	300K
TBA221A-741	SIW	GPK	INT	.	0.2V/uS	+18V	-18V	70C	86dB	6MV	500NA	200NA	375MWF	5MA	12V	15V	30V	.	85MW	3MA	70dB	76dB	300K
TBA221B	SIW	GPK	INT	.	0.2V/uS	+18V	-18V	70C	86dB	6MV	500NA	200NA	320MWF	5MA	12V	15V	30V	.	85MW	3MA	70dB	76dB	300K
TBA221B-741	SIW	GPK	INT	.	0.2V/uS	+18V	-18V	70C	86dB	6MV	500NA	200NA	320MWF	5MA	12V	15V	30V	.	85MW	3MA	70dB	76dB	300K
TBA221D	MJG	GPK	INT	.	0.2V/uS	+18V	-18V	70C	86dB	6MV	500NA	200NA	225MWF	5MA	12V	15V	30V	.	85MW	3MA	70dB	76dB	300K
TBA221G	SIW	GPK	INT	.	0.2V/uS	+18V	-18V	70C	86dB	6MV	500NA	200NA	225MWF	5MA	12V	15V	30V	.	85MW	3MA	70dB	76dB	300K
TBA221G-741	SIW	GPK	INT	.	0.2V/uS	+18V	-18V	70C	86dB	6MV	500NA	200NA	225MWF	5MA	12V	15V	30V	.	85MW	3MA	70dB	76dB	300K
TBA221N	SIW	LNA	INT	.	0.2V/uS	+18V	-18V	70C	86dB	6MV	500NA	200NA	225MWF	5MA	12V	15V	30V	.	85MW	3MA	70dB	76dB	300K
TBA221W	SIW	GPK	INT	.	0.2V/uS	+18V	-18V	70C	86dB	6MV	500NA	200NA	225MWF	5MA	12V	15V	30V	.	85MW	3MA	70dB	76dB	300K
TBA221W-741	SIW	GPK	INT	.	0.2V/uS	+18V	-18V	70C	86dB	6MV	500NA	200NA	225MWF	5MA	12V	15V	30V	.	85MW	3MA	70dB	76dB	300K
TBA222	SIW	GPK	INT	.	0.3V/uS	+22V	-22V	125C	94dB	5MV	350NA	100NA	520MWF	6MA	13V	15V	30V	15uV/dT	85MW	3MA	80dB	76dB	300K
TBA222-741	SIW	GPK	INT	.	0.3V/uS	+22V	-22V	125C	94dB	5MV	350NA	100NA	520MWF	6MA	13V	15V	30V	15uV/C	85MW	3MA	80dB	76dB	300K
TBA222Q1	SIW	LNA	INT	.	0.3V/uS	+22V	-22V	125C	94dB	5MV	350NA	100NA	520MWF	6MA	13V	15V	30V	15uV/C	85MW	3MA	80dB	76dB	300K
TBA222Q1-741	SIW	LNA	INT	.	0.3V/uS	+22V	-22V	125C	94dB	5MV	350NA	100NA	520MWF	6MA	13V	15V	30V	15uV/C	85MW	3MA	80dB	76dB	300K
TBA222Q2	SIW	LNA	INT	.	0.3V/uS	+22V	-22V	125C	94dB	5MV	350NA	100NA	520MWF	6MA	13V	15V	30V	15uV/C	85MW	3MA	80dB	76dB	300K
TBA222Q2-741	SIW	LNA	INT	.	0.3V/uS	+22V	-22V	125C	94dB	5MV	350NA	100NA	520MWF	6MA	13V	15V	30V	15uV/C	85MW	3MA	80dB	76dB	300K
TBA222S1	SIW	LNA	INT	.	0.3V/uS	+22V	-22V	125C	94dB	5MV	350NA	100NA	520MWF	6MA	13V	15V	30V	15uV/C	85MW	3MA	80dB	76dB	300K
TBA222S1-741	SIW	GPK	INT	.	0.3V/uS	+22V	-22V	125C	94dB	5MV	350NA	100NA	520MWF	6MA	13V	15V	30V	15uV/C	85MW	3MA	80dB	76dB	300K
TBA231	SGI	DLN	EXT	.	.	+18V	-18V	70C	76dB	6MV	2uA	1uA	500MWF	.2MA	12V	15V	5V	.	420MW	14MA	70dB	74dB	37K
TBB0747	THF	DGK	INT	.	0.2V/uS	+18V	-18V	70C	86dB	6MV	500NA	200NA	237MWF	5MA	12V	15V	30V	.	85MW	3MA	70dB	76dB	300K
TBB0747-747	SIW	DGK	INT	.	0.2V/uS	+18V	-18V	70C	86dB	6MV	500NA	200NA	237MWF	5MA	12V	15V	30V	15uV/C	85MW	3MA	70dB	76dB	300K
TBB0747A	SIW	DGK	INT	.	0.2V/uS	+18V	-18V	70C	86dB	6MV	500NA	200NA	410MWF	5MA	12V	15V	30V	15uV/C	85MW	3MA	70dB	76dB	300K
TBB0747A-747	SIW	DGK	INT	.	0.2V/uS	+18V	-18V	70C	86dB	6MV	500NA	200NA	237MWF	5MA	12V	15V	30V	15uV/C	85MW	3MA	70dB	76dB	300K
TBB0748	SIW	GPU	EXT	.	.25V/uS	+18V	-18V	70C	86dB	6MV	500NA	200NA	525MWF	5MA	12V	15V	30V	.	85MW	3MA	70dB	76dB	300K
TBB0748-748	SIW	GPU	EXT	.	.25V/uS	+18V	-18V	70C	86dB	6MV	500NA	200NA	525MWF	5MA	12V	15V	30V	.	85MW	3MA	70dB	76dB	300K
TBB0748B	SIW	GPU	EXT	.	.25V/uS	+18V	-18V	70C	86dB	6MV	500NA	200NA	410MWF	5MA	12V	15V	30V	.	85MW	3MA	70dB	76dB	300K
TBB0748B-748	SIW	GPU	EXT	.	.25V/uS	+18V	-18V	70C	86dB	6MV	500NA	200NA	410MWF	5MA	12V	15V	30V	.	85MW	3MA	70dB	76dB	300K
TBB1458	SIW	DGK	INT	.5MHZ	0.3V/uS	+18V	-18V	70C	86dB	6MV	500NA	200NA	237MWF	5MA	12V	15V	30V	.	125MW	6MA	70dB	76dB	300K
TBB1458-1458	SIW	DGK	INT	.5MHZ	0.3V/uS	+18V	-18V	70C	86dB	6MV	500NA	200NA	237MWF	5MA	12V	15V	30V	.	125MW	6MA	70dB	76dB	300K
TBB1458B	SIW	DGK	INT	.5MHZ	0.3V/uS	+18V	-18V	70C	86dB	6MV	500NA	200NA	320MWF	5MA	12V	15V	30V	.	125MW	6MA	70dB	76dB	300K
TBB1458B1458	SIW	DGK	INT	.5MHZ	0.3V/uS	+18V	-18V	70C	86dB	6MV	500NA	200NA	320MWF	5MA	12V	15V	30V	.	125MW	6MA	70dB	76dB	300K
TBB2331	SIW	DGK	INT	.	3V/uS	+15V	-15V	70C	75dB	15MV	50NA	20NA	235MWF	24MA	15V	13V	15V	50uV/C	.	2MA	65dB	80dB	1M
TBB2331B	SIW	DGK	INT	.	3V/uS	+15V	-15V	70C	75dB	15MV	50NA	20NA	320MWF	24MA	15V	13V	15V	50uV/C	.	2MA	65dB	80dB	1M
TBB4331A	SIW	DGK	INT	.	.	+15V	-15V	70C	70dB	15MV	50NA	20NA	300MWF	25MA	14V	.	13V	50uV/C	.	3MA	65dB	80dB	1M
TBC0747	SIW	DGK	INT	.	0.2V/uS	+22V	-22V	125C	94dB	4MV	350NA	100NA	526MWF	5MA	12V	15V	30V	15uV/C	85MW	3MA	80dB	76dB	300K
TBC0747-747	SIW	DGK	INT	.	0.2V/uS	+22V	-22V	125C	94dB	4MV	350NA	100NA	526MWF	5MA	12V	15V	30V	15uV/C	85MW	3MA	80dB	76dB	300K
TBC0748	SIW	GPU	EXT	.	.25V/uS	+22V	-22V	125C	94dB	4MV	350NA	100NA	525MWF	5MA	12V	15V	30V	15uV/C	85MW	3MA	80dB	76dB	300K
TBC0748-748	SIW	GPU	EXT	.	.25V/uS	+22V	-22V	125C	94dB	4MV	350NA	100NA	525MWF	5MA	12V	15V	30V	15uV/C	85MW	3MA	80dB	76dB	300K
TBC1458	SIW	DGK	INT	.5MHZ	0.3V/uS	+22V	-22V	125C	94dB	4MV	350NA	100NA	525MWF	5MA	12V	15V	30V	.	150MW	6MA	80dB	76dB	300K
TBC1458-1558	SIW	DGK	INT	.5MHZ	0.3V/uS	+22V	-22V	125C	94dB	4MV	350NA	100NA	525MWF	5MA	12V	15V	30V	.	150MW	6MA	80dB	76dB	300K
TBC2332	SIW	DGK	INT	.	3V/uS	+15V	-15V	125C	80dB	10MV	30NA	15NA	525MWF	24MA	15V	13V	15V	50uV/C	.	2MA	70dB	80dB	1M
TBE2335	SIW	DGK	INT	.	3V/uS	+15V	-15V	85C	75dB	15MV	50NA	25NA	315MWF	15V	13V	15V	50uV/C	.	2MA	65dB	80dB	1M
TBE2335B	SIW	DGK	INT	.	3V/uS	+15V	-15V	85C	75dB	15MV	50NA	25NA	425MWF	24MA	15V	13V	15V	50uV/C	.	2MA	65dB	80dB	1M
TBE4335A	SIW	OGK	INT	.	.	+15V	-15V	85C	70dB	15MV	50NA	20NA	430MWF	25MA	14V	.	13V	50uV/C	.	3MA	65dB	80dB	1M
TCA220B	MJG	TGU	EXT	2MHZ	0.2V/uS	+9V	-9V	70C	64dB	10MV	2uA	0.2uA	.	.	4V	9V	5V	.	.	2MA	80dB	56dB	.
TCA220N(16)	MJG	TGU	EXT	2MHZ	0.2V/uS	+9V	-9V	70C	64dB	10MV	2uA	0.2uA	.	.	4V	9V	5V	.	.	2MA	80dB	56dB	.

For detailed explanations of column heading notations, see App. A.

Also for ready references the more important abbreviations used in the column headings are listed below:

LEFT HAND PAGE
APP = application (codes at APP.E.)
CMRR = common mode rejection ratio
CMP = compensation (frequency)
dV_{IO}/dT = input offset voltage temperature drift
GBP = gain bandwidth product
I_B = input bias current
I_{IO} = input bias offset current
I_Q = quiescent supply current
MFR = manufacturer (codes at App.C.)
P_Q = quiescent power consumer
PSRR = power supply rejection ratio
V_{ICM} = common mode input voltage rating
V_{IDF} = differential input voltage rating
V_{IO} = input offset voltage
V_S = dc supply voltage

RIGHT HAND PAGE
Lead out coding summary (details at APP.G.) for different cases (APP.F.)
A = gain adjust
B = bias adjust
C = case
E— = inverting input
E+ = non-inverting input
F,F* = input frequency compensation
G = ground
J = high level input
K = output, open collector
L = output, open emitter
M = metal case
N = not connected
Q = special terminal
R,R* = outputs
S = strobe
T,T* = offset balance
V+ = +ve dc supply
V- = —ve dc supply
W = guard ring
X = blank position, no lead
+ + = +ve supplementary dc supply
— — = —ve supplementary dc supply
ø,ø* = output frequency compensation

CASE (APP F)	LD 1	LD 2	LD 3	LD 4	LD 5	LD 6	LD 7	LD 8	LD 9	LD 10	LD 11	LD 12	LD 13	LD 14	LD 15	LD 16	EUROPE SUBSTITUTE	USA SUBSTITUTE	IS	TYPE NUMBER
FLP-10/3P	V+	N	E+	N	E-	V-	N	K	N	ø	.	.	.	.					0	TAA862F
T05-8/4M	X	V+	E+	E-	X	V-	K	ø	.	.								TAA862	0	TAA865
DIL-6/1P	V+	E+	E-	V-	K	ø	.	.											0	TAA865A
FLP-6/2P	V+	E+	E-	V-	K	ø	.												0	TAA865W
T05-8/1M	E+1	E-1	V+	E-2	E+2	R2	V-	R1	.	.								TAA2765	0	TAA2761
DIL-8/1P	E+1	E-1	V+	E-2	E+2	R2	V-	R1	.									TAA2765A	0	TAA2761A
T05-8/1M	E+1	E-1	V+	E-2	E+2	R2	V-	R1	.										0	TAA2762
T05-8/1M	E+1	E-1	V+	E-2	E+2	R2	V-	R1	.									TAA2762	0	TAA2765
DIL-8/1P	E+1	E-1	V+	E-2	E+2	R2	V-	R1	.										0	TAA2765A
DIL-14/1C	V-	R3	E+3	E-3	E-4	E+4	R4	R1	E+1	E-1	V+	E-2	E+2	R2	.				0	TAA4761A
DIL-14/1C	V-	R3	E+3	E-3	E-4	E+4	R4	R1	E+1	E-1	V+	E-2	E+2	R2	.				0	TAA4765A
T05-8/1M	T	E-	E+	V-M	T*	R	V+	N	.								LM741CH	UA741HC	0	TBA221
T05-8/1M	T	E-	E+	V-M	T*	R	V+	N	.								LM741CH	UA741HC	0	TBA221-741
DIL-14/1P	N	N	T	E-	E+	V-	N	N	T*	R	V+	N	N	N	.		LM741CD	UA741DC	0	TBA221A
DIL-14/1P	N	N	T	E-	E+	V-	N	N	T*	R	V+	N	N	N	.		LM741CD	UA741DC	0	TBA221A-741
DIL-8/1P	T	E-	E+	V-	T*	R	V+	N	.								LM741CJ	UA741TC	0	TBA221B
DIL-8/1P	T	E-	E+	V-	T*	R	V+	N	.								LM741CJ	UA741TC	0	TBA221B-741
MDL-8/2P	T	E-	E+	V-	T*	R	V+	N	.									TBA221G	0	TBA221D
MDL-8/2P	T	E-	E+	V-	T*	R	N	V+	.										0	TBA221G
MDL-8/2P	T	E-	E+	V-	T*	R	N	V+	.										0	TBA221G-741
FLP-8/2P	T	E-	E+	V-	T*	R	N	V+											0	TBA221N
FLP-8/2P	T	E-	E+	V-	T*	R	N	V+											0	TBA221W
FLP-8/2P	T	E-	E+	V-	T*	R	N	V+											0	TBA221W-741
T05-8/1M	T	E-	E+	V-M	T*	R	V+	N									LM741H	UA741HM	0	TBA222
T05-8/1M	T	E-	E+	V-M	T*	R	V+	N									LM741H	UA741HM	0	TBA222-741
T05-8/1M	T	E-	E+	V-M	T*	R	V+	N										ICL741LNTY	0	TBA222Q1
T05-8/1M	T	E-	E+	V-M	T*	R	V+	N										ICL741LNTY	0	TBA222Q1-741
T05-8/1M	T	E-	E+	V-M	T*	R	V+	N										ICL741LNTY	0	TBA222Q2
T05-8/1M	T	E-	E+	V-M	T*	R	V+	N										ICL741LNTY	0	TBA222Q2-741
T05-8/1M	T	E-	E+	V-M	T*	R	V+	N										ICL741LNTY	0	TBA222S1
T05-8/1M	T	E-	E+	V-M	T*	R	V+	N										ICL741LNTY	0	TBA222S1-741
DIL-14/1P	R1	ø1	F1	F*1	E+1	E-1	V-	E-2	E+2	F2	F*2	ø2	R2	V+			RC4739DB	UA739DC	0	TBA231
T05-10/1M	R1	V+	E-1	E+1	V-	E+2	E-2	V+	R2	N	.						SN72747L	UA747HC	0	TBB0747
T05-10/1M	R1	V+	E-1	E+1	V-	E+2	E-2	V+	R2	N	.						SN72747L	UA747HC	0	TBB0747-747
DIL-14/1C	E-1	E+1	T1	V-	T2	E+2	E-2	T*2	V+	R2	N	R1	V+	T*1			SN72747N	UA747DC	0	TBB0747A
DIL-14/1C	E-1	E+1	T1	V-	T2	E+2	E-2	T*2	V+	R2	N	R1	V+	T*1			SN72747L	UA747HC	0	TBB0747A-747
T05-8/1M	T	E-	E+	V-M	T*	R	V+	F*	.								SN72748L	UA748HC	0	TBB0748
T05-8/1M	T	E-	E+	V-M	T*	R	V+	F*	.								SN72748L	UA748HC	0	TBB0748-748
DIL-8/1P	T	E-	E+	V-	T*	R	V+	F*	.								SN72748P	UA748TC	0	TBB0748B
DIL-8/1P	T	E-	E+	V-	T*	R	V+	F*	.								SN72748L	UA748HC	0	TBB0748B-748
T05-8/1M	R1	E-1	E+1	V-	E+2	E-2	R2	V+	.								LM1458H	MC1458G	0	TBB1458
T05-8/1M	R1	E-1	E+1	V-	E+2	E-2	R2	V+	.								LM1458H	MC1458G	0	TBB1458-1458
DIL-8/1P	R1	E-1	E+1	V-	E+2	E-2	R2	V+	.								LM1458J	MC1458U	0	TBB1458B
DIL-8/1P	R1	E-1	E+1	V-	E+2	E-2	R2	V+	.								LM1458J	MC1458U	.	TBB1458B1458
T05-8/1M	E+1	E-1	V+	E-2	E+2	R2	V-	R1	.									TBE2335	0	TBB2331
DIL-8/1P	E+1	E-1	V+	E-2	E+2	R2	V-	R1	.									TBE2335B	0	TBB2331B
DIL-14/1P	V-	R3	E+3	E-3	E-4	E+4	R4	R1	E+1	E-1	V+	E-2	E+2	R2				TBE4335A	0	TBB4331A
T05-10/1M	R1	V+	E-1	E+1	V-	E+2	E-2	V+	R2	N	.						SFC2747KM	UA747HM	0	TBC0747
T05-10/1M	R1	V+	E-1	E+1	V-	E+2	E-2	V+	R2	N	.						SFC2747KM	UA747HM	0	TBC0747-747
T05-8/1M	T	E-	E+	V-M	T*	R	V+	F*	.								SN52748L	UA748HM	0	TBC0748
T05-8/1M	T	E-	E+	V-M	T*	R	V+	F*	.								SN52748L	UA748HM	0	TBC0748-748
T05-8/1M	R1	E-1	E+1	V-	E+2	E-2	R2	V+	.								LM1558H	MC1558G	0	TBC1458
T05-8/1M	R1	E-1	E+1	V-	E+2	E-2	R2	V+	.								LM1558H	MC1558G	.	TBC1458-1558
T05-8/1M	E+1	E-1	V+	E-2	E+2	R2	V-	R1	.										0	TBB2332
T05-8/1M	E+1	E-1	V+	E-2	E+2	R2	V-	R1	.									TBC2332	0	TBE2335
DIL8/1P	E+1	E-1	V+	E-2	E+2	R2	V-	R1	.										0	TBB2335B
DIL-14/1P	V-	R3	E+3	E-3	E-4	E+4	R4	R1	E+1	E-1	V+	E-2	E+2	R2					0	TBE4335A
DIL-16/1P	E-1	E+1	B	E+2	E-2	V+	E+3	E-3	N	F3	R3	F2	R2	R1	F1	V-			0	TCA220B
DIL-16/1P	E-1	E+1	B	E+2	E-2	V+	E+3	E-3	N	F3	R3	F2	R2	R1	F1	V-		TCA220B	0	TCA220N(16)
DIL-14/1P	R1	ø1	F1	F*1	E+1	E-1	V-	E+1	E+2	F2	F*2	ø2	R2	V+			TBA231	UA739PC	0	TCA250

TYPE NUMBER	M F R	A P D	C M P	GBP MIN	SLEW RATE MIN	$V_S{}^+$ MAX	V_S MAX	T_{op} MAX	A_{VOL} MIN	V_{IO} MAX	I_B MAX	I_{IO} MAX	P_{TOT} MAX	I_{OUT} MIN	V_{OUT} MIN	V_{ICM} MAX	V_{IDF} MAX	dV_{IO}/dT MAX	P_Q MAX	I_Q MAX	CM RR MIN	PS RR MIN	R_{IN} MIN
TCA250	ITG	DLN	EXT	.	.	+12V	-12V	60C	72dB	10MV	5uA	1uA	670MWF	4MA	4V	.	5V	.	420MW	14MA	70dB	70dB	25K
TCA311	SIW	HIR	EXT	.	3V/uS	+15V	-15V	70C	75dB	20MV	50NA	25NA	235MWF	24MA	15V	13V	15V	50uV/C	.	3MA	60dB	74dB	1M
TCA311A	SIW	HIR	EXT	.	3V/uS	+15V	-15V	70C	75dB	20MV	50NA	25NA	320MWF	24MA	15V	13V	15V	50uV/C	.	3MA	60dB	74dB	1M
TCA311W	SIW	HIR	EXT	.	3V/uS	+15V	-15V	70C	75dB	20MV	50NA	25NA	225MWF	24MA	15V	13V	15V	50uV/C	.	3MA	60dB	74dB	1M
TCA312	SIW	HIR	EXT	.	3V/uS	+15V	-15V	125C	80dB	14MV	30NA	15NA	525MWF	24MA	15V	13V	15V	50uV/C	.	3MA	65dB	74dB	1M
TCA315	SIW	HIR	EXT	.	3V/uS	+15V	-15V	85C	75dB	20MV	50NA	25NA	315MWF	24MA	15V	13V	15V	50uV/C	.	3MA	60dB	74dB	1M
TCA315A	SIW	HIR	EXT	.	3V/uS	+15V	-15V	85C	75dB	20MV	50NA	25NA	425MWF	24MA	15V	13V	15V	50uV/C	.	3MA	60dB	74dB	1M
TCA315W	SIW	HIR	EXT	.	3V/uS	+15V	-15V	85C	75dB	20MV	50NA	20NA	300MWF	24MA	15V	13V	15V	50uV/C	.	3MA	60dB	74dB	1M
TCA321	SIW	HCO	EXT	.	3V/uS	+15V	-15V	70C	75dB	7.5MV	1uA	0.3uA	235MWF	25MA	15V	13V	15V	50uV/C	.	3MA	60dB	74dB	50K
TCA321A	SIW	HCO	EXT	.	3V/uS	+15V	-15V	70C	75dB	7.5MV	1uA	0.3uA	320MWF	25MA	15V	13V	15V	50uV/C	.	3MA	60dB	74dB	50K
TCA321W	SIW	HCO	EXT	.	3V/uS	+15V	-15V	70C	75dB	7.5MV	1uA	0.3uA	225MWF	25MA	15V	13V	15V	50uV/C	.	3MA	60dB	74dB	50K
TCA322	SIW	HCO	EXT	.	3V/uS	+15V	-15V	125C	80dB	5MV	0.7uA	0.1uA	525MWF	25MA	15V	13V	15V	30uV/C	.	3MA	65dB	74dB	50K
TCA325	SIW	HCO	EXT	.	3V/uS	+15V	-15V	85C	75dB	7.5MV	1uA	0.3uA	315MWF	25MA	15V	13V	15V	50uV/C	.	3MA	60dB	74dB	50K
TCA325A	SIW	HCO	EXT	.	3V/uS	+15V	-15V	85C	75dB	7.5MV	1uA	0.3uA	425MWF	25MA	15V	13V	15V	50uV/C	.	3MA	60dB	74dB	50K
TCA325W	SIW	HCO	EXT	.	3V/uS	+15V	-15V	85C	75dB	7.5MV	1uA	0.3uA	300MWF	25MA	15V	13V	15V	50uV/C	.	3MA	60dB	74dB	50K
TCA331	SIW	HIR	EXT	.	3V/uS	+15V	-15V	70C	75dB	20MV	50NA	25NA	235MWF	25MA	14V	13V	15V	50uV/C	.	3MA	60dB	74dB	1M
TCA331A	SIW	HIR	EXT	.	3V/uS	+15V	-15V	70C	75dB	20MV	50NA	25NA	320MWF	25MA	14V	13V	15V	50uV/C	.	3MA	60dB	74dB	1M
TCA331W	SIW	HIR	EXT	.	3V/uS	+15V	-15V	70C	75dB	20MV	50NA	25NA	225MWF	25MA	14V	13V	15V	50uV/C	.	3MA	60dB	74dB	1M
TCA332	SIW	HIR	EXT	.	3V/uS	+15V	-15V	125C	80dB	14MV	30NA	15NA	525MWF	25MA	14V	13V	15V	50uV/C	.	3MA	65dB	74dB	1M
TCA335	SIW	HIR	EXT	.	3V/uS	+15V	-15V	85C	75dB	20MV	50NA	25NA	315MWF	25MA	14V	13V	15V	50uV/C	.	3MA	60dB	74dB	1M
TCA335A	SIW	HIR	EXT	.	3V/uS	+15V	-15V	85C	75dB	20MV	50NA	25NA	425MWF	25MA	14V	13V	15V	50uV/C	.	3MA	60dB	74dB	1M
TCA335W	SIW	HIR	EXT	.	3V/uS	+15V	-15V	85C	75dB	20MV	50NA	25NA	300MWF	25MA	14V	13V	15V	50uV/C	.	3MA	60dB	74dB	1M
TCA410A	MUG	VFA	INT	.	1V/uS	+18V	-18V	70C	0dB	10MV	1NA	.	200MWF	2MA	13V	15V	.	.	100MW	4MA	.	.	.
TCA410B	MUG	VFA	T	.	1V/uS	+18V	-18V	70C	0dB	10MV	3NA	.	200MWF	2MA	13V	15V	.	.	100MW	4MA	.	.	.
TCA410D	MUG	VFA	INT	.	1V/uS	+18V	-18V	70C	0dB	10MV	3NA	.	.	2MA	13V	15V	.	.	100MW	4MA	.	.	.
TCA410DE	MUG	VFA	INT	.	1V/uS	+18V	-18V	70C	0dB	10MV	1NA	.	200MWF	2MA	13V	15V	.	.	100MW	4MA	.	.	.
TCA490	MUG	DLN	EXT	.	1V/uS	+18V	-18V	70C	84dB	6MV	1.5uA	750NA	400MWF	2MA	12V	15V	5V	10uV/C	330MW	14MA	70dB	70dB	50K
TCA490A	MUG	DLN	EXT	.	1V/uS	+18V	-18V	70C	84dB	6MV	1.5uA	750NA	400MWF	2MA	12V	15V	5V	10uV/C	330MW	14MA	70dB	70dB	50K
TCA490B	MUG	DLN	EXT	.	1V/uS	+18V	-18V	70C	84dB	6MV	1.5uA	750NA	400MWF	2MA	12V	15V	5V	10uV/C	330MW	14MA	70dB	70dB	50K
TCA490C	MUG	DLN	EXT	.	1V/uS	+18V	-18V	70C	84dB	6MV	1.5uA	750NA	400MWF	2MA	12V	15V	5V	10uV/C	330MW	14MA	70dB	70dB	50K
TCA520B	MUG	GPU	EXT	.3MHZ	0.1V/uS	+10V	-10V	70C	84dB	6MV	100NA	30NA	400MWF	12MA	10V	.	6V	.	.	2MA	.	.	.
TCA520D	MUG	GPU	EXT	.3MHZ	0.1V/uS	+10V	-10V	70C	84dB	6MV	100NA	30NA	.	12MA	10V	.	6V	.	.	2MA	.	.	.
TCA520N(8)	MUG	GPU	EXT	.3MHZ	0.1V/uS	+10V	-10V	70C	84dB	6MV	100NA	30NA	400MWF	12MA	10V	.	6V	.	.	2MA	.	.	.
TCA520V	MUG	GPU	EXT	.3MHZ	0.1V/uS	+10V	-10V	70C	84dB	6MV	100NA	30NA	400MWF	12MA	10V	.	6V	.	.	2MA	.	.	.
TCA680	MUG	HSR	INT	2MHZ	5V/uS	+18V	-18V	70C	90dB	8MV	100NA	30NA	500MWF	.	13V	13V	.	.	.	2MA	80dB	80dB	.
TCA680B	MUG	HSR	INT	2MHZ	5V/uS	+18V	-18V	70C	90dB	8MV	100NA	30NA	400MWF	.	13V	13V	.	.	.	2MA	80dB	80dB	.
TCA680D	MUG	HSR	INT	2MHZ	5V/uS	+18V	-18V	70C	90dB	8MV	100NA	30NA	.	.	13V	13V	.	.	.	2MA	80dB	80dB	.
TCA680N(8)	MUG	HSR	INT	2MHZ	5V/uS	+18V	-18V	70C	90dB	8MV	100NA	30NA	400MWF	.	13V	13V	.	.	.	2MA	80dB	80dB	.
TCA680T	MUG	HSR	INT	2MHZ	5V/uS	+18V	-18V	70C	90dB	8MV	100NA	30NA	500MWF	.	13V	13V	.	.	.	2MA	80dB	80dB	.
TCA680V	MUG	HSR	INT	2MHZ	5V/uS	+18V	-18V	70C	90dB	8MV	100NA	30NA	400MWF	.	13V	13V	.	.	.	2MA	80dB	80dB	.
TDA0301D	MUG	GPU	EXT	.	.	+18V	-18V	70C	88dB	7.5MV	250NA	50NA	.	5MA	12V	15V	30V	30uV/C	.	3MA	70dB	70dB	500K
TDA0319D	MUG	DCP	INT	.	.	+18V	-18V	70C	78dB	8MV	1uA	0.2uA	500MWF	.	.	15V	5V	.	.	12MA	.	.	.
TDAG324D	MUG	OGK	INT	.	.	+16V	-16V	70C	88dB	7MV	250NA	50NA	.	.	.	16V	16V	35uV/C	.	2MA	65dB	65dB	.
TDA0358D	MUG	DGK	INT	.	.	+16V	-16V	70C	88dB	7MV	250NA	50NA	.	10MA	.	16V	32V	30uV/C	.	3MA	65dB	65dB	.
TDA0741D	MUG	GPK	INT	.	.25V/uS	+18V	-18V	70C	86dB	6MV	500NA	200NA	.	3MA	10V	15V	30V	.	85MW	3MA	70dB	76dB	300K
TDA0748D	MUG	GPU	EXT	.	.25V/uS	+18V	-18V	70C	94dB	6MV	500NA	200NA	.	50MA	10V	15V	30V	.	85MW	3MA	70dB	96dB	300K
TDA1034D	MUG	LNA	INT	3MHZ	4V/uS	+20V	-20V	85C	90dB	4MV	1.5uA	0.5uA	.	6MA	12V	13V	.	.	.	7MA	86dB	86dB	.
TDA1034N(8)	MUG	LNA	INT	3MHZ	4V/uS	+20V	-20V	85C	90dB	4MV	1.5uA	0.5uA	.	6MA	12V	13V	.	.	.	7MA	86dB	86dB	.
TDA1034NT	MUG	LNA	INT	3MHZ	4V/uS	+20V	-20V	85C	90dB	4MV	1.5uA	0.5uA	.	6MA	12V	13V	.	.	.	7MA	86dB	86dB	.
TDA1034T	MUG	LNA	INT	3MHZ	4V/uS	+20V	-20V	85C	90dB	4MV	1.5uA	0.5uA	.	6MA	12V	13V	.	.	.	7MA	86dB	86dB	.
TDA1034NV	MUG	LNA	INT	3MHZ	4V/uS	+20V	-20V	85C	90dB	4MV	1.5uA	0.5uA	.	6MA	12V	13V	.	.	.	7MA	86dB	86dB	.
TDA1034V	MUG	LNA	INT	3MHZ	4V/uS	+20V	-20V	85C	90dB	4MV	1.5uA	0.5uA	.	6MA	12V	13V	.	.	.	7MA	86dB	86dB	.
TDA1458D	MUG	DGK	INT	.5MHZ	0.3V/uS	+18V	-18V	75C	88dB	6MV	0.5uA	0.2uA	.	5MA	12V	15V	30V	.	125MW	6MA	70dB	76dB	300K
TDA4250CD	MUG	PRA	INT	.	.	+18V	-18V	70C	95dB	6MV	75NA	20NA	.	1MA	12V	15V	30V	.	3MW	.1MA	70dB	74dB	.
TDA4250CN(8)	MUG	PRA	INT	.	.	+18V	-18V	70C	95dB	6MV	75NA	20NA	500MWF	1MA	12V	15V	30V	.	3MW	.1MA	70dB	74dB	.
TDA4250CV	MUG	PRA	INT	.	.	+18V	-18V	70C	95dB	6MV	75NA	20NA	500MWF	1MA	12V	15V	30V	.	3MW	.1MA	70dB	74dB	.
TDA4250T	MUG	PRA	INT	.	.	+18V	-18V	125C	100dB	5MV	50NA	10NA	500MWF	1MA	12V	15V	30V	.	2.7MW	90UA	70dB	76dB	.
TDB0118-CM	THF	XSR	INT	.	50V/uS	+18V	-18V	70C	88dB	10MV	600NA	200NA	500MWF	5MA	12V	15V	1V	.	.	10MA	70dB	65dB	1M
TDB0119-CM	THF	DCP	INT	.	.	+18V	-18V	70C	78dB	8MV	1uA	0.2uA	500MWF	6MA	.	15V	5V	.	.	12MA	.	.	.
TDB0119-DP	THF	DCP	INT	.	.	+18V	-18V	70C	78dB	8MV	1uA	0.2uA	500MWF	6MA	.	15V	5V	.	.	12MA	.	.	.

154

For detailed explanations of column heading notations, see App. A.

Also for ready references the more important abbreviations used in the column headings are listed below:

LEFT HAND PAGE
APP = application (codes at APP.E.)
CMRR = common mode rejection ratio
CMP = compensation (frequency)
dV_{IO}/dT = input offset voltage temperature drift
GBP = gain bandwidth product
I_B = input bias current
I_{IO} = input bias offset current
I_Q = quiescent supply current
MFR = manufacturer (codes at App.C.)
P_Q = quiescent power consumer
PSRR = power supply rejection ratio
V_{ICM} = common mode input voltage rating
V_{IDF} = differential input voltage rating
V_{IO} = input offset voltage
V_S = dc supply voltage

RIGHT HAND PAGE
Lead out coding summary (details at APP.G.) for different cases (APP.F.)
A = gain adjust
B = bias adjust
C = case
E— = inverting input
E+ = non-inverting input
F,F* = input frequency compensation
G = ground
J = high level input
K = output, open collector
L = output, open emitter
M = metal case
N = not connected
Q = special terminal
R,R* = outputs
S = strobe
T,T* = offset balance
V+ = +ve dc supply
V— = —ve dc supply
W = guard ring
X = blank position, no lead
+ + = +ve supplementary dc supply
— — = —ve supplementary dc supply
$\phi,\phi*$ = output frequency compensation

CASE (APP F)	LD1	LD2	LD3	LD4	LD5	LD6	LD7	LD8	LD9	LD10	LD11	LD12	LD13	LD14	LD15	LD16	EUROPE SUBSTITUTE	USA SUBSTITUTE	IS	TYPE NUMBER
T05-8/4M	X	V+	E+	E-	X	V-	K	B										TCA315	0	TCA311
DIL-6/1P	V+	E+	E-	V-	K	B												TCA315A	0	TCA311A
FLP-6/2P	E+	E-	V-	K	B	V+												TCA315W	0	TCA311W
T05-8/4M	X	V+	E+	E-	X	V-	K	B											0	TCA312
T05-8/4M	X	V+	E+	E-	X	V-	K	B										TCA312	0	TCA315
DIL-6/1P	V+	E+	E-	V-	K	B													0	TCA315A
FLP-6/2P	E+	E-	V-	K	B	V+													0	TCA315W
T05-8/4M	X	V+	E+	E-	X	V-	K	B										TCA325	0	TCA321
DIL-6/1P	V+	E+	E-	V-	K	B												TCA325A	0	TCA321A
FLP-6/2P	E+	E-	V-	K	B	V+												TCA325W	0	TCA321W
T05-8/4M	X	V+	E+	E-	X	V-	K	B											0	TCA322
T05-8/4M	X	V+	E+	E-	X	V-	K	B										TCA322	0	TCA325
DIL-6/1P	V+	E+	E-	V-	K	B													0	TCA325A
FLP-6/2P	E+	E-	V-	K	B	V+													0	TCA325W
T05-8/4M	X	V+	E+	E-	X	V-	K	ø										TCA335	G	TCA331
DIL-6/1P	V+	E+	E-	V-	K	ø												TCA335A	0	TCA331A
FLP-6/2P	E+	E-	V-	K	ø	V+												TCA335W	0	TCA331W
T05-8/4M	X	V+	E+	E-	X	V-	K	ø											0	TCA332
T05-8/4M	X	V+	E+	E-	X	V-	K	ø										TCA332	0	TCA335
DIL-6/1P	V+	E+	E-	V-	K	ø													0	TCA335A
FLP-6/2P	E+	E-	V-	K	ø	V+													0	TCA335W
T18-4/2M	R	V+	E	V-														TCA410DE	0	TCA410A
T18-4/2M	R	V+	E	V-														TCA410A	0	TCA410B
MDL-6/1P	N	E	V-	N	R	V+													0	TCA410D
T18-4/2M	R	V+	E	V-														TCA410A	0	TCA410DE
DIL-14/1P	R1	ø1	F1	F*1	E+1	E-1	V-	E-2	E+2	F2	F*2	ø2	R2	V+			TCA490A	UA749PC	0	TCA490
DIL-14/1P	R1	ø1	F1	F*1	E+1	E-1	V-	E-2	E+2	F2	F*2	ø2	R2	V+				TCA490B	0	TCA490A
DIL-14/1P	R1	ø1	F1	F*1	E+1	E-1	V-	E-2	E+2	F2	F*2	ø2	R2	V+				TCA490C	0	TCA490B
DIL-14/1P	R1	ø1	F1	F*1	E+1	E-1	V-	E-2	E+2	F2	F*2	ø2	R2	V+				UA749PC	0	TCA490C
DIL-8/1P	T	E-	E+	V-	F	R	V+	T*										TCA520V	0	TCA520B
MDL-8/2P	T	E-	E+	V-	F	R	V+	T*											0	TCA520D
DIL-8/1P	T	E-	E+	V-	F	R	V+	T*										TCA520V	0	TCA520N(8)
DIL-8/1P	T	E-	E+	V-	F	R	V+	T*										TCA520B	0	TCA520V
T05-8/1M	T	E-	E+	V-	N	R	V+	T*										TCA680T	0	TCA680
DIL-8/1P	T	E-	E+	V-	F	R	V+	T*										TCA680V	0	TCA680B
MDL-8/2P	T	E-	E+	V-	N	R	V+	T*											0	TCA680D
DIL-8/1P	T	E-	E+	V-	N	R	V+	T*										TCA680V	0	TCA680N(8)
T05-8/1M	T	E-	E+	V-	N	R	V+	T*										TCA680	0	TCA680T
DIL-8/1P	T	E-	E+	V-	N	R	V+	T*										TCA680N(8)	0	TCA680V
MDL-8/2P	FT	E-	E+	V-	T*	R	V+	F*										LM301AD	0	TDA301D
MDL-14/4P	N	N	G1	E+1	E-1	V-	R2	G2	E+2	E-2	V+	R1	N	N					0	TDA0319D
MDL-14/4P	R1	E-1	E+1	V+	E+2	E-2	R2	R3	E-3	E+3	G	E+4	E-4	R4				LM324D	0	TDA0324D
MDL-14/4P	R1	E-1	E+1	G	E+2	E-2	R2	V+										LM358D	0	TDA0358D
MDL-8/2P	T	E-	E+	V-	T*	R	V+	N										UA741CD	0	TDA0741D
MDL-8/2P	FT	E-	E+	V-	T*	R	V+	F*										UA748CD	0	TDA0748D
MDL-8/2P	T	E-	E+	V-	F	R	V+	T*											0	TDA1034D
DIL-8/1P	T	E-	E+	V-	F	R	V+	T*										TDA1034	0	TDA1034N(8)
T05-8/1M	T	E-	E+	V-	F	R	V+	T*											0	TDA1034NT
T05-8/1M	T	E-	E+	V-	F	R	V+	T*											0	TDA1034T
DIL-8/1P	T	E-	E+	V-	F	R	V+	T*											0	TDA1034NV
DIL-8/1P	T	E-	E+	V-	F	R	V+	T*										TDA1034N8	0	TDA1034V
MDL-8/2P	R1	E-1	E+1	V-	E+2	E-2	R2	V+										MC1458D	0	TDA1458D
MDL-8/2P	T	E-	E+	V-	T*	R	V+	B											0	TDA4250CD
DIL-8/1P	T	E-	E+	V-	T*	R	V+	B									LM4250J	LM4250CN	0	TDA4250CN(8)
DIL-8/1P	T	E-	E+	V-	T*	R	V+	B									LM4250J	LM4250CN	0	TDA4250CV
T05-8/1M	T	E-	E+	V-	T*	R	V+	B									SG4250T	LM4250H	0	TDA4250T
T05-8/1M	T*F	E-	E+	V-	F*T	R	V+	ø									SF.C2138	LM318H	0	TDB0118-CM
T05-10/1M	R1	G1	E+1	E-1	V-	R2	G2	E+2	E-2	V+								LM319H	0	TDB0119-CM
DIL-14/1P	N	N	G1	E+1	E-1	V-	R2	G2	E+2	E-2	V+	R1	N	N				LM319D	0	TDB0119-DP
DIL-14/1P	R1	E-1	E+1	V+	E+2	E-2	R2	R3	E-3	E+3	G	E+4	E-4	R4			MLM324J	LM324D	0	TDB0124

TYPE NUMBER	M F R	A P P	C M P	GBP MIN	SLEW RATE MIN	V$_S^+$ MAX	V$_S^-$ MAX	T$_{OP}$ MAX	A$_{VOL}$ MIN	V$_{IO}$ MAX	I$_B$ MAX	I$_{IO}$ MAX	P$_{TOT}$ MAX	I$_{OUT}$ MIN	V$_{OUT}$ MIN	V$_{ICM}$ MAX	V$_{IDF}$ MAX	dV$_{IO}$/dT MAX	P$_Q$ MAX	I$_O$ MAX	CM RR MIN	PS RR MIN	R$_{IN}$ MIN
TDB0124	THF	QGK	INT	.	.	+16V	-16V	70C	88dB	5MV	250NA	50NA	900MWF	.	.	16V	16V	35uV/C	.	2MA	65dB	65dB	.
TDB0791-DP	THF	HPO	INT	.	.	+18V	-18V	70C	86dB	6MV	500NA	200NA	500MWF	1A	11V	15V	30V	.	.	25MA	70dB	76dB	300K
TDB0791-EP	THF	HPO	INT	.	.	+18V	-18V	70C	86dB	6MV	500NA	200NA	15WH	1A	11V	15V	30V	.	.	25MA	70dB	76dB	300K
TDB0791-KM	THF	HPO	INT	.	.	+18V	-18V	70C	86dB	6MV	500NA	200NA	15WC	1A	11V	15V	30V	.	.	25MA	70dB	76dB	300K
TDC0118-CM	TFH	XSR	INT	.	50V/uS	+18V	-18V	125C	94dB	6MV	300NA	50NA	500MWF	7MA	12V	15V	1V	.	.	8MA	80dB	70dB	1M
TDC0119-CM	THF	DCP	INT	.	.	+18V	-18V	125C	80dB	4MV	500NA	75NA	500MWF	6MA	.	15V	5V	.	.	12MA	.	.	.
TDC0119-DC	THF	DCP	INT	.	.	+18V	-18V	125C	80dB	4MV	500NA	75NA	500MWF	6MA	.	15V	5V	.	.	12MA	.	.	.
TDC0124	THF	QGK	INT	.	.	+16V	-16V	125C	94dB	7MV	150NA	30NA	900MWF	.	.	16V	16V	35uV/C	.	2MA	70dB	65dB	.
TDC0791-KM	THF	HPO	INT	.	.	+22V	-22V	125C	94dB	5MV	500NA	200NA	15WC	1A	11V	15V	30V	.	.	25MA	70dB	76dB	300K
TDC5711	TRU	DCP	EXT	.	.	+14V	-7V	70C	58dB	5MV	100uA	15uA	300MWF	5MA	1.7V	7V	5V	.	300MW	11MA	.	.	.
TDC5711F	TRU	DCP	EXT	.	.	+14V	-7V	70C	58dB	5MV	100uA	15uA	300MWF	5MA	1.7V	7V	5V	.	300MW	11MA	.	.	.
TDC5711P	TRU	DCP	EXT	.	.	+14V	-7V	70C	58dB	5MV	100uA	15uA	300MWF	5MA	1.7V	7V	5V	.	300MW	11MA	.	.	.
TDE0118-CM	TFH	XSR	INT	.	50V/uS	+18V	-18V	85C	88dB	10MV	600NA	200NA	500MW	5MA	12V	15V	1V	.	.	10MA	70dB	65dB	1M
TDE0119-CM	THF	DCP	INT	.	.	+18V	-18V	85C	80dB	4MV	500NA	75NA	500MWF	6MA	.	15V	5V	.	.	12MA	.	.	.
TDE0119-DP	THF	DCP	INT	.	.	+18V	-18V	85C	80dB	4MV	500NA	75NA	500MWF	6MA	.	15V	5V	.	.	12MA	.	.	.
TDE0124	THF	QGK	INT	.	.	+16V	-16V	85C	94dB	5MV	150NA	30NA	900MWF	.	.	16V	16V	35uV/C	.	2MA	70dB	65dB	.
TDF2902DP	THG	QGK	INT	.	.	+16V	-16V	85C	88dB	7MV	250NA	50NA	570MWF	.	.	16V	16V	35uV/C	.	2MA	50dB	50dB	.
TOA101AE	TRU	GPU	EXT	.	.	+22V	-22V	125C	94dB	2MV	75NA	10NA	500MWF	5MA	12V	15V	30V	15uV/C	.	3MA	80dB	80dB	1.5M
TOA101AF	TRU	GPU	EXT	.	.	+22V	-22V	125C	94dB	2MV	75NA	10NA	500MWF	5MA	12V	15V	30V	15uV/C	.	3MA	80dB	80dB	1.5M
TOA101AJ	TRU	GPU	EXT	.	.	+22V	-22V	125C	94dB	2MV	75NA	10NA	500MWF	5MA	12V	15V	30V	15uV/C	.	3MA	80dB	80dB	1.5M
TOA101AV	TRU	GPU	EXT	.	.	+22V	-22V	125C	94dB	2MV	75NA	10NA	500MWF	5MA	12V	15V	30V	15uV/C	.	3MA	80dB	80dB	1.5M
TOA201AE	TRU	GPU	EXT	.	.	+22V	-22V	85C	94dB	2MV	75NA	10NA	500MWF	5MA	12V	15V	30V	15uV/C	.	3MA	80dB	80dB	500K
TOA201AF	TRU	GPU	EXT	.	.	+22V	-22V	85C	94dB	2MV	75NA	10NA	500MWF	5MA	12V	15V	30V	15uV/C	.	3MA	80dB	80dB	500K
TOA201AJ	TRU	GPU	EXT	.	.	+22V	-22V	85C	94dB	2MV	75NA	10NA	500MWF	5MA	12V	15V	30V	15uV/C	.	3MA	80dB	80dB	500K
TOA201AV	TRU	GPU	EXT	.	.	+22V	-22V	85C	94dB	2MV	75NA	10NA	500MWF	5MA	12V	15V	30V	15uV/C	.	3MA	80dB	80dB	500K
TOA301AE	TRU	GPU	EXT	.	.	+18V	-18V	70C	88dB	7.5MV	250NA	50NA	500MWF	5MA	12V	15V	30V	30uV/C	.	3MA	70dB	70dB	500K
TOA301AF	TRU	GPU	EXT	.	.	+18V	-18V	70C	88dB	7.5MV	250NA	50NA	.	5MA	12V	15V	30V	30uV/C	.	3MA	70dB	70dB	500K
TOA301AJ	TRU	GPU	EXT	.	.	+18V	-18V	70C	88dB	7.5MV	250NA	50NA	500MWF	5MA	12V	15V	30V	30uV/C	.	3MA	70dB	70dB	500K
TOA301AV	TRU	GPU	EXT	.	.	+18V	-18V	70C	88dB	7.5MV	250NA	50NA	500MWF	5MA	12V	15V	30V	30uV/C	.	3MA	70dB	70dB	500K
TOA1709E	TRU	GPU	EXT	.	.	+18V	-18V	125C	88dB	5MV	500NA	200NA	330MWF	5MA	12V	10V	5V	15uV/C	165MW	.	70dB	76dB	150K
TOA1709F	TRU	GPU	EXT	.	.	+18V	-18V	125C	88dB	5MV	500NA	200NA	330MWF	5MA	12V	10V	5V	15uV/C	165MW	.	70dB	76dB	150K
TOA1709J	TRU	GPU	EXT	.	.	+18V	-18V	125C	88dB	5MV	500NA	200NA	330MWF	5MA	12V	10V	5V	15uV/C	165MW	.	70dB	76dB	150K
TOA1709P	TRU	GPU	EXT	.	.	+18V	-18V	125C	88dB	5MV	500NA	200NA	330MWF	5MA	12V	10V	5V	15uV/C	165MW	.	70dB	76dB	150K
TOA1709V	TRU	GPU	EXT	.	.	+18V	-18V	125C	88dB	5MV	500NA	200NA	300MWF	5MA	12V	10V	5V	15uV/C	165MW	.	70dB	76dB	150K
TOA1741E	TRU	GPK	INT	.	0.1V/uS	+22V	-22V	125C	94dB	5MV	500NA	200NA	500MWF	5MA	12V	15V	30V	.	85MW	3MA	70dB	76dB	300K
TOA1741F	TRU	GPK	INT	.	0.1V/uS	+22V	-22V	125C	94dB	5MV	500NA	200NA	500MWF	5MA	12V	15V	30V	.	85MW	3MA	70dB	76dB	300K
TOA1741J	TRU	GPK	INT	.	0.1V/uS	+22V	-22V	125C	94dB	5MV	500NA	200NA	500MWF	5MA	12V	15V	30V	.	85MW	3MA	70dB	76dB	300K
TOA1741P	TRU	GPK	INT	.	0.1V/uS	+22V	-22V	125C	94dB	5MV	500NA	200NA	500MWF	5MA	12V	15V	30V	.	85MW	3MA	70dB	76dB	300K
TOA1741V	TRU	GPK	INT	.	0.1V/uS	+22V	-22V	125C	94dB	5MV	500NA	200NA	500MWF	5MA	12V	15V	30V	.	85MW	3MA	70dB	76dB	300K
TOA1741WF	TRU	GPU	EXT	.	0.1V/uS	+22V	-22V	125C	94dB	5MV	500NA	200NA	500MWF	5MA	12V	15V	30V	.	85MW	.	70dB	76dB	300K
TOA1741WP	TRU	GPU	EXT	.	0.1V/uS	+22V	-22V	125C	94dB	5MV	500NA	200NA	500MWF	5MA	12V	15V	30V	.	85MW	.	70dB	76dB	300K
TOA1741WV	TRU	GPU	EXT	.	0.1V/uS	+22V	-22V	125C	94dB	5MV	500NA	200NA	500MWF	5MA	12V	15V	30V	.	85MW	.	70dB	76dB	300K
TOA1747AV	TRU	DGK	INT	.	0.1V/uS	+22V	-22V	125C	94dB	5MV	500NA	200NA	800MWF	5MA	12V	15V	30V	.	85MW	3MA	70dB	76dB	300K
TOA1747BV	TRU	DGK	INT	.	0.1V/uS	+22V	-22V	125C	94dB	5MV	500NA	200NA	800MWF	5MA	12V	15V	30V	.	85MW	3MA	70dB	76dB	300K
TOA1747J	TRU	DGK	INT	.	0.1V/uS	+22V	-22V	125C	94dB	5MV	500NA	200NA	800MWF	5MA	12V	15V	30V	.	85MW	3MA	70dB	76dB	300K
TOA1748E	TRU	GPU	EXT	.	0.1V/uS	+22V	-22V	125C	94dB	5MV	500NA	200NA	500MWF	5MA	12V	15V	30V	.	85MW	3MA	70dB	76dB	300K
TOA1748F	TRU	GPU	EXT	.	0.1V/uS	+22V	-22V	125C	94dB	5MV	500NA	200NA	500MWF	5MA	12V	15V	30V	.	85MW	3MA	70dB	76dB	300K
TOA1748J	TRU	GPU	EXT	.	0.1V/uS	+22V	-22V	125C	94dB	5MV	500NA	200NA	500MWF	5MA	12V	15V	30V	.	85MW	3MA	70dB	76dB	300K
TOA1748P	TRU	GPU	EXT	.	0.1V/uS	+22V	-22V	125C	94dB	5MV	500NA	200NA	500MWF	5MA	12V	15V	30V	.	85MW	3MA	70dB	76dB	300K
TOA1748V	TRU	GPU	EXT	.	0.1V/uS	+22V	-22V	125C	94dB	5MV	500NA	200NA	500MWF	5MA	12V	15V	30V	.	85MW	3MA	70dB	76dB	300K
TOA1809E	TRU	DGU	EXT	.	.	+18V	-18V	125C	88dB	5MV	500NA	200NA	415MWF	5MA	12V	10V	5V	15uV/C	160MW	.	70dB	76dB	150K
TOA1809J	TRU	DGU	EXT	.	.	+18V	-18V	125C	88dB	5MV	500NA	200NA	415MWF	5MA	12V	10V	5V	15uV/C	160MW	.	70dB	76dB	150K
TOA2709E	TRU	GPU	EXT	.	.	+18V	-18V	70C	84dB	7.5MV	1.5uA	0.5uA	115MWF	5MA	12V	10V	5V	.	200MW	.	65dB	74dB	50K
TOA2709F	TRU	GPU	EXT	.	.	+18V	-18V	70C	84dB	7.5MV	1.5uA	0.5uA	115MWF	5MA	12V	10V	5V	.	200MW	.	65dB	74dB	50K
TOA2709J	TRU	GPU	EXT	.	.	+18V	-18V	70C	84dB	7.5MV	1.5uA	0.5uA	115MWF	5MA	12V	10V	5V	.	200MW	.	65dB	74dB	50K
TOA2709P	TRU	GPU	EXT	.	.	+18V	-18V	70C	84dB	7.5MV	1.5uA	0.5uA	115MWF	5MA	12V	10V	5V	.	200MW	.	65dB	74dB	50K
TOA2709V	TRU	GPU	EXT	.	.	+18V	-18V	70C	84dB	7.5MV	1.5uA	0.5uA	300MWF	5MA	12V	10V	5V	.	200MW	.	65dB	74dB	50K
TOA2741E	TRU	GPK	INT	.	0.1V/uS	+18V	-18V	70C	86dB	6MV	500NA	200NA	500MW	5MA	12V	15V	30V	.	85MW	3MA	70dB	76dB	300K
TOA2741F	TRU	GPK	INT	.	0.1V/uS	+18V	-18V	70C	86dB	6MV	500NA	200NA	500MWF	5MA	12V	15V	30V	.	85MW	3MA	70dB	76dB	300K
TOA2741J	TRU	GPK	INT	.	0.1V/uS	+18V	-18V	70C	86dB	6MV	500NA	200NA	500MWF	5MA	12V	15V	30V	.	85MW	3MA	70dB	76dB	300K

For detailed explanations of column heading notations, see App. A.

Also for ready references the more important abbreviations used in the column headings are listed below:

LEFT HAND PAGE
- APP = application (codes at APP.E.)
- CMRR = common mode rejection ratio
- CMP = compensation (frequency)
- dV_{IO}/dT = input offset voltage temperature drift
- GBP = gain bandwidth product
- I_B = input bias current
- I_{IO} = input bias offset current
- I_Q = quiescent supply current
- MFR = manufacturer (codes at App.C.)
- P_Q = quiescent power consumer
- PSRR = power supply rejection ratio
- V_{ICM} = common mode input voltage rating
- V_{IDF} = differential input voltage rating
- V_{IO} = input offset voltage
- V_S = dc supply voltage

RIGHT HAND PAGE
Lead out coding summary (details at APP.G.) for different cases (APP.F.)
- A = gain adjust
- B = bias adjust
- C = case
- E− = inverting input
- E+ = non-inverting input
- F,F* = input frequency compensation
- G = ground
- J = high level input
- K = output, open collector
- L = output, open emitter
- M = metal case
- N = not connected
- Q = special terminal
- R,R* = outputs
- S = strobe
- T,T* = offset balance
- V+ = +ve dc supply
- V− = −ve dc supply
- W = guard ring
- X = blank position, no lead
- ++ = +ve supplementary dc supply
- −− = −ve supplementary dc supply
- ø,ø* = output frequency compensation

CASE (APP F)	LD1	LD2	LD3	LD4	LD5	LD6	LD7	LD8	LD9	LD10	LD11	LD12	LD13	LD14	LD15	LD16	EUROPE SUBSTITUTE	USA SUBSTITUTE	I S	TYPE NUMBER
DIL-14/1P	F	Q	V+	V+	R	N	V-	V-	ø	T	N	T*	E-	N					O	TDB0791-DP
HIL-14/1P	F	Q	V+	V+	R	N	V-	V-	ø	T	N	T*	E-	N					O	TDB0791-EP
T03-10/2M	R	V+	Q	F	E+	E-	T	T*	ø	V-M								UA791KC	O	TDB0791-KM
T05-8/1M	T*F	E-	E+	V-	F*T	R	V+	ø									SF.C2118M	LM218H	O	TDC0118-CM
T05-10/1M	R1	G1	E+1	E-1	V-	R2	G2	E+2	E-2	V+								LM119H	O	TDC0119-CM
DIL-14/1C	N	N	G1	E+1	E-1	V-	R2	G2	E+2	E-2	V+	R1	N	N				LM119D	O	TDC0119-DC
DIL-14/1P	R1	E-1	E+1	V+	E+2	E-2	R2	R3	E-3	E+3	G	E+4	E-4	R4			MLM124D	LM124D	O	TDC0124
T03-10/2M	R	V+	Q	F	E+	E-	T	T*	ø	V-M								UA791KM	O	TDC0791-KM
T05-10/1M	G	S1	E-1	E+1	V-	E+2	E-2	S2	R	V+							SFC2711C	UA711HC	O	TDC5711
FLP-10/3C	E-1	E+2	V-	E+2	E-2	S2	R	V+	G	S1							SFC2711PM	UA711FM	O	TDC5711F
DIL-14/1P	N	E-2	E+2	V-	E+2	E-2	N	N	S2	R	V+	G	S1	N			SFC2711EC	UA711DC	O	TDC5711P
T05-8/1M	T*F	E-	E+	V-	F*T	R	V+	ø									SF.C2218	LM118H	O	TDE0118-CM
T05-10/1M	R1	G1	E+1	E-1	V-	R2	G2	E+2	E-2	V+								LM219H	O	TDE0119-CM
DIL-14/1C	N	N	G1	E+1	E-1	V-	R2	G2	E+2	E-2	V+	R1	N	N				LM219D	O	TDE0119-DP
DIL-14/1P	R1	E-1	E+1	V+	E+2	E-2	R2	R3	E-3	E+3	G	E+4	E-4	R4			SG224J	LM224D	O	TDE0124
DIL-14/1P	R1	E-1	E+1	V+	E+2	E-2	R2	R3	E-3	E+3	G	E+4	E-4	R4			MLM2902P	LM2902J	O	TDF2902DP
DIL-14/1P	N	N	FT	E-	E+	V-	N	N	T*	R	V+	F*	N	N			UA101AD	LM101AJ14	O	T0A101AE
FLP-10/3C	N	FT	E-	E+	V-	T*	R	V+	F*	N							SFC2101APM	LM101AF	O	T0A101AF
DIL-14/1C	N	N	FT	E-	E+	V-	N	N	T*	R	V+	F*	N	N			UA101AD	LM101AJ14	O	T0A101AJ
T05-8/1M	FT	E-	E+	V-M	T*	R	V+	F*									SFC2101A	LM101AH	O	T0A101AV
DIL-14/1P	N	N	FT	E-	E+	V-	N	N	T*	R	V+	F*	N	N			UA201D	LM201AJ14	O	T0A201AE
FLP-10/3C	N	FT	E-	E+	V-	T*	R	V+	F*	N							SFC2201APT	LM201AF	O	T0A201AF
DIL-14/1C	N	N	FT	E-	E+	V-	N	N	T*	R	V+	F*	N	N			UA201AD	LM201AJ14	O	T0A201AJ
T05-8/1M	FT	E-	E+	V-M	T*	R	V+	F*									SFC2101A	LM201AH	O	T0A201AV
DIL-14/1P	N	N	FT	E-	E+	V-	N	N	T*	R	V+	F*	N	N			UA301AD	LM301AJ14	O	T0A301AE
FLP-10/3C	N*	FT	E-	E+	V-	T*	R	V+	F*	N							SFC2201APM	LM201AF	O	T0A301AF
DIL-14/1C	N	N	FT	E-	E+	V-	N	N	T*	R	V+	F*	N	N			UA301AD	LM301AJ14	O	T0A301AJ
T05-8/1M	FT	E-	E+	V-M	T*	R	V+	F*									SFC2301AH	LM301AH	O	T0A301AV
DIL-14/1P	N	N	F*	E-	E+	V-	N	N	ø	R	V+	F	N	N			LM709D	UA709DM	O	T0A1709E
FLP-10/3C	N	F*	E-	E+	V-	ø	R	V+	F	N								UA709FM	O	T0A1709F
DIL-14/1C	N	N	F*	E-	E+	V-	N	N	ø	R	V+	F	N	N			LM709D	UA709DM	O	T0A1709J
DIL-14/1P	N	N	F*	E-	E+	V-	N	N	ø	R	V+	F+	N	N			LM709D	UA709DM	O	T0A1709P
T05-8/1M	F*	E-	E+	V-M	ø	R	V+	F									TAA522	UA709HM	O	T0A1709V
DIL-14/1P	N	N	T	E-	E+	V-	N	N	T*	R	V+	N	N				LM741D	UA741DM	O	T0A1741E
FLP-10/3C	N	T	E-	E+	V-	T*	R	V+	N	N							LM741F	UA741FM	O	T0A1741F
DIL-14/1C	N	N	T	E-	E+	V-	N	N	T*	R	V+	N	N				LM741D	UA741DM	O	T0A1741J
DIL-14/1P	N	N	T	E-	E+	V-	N	N	T*	R	V+	N	N				LM741D	UA741DM	O	T0A1741P
T05-8/1M	T	E-	E+	V-M	T*	R	V+	N									TBA222	UA741HM	O	T0A1741V
FLP-10/3C	N	T	E-	E+	V-	T*	R	V+	F	N							SN52748FA	UA748FM	O	T0A1741WF
DIL-14/1C	N	N	T	E-	E+	V-	N	N	T*	R	V+	F	N	N			SN52748JA	UA748DM	O	T0A1741WP
T05-8/1M	T	E-	E+	V-	T*	R	V+	F									TBC0748	UA748HM	O	T0A1741WV
T05-10/1M	R1	V+1	E-1	E+1	V-	E+2	V+2	R2	N								SFC2747M	UA747HM	O	T0A1747AV
T05-8/1M	R1	E-1	E+1	V-	E+2	E-2	R2	V+									TBC1458	MC1558G	O	T0A1747BV
DIL-14/1C	E-1	E+1	T1	V-	T2	E+2	E-2	T*2	V+2	R2	N	R1	V+1	T*1			SFC2747KM	UA747DM	O	T0A1747J
DIL-14/1P	N	N	TF	E-	E+	V-	N	N	T*	R	V+	F*	N	N			SN52748JA	UA748DM	O	T0A1748E
FLP-10/3C	N	TF	E-	E+	V-	T*	R	V+	F*	N							SN52748FA	UA748FM	O	T0A1748F
DIL-14/1C	N	N	TF	E-	E+	V-	N	N	T*	R	V+	F*	N	N			SN52748JA	UA748DM	O	T0A1748J
DIL-14/1P	N	N	TF	E-	E+	V-	N	N	T*	R	V+	F*	N	N			SN52748JA	UA748DM	O	T0A1748P
T05-8/1M	TF	E-	E+	V-M	T*	R	V+	F*									TBC0748	UA748HM	O	T0A1748V
DIL-14/1P	ø2	R	F2	F*2	E-2	E+2	V-	E+1	E-1	F*1	F1	R1	ø1	V+			RM1537DC	MC1537L	O	T0A1809E
DIL-14/1C	ø2	R	F2	F*2	E-2	E+2	V-	E+1	E-1	F*1	F1	R1	ø1	V+			RM1537DC	MC1537L	O	T0A1809J
DIL-14/1P	N	N	F*	E-	E+	V-	N	N	ø	R	V+	F	N	N			TAA521A	UA709DC	O	T0A2709E
FLP-10/1C	N	F*	E-	E+	V-	ø	R	V+	F	N								MC1709F	O	T0A2709F
DIL-14/1C	N	N	F*	E-	E+	V-	N	N	ø	R	V+	F	N	N			TAA521A	UA709DC	O	T0A2709J
DIL-14/1P	N	N	F*	E-	E+	V-	N	N	ø	R	V+	F	N	N			TAA521A	UA709DC	O	T0A2709P
T05-8/1M	F*	E-	E+	V-M	ø	R	V+	F									TAA521	UA709HC	O	T0A2709V
DIL-14/1P	N	N	T	E-	E+	V-	N	N	T*	R	V+	N	N				TBA221A	UA741DC	O	T0A2741E
FLP-10/3C	N	T	E-	E+	V-	T*	R	V+	N	N							LM741F	UA741FM	O	T0A2741F
DIL-14/1C	N	N	T	E-	E+	V-	N	N	T*	R	V+	N	N				TBA221A	UA741DC	O	T0A2741J
DIL-14/1P	N	N	T	E-	E+	V-	N	N	T*	R	V+	N	N				TBA221A	UA741DC	O	T0A2741P

TYPE NUMBER	MFR	APP	CMP	GBP MIN	SLEW RATE MIN	Vs+ MAX	Vs− MAX	Topp MAX	AVUL MIN	VIU MAX	IB MAX	IIU MAX	PTOT MAX	IOUT MIN	VOUT MIN	VICM MAX	VIDF MAX	dVIO/dT MAX	PQ MAX	IQ MAX	CMRR MIN	PSRR MIN	RIN MIN
TOA2741P	TRU	GPK	INT	.	0.1V/uS	+18V	−18V	70C	86dB	6MV	500NA	200NA	500MWF	5MA	12V	15V	30V	.	85MW	3MA	70dB	76dB	300K
TOA2741V	TRU	GPK	INT	.	0.1V/uS	+18V	−18V	70C	86dB	6MV	500NA	200NA	500MWF	5MA	12V	15V	30V	.	85MW	3MA	70dB	76dB	300K
TOA2741WF	TRU	GPU	EXT	.	0.1V/uS	+18V	−18V	70C	86dB	6MV	500NA	200NA	500MWF	5MA	12V	15V	30V	.	85MW	.	70dB	76dB	300K
TOA2741WP	TRU	GPU	EXT	.	0.1V/uS	+18V	−18V	70C	86dB	6MV	500NA	200NA	500MWF	5MA	12V	15V	30V	.	85MW	.	70dB	76dB	300K
TOA2741WV	TRU	GPU	EXT	.	0.1V/uS	+18V	−18V	70C	86dB	6MV	500NA	200NA	500MWF	5MA	12V	15V	30V	.	85MW	.	70dB	76dB	300K
TOA274AV	TRU	DGK	INT	.	0.1V/uS	+18V	−18V	70C	86dB	6MV	500NA	200NA	800MWF	5MA	12V	15V	30V	.	85MW	3MA	70dB	76dB	300K
TOA2747BV	TRU	DGK	INT	.	0.1V/uS	+18V	−18V	70C	86dB	6MV	500NA	200NA	800MWF	5MA	12V	15V	30V	.	85MW	3MA	70dB	76dB	300K
TOA2747J	TRU	DGK	INT	.	0.1V/uS	+18V	−18V	70C	86dB	6MV	500NA	200NA	800MWF	5MA	12V	15V	30V	.	85MW	3MA	70dB	76dB	300K
TOA2748E	TRU	GPU	EXT	.	0.1V/uS	+18V	−18V	70C	86dB	6MV	500NA	200NA	500MWF	5MA	12V	15V	30V	.	85MW	3MA	70dB	76dB	300K
TOA2748F	TRU	GPU	EXT	.	0.1V/uS	+18V	−18V	70C	86dB	6MV	500NA	200NA	500MWF	5MA	12V	15V	30V	.	85MW	3MA	70dB	76dB	300K
TOA2748J	TRU	GPU	EXT	.	0.1V/uS	+18V	−18V	70C	86dB	6MV	500NA	200NA	500MWF	5MA	12V	15V	30V	.	85MW	3MA	70dB	76dB	300K
TOA2748P	TRU	GPU	EXT	.	0.1V/uS	+18V	−18V	70C	86dB	6MV	500NA	200NA	500MWF	5MA	12V	15V	30V	.	85MW	3MA	70dB	76dB	300K
TOA2748V	TRU	GPU	EXT	.	0.1V/uS	+18V	−18V	70C	86dB	6MV	500NA	200NA	500MWF	5MA	12V	15V	30V	.	85MW	3MA	70dB	76dB	300K
TOA2809E	TRU	DGU	EXT	.	.	+18V	−18V	70C	84dB	7.5MV	1.5uA	0.5uA	415MWF	5MA	12V	10V	5V	.	200MW	.	65dB	74dB	50K
TOA2809J	TRU	DGU	EXT	.	.	+18V	−18V	70C	84dB	7.5MV	1.5uA	0.5uA	500MWF	5MA	12V	10V	5V	.	200MW	.	65dB	74dB	50K
TUA3709E	TRU	GPU	EXT	.	.	+18V	−18V	70C	88dB	5MV	500NA	200NA	115MWF	5MA	12V	10V	5V	15uV/C	165MW	.	70dB	76dB	150K
TOA3709F	TRU	GPU	EXT	.	.	+18V	−18V	70C	88dB	5MV	500NA	200NA	115MWF	5MA	12V	10V	5V	15uV/C	165MW	.	70dB	76dB	150K
TOA3709J	TRU	GPU	EXT	.	.	+18V	−18V	70C	88dB	5MV	500NA	200NA	115MWF	5MA	12V	10V	5V	15uV/C	165MW	.	70dB	76dB	150K
TOA3709P	TRU	GPU	EXT	.	.	+18V	−18V	70C	88dB	5MV	500NA	200NA	115MWF	5MA	12V	10V	5V	15uV/C	165MW	.	70dB	76dB	150K
TOA3709V	TRU	GPU	EXT	.	.	+18V	−18V	70C	88dB	5MV	500NA	200NA	300MWF	5MA	12V	10V	5V	15uV/C	165MW	.	70dB	76dB	150K
TOA3741E	TRU	GPK	INT	.	0.2V/uS	+22V	−22V	125C	94dB	2MV	10NA	5NA	500MWF	5MA	12V	15V	30V	.	85MW	3MA	70dB	76dB	10M
TOA3741F	TRU	GPK	INT	.	0.2V/uS	+22V	−22V	125C	94dB	2MV	10NA	5NA	500MWF	5MA	12V	15V	30V	.	85MW	3MA	70dB	76dB	10M
TOA3741J	TRU	GPK	INT	.	0.2V/uS	+22V	−22V	125C	94dB	2MV	10NA	5NA	500MWF	5MA	12V	15V	30V	.	85MW	3MA	70dB	76dB	10M
TOA3741P	TRU	GPK	INT	.	0.2V/uS	+22V	−22V	125C	94dB	2MV	10NA	5NA	500MWF	5MA	12V	15V	30V	.	85MW	3MA	70dB	76dB	10M
TOA3741V	TRU	GPK	INT	.	0.2V/uS	+22V	−22V	125C	94dB	2MV	10NA	5NA	500MWF	5MA	12V	15V	30V	.	85MW	3MA	70dB	76dB	10M
TOA3748E	TRU	GPU	EXT	.	0.2V/uS	+22V	−22V	125C	94dB	2MV	75NA	10NA	500MWF	5MA	12V	15V	30V	30uV/C	85MW	3MA	80dB	80dB	1.5M
TOA3748F	TRU	GPU	EXT	.	0.2V/uS	+22V	−22V	125C	94dB	2MV	75NA	10NA	500MWF	5MA	12V	15V	30V	30uV/C	85MW	3MA	80dB	80dB	1.5M
TOA3748J	TRU	GPU	EXT	.	0.2V/uS	+22V	−22V	125C	94dB	2MV	75NA	10NA	500MWF	5MA	12V	15V	30V	30uV/C	85MW	3MA	80dB	80dB	1.5M
TOA3748V	TRU	GPU	EXT	.	0.2V/uS	+22V	−22V	125C	94dB	2MV	75NA	10NA	500MWF	5MA	12V	15V	30V	30uV/C	85MW	3MA	80dB	80dB	1.5M
TOA4709E	TRU	GPU	EXT	.	.	+18V	−18V	125C	88dB	2MV	200NA	50NA	330MWF	5MA	12V	10V	5V	10uV/C	108MW	.	80dB	80dB	350K
TOA4709F	TRU	GPU	EXT	.	.	+18V	−18V	125C	88dB	2MV	200NA	50NA	330MWF	5MA	12V	10V	5V	10uV/C	108MW	.	80dB	80dB	350K
TOA4709J	TRU	GPU	EXT	.	.	+18V	−18V	125C	88dB	2MV	200NA	50NA	330MWF	5MA	12V	10V	5V	10uV/C	108MW	.	80dB	80dB	350K
TOA4709P	TRU	GPU	EXT	.	.	+18V	−18V	125C	88dB	2MV	200NA	50NA	330MWF	5MA	12V	10V	5V	10uV/C	108MW	.	80dB	80dB	350K
TOA4709V	TRU	GPU	EXT	.	.	+18V	−18V	125C	88dB	2MV	200NA	50NA	300MWF	5MA	12V	10V	5V	10uV/C	108MW	.	80dB	80dB	350K
TOA7741E	TRU	GPK	INT	.	0.2V/uS	+22V	−22V	125C	94dB	5MV	30NA	10NA	500MWF	5MA	12V	15V	30V	.	85MW	3MA	70dB	76dB	3M
TOA7741F	TRU	GPK	INT	.	0.2V/uS	+22V	−22V	125C	94dB	5MV	30NA	10NA	500MWF	5MA	12V	15V	30V	.	85MW	3MA	70dB	76dB	3M
TOA7741J	TRU	GPK	INT	.	0.2V/uS	+22V	−22V	125C	94dB	5MV	30NA	10NA	500MWF	5MA	12V	15V	30V	.	85MW	3MA	70dB	76dB	3M
TOA7741P	TRU	GPK	INT	.	0.2V/uS	+22V	−22V	125C	94dB	5MV	30NA	10NA	500MWF	5MA	12V	15V	30V	.	85MW	3MA	70dB	76dB	3M
TOA7741V	TRU	GPK	INT	.	0.2V/uS	+22V	−22V	125C	94dB	5MV	30NA	10NA	500MWF	5MA	12V	15V	30V	.	85MW	3MA	70dB	76dB	3M
TOA7747AV	TRU	DGK	INT	.	0.2V/uS	+22V	−22V	125C	94dB	5MV	30NA	10NA	800MWF	5MA	12V	15V	30V	.	85MW	3MA	70dB	76dB	3M
TOA7747BV	TRU	DGK	INT	.	0.2V/uS	+22V	−22V	125C	94dB	5MV	30NA	10NA	800MWF	5MA	12V	15V	30V	.	85MW	3MA	70dB	76dB	3M
TOA7748E	TRU	GPU	EXT	.	0.2V/uS	+22V	−22V	125C	94dB	5MV	30NA	10NA	500MWF	5MA	12V	15V	30V	.	85MW	3MA	70dB	76dB	3M
TOA7748F	TRU	GPU	EXT	.	0.2V/uS	+22V	−22V	125C	94dB	5MV	30NA	10NA	500MWF	5MA	12V	15V	30V	.	85MW	3MA	70dB	76dB	3M
TOA7748J	TRU	GPU	EXT	.	0.2V/uS	+22V	−22V	125C	94dB	5MV	30NA	10NA	500MWF	5MA	12V	15V	30V	.	85MW	3MA	70dB	76dB	3M
TOA7748V	TRU	GPU	EXT	.	0.2V/uS	+22V	−22V	125C	94dB	5MV	30NA	10NA	500MWF	5MA	12V	15V	30V	.	85MW	3MA	70dB	76dB	3M
TOA8741E	TRU	GPK	INT	.	0.2V/uS	+18V	−18V	70C	86dB	6MV	60NA	20NA	500MWF	5MA	12V	15V	30V	.	85MW	3MA	70dB	76dB	1M
TOA8741F	TRU	GPK	INT	.	0.2V/uS	+18V	−18V	70C	86dB	6MV	60NA	20NA	500MWF	5MA	12V	15V	30V	.	85MW	3MA	70dB	76dB	1M
TOA8741J	TRU	GPK	INT	.	0.2V/uS	+18V	−18V	70C	86dB	6MV	60NA	20NA	500MWF	5MA	12V	15V	30V	.	85MW	3MA	70dB	76dB	1M
TOA8741P	TRU	GPK	INT	.	0.2V/uS	+18V	−18V	70C	86dB	6MV	60NA	20NA	500MWF	5MA	12V	15V	30V	.	85MW	3MA	70dB	76dB	1M
TOA8741V	TRU	GPK	INT	.	0.2V/uS	+18V	−18V	70C	86dB	6MV	60NA	20NA	500MWF	5MA	12V	15V	30V	.	85MW	3MA	70dB	76dB	1M
TOA8747AV	TRU	DGK	INT	.	0.2V/uS	+18V	−18V	70C	86dB	6MV	60NA	20NA	800MWF	5MA	12V	15V	30V	.	85MW	3MA	70dB	76dB	1M
TOA8747BV	TRU	DGK	INT	.	0.2V/uS	+18V	−18V	70C	86dB	6MV	60NA	20NA	800MWF	5MA	12V	15V	30V	.	85MW	3MA	70dB	76dB	1M
TOA8748E	TRU	GPU	EXT	.	0.2V/uS	+18V	−18V	70C	86dB	6MV	60NA	20NA	500MWF	5MA	12V	15V	30V	.	85MW	3MA	70dB	76dB	1M
TOA8748F	TRU	GPU	EXT	.	0.2V/uS	+18V	−18V	70C	86dB	6MV	60NA	20NA	500MWF	5MA	12V	15V	30V	.	85MW	3MA	70dB	76dB	1M
TOA8748J	TRU	GPU	EXT	.	0.2V/uS	+18V	−18V	70C	86dB	6MV	60NA	20NA	500MWF	5MA	12V	15V	30V	.	85MW	3MA	70dB	76dB	1M
TOA8748V	TRU	GPU	EXT	.	0.2V/uS	+18V	−18V	70C	86dB	6MV	60NA	20NA	500MWF	5MA	12V	15V	30V	.	85MW	3MA	70dB	76dB	1M
TSC1225E	TRU	DCP	EXT	.	.	+14V	−7V	125C	84dB	3.5MV	30uA	10uA	300MWF	.	1.3V	7V	5V	.	.	20MA	.	.	.
TSC1225F	TRU	DCP	EXT	.	.	+14V	−7V	125C	84dB	3.5MV	30uA	10uA	300MWF	.	1.3V	7V	5V	.	.	20MA	.	.	.
TSC1225J	TRU	DCP	EXT	.	.	+14V	−7V	125C	84dB	3.5MV	30uA	10uA	300MWF	.	1.3V	7V	5V	.	.	20MA	.	.	.
TSC1225V	TRU	DCP	EXT	.	.	+14V	−7V	125C	84dB	3.5MV	30uA	10uA	300MWF	.	1.3V	7V	5V	.	.	20MA	.	.	.

158

For detailed explanations of column heading notations, see App. A.

Also for ready references the more important abbreviations used in the column headings are listed below.

LEFT HAND PAGE

APP = application (codes at APP.E.)

CMRR = common mode rejection ratio

CMP = compensation (frequency)

dV_{io}/dT = input offset voltage temperature drift

GBP = gain bandwidth product

I_B = input bias current

I_{io} = input bias offset current

I_Q = quiescent supply current

MFR = manufacturer (codes at App.C.)

P_Q = quiescent power consumer

PSRR = power supply rejection ratio

V_{ICM} = common mode input voltage rating

V_{IDF} = differential input voltage rating

V_{io} = input offset voltage

V_S = dc supply voltage

RIGHT HAND PAGE

Lead out coding summary (details at APP.G.) for different cases (APP.F.)

A = gain adjust
B = bias adjust
C = case
E– = inverting input
E+ = non-inverting input
F,F* = input frequency compensation
G = ground
J = high level input
K = output, open collector
L = output, open emitter
M = metal case
N = not connected
Q = special terminal
R,R* = outputs
S = strobe
T,T* = offset balance
V+ = +ve dc supply
V– = –ve dc supply
W = guard ring
X = blank position, no lead
++ = +ve supplementary dc supply
–– = –ve supplementary dc supply
ø,ø* = output frequency compensation

CASE (APP F)	LD1	LD2	LD3	LD4	LD5	LD6	LD7	LD8	LD9	LD10	LD11	LD12	LD13	LD14	LD15	LD16	EUROPE SUBSTITUTE	USA SUBSTITUTE	I/S	TYPE NUMBER
T05-8/1M	T	E–	E+	V-M	T*	R	V+	N									TBA221	UA741HC	0	T0A2741V
FLP-10/3C	N	T	E–	E+	V–	T*	R	V+	F	N							LM741F	UA741FM	0	T0A2741WF
DIL-14/1C	N	N	T	E–	E+	V–	N	T*	R	V+	F	N					TBA221A	UA741DC	0	T0A2741WP
T05-8/1M	T	E–	E+	V-M	T*	R	V+	F									TSB0748	UA748HC	0	T0A2741WV
T05-10/1M	R1	V+1	E-1	E+1	V–		E+2	E-2	V+2	R2	N						TBB0747	UA747HC	0	T0A2747AV
T05-8/1M	R1	E-1	E+1	V–	E+2	E-2	R2	V+									TBB1458	MC1458	0	T0A2747BV
DIL-14/1C	E-1	E+1	T1	V–	T2	E+2	E-2	T*2	V+2	R2	N	R1	V+	T*1			TBB0747A	UA747DC	0	T0A2747J
DIL-14/1P	N	N	TF	E–	E+	V–	N	N	T*	R	V+	F*	N	N			SN72748J	UA748DC	0	T0A2748E
FLP-10/3C	N	TF	E–	E+	V–	T*	R	V+	F*	N							SN72748FA	UA748FM	0	T0A2748F
DIL-14/1C	N	N	TF	E–	E+	V–	N	N	T*	R	V+	F*	N	N			SN72748J	UA748DC	0	T0A2748J
DIL-14/1P	N	N	TF	E–	E+	V–	N	N	T*	R	V+	F*	N	N			SN72748J	UA748DC	0	T0A2748P
T05-8/1M	TF	E–	E+	V-M	T*	R	V+	F*									TBB0748	UA748HC	0	T0A2748V
DIL-14/1P	ø2	R	F2	F*2	E-2	E+2	V–	E+1	E-1	F*1	F1	R1	ø1	V+			RC1437DC	MC1437L	0	T0A2809E
DIL-14/1C	ø2	R	F2	F*2	E-2	E+2	V–	E+1	E-1	F*1	F1	R1	ø1	V+			RC1437DC	MC1437L	0	T0A2809J
DIL-14/1P	N	N	F*	E–	E+	V–	N	N	ø	R	V+	N	N					709BE	0	T0A3709E
FLP-10/3C	N	F*	E–	E+	V–	ø	R	V+	F	N								709BH	0	T0A3709H
DIL-14/1P	N	N	F*	E–	E+	V–	N	N	ø	R	V+	F	N	N				709BL	0	T0A3709J
DIL-14/1P	N	N	F*	E–	E+	V–	N	N	ø	R	V+	F	N	N				709BL	0	T0A3709P
T05-8/1M	F*	E–	E+	V-M	ø	R	V+	F										709BE	0	T0A3709V
DIL-14/1P	N	N	T	E–	E+	V–	N	N	T*	R	V+	N	N					MC1556L	0	T0A3741E
FLP-10/3C	N	T	E–	E+	V–	T*	R	V+	N	N								LM143F	0	T0A3741F
DIL-14/1C	N	N	T	E–	E+	V–	N	N	T*	R	V+	N	N					MC1556L	0	T0A3741J
DIL-14/1P	N	N	T	E–	E+	V–	N	N	T*	R	V+	N	N					MC1556L	0	T0A3741P
T05-8/1M	T	E–	E+	V-M	T*	R	V+	N									S5556T	MC1556G	0	T0A3741V
DIL-14/1P	N	N	TF	E–	E+	V–	N	N	T*	R	V+	F*	N	N				UA748ADM	0	T0A3748E
FLP-10/3C	N	TF	E–	E+	V–	T*	R	V+	F*	N								UA748AFM	0	T0A3748F
DIL-14/1C	N	N	TF	E–	E+	V–	N	N	T*	R	V+	F*	N	N				UA748ADM	0	T0A3748J
T05-8/1M	TF	E–	E+	V-M	T*	R	V+	F*										UA748AHM	0	T0A3748V
DIL-14/1P	N	N	F*	E–	E+	V–	N	N	ø	R	V+	F	N	N			LM709AJ	UA709ADM	0	T0A4709E
FLP-10/3C	N	F*	E–	E+	V–	ø	R	V+	F	N							SN52709AFA	UA709AFM	0	T0A4709F
DIL-14/1C	N	N	F*	E–	E+	V–	N	N	ø	R	V+	F	N	N			LM709AJ	UA709ADM	0	T0A4709J
DIL-14/1P	N	N	F*	E–	E+	V–	N	N	ø	R	V+	F	N	N			LM709AJ	UA709ADM	0	T0A4709P
T05-8/1M	F*	E–	E+	V-M	ø	R	V+	F									MC1709AG	UA709AHM	0	T0A4709V
DIL-14/1P	N	N	T	E–	E+	V–	N	N	T*	R	V+	N	N					MC1556L	0	T0A7741E
FLP-10/3C	N	T	E–	E+	V–	T*	R	V+	N	N								LM143F	0	T0A7741F
DIL-14/1C	N	N	T	E–	E+	V–	N	N	T*	R	V+	N	N					MC1556L	0	T0A7741J
DIL-14/1P	N	N	T	E–	E+	V–	N	N	T*	R	V+	N	N					MC1556L	0	T0A7741P
T05-8/1M	T	E–	E+	V-M	T*	R	V+	N										MC1556G	0	T0A7741V
T05-10/1M	R1	V+1	E-1	E+1	V–		E+2	E-2	V+2	R2	N						TBC0747	UA747AHM	0	T0A7747AV
T05-8/1M	R1	E-1	E+1	V–	E+2	E-2	R2	V+										UA798HM	0	T0A7747BV
DIL-14/1P	N	N	TF	E–	E+	V–	N	N	T*	R	V+	F*	N	N			UA101AD	LM101AJ14	0	T0A7748E
FLP-10/3C	N	TF	E–	E+	V–	T*	R	V+	F*	N							SFC2101APM	LM101AF	0	T0A7748F
DIL-14/1C	N	N	TF	E–	E+	V–	N	N	T*	R	V+	F*	N	N			UA101AD	LM1C1AJ14	0	T0A7748J
T05-8/1M	TF	E–	E+	V-M	T*	R	V+	F*									SFC2101A	LM101AH	0	T0A7748V
DIL-14/1P	N	N	T	E–	E+	V–	N	N	T*	R	V+	N	N					T0A7741E	0	T0A8741E
FLP-10/3C	N	T	E–	E+	V–	T*	R	V+	N	N								T0A7741F	0	T0A8741F
DIL-14/1C	N	N	T	E–	E+	V–	N	N	T*	R	V+	N	N					T0A7741J	0	T0A8741J
DIL-14/1P	N	N	T	E–	E+	V–	N	N	T*	R	V+	N	N					T0A7741P	0	T0A8741P
T05-8/1M	T	E–	E+	V-M	T*	R	V+	N										T0A7741V	0	T0A8741V
T05-10/1M	R1	V+1	E-1	E+1	V–		E+2	E-2	V+2	R2	N						TBB0747	UA747HC	0	T0A8747AV
T05-8/1M	R1	E-1	E+1	V–	E+2	E-2	R2	V+										UA798HM	0	T0A8747BV
DIL-14/1P	N	N	TF	E–	E+	V–	N	N	T*	R	V+	F*	N	N			UA201AD	LM201AJ14	0	T0A8748E
FLP-10/3C	N	TF	E–	E+	V–	T*	R	V+	F*	N							SFC2201APM	LM201AF	0	T0A8748F
DIL-14/1C	N	N	TF	E–	E+	V–	N	N	T*	R	V+	F*	N	N			UA201AD	LM201AJ14	0	T0A8748J
T05-8/1M	TF	E–	E+	V-M	T*	R	V+	F*									SFC2101A	LM201AH	0	T0A8748V
DIL-14/1P	N	E-1	E+1	V–	E+2	E-2	N	N	S2	R	V+	G	S1	N					0	TSC1225E
FLP-10/3C	E-1	E+1	V–	E+2	E-2	S2	R	V+	G	S1									0	TSC1225F
DIL-14/1C	N	E-1	E+1	V–	E+2	E-2	N	N	S2	R	V+	G	S1	N					0	TSC1225J
T05-10/1M	G	S1	E-1	E+1	V-M	E+2	E-2	S2	R	V+									0	TSC1225V
DIL-14/1P	N	E-1	E+1	V–	E+2	E-2	N	N	S2	R	V+	G	S1	N			SFC2711KM	UA711DM	0	TSC1711E

TYPE NUMBER	MFR	APP	CMP	GBP MIN	SLEW RATE MIN	V_S^+ MAX	V_S^- MAX	T_{OP} MAX	A_{VOL} MIN	V_{IO} MAX	I_B MAX	I_{IO} MAX	P_{TOT} MAX	I_{OUT} MIN	V_{OUT} MIN	V_{ICM} MAX	V_{IDF} MAX	dV_{IO}/dT MAX	P_Q MAX	I_O MAX	CMRR MIN	PSRR MIN	R_{IN} MIN
TSC1711E	TRU	DCP	EXT	.	.	+14V	-7V	125C	54dB	3.5MV	75uA	10uA	670MWF	.5MA	1.2V	7V	5V	25uV/C	200MW	9MA	.	.	.
TSC1711F	TRU	DCP	EXT	.	.	+14V	-7V	125C	54dB	3.5MV	75uA	10uA	300MWF	.5MA	1.2V	7V	5V	25uV/C	200MW	9MA	.	.	.
TSC1711J	TRU	DCP	EXT	.	.	+14V	-7V	125C	54dB	3.5MV	75uA	10uA	300MWF	.5MA	1.2V	7V	5V	25uV/C	200MW	9MA	.	.	.
TSC1711V	TRU	DCP	EXT	.	.	+14V	-7V	125C	54dB	3.5MV	75uA	10uA	300MWF	.5MA	1.2V	7V	5V	25uV/C	200MW	9MA	.	.	.
TSC2225E	TRU	DCP	EXT	.	.	+14V	-7V	70C	.	3.5MV	30uA	10uA	300MWF	.	1.3V	7V	5V	.	.	20MA	.	.	.
TSC2225F	TRU	DCP	EXT	.	.	+14V	-7V	70C	.	3.5MV	30uA	10uA	300MWF	.	1.3V	7V	5V	.	.	20MA	.	.	.
TSC2225J	TRU	DCP	EXT	.	.	+14V	-7V	70C	.	3.5MV	30uA	10uA	300MWF	.	1.3V	7V	5V	.	.	20MA	.	.	.
TSC2225V	TRU	DCP	EXT	.	.	+14V	-7V	70C	.	3.5MV	30uA	10uA	300MWF	.	1.3V	7V	5V	.	.	20MA	.	.	.
TSC2711E	TRU	DCP	EXT	.	.	+14V	-7V	70C	54dB	5MV	100uA	15uA	300MWF	.5MA	1.2V	7V	5V	25uV/C	230MW	9MA	.	.	.
TSC2711F	TRU	DCP	EXT	.	.	+14V	-7V	70C	54dB	5MV	100uA	15uA	300MWF	.5MA	1.2V	7V	5V	25uV/C	230MW	9MA	.	.	.
TSC2711J	TRU	DCP	EXT	.	.	+14V	-7V	70C	54dB	5MV	100uA	15uA	300MWF	.5MA	1.2V	7V	5V	25uV/C	230MW	9MA	.	.	.
TSC2711V	TRU	DCP	EXT	.	.	+14V	-7V	70C	54dB	5MV	100uA	15uA	300MWF	.5MA	1.2V	7V	5V	25uV/C	230MW	9MA	.	.	.
TSC3225E	TRU	DCP	EXT	.	.	+14V	-7V	70C	80dB	5MV	50uA	15uA	300MWF	.	1.3V	7V	5V	.	.	20MA	.	.	.
TSC3225F	TRU	DCP	EXT	.	.	+14V	-7V	70C	80dB	5MV	50uA	15uA	300MWF	.	1.3V	7V	5V-	.	.	20MA	.	.	.
TSC3225J	TRU	DCP	EXT	.	.	+14V	-7V	70C	80dB	5MV	50uA	15uA	300MWF	.	1.3V	7V	5V	.	.	20MA	.	.	.
TSC3225V	TRU	DCP	EXT	.	.	+14V	-7V	70C	80dB	5MV	50uA	15uA	300MWF	.	1.3V	7V	5V	.	.	20MA	.	.	.
TSC4711E	TRU	DCP	EXT	.	.	+14V	-7V	125C	54dB	3.5MV	75uA	10uA	300MWF	5MA	1.2V	7V	5V	25uV/C	300MW	11MA	.	.	.
TSC4711F	TRU	DCP	EXT	.	.	+14V	-7V	125C	54dB	3.5MV	75uA	10uA	300MWF	5MA	1.2V	7V	5V	25uV/C	300MW	11MA	.	.	.
TSC4711J	TRU	DCP	EXT	.	.	+14V	-7V	125C	54dB	3.5MV	75uA	10uA	300MWF	5MA	1.2V	7V	5V	25uV/C	300MW	11MA	.	.	.
TSC4711V	TRU	DCP	EXT	.	.	+14V	-7V	125C	54dB	3.5MV	75uA	10uA	300MWF	5MA	1.2V	7V	5V	25uV/C	300MW	11MA	.	.	.
TSC5711E	TRU	DCP	EXT	.	.	+14V	-7V	70C	54dB	5MV	100uA	15uA	300MWF	5MA	1.2V	7V	5V	25uV/C	300MW	11MA	.	.	.
TSC5711F	TRU	DCP	EXT	.	.	+14V	-7V	70C	54dB	5MV	100uA	15uA	300MWF	5MA	1.2V	7V	5V	25uV/C	300MW	11MA	.	.	.
TSC5711J	TRU	DCP	EXT	.	.	+14V	-7V	70C	54dB	5MV	100uA	15uA	300MWF	5MA	1.2V	7V	5V	25uV/C	300MW	11MA	.	.	.
TSC5711V	TRU	DCP	EXT	.	.	+14V	-7V	70C	54dB	5MV	100uA	15uA	300MWF	5MA	1.2V	7V	5V	25uV/C	300MW	11MA	.	.	.
uA101AD	FAU	GPU	EXT	.	*V/uS	+22V	-22V	125C	94dB	2MV	75NA	10NA	670MWF	5MA	12V	15V	30V	15uV/dT	.	3MA	80dB	80dB	1.5M
uA101AF	FAU	GPU	EXT	.	.	+22V	-22V	125C	94dB	2MV	75NA	10NA	670MWF	5MA	12V	15V	30V	15uV/C	.	3MA	80dB	80dB	1.5M
uA101AH	FAU	GPU	EXT	.	.	+22V	-22V	125C	94dB	2MV	75NA	10NA	500MWF	5MA	12V	15V	30V	15uV/C	.	3MA	80dB	80dB	1.5M
uA101D	FAU	GPU	EXT	.	.	+22V	-22V	125C	94dB	5MV	500NA	200NA	670MWF	5MA	12V	15V	30V	15uV/C	.	3MA	70dB	70dB	300K
uA101H	FAU	GPU	EXT	.	.	+22V	-22V	125C	94dB	5MV	500NA	200NA	500MWF	5MA	12V	15V	30V	15uV/C	.	3MA	70dB	70dB	300K
uA102M	FAU	VFA	INT	.	.	+18V	-18V	125C	0dB	5MV	10NA	.	500MWF	1MA	10V	15V	.	20uV/C	.	6MA	.	60dB	10G
uA107H	FAU	GPK	INT	.	.	+22V	-22V	125C	94dB	2MV	75NA	10NA	500MWF	5MA	12V	15V	30V	15uV/C	.	3MA	80dB	80dB	1.5M
uA108AD	FAU	SBA	EXT	.	.	+20V	-20V	125C	98dB	0.5MV	2NA	0.2NA	500MWF	1MA	13V	15V	0.5V	5uV/C	.	1MA	96dB	96dB	30M
uA108AF	FAU	SBA	EXT	.	.	+20V	-20V	125C	98dB	0.5MV	2NA	0.2NA	500MWF	1MA	13V	15V	1V	5uV/C	.	1MA	96dB	96dB	30M
uA108AH	FAU	SBA	EXT	.	.	+20V	-20V	125C	98dB	0.5MV	2NA	0.2NA	500MWF	1MA	13V	15V	1V	5uV/C	.	1MA	96dB	96dB	30M
uA108D	FAU	SBA	EXT	.	.	+20V	-20V	125C	94dB	2MV	2NA	0.2NA	500MWF	1MA	13V	15V	1V	15uV/C	.	1MA	80dB	80dB	30M
uA108F	FAU	SBA	EXT	.	.	+20V	-20V	125C	94dB	2MV	2NA	0.2NA	500MWF	1MA	13V	15V	1V	15uV/C	.	1MA	80dB	80dB	30M
uA108H	FAU	SBA	EXT	.	.	+20V	-20V	125C	94dB	2MV	2NA	0.2NA	500MWF	1MA	13V	15V	1V	15uV/C	.	1MA	80dB	80dB	30M
uA110M	FAU	VFA	INT	.	.	+18V	-18V	125C	0dB	4MV	3NA	.	500MWF	1MA	10V	15V	.	20uV/C	.	6MA	.	70dB	10G
uA111H	FAU	CPR	EXT	.	.	+18V	-18V	125C	100dB	3MV	100NA	10NA	500MWF	8MA	.	15V	30V	.	.	6MA	.	.	.
uA111R	FAU	CPR	EXT	.	.	+18V	-18V	125C	100dB	3MV	100NA	10NA	500MWF	8MA	.	15V	30V	.	.	6MA	.	.	.
uA201AD	FAU	GPU	EXT	.	.	+22V	-22V	85C	94dB	2MV	75NA	10NA	670MWF	5MA	12V	15V	30V	15uV/C	.	3MA	80dB	80dB	1.5M
uA201AF	FAU	GPU	EXT	.	.	+22V	-22V	85C	94dB	2MV	75NA	10NA	670MWF	5MA	12V	15V	30V	15uV/C	.	3MA	80dB	80dB	1.5M
uA201AH	FAU	GPU	EXT	.	.	+22V	-22V	85C	94dB	2MV	75NA	10NA	500MWF	5MA	12V	15V	30V	15uV/C	.	3MA	80dB	80dB	1.5M
uA201D	FAU	GPU	EXT	.	.	+22V	-22V	70C	86dB	7.5MV	1.5uA	500NA	670MWF	5MA	12V	15V	30V	30uV/C	.	3MA	65dB	70dB	100K
uA201H	FAU	GPU	EXT	.	.	+22V	-22V	70C	86dB	7.5MV	1.5uA	500NA	500MWF	5MA	12V	15V	30V	30uV/C	.	3MA	65dB	70dB	100K
uA207H	FAU	GPK	INT	.	.	+22V	-22V	85C	94dB	2MV	75NA	10NA	500MWF	5MA	12V	15V	30V	15uV/C	.	3MA	80dB	80dB	1.5M
uA208AD	FAU	SBA	EXT	.	.	+20V	-20V	85C	98dB	0.5MV	2NA	0.2NA	500MWF	1MA	13V	15V	1V	5uV/C	.	1MA	96dB	96dB	30M
uA208AF	FAU	SBA	EXT	.	.	+20V	-20V	85C	98dB	0.5MV	2NA	0.2NA	500MWF	1MA	13V	15V	1V	5uV/C	.	1MA	96dB	96dB	30M
uA208AH	FAU	SBA	EXT	.	.	+20V	-20V	85C	98dB	0.5MV	2NA	0.2NA	500MWF	1MA	13V	15V	1V	5uV/C	.	1MA	96dB	96dB	30M
uA208D	FAU	SBA	EXT	.	.	+20V	-20V	85C	94dB	2MV	2NA	0.2NA	500MWF	1MA	13V	15V	1V	15uV/C	.	1MA	80dB	80dB	30M
uA208F	FAU	SBA	EXT	.	.	+20V	-20V	85C	94dB	2MV	2NA	0.2NA	500MWF	1MA	13V	15V	1V	15uV/C	.	1MA	80dB	80dB	30M
uA208H	FAU	SBA	EXT	.	.	+20V	-20V	85C	94dB	2MV	2NA	0.2NA	500MWF	1MA	13V	15V	1V	15uV/C	.	1MA	80dB	80dB	30M
uA301AD	FAU	GPU	EXT	.	.	+18V	-18V	70C	88dB	7.5MV	250NA	50NA	670MWF	5MA	12V	15V	30V	30uV/C	.	3MA	70dB	70dB	0.5M
uA301AH	FAU	GPU	EXT	.	.	+18V	-18V	70C	88dB	7.5MV	250NA	50NA	500MWF	5MA	12V	15V	30V	30uV/C	.	3MA	70dB	70dB	0.5M
uA301AT	FAU	GPU	EXT	.	.	+18V	-18V	70C	88dB	7.5MV	250NA	50NA	310MWF	5MA	12V	15V	30V	30uV/C	.	3MA	70dB	70dB	0.5M
uA302C	FAU	VFA	INT	.	.	+18V	-18V	70C	0dB	15MV	30NA	.	500MWF	1MA	10V	15V	.	50uV/C	.	6MA	.	60dB	10G
uA307H	FAU	GPK	INT	.	.	+18V	-18V	70C	88dB	7.5MV	250NA	50NA	500MWF	5MA	12V	15V	30V	30uV/C	.	3MA	70dB	70dB	0.5M
uA307T	FAU	GPK	INT	.	.	+18V	-18V	70C	88dB	7.5MV	250NA	50NA	310MWF	5MA	12V	15V	30V	30uV/C	.	3MA	70dB	70dB	0.5M
uA308AD	FAU	SBA	EXT	.	.	+18V	-18V	70C	98dB	0.5MV	7NA	1NA	500MWF	1MA	13V	15V	1V	5uV/C	.	1MA	96dB	96dB	10M
uA308AH	FAU	SBA	EXT	.	.	+18V	-18V	70C	98dB	0.5MV	7NA	1NA	500MWF	1MA	13V	15V	1V	5uV/C	.	1MA	96dB	96dB	10M

For detailed explanations of column heading notations, see App. A.

Also for ready references the more important abbreviations used in the column headings are listed below:

LEFT HAND PAGE

APP = application (codes at APP.E.)

CMRR = common mode rejection ratio

CMP = compensation (frequency)

dV_{io}/dT = input offset voltage temperature drift

GBP = gain bandwidth product

I_B = input bias current

I_{io} = input bias offset current

I_Q = quiescent supply current

MFR = manufacturer (codes at App.C.)

P_Q = quiescent power consumer

PSRR = power supply rejection ratio

V_{ICM} = common mode input voltage rating

V_{DI} = differential input voltage rating

V_{io} = input offset voltage

V_S = dc supply voltage

RIGHT HAND PAGE

Lead out coding summary (details at APP.G.) for different cases (APP.F.)

A = gain adjust

B = bias adjust

C = case

E− = inverting input

E+ = non-inverting input

F,F* = input frequency compensation

G = ground

J = high level input

K = output, open collector

L = output, open emitter

M = metal case

N = not connected

Q = special terminal

R,R* = outputs

S = strobe

T,T* = offset balance

V+ = +ve dc supply

V− = −ve dc supply

W = guard ring

X = blank position, no lead

+ + = +ve supplementary dc supply

− − = −ve supplementary dc supply

◊,◊* = output frequency compensation

CASE (APP F)	LD 1	LD 2	LD 3	LD 4	LD 5	LD 6	LD 7	LD 8	LD 9	LD 10	LD 11	LD 12	LD 13	LD 14	LD 15	LD 16	EUROPE SUBSTITUTE	USA SUBSTITUTE	!SS	TYPE NUMBER
FLP-10/1C	E-1	E+1	V-	E+2	E-2	S2	R	V+	G	S1	.	.	.	.			SFC2711PM	UA711FM	0	TSC1711F
DIL-14/1C	N	E-1	E+1	V-	E+2	E-2	N	N	S2	R	V+	G	S1	N	.	.	SFC2711KM	UA711DM	0	TSC1711J
TO5-10/1M	G	S1	E-1	E+1	V-	E+2	E-2	S2	R	V+	.	.	.	.	.	.	SFC2711M	UA711HM	0	TSC1711V
DIL-14/1P	N	E-1	E+1	V-	E+2	E-2	N	N	S2	R	V+	G	S1	N	.	.			0	TSC2225E
FLP-10/3C	E-1	E+1	V-	E+2	E-2	S2	R	V+	G	S1	.	.	.	.					0	TSC2225F
DIL-14/1C	N	E-1	E+1	V-	E+2	E-2	N	N	S2	R	V+	G	S1	N	.	.			0	TSC2225J
TO5-10/1M	G	S1	E-1	E+1	V-M	E+2	E-2	S2	R	V+	.	.	.	.	.	.			0	TSC2225V
DIL-14/1P	N	E-1	E+1	V-	E+2	E-2	N	N	S2	R	V+	G	S1	N	.	.	SFC2711EC	UA711DC	0	TSC2711E
FLP-10/3C	E-1	E+1	V-	E+2	E-2	S2	R	V+	G	S1	.	.	.	.			SFC2711PM	UA711FM	0	TSC2711F
DIL-14/1C	N	E-1	E+1	V-	E+2	E-2	N	N	S2	R	V+	G	S1	N	.	.	SFC2711EC	UA711DC	0	TSC2711J
TO5-10/1M	G	S1	E-1	E+1	V-	E+2	E-2	S2	R	V+	.	.	.	.	.	.	SFC2711C	UA711HC	0	TSC2711V
DIL-14/1P	N	E-1	E+1	V-	E+2	E-2	N	N	S2	R	V+	G	S1	N	.	.			0	TSC3225E
FLP-10/3C	E-1	E+1	V-	E+2	E-2	S2	R	V+	G	S1	.	.	.	.					0	TSC3225F
DIL-14/1C	N	E-1	E+1	V-	E+2	E-2	N	N	S2	R	V+	G	S1	N	.	.			0	TSC3225J
TO5-10/1M	G	S1	E-1	E+1	V-M	E+2	E-2	S2	R	V+	.	.	.	.	.	.			0	TSC3225V
DIL-14/1P	N	E-1	E+1	V-	E+2	E-2	N	N	S2	R	V+	G	S1	N	.	.	SFC2711KM	UA711DM	0	TSC4711E
FLP-10/3C	E-1	E+1	V-	E+2	E-2	S2	R	V+	G	S1	.	.	.	.			SFC2711PM	UA711FM	0	TSC4711F
DIL-14/1C	N	E-1	E+1	V-	E+2	E-2	N	N	S2	R	V+	G	S1	N	.	.	SFC2711KM	UA711DM	0	TSC4711J
TO5-10/1M	G	S1	E-1	E+1	V-	E+2	E-2	S2	R	V+	.	.	.	.	.	.	SFC2711M	UA711HM	0	TSC4711V
DIL-14/1P	N	E-1	E+1	V-	E+2	E-2	N	N	S2	R	V+	G	S1	N	.	.	SFC2711EC	UA711DC	0	TSC5711E
FLP-10/3C	E-1	E+1	V-	E+2	E-2	S2	R	V+	G	S1	.	.	.	.			SFC2711PM	UA711FM	0	TSC5711F
DIL-14/1C	N	E-1	E+1	V-	E+2	E-2	N	N	S2	R	V+	G	S1	N	.	.	SFC2711EC	UA711DC	0	TSC5711J
TO5-10/1M	G	S1	E-1	E+1	V-	E+2	E-2	S2	R	V+	.	.	.	.	.	.	SFC2711C	UA711HC	0	TSC5711V
DIL-14/1C	N	N	TF	E-	E+	V-	N	N	T*	R	V+	F*	N	N	.	.		LM101AJ14	0	uA101AD
FLP-10/3C	N	TF	E-	E+	V-	T*	R	V+	F*	N	.	.	.	.			SFC2101APM	LM101AF	0	uA101AF
TO5-8/1M	TF	E-	E+	V-M	T*	R	V+	F*	.	.	.	.	.	.	.	.	SFC2101A	LM101AH	0	uA101AH
DIL-14/1C	N	N	TF	E-	E+	V-	N	N	T*	R	V+	F*	N	N	.	.		LM101AJ14	0	uA101D
TO5-8/1M	TF	E-	E+	V-M	T*	R	V+	F*	.	.	.	.	.	.	.	.	SFC2101APM	LM101AH	0	uA101H
TO5-8/1M	T	N	E+	V-	B	R	V+	T*	.	.	.	.	.	.	.	.		LM102H	0	uA102M
TO5-8/1M	N	E-	E+	V-M	N	R	V+	N	.	.	.	.	.	.	.	.	SFC2107M	LM107H	0	uA107H
DIL-14/1C	N	F	W	E-	E+	W	V-	N	N	R	V+	F*	N	N	.	.		LM108AD	0	uA108AD
FLP-10/3C	N	W	E-	E+	W	V-	R	V+	F*	F	.	.	.	.			SFC2108PM	LM108AF	0	uA108AF
TO5-8/1M	F	E-	E+	V-	N	R	V+	F*	.	.	.	.	.	.	.	.	SFC2108A	LM108AH	0	uA108AH
DIL-14/1C	N	F	W	E-	E+	W	V	N	N	R	V+	F*	N	N	.	.		LM108AD	0	uA108D
FLP-10/3C	N	W	E-	E+	W	V-	R	V+	F*	F	.	.	.	.			SFC2108PM	LM108AF	0	uA108F
TO5-8/1M	F	E-	E+	V-	N	R	V+	F*	.	.	.	.	.	.	.	.	SFC2108A	LM108AH	0	uA108H
TO5-8/1M	T	N	E+	V-	B	R	V+	T*	.	.	.	.	.	.	.	.	SFC2110M	LM110H	0	uA110M
TO5-8/1M	G	E+	E-	V-	T	T*S	R	V+	.	.	.	.	.	.	.	.	SFC2111M	LM111H	0	uA111H
DIL-8/1P	G	E+	E-	V-	T	T*S	R	V+	.	.	.	.	.	.	.	.		SN52111JP	0	uA111R
DIL-14/1C	N	N	TF	E-	E+	V-	N	N	T*	R	V+	F*	N	N	.	.		LM201AJ14	0	uA201AD
FLP-10/3C	N	TF	E-	E+	V-	T*	R	V+	F*	N	.	.	.	.			SFC2201APT	LM201AF	0	uA201AF
TO5-8/1M	TF	E-	E+	V-M	T*	R	V+	F*	.	.	.	.	.	.	.	.	SFC2101A	LM201AH	0	uA201AH
DIL-14/1C	N	N	TF	E-	E+	V-	N	N	T*	R	V+	F*	N	N	.	.		LM201AJ14	0	uA201D
TO5-8/1M	TF	E-	E+	V-M	T*	R	V+	F*	.	.	.	.	.	.	.	.	SFC2201A	LM201AH	0	uA201H
TO5-8/1M	N	E-	E+	V-M	N	R	V+	N	.	.	.	.	.	.	.	.	SFC2207M	LM207H	0	uA207H
DIL-14/1C	N	F	W	E-	E+	W	V-	N	N	R	V+	F*	N	N	.	.		LM208AD	0	uA208AD
FLP-10/3C	N	W	E-	E+	W	V-	R	V+	F*	F	.	.	.	.			SFC2208PT	LM208AF	0	uA208AF
TO5-8/1M	F	E-	E+	V-	N	R	V+	F*	.	.	.	.	.	.	.	.	SFC2208A	LM208AH	0	uA208AH
DIL-14/1C	N	F	W	E-	E+	W	V-	N	N	R	V+	F*	N	N	.	.		LM208D	0	uA208D
FLP-10/3C	N	W	E-	E+	W	V-	R	V+	F*	F	.	.	.	.			SFC2208PT	LM208F	0	uA208F
TO5-8/1M	F	E-	E+	V-	N	R	V+	F*	.	.	.	.	.	.	.	.	SFC2208	LM208H	0	uA208H
DIL-14/1C	N	N	TF	E-	E+	V-	N	N	T*	R	V+	F*	N	N	.	.		LM301AJ14	0	uA301AD
TO5-8/1M	TF	E-	E+	V-M	T*	R	V+	F*	.	.	.	.	.	.	.	.	SFC2301A	LM301AH	0	uA301AH
DIL-8/1P	TF	E-	E+	V-	T*	R	V+	F*	.	.	.	.	.	.	.	.	SFC301ADC	LM301AJ	0	uA301AT
TO5-8/1M	T	N	E+	V-	B	R	V+	T*	.	.	.	.	.	.	.	.		LM302H	0	uA302C
TO5-8/1M	N	E-	E+	V-M	N	R	V+	N	.	.	.	.	.	.	.	.	SFC2307	LM307H	0	uA307H
DIL-8/1P	N	E-	E+	V-	N	R	V+	N	.	.	.	.	.	.	.	.	SFC2307DC	LM307J	0	uA307T
DIL-14/1C	N	F	W	E-	E+	W	V-	N	N	R	V+	F*	N	N	.	.		LM308AD	0	uA308AD
TO5-8/1M	F	E-	E+	V-	N	R	V+	F*	.	.	.	.	.	.	.	.	SFC2308A	LM308AH	0	uA308AH
DIL-14/1C	N	F	W	E-	E+	W	V-	N	N	R	V+	F*	N	N	.	.		LM308AD	0	uA308D

TYPE NUMBER	M F R	A P P	C M P	GBP MIN	SLEW RATE MIN	V_S^+ MAX	V_S^- MAX	T_{op} MAX	A_{VOL} MIN	V_{IO} MAX	I_B MAX	I_{IO} MAX	P_{TOT} MAX	I_{OUT} MIN	V_{OUT} MIN	V_{ICM} MAX	V_{IDF} MAX	dV_{IO}/dT MAX	P_Q MAX	I_Q MAX	CMRR MIN	PSRR MIN	R_{IN} MIN
uA308D	FAU	SBA	EXT	.	.	+18V	-18V	70C	88dB	7.5MV	7NA	1NA	500MWF	1MA	13V	15V	1V	30uV/C	.	1MA	80dB	80dB	10M
uA308H	FAU	SBA	EXT	.	.	+18V	-18V	70C	88dB	7.5MV	7NA	1NA	500MWF	1MA	13V	15V	1V	30uV/C	.	1MA	80dB	80dB	10M
uA310C	FAU	VFA	INT	.	.	+18V	-18V	70C	0dB	7.5MV	7NA	.	500MWF	1MA	10V	15V	.	30uV/C	.	6MA	.	70dB	10G
uA311H	FAU	CPR	EXT	.	.	+18V	-18V	70C	100dB	7.5MV	250NA	50NA	500MWF	8MA	.	15V	30V	.	.	8MA	.	.	.
uA311R	FAU	CPR	EXT	.	.	+18V	-18V	70C	100dB	7.5MV	250NA	50NA	500MWF	8MA	.	15V	30V	.	.	8MA	.	.	.
uA311T	FAU	CPR	EXT	.	.	+18V	-18V	70C	100dB	7.5MV	250NA	50NA	500MWF	8MA	.	15V	30V	.	.	8MA	.	.	.
uA702A	SGG	WBA	EXT	3MHZ	.	+14V	-7V	125C	68dB	2MV	5uA	0.5uA	500MWF	.3MA	5V	1.5V	5V	10uV/C	120MW	7MA	80dB	74dB	16K
uA702DC	FAU	WBA	EXT	3MHZ	.	+13V	-8V	70C	66dB	5MV	7.5uA	2uA	670MWF	.3MA	5V	5V	5V	20uV/C	120MW	7MA	70dB	70dB	10K
uA702DM	FAU	WBA	EXT	3MHZ	.	+13V	-8V	125C	68dB	2MV	5uA	0.5uA	670MWF	.3MA	5V	5V	5V	10uV/C	120MW	7MA	80dB	74dB	16K
uA702FM	FAU	WBA	EXT	3MHZ	.	+13V	-8V	125C	68dB	2MV	5uA	0.5uA	570MWF	.3MA	5V	5V	5V	10uV/C	120MW	7MA	80dB	74dB	16K
uA702HC	FAU	WBA	EXT	3MHZ	.	+13V	-8V	70C	66dB	5MV	7.5uA	2uA	500MWF	.3MA	5V	5V	5V	20uV/C	120MW	7MA	70dB	70dB	10K
uA702HM	FAU	WBA	EXT	3MHZ	.	+13V	-8V	125C	68dB	2MV	5uA	0.5uA	500MWF	.3MA	5V	5V	5V	10uV/C	120MW	7MA	80dB	74dB	16K
uA709	SGG	GPU	EXT	.3MHZ	.15V/uS	+18V	-18V	125C	88dB	5MV	500NA	200NA	500MWF	5MA	12V	10V	5V	15uV/C	165MW	.	70dB	76dB	150K
uA709A	SJU	GPU	EXT	.	.	+18V	-18V	125C	85dB	5MV	500NA	200NA	300MWF	10MA	12V	10V	5V	6uV/C	165MW	.	70dB	100dB	150K
uA709ADM	FAU	GPU	EXT	.3MHZ	.15V/uS	+18V	-18V	125C	88dB	2MV	200NA	50NA	670MWF	5MA	12V	10V	5V	10uV/C	108MW	.	80dB	80dB	350K
uA709AFM	FAU	GPU	EXT	.3MHZ	.15V/uS	+18V	-18V	125C	88dB	2MV	200NA	50NA	570MWF	5MA	12V	10V	5V	10uV/C	108MW	.	80dB	80dB	350K
uA709AHM	FAU	GPU	EXT	.3MHZ	.15V/uS	+18V	-18V	125C	88dB	2MV	200NA	50NA	500MWF	5MA	12V	10V	5V	10uV/C	108MW	.	80dB	80dB	350K
uA709CA	SJU	GPU	EXT	.	.	+18V	-18V	75C	84dB	7.5MV	1.5uA	500NA	250MWF	10MA	12V	10V	5V	.	200MW	.	65dB	100dB	50K
uA709CN(8)	MUG	GPU	EXT	.	.	+18V	-18V	75C	84dB	7.5MV	1.5uA	500NA	250MWF	10MA	12V	10V	5V	.	200MW	.	65dB	100dB	50K
uA709CN(14)	SJU	GPU	EXT	.	.	+18V	-18V	75C	84dB	7.5MV	1.5uA	500NA	250MWF	10MA	12V	10V	5V	.	200MW	.	65dB	100dB	50K
uA709CT	SJU	GPU	EXT	.	.	+18V	-18V	75C	84dB	7.5MV	1.5uA	500NA	250MW	10MA	12V	10V	5V	.	200MW	.	65dB	100dB	50K
uA709CV	MUG	GPU	EXT	.	.	+18V	-18V	75C	84dB	7.5MV	1.5uA	500NA	250MWF	10MA	12V	10V	5V	.	200MW	.	65dB	100dB	50K
uA709DC	FAU	GPU	EXT	.3MHZ	.15V/uS	+18V	-18V	70C	84dB	7.5MV	1.5uA	500NA	670MWF	5MA	12V	10V	5V	.	200MW	.	65dB	74dB	50K
uA709DM	FAU	GPU	EXT	.3MHZ	.15V/uS	+18V	-18V	125C	88dB	5MV	500NA	200NA	670MWF	5MA	12V	10V	5V	15uV/C	165MW	.	70dB	76dB	150K
uA709F	MUG	GPU	EXT	.	.	+18V	-18V	125C	.	5MV	500NA	200NA	300MWF	10MA	12V	10V	5V	6uV/C	165MW	.	70dB	100dB	150K
uA709FM	FAU	GPU	EXT	.3MHZ	.15V/uS	+18V	-18V	125C	88dB	5MV	500NA	200NA	570MWF	5MA	12V	10V	5V	15uV/C	165MW	.	70dB	76dB	150K
uA709HC	FAU	GPU	EXT	.3MHZ	.15V/uS	+18V	-18V	70C	84dB	7.5MV	1.5uA	500NA	500MWF	5MA	12V	10V	5V	.	200MW	.	.65dB	74dB	50K
uA709HM	FAU	GPU	EXT	.3MHZ	.15V/uS	+18V	-18V	125C	88dB	5MV	500NA	200NA	500MWF	5MA	12V	10V	5V	15uV/C	165MW	.	70dB	76dB	150K
uA709N(8)	MUG	GPU	EXT	.	.	+18V	-18V	125C	85dB	5MV	500NA	200NA	300MWF	10MA	12V	10V	5V	6uV/C	165MW	.	70dB	100dB	150K
uA709N(14)	MUG	GPU	EXT	.	.	+18V	-18V	125C	85dB	5MV	500NA	200NA	300MWF	10MA	12V	10V	5V	6uV/C	165MW	.	70dB	100dB	150K
uA709PC	FAU	GPU	EXT	.3MHZ	.15V/uS	+18V	-18V	70C	84dB	7.5MV	1.5uA	500NA	670MWF	5MA	12V	10V	5V	.	200MW	.	65dB	74dB	50K
uA709T	SJU	GPU	EXT	.	.	+18V	-18V	125C	88dB	5MV	500NA	200NA	300MWF	10MA	12V	10V	5V	6uV/C	165MW	.	70dB	76dB	150K
uA709TC	FAU	GPU	EXT	.3MHZ	.15V/uS	+18V	-18V	70C	84dB	7.5MV	1.5uA	500NA	670MWF	5MA	12V	10V	5V	.	200MW	.	65dB	74dB	.
uA709V	MUG	GPU	EXT	.	.	+18V	-18V	125C	85dB	5MV	500NA	200NA	300MWF	10MA	12V	10V	5V	6uV/C	165MW	.	70dB	100dB	150K
uA710DC	FAU	CPR	EXT	.	.	+14V	-6V	70C	60dB	5MV	25uA	5uA	670MWF	5MA	2.5V	7V	5V	20uV/C	150MW	.	70dB	.	.
uA710DM	FAU	CPR	EXT	.	.	+14V	-6V	125C	62dB	3MV	20uA	3uA	670MWF	5MA	2.5V	7V	5V	10uV/C	150MW	.	80dB	.	.
uA710F	MUG	CPR	EXT	.	.	+14V	-7V	125C	62dB	2MV	16uA	3uA	200MWF	5MA	2.5V	5V	5V	10uV/C	150MW	9MA	80dB	.	.
uA710FM	FAU	CPR	EXT	.	.	+14V	-6V	125C	62dB	2MV	20uA	3uA	570MWF	5MA	2.5V	7V	5V	10uV/C	150MW	.	80dB	.	.
uA710HC	FAU	CPR	EXT	.	.	+14V	-6V	70C	60dB	5MV	25uA	5uA	500MWF	5MA	2.5V	7V	5V	20uV/C	150MW	.	70dB	.	.
uA710HM	FAU	CPR	EXT	.	.	+14V	-7V	125C	62dB	2MV	20uA	3uA	500MWF	5MA	2.5V	7V	5V	10uV/C	150MW	.	80dB	.	.
uA710A	SJU	CPR	EXT	.	.	+14V	-7V	125C	62dB	2MV	16uA	3uA	200MWF	5MA	2.5V	5V	5V	10uV/C	150MW	9MA	80dB	.	.
uA710CA	SJU	CPR	EXT	.	.	+14V	-7V	75C	60dB	5MV	25uA	5uA	300MWF	5MA	2.5V	5V	5V	20uV/C	150MW	9MA	70dB	.	.
uA710CF	MUG	CPR	EXT	.	.	+14V	-7V	75C	60dB	5MV	25uA	5uA	300MWF	5MA	2.5V	5V	5V	20uV/C	150MW	9MA	70dB	.	.
uA710CN(8)	MUG	CPR	EXT	.	.	+14V	-7V	75C	60dB	5MV	25uA	5uA	300MWF	5MA	2.5V	5V	5V	20uV/C	150MW	9MA	70dB	.	.
uA710CN(14)	MUG	CPR	EXT	.	.	+14V	-7V	75C	60dB	5MV	25uA	5uA	300MWF	5MA	2.5V	5V	5V	20uV/C	150MW	9MA	70dB	.	.
uA710CT	SJU	CPR	EXT	.	.	+14V	-7V	75C	60dB	5MV	25uA	5uA	200MWF	5MA	2.5V	5V	5V	20uV/C	150MW	9MA	70dB	.	.
uA710CV	MUG	CPR	EXT	.	.	+14V	-7V	75C	60dB	5MV	25uA	5uA	300MWF	5MA	2.5V	5V	5V	20uV/C	150MW	9MA	70dB	.	.
uA710PC	FAU	CPR	EXT	.	.	+14V	-7V	70C	60dB	5MV	25uA	5uA	670MWF	5MA	2.5V	7V	5V	20uV/C	150MW	.	70dB	.	.
uA710T	SJU	CPR	EXT	.	.	+14V	-7V	125C	62dB	2MV	16uA	3uA	300MWF	5MA	2.5V	5V	5V	10uV/C	150MW	9MA	80dB	.	.
uA711A	SJU	DCP	EXT	.	.	+14V	-7V	125C	58dB	3.5MV	75uA	10uA	300MWF	5MA	2.5V	5V	5V	10uV/C	200MW	15MA	.	.	.
uA711CA	SJU	DCP	EXT	.	.	+14V	-7V	75C	57dB	5MV	100uA	15uA	300MWF	5MA	2.5V	5V	5V	10uV/C	200MW	15MA	.	.	.
uA711CF	MUG	DCP	EXT	.	.	+14V	-7V	75C	57dB	5MV	100uA	15uA	300MWF	5MA	2.5V	5V	5V	10uV/C	200MW	15MA	.	.	.
uA711CK	SJU	DCP	EXT	.	.	+14V	-7V	75C	57dB	5MV	100uA	15uA	300MWF	5MA	2.5V	5V	5V	10uV/C	200MW	15MA	.	.	.
uA711CN(14)	SJU	DCP	EXT	.	.	+14V	-7V	75C	57dB	5MV	100uA	15uA	300MWF	5MA	2.5V	5V	5V	10uV/C	200MW	15MA	.	.	.
uA711DC	FAU	DCP	EXT	.	.	+14V	-7V	70C	57dB	5MV	100uA	15uA	670MWF	5MA	2.5V	7V	5V	20uV/C	230MW	.	.	.	.
uA711DM	FAU	DCP	EXT	.	.	+14V	-7V	125C	58dB	3.5MV	75uA	10uA	670MWF	5MA	2.5V	7V	5V	20uV/C	200MW	.	.	.	.
uA711F	MUG	DCP	EXT	.	.	+14V	-7V	125C	58dB	3.5MV	75uA	10uA	300MWF	5MA	2.5V	5V	5V	10uV/C	200MW	15MA	.	.	.
uA711FM	FAU	DCP	EXT	.	.	+14V	-7V	125C	58dB	3.5MV	75uA	10uA	570MWF	5MA	2.5V	7V	5V	20uV/C	200MW	.	.	.	.
uA711HC	FAU	DCP	EXT	.	.	+14V	-7V	70C	57dB	5MV	100uA	15uA	500MWF	5MA	2.5V	7V	5V	20uV/C	230MW	.	.	.	.
uA711HM	FAU	DCP	EXT	.	.	+14V	-7V	125C	58dB	3.5MV	75uA	10uA	500MWF	5MA	2.5V	7V	5V	20uV/C	200MW	.	.	.	.

For detailed explanations of column heading notations, see App. A.

Also for ready references the more important abbreviations used in the column headings are listed below:

LEFT HAND PAGE

APP = application (codes at APP.E.)
CMRR = common mode rejection ratio
CMP = compensation (frequency)
dV_{io}/dT = input offset voltage temperature drift
GBP = gain bandwidth product
I_B = input bias current
I_{io} = input bias offset current
I_Q = quiescent supply current
MFR = manufacturer (codes at App.C.)
P_Q = quiescent power consumer
PSRR = power supply rejection ratio
V_{ICM} = common mode input voltage rating
V_{IDF} = differential input voltage rating
V_{io} = input offset voltage
V_S = dc supply voltage

RIGHT HAND PAGE

Lead out coding summary (details at APP.G.) for different cases (APP.F.)

A = gain adjust
B = bias adjust
C = case
E— = inverting input
E+ = non-inverting input
F,F* = input frequency compensation
G = ground
J = high level input
K = output, open collector
L = output, open emitter
M = metal case
N = not connected
Q = special terminal
R,R* = outputs
S = strobe
T,T* = offset balance
V+ = +ve dc supply
V— = —ve dc supply
W = guard ring
X = blank position, no lead
+ + = +ve supplementary dc supply
— — = —ve supplementary dc supply
ø,ø* = output frequency compensation

CASE (APP F)	LD 1	LD 2	LD 3	LD 4	LD 5	LD 6	LD 7	LD 8	LD 9	LD 10	LD 11	LD 12	LD 13	LD 14	LD 15	LD 16	EUROPE SUBSTITUTE	USA SUBSTITUTE	IS	TYPE NUMBER
TO5-8/1M	F	E-	E+	V-	N	R	V+	F*									SFC2308A	LM308AH	0	uA308H
TO5-8/1M	T	N	E+	V-	B	R	V+	T*									SFC2310EC	LM310H	0	uA310C
TO5-8/1M	G	E+	E-	V-	T	T*S	R	V+									SFC2311	LM311H	0	uA311H
DIL-8/1P	G	E+	E-	V-	T	T*S	R	V+									SFC2311DC	LM311N	0	uA311R
DIL-8/1P	G	E+	E-	V-	T	T*S	R	V+									SFC2311DC	LM311N	0	uA311T
TO5-8/1M	G	E-	E+	V-M	F	ø	R	V+										SN52702AL	0	uA702A
DIL-14/1C	N	N	G	E-	E+	V-	N	N	F	ø	R	N	V+	N			U6E7712393	SN72702J	0	uA702DC
DIL-14/1C	N	N	G	E-	E+	V-	N	N	F	ø	R	N	V+	N				SN52702AJ	0	uA702DM
FLP-10/3C	N	G	E-	E+	V-	F	ø	R	N	V+								SN52702AFA	0	uA702FM
TO5-8/1M	G	E-	E+	V-M	F	ø	R	V+									U5B771239X	SN72702L	0	uA702HC
TO5-8/1M	G	E-	E+	V-M	F	ø	R	V+									U5B771231X	SN52702AL	0	uA702HM
TO5-8/1M	F	E-	E+	V-M	ø	ø*R	V+	F*									TAA522	UA709HM	0	uA709
DIL-14/1P	N	N	F	E-	E+	V-	N	N	ø	R	V+	F*	N	N			LM709J	UA709DM	0	uA709A
DIL-14/1C	N	N	F	E-	E+	V-	N	N	ø	R	V+	F*	N	N			SN52709AJ	LM709AJ	0	uA709ADM
FLP-10/3C	N	F	E-	E+	V-	ø	R	V+	F*	N								SN52709AFA	0	uA709AFM
TO5-8/1M	F	E-	E+	V-	ø	ø*R	V+	F*									TAA522	SN52709AL	0	uA709AHM
DIL-14/1P	N	N	F	E-	E+	V-	N	N	ø	R	V+	F*	N	N			TAA521A	UA709DC	0	uA709CA
DIL-8/1P	F	E-	E+	V-	ø	ø*R	V+	F*									LM709CN8	UA709TC	0	uA709CN(8)
DIL-14/1P	N	N	F	E-	E+	V-	N	N	ø	R	V+	F*	N	N			TAA521A	UA709DC	0	uA709CN(14)
TO5-8/1M	F	E-	E+	V-M	ø	ø*R	V+	F*									TAA521	UA709HC	0	uA709CT
DIL-8/1P	F	E-	E+	V-	ø	ø*R	V+	F*									LM709CN8	UA709TC	0	uA709CV
DIL-14/1C	N	N	F	E-	E+	V-	N	N	ø	R	V+	F*	N	N			TAA521A	LM709CN	0	uA709DC
DIL-14/1C	N	N	F	E-	E+	V-	N	N	ø	R	V+	F*	N	N			SN52709AJ	LM709J	0	uA709DM
DIL-14/1C	N	N	F	E-	E+	V-	N	N	ø	R	V+	F*	N	N			LM709J	UA709DM	0	uA709F
FLP-10/3C	N	F	E-	E+	V-	ø	R	V+	F*	N								SN52709AFA	0	uA709FM
TO5-8/1M	F	E-	E+	V-	ø	ø*R	V	F*									TAA521	LM709CH	0	uA709HC
TO5-8/1M	F	E-	E+	V-	ø	ø*R		F*									TAA522	LM709H	0	uA709HM
DIL-8/1P	F	E-	E+	V-	ø	ø*R	V+	F*									SN52709AJP	MC1709U	0	uA709N(8)
DIL-14/1P	N	N	F	E-	E+	V-	N	N	ø	R	V+	F*	N	N			LM709J	UA709DM	0	uA709N(14)
DIL-14/1P	N	N	F	E-	E+	V-	N	N	ø	R	V+	F*	N	N			TAA521A	LM709CN	0	uA709PC
TO5-8/1M	F	E-	E+	V-M	ø	ø*R	V+	F*									TAA522	UA709HM	0	uA709T
DIL-8/1P	F	E-	E+	V-	ø	ø*R	V+	F*									SN72709CJP	LM709CN8	0	uA709TC
DIL-8/1P	F	E-	E+	V-	ø	ø*R	V+	F*									SN52709AJP	MC1709U	0	uA709V
DIL-14/1C	N	G	E+	E-	N	V-	N	N	R	N	V+	N	N	N			SFC2710EC	UA710CF	0	uA710DC
DIL-14/1C	N	G	E+	E-	N	V-	N	N	R	N	V+	N	N	N			SFC2710KM	UA710F	0	uA710DM
DIL-14/1C	N	G	E-	E+	N	V-	N	N	R	N	V+	N	N	N			SFC2710KM	UA710DM	0	uA710F
FLP-10/3C	G	E+	E-	N	V-	R	N	V+	N	N							SFC2710PM	UA710FM	0	uA710FM
TO5-8/1M	G	E+	E-	V-	N	N	R	V+									SFC2710C	LM710CH	0	uA710HC
TO5-8/1M	G	E+	E-	V-	N	N	R	V+									SFC2710M	LM710H	0	uA710HM
DIL-14/1P	N	G	E-	E+	N	V-	N	N	R	N	V+	N	N	N			SFC2710KM	UA710DM	0	uA710A
DIL-14/1P	N	G	E-	E+	N	V-	N	N	R	N	V+	N	N	N			SFC2710EC	UA710DC	0	uA710CA
DIL-14/1C	N	G	E-	E+	N	V-	N	N	R	N	V+	N	N	N			SFC2710KM	UA710DC	0	uA710CF
DIL-8/1P	G	E+	E-	V-	N	N	R	V+										UA710CV	0	uA710CN(8)
DIL-14/1P	N	G	E-	E+	N	V-	N	N	R	N	V+	N	N	N			SFC2710EC	UA710DC	0	uA710CN(14)
TO5-8/1M	G	E+	E-	V-	N	N	R	V+									SFC2710C	UA710HC	0	uA710CT
DIL-8/1P	G	E+	E-	V-	N	N	R	V+										UA710CN8	0	uA710CV
DIL-14/1P	N	G	E+	E-	N	V-	N	N	R	N	V+	N	N	N			SFC2710EC	UA710DC	0	uA710PC
TO5-8/1M	G	E+	E-	V-	N	N	R	V+									SFC2710M	UA710HM	0	uA710T
DIL-14/1P	N	E-1	E+1	V-	E+2	E-2	N	N	S2	R	V+	G	S1	N			SFC2711KM	UA711DM	0	uA711A
DIL-14/1P	N	E-1	E+1	V-	E+2	E-2	N	N	S2	R	V+	G	S1	N			SFC2711EC	UA711DC	0	uA711CA
DIL-14/1C	N	E-1	E+1	V-	E+2	E-2	N	N	S2	R	V-	G	S1	N			SFC2711EC	UA711DC	0	uA711CF
TO5-10/1M	G	S1	E-1	E+1	V-M	E+2	E-2	S2	R	V+							SFC2711C	UA711HC	0	uA711CK
DIL-14/1P	N	E-1	E+1	V-	E+2	E-2	N	N	S2	R	V+	G	S1	N			SFC2711EC	UA711DC	0	uA711CN(14)
DIL-14/1C	N	E-1	E+1	V-	E+2	E-2	N	N	S2	R	V+	G	S1	N			SFC2711EC	LM711CN	0	uA711DC
DIL-14/1C	N	E-1	E+1	V-	E+2	E-2	N	N	S2	R	V+	G	S1	N			SFC2711KM	UA711N14	0	uA711DM
DIL-14/1C	N	E-1	E+1	V-	E+2	E-2	N	N	S2	R	V+	G	S1	N			SFC2711KM	UA711DM	0	uA711F
FLP-10/3C	E-1	E+1	V-	E+2	E-2	S2	R	V+	G	S1							SFC2711PM	711BH	0	uA711FM
TO5-10/1M	G	S1	E-1	E+1	V-	E+2	E-2	S2	R	V+							SFC2711C	LM711CH	0	uA711HC
TO5-10/1M	G	S1	E-1	E+1	V-	E+2	E-2	S2	R	V+							SFC2711M	LM711H	0	uA711HM
TO5-10/1M	G	S1	E-1	E+1	V-	E+2	E-2	S2	R	V+							SFC2711M	UA711HM	0	uA711K

TYPE NUMBER	M F R	A P P	C M P	GBP MIN	SLEW RATE MIN	V$_S^+$ MAX	V$_S^-$ MAX	T$_{op}$ MAX	A$_{VOL}$ MIN	V$_{IO}$ MAX	I$_B$ MAX	I$_{IO}$ MAX	P$_{TOT}$ MAX	I$_{OUT}$ MIN	V$_{OUT}$ MIN	V$_{ICM}$ MAX	V$_{IDF}$ MAX	dV$_{IO}$/dT MAX	P$_O$ MAX	I$_O$ MAX	CM RR MIN	PS RR MIN	R$_{IN}$ MIN
uA711K	SJU	DCP	EXT	.	.	+14V	-7V	125C	58dB	3.5MV	75uA	10uA	300MWF	5MA	2.5V	5V	5V	10uV/C	200MW	15MA		.	.
uA711N(14)	MUG	DCP	EXT	.	.	+14V	-7V	125C	58dB	3.5MV	75uA	10uA	300MWF	5MA	2.5V	5V	5V	10uV/C	200MW	15MA			
uA711PC	FAU	DCP	EXT	.	.	+14V	-7V	70C	57dB	5MV	100uA	15uA	670MWF	5MA	2.5V	7V	5V	20uV/C	230MW			.	.
uA715DC	FAU	HSR	EXT	.	10V/uS	+18V	-18V	70C	80dB	7.5MV	1.5uA	250NA	670MWF	5MA	10V	15V	15V		300MW	10MA	74dB	68dB	300K
uA715DM	FAU	HSR	EXT	.	15V/uS	+18V	-18V	125C	84dB	5MV	750NA	250NA	670MWF	5MA	10V	15V	15V	.	210MW	7MA	74dB	70dB	300K
uA715HC	FAU	HSR	EXT	.	10V/uS	+18V	-18V	70C	80dB	7.5MV	1.5uA	250NA	500MWF	5MA	10V	15V	15V		300MW	10MA	74dB	68dB	300K
uA715HM	FAU	HSR	EXT	.	15V/uS	+18V	-18V	125C	84dB	5MV	750NA	250NA	500MWF	5MAA	10V	15V	15V		210MW	7MA	74dB	70dB	300K
uA725AHM	FAU	PIA	EXT	.	.	+22V	-22V	125C	120dB	0.5MV	75NA	5NA	500MWF	5MA	12V	22V	5V	2uV/C	120MW	.	110dB	102dB	500K
uA725EHC	FAU	PIA	EXT	.	.	+22V	-22V	70C	120dB	0.5MV	75NA	5NA	500MWF	5MA	12V	22V	5V	2uV/C	150MW	.	120dB	106dB	500K
uA725HC	FAU	PIA	EXT	.	.	+22V	-22V	70C	108dB	2.5MV	125NA	35NA	500MWF	5MA	12V	22V	5V	5uV/C	150MW	.	94dB	90dB	500K
uA725HM	FAU	PIA	EXT	.	.	+22V	-22V	125C	120dB	1MV	100NA	20NA	500MWF	5MA	12V	22V	5V	5uV/C	150MW	.	110dB	100dB	500K
uA727HC	FAU	BDO	EXT	.	.	+18V	-18V	85C	34dB	10MV	75NA	25NA	500MWF	.	5V	15V	10V	3uV/C	.	17MA	70dB	74dB	100M
uA727HM	FAU	BDO	EXT	.	.	+18V	-18V	125C	36dB	10MV	40NA	15NA	500MWF	.	5V	15V	10V	1.5uV/C	.	17MA	80dB	74dB	100M
uA730HC	FAU	BDO	EXT	1MHZ	.	+15V	.	70C	40dB	5MV	16uA	3uA	500MWF	.	2V	4V	5V		156MW	13MA	60dB	.	2.5K
uA730HM	FAU	BDO	EXT	1MHZ	.	+15V	.	125C	40dB	2.5MV	7.5uA	1.5uA	500MWF	.	2V	4V	5V		155MW	13MA	70dB	.	5K
uA733A	SJU	BDO	EXT	20MHZ	.	+8V	-8V	125C	50dB	5MV	20uA	3uA	600MWF	2MA	1.5V	6V	5V	.		24MA	60dB	50dB	2K
uA733CA	SJU	BDO	EXT	20MHZ	.	+8V	-8V	75C	48dB	6MV	30uA	5uA	300MWF	2MA	1.5V	6V	5V	.		24MA	60dB	50dB	2K
uA733CF	MUG	BDO	EXT	20MHZ	.	+8V	-8V	75C	48dB	6MV	30uA	5uA	500MWF	2MA	1.5V	6V	5V	.		24MA	60dB	50dB	2K
uA733CI	FAU	BDO	EXT	20MHZ	.	+8V	-8V	75C	48dB	6MV	30uA	5uA	500MWF	2MA	1.5V	6V	5V	.		24MA	60dB	50dB	2K
uA733CK	SJU	BDO	EXT	20MHZ	.	+8V	-8V	75C	48dB	6MV	30uA	5uA	350MWF	2MA	1.5V	6V	5V	.		24MA	60dB	50dB	2K
uA733CN(14)	MUG	BDO	EXT	20MHZ	.	+8V	-8V	75C	48dB	6MV	30uA	5uA	300MWF	2MA	1.5V	6V	5V	.		24MA	60dB	50dB	2K
uA733DC	FAU	BDO	EXT	20MHZ	.	+8V	-8V	70C	48dB	6MV	30uA	5uA	670MWF	2MA	1.5V	6V	5V	.		24MA	60dB	50dB	2K
uA733DM	FAU	BDO	EXT	20MHZ	.	+8V	-8V	125C	50dB	5MV	20uA	3uA	670MWF	2MA	1.5V	6V	5V	.		24MA	60dB	50dB	2K
uA733F	MUG	BDO	EXT	20MHZ	.	+8V	-8V	125C	50dB	5MV	20uA	3uA	1WF	2MA	1.5V	6V	5V	.		24MA	60dB	50dB	2K
uA733FM	FAU	BDO	EXT	20MHZ	.	+8V	-8V	125C	50dB	5MV	20uA	3uA	570MWF	2MA	1.5V	6V	5V	.		24MA	60dB	50dB	2K
uA733HC	FAU	BDO	EXT	20MHZ	.	+8V	-8V	70C	48dB	6MV	30uA	5uA	500MWF	2MA	1.5V	6V	5V	.		24MA	60dB	50dB	2K
uA733HM	FAU	BDO	EXT	20M	.	+8V	-8V	125C	50dB	5MV	20uA	3uA	500MWF	2MA	1.5V	6V	5V	.		24MA	60dB	50dB	2K
uA733I	SJU	BDO	EXT	20MHZ	.	+8V	-8V	125C	50dB	5MV	20uA	3uA	500MWF	2MA	1.5V	6V	5V	.		24MA	60dB	50dB	2K
uA733K	SJU	BDO	EXT	20MHZ	.	+8V	-8V	125C	50dB	5MV	20uA	3uA	700MWF	2MA	1.5V	6V	5V	.		24MA	..dB	50dB	2K
uA734DC	FAU	CPR	EXT	.	.	+18V	-18V	70C	91dB	5MV	100NA	25NA	670MWF	3MA	.	13V	10V	20uV/C	145MW	.	70dB	80dB	7M
uA734DM	FAU	CPR	EXT	.	.	+18V	-18V	125C	91dB	3MV	50NA	10NA	670MWF	3MA	.	13V	10V	15uV/C	145MW	.	70dB	80dB	20M
uA734HC	FAU	CPR	EXT	.	.	+18V	-18V	70C	91dB	5MV	100NA	25NA	500MWF	3MA	.	13V	10V	20uV/C	145MW	.	70dB	80dB	7M
uA734HM	FAU	CPR	EXT	.	.	+18V	-18V	125C	91dB	3MV	50NA	10NA	500MWF	3MA	.	13V	10V	15uV/C	145MW	.	70dB	80dB	20M
uA739DC	FAU	DLN	EXT	.	.	+18V	-18V	70C	76dB	6MV	2uA	1uA	670MWF	.2MA	12V	15V	5V		420MW	14MA	70dB	74dB	37K
uA739PC	FAU	DLN	EXT	.	.	+18V	-18V	70C	76dB	6MV	2uA	1uA	670MWF	5MA	12V	15V	5V		420MW	14MA	70dB	74dB	37K
uA740CT	SJU	FET	INT	.5MHZ	3V/uS	+22V	-22V	70C	88dB	60MV	2NA	120pA	500MWF	40MA	12V	22V	30V	.	240MW	3MA	74dB	77dB	500K
uA740HC	FAU	FET	INT	1MHZ	2V/uS	+22V	-22V	70C	86dB	20MV	2NA	300pA	500MWF	5MA	12V	15V	30V	.	240MW	8MA	55dB	66dB	300K
uA740HM	FAU	FET	INT	1MHZ	2V/uS	+22V	-22V	125C	94dB	110MV	200pA	150pA	500MWF	5MA	12V	15V	30V	.	156MW	5MA	64dB	70dB	300K
uA740T	SJU	FET	INT	.5MHZ	3V/uS	+22V	-22V	70C	88dB	30MV	2NA	120pA	500MWF	40MA	12V	22V	30V	.	240MW	3MA	74dB	77dB	500K
uA741A	MUG	GPK	INT	.	.25V/uS	+22V	-22V	125C	94dB	5MV	500NA	200NA	500MWF	5MA	10V	12V	30V	.	85MW	3MA	70dB	76dB	300K
uA741ADM	FAU	GPK	INT	.4MHZ	0.3V/uS	+22V	-22V	125C	94dB	3MV	80NA	30NA	670MWF	10MA	16V	15V	30V	15uV/C	150MW	.	80dB	86dB	1M
uA741AFM	FAU	GPK	INT	.4MHZ	0.3V/uS	+22V	-22V	125C	94dB	3MV	80NA	30NA	570MWF	10MA	16V	15V	30V	15uV/C	150MW	.	80dB	86dB	1M
uA741AHM	FAU	GPK	INT	.4MHZ	0.3V/uS	+22V	-22V	125C	94dB	3MV	80NA	30NA	500MWF	10MA	16V	15V	30V	15uV/C	150MW	.	80dB	86dB	1M
uA741CA	SJU	GPK	INT	.	.25V/uS	+18V	-18V	70C	86dB	6MV	500NA	200NA	500MWF	5MA	10V	15V	30V	.	85MW	3MA	70dB	76dB	300K
uA741CD	MUG	GPK	INT	.	.25V/uS	+18V	-18V	70C	86dB	6MV	500NA	200NA		5MA	10V	15V	30V	.	85MW	3MA	70dB	76dB	300K
uA741CN(8)	SJU	GPK	INT	.	.25V/uS	+18V	-18V	70C	86dB	6MV	500NA	200NA	500MWF	3MA	10V	15V	30V	.	85MW	3MA	70dB	76dB	300K
uA741CN(14)	MUG	GPK	INT	.	.25V/uS	+18V	-18V	70C	86dB	6MV	500NA	200NA	500MWF	5MA	10V	15V	30V	.	85MW	3MA	70dB	76dB	300K
uA741CT	SJU	GPK	INT	.	.25V/uS	+18V	-18V	70C	86dB	6MV	500NA	200NA	500MWF	5MA	10V	15V	30V	.	85MW	3MA	70dB	76dB	300K
uA741CV	SJU	GPK	INT	.	.25V/uS	+18V	-18V	70C	86dB	6MV	500NA	200NA	500MWF	3MA	10V	15V	30V	.	85MW	3MA	70dB	76dB	300K
uA741DC	FAU	GPK	INT	.	0.2V/uS	+18V	-18V	70C	86dB	6MV	500NA	200NA	670MWF	5MA	12V	15V	30V	.	85MW	3MA	70dB	76dB	300K
uA741DM	FAU	GPK	INT	.4MHZ	0.3V/uS	+22V	-22V	125C	94dB	3MV	80NA	30NA	670MWF	5MA	12V	15V	30V	15uV/C	150MW	.	80dB	86dB	1M
uA741EDC	FAU	GPK	INT	.4MHZ	0.3V/uS	+22V	-22V	70C	94dB	3MV	80NA	30NA	670MWF	7MA	16V	15V	30V	15uV/C	150MW	.	80dB	86dB	1M
uA741EHC	FAU	GPK	INT	.4MHZ	0.3V/uS	+22V	-22V	70C	94dB	3MV	80NA	30NA	500MWF	7MA	16V	15V	30V	15uV/C	150MW	.	80dB	86dB	1M
uA741F	MUG	GPK	INT	.	.25V/uS	+22V	-22V	125C	94dB	5MV	500NA	200NA	500MWF	5MA	10V	12V	30V	.	85MW	3MA	70dB	76dB	300K
uA741FM	FAU	GPK	INT	.	0.2V/uS	+22V	-22V	125C	94dB	5MV	500NA	200NA	570MWF	5MA	12V	15V	30V	.	85MW	3MA	70dB	76dB	300K
uA741HC	FAU	GPK	INT	.	0.2V/uS	+18V	-18V	70C	86dB	6MV	500NA	200NA	500MWF	5MA	12V	15V	30V	.		3MA	70dB	76dB	300K
uA741HM	FAU	GPK	INT	.	0.2V/uS	+22V	-22V	125C	94dB	5MV	500NA	200NA	500MWF	7MA	12V	15V	30V	.	85MW	3MA	70dB	76dB	300K
uA741N(8)	FAU	GPK	INT	.	.25V/uS	+22V	-22V	125C	94dB	5MV	500NA	200NA	310MWF	5MA	10V	12V	30V	.	85MW	3MA	70dB	76dB	300K
uA741N(14)	MUG	GPK	INT	.	.25V/uS	+22V	-22V	125C	94dB	5MV	500NA	200NA	500MWF	5MA	10V	12V	30V	.	85MW	3MA	70dB	76dB	300K
uA741PC	FAU	GPK	INT	.	0.2V/uS	+18V	-18V	70C	86dB	6MV	500NA	200NA	670MWF	5MA	12V	15V	30V	.	150MW	3MA	70dB	76dB	300K

For detailed explanations of column heading notations, see App. A.
Also for ready references the more important abbreviations used in the column headings are listed below:

LEFT HAND PAGE

APP = application (codes at APP.E.)
CMRR = common mode rejection ratio
CMP = compensation (frequency)
dV_{io}/dT = input offset voltage temperature drift
GBP = gain bandwidth product
I_B = input bias current
I_{io} = input bias offset current
i_Q = quiescent supply current
MFR = manufacturer (codes at App.C.)
P_Q = quiescent power consumer
PSRR = power supply rejection ratio
V_{icm} = common mode input voltage rating
V_{inf} = differential input voltage rating
V_{io} = input offset voltage
V_S = dc supply voltage

RIGHT HAND PAGE

Lead out coding summary (details at APP.G.) for different cases (APP.F.)

A = gain adjust
B = bias adjust
C = case
E− = inverting input
E+ = non-inverting input
F,F* = input frequency compensation
G = ground
J = high level input
K = output, open collector
L = output, open emitter
M = metal case
N = not connected
Q = special terminal
R,R* = outputs
S = strobe
T,T* = offset balance
V+ = +ve dc supply
V− = −ve dc supply
W = guard ring
X = blank position, no lead
+ + = +ve supplementary dc supply
− − = −ve supplementary dc supply
ø,ø* = output frequency compensation

CASE (APP.F)	LD1	LD2	LD3	LD4	LD5	LD6	LD7	LD8	LD9	LD10	LD11	LD12	LD13	LD14	LD15	LD16	EUROPE SUBSTITUTE	USA SUBSTITUTE	IS	TYPE NUMBER
DIL-14/1P	N	E-1	E+1	V-	E+2	E-2	N	N	S2	R	V+	G	S1	N	.	.	SFC2711KM	UA711DM	0	uA711N(14)
DIL-14/1P	N	E-1	E+1	V-	E+2	E-2	N	N	S2	R	V+	G	S1	N	.	.	SFC2711EC	LM711CN	0	uA711PC
DIL-14/1C	F	F*	Q	E-	E+	N	N	N	N	V-	R	ø	V+	ø*	.	.		UA715DM	0	uA715DC
DIL-14/1C	F	F*	Q	E-	E+	N	N	N	N	V-	R	ø	V+	ø*	.	.			0	uA715DM
TO5-10/1M	F	Q	E-	E+	V-	R	ø	V+	ø*	F*	.							UA715HM	0	uA715HC
TO5-10/1M	F	Q	E-	E+	V-	R	ø	V+	ø*	F*									0	uA715HM
TO5-8/1M	T	E-	E+	V-	ø	ø*	V+	T*									RM725T	SSS725AJ	0	uA725AHM
TO5-8/1M	T	E-	E+	V-	ø	ø*	V+	T*									RC725J	SSS725EJ	G	uA725EHC
TO5-8/1M	T	E-	E+	V-	ø	ø*	V+	T*									PC725T	SSS725EJ	0	uA725HC
TO5-8/1M	T	E-	E+	V-	ø	ø*	V+	T*									RM725T	SSS725AJ	0	uA725HM
TO5-10/1M	V+	E-	E+	F*	V-	B	++	R	R*	F									0	uA727HC
TO5-10/1M	V+	E-	E+	F*	V-	B	++	R	R*	F									0	uA727HM
TO5-8/1M	R*1	E-	E+	G	R.	R2	V+	R*2											0	uA730HC
TO5-8/1M	R*1	E-	E+	G	R1	R2	V+	R*2											0	uA730HM
DIL-14/1P	E+	N	A2	A*2	V-	N	R	R*	N	V+	A1	A*1	N	E-			SN52733J	UA733DM	0	uA733A
DIL-14/1P	E+	N	A2	A*2	V-	N	R	R*	N	V+	A1	A*1	N	E-			SN72733J	UA733DC	0	uA733CA
DIL-14/1C	E+	N	A2	A*2	V-	N	R	R*	N	V+	A1	A*1	N	E-			SN72733J	UA733DC	0	uA733CF
DIL-14/1C	E+	N	A2	A*2	V-	N	R	R*	N	V+	A1	A*1	N	E-			SN72733J	UA733DC	0	uA733CI*
TO5-10/1M	E-	E+	A2	A*2	V-	R	R*	V+	A1	A*1							SN72733L	UA733HC	0	uA733CK
DIL-14/1P	E+	N	A2	A*2	V-	N	R	R*	N	V+	A1	A*1	N	E-			SN72733J	UA733DC	0	uA733CN(14)
DIL-14/1C	E+	N	A2	A*2	V-	N	R	R*	N	V+	A1	A*1	N	E-			SN72733J	LM733CN	0	uA733DC
DIL-14/1C	E+	N	A2	A*2	V-	N	R	R*	N	V+	A1	A*1	N	E-			SN52733J	LM733D	0	uA733DM
DIL-14/1C	E+	N	A2	A*2	V-	N	R	R*	N	V+	A1	A*1	N	E-			SN52733J	LM733D	0	uA733F
FLP-10/3C	E+	A2	A*2	V-	R	R*	V+	A1	A*1	E-								SN52733FA	0	uA733FM
TO5-10/1M	E-	E+	A2	A*2	V-	R	R*	V+	A1	A*1							SN72733L	LM733CH	0	uA733HC
TO5-10/1M	E-	E+	A2	A*2	V-	R	R*	V+	A1	A*1							SN52733L	LM733H	0	uA733HM
DIL-14/1C	E+	N	A2	A*2	V-	N	R	R*	N	V+	A1	A*1	N	E-			SN52733J	UA733DM	0	uA733I
TO5-10/1M	E-	E+	A2	A*2	V-	R	R*	V+	A1	A*1							SN52733L	LM733HM	0	uA733K
DIL-14/1C	N	N	V+	N	E+	N	E-	T	N	T*	V-	G	R	R*				UA734DM	0	uA734DC
DIL-14/1C	N	N	V+	N	E+	N	E-	T	N	T*	V-	G	R	R*					0	uA734DM
TO5-10/1M	E-	N	T	T*	V-	G	R	R*	V+	E+								ICL8001C	0	uA734HC
TC5-10/1M	E-	N	T	T*	V-	G	R	R*	V+	E+								ICL8001M	0	uA734HM
DIL-14/1C	R1	ø1	F1	F*1	E+1	E-1	V-	E-2	E+2	F2	F*2	ø2	R2	V+			TBA231	UA739PC	0	uA739DC
DIL-14/1P	R1	ø1	F1	F*1	E+1	E-1	V-	E-2	E+2	F2	F*2	ø2	R2	V+			TBA231	UA739DC	0	uA739PC
TO5-8/1M	T	E-	E+	V-	T*	R	V+	N									LF13741H	UA740HC	0	uA740CT
TO5-8/1M	T	E-	E+	V-	T*	R	V+	N										NE536	0	uA740HC
TO5-8/1M	T	E-	E+	V-	T*	R	V+	N										SU536	0	uA740HM
TO5-8/1M	T	E-	E+	V-	T*	R	V+	N									LF13741H	UA740HM	0	uA740T
DIL-14/1P	N	N	T	E-	E+	V-	N	N	T*	R	V+	N	N	N			LM741D	UA741DM	0	uA741A
DIL-14/1C	N	N	T	E-	E+	N	N	N	T*	R	V+	N	N	N				LM741AD	0	uA741ADM
FLP-10/3C	N	T	E-	E+	V-	T*	R	V+	N	N							SFC2741PM	UA741AF	0	uA741AFM
TO5-8/1M	T	E-	E+	V-M	T*	R	V+	N									TBA222	LM741AH	0	uA741AHM
DIL-14/1P	N	N	T	E-	E+	V-	N	N	T*	R	V+	N	N	N			TBA221A	UA741DC	0	uA741CA
MDL-8/2P	T	E-	E+	V-	T*	R	V+	N											0	uA741CD
DIL-8/1P	T	E-	E+	V-	T*	R	V+	N									TBA221B	UA741TC	0	uA741CN(8)
DIL-14/1P	N	N	T	E-	E+	V-	N	N	T*	R	V+	N	N	N			TBA221A	UA741DC	0	uA741CN(14)
TO5-8/1M	T	E-	E+	V-	T*	R	V+	N									TBA221	UA741HC	0	uA741CT
DIL-8/1P	T	E-	E+	V-	T*	R	V+	N									TBA221B	UA741TC	0	uA741CV
DIL-14/1C	N	N	T	E-	E+	V-	N	N	T*	R	V+	N	N	N			TBA221A	LM741CJ14	0	uA741DC
DIL-14/1C	N	N	T	E-	E+	V-	N	N	T*	R	V+	N	N	N				LM741D	0	uA741DM
DIL-14/1C	N	N	T	E-	E+	V-	N	N	T*	R	V+	N	N	N				LM741EJ14	0	uA741EDC
TO5-8/1M	T	E-	E+	V-M	T*	R	V+	N										LM741EH	0	uA741EHC
DIL-14/1C	N	N	T	E-	E+	V-	N	N	T*	R	V+	N	N	N			LM741D	UA741DM	0	uA741F
FLP-10/3C	N	T	E-	E+	V-	T*	R	V+	N	N							SFC2741PM	LM741F	0	uA741FM
TO5-8/1M	T	E-	E+	V-M	T*	R	V+	N									TBA221	LM741CH	0	uA741HC
TO5-8/1M	T	E-	E+	V-M	T*	R	V+	N									TBA222	LM741H	0	uA741HM
DIL-8/1P	T	E-	E+	V-	T*	R	V+	N										UA741V	0	uA741N(8)
DIL-14/1P	N	N	T	E-	E+	V-	N	N	T*	R	V+	N	N	N			LM741D	UA741DM	0	uA741N(14)
DIL-14/1P	N	N	T	E-	E+	V-	N	N	T*	R	V+	N	N	N			TBA221A	LM741CD	0	uA741PC
DIL-8/1P	T	E-	E+	V-	T*	R	V+	N									TBA221B	LM741CN	0	uA741RC

TYPE NUMBER	MFR	APP	CMP	GBP MIN	SLEW RATE MIN	V_S^+ MAX	V_S^- MAX	T_{op} MAX	A_{VOL} MIN	V_{IO} MAX	I_B MAX	I_{IO} MAX	P_{TOT} MAX	I_{OUT} MIN	V_{OUT} MIN	V_{ICM} MAX	V_{IDR} MAX	dV_{IO}/dT MAX	P_O MAX	I_O MAX	CMRR MIN	PSRR MIN	R_{IN} MIN
uA741RC	FAU	GPK	INT	.	0.2V/uS	+18V	-18V	70C	86dB	6MV	500NA	200NA	310MWF	5MA	12V	15V	30V	.	.	3MA	70dB	76dB	300K
uA741T	SJU	GPK	INT	.	.25V/uS	+22V	-22V	125C	94dB	5MV	500NA	200NA	500MWF	5MA	10V	12V	30V	.	85MW	3MA	70dB	76dB	300K
uA741TC	FAU	GPK	INT	.	0.2V/uS	+18V	-18V	70C	86dB	6MV	500NA	200NA	310MWF	5MA	12V	15V	30V	.	.	3MA	70dB	76dB	300K
uA741V	MUG	GPK	INT	.	.25V/uS	+22V	-22V	125C	94dB	5MV	500NA	200NA	310MWF	5MA	10V	12V	30V	.	85MW	3MA	70dB	76dB	300K
uA747A	SJU	DGK	INT	.5MHZ	.25V/uS	+22V	-22V	125C	94dB	6MV	500NA	300NA	670MWF	7MA	10V	15V	30V	.	85MW	3MA	70dB	76dB	300K
uA747ADM	FAU	DGK	INT	.4MHZ	0.3V/uS	+22V	-22V	125C	94dB	3MV	80NA	30NA	670MWF	10MA	16V	15V	30V	15uV/C	150MW	3MA	80dB	86dB	1M
uA747AHM	FAU	DGK	INT	.4MHZ	0.3V/uS	+22V	-22V	125C	94dB	3MV	80NA	30NA	500MWF	10MA	16V	15V	30V	15uV/C	150MW	3MA	80dB	86dB	1M
uA747CA	SJU	DGK	INT	.5MHZ	.25V/uS	+18V	-18V	70C	88dB	7.5MV	800NA	200NA	670MWF	5MA	10V	15V	30V	.	85MW	3MA	70dB	76dB	300K
uA747CF	MUG	DGK	INT	.5MHZ	.25V/uS	+18V	-18V	70C	88dB	7.5MV	800NA	200NA	670MWF	5MA	10V	15V	30V	.	85MW	3MA	70dB	76dB	300K
uA747CK	SJU	DGK	INT	.5MHZ	.25V/uS	+18V	-18V	70C	88dB	7.5MV	800NA	200NA	500MWF	5MA	10V	15V	30V	.	85MW	3MA	70dB	76dB	300K
uA747CN(14)	MUG	DGK	INT	.5MHZ	.25V/uS	+18V	-18V	70C	88dB	7.5MV	800NA	200NA	670MWF	5MA	10V	15V	30V	.	85MW	3MA	70dB	76dB	300K
uA747DC	FAU	DGK	INT	.	0.2V/uS	+18V	-18V	70C	88dB	6MV	500NA	200NA	670MWF	5MA	12V	15V	30V	.	85MW	3MA	70dB	76dB	300K
uA747DM	FAU	DGK	INT	.	0.2V/uS	+22V	-22V	125C	94dB	5MV	500NA	200NA	500MWF	5MA	12V	15V	30V	.	85MW	3MA	70dB	76dB	300K
uA747EDC	FAU	DGK	INT	.4MHZ	0.3V/uS	+22V	-22V	70C	94dB	3MV	80NA	30NA	670MWF	10MA	16V	15V	30V	15uV/C	150MW	3MA	80dB	86dB	1M
uA747EHC	FAU	DGK	INT	.4MHZ	0.3V/uS	+22V	-22V	70C	94dB	3MV	80NA	30NA	500MWF	10MA	16V	15V	30V	15uV/C	150MW	3MA	80dB	86dB	1M
uA747F	MUG	DGK	INT	.5MHZ	.25V/uS	+22V	-22V	125C	94dB	6MV	500NA	300NA	670MWF	5MA	10V	15V	30V	.	85MW	3MA	70dB	76dB	300K
uA747HC	FAU	DGK	INT	.	0.2V/uS	+18V	-18V	70C	88dB	6MV	500NA	200NA	500MWF	5MA	12V	15V	30V	.	85MW	3MA	70dB	76dB	300K
uA747HM	FAU	DGK	INT	.	0.2V/uS	+22V	-22V	125C	94dB	5MV	500NA	200NA	500MWF	5MA	12V	15V	30V	.	85MW	3MA	70dB	76dB	300K
uA747-IDC	FAU	DGK	INT	.	0.2V/uS	+18V	-18V	70C	88dB	6MV	500NA	200NA	670MWF	5MA	12V	15V	30V	.	85MW	3MA	70dB	76dB	300K
uA747-IDM	FAU	DGK	INT	.	0.2V/uS	+22V	-22V	125C	94dB	5MV	500NA	200NA	670MWF	5MA	12V	15V	30V	.	85MW	3MA	70dB	76dB	300K
uA747-IHC	FAU	DGK	INT	.	0.2V/uS	+18V	-18V	70C	88dB	6MV	500NA	200NA	500MWF	5MA	12V	15V	30V	.	85MW	3MA	70dB	76dB	300K
uA747-IHM	FAU	DGK	INT	.	0.2V/uS	+22V	-22V	125C	94dB	5MV	500NA	200NA	500MWF	5MA	12V	15V	30V	.	85MW	3MA	70dB	76dB	300K
uA747K	SJU	DGK	INT	.5MHZ	.25V/uS	+22V	-22V	125C	94dB	6MV	500NA	300NA	500MWF	7MA	10V	15V	30V	.	85MW	3MA	70dB	76dB	300K
uA747N(14)	MUG	DGK	INT	.5MHZ	.25V/uS	+22V	-22V	125C	94dB	6MV	500NA	300NA	670MWF	7MA	10V	15V	30V	.	85MW	3MA	70dB	76dB	300K
uA747PC	FAU	DGK	INT	.	0.2V/uS	+18V	-18V	70C	88dB	6MV	500NA	200NA	670MWF	5MA	12V	15V	30V	.	85MW	3MA	70dB	76dB	300K
uA748A	SJU	GPU	EXT	.	.25V/uS	+22V	-22V	125C	94dB	5MV	500NA	200NA	500MWF	5MA	10V	15V	30V	.	85MW	3MA	70dB	96dB	300K
uA748ADM	FAU	GPU	EXT	.	0.2V/uS	+22V	-22V	125C	94dB	2MV	75NA	10NA	670MWF	5MA	12V	15V	30V	15uV/C	85MW	3MA	80dB	80dB	2M
uA748AFM	FAU	GPU	EXT	.	0.2V/uS	+22V	-22V	125C	94dB	2MV	75NA	10NA	570MW	5MA	12V	15V	30V	15uV/C	85MW	3MA	80dB	80dB	2M
uA748AHM	FAU	GPU	EXT	.	0.2V/uS	+22V	-22V	125C	94dB	2MV	75NA	10NA	570MW	5MA	12V	15V	30V	15uV/C	85MW	3MA	80dB	80dB	2M
uA748CA	SJU	GPU	EXT	.	.25V/uS	+18V	-18V	70C	94dB	6MV	500NA	200NA	500MWF	5MA	10V	15V	30V	.	85MW	3MA	70dB	96dB	300K
uA748CD	MUG	GPU	EXT	.	.25V/uS	+18V	-18V	70C	94dB	6MV	500NA	200NA	.	5MA	10V	15V	30V	.	85MW	3MA	70dB	96dB	300K
uA748CN(8)	MUG	GPU	EXT	.	.25V/uS	+18V	-16V	70C	94dB	6MV	500NA	200NA	500MWF	5MA	10V	15V	30V	.	85MW	3MA	70dB	96dB	300K
uA748CN(14)	MUG	GPU	EXT	.	.25V/uS	+18V	-18V	70C	94dB	6MV	500NA	200NA	500MWF	5MA	10V	15V	30V	.	85MW	3MA	70dB	96dB	300K
uA748CT	SJU	GPU	EXT	.	.25V/uS	+18V	-18V	70C	94dB	6MV	500NA	200NA	500MWF	5MA	10V	15V	30V	.	85MW	3MA	70dB	96dB	300K
uA748CV	SJU	GPU	EXT	.	.25V/uS	+18V	-18V	70C	94dB	6MV	500NA	200NA	500MWF	5MA	10V	15V	30V	.	85MW	3MA	70dB	96dB	300K
uA748DC	FAU	GPU	EXT	.	0.2V/uS	+22V	-22V	70C	86dB	6MV	500NA	200NA	670MWF	5MA	12V	15V	30V	.	85MW	3MA	70dB	76dB	300K
uA748DM	FAU	GPU	EXT	.	0.2V/uS	+22V	-22V	125C	94dB	5MV	500NA	200NA	670MWF	5MA	12V	15V	30V	.	85MW	3MA	70dB	76dB	300K
uA748FM	FAU	GPU	EXT	.	0.2V/uS	+22V	-22V	125C	94dB	5MV	500NA	200NA	570MWF	5MA	12V	15V	30V	.	85MW	3MA	70dB	76dB	300K
uA748HC	FAU	GPU	EXT	.	0.2V/uS	+18V	-18V	70C	86dB	6MV	500NA	200NA	500MWF	5MA	12V	15V	30V	.	85MW	3MA	70dB	76dB	300K
uA748HM	FAU	GPU	EXT	.	0.2V/uS	+22V	-22V	125C	94dB	5MV	500NA	200NA	500MWF	5MA	12V	15V	30V	.	85MW	3MA	70dB	76dB	300K
uA748N(8)	MUG	GPU	EXT	.	.25V/uS	+22V	-22V	125C	94dB	5MV	500NA	200NA	310MWF	5MA	10V	15V	30V	.	85MW	3MA	70dB	96dB	300K
uA748N(14)	MUG	GPU	EXT	.	.25V/uS	+22V	-22V	125C	94dB	5MV	500NA	200NA	500MWF	5MA	10V	15V	30V	.	85MW	3MA	70dB	96dB	300K
uA748T	SJU	GPU	EXT	.	.25V/uS	+22V	-22V	125C	94dB	5MV	500NA	200NA	500MWF	5MA	10V	15V	30V	.	85MW	3MA	70dB	96dB	300K
uA748TC	FAU	GPU	EXT	.	0.2V/uS	+18V	-18V	70C	86dB	6MV	500NA	200NA	310MWF	5MA	12V	15V	30V	.	85MW	3MA	70dB	76dB	300K
uA748V	MUG	GPU	EXT	.	.25V/uS	+22V	-22V	125C	94dB	5MV	500NA	200NA	310MWF	5MA	10V	15V	30V	.	85MW	3MA	70dB	96dB	300K
uA749DC	FAU	DLN	EXT	.	1V/uS	+18V	-18V	70C	84dB	6MV	1.5uA	750NA	650MWF	2MA	12V	15V	5V	10uV/C	330MW	14MA	70dB	70dB	50K
uA749DHC	FAU	DLN	EXT	.	.	+12V	-12V	70C	80dB	10MV	1.5uA	0.6MA	500MWF	2MA	12V	12V	5V	10uV/C	54MW	5MA	70dB	80dB	50K
uA749DM	FAU	DLN	EXT	.	1V/uS	+18V	-18V	125C	86dB	3MV	750NA	400NA	650MWF	2MA	12V	15V	5V	10uV/C	220MW	11MA	70dB	74dB	100K
uA749PC	FAU	DLN	EXT	.	1V/uS	+18V	-18V	70C	84dB	6MV	1.5uA	750NA	650MWF	2MA	12V	15V	5V	10uV/C	330MW	14MA	70dB	70dB	50K
uA760DC	FAU	CPR	EXT	.	.	+8V	-8V	70C	.	6MV	60uA	7.5uA	570MWF	5MA	.	4V	4V	10uV/C	.	32MA	.	.	5K
uA760DM	FAU	CPR	EXT	.	.	+8V	-8V	125C	.	6MV	60uA	7.5uA	670MWF	5MA	.	4V	4V	10uV/C	.	34MA	.	.	5K
uA760HC	FAU	CPR	EXT	.	.	+8V	-8V	70C	.	6MV	60uA	7.5uA	500MWF	5MA	.	4V	4V	10uV/C	.	32MA	.	.	5K
uA760HM	FAU	CPR	EXT	.	.	+8V	-8V	125C	.	6MV	60uA	7.5uA	500MWF	5MA	.	4V	4V	10uV/C	.	34MA	.	.	5K
uA775DC	FAU	QCP	EXT	.	.	+18V	-18V	70C	100dB	5MV	250NA	50NA	570MWF	6MA	.	+18V	-18V	30uV/C	.	2MA	70dB	70dB	150K
uA775DM	FAU	QCP	EXT	.	.	+18V	-18V	70C	100dB	5MV	100NA	25NA	670MWF	6MA	.	+18V	-18V	30uV/C	.	2MA	70dB	70dB	400K
uA775PC	FAU	QCP	EXT	.	.	+18V	-18V	70C	100dB	5MV	250NA	50NA	670MWF	6MA	.	+18V	18V	30uV/C	.	2MA	70dB	70dB	150K
uA776DC	FAU	PRA	INT	.	0.3V/uS	+18V	-18V	70C	94dB	6MV	50NA	25NA	670MWF	2MA	10V	15V	30V	.	6MW	.2MA	70dB	74dB	2M
uA776DM	FAU	PRA	INT	.	0.3V/uS	+18V	-18V	125C	106dB	5MV	50NA	15NA	670MWF	2MA	10V	15V	30V	.	6MW	.2MA	70dB	76dB	2M
uA776HC	FAU	PRA	INT	.	0.3V/uS	+18V	-18V	70C	94dB	6MV	50NA	25NA	500MWF	2MA	10V	15V	30V	.	6MW	.2MA	70dB	74dB	2M
uA776HM	FAU	PRA	INT	.	0.3V/uS	+18V	-18V	125C	106dB	5MV	50NA	15NA	500MWF	2MA	10V	15V	30V	.	6MW	.2MA	70dB	76dB	2M

For detailed explanations of column heading notations, see App. A.

Also for ready references the more important abbreviations used in the column headings are listed below:

LEFT HAND PAGE
- APP = application (codes at APP.E.)
- CMRR = common mode rejection ratio
- CMP = compensation (frequency)
- dV_{io}/dT = input offset voltage temperature drift
- GBP = gain bandwidth product
- I_B = input bias current
- I_{io} = input bias offset current
- I_Q = quiescent supply current
- MFR = manufacturer (codes at App.C.)
- P_Q = quiescent power consumer
- PSRR = power supply rejection ratio
- V_{ICM} = common mode input voltage rating
- V_{inf} = differential input voltage rating
- V_{io} = input offset voltage
- V_S = dc supply voltage

RIGHT HAND PAGE
Lead out coding summary (details at APP.G.) for different cases (APP.F.)
- A = gain adjust
- B = bias adjust
- C = case
- E− = inverting input
- E+ = non-inverting input
- F,F* = input frequency compensation
- G = ground
- J = high level input
- K = output, open collector
- L = output, open emitter
- M = metal case
- N = not connected
- Q = special terminal
- R,R* = outputs
- S = strobe
- T,T* = offset balance
- V+ = +ve dc supply
- V− = −ve dc supply
- W = guard ring
- X = blank position, no lead
- + + = +ve supplementary dc supply
- − − = −ve supplementary dc supply
- b.b* = output frequency compensation

CASE (APP F)	LD1	LD2	LD3	LD4	LD5	LD6	LD7	LD8	LD9	LD10	LD11	LD12	LD13	LD14	LD15	LD16	EUROPE SUBSTITUTE	USA SUBSTITUTE	IS	TYPE NUMBER
TO5-8/1M	T	E-	E+	V-	T*	R	V+	N									TBA222	UA741HM	0	uA741T
DIL-8/1C	T	E-	E+	V-	T*	R	V+	N									TBA221B	LM741CN	0	uA741TC
DIL-8/1P	T	E-	E+	V-	T*	R	V+	N										UA741N8	0	uA741V
DIL-14/1P	E-1	E+1	T1	V-	T2	E+2	E-2	T*2	V+2	R2	N	R1	V+1	T*1			SFC2747KM	UA747DM	0	uA747A
DIL-14/1C	E-1	E+1	T1	V-	T2	E+2	E-2	T*2	V+2	R2	N	R1	V+	T*1			SFC2747KM	LM747AD	0	uA747ADM
TO5-10/1M	R1	V+1	E-1	E+1	V-	E+2	E-2	V+2	R2	N							TBC0747	LM747AH	0	uA747AH
DIL-14/1P	E-1	E+1	T1	V-	T2	E+2	E-2	T*2	V+2	R2	N	R1	V+1	T*1			TBB0747A	UA747DC	0	uA747CA
DIL-14/1C	E-1	E+1	T1	V-	T2	E+2	E-2	T*2	V+2	R2	N	R1	V+1	T*1			TBB0747A	UA747CF	0	uA747CF
TO5-10/1M	R1	V+1	E-1	E+1	V-	E+2	E-2	V+2	R2	N							TBB0747	UA747HC	0	uA747CK
DIL-14/1P	E-1	E+1	T1	V-	T2	E+2	E-2	T*2	V+2	R2	N	R1	V+1	T*1			TBB0747A	UA747DC	0	uA747CN(14)
DIL-14/1C	E-1	E+1	T1	V-	T2	E+2	E-2	T*2	V+2	R2	N	R1	V+1	T*1			TBB0747A	LM747CD	0	uA747DC
DIL-14/1C	E-1	E+1	T1	V-	T2	E+2	E-2	T*2	V+2	R2	N	R1	V+1	T*1			SFC2747KM	LM747D	0	uA747DM
DIL-14/1C	E-1	E+1	T1	V-	T2	E+2	E-2	T*2	V+2	R2	N	R1	V+1	T*1			SFC2747KM	LM747ED	0	uA747EDC
TO5-10/1M	R1	V+1	E-1	E+1	V-	E+2	E-2	V+2	R2	N							TBB0747	LM747EH	0	uA747EHC
DIL-14/1C	E-1	E+1	T1	V-	T2	E+2	E-2	T*2	V+2	R2	N	R1	V+1	T*1			SFC2747KM	UA747DM	0	uA747F
TO5-10/1M	R1	V+1	E-1	E+1	V-	E+2	E-2	V+2	R2	N							TBB0747	LM747CH	0	uA747CH
TO5-10/1M	R1	V+1	E-1	E+1	V-	E+2	E-2	V+2	R2	N							TBC0747	LM747H	0	uA747HM
DIL-14/1C	E-1	E+1	T1	V-	T2	E+2	E-2	T*2	V+2	R2	N	R1	V+	T*1				LM747-ICJ	0	uA747IDC
DIL-14/1C	E-1	E+1	T1	V-	T2	E+2	E-2	T*2	V+2	R2	N	R1	V+	T*1				LM747-IJ	0	uA747IDM
TO5-10/1M	R1	V+1	E-1	E+1	V-	E+2	E-2	V+2	R2	N								LM747-ICH	0	uA747IHC
TO5-10/1M	R1	V+1	E-1	E+1	V-	E+2	E-2	V+2	R2	N								LM747-IH	0	uA747IHM
TO5-10/1M	R1	V+1	E-1	E+1	V-	E+2	E-2	V+2	R2	N							SFC2747M	UA747HM	0	uA747K
DIL-14/1P	E-1	E+1	T1	V-	T2	E+2	E-2	T*2	V+2	R2	N	R1	V+1	T*1			SFC2747KM	UA747DM	0	uA747N(14)
DIL-14/1P	E-1	E+1	T1	V-	T2	E+2	E-2	T*2	V+2	R2	N	R1	V+1	T*1			TBB0747A	LM747CD	0	uA747PC
DIL-14/1P	N	N	FT	E-	E+	V-	N	N	T*	R	V+	F*	N	N			SN52748JA	UA748DM	0	uA748A
DIL-14/1C	N	N	FT	E-	E+	V-	N	N	T*	R	V+	F*	N	N				SN52748JA	0	uA748ADM
FLP-10/3C	N	FT	E-	E+	V-		T*	R	V+	F*	N							SN52748FA	0	uA748AFM
TO5-8/1M	FT	E-	E+	V-	T*	R	V+	F*									TBC0748	LM748H	0	uA748AHM
DIL-8/1P	N	N	FT	E-	E+	V-	N	N	T*	R	V+	F*	N	N			SN72748J	UA748DC	0	uA748CA
MDL-8/2P	FT	E-	E+	V-	T*	R	V+	F*											0	uA748CD
DIL-8/1P	FT	E-	E+	V-	T*	R	V+	F*									TBB0748B	UA748TC	0	uA748CN(8)
DIL-14/1P	N	N	FT	E-	E+	V-	N	N	T*	R	V+	F*	N	N			SN72748J	UA748DC	0	uA748CN(14)
TO5-8/1M	FT	E-	E+	V-	T*	R	V+	F*									TBB0748	UA748HC	0	uA748CT
DIL-8/1P	FT	E-	E+	V-	T*	R	V+	F*									TBB0748B	UA748TC	0	uA748CV
DIL-14/1C	N	N	FT	E-	E+	V-	N	N	T*	R	V+	F*	N	N				SN72748J	0	uA748DC
DIL-14/1C	N	N	FT	E-	E+	V-	N	N	T*	R	V+	F*	N	N				SN52748JA	0	uA748DM
FLP-10/3C	N	FT	E-	E+	V-		T*	R	V+	F*	N							SN52748FA	0	uA748FM
TO5-8/1M	FT	E-	E+	V-	T*	R	V+	F*									TBB0748	LM748CH	0	uA748HC
TO5-8/1M	FT	E-	E+	V-	T*	R	V+	F*									TBC0748	LM748H	0	uA748HM
TO5-8/1M	FT	E-	E+	V-	T*	R	V+	F*									SN52748JP	LM748J	0	uA748N(8)
DIL-14/1P	N	N	FT	E-	E+	V-	N	N	T*	R	V+	F*	N	N			SN52748JA	UA748DM	0	uA748N(14)
TO5-8/1M	FT	E-	E+	V-	T*	R	V+	F*									TBC0748	UA748HM	0	uA748T-
DIL-8/1C	FT	E-	E+	V-	T*	R	V+	F*									TBB0748	LM748CJ	0	uA748TC
DIL-8/1P	FT	E-	E+	V-	T*	R	V+	F*									SN52748JP	LM748J	0	uA748V
DIL-14/1C	R1	ø1	F1	F*1	E+1	E-1	V-	E-2	E+2	F2	F*2	ø2	R2	V+			TBA231	UA749DC	0	uA749DC
TO5-8/1M	R1	E+1	E-1	V-	E-2	E+2	R2	V+											0	uA749DHC
DIL-14/1C	R1	ø1	F1	F*1	E+1	E-1	V-	E-2	E+2	F2	F*2	ø2	R2	V+					0	uA749DM
DIL-14/1P	R1	ø1	F1	F*1	E+1	E-1	V-	E-2	E+2	F2	F*2	ø2	R2	V+					0	uA749PC
DIL-14/1C	N	N	N	E2	E1	V-	N	N	G	R2	R1	V+	N	N				LM360D	0	uA760DC
DIL-14/1C	N	N	N	E2	E1	V-	N	N	G	R2	R1	V+	N	N				LM160D	0	uA760DM
TO5-8/1M	N	E2	E1	V-	G	R2	R1	V+										LM360H	0	uA760HC
TO5-8/1M	N	E2	E1	V-	G	R2	R1	V+										LM160H	0	uA760HM
DIL-14/1C	R1	R2	V+	E-1	E+1	E-2	E+2	E-3	E+3	E-4	E+4	G	R4	R3			MLM339AL	LM339AJ	0	uA775DC
DIL-14/1C	R1	R2	V+	E-1	E+1	E-2	E+2	E-3	E+3	E-4	E+4	G	R4	R3			MLM139AL	LM139AD	0	uA775DM
DIL-14/1P	R1	R2	V+	E-1	E+1	E-2	E+2	E-3	E+3	E-4	E+4	G	R4	R3			MLM339AL	LM339AJ	0	uA775PC
DIL-14/1C	N	N	T	E-	E+	V-	N	N	T*	R	V+	B	N	N			SF2776KM	MC1776L	0	uA776DC
DIL-14/1C	N	N	T	E-	E+	V-	N	N	T*	R	V+	B	N	N			SFC2776KM	MC1776L	0	uA776DM
TO5-8/1M	T	E-	E+	V-	T*	R	V+	B									SFC2776C	MC1776CG	0	uA776HC
TO5-8/1M	T	E-	E+	V-	T*	R	V+	B									SFC2776M	MC1776G	0	uA776HM
DIL-8/1P	T	E-	E+	V-	T*	R	V+	B									SFC2776DC		0	uA776TC

TYPE NUMBER	MFR	APP	CMP	GBP MIN	SLEW RATE MIN	V_S^+ MAX	V_S^- MAX	T_{op} MAX	A_{VOL} MIN	V_{IO} MAX	I_B MAX	I_{IO} MAX	P_{TOT} MAX	I_{OUT} MIN	V_{OUT} MIN	V_{ICM} MAX	V_{IDF} MAX	dV_{IO}/dT MAX	P_O MAX	I_O MAX	CMRR MIN	PSRR MIN	R_{IN} MIN
uA776TC	FAU	PRA	INT		0.3V/uS	+18V	-18V	70C	94dB	6MV	50NA	25NA	310MWF	2MA	10V	15V	30V	.	6MW	.2MA	70dB	74dB	2M
uA777DC	FAU	PIA	EXT	.	0.2V/uS	+22V	-22V	70C	88dB	5MV	100NA	20NA	670MWF	5MA	12V	15V	30V	30uV/C	85MW	3MA	70dB	76dB	1M
uA777HC	FAU	PIA	EXT	.	0.2V/uS	+22V	-22V	70C	88dB	5MV	100NA	20NA	500MWF	5MA	12V	15V	30V	30uV/C	85MW	3MA	70dB	76dB	1M
uA777TC	FAU	PIA	EXT	.	0.2V/uS	+22V	-22V	70C	88dB	5MV	100NA	20NA	310MWF	5MA	12V	15V	30V	30MW	85MW	3MA	70dB	76dB	1M
uA791KC	FAU	HPO	INT			+18V	-18V	125C	86dB	6MV	500NA	200NA	16WC	1A	12V	15V	30V		.	30MA	70dB	76dB	300K
uA791KM	FAU	HPO	INT			+22V	-22V	150C	94dB	5MV	500NA	200NA	20WC	1A	12V	15V	30V		.	25MA	70dB	76dB	300K
uA791P5	FAU	HPO	INT			+18V	-18V	125C	86dB	6MV	500NA	200NA	10WC	1A	12V	15V	30V		.	30MA	70dB	76dB	300K
uA798HC	FAU	DGK	T	.3MHZ	0.2V/uS	+18V	-18V	70C	86dB	6MV	250NA	75NA	500MWF	6MA	13V	18V	30V	30uV/C	.	4MA	70dB	76dB	300K
uA798HM	FAU	DGK	T	.3MHZ	0.2V/uS	+18V	-18V	125C	94dB	5MV	100NA	25NA	500MWF	6MA	13V	18V	30V	30uV/C	.	3MA	70dB	76dB	300K
uA798RC	FAU	DGK	T	.3MHZ	0.2V/uS	+18V	-18V	70C	86dB	6MV	250NA	75NA	310MWF	6MA	13V	18V	30V	30uV/C	.	4MA	70dB	76dB	300K
uA798RM	FAU	DGK	INT	.3MHZ	0.2V/uS	+18V	-18V	125C	94dB	5MV	100NA	25NA	310MWF	6MA	13V	18V	30V	30uV/C	.	3MA	70dB	76dB	300K
uA798TC	FAU	DGK	T	.3MHZ	0.2V/uS	+18V	-18V	70C	86dB	6MV	250NA	75NA	310MWF	6MA	13V	18V	30V	30uV/C	.	4MA	70dB	76dB	300K
uA799HC	FAU	GPK	INT	.3MHZ	0.2V/uS	+18V	-18V	70C	86dB	6MV	250NA	75NA	500MWF	6MA	12V	18V	30V	30uV/C	.	4MA	70dB	76dB	300K
uA799HM	FAU	GPK	INT	.3MHZ	0.2V/uS	+18V	-18V	125C	94dB	5MV	100NA	25NA	500MWF	6MA	12V	18V	30V	30uV/C	.	4MA	70dB	76dB	300K
uA799RC	FAU	GPK	INT	.3MHZ	0.2V/uS	+18V	-18V	70C	86dB	6MV	250NA	75NA	310MWF	6MA	12V	18V	30V	30uV/C	.	4MA	70dB	76dB	300K
uA799RM	FAU	GPK	INT	.3MHZ	0.2V/uS	+18V	-18V	125C	94dB	5MV	100NA	25NA	500MWF	6MA	12V	18V	30V	30uV/C	.	4MA	70dB	76dB	300K
uA799TC	FAU	GPK	INT	.3MHZ	0.2V/uS	+18V	-18V	70C	86dB	6MV	250NA	75NA	310MWF	6MA	12V	18V	30V	30uV/C	.	4MA	70dB	76dB	300K
uA1458CHC	FAU	DGK	INT	.3MHZ	0.3V/uS	+18V	-18V	70C	86dB	10MV	700NA	300NA	800MWF	4MA	11V	15V	30V	50uV/C	240MW	10MA	60dB	.	100M
uA1458CRC	FAU	DGK	INT	.3MHZ	0.3V/uS	+18V	-18V	70C	86dB	10MV	700NA	300NA	560MWF	4MA	11V	15V	30V	50uV/C	240MW	10MA	60dB	.	100M
uA1458CTC	FAU	DGK	INT	.3MHZ	0.3V/uS	+18V	-18V	70C	86dB	10MV	700NA	300NA	560MWF	4MA	11V	15V	30V	50uV/C	240MW	10MA	60dB	.	100M
uA1458HC	FAU	DGK	INT	.3MHZ	0.3V/uS	+18V	-18V	70C	86dB	6MV	500NA	200NA	800MWF	5MA	12V	15V	30V	50uV/C	170MW	6MA	70dB	76dB	100M
uA1458RC	FAU	DGK	INT	.3MHZ	0.3V/uS	+18V	-18V	70C	86dB	6MV	500NA	200NA	560MWF	5MA	12V	15V	30V	50uV/C	170MW	6MA	70dB	76dB	100M
uA1458TC	FAU	DGK	INT	.3MHZ	0.3V/uS	+18V	-18V	70C	86dB	6MV	500NA	200NA	560MWF	5MA	12V	15V	30V	50uV/C	170MW	6MA	70dB	76dB	100M
uA1558HC	FAU	DGK	INT	.3MHZ	0.3V/uS	+22V	-22V	125C	94dB	5MV	500NA	200NA	800MWF	5MA	12V	15V	30V	50uV/C	150MW	5MA	70dB	76dB	100M
uA3301P	FAU	QCD	INT	2MHZ		+28V		85C	60dB			300NA		670MWF	5MA	5V	.	.	.	.	3MA	.	70dB	100K
uA3302P	FAU	QCP	EXT		50V/uS	+18V		85C	66dB	20MV	500NA	15NA	670MWF	2MA	.	+18V	-18V		.	20MA	70dB		
uA3303P	FAU	QCP	INT	.3MHZ	0.2V/uS	+18V	-18V	85C	86dB	8MV	500NA	75NA	670MWF	10MA	12V	18V	30V	30uV/C	.	7MA	70dB	76dB	300K
LA3401P	FAU	QCD	INT	2MHZ		+18V		70C	60dB			300NA		670MWF	5MA	5V	.	.	.	.	3MA	.	70dB	100K
uA3403D	FAU	QGK	INT	.3MHZ	0.2V/uS	+18V	-18V	70C	86dB	8MV	500NA	50NA	670MWF	10MA	12V	18V	30V	30uV/C	.	7MA	70dB	76dB	300K
uA3403P	FAU	QGK	INT	.3MHZ	0.2V/uS	+18V	-18V	70C	86dB	8MV	500NA	50NA	670MWF	10MA	12V	18V	30V	30uV/C	.	7MA	70dB	76dB	300K
uA3503D	FAU	QGK	INT	.3MHZ	0.2V/uS	+18V	-18V	125C	94dB	5MV	500NA	50NA	670MWF	20MA	12V	18V	30V	30uV/C	.	4MA	70dB	76dB	300K
uA4136DC	FAU	QGK	INT	1MHZ	0.3V/uS	+18V	-18V	70C	86dB	6MV	500NA	200NA	670MWF	5MA	12V	15V	30V		340MW		70dB	76dB	300K
uA4136DM	FAU	QGK	INT	2MHZ	0.5V/uS	+22V	-22V	125C	94dB	5MV	500NA	200NA	670MWF	5MA	12V	15V	30V		340MW		70dB	76dB	300K
uA4136PC	FAU	QGK	INT	1MHZ	0.3V/uS	+18V	-18V	70C	86dB	6MV	500NA	200NA	670MWF	5MA	12V	15V	30V		340MW	.	70dB	76dB	300K
uAF111D	FAU	CPR	EXT	.		+18V	-18V	125C	100dB	4MV	50pA	25pA	500MWF	8MA		15V	30V		.	6MA	.	.	.
uAF111H	FAU	CPR	EXT	.		+18V	-18V	125C	100dB	4MV	50pA	25pA	500MWF	8MA		15V	30V		.	6MA	.	.	.
uAF155AHM	FAU	FET	INT	1MHZ	3V/uS	+22V	-22V	125C	94dB	2MV	50pA	10pA	500MWF	5MA	12V	20V	40V	5uV/C	.	4MA	85dB	85dB	0.1T
uAF155HM	FAU	FET	INT	1MHZ	2V/uS	+18V	-18V	125C	94dB	5MV	100pA	20pA	500MWF	5MA	12V	20V	40V	15uV/C	.	4MA	85dB	85dB	0.1T
uAF156AHM	FAU	HSR	INT	4MHZ	10V/uS	+22V	-22V	125C	94dB	2MV	50pA	10pA	500MWF	5MA	12V	20V	40V	5uV/C	.	7MA	85dB	85dB	0.1T
uAF156HM	FAU	HSR	INT	3MHZ	7.5V/uS	+18V	-18V	125C	94dB	5MV	100pA	20pA	500MWF	5MA	12V	20V	40V	15uV/C	.	7MA	85dB	85dB	0.1T
uAF157AHM	FAU	XSR	INT	15MHZ	8V/uS	+22V	-22V	125C	94dB	2MV	50pA	10pA	500MWF	5MA	12V	20V	40V	5uV/C	.	7MA	85dB	85dB	0.1T
uAF157HM	FAU	XSR	INT	15MHZ	6V/uS	+18V	-18V	125C	94dB	5MV	100pA	20pA	500MWF	5MA	12V	20V	40V	15uV/C	.	7MA	85dB	85dB	0.1T
uAF311D	FAU	CPR	EXT			+18V	-18V	125C	100dB	10MV	150pA	75pA		8MA		15V	30V		.	7MA	.	.	.
uAF311H	FAU	CPR	EXT			+18V	-18V	125C	100dB	10MV	150pA	75pA	500MWF	8MA		15V	30V		.	7MA	.	.	.
uAF355AHC	FAU	FET	INT	1MHZ	3V/uS	+22V	-22V	70C	94dB	2MV	50pA	10pA	500MWF	5MA	12V	20V	40V	5uV/C	.	4MA	85dB	85dB	0.1T
uAF355HC	FAU	FET	INT	1MHZ	2V/uS	+18V	-18V	70C	94dB	10MV	200pA	50pA	500MWF	5MA	12V	16V	30V	15uV/C	.	4MA	85dB	85dB	0.1T
uAF356AHC	FAU	HSR	INT	4MHZ	10V/uS	+22V	-22V	70C	94dB	2MV	50pA	10pA	500MWF	5MA	12V	20V	40V	5uV/C	.	7MA	85dB	85dB	0.1T
uAF356HC	FAU	HSR	INT	3MHZ	7V/uS	+18V	-18V	70C	94dB	10MV	200pA	50pA	500MWF	5MA	12V	15V	30V	15uV/C	.	10MA	85dB	85dB	0.1T
uAF357AHC	FAU	XSR	INT	15MHZ	8V/uS	+22V	-22V	70C	94dB	2MV	50pA	10pA	500MWF	5MA	12V	20V	40V	5uV/C	.	7MA	85dB	85dB	0.1T
uAF357HC	FAU	XSR	INT	15MHZ	6V/uS	+18V	-18V	70C	94dB	10MV	200pA	50pA	500MWF	5MA	12V	16V	30V	15uV/C	.	10MA	85dB	85dB	0.1T
U3F7702312	OBS	WBA	EXT	3MHZ		+13V	-8V	125C	68dB	2MV	5uA	0.5uA	570MWF	.3MA	5V	5V	5V	10uV/C	120MW	7MA	80dB	74dB	16K
U3F7702313	OBS	WBA	EXT	3MHZ	.	+13V	-8V	125C	68dB	2MV	5uA	0.5uA	570MWF	.3MA	5V	5V	5V	10uV/C	120MW	7MA	80dB	74dB	16K
U3F7709311	OBS	GPU	EXT	.3MHZ	.15V/uS	+18V	-18V	125C	88dB	5MV	500NA	200NA	570MWF	5MA	12V	10V	5V	15uV/C	165MW	.	70dB	76dB	150K
U3F7709312	OBS	GPU	EXT	.3MHZ	.15V/uS	+18V	-18V	125C	88dB	5MV	500NA	200NA	570MWF	5MA	12V	10V	5V	15uV/C	165MW	.	70dB	76dB	150K
U3F7709313	OBS	GPU	EXT	.3MHZ	.15V/uS	+18V	-18V	125C	88dB	5MV	500NA	200NA	570MWF	5MA	12V	10V	5V	15uV/C	165MW	.	70dB	76dB	150K
U3F770931X	OBS	GPU	EXT	.3MHZ	.15V/uS	+18V	-18V	125C	88dB	5MV	500NA	200NA	570MWF	5MA	12V	10V	5V	15uV/C	165MW	.	70dB	76dB	150K
U3F7710312	OBS	CPR	EXT	.		+14V	-6V	125C	62dB	2MV	20uA	31A	570MWF	5MA	2.5V	7V	5V	10uV/C	150MW	.	80dB	.	.
U3F7710313	OBS	CPR	EXT	.		+14V	-6V	125C	62dB	2MV	20uA	3uA	570MWF	5MA	2.5V	7V	5V	10uV/C	150MW	.	80dB	.	.
U3F771031X	OBS	CPR	EXT	.		+14V	-6V	125C	62dB	2MV	20uA	3uA	570MWF	5MA	2.5V	7V	5V	10uV/C	150MW	.	80dB	.	.
U3F7711312	OBS	DCP	EXT	.		+14V	-7V	125C	58dB	3.5MV	75uA	10uA	570MWF	5MA	2.5V	7V	5V	20uV/C	200MW	.	.	.	.

For detailed explanations of column heading notations, see App. A.

Also for ready references the more important abbreviations used in the column headings are listed below:

LEFT HAND PAGE

APP = application (codes at APP.E.)

CMRR = common mode rejection ratio

CMP = compensation (frequency)

dV_{in}/dT = input offset voltage temperature drift

GBP = gain bandwidth product

I_B = input bias current

I_{IO} = input bias offset current

I_Q = quiescent supply current

MFR = manufacturer (codes at App.C.)

P_Q = quiescent power consumer

PSRR = power supply rejection ratio

V_{ICM} = common mode input voltage rating

V_{IDF} = differential input voltage rating

V_{IO} = input offset voltage

V_S = dc supply voltage

RIGHT HAND PAGE

Lead out coding summary (details at APP.G.) for different cases (APP.F.)

A = gain adjust

B = bias adjust

C = case

E- = inverting input

E+ = non-inverting input

F,F* = input frequency compensation

G = ground

J = high level input

K = output, open collector

L = output, open emitter

M = metal case

N = not connected

Q = special terminal

R,R* = outputs

S = strobe

T,T* = offset balance

V+ = +ve dc supply

V- = -ve dc supply

W = guard ring

X = blank position, no lead

++ = +ve supplementary dc supply

-- = -ve supplementary dc supply

ø,ø* = output frequency compensation

CASE (APP.F)	LD 1	LD 2	LD 3	LD 4	LD 5	LD 6	LD 7	LD 8	LD 9	LD 10	LD 11	LD 12	LD 13	LD 14	LD 15	LD 16	EUROPE SUBSTITUTE	USA SUBSTITUTE	S	TYPE NUMBER
DIL-14/1C	N	N	TF	E-	E+	V-	N	N	T*	R	V+	F*	N	N				SN72777JA	0	uA777DC
TO5-8/1M	TF	E-	E+	V-	T*	R	R	V+	F*									SN72777L	0	uA777HC
DIL-8/1P	TF	E-	E+	V-	T*	R	R	V+	F*									SN72777P	0	uA777TC
TO3-10/2M	R	V+	Q	F	E+	E-	T	T*	ø	V-M							TDB0791KM		0	uA791KC
TO3-10/2M	R	V+	Q	F	E+	E-	T	T*	ø	V-M							TDC0791KM		0	uA791KM
HIL-12/1C	V-	V-	N	R	V+	Q	F	E+	E-	T	T*	ø					TDB0791EP		0	uA791P5
TO5-8/1M	R1	E-1	E+1	V-	E+2	E-2	R2	V+									TBB1458	MC1458G	0	uA798HC
TO5-8/1M	R1	E-1	E+1	V-	E+2	E-2	R2	V+									TBC1458	MC1558G	0	uA798HM
DIL-8/1C	R1	E-1	E+1	V-	E+2	E-2	R2	V+									TBB1458B	MC1458U	0	uA798RC
DIL-8/1C	R1	E-1	E+1	V-	E+2	E-2	R2	V+									LM1558J	MC1558U	0	uA798RM
DIL-8/1P	R1	E-1	E+1	V-	E+2	E-2	R2	V+									TBB1458B	MC1458U	0	uA798TC
TO5-8/1M	T	E-	E+	V-	T*	R	V+	N									UA741EHC	LM741EH	0	uA799HC
TO5-8/1M	T	E-	E+	V-	T*	R	V+	N									TBA222	LM741H	0	uA799HM
DIL-8/1C	T	E-	E+	V-	T*	R	V+	N									TBA221B	LM741EJ	0	uA799RC
DIL-8/1C	T	E-	E+	V-	T*	R	V+	N											0	uA799RM
DIL-8/1C	T	E-	E+	V-	T*	R	V+	N									TBA221B	LM741EJ	0	uA799TC
TO5-8/1M	R1	E-1	E+1	V-	E+2	E-2	R2	V+									SFC2458	MC1458G	0	uA1458CH
DIL-8/1C	R1	E-1	E+1	V-	E+2	E-2	R2	V+									SFC2458DC	MC1458U	0	uA1458CR
DIL-8/1P	R1	E-1	E+1	V-	E+2	E-2	R2	V+									SFC2458DC	MC1458U	0	uA1458CT
TO5-8/1M	R1	E-1	E+1	V-	E+2	E-2	R2	V+									SFC2458	MC1458G	0	uA1458HC
DIL-8/1C	R1	E-1	E+1	V-	E+2	E-2	R2	V+									SFC2458DC	MC1458U	0	uA1458RC
DIL-8/1P	R1	E-1	E+1	V-	E+2	E-2	R2	V+									SFC2458DC	MC1458U	0	uA1458TC
TO5-8/1M	R1	E-1	E+1	V-	E+2	E-2	R2	V+									SFC2458M	MC1558G	0	uA1558HC
DIL-14/1P	E+2	E+1	E-1	R1	R2	E-2	G	E-3	R3	R4	E-4	E+4	E+3	V+			LM3301N	MC3301P	0	uA3301P
DIL-14/1P	R2	R1	V+	E-1	E+1	E-2	E+2	E-3	E+3	E-4	E+4	G	R4	R3			MC3302L	LM3302N	0	uA3302P
DIL-14/1P	R1	E-1	E+1	V+	E+2	E-2	R2	R3	E-3	G	E+4	E-4	R4				RV3403ADC	MC3303P	0	uA3303P
DIL-14/1P	E+2	E+1	E-1	R1	R2	E-2	G	E-3	R3	R4	E-4	E+4	E+3	V+			LM3401N	MC3401P	0	uA3401P
DIL-14/1C	R1	E-1	E+1	V+	E+2	E-2	R2	R3	E-3	E+3	G	E+4	E-4	R4			RC3403ADC	MC3403L	0	uA3403D
DIL-14/1P	R1	E-1	E+1	V+	E+2	E-2	R2	R3	E-3	E+3	G	E+4	E-4	R4			RC3403ADC	MC3403L	0	u3403P
DIL-14/1C	R1	E-1	E+1	V+	E+2	E-2	R2	R3	E-3	E+3	G	E+4	E-4	R4			RM3503ADC	MC3503L	0	uA3503D
DIL-14/1C	E-1	E+1	R1	R2	E+2	E-2	V-	E-3	E+3	R3	V+	R4	E+4	E-4			RC4136DB		0	uA4136DC
DIL-14/1C	E-1	E+1	R1	R2	E+2	E-2	V-	E-3	E+3	R3	V+	R4	E+4	E-4			RM4136DC		0	uA4136DM
DIL-14/1P	E-1	E+1	R1	R2	E+2	E-2	V-	E-3	E+3	R3	V+	R4	E+4	E-4			RC4136DC		0	uA4136PC
DIL-14/1C	N	G	E+	E-	N	V-	T	T*S	R	N	V+	N	N	N			LF111D		0	uAF111D
TO5-8/1M	G	E+	E-	V-	T	T*S	R	V+									LF111H		0	uAF111H
TO5-8/1M	T	E-	E+	V-M	T*								R	V+	N		LF155AH		0	uAF155AHM
TO5-8/1M	T	E-	E+	V-M	T*	R		N									LF155H		0	uAF155HM
TO5-8/1M	T	E-	E+	V-M	T*	R	V+	N									LF156AH		0	uAF156AHM
TO5-8/1M	T	E-	E+	V-M	T*	R	V+	N									LF156H		0	uAF156HM
TO5-8/1M	T	E-	E+	V-M	T*	R	V+	N									LF157AH		0	uAF157AHM
TO5-8/1M	T	E-	E+	V-M	T*	R	V+	N									LF157H		0	uA157HM
DIL-14/1C	N	G	E+	E-	N	V-	T	T*S	R	N	V+	N	N	N			LF311D		0	uAF311D
TO5-8/1M	G	E+	E-	V-	T	T*S	R	V+									LF311H		0	uAF311H
TO5-8/1M	T	E-	E+	V-M	T*	R	V+	N									LF355AH		0	uAF355AHC
TO5-8/1M	T	E-	E+	V-M	T*	R	V+	N									LF355H		0	uAF355HC
TO5-8/1M	T	E-	E+	V-M	T*	R	V+	N									LF356AH		0	uAF356AHC
TO5-8/1M	T	E-	E+	V-M	T*	R	V+	N									LF356H		0	uAF356HC
TO5-8/1M	T	E-	E+	V-M	T*	R	V+	N									LF357AH		0	uAF357AHC
TO5-8/1M	T	E-	E+	V-M	T*	R	V+	N									LF357H		0	uAF357HC
FLP-10/3C	N	G	E-	E+	V-	F	ø	R	N	V+							MC1712F	UA702FM	0	U3F7702312
FLP-10/3C	N	G	E-	E+	V-	F	ø	R	N	V+							MC1712F	UA702FM	0	U3F7702313
FLP-10/3C	N	F	E-	E+	V-	ø	R	V+	F*	N							MC1709F	UA709FM	0	U3F7709311
FLP-10/3C	N	F	E-	E+	V-	ø	R	V+	F*	N							MC1709F	UA709FM	0	U3F7709312
FLP-10/3C	N	F	E-	E+	V-	ø	R	V+	F*	N							MC1709F	UA709FM	0	U3F7709313
FLP-10/3C	N	F	E-	E+	V-	ø	R	V+	F*	N							MC1709F	UA709FM	0	U3F770931X
FLP-10/3C	G	E+	E-	N	V-	R	N	V+	N	N							SFC2710PM	UA710FM	0	U3F7710312
FLP-10/3C	G	E+	E-	N	V-	R	N	V+	N	N							SFC2710PM	UA710FM	0	U3F7710313
FLP-10/3C	G	E+	E-	N	V-	R	N	V+	N	N							SFC2710PM	UA710FM	0	U3F771031X
FLP-10/3C	E-1	E+1	V-	E+2	E-2	S2	R	V+	G	S1							SFC2711PM	UA711FM	0	U3F7711312
FLP-10/3C	E-1	E+1	V-	E+2	E-2	S2	R	V+	G	S1							SFC2711PM	UA711FM	0	U3F7711313

TYPE NUMBER	M F R	A P P	C M P	GBP MIN	SLEW RATE MIN	V_S^+ MAX	V_S^- MAX	T_{op} MAX	A_{VOL} MIN	V_{IO} MAX	I_B MAX	I_{IO} MAX	P_{TOT} MAX	I_{OUT} MIN	V_{OUT} MIN	V_{ICM} MAX	V_{IDF} MAX	dV_{IO}/dT MAX	P_Q MAX	I_Q MAX	CM RR MIN	PS RR MIN	R_{IN} MIN
U3F7711313	OBS	DCP	EXT	.	.	+14V	-7V	125C	58dB	3.5MV	75uA	10uA	570MWF	5MA	2.5V	7V	5V	20uV/C	200MW	.	.	.	2K
U3F7733312	OBS	BDO	EXT	20MHZ	.	+8V	-8V	125C	50dB	5MV	20uA	3uA	570MWF	2MA	1.5V	6V	5V	.	.	24MA	60dB	50dB	2K
U3F7733313	OBS	BDO	EXT	20MHZ	.	+8V	-8V	125C	50dB	5MV	20uA	3uA	570MWF	2MA	1.5V	6V	5V	.	.	24MA	60dB	50dB	2K
U3F7741312	OBS	GPK	INT	.	0.2V/uS	+22V	-22V	125C	94dB	5MV	500NA	200NA	570MWF	5MA	12V	15V	30V	.	85MW	3MA	70dB	76dB	300K
U3F7741313	OBS	GPK	INT	.	0.2V/uS	+22V	-22V	125C	94dB	5MV	500NA	200NA	570MWF	5MA	12V	15V	30V	.	85MW	3MA	70dB	76dB	300K
U3F7748312	OBS	GPU	EXT	.	0.2V/uS	+22V	-22V	125C	94dB	5MV	500NA	200NA	570MWF	5MA	12V	15V	30V	.	85MW	3MA	70dB	76dB	300K
U3F7748313	OBS	GPU	EXT	.	0.2V/uS	+22V	-22V	125C	94dB	5MV	500NA	200NA	570MWF	5MA	12V	15V	30V	.	85MW	5MA	70dB	76dB	300K
U3H7702313	OBS	WBA	EXT	3MHZ	.	+13V	-8V	125C	68dB	2MV	5uA	0.5uA	570MWF	.3MA	5V	5V	5V	10uV/C	120MW	7MA	80dB	74dB	16K
U3H770231X	OBS	WBA	EXT	3MHZ	.	+13V	-8V	125C	68dB	2MV	5uA	0.5uA	570MWF	.3MA	5V	5V	5V	10uV/C	120MW	7MA	80dB	74dB	16K
U3H771231X	OBS	WBA	EXT	3MHZ	.	+13V	-8V	125C	68dB	2MV	5uA	0.5uA	570MWF	.3MA	5V	5V	5V	10uV/C	120MW	7MA	80dB	74dB	16K
U3T771131X	OBS	DCP	EXT	.	.	+14V	-7V	125C	58dB	3.5MV	75uA	10uA	570MWF	5MA	2.5V	7V	5V	20uV/C	200MW	.	.	.	2K
U3T771139X	OBS	DCP	EXT	.	.	+14V	-7V	70C	57dB	5MV	100uA	15uA	300MWF	5MA	2.5V	7V	5V	20uV/C	230MW	.	65dB	.	.
U5B101A312	OBS	GPU	EXT	.	.	+22V	-22V	125C	94dB	2MV	75NA	10NA	500MWF	5MA	12V	15V	30V	15uV/C	.	3MA	80dB	80dB	1.5M
U5B201A333	OBS	GPU	EXT	.	.	+22V	-22V	85C	94dB	2MV	75NA	10NA	500MWF	5MA	12	15V	30V	15uV/C	.	3MA	80dB	80dB	500K
U5B301A393	OBS	GPU	EXT	.	.	+18V	-18V	70C	88dB	7.5MV	250NA	50NA	500MWF	5MA	12V	15V	30V	30uV/C	.	3MA	70dB	70dB	500K
U5B7101312	OBS	GPU	EXT	.	.	+22V	-22V	125C	94dB	5MV	500NA	200NA	500MWF	5MA	12V	15V	30V	15uV/C	.	3MA	70dB	70dB	300K
U5B7201393	OBS	GPU	EXT	.	.	+22V	-22V	70C	86dB	7.5MV	1.5uA	0.5uA	500MWF	5MA	12V	15V	30V	30uV/C	.	3MA	65dB	70dB	100K
U5B7702312	OBS	WBA	EXT	3MHZ	.	+13V	-8V	125C	68dB	2MV	5uA	0.5uA	500MWF	.3MA	5V	5V	5V	10uV/C	120MW	7MA	80dB	74dB	16K
U5B770231X	OBS	WBA	EXT	3M	.	+13V	-8V	125C	68dB	2MV	5uA	0.5uA	500MWF	.3MA	5V	5V	5V	10uV/C	120MW	7MA	80dB	74dB	16K
U5B7702393	OBS	WBA	EXT	3MHZ	.	+13V	-8V	70C	65dB	5MV	7.5uA	2uA	500MWF	.3MA	5V	5V	5V	20uV/C	120MW	7MA	70dB	70dB	10K
U5B770239X	OBS	WBA	EXT	3MHZ	.	+13V	-8V	70C	66dB	5MV	7.5uA	2uA	500MWF	.3MA	5V	5V	5V	20uV/C	120MW	7MA	70dB	70dB	10K
U5B7709311	SGG	GPU	EXT	.3MHZ	.15V/uS	+18V	-18V	125C	88dB	2MV	200NA	50NA	500MWF	5MA	12V	10V	5V	10uV/C	108MW	.	80dB	80dB	350K
U5B7709312	OBS	GPU	EXT	.3MHZ	.15V/uS	+18V	-18V	125C	88dB	5MV	500NA	200NA	500MWF	5MA	12V	10V	5V	15uV/C	165MW	.	70dB	76dB	150K
U5B770931X	SGG	GPU	EXT	.3MHZ	.15V/uS	+18V	-18V	125C	88dB	5MV	500NA	200NA	500MWF	5MA	12V	10V	5V	15uV/C	165MW	.	70dB	76dB	150K
U5B7709393	OBS	GPU	EXT	.3MHZ	.15V/uS	+18V	-18V	70C	84dB	7.5MV	1.5uA	500NA	500MWF	5MA	12V	10V	5V	.	200MW	.	65dB	74dB	50K
U5B770939X	SGG	GPU	EXT	.3MHZ	.15V/uS	+18V	-18V	70C	84dB	7.5MV	1.5uA	500NA	500MWF	5MA	12V	10V	5V	.	200MW	.	65dB	74dB	50K
U5B7710312	OBS	CPR	EXT	.	.	+14V	-7V	125C	62dB	2MV	20uA	3uA	500MWF	5MA	2.5V	7V	5V	10uV/C	150MW	.	80dB	.	.
U5B771031X	SGG	CPR	EXT	.	.	+14V	-7V	125C	62dB	2MV	20uA	3uA	500MWF	5MA	2.5V	7V	5V	10uV/C	150MW	.	80dB	.	.
U5B7710393	OBS	CPR	EXT	.	.	+14V	-6V	70C	60dB	5MV	25uA	5uA	500MWF	5MA	2.5V	7V	5V	20uV/C	150MW	.	70dB	.	.
U5B771039X	SGG	CPR	EXT	.	.	+14V	-6V	70C	60dB	5MV	25uA	5uA	500MWF	5MA	2.5V	7V	5V	20uV/C	150MW	.	70dB	.	.
U5B771231X	SGG	WBA	EXT	3MHZ	.	+13V	-8V	125C	68dB	2MV	5uA	0.5uA	500MWF	.3MA	5V	5V	5V	10uV/C	120MW	7MA	80dB	74dB	16K
U5B771239X	OBS	WBA	EXT	3MHZ	.	+13V	-8V	70C	66dB	5MV	7.5uA	2uA	500MWF	.3MA	5V	5V	5V	20uV/C	120MW	7MA	70dB	70dB	10K
U5B7730312	OBS	BDO	EXT	1MHZ	.	+15V	.	125C	40dB	2.5MV	7.5uA	1.5uA	500MWF	.	2V	4V	5V	.	156MW	13MA	70dB	.	5K
U5B773031X	OBS	BDO	EXT	1MHZ	.	+15V	.	125C	40dB	2.5MV	7.5uA	1.5uA	500MWF	.	2V	4V	5V	.	156MW	13MA	70dB	.	5K
U5B7730393	OBS	BDO	EXT	1MHZ	.	+15V	.	70C	40dB	5MV	16uA	3uA	500MWF	.	2V	4V	5V	.	156MW	13MA	60dB	.	2.5K
U5B773039X	OBS	BDO	EXT	1MHZ	.	+15V	.	70C	40dB	5MV	16uA	3uA	500MWF	.	2V	4V	5V	.	156MW	13MA	60dB	.	2.5K
U5B7740312	OBS	FET	INT	1MHZ	2V/uS	+22V	-22V	125C	94dB	110MV	200pA	150pA	500MWF	5MA	12V	15V	30V	.	156MW	5MA	64dB	70dB	300K
U5B7740393	OBS	FET	INT	1MHZ	2V/uS	+22V	-22V	70C	86dB	20MV	2NA	300pA	500MWF	5MA	12V	15V	30V	.	240MW	8MA	55dB	66dB	300K
U5B7741312	ING	GPK	INT	.	0.2V/uS	+22V	-22V	125C	94dB	5MV	500NA	200NA	500MWF	7MA	12V	15V	30V	.	85MW	3MA	70dB	76dB	300K
U5B7741392	OBS	GPK	INT	.	0.2V/uS	+18V	-18V	70C	86dB	6MV	500NA	200NA	500MWF	5MA	12V	15V	30V	.	.	3MA	70dB	76dB	300K
U5B7741393	ING	GPK	INT	.	0.2V/uS	+18V	-18V	70C	86dB	6MV	500NA	200NA	500MWF	5MA	12V	15V	30V	.	.	3MA	70dB	76dB	300K
U5B7748312	OBS	GPU	EXT	.	0.2V/uS	+22V	-22V	125C	94dB	5MV	500NA	200NA	500MWF	5MA	12V	15V	30V	.	85MW	3MA	70dB	76dB	300K
U5B7748393	OBS	GPU	EXT	.	0.2V/uS	+18V	-18V	70C	86dB	6MV	500NA	200NA	500MWF	5MA	12V	15V	30V	.	85MW	3MA	70dB	76dB	300K
U5B7777312	OBS	PIA	EXT	.	0.2V/uS	+22V	-22V	125C	94dB	2MV	25NA	3NA	500MWF	5MA	12V	15V	30V	15uV/C	.	4MA	80dB	80dB	2M
U5B7777393	OBS	PIA	EXT	.	0.2V/uS	+22V	-22V	70C	88dB	5MV	100NA	20NA	500MWF	5MA	12V	15V	30V	30uV/C	85MW	3MA	70dB	76dB	1M
U5F7711312	OBS	DCP	EXT	.	.	+14V	-7V	125C	58dB	3.5MV	75uA	10uA	500MWF	5MA	2.5V	7V	5V	20uV/C	200MW	.	.	.	.
U5F771131X	SGG	DCP	EXT	.	.	+14V	-7V	125C	58dB	3.5MV	75uA	10uA	500MWF	5MA	2.5V	7V	5V	20uV/C	200MW	.	.	.	.
U5F7711393	OBS	DCP	EXT	.	.	+14V	-7V	70C	57dB	5MV	100uA	15uA	500MWF	5MA	2.5V	7V	5V	20uV/C	230MW	.	.	.	.
U5F771139X	SGG	DCP	EXT	.	.	+14V	-7V	70C	57dB	5MV	100uA	15uA	500MWF	5MA	2.5V	7V	5V	20uV/C	230MW	.	.	.	.
U5F7733312	OBS	BDO	EXT	20MHZ	.	+8V	-8V	125C	50dB	5MV	20uA	3uA	500MWF	2MA	1.5V	6V	5V	.	.	24MA	60dB	50dB	2K
U5F7733393	OBS	BDO	EXT	20MHZ	.	+8V	-8V	70C	48dB	6MV	30uA	5uA	500MWF	2MA	1.5V	6V	5V	.	.	24MA	60dB	50dB	2K
U5F7747393	FAU	DGK	INT	.	0.2V/uS	+18V	-18V	70C	88dB	6MV	500NA	200NA	500MWF	5MA	12V	15V	30V	.	85MW	3MA	70dB	76dB	300K
U5T7725311	OBS	PIA	EXT	.	.	+22V	-22V	125C	120dB	0.5MV	75NA	5NA	500MWF	5MA	12V	22V	5V	2uV/C	120MW	.	110dB	102dB	500K
U5T7725312	OBS	PIA	EXT	.	.	+22V	-22V	125C	120dB	1MV	100NA	20NA	500MWF	5MA	12V	22V	5V	5uV/C	150MW	.	110dB	100dB	500K
U5T7725333	OBS	PIA	EXT	.	.	+22V	-22V	85C	120dB	0.5MV	75NA	5NA	500MWF	5MA	12V	22V	5V	2uV/C	150MW	.	120dB	106dB	500K
U5T7725393	OBS	PIA	EXT	.	.	+22V	-22V	70C	108dB	2.5MV	125NA	35NA	500MWF	5MA	12V	22V	5V	5uV/C	150MW	.	94dB	90dB	500K
U6A7101312	OBS	GPU	EXT	.	.	+22V	-22V	125C	94dB	5MV	1.5uA	0.5uA	500MWF	5MA	12V	15V	30V	15uV/C	.	3MA	70dB	70dB	300K
U6A7702312	FAU	WBA	EXT	3MHZ	.	+13V	-13V	125C	68dB	2MV	5uA	0.5uA	670MWF	.3MA	5V	5V	5V	10uV/C	120MW	7MA	80dB	74dB	16K
U6A7702393	OBS	WBA	EXT	3MHZ	.	+13V	-8V	70C	66dB	5MV	7.5uA	2uA	670MWF	.3MA	5V	5V	5V	20uV/C	120MW	7MA	70dB	70dB	10K
U6A7709311	FAU	GPU	EXT	.3MHZ	.15V/uS	+18V	-18V	125C	88dB	2MV	200NA	50NA	670MWF	5MA	12V	10V	5V	10uV/C	108MW	.	80dB	80dB	350K

For detailed explanations of column heading notations, see App. A.
Also for ready references the more important abbreviations used in the column headings are listed below:

LEFT HAND PAGE

APP = application (codes at APP.E.)
CMRR = common mode rejection ratio
CMP = compensation (frequency)
dV_{io}/dT = input offset voltage temperature drift
GBP = gain bandwidth product
I_B = input bias current
I_{io} = input bias offset current
I_o = quiescent supply current
MFR = manufacturer (codes at App.C.)
P_o = quiescent power consumer
PSRR = power supply rejection ratio
V_{ICM} = common mode input voltage rating
V_{IDF} = differential input voltage rating
V_{io} = input offset voltage
V_S = dc supply voltage

RIGHT HAND PAGE

Lead out coding summary (details at APP.G.) for different cases (APP.F.)

A = gain adjust
B = bias adjust
C = case
E— = inverting input
E+ = non-inverting input
F,F* = input frequency compensation
G = ground
J = high level input
K = output, open collector
L = output, open emitter
M = metal case
N = not connected
Q = special terminal
R,R* = outputs
S = strobe
T,T* = offset balance
V+ = +ve dc supply
V— = —ve dc supply
W = guard ring
X = blank position, no lead
++ = +ve supplementary dc supply
—— = —ve supplementary dc supply
φ,φ* = output frequency compensation

CASE (APP F)	LD 1	LD 2	LD 3	LD 4	LD 5	LD 6	LD 7	LD 8	LD 9	LD 10	LD 11	LD 12	LD 13	LD 14	LD 15	LD 16	EUROPE SUBSTITUTE	USA SUBSTITUTE	IS	TYPE NUMBER
FLP-10/3C	E+	A2	A*2	V-	R	R*	V+	A1	A*1	E-							SN52733FA	UA733FM	0	U3F7733312
FLP-10/3C	E+	A2	A*2	V-	R	R*	V+	A1	A*1	E-							SN52733FA	UA733FM	0	U3F7733313
FLP-10/3C	N	T	E-	E+	V-	T*	R	V+	N	N							SFC2741PM	UA741FM	0	U3F7741312
FLP-10/3C	N	T	E-	E+	V-	T*	R	V+	N	N							SFC2741PM	UA741FM	0	U3F7741313
FLP-10/3C	N	FT	E-	E+	V-	T*	R	V+	F*	N							SN52748FA	UA748FM	0	U3F7748312
FLP-10/3C	N	FT	E-	E+	V-	T*	R	V+	F*	N							SN52748FA	UA748FM	0	U3F7748313
FLP-10/3G	N	G	E-	E+	V-	F	ø	R	N	V+							MC1712F	UA702FM	0	U3H7702313
FLP-10/3G	N	G	E-	E+	V-	F	ø	R	N	V+							MC1712F	UA702FM	0	U3H770231X
FLP-10/3G	N	G	E-	E+	V-	F	ø	R	N	V+							MC1712F	UA702FM	0	U3H771231X
FLP-10/1C	E-1	E+1	V-	E+2	E-2	S2	R	V+	G	S1							SFC2711PM	UA711PM	0	U3T771131X
FLP-10/1C	E-1	E+1	V-	E+2	E-2	S2	R	V+	G	S1								SN72711FA	0	U3T771139X
T05-8/1M	FT	E-	E+	V-M	T*	R	V+	F*									SFC2101A	LM101AH	0	U5B101A312
T05-8/1M	FT	E-	E+	V-M	T*	R	V+	F*									UA201AH	LM201AH	0	U5B201A393
T05-8/1M	FT	E-	E+	V-M	T*	R	V+	F*									UA301AH	LM301AH	0	U5B301A393
T05-8/1M	FT	E-	E+	V-M	T*	R	V+	F*									SFC2101A	LM101H	0	U5B7101312
T05-8/1M	FT	E-	E+	V-M	T*	R	V+	F*									UA201H	LM201H	0	U5B7201393
T05-8/1M	G	E-	E+	V-M	F	ø	R	V+									SN52702AL	UA702HM	0	U5B7702312
T05-8/1M	G	E-	E+	V-M	F	ø	R	V+									SN52702AL	UA702HM	0	U5B770231X
T05-8/1M	G	E-	E+	V-M	F	ø	R	V+									SN72702L	UA702HC	0	U5B7702393
T05-8/1M	G	E-	E+	V-M	F	ø	R	V+									SN72702L	UA702HC	0	U5B770239X
T05-8/1M	F	E-	E+	V-M	ø	ø*R	V+	F*									TAA522	UA709AHM	0	U5B7709311
T05-8/1M	F	E-	E+	V-M	ø	ø*R	V+	F*									TAA522	UA709HM	0	U5B7709312
T05-8/1M	F	E-	E+	V-M	ø	ø*R	V+	F*									TAA522	UA709HM	0	U5B770931X
T05-8/1M	F	E-	E+	V-M	ø	ø*R	V+	F*									TAA521	UA709HC	0	U5B7709393
T05-8/1M	F	E-	E+	V-M	ø	ø*R	V+	F*									TAA521	UA709HC	0	U5B770939X
T05-8/1M	G	E+	E-	V-M	N	N	R	V+									SFC2710M	UA710HM	0	U5B7710312
T05-8/1M	G	E+	E-	V-M	N	N	R	V+									SFC2710M	UA710HM	0	U5B771031X
T05-8/1M	G	E+	E-	V-M	N	N	R	V+									SFC2710C	UA710HC	0	U5B7710393
T05-8/1M	G	E+	E-	V-M	N	N	R	V+									SFC2710C	UA710HC	0	U5B771039X
T05-8/1M	G	E-	E+	V-M	F	ø	R	V+									SN52702AL	UA702HM	0	U5B771231X
T05-8/1M	G	E-	E+	V-M	F	ø	R										SN72702L	UA702HC	0	U5B771239X
T05-8/1M	R*1	E-	E+	G-M	R1	R2	V+	R*2										UA730HM	0	U5B7730312
T05-8/1M	R*1	E-	E+	G-M	R1	R2	V+	R*2										UA730HM	0	U5B773031X
T05-8/1M	R*1	E-	E+	G-M	R1	R2	V+	R*2										UA730HC	0	U5B7730393
T05-8/1M	R*1	E-	E+	G-M	R1	R2	V+	R*2										UA730HC	0	U5B773039X
T05-8/1M	T	E-	E+	V-M	T*	R	V+	N									UA740T	UA740HM	0	U5B7740312
T05-8/1M	T	E-	E+	V-M	T*	R	V+	N									UA740CT	UA740HC	0	U5B7740393
T05-8/1M	T	E-	E+	V-M	T*	R	V+	N									TBA222	UA741HM	0	U5B7741312
T05-8/1M	T	E-	E+	V-M	T*	R	V+	N									TBA221	UA741HC	0	U5B7741392
T05-8/1M	T	E-	E+	V-M	T*	R	V+	N									TBA221	UA741HC	0	U5B7741393
T05-8/1M	FT	E-	E+	V-M	T*	R	V+	F*									TBC0748	UA748HM	0	U5B7748312
T05-8/1M	FT	E-	E+	V-M	T*	R	V+	F*									TBB0748	UA748HC	0	U5B7748393
T05-8/1M	TF	E-	E+	V-M	T*	R	V+	F*										SN52777L	0	U5B7777312
T05-8/1M	TF	E-	E+	V-M	T*	R	V+	F*									SN72777L	UA777HC	0	U5B7777393
T05-10/1M	G	S1	E-1	E+1	V-M	E+2	E-2	S2	R	V+							SFC2711M	UA711HM	0	U5F7711312
T05-10/1M	G	S1	E-1	E+1	V-	E+2	E-2	S2	R	V+							SFC2711M	UA711HM	0	U5F771131X
T05-10/1M	G	S1	E-1	E+1	V-M	E+2	E-2	S2	R	V+							SFC2711C	UA711HC	0	U5F7711393
T05-10/1M	G	S1	E-1	E+1	V-M	E+2	E-2	S2	R	V+							SFC2711C	UA711HC	0	U5F771139X
T05-10/1M	E-	E+	A2	A*2	V-M	R	R*	V+	A1	A*1							SN52733L	UA733HM	0	U5F7733312
T05-10/1M	E-	E+	A2	A*2	V-M	R	R*	V+	A1	A*1							SN72733L	UA733HC	0	U5F7733393
T05-10/1M	R1	V+1	E-1	E+1	V-M	E+2	E-2	V+2	R2	N							TBB0747	UA747HC	0	U5F7747393
T05-8/1M	T	E-	E+	V-	ø	ø*	V+	T*									LM725AH	UA725AHM	0	U5T7725311
T05-8/1M	T	E-	E+	V-	ø	ø*	V+	T*									LM725H	UA725HM	0	U5T7725312
T05-8/1M	T	E-	E+	V-	ø	ø*	V+	T*										UA725EHC	0	U5T7725333
T05-8/1M	T	E-	E+	V-	ø	ø*	V+	T*									LM725CH	UA725HC	0	U5T7725393
DIL-14/1C	N	N	FT	E-	E+	V-	N	N	T*	R	V+	F*	N	N				LM101J14	0	U6A7101312
DIL-14/1C	N	N	G	E-	E+	V-	N	N	F	ø	R	N	V+	N			SN52702J	UA702DM	0	U6A7702312
DIL-14/1C	N	N	G	E-	E+	V-	N	N	F	ø	R	N	V+	N			SN72702J	UA702DC	0	U6A7702393
DIL-14/1C	N	N	F	E-	E+	V-	N	N	N	ø	R	V+	F*	N			LM709AJ	UA709ADM	0	U6A7709311
DIL-14/1C	N	N	F	E-	E+	V-	N	N	N	ø	R	V+	F*	N			LM709J	UA709DM	0	U6A7709312

TYPE NUMBER	MFR	APP	CMP	GBP MIN	SLEW RATE MIN	Vs+ MAX	Vs- MAX	Top MAX	Avol MIN	Vio MAX	Ib MAX	Iio MAX	Ptot MAX	Iout MIN	Vout MIN	Vicm MAX	Vidf MAX	dVio/dT MAX	Pq MAX	Iq MAX	CMRR MIN	PSRR MIN	Rin MIN
U6A7709312	FAU	GPU	EXT	.3MHZ	.15V/uS	+18V	-18V	125C	88dB	5MV	500NA	200NA	670MWF	5MA	12V	10V	5V	15uV/C	165MW	.	70dB	76dB	150K
U6A7709393	FAU	GPU	EXT	.3MHZ	.15V/uS	+18V	-18V	70C	84dB	7.5MV	1.5uA	500NA	670MWF	5MA	12V	10V	5V	.	200MW	.	65dB	74dB	50K
U6A7710312	FAU	CPR	EXT	.	.	+14V	-6V	125C	62dB	2MV	20uA	3uA	670MWF	5MA	2.5V	7V	5V	10uV/C	150MW	.	80dB	.	.
U6A7710393	FAU	CPR	EXT	.	.	+14V	-6V	70C	60dB	5MV	25uA	5uA	670MWF	5MA	2.5V	7V	5V	20uV/C	150MW	.	70dB	.	.
U6A7711312	OBS	DCP	EXT	.	.	+14V	-7V	125C	58dB	3.5MV	75uA	10uA	670MWF	5MA	2.5V	7V	5V	20uV/C	200MW	.	.	.	.
U6A7711393	OBS	DCP	EXT	.	.	+14V	-7V	70C	57dB	5MV	100uA	15uA	670MWF	5MA	2.5V	7V	5V	20uV/C	230MW	.	.	.	.
U6A7733312	OBS	BDO	EXT	20MHZ	.	+8V	-8V	125C	50dB	5MV	20uA	3uA	670MWF	2MA	1.5V	6V	5V	.	.	24MA	60dB	50dB	2K
U6A7733393	ADU	BDO	EXT	20MHZ	.	+8V	-8V	70C	48dB	6MV	30uA	5uA	670MWF	2MA	1.5V	6V	5V	.	.	24MA	60dB	50dB	2K
U6A7739393	OBS	DLN	EXT	.	.	+18V	-18V	70C	76dB	6MV	2uA	1uA	670MWF	.2MA	12V	15V	5V	.	420MW	14MA	70dB	74dB	37K
U6A7741312	ADU	GPK	INT	.4MHZ	0.3V/uS	+22V	-22V	125C	94dB	3MV	80NA	30NA	670MWF	5MA	12V	15V	30V	15uV/C	150MW	.	80dB	86dB	1M
U6A7741393	OBS	GPK	INT	.	0.2V/uS	+18V	-18V	70C	86dB	6MV	500NA	200NA	670MWF	5MA	12V	15V	30V	.	85MW	3MA	70dB	76dB	300K
U6A7748312	OBS	GPU	EXT	.	0.2V/uS	+22V	-22V	125C	94dB	5MV	500NA	200NA	670MWF	5MA	12V	15V	30V	.	85MW	3MA	70dB	76dB	300K
U6A7748393	OBS	GPU	EXT	.	0.2V/uS	+22V	-22V	70C	86dB	6MV	500NA	200NA	670MWF	5MA	12V	15V	30V	.	85MW	3MA	70dB	76dB	300K
U6A7749312	OBS	DLN	EXT	.	1V/uS	+18V	-18V	125C	86dB	3MV	750NA	400NA	650MWF	2MA	12V	15V	5V	10uV/C	220MW	11MA	70dB	74dB	100K
U6A7749393	OBS	DLN	EXT	.	1V/uS	+18V	-18V	70C	84dB	6MV	1.5uA	750NA	650MWF	2MA	12V	15V	5V	10uV/C	330MW	14MA	70dB	70dB	50K
U6E7201393	ADU	GPU	EXT	.	.	+22V	-22V	70C	86dB	7.5MV	1.5uA	0.5uA	500MWF	5MA	12V	15V	30V	30uV/C	.	3MA	65dB	70dB	100K
U6E7709393	SGG	GPU	EXT	.3MHZ	.15V/uS	+18V	-18V	70C	84dB	7.5MV	1.5uA	500NA	670MWF	5MA	12V	10V	5V	.	200MW	.	65dB	74dB	50K
U6E7710393	SGG	CPR	EXT	.	.	+14V	-6V	70C	60dB	5MV	25uA	5uA	670MWF	5MA	2.5V	7V	5V	20uV/C	150MW	.	70dB	.	.
U6E7711393	SGG	DCP	EXT	.	.	+14V	-7V	70C	57dB	5MV	100uA	15uA	670MWF	5MA	2.5V	7V	5V	20uV/C	230MW	.	.	.	.
U6E7712393	OBS	WBA	EXT	3MHZ	.	+13V	-18V	70C	66dB	5MV	7.5uA	2uA	670MWF	.3MA	5V	5V	5V	20uV/C	120MW	7MA	70dB	70dB	10K
U6E7739393	OBS	DLN	EXT	.	.	+18V	-18V	70C	76dB	6MV	2uA	1uA	670MWF	.2MA	12V	15V	5V	.	420MW	14MA	70dB	74dB	37K
U6E7741393	OBS	GPK	INT	.	0.2V/uS	+18V	-18V	70C	86dB	6MV	500NA	200NA	670MWF	5MA	12V	15V	30V	.	85MW	3MA	70dB	76dB	300K
U6E7748393	OBS	GPU	EXT	.	0.2V/uS	+22V	-22V	70C	86dB	6MV	500NA	200NA	670MWF	5MA	12V	15V	30V	.	85MW	3MA	70dB	76dB	300K
U6T7201393	ADU	GPU	EXT	.	.	+22V	-22V	70C	86dB	7.5MV	1.5uA	0.5uA	500MWF	5MA	12V	15V	30V	30uV/C	.	3MA	65dB	70dB	100K
U6T7741393	OBS	GPK	INT	.	0.2V/uS	+18V	-18V	70C	86dB	6MV	500NA	200NA	310MWF	5MA	12V	15V	30V	.	.	3MA	70dB	76dB	300K
U6T7748393	OBS	GPU	EXT	.	0.2V/uS	+18V	-18V	70C	86dB	6MV	500NA	200NA	310MWF	5MA	12V	15V	30V	.	85MW	3MA	70dB	76dB	300K
U6W7747312	OBS	DGK	INT	.	0.2V/uS	+22V	-22V	125C	94dB	5MV	500NA	200NA	670MWF	5MA	12V	15V	30V	.	85MW	3MA	70dB	76dB	300K
U6W7747392	OBS	DGK	INT	.	0.2V/uS	+18V	-18V	70C	88dB	6MV	500NA	200NA	670MWF	5MA	12V	15V	30V	.	85MW	3MA	70dB	76dB	300K
U6W7747393	OBS	DGK	INT	.	0.2V/uS	+18V	-18V	70C	88	6MV	500NA	200NA	670MWF	5MA	12V	15V	30V	.	85MW	3MA	70dB	76dB	300K
U7A7747312	OBS	DGK	INT	.	0.2V/uS	+22V	-22V	125C	94o.	5MV	500NA	200NA	670MWF	5MA	12V	15V	30V	.	85MW	3MA	70dB	76dB	300K
U7A7747393	OBS	DGK	INT	.	0.2V/uS	+18V	-18V	70C	88dB	6MV	500NA	200NA	670MWF	5MA	12V	15V	30V	.	85MW	3MA	70dB	76dB	300K
U9T7741393	ING	GPK	INT	.	0.2V/uS	+18V	-18V	70C	86dB	6MV	500NA	200NA	670MWF	5MA	12V	15V	30V	.	85MW	3MA	70dB	76dB	300K
UC4250CTY	ING	PRA	INT	.	.	+18V	-18V	70C	95dB	6MV	75NA	20NA	500MWF	1MA	12V	15V	30V	.	3MW	.1MA	70dB	74dB	.
UC4250TY	ING	PRA	INT	.	.	+18V	-18V	125C	100dB	5MV	50NA	10NA	500MWF	1MA	12V	15V	30V	.	2.7MW	90UA	70dB	76dB	.
ULN2139D	SPU	GPU	EXT	.3MHZ	0.8V/uS	+18V	-18V	100C	86dB	7.5MV	1uA	100NA	.	5MA	10V	18V	18V	.	200MW	.	80dB	75dB	100K
ULN2139G	SPU	GPU	EXT	.3MHZ	0.8V/uS	+18V	-18V	100C	86dB	7.5MV	1uA	100NA	.	5MA	10V	18V	18V	.	200MW	.	80dB	75dB	100K
ULN2139H	SPU	GPU	EXT	.3MHZ	0.8V/uS	+18V	-18V	100C	86dB	7.5MV	1uA	100NA	.	5MA	10V	18V	18V	.	200MW	.	80dB	75dB	100K
ULN2139M	SPU	GPU	EXT	.3MHZ	0.8V/uS	+18V	-18V	100C	86dB	7.5MV	1uA	100NA	.	5MA	10V	18V	18V	.	200MW	.	80dB	75dB	100K
ULN2151D	SPU	GPK	INT	.	0.4V/uS	+20V	-20V	100C	88dB	5MV	250NA	25NA	.	5MA	10V	15V	30V	.	85MW	.	75dB	75dB	0.5M
ULN2151G	SPU	GPK	INT	.	0.4V/uS	+20V	-20V	100C	88dB	5MV	250NA	25NA	.	5MA	10V	15V	30V	.	85MW	.	75dB	75dB	0.5M
ULN2151H	SPU	GPK	C	.	0.4V/uS	+20V	-20V	100C	88dB	5MV	250NA	25NA	.	5MA	10V	15V	30V	.	85MW	.	75dB	75dB	0.5M
ULN2151M	SPU	GPK	INT	.	0.4V/uS	+20V	-20V	100C	88dB	5MV	250NA	25NA	.	5MA	10V	15V	30V	.	85MW	.	75dB	75dB	0.5M
ULN2156D	OBS	HSR	INT	.	1V/uS	+18V	-18V	100C	97dB	10MV	30NA	10NA	.	.	.	15V	30V	.	90MW	.	.	.	1M
ULN2156M	OBS	HSR	INT	.	1V/uS	+18V	-18V	100C	97dB	10MV	30NA	10NA	.	.	.	15V	30V	.	90MW	.	.	.	1M
ULN2157A	OBS	DGK	INT	.	0.4V/uS	+20V	-20V	100C	88dB	5MV	250NA	25NA	.	5MA	10V	15V	30V	.	85MW	.	75dB	75dB	0.5M
ULN2157H	OBS	DGK	INT	.	0.4V/uS	+20V	-20V	100C	88dB	5MV	250NA	25NA	.	5MA	10V	15V	30V	.	85MW	.	75dB	75dB	0.5M
ULN2157K	OBS	DGK	INT	.	0.4V/uS	+20V	-20V	100C	88dB	5MV	250NA	25NA	.	5MA	10V	15V	30V	.	85MW	.	75dB	75dB	0.5M
ULN2158D	OBS	GPU	EXT	.	0.4V/uS	+20V	-20V	100C	88dB	5MV	250NA	25NA	.	5MA	10V	15V	30V	.	85MW	.	75dB	75dB	0.5M
ULN2158M	OBS	GPU	EXT	.	0.4V/uS	+20V	-20V	100C	88dB	5MV	250NA	25NA	.	5MA	10V	15V	30V	.	85MW	.	75dB	75dB	0.5M
ULN2171D	SPU	GPK	INT	.	0.8V/uS	+20V	-20V	100C	88dB	5MV	50NA	20NA	.	5MA	10V	15V	30V	.	95MW	.	80dB	80dB	2M
ULN2171G	SPU	GPK	INT	.	0.8V/uS	+20V	-20V	100C	88dB	5MV	50NA	20NA	.	5MA	10V	15V	30V	.	95MW	.	80dB	80dB	2M
ULN2171H	SPU	GPK	INT	.	0.8V/uS	+20V	-20V	100C	88dB	5MV	50NA	20NA	.	5MA	10V	15V	30V	.	95MW	.	80dB	80dB	2M
ULN2171M	SPU	GPK	INT	.	0.8V/uS	+20V	-20V	100C	88dB	5MV	50NA	20NA	.	5MA	10V	15V	30V	.	95MW	.	80dB	80dB	2M
ULN2172D	OBS	GPU	EX	.	0.8V/uS	+20V	-20V	100C	88dB	5MV	50NA	20NA	.	5MA	10V	15V	30V	.	95MW	.	80dB	80dB	2M
ULN2172M	OBS	GPU	EXT	.	0.8V/uS	+20V	-20V	100C	88dB	5MV	50NA	20NA	.	5MA	10V	15V	30V	.	95MW	.	80dB	80dB	2M
ULN2173D	OBS	LBC	INT	.	0.1V/uS	+20V	-20V	100C	94dB	5MV	10NA	5NA	.	.	.	15V	30V	.	45MW	.	.	.	3M
ULN2173M	OBS	LBC	INT	.	0.1V/uS	+20V	-20V	100C	94dB	5MV	10NA	5NA	.	.	.	15V	30V	.	45MW	.	.	.	3M
ULN2174D	OBS	LBC	EXT	.	0.1V/uS	+20V	-20V	100C	94dB	5MV	10NA	5NA	.	.	.	15V	30V	.	45MW	.	.	.	3M
ULN2174M	OBS	LBC	EXT	.	0.1V/uS	+20V	-20V	100C	94dB	5MV	10NA	5NA	.	.	.	15V	30V	.	45MW	.	.	.	3M
ULN2741D	OBS	GPK	INT	.4MHZ	0.3V/uS	+18V	-18V	70C	94dB	5MV	80NA	30NA	.	6MA	12V	15V	30V	15uV/C	85MW	3MA	70dB	76dB	300K

For detailed explanations of column heading notations, see App. A.

Also for ready references the more important abbreviations used in the column headings are listed below:

LEFT HAND PAGE

APP = application (codes at APP.E.)
CMRR = common mode rejection ratio
CMP = compensation (frequency)
dV_{IO}/dT = input offset voltage temperature drift
GBP = gain bandwidth product
I_B = input bias current
I_{IO} = input bias offset current
I_Q = quiescent supply current
MFR = manufacturer (codes at App.C.)
P_Q = quiescent power consumer
PSRR = power supply rejection ratio
V_{ICM} = common mode input voltage rating
V_{IDF} = differential input voltage rating
V_{IO} = input offset voltage
V_S = dc supply voltage

RIGHT HAND PAGE

Lead out coding summary (details at APP.G.) for different cases (APP.F.)

A = gain adjust
B = bias adjust
C = case
E— = inverting input
E+ = non-inverting input
F,F* = input frequency compensation
G = ground
J = high level input
K = output, open collector
L = output, open emitter
M = metal case
N = not connected
Q = special terminal
R,R* = outputs
S = strobe
T,T* = offset balance
V+ = +ve dc supply
V— = —ve dc supply
W = guard ring
X = blank position, no lead
+ + = +ve supplementary dc supply
— — = —ve supplementary dc supply
ø,ø* = output frequency compensation

CASE (APP F)	LD 1	LD 2	LD 3	LD 4	LD 5	LD 6	LD 7	LD 8	LD 9	LD 10	LD 11	LD 12	LD 13	LD 14	LD 15	LD 16	EUROPE SUBSTITUTE	USA SUBSTITUTE	I S	TYPE NUMBER
DIL-14/1C	N	N	F	E-	E+	V-	N	N	ø	R	V+	F*	N	N	.	.	LM709CJ	UA709DC	O	U6A7709393
DIL-14/1C	N	G	E+	E-	N	V-	N	N	R	R	V+	N	N	N	.	.	SFC2710KM	UA710DM	O	U6A7710312
DIL-14/1C	N	G	E+	E-	N	V-	N	N	R	R	V+	N	N	N	.	.	SFC2710EC	UA710DC	O	U6A7710393
DIL-14/1C	N	E-1	E+1	V-	E+2	E-2	N	N	S2	R	V+	G	S1	N	.	.	SFC2711KM	UA711DM	O	U6A7711312
DIL-14/1C	N	E-1	E+1	V-	E+2	E-2	N	N	S2	R	V+	G	S1	N	.	.	SFC2711EC	UA711DC	O	U6A7711393
DIL-14/1C	E+	N	A2	A*2	V-	R	R	R*	N	V+	A1	A*1	N	E-	.	.	SN52733J	UA733DM	O	U6A7733312
DIL-14/1C	E+	N	A2	A*2	V-	N	R	R*	N	V+	A1	A*1	N	E-	.	.	SN72733J	UA733DC	O	U6A7733393
DIL-14/1C	R1	ø1	F1	F*1	E+1	E-1	V-	E-2	E+2	F2	F*2	ø2	R2	V+	.	.	TBA231	UA739DC	O	U6A7739393
DIL-14/1C	N	N	T	E-	E+	V-	N	N	T*	R	V+	N	N	N	.	.	LM741D	UA741DM	O	U6A7741312
DIL-14/1C	N	N	T	E-	E+	V-	N	N	T*	R	V+	N	N	N	.	.	TBA221A	UA741DC	O	U6A7741393
DIL-14/1C	N	N	FT	E-	E+	V-	N	N	T*	R	V+	F*	N	N	.	.	SN52748JA	UA748DM	O	U6A7748312
DIL-14/1C	N	N	FT	E-	E+	V-	N	N	T*	R	V+	F*	N	N	.	.	SN72748J	UA748DC	O	U6A7748393
DIL-14/1C	R1	ø1	F1	F*1	E+1	E-1	V-	E-2	E+2	F2	F*2	ø2	R2	V+	.	.		UA749DM	O	U6A7749312
DIL-14/1C	R1	ø1	F1	F*1	E+1	E-1	V-	E-2	E+2	F2	F*2	ø2	R2	V+	.	.		UA749DC	O	U6A7749393
DIL-14/1C	N	N	FT	E-	E+	V-	N	N	T*	R	V+	F*	N	N	.	.	MLM201D	LM201D	O	U6E7201393
DIL-14/1P	N	N	F	E-	E+	V-	N	N	ø	R	V+	F*	N	N	.	.	TAA521A	UA709DC	O	U6E7709393
DIL-14/1P	N	G	E+	E-	N	V-	N	N	R	R	V+	N	N	N	.	.	SFC2710EC	UA710DC	O	U6E7710393
DIL-14/1P	N	E-1	E+1	V-	E+2	E-2	N	N	S2	R	V+	G	S1	N	.	.	SFC2711EC	UA711DC	O	U6E7711393
DIL-14/1P	N	N	G	E-	E+	V-	N	N	F	ø	R	N	V+	N	.	.	SN72702J	UA702DC	O	U6E7712393
DIL-14/1C	R1	ø1	F1	F*1	E+1	E-1	V-	E-2	E+2	F2	F*2	ø2	R2	V+	.	.	TBA231	UA739DC	O	U6E7739393
DIL-14/1C	N	N	T	E-	E+	V-	N	N	T*	R	V+	N	N	N	.	.	TBA221A	UA741DC	O	U6E7741393
DIL-14/1C	N	N	FT	E-	E+	V-	N	N	T*	R	V+	F*	N	N	.	.	SN72748J	UA748DC	O	U6E7748393
DIL-8/1P	FT	E-	E+	V-	T*	R	V+	F*	.								MLM201J	LM201J	O	U6T7201393
DIL-8/1P	T	E-	E+	V-	T*	R	V+	N	.								TBA221B	UA741TC	O	U6T7741393
DIL-8/1P	FT	E-	E+	V-	T*	R	V+	F*	.								TBB0748	UA748TC	O	U6T7748393
DIL-14/1M	E-1	E+1	T1	V-	T2	E+2	E-2	T*2	V+2	R2	N	R1	V+	T*1	.	.	SFC2747KM	UA747DM	O	U6W7747312
DIL-14/1M	E-1	E+1	T1	V-	T2	E+2	E-2	T*2	V+2	R2	N	R1	V+1	T*	.	.	TBB0747A	UA747DC	O	U6W7747392
DIL-14/1M	E-1	E+1	T1	V-	T2	E+2	E-2	T*2	V+2	R2	N	R1	V+1	T*1	.	.	TBB0747A	UA747DC	O	U6W7747393
DIL-14/1C	E-1	E+1	T1	V-	T2	E+2	E-2	T*2	V+2	R2	N	R1	V+	T*1	.	.	SFC2747KM	UA747DM	O	U7A7747312
DIL-14/1C	E-1	E+1	T1	V-	T2	E+2	E-2	T*2	V+2	R2	N	R1	V+1	T*1	.	.	TBB0747A	UA747DC	O	U7A7747393
DIL-8/1P	T	E-	E+	V-	T*	R	V+	N	.								TBA221B	UA741TC	O	U9T7741393
TO5-8/1M	T	E-	E+	V-	T*	R	V+	B	.								SG4250CT	LM4250CH	O	UC4250CTY
TO5-8/1M	T	E-	E+	V-	T*	R	V+	B	.								SG4250T	LM4250H	O	UC4250TY
TO5-8/1M	F	E-	E+	V-	ø	R	V+	F*	.									MC1439G	O	ULN2139D
FLP-10/3C	N	F	E-	E+	V-	ø	R	V-	F*	N	.								O	ULN2139G
DIL-14/1C	N	F	E-	E+	V-	N	N	N	N	ø	R	V+	F*	N	.	.	MC1439L		O	ULN2139H
DIL-8/1P	F	E-	E+	V-	ø	R	V+	F*	.									MC1439P1	O	ULN2139M
TO5-8/1M	T	E-	E+	V-	T*	R	V+	N	.								LM741EH	UA741EHC	O	ULN2151D
FLP-10/3C	N	T	E-	E+	V-	T*	R	V+	N	N	.						LM741F	UA741FM	O	ULN2151G
DIL-14/1C	N	T	E-	E+	V-	N	N	N	T*	R	V+	N	N		.	.	LM741ED	UA741EDC	O	ULN2151H
DIL-8/1P	T	E-	E+	V-	T*	R	V+	N	.								TBA221B	LM741EJ	O	ULN2151M
TO5-8/1P	T	E-	E+	V-	T*	R	V+	N	.									RC4131T	O	ULN2156D
DIL-8/1P	T	E-	E+	V-	T*	R	V+	N	.									RC4131NB	O	ULN2156M
DIL-14/1P	E-1	E+1	T1	V-	T2	E+2	E-2	T*2	V+2	R2	N	R1	V+1	T*1	.	.	SFC2747KM	UA747DM	O	ULN2157A
DIL-14/1C	E-1	E+1	T1	V-	T2	E+2	E-2	T*2	V+2	R2	N	R1	V+1	T*1	.	.	SFC2747KM	UA747DM	O	ULN2157H
TO5-10/1M	R1	V+1	E-1	E+1	V-	E+2	E-2	V+2	R2	N	.						SFC2747M	UA747HM	O	ULN2157K
TO5-8/1M	TF	E-	E+	V-	T*	R	V+	F*	.								TBC0748	UA748HM	O	ULN2158D
DIL-8/1P	T	E-	E+	V-	T*	R	V+	F	.								SN52748JP	LM748J	O	ULN2158M
TO5-8/1M	T	E-	E+	V-	T*	R	V+	N	.								N5556T	MC1456G	O	ULN2171D
FLP-10/3C	N	T	E-	E+	V-	T*	R	V+	N	N	.								O	UL2171G
DIL-14/1C	N	T	E-	E+	V-	N	N	N	T*	R	V+	N	N		.	.			O	ULN2171H
DIL-8/1P	T	E-	E+	V-	T*	R	V+	N	.								RC1556NB	RC1556NB	O	ULN2171M
TO5-8/1M	TF	E-	E+	V-	T*	R	V+	F*	.								SFC2101A	LM201AH	O	ULN2172D
DIL-8/1P	T	E-	E+	V-	T*	R	V+	ø	.								SFC2301ADC	LM301AJ	O	ULN2172M
TO5-8/1M	T	E-	E+	V-	T*	R	V+	N	.									RC1556AT	O	ULN2173D
DIL-8/1P	T	E-	E+	V-	T*	R	V+	N	.									RC1556ANB	O	ULN2173M
TO5-8/1M	TF	E-	E+	V-	T*	R	V+	F*	.										O	ULN2174D
DIL-8/1P	T	E-	E+	V-	T*	R	V+	ø	.										O	ULN2174M
TO5-8/1M	T	E-	E+	V-	T*	R	V+	N	.								TBA222	UA741HM	O	ULN2741D
DIL14-1P	E-1	E+1	T1	V-	T2	E+2	E-2	T*2	V+2	R2	N	R1	V+1	T*1	.		TBB0747A	UA747DC	O	ULN2747A

TYPE NUMBER	M F R	A P P	C M P	GBP MIN	SLEW RATE MIN	V_S^+ MAX	V_S^- MAX	T_{op} MAX	A_{VOL} MIN	V_{IO} MAX	I_B MAX	I_{IO} MAX	P_{TOT} MAX	I_{OUT} MIN	V_{OUT} MIN	V_{ICM} MAX	V_{IDF} MAX	dV_{IO}/dT MAX	P_O MAX	I_O MAX	CM RR MIN	PS RR MIN	R_{IN} MIN
ULN2747A	OBS	DGK	INT	.4MHZ	0.3V/uS	+18V	-18V	70C	88dB	6MV	500NA	200NA	.	5MA	12V	15V	30V	.	85MW	3MA	70dB	76dB	300K
ULS2139D	SPU	GPU	EXT	.3MHZ	1V/uS	+18V	-18V	125C	94dB	3MV	500MA	60NA	.	10MA	10V	18V	18V	.	150MW	.	80dB	75dB	150K
ULS2139G	SPU	GPU	EXT	.3MHZ	1V/uS	+18V	-18V	125C	94dB	3MV	500MA	60NA	.	10MA	10V	18V	18V	.	150MW	.	80dB	75dB	150K
ULS2139H	SPU	GPU	EXT	.3MHZ	1V/uS	+18V	-18V	125C	94dB	3MV	500MA	60NA	.	10MA	10V	18V	18V	.	150MW	.	80dB	75dB	150K
ULS2139M	SPU	GPU	EXT	.3MHZ	1V/uS	+18V	-18V	125C	94dB	3MV	500MA	60NA	.	10MA	10V	18V	18V	.	150MW	.	80dB	75dB	150K
ULS2151D	SPU	GPK	INT	.	0.5V/uS	+22V	-22V	125C	94dB	2MV	50NA	5NA	.	5MA	10V	15V	30V	.	85MW	.	85dB	85dB	1.5M
ULS2151G	SPU	GPK	INT	.	0.5V/uS	+22V	-22V	125C	94dB	2MV	50NA	5NA	.	5MA	10V	15V	30V	.	85MW	.	85dB	85dB	1.5M
ULS2151H	SPU	GPK	INT	.	0.5V/uS	+22V	-22V	125C	94dB	2MV	50NA	5NA	.	5MA	10V	15V	30V	.	85MW	.	85dB	85dB	1.5M
ULS2151M	SPU	GPK	INT	.	0.5V/uS	+22V	-22V	125C	94dB	2MV	50NA	5NA	.	5MA	10V	15V	30V	.	85MW	.	85dB	85dB	1.5M
ULS2156D	OBS	HSR	INT	.	1V/uS	+22V	-22V	125C	100dB	4MV	15NA	2NA	.	.	.	15V	30V	.	45MW	.	.	.	1M
ULS2157H	OBS	DGK	INT	94HZ	0.5V/uS	+22V	-22V	125C	94dB	2MV	50NA	5NA	.	5MA	10V	15V	30V	.	85MW	.	85dB	85dB	1.5M
ULS2157K	OBS	DGK	INT	.	0.5V/uS	+22V	-22V	125C	94dB	2MV	50NA	5NA	.	5MA	10V	15V	30V	.	85MW	.	85dB	85dB	1.5M
ULS2158D	OBS	GPU	EXT	.	0.5V/uS	+22V	-22V	125C	94dB	2MV	50NA	5NA	.	5MA	10V	15V	30V	.	85MW	.	85dB	85dB	1.5M
ULS2171D	SPU	GPK	INT	.	1V/uS	+22V	-22V	125C	94dB	2MV	15NA	7NA	.	5MA	10V	15V	30V	.	90MW	.	85dB	85dB	8M
ULS2171G	SPU	GPK	INT	.	1V/uS	+22V	-22V	125C	94dB	2MV	15NA	7NA	.	5MA	10V	15V	30V	.	90MW	.	85dB	85dB	8M
ULS2171H	SPU	GPK	INT	.	1V/uS	+22V	-22V	125C	94dB	2MV	15NA	7NA	.	5MA	10V	15V	30V	.	90MW	.	85dB	85dB	8M
ULS2171M	SPU	GPK	INT	.	1V/uS	+22V	-22V	125C	94dB	2MV	15NA	7NA	.	5MA	10V	15V	30V	.	90MW	.	85dB	85dB	8M
ULS2172D	OBS	GPU	EXT	.	1V/uS	+22V	-22V	125C	94dB	2MV	15NA	7NA	.	5MA	10V	15V	30V	.	90MW	.	85dB	85dB	8M
ULS2173D	OBS	LBC	INT	.	0.1V/uS	+22V	-22V	125C	100dB	2MV	3NA	1.5NA	.	.	.	15V	30V	.	35MW	.	.	.	7M
ULS2174D	OBS	LBC	EXT	.	0.1V/uS	+22V	-22V	125C	100dB	2MV	3NA	1.5NA	.	.	.	15V	30V	.	35MW	.	.	.	7M
ULS2741D	OBS	GPK	INT	.4MHZ	0.3V/uS	+22V	-22V	125C	94dB	3MV	80MA	30NA	.	10MA	16V	15V	30V	15uV/C	150MW	3MA	80dB	86dB	1M
ZA702M1	ZEU	PIA	INT	.	0.1V/uS	+18V	-18V	70C	.	4MV	50NA	.	.	5MA	10V	10V	10V	50uV/C	.	10MA	80dB	.	100M
ZA703M1	ZEU	PIA	INT	.	0.1V/uS	+18V	-18V	70C	.	4MV	5pA	.	.	5MA	10V	10V	10V	60uV/C	.	10MA	80dB	.	10G
ZA801M1	ZEU	GPK	INT	2MHZ	6V/uS	+18V	-18V	85C	100dB	2MV	25pA	.	.	10MA	10V	15V	15V	50uV/C	.	3MA	70dB	.	10G
ZA801M2	ZEU	GPK	INT	2MHZ	6V/uS	+18V	-18V	85C	100dB	2MV	25pA	.	.	10MA	10V	15V	15V	20uV/C	.	3MA	70dB	.	10G
ZA801M3	ZEU	GPK	INT	2MHZ	6V/uS	+18V	-18V	85C	100dB	2MV	25pA	.	.	10MA	10V	15V	15V	10uV/C	.	3MA	70dB	.	10G
ZA801D1	ZEU	GPK	INT	2MHZ	6V/uS	+18V	-18V	85C	100dB	2MV	25pA	.	.	10MA	10V	15V	15V	50uV/C	.	3MA	70dB	.	10G
ZA801E1	ZEU	GPK	INT	2MHZ	6V/uS	+18V	-18V	85C	100dB	2MV	25pA	.	.	10MA	10V	15V	15V	50uV/C	.	3MA	70dB	.	10G
ZA802M1	ZEU	PIA	INT	3MHZ	6V/uS	+18V	-18V	85C	100dB	2MV	5pA	.	.	10MA	10V	15V	15V	50uV/C	.	3MA	90dB	.	10G
ZA804M1	ZEU	LNA	INT	2MHZ	6V/uS	+18V	-18V	85C	100dB	2MV	25pA	.	.	10MA	10V	15V	15V	50uV/C	.	3MA	70dB	.	10G
ZA804M2	ZEU	LNA	INT	2MHZ	6V/uS	+18V	-18V	85C	100dB	2MV	25pA	.	.	10MA	10V	15V	15V	20uV/C	.	3MA	70dB	.	10G
ZA903M1	ZEU	LVD	INT	2MHZ	6V/uS	+18V	-18V	60C	100dB	0.5MV	10pA	.	.	10MA	10V	15V	15V	3uV/C	.	3MA	78dB	.	10G
ZA903M2	ZEU	LVD	INT	2MHZ	6V/uS	+18V	-18V	60C	100dB	0.5MV	10pA	.	.	10MA	10V	15V	15V	1uV/C	.	3MA	78dB	.	10G
ZEL-1	ZEU	PIA	INT	.5MHZ	2.5V/uS	+18V	.	85C	114dB	.	50NA	15NA	.	5MA	10V	10V	15V	20uV/C	.	5MA	80dB	.	300K
ZEL-1/02	ZEU	PIA	INT	.5MHZ	2.5V/uS	+18V	.	85C	114dB	.	50NA	15NA	.	5MA	10V	10V	15V	2.5uV/C	.	5MA	80dB	.	300K
ZEL-1/03	ZEU	PIA	INT	.5MHZ	2.5V/uS	+18V	.	85C	114dB	.	50NA	15NA	.	5MA	10V	10V	15V	5uV/C	.	5MA	80dB	.	300K
ZEL-1/04	ZEU	PIA	INT	.5MHZ	2.5V/uS	+18V	.	85C	114dB	.	50NA	15NA	.	5MA	10V	10V	15V	10uV/C	.	5MA	80dB	.	300K
ZEL-1AC	ZEU	PIA	INT	.5MHZ	2V/uS	+18V	-18V	85C	114dB	.	50NA	5NA	.	20MA	10V	10V	15V	20uV/C	.	8MA	80dB	.	300K
ZEL-1C	ZEU	PIA	INT	.5MHZ	2V/uS	+18V	-18V	85C	114dB	.	50NA	5NA	.	20MA	10V	10V	15V	20uV/C	.	8MA	80dB	.	300K
ZEL-1E	ZEU	PIA	INT	.5MHZ	2.5V/uS	+25V	-25V	85C	114dB	.	50NA	5NA	.	4MA	20V	20V	25V	20uV/C	.	5MA	80dB	.	300K
ZLD709	FEG	GPU	EXT	.	.	+18V	-18V	125C	88dB	5MV	500NA	200NA	300MWF	.	12V	10V	5V	20uV/C	165MW	.	80dB	76dB	150K
ZLD709C	FEG	GPU	EXT	.	.	+18V	-18V	75C	84dB	7.5MV	1.5uA	500NA	300MWF	.	10V	10V	5V	20uV/C	200MW	.	65dB	74dB	50K
ZLD709CE	FEG	GPU	EXT	.	.	+18V	-18V	75C	84dB	7.5MV	1.5uA	500NA	300MWF	.	10V	10V	5V	20uV/C	200MW	.	65dB	74dB	50K
ZLD709CF	FEG	GPU	EX	.	.	+18V	-18V	75C	84dB	7.5MV	1.5uA	500NA	300MWF	.	10V	10V	5V	20uV/C	200MW	.	65dB	74dB	50K
ZLD709F	FEG	GPU	EXT	.	.	+18V	-18V	125C	88dB	5MV	500NA	200NA	300MWF	.	12V	10V	5V	20uV/C	165MW	.	80dB	76dB	150K
ZLD741	FEG	GPK	INT	.	.25V/uS	+22V	-22V	125C	94dB	5MV	500NA	200NA	500MWF	.	12V	15V	30V	.	85MW	3MA	70dB	76dB	300K
ZLD741C	FEG	GPK	INT	.	.25V/uS	+18V	-18V	70C	86dB	6MV	500NA	200NA	500MWF	.	12V	15V	30V	.	85MW	3MA	70dB	76dB	300K
ZLD741CE	FEG	GPK	INT	.	.25V/uS	+18V	-18V	70C	86dB	6MV	500NA	200NA	500MWF	.	12V	15V	30V	.	85MW	3MA	70dB	76dB	300K
ZLD741CP	FEG	GPK	INT	.	.25V/uS	+18V	-18V	70C	86dB	6MV	500NA	200NA	300MWF	.	12V	15V	30V	.	85MW	3MA	70dB	76dB	300K
ZN402E	FEG	PRA	EXT	2MHZ	6V/uS	+18V	-18V	70C	80dB	6MV	1.2uA	0.5uA	250MWF	.	10V	10V	5V	15uV/C	.	7MA	70dB	80dB	100K
ZN402T	FEG	PRA	T	2MHZ	6V/uS	+18V	-18V	70C	80dB	6MV	1.2uA	0.5uA	300MWF	.	10V	10V	5V	15uV/C	.	7MA	70dB	80dB	100K
ZN402P	FEG	PRA	EXT	2MHZ	6V/uS	+18V	-18V	70C	80dB	6MV	1.2uA	0.5uA	250MWF	.	10V	10V	5V	15uV/C	.	7MA	70dB	80dB	100K
ZN424E	FEG	PRA	EXT	2MHZ	6V/uS	+18V	-18V	70C	80dB	6MV	1.2uA	0.5uA	250MWF	.	10V	10V	5V	15uV/C	.	7MA	70dB	80dB	100K
ZN424P	FEG	PRA	EXT	2MHZ	6V/uS	+18V	-18V	70C	80dB	6MV	1.2uA	0.5uA	250MWF	.	10V	10V	5V	15uV/C	.	7MA	70dB	80dB	100K
ZN424T	FEG	PRA	EXT	2MHZ	6V/uS	+18V	-18V	70C	80dB	6MV	1.2uA	0.5uA	300MWF	.	10V	10V	5V	15uV/C	.	7MA	70dB	80dB	100K

For detailed explanations of column heading notations, see App. A.

Also for ready references the more important abbreviations used in the column headings are listed below:

LEFT HAND PAGE

APP = application (codes at APP.E.)

CMRR = common mode rejection ratio

CMP = compensation (frequency)

dV_{I0}/dT = input offset voltage temperature drift

GBP = gain bandwidth product

I_B = input bias current

I_{I0} = input bias offset current

I_Q = quiescent supply current

MFR = manufacturer (codes at App.C.)

P_Q = quiescent power consumer

PSRR = power supply rejection ratio

V_{ICM} = common mode input voltage rating

V_{IDF} = differential input voltage rating

V_{I0} = input offset voltage

V_S = dc supply voltage

RIGHT HAND PAGE

Lead out coding summary (details at APP.G.) for different cases (APP.F.)

A = gain adjust

B = bias adjust

C = case

E− = inverting input

E+ = non-inverting input

F,F* = input frequency compensation

G = ground

J = high level input

K = output, open collector

L = output, open emitter

M = metal case

N = not connected

Q = special terminal

R,R* = outputs

S = strobe

T,T* = offset balance

V+ = +ve dc supply

V− = −ve dc supply

W = guard ring

X = blank position, no lead

+ + = +ve supplementary dc supply

− − = −ve supplementary dc supply

$\emptyset,\emptyset^*$ = output frequency compensation

CASE (APP F)	LD 1	LD 2	LD 3	LD 4	LD 5	LD 6	LD 7	LD 8	LD 9	LD 10	LD 11	LD 12	LD 13	LD 14	LD 15	LD 16	EUROPE SUBSTITUTE	USA SUBSTITUTE	I S	TYPE NUMBER
TO5-8/1M	F	E-	E+	V-	∅	R	V+	F*	.	.	.	.	.	.			.	MC1539G	0	ULS2139D
FLP-10/3C	N	F	E-	E+	V-	∅	R	V-	F*	N							.	.	0	ULS2139G
DIL-14/1C	N	F	E-	E+	V-	N	N	∅	R	V+	F*	N					.	.	0	ULS2139H
DIL-8/1P	F	E-	E+	V-	∅	R	V+	F*	.	.							.	.	0	ULS2139M
TO5-8/1M	T	E-	E+	V-	T*	R	V+	N	.	.							LM741AH	UA741AHM	0	ULS2151D
FLP-10/3C	N	T	E-	E+	V-	T*	R	V+	N	N							LM741AF	UA741AFM	0	ULS2151G
DIL-14/1C	N	T	E-	E+	V-	N	N	N	N	T*	R	V+	N	N			LM741AD	UA741ADM	0	ULS2151H
DIL-8/1P	T	E-	E+	V-	T*	R	V+	N	.								LM741EJ	.	0	ULS2151M
TO5-8/1M	T	E-	E+	V-	T*	R	V+	N	.								.	RM4131T	0	ULS2156D
DIL-14/1C	E-1	E+1	T1	V-	T2	E+2	E-2	T*2	V+2	R2	N	R1	V+1	T*1			LM747AD	UA747ADM	0	ULS2157H
TO5-10/1M	R1	V+1	E-1	E+1	V-	E+2	E-2	V+2	R2	N	.						TBC0747	UA747AHM	0	ULS2157K
TO5-8/1M	TF	E-	E+	V-	T*	R	V+	F*	.								.	UA748AHM	0	ULS2158D
TO5-8/1M	T	E-	E+	V-	T*	R	V+	N	.								S5556T	MC1556G	0	ULS2171D
FLP-10/3G	N	T	E-	E+	V-	T*	R	V+	N	N							.	.	0	ULS2171G
DIL-14/1C	N	T	E-	E+	V-	N	N	N	N	T*	R	V+	N	N			.	.	0	ULS2171H
DIL-8/1P	T	E-	E+	V-	T*	R	V+	N	.								.	S5556V	0	ULS2171M
TO5-8/1M	TF	E-	E+	V-	T*	R	V+	F*	.								SFC2101A	LM101AH	0	ULS2172D
TO5-8/1M	T	E-	E+	V-	T*	R	V+	N	.								.	RM1556AT	0	ULS2173D
TO5-8/1M	TF	E-	E+	V-	T*	R	V+	F*	.								.	.	0	ULS2174D
TO5-8/1M	T	E-	E+	V-	T*	R	V+	N	.								LM741AH	UA741AHM	0	ULS2741D
DIM-9/5P	A	E+	E-	T	E+	Q	V-	R	A*	.							.	.	0	ZA702M1
DIM-9/5P	A	E+	E-	T	E+	Q	V-	R	A*	.							.	.	0	ZA703M1
DIM-7/5P	E+	E-	V+	G	V-	R	T	.									ZA801M2	.	0	ZA801M1
DIM-7/5P	E+	E-	V+	G	V-	R	T	.									ZA801M3	.	0	ZA801M2
DIM-7/5P	E+	E-	V+	G	V-	R	T	.									.	.	0	ZA801M3
DIM-14/1P	N	N	N	E-	E+	V-	V-	N	N	R	V+	N	N	T			.	.	0	ZA801D1
DIM-14/1M	N	N	N	E-	E+	V-	GC	N	R	V+	N	N	T				.	.	0	ZA801E1
DIM-7/5P	E+	E-	V+	G	V-	R	T	.									.	.	0	ZA802M1
DIM-7/5P	E+	E-	V+	G	V-	R	T	.									.	ZA804M2	0	ZA804M1
DIM-7/5P	E+	E-	V+	G	V-	R	T	.									.	.	0	ZA804M2
DIM-9/5P	T	E+	E-	T*	V+	G	V-	R	T1	.							.	ZA903M2	0	ZA903M1
DIM-9/5P	T	E+	E-	T*	V+	G	V-	R	T1	.							.	.	0	ZA903M2
DIM-7/5P	E+	E-	V+	G	V-	R	T	.									.	ZEL-1/04	0	ZEL-1
DIM-7/5P	E+	E-	V+	G	V-	R	T	.									.	.	0	ZEL-1/02
DIM-7/5P	E+	E-	V+	G	V-	R	T.	.									.	ZEL-1/02	0	ZEL-1/03
DIM-7/5P	E+	E-	V+	G	V-	R	T	.									ZEL-1/03	.	0	ZEL-1/04
DIM-7/5P	E+	E-	V+	G	V-	R	T	.									ZEL-1C	.	0	ZEL-1AC
DIM-7/5P	E+	E-	V+	G	V-	R	T	.									.	ZEL-1C	0	ZEL-1C
DIM-7/5P	E+	E-	V+	G	V-	R	T	.									ZEL-1C	.	0	ZEL-1E
TO5-8/1M	F	E-	E+	V-M	∅	∅*R	V+	F*	.								TAA522	UA709AHM	0	ZLD709
TO5-8/1M	F	E-	E+	V-M	∅	∅*R	V+	F*	.								TAA521	UA709HC	0	ZLD709C
DIL-8/1P	F	N	F	E-	E+	V-	N	N	∅	R	V+	F*	N	N			TAA521A	UA709CDL14	0	ZLD709CE
FLP-10/3C	N	F	E-	E+	V-	∅	R∅*	V+	F*	N							SF.C2709PT	UA709CFP10	0	ZLD709CF
FLP-10/3C	N	F	E-	E+	V-	∅	R∅*	V+	F*	N							SF.C2709AP	UA709AFP10	0	ZLD709F
TO5-8/1M	T	E-	E+	V-	T*	R	V+	N	.								TBA222	UA741T05	0	ZLD741
TO5-8/1M	T	E-	E+	V-	T*	R	V+	N	.								TBA221	UA741CT05	0	ZLD741C
DIL-14/1P	N	N	T	E-	E+	V-	N	N	T*	R	V+	N	N	N			TBA221A	UA741CDL14	0	ZLD741CE
DIL-14/1P	N	N	T	E-	E+	V-	N	N	T*	R	V+	N	N	G			TBA221A	UA741CDL14	0	ZLD741CP
DIL-14/1P	T	E-	N	E+	T*∅	R	V-	N	N	S	N	N	V+				ZN424E	.	0	ZN402E
TO5-8/1M	E-	E+	T∅	R	V-	S	V+	T*	.								ZN424T	.	0	ZN402T
DIL-8/1P	V+	T	E-	E+	T*∅	R	V-	S	.								ZN424P	.	0	ZN402P
DIL-14/1P	T	E-	N	E+	T*∅	R	V-	N	N	S	N	N	V+				ZN402E	.	0	ZN424E
DIL-8/1P	V+	T	E-	E+	T*∅	R	V-	S	.								ZN402P	.	0	ZN424P
TO5-8/1M	E-	E+	T∅	R	V-	S	V+	T*	.								ZN402T	.	0	ZN424T

Appendix A

Explanatory notes to tabulations

The general layout plan of the information in the tables of this compendium should be immediately evident from the data tabulation explanatory chart set out overleaf.

Supporting Appendices with additional information are:

App. B   Glossary of *Opamp Terms*
App. C   Tabulation *Codes for Manufacturers*
App. D   IC Manufacturers' *House Numbers*
App. E   Tabulation *Codes for Applications*
App. F   *Case Outline and Leadout Diagrams*
App. G   Codes for *Leadout Connections*

Unit symbols used in the tables are:

A   = amperes
C   = ° centigrade
dB   = decibels
G   = gigaohms (megohms × 10^3)
GHZ = gigahertz (megahertz × 10^3)
K   = kilohms
KHZ = kilohertz
M   = megohms
MA   = milliamperes, mA
MAX = maximum
MHZ = megahertz
MIN   = minimum
MV   = millivolts
MWC = milliwatts, case at 25C
MWF = milliwatts, free air at 25C
MWH = milliwatts, heat sink, 25C
NA   = nanoamps (microamps × 10^{-3})
NV   = nanovolts (microvolts × 10^{-3})
PA   = picoamps (microamps × 10^{-12})
R   = ohms
T   = teraohms (megohms × 10^6)
V   = volts
WC   = watts, case at 25C
WF   = watts, free air at 25C
WH   = watts, heatsink, 25C
μA   = microamps
μS   = microseconds
μV   = microvolts
μW   = microwatts
μWF = microwatts, free air at 25C

Where a unit symbol appears in the middle of a value, it indicates the position of the decimal point, e.g. 3K3 = 3·3K.

TYPE NUMBER	M F R	A P P	C M P	GBP MIN	SLEW RATE MIN	V_S^+ MAX	V_S^- MAX	T_{op} MAX	A_{VOL} MIN	V_{IO} MAX	I_B MAX	I_{IO} MAX	P_{TOT} MAX	I_{OUT} MIN	V_{OUT} MIN	V_{ICM} MAX	V_{IDF} MAX	dV_{IO}/dT MAX	P_Q MAX	I_Q MAX	CM RR MIN	PS RR MIN	R_{IN} MIN
(EXAMPLE) LH0022CH	NAU	FET	INT	3MHZ	1V/uS	+22V	-22V	85C	97dB	6MV	25pA	5pA	500MWF	10MA	10V	15V	30V	15uV/C	85MW	3MA	70dB	70dB	0.1T

TYPE No. NUMERO-ALPHABETIC LISTING

MFR = MANUFACTURER CODED AS APP. C

APP = APPLICATION CODED AS APP. E

CMP = FREQUENCY COMPENSATION WITH INT = INTERNAL EXT = EXTERNAL

GBP MIN = UNITY GAIN BANDWIDTH PRODUCT, MIN; IN KHZ, MHZ, or GHZ

SLEW RATE, MIN. IN VOLTS PER MICROSECOND, V/µS

V_S^+ MAX = MAX. PERMISSIBLE +VE DC SUPPLY VOLTAGE IN VOLTS, V

V_S^- MAX = MAX PERMISSIBLE −VE DC SUPPLY VOLTAGE IN VOLTS, V

T_{OP} MAX = MAX. PERMISSIBLE OPERATIONAL AMBIENT TEMPERATURE IN °C.

A_{VOL} MIN = MIN. OPEN-LOOP VOLTAGE GAIN IN DB

V_{IO} MAX = MAX INPUT OFFSET VOLTAGE AT 25°C IN MV or µV

I_B MAX = MAX. INPUT BIAS CURRENT AT 25°C IN MA, µA, nA or pA

R_{IN} MIN = MIN INPUT RESISTANCE

PSRR MIN = MIN. POWER SUPPLY REJECTION RATIO IN DB

CMRR MIN = MIN. COMMON MODE REJECTION RATIO IN DB

I_Q MAX = MAX. QUIESCENT (NO SIGNAL, NO LOAD) CURRENT CONSUMPTION IN MA

P_Q MAX = MAX. QUIESCENT (NO SIGNAL, NO LOAD) POWER CONSUMPTION IN MW

dV_{IO}/dT MAX = MAX. INPUT OFFSET VOLTAGE TEMPERATURE DRIFT IN µV/C OR MV/C

V_{IDF} MAX = MAX. PERMISSIBLE DIFFERENTIAL INPUT VOLTAGE IN V.

V_{ICM} MAX = MAX. PERMISSIBLE COMMON-MODE INPUT VOLTAGE IN VOLTS, V

V_{OUT} MIN = GUARANTEED MIN. OUTPUT VOLTAGE, PEAK VALUE, IN VOLTS, V

I_{OUT} MIN = GUARANTEED MINIMUM OUTPUT CURRENT, PEAK VALUE, IN MA OR µA.

P_{TOT} MAX = MAX. PERMISSIBLE POWER DISSIPATION IN W, mW, µW WITH F = FREE AIR 25°C, C = CASE 25°C. H = HEATSINK 25°C.

I_{IO} MAX = MAX. INPUT OFFSET CURRENT AT 25°C IN MA, µA, nA, OR pA

• R_{IN} EXPRESSED AS OHMS (R), KILOHMS (K), MEGOHMS (M), GIGAOHMS (G) OR TERAOHMS (T)

[NOTE: FOR FURTHER EXPLANATION OF SPECIAL TERMS SEE APP. B]

LEFT HAND PAGE

For detailed explanations of column heading notations, see App. A.

Also for ready references the more important abbreviations used in the column headings are listed below:

APP = application (codes at APP.E.)

CMRR = common mode rejection ratio

CMP = compensation (frequency)

dV_{10}/dT = input offset voltage temperature drift

GBP = gain bandwidth product

I_B = input bias current

I_{10} = input bias offset current

I_0 = quiescent supply current

MFR = manufacturer (codes at App.C.)

P_0 = quiescent power consumer

PSRR = power supply rejection ratio

V_{ICM} = common mode input voltage rating

V_{IDF} = differential input voltage rating

V_{10} = input offset voltage

V_S = dc supply voltage

RIGHT HAND PAGE

Lead out coding summary (details at APP.G.) for different cases (APP.F.)

A = gain adjust

B = bias adjust

C = case

E— = inverting input

E+ = non-inverting input

F,F* = input frequency compensation

G = ground

J = high level input

K = output, open collector

L = output, open emitter

M = metal case

N = not connected

Q = special terminal

R,R* = outputs

S = strobe

T,T* = offset balance

V+ = +ve dc supply

V— = —ve dc supply

W = guard ring

X = blank position, no lead

+ + = +ve supplementary dc supply

— — = —ve supplementary dc supply

◊,◊* = output frequency compensation

CASE (APP F)	LD 1	LD 2	LD 3	LD 4	LD 5	LD 6	LD 7	LD 8	LD 9	LD 10	LD 11	LD 12	LD 13	LD 14	LD 15	LD 16	EUROPE SUBSTITUTE	USA SUBSTITUTE	ISS	TYPE NUMBER
TO5-8/1M	T	E—	E+	V—	T*	R	V+	N	.	.	.	.	.	.	.	.		LH0022H	0	LH0022CH

CASE = PACKAGE OF DIFFERENT TYPES CODED ACCORDING TO APP. F – FIRST NUMBER INDICATES NUMBER OF LEAD POSITIONS EG DIL-14 = 14-LEAD DUAL-IN-LINE PACKAGE

LD1, LD2, ETC = LEAD NUMBERS WITH CONNECTIONS ACCORDING TO PAGE FOOTNOTE OR APP. G.

TYPE No. REPEATED ON R.H. MARGIN

ISS = ISSUE NUMBER OF DATA ENTRY

USA SUBSTITUTE = SUGGESTED ALTERNATIVE AVAILABLE IN USA

EURO SUBSTITUTE = PROELECTRON STANDARD OR OTHER TYPE AVAILABLE IN EUROPE

Appendix B
Glossary of Opamp Terms

(General Note: All voltage values, unless otherwise noted are normally stated with respect to ground as the zero reference level, or to the mid-point of the two dc supply voltages, if no ground terminal is provided in the amplifier. For single-supply opamps, the reference level is taken as half the supply voltage.)

AVERAGE BIAS CURRENT DRIFT

The ratio of the change in the input bias current to the change in temperature producing it.

COMMON-MODE INPUT RESISTANCE

The ratio of the input voltage range to the change in the input bias current over this range.

COMMON-MODE INPUT VOLTAGE RANGE (V_{ICM})

The maximum (+ve or −ve peak value) voltage that can safely be applied between the input terminals connected together and ground for the amplifier to operate in linear fashion. (Note that the amplifier specifications are not usually guaranteed over the full common-mode voltage range, unless specifically so stated.)

COMMON-MODE REJECTION RATIO (CMRR)

The ratio of the peak-to-peak input common-mode voltage range to the peak-to-peak change in input offset voltage over this range.

DIFFERENTIAL INPUT VOLTAGE RANGE (V_{IDF})

The maximum voltage (+ve or −ve) that may safely be applied between the two input terminals without excessive current flow.

INPUT BIAS CURRENT (I_B)

The average of the currents into the two input terminals for the output at zero volts with no load.

INPUT OFFSET CURRENT (I_{IO})

The difference between the bias currents into the two input terminals when the output is at zero volts.

INPUT OFFSET VOLTAGE (V_{IO})

The voltage which must be applied between the input terminals through two equal (or zero) resistances to obtain zero output voltage.

INPUT OFFSET VOLTAGE DRIFT, OR TEMPERATURE COEFFICIENT (dV_{IO}/dT)

The ratio of the change in input offset voltage to the change of ambient temperature for a constant output voltage (usually zero).

INPUT RESISTANCE (R_{IN})

The ratio of the change in input voltage to the change in input current on either input terminal with the other grounded.

INPUT VOLTAGE RANGE

(Usually taken to mean Common-mode Input Voltage Range, and distinguished from differential Input Voltage Range.)

LARGE-SIGNAL VOLTAGE GAIN

The ratio of the output voltage swing to the change in input differential voltage required to drive the output from zero to this value. For most practical purposes, same as Open Loop Voltage Gain.

OFFSET VOLTAGE TEMPERATURE DRIFT

(See Input Offset Voltage Temperature Drift)

OPEN-LOOP VOLTAGE GAIN (A_{VOL})

The ratio of the output signal voltage to the corresponding differential input signal voltage, with no feedback applied.

OPERATING TEMPERATURE (T_{op})

The range of ambient temperature over which the amplifier will perform in linear fashion.

OUTPUT CURRENT (I_{OUT})

The output current, peak value, that the amplifier can supply into a specified load, usually specified as a minimum guaranteed value.

OUTPUT RESISTANCE

The small-signal driving-point resistance of the output terminal with respect to ground at a specified quiescent dc output voltage and current, with measurements made under open-loop conditions.

OUTPUT VOLTAGE SWING (V_{OUT})

The peak output voltage swing referred to zero that can be obtained without clipping.

POWER CONSUMPTION (P_Q)

The power drawn from the dc supply to permit the amplifier to operate with the output at zero and with no load current.

POWER DISSIPATION (P_{TOT})

The power that can be dissipated safely on a continuous basis in the amplifier while operating over a specified temperature range.

POWER SUPPLY REJECTION RATIO (PSRR)

The inverse ratio of the change in input offset voltage

to the change in power supply voltage producing it, with the supplies varying symmetrically.

SLEW RATE (SR)

The ratio of a change in output voltage to the time required to effect this change under large-signal drive conditions. Slew rate may be specified separately for positive- and negative-going changes.

SUPPLY CURRENT (I_Q)

The current required from the dc power supply to operate the amplifier with no load and the output at zero volts.

SUPPLY VOLTAGE (V_S)

Maximum permissible (rated) dc supply voltages (+ve and −ve) that can safely be used with the amplifier.

UNITY-GAIN BANDWIDTH (GBP)

The frequency range from dc to the frequency at which the amplifier open-loop, small-signal voltage gain falls off to unity, with the amplifier compensated for unity-gain stability.

UNITY-GAIN CLOSED-LOOP BANDWIDTH

The frequency at which the magnitude of the small-signal voltage gain of the amplifier, operated closed-loop as a unity-gain voltage-follower, is 3dB below unity.

VOLTAGE GAIN (A_V)

The ratio of the change in output voltage to the change in voltage between the input terminals producing it.

Appendix C
Tabulation Codes for Manufacturers

ADU   **Advanced Micro Devices Inc.,**
901 Thompson Pl., Sunnyvale, CA 94086, USA

ANG   **Analog Devices Ltd,**
Central Ave., East Molesey, KT8 9BR, Surrey, UK

ANU   **Analog Devices Inc.,**
P.O. Box 280, Norwood, Mass., 02062

BLG   **Bell & Howell Ltd,**
Lennox Road, Basingstoke, Hants, UK

BLU   **Bell & Howell** (Control Products Divison),
706 Bostwick Ave, Bridgeport, Conn. 06605, USA

BUG   **Burr-Brown International Ltd,**
17 Exchange Rd, Watford, WQD1 7EB, Herts., UK

BUU   **Burr-Brown Research Corp.,**
P.O. Box 11400, Tucson, AZ. 85734, USA

CMG   **Computing Techniques Ltd,**
Brookers Rd, Billingshurst, Sussex, RH14 9RZ, UK

DAG   **Datel UK Ltd,**
Stephenson Close, Portway Ind. Estate, Andover, Hants, UK

DAU   **Datel Systems Inc.,**
1020 Turnpike St., Canton, MA 02021, USA

FAG   **Fairchild Camera & Instrument (UK) Ltd,**
230 High St., Potters Bar, Herts., UK

FAU   **Fairchild Semiconductor**
464 Ellis St., Mountain View, CA 94042, USA

FEG   **Ferranti Ltd,** (Electronic Department),
Gem Mill, Chadderton, Oldham, Lancs., OL9 8NP, UK

FUJ   **Fujitsu Ltd,**
1015 Kamikodanaka, Kawasaki, Japan

HAG   **Harris Semiconductor (Memec) Ltd,**
The Firs, Whitchurch, Nr. Aylesbury, Bucks., HP22 4JU, UK

HAU   **Harris Semiconductor**
P.O. Box 883, Melbourne, FL, 32901, USA

HIJ   **Hitachi Ltd** (Semiconductor and IC Div.),
1450 Josuihonimachi, Kodaira City, Tokyo, Japan

ING   **Intersil Inc.,**
8 Tessa Rd, Richfield Trading Estate, Reading, Berks., UK

INU   **Intersil Inc.,**
10900 N. Tantau Ave, Cupertino, CA, 95014, USA

ITG   **ITT Semiconductors**
Maidstone Rd, Foots Cray, Sidcup, Kent, DA14 5HT, UK

ITU   **ITT Semiconductors**
74 Commerce Way, Woburn, MA, 01801, USA

MNG   **Mitsubishi Shoji Kaisha Ltd,**
Bow Bells House, Bread St., London, EC4, UK

MNJ   **Mitsubishi Electric Corp.,**
2–12 Marunouchi, Chiyoda-ku, Tokyo, Japan

MTG   **Motorola Ltd** (Semiconductor Products Div.)
York House, Empire Way, Wembley, Middlesex, HA9 0PR, UK

MTU   **Motorola Semiconductor Products Inc.,**
5005 E. McDowell Road, Phoenix, AZ, 85008 USA

MUG   **Mullard Ltd,**
Mullard House, Torrington Place, London, WC1E 7HD, UK

NAG   **National Semiconductor (UK) Ltd,**
Harpur Centre, Bedford, MK40 3LF, UK

NAU   **National Semiconductor Corp.,**
2900 Semiconductor Drive, Santa Clara, CA, 95051, USA

NIJ   **Nippon Electric Co. Ltd,**
1753 Shimonumabe, Nakahara-ku, Kawasaki, Japan

OAU   **Opamp Labs Inc.,**
1033 N. Sycamore Ave., Los Angeles, CA 90038, USA

OBS   Obsolete – no longer commercially available.

OTU   **Optical Electronics Inc.,**
P.O. Box 11140, Tucson, AZ, 85734, USA

PLG   **Plessey Semiconductors,**
Cheney Manor, Swindon, Wilts., SN2 2QW, UK

PRG   **Precision Monolithics** (Bourns Trimpot Ltd)
17/27 High St., Hounslow, Middlesex, UK

PRU   **Precision Monolithics (Bourns) Inc.,**
1500 Space Park Drive, Santa Clara, CA, 95050, USA

RAG   **Raytheon Semiconductor**
The Pinnacles, Harlow, Essex, CM19 5BB, UK

RAU   **Raytheon Semiconductor,**
350 Ellis Street, Mountain View, CA, 94042, USA

RCG   **RCA (Great Britain) Ltd,**
Lincoln Way, Windmill Road, Sunbury-on-Thames, Middlesex, UK

RCU   **RCA Solid State Division**
Route 202, Somerville, NJ, 08876, USA

SAJ   **Sanken Electric Co. Ltd,**
1-22-8 Nishi-Ikebukuro, Toshima-Ku, Tokyo, Japan

SGG	**SGS-ATES (UK) Ltd,** Planar House, Walton Street, Aylesbury, Bucks., UK
SGI	**SGS-ATES Componenti Spa,** Via Olivetti, 2 Agrate Brianza, 20041, Milan, Italy
SHG	**Shindengen Hyokuto Boeki Haisha Ltd,** St. Alphage House, Fore St., London, EC2Y 5DA, UK
SHJ	**Shindengen Electric Mfg Co., Ltd,** New Ohtemachi Bldng, 2–1, 2-chome, Ohtemachi, Chiyoda-ku, Tokyo, Japan
SIG	**Siemens Ltd,** Great West Road, Brentford, Middlesex, TW8 9DG, UK
SIW	**Siemens Aktiengesellschaft,** Richard-Strauss-Strasse 76, D-8000 Munchen 2, Postfach 202109, W. Germany
SJG	**Signetics International Corporation** Yeoman House, 63 Croydon Rd, London, SE20, UK
SJU	**Signetics Corp.,** 811 East Arques Ave, Sunnydale, CA. 94086, USA
SKU	**Silicon General Inc.,** 7382 Bolsa Avenue, Westminster, CA, 92683, USA
SLG	**Siliconix Ltd,** 30A High St., Thatcham, Newbury, Berks., RG13 4JG, UK
SLU	**Siliconix Incorporated,** 2201 Laurelwood Road, Santa Clara, CA, 95054, USA
SOJ	**Sony Semiconductor Corp.,** 14–1, Asa hi-sho 4, Atsuigi-shi, Kanagawa-ken, 243, Japan
SPG	**Sprague Electric (UK) Ltd,** 159 High St., Yiewsley, W. Drayton, Middlesex, UB7 7RY, UK
SPU	**Sprague Electric Company** (Semiconductor Div.), 115 Northeast Cutoff, Worcester, MA, 01606, USA
TDG	**Teledyne Semiconductor,** Heathrow House, Bath Road, Cranford, Hounslow, Middlesex, TW5 9QP, UK
TDU	**Teledyne (Amelco) Semiconductor,** 1300 Terra Bella Ave, Mountain View, CA, 94032, USA
TEB	**Teledyne-Philbrick,** Heathrow House, Bath Road, Cranford, Hounslow, Middlesex, TW5 9QP, UK
TEU	**Teledyne-Philbrick,** Allied Drive at Route 128, Dedham, MA, 02026, USA
TGG	**Texas Instruments Ltd,** Manton Lane, Bedford, UK
TGU	**Texas Instruments Inc.** (Components Group), P.O. Box 5012, Dallas, Texas, 75222, USA
THF	**Thomson-CSF (Sescosem),** 50 Rue Jean Pierre Timbaud, BP 120, 92403, Courbevoie, France
THG	**Thomson-CSF (UK) Ltd,** Ringway House, Bell Rd, Daneshill, Basingstoke, Hants., RG24 0QG, UK.
TKJ	**Tokyo Sanyo Electric Co. Ltd** (Semiconductor Div.), Oizumachi, Oragun, Gumma, Japan
TOG	**Toshiba (UK) Ltd,** Toshiba House, Great South West Rd, Feltham, Middlesex, UK
TOJ	**Toshiba (Tokyo Shibaura) Electric Co.,** 2–1, 5-chome, Ginza Chuo-ku, Tokyo, Japan
TRU	**Transitron Electronic Corp.,** 168 Albion St., Wakefield, MA, 01881, USA
ZEU	**Zeltex Inc.,** 940 Detrolt Ave, Concord, CA, 94518, USA

Appendix D
IC Manufacturers' House Numbers

(General Note: Manufacturers often adopt their own 'in-house' serial numbering for their ICs. Listed below are the initial letters of numerical series used by different manufacturers.)

AD	Analog Devices
ADO	Analog Devices
AM	Advanced Micro Devices; Datel
AMD	Advanced Micro Devices
AMLM	Advanced Micro Devices
AMSSS	Advanced Micro Devices
AMU	Advanced Micro Devices
C	Bell & Howell
CA	RCA
CIA	Teledyne-Philbrick
CMP	Precision Monolithics
CN	Ferranti
DA	Teledyne-Philbrick
EP	Teledyne-Philbrick
ESL	Teledyne-Philbrick
FSL	Teledyne-Philbrick
FSS	Ferranti
HA	Harris
HEPC	Motorola
ICH	Intersil
ICL	Intersil
JM	Fairchild
JSF	Thomson-CSF
L	Analog Devices; SGS-ATES
LA	Teledyne-Philbrick
LF	National Semiconductor
LH	National Semiconductor
LM	National Semiconductor
M	Mitsubishi
MC	Motorola Semiconductors
MCC	Motorola Semiconductors
MCCF	Motorola Semiconductors
MCE	Motorola Semiconductors
MCH	Motorola Semiconductors
MIC	ITT Semiconductors
MLF	Motorola; Teledyne-Philbrick
MLM	Motorola Semiconductors
MLMC	Motorola Semiconductors
MONO-OP	Precision Monolithics
N	Signetics; Mullard
NC	General Instruments (obs.)
NE	Signetics; Mullard
NH	National Semiconductor

OP	Precision Monolithics
P	Teledyne-Philbrick
PF	Teledyne-Philbrick
PG	General Instruments (obs.)
PP	Teledyne-Philbrick
RA	Radiation (now Harris)
RC	Raytheon
RL	Raytheon
RM	Raytheon
RSN	Raytheon
RV	Raytheon
S	Signetics
SA	Teledyne-Philbrick
SE	Signetics; Mullard
SFC	Thomson-CSF
SG	Silicon General
SH	Fairchild
SK	RCA
SL	Plessey; Teledyne-Philbrick
SN	Texas Instruments
SP	Teledyne-Philbrick
SQ	Teledyne-Philbrick
SSS	Precision Monolithics
SU	Signetics; Mullard
T	Teledyne-Philbrick Transitron
TA	AEG-Telefunken
TAA	Proelectron Standard
TBA	Proelectron Standard
TBB	Proelectron Standard
TBC	Proelectron Standard
TBE	Proelectron Standard
TCA	Proelectron Standard
TDA	Proelectron Standard
TDB	Proelectron Standard
TDC	Proelectron Standard
TDE	Proelectron Standard
TL	AEG-Telefunken
TOA	Transitron
TSC	Transitron
U	Fairchild
ULN	Sprague
ULS	Sprague
USL	Teledyne-Philbrick
ZA	Zeltex
ZEL	Zeltex
ZLD	Ferranti
ZN	Ferranti
μA	Fairchild

Appendix E
Tabulation Codes for Applications

BDO	Balanced differential-output amplifier		**PAA**	Parametric amplifier
CDA	Current-difference amplifier		**PIA**	Precision instrumentation amplifier
CHP	Chopper-stabilized amplifier		**PRA**	Programmable opamp
CPR	DC comparator		**QCD**	Quad current-difference amplifier
DBD	Dual balanced differential-output amplifier		**QCP**	Quad comparator
DCP	Dual Comparator		**QFE**	Quad fet-input opamp
DFE	Dual fet-input opamp		**QGK**	Quad general-purpose, internally-compensated, opamp
DGK	Dual general purpose opamp			
DGU	Dual general-purpose uncompensated opamp		**QGU**	Quad general-purpose, uncompensated, opamp
DHS	Dual high-slew-rate opamp		**QLQ**	Quad low-quiescent-power opamp
DLN	Dual low-noise opamp		**QPI**	Quad precision instrumentation amplifier
DPI	Dual precision instrumentation amplifier		**QPR**	Quad programmable opamp
DPR	Dual programmable opamp		**QSB**	Quad super-beta opamp
DSB	Dual super-beta opamp		**SBA**	Super-beta opamp
FET	Fet-input opamp		**TCP**	Triple comparator
GPK	General-purpose, internally-compensated, opamp		**TFE**	Triple fet-input opamp
			TGK	Triple general-purpose, internally compensated, opamp
GPU	General-purpose, uncompensated, opamp			
HCO	High current output opamp		**TGU**	Triple general-purpose, uncompensated, opamp
HIR	High input resistance opamp		**TLN**	Triple low-noise opamp
HPO	High power output opamp		**TLP**	Triple low-quiescent-power opamp
HSR	High slew rate opamp		**TOT**	Triple operational transconductance amplifier
HVO	High voltage output opamp		**TPI**	Triple precision instrumentation amplifier
LBC	Low input bias current opamp		**TPR**	Triple programmable opamp
LCD	Low input offset current drift opamp		**TSB**	Triple super-beta opamp
LNA	Low noise opamp		**VFA**	Voltage-follower amplifier
LOC	Low input offset current opamp		**WBA**	Wide-band opamp
LOV	Low input offset voltage opamp		**XHG**	Extra-high-gain opamp
LQP	Low quiescent power opamp		**XLP**	Extra-low quiescent power opamp
LVD	Low input offset voltage drift opamp		**XSR**	Extra-high slew rate opamp
MWB	Medium-wideband opamp		**XWB**	Extra-wide-band opamp
OTA	Operational transconductance amplifie			

Appendix F
Case Outline and
Leadout Diagrams

In the data tabulations, the opamp packages (cases) are specified by a four-part coding system. Diagrams for each of the codings are set out in this appendix.

As an illustration,

(1)	(2)	(3)	(4)
DIL	−12	/1	P

(1) (Casing)	(2) (Number of lead positions, including blanks)	(3) (Serial number of package variants)	(4) (Main body material)
BML = Beam lead **CFL** = Flip chip **CHP** = Chip or dice **DIL** = Dual-in-line **DIM** = Modified DIL **FLP** = Flat pack **HIL** = Heat-sinked DIL **MDL** = Miniature DIL **QIL** = Quad-in-line **SIH** = Heatsinked SIL **SIL** = Single-in-line **T66** = TO66 can **TIL** = Triple-in-line **T03** = TO3 can **T05** = TO5 can **T08** = TO8 can **XTR** = Special			B = Beryllium oxide C = Ceramic G = Glass M = Metal P = Plastic

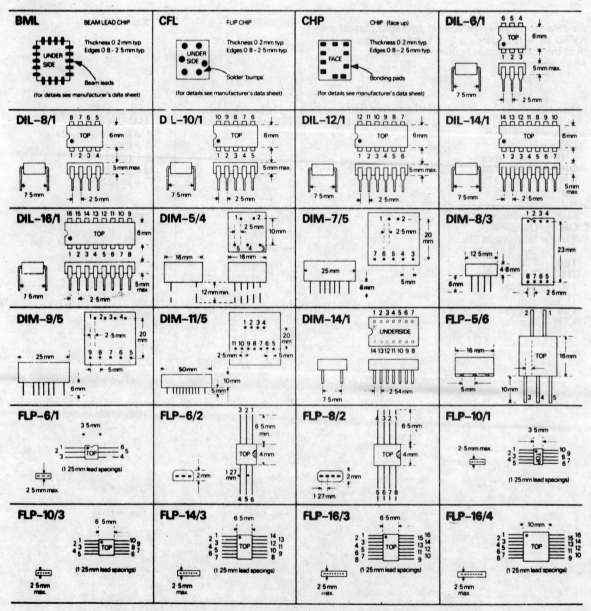

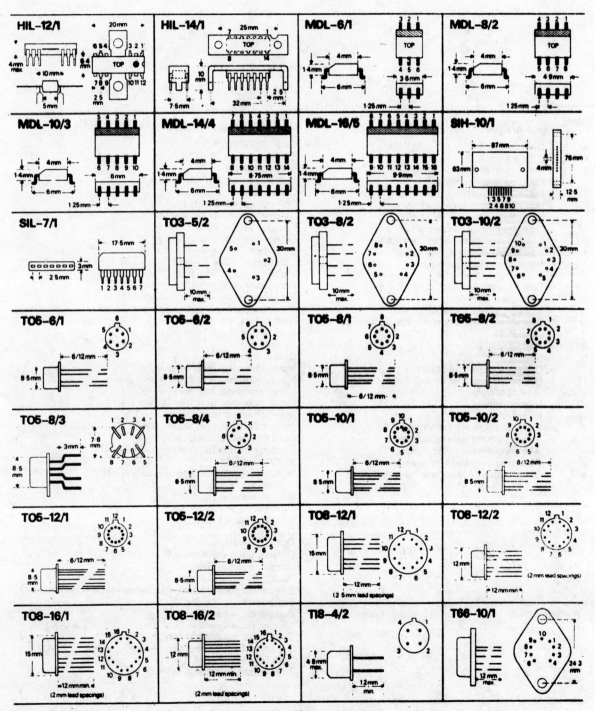

Appendix G
Codes for Leadout Connections

I: Connection Codes in Serial Order

A	=	Gain adjust, 1
A*	=	Gain adjust, 2
B	=	Bias adjust or set
C	=	Case, package, screen
E +	=	Input, non-inverting, low-level
E −	=	Input, inverting, low-level
F	=	Input frequency compensation, 1
F*	=	Input frequency compensation, 2
G	=	Ground, common, earth, zero volts
J +	=	Input, non-inverting, high-level
J −	=	Input, inverting, high-level
K	=	Output, open collector
L	=	Output, open emitter
M	=	Metal casing
N	=	Not connected, i.e. isolated lead
Q	=	Special terminal (consult manufacturer's data)
R	=	Output, 1
R*	=	Output, 2
S	=	Strobe
T	=	Offset balance, trim or null, 1
T*	=	Offset balance, trim or null, 2
V +	=	+ ve dc supply
V −	=	− ve dc supply
W	=	Guard ring
X	=	Blank position, lead omitted
+ +	=	+ ve supplementary dc supply
− −	=	− ve supplementary dc supply
φ	=	Output frequency compensation, 1
φ*	=	Output frequency compensation, 2

II: Lead Assignments in Alphabetical Order

Balance, offset, 1 = T
Balance, offset, 2 = T*
Bias adjust = B
Blank position, without lead = X
Case = C
Compensation, input, 1 = F
Compensation, input, 2 = F*
Compensation, output, 1 = φ
Compensation, output, 2 = φ*
DC supply, + ve = V +
DC supply, − ve = V −
Frequency compensation, input, 1 = F
Frequency compensation, input, 2 = F*
Frequency compensation, output, 1 = φ
Frequency compensation, output, 2 = φ*
Gain adjust, 1 = A
Gain adjust, 2 = A*
Ground = G
Guard ring = W
Input, inverting, high-level = J
Input, non-inverting, high-level = J +
Input, inverting, low-level = E −
Input, non-inverting, low-level = E +
Input offset voltage, adjust, 1 = T
Input offset voltage, adjust, 2 = T*
Lead omitted, blank position = X
Lead in position but not connected = N
Metal case = M
Not connected, but lead in position = N
Null, offset, 1 = T
Null, offset, 2 = T*
Offset voltage adjust, 1 = T
Offset voltage adjust, 2 = T*
Output, 1 = R
Output, 2 = R*
Output, open-collector = K
Output, open-emitter = L
Package = C
Special purpose terminal (data sheet to be consulted) = Q
Strobe = S
Supply, dc, + ve = V +
Supply, dc, − ve = V −
Supply, dc, supplementary, + ve = + +
Supply, dc, supplementary, − ve = − −
Trim (offset voltage), 1 = T
Trim (offset voltage), 2 = T*